PROGRESS IN BRAIN RESEARCH

VOLUME 195

EVOLUTION OF THE PRIMATE BRAIN: FROM NEURON TO BEHAVIOR

SERIES EDITORS

PROGRESS IN BRAIN RESEARCH

VOLUME 195

EVOLUTION OF THE PRIMATE BRAIN: FROM NEURON TO BEHAVIOR

EDITED BY

MICHEL A. HOFMAN

Netherlands Institute for Neuroscience
Royal Netherlands Academy of Arts and Sciences
Amsterdam, The Netherlands

DEAN FALK

School for Advanced Research
Santa Fe, New Mexico, USA
and
Department of Anthropology
Florida State University
Tallahassee, Florida, USA

ELSEVIER

AMSTERDAM – BOSTON – HEIDELBERG – LONDON – NEW YORK – OXFORD
PARIS – SAN DIEGO – SAN FRANCISCO – SINGAPORE – SYDNEY – TOKYO

Elsevier
Radarweg 29, PO Box 211, 1000 AE Amsterdam, The Netherlands
Linacre House, Jordan Hill, Oxford OX2 8DP, UK
360 Park Avenue South, New York, NY 10010-1710

First edition 2012

Library of Congress Cataloging-in-Publication Data
A catalog record for this book is available from the Library of Congress

British Library Cataloguing in Publication Data
A catalogue record for this book is available from the British Library

ISBN: 978-0-444-53860-4
ISSN: 0079-6123

Printed and bound by CPI Group (UK) Ltd, Croydon, CR0 4YY

Transferred to Digital Print 2012

List of Contributors

F. Aboitiz, Departamento de Psiquiatría, Centro Interdisciplinario de Neurociencias Escuela de Medicina, Pontificia Universidad Católica de Chile, Santiago, Chile
A.L. Bauernfeind, Department of Anthropology, The George Washington University, Washington, DC, USA
S. Bianchi, Department of Anthropology, The George Washington University, Washington, DC, USA
D.P. Buxhoeveden, College of Social Work and Department of Anthropology, University of South Carolina, Columbia, SC, USA
C.J. Charvet, Behavioral and Evolutionary Neuroscience Group, Department of Psychology, Cornell University, Ithaca NY, USA
C. Cherniak, Committee for Philosophy and the Sciences, Department of Philosophy, University of Maryland, College Park, MD, USA
G. Clowry, Institute of Neuroscience, Newcastle University, Newcastle upon Tyne, United Kingdom
M.C. Corballis, Department of Psychology, University of Auckland, Auckland, New Zealand
E. Cunha, Department of Life Sciences, Forensic Sciences Center, University of Coimbra, Coimbra, Portugal
A. de Sousa, Department of Life Sciences, Forensic Sciences Center, University of Coimbra, Coimbra, Portugal
C. Dehay, Université de Lyon, Université Lyon I, Lyon, France
U. Dicke, Brain Research Institute, University of Bremen, Bremen, Germany
D. Falk, School for Advanced Research, Santa Fe, NM, USA, and Department of Anthropology, Florida State University, Tallahassee, FL, USA
B.L. Finlay, Behavioral and Evolutionary Neuroscience Group, Department of Psychology, Cornell University, Ithaca NY, USA
S. Herculano-Houzel, Instituto de Ciências Biomédicas, Universidade Federal do Rio de Janeiro, Brasil and Instituto Nacional de Neurociência Translacional, Rio de Janeiro, Brazil
P.R. Hof, Fishberg Department of Neuroscience and Friedman Brain Institute, Mount Sinai School of Medicine, New York, NY, USA, and New York Consortium in Evolutionary Primatology, New York, NY, USA
M.A. Hofman, Netherlands Institute for Neuroscience, Royal Netherlands Academy of Arts and Sciences, Amsterdam, The Netherlands
J.H. Kaas, Department of Psychology, Vanderbilt University, Nashville, TN, USA
H. Kennedy, Inserm U846, Stem cell and Brain Research Institute, Bron, France
J.E. LeDoux, Center for Neural Science, New York University, New York, NY, USA
L. Lefebvre, Department of Biology, McGill University, Montréal, QC, Canada
C. MacLeod, Department of Anthropology, Langara College, Vancouver, BC, Canada
Z. Molnár, Department of Physiology, Anatomy and Genetics, University of Oxford, Oxford, United Kingdom

J.F. Montiel, Departamento de Psiquiatría, Centro Interdisciplinario de Neurociencias Escuela de Medicina, Pontificia Universidad Católica de Chile, Santiago, Chile, and Facultad de Medicina, Centro de Investigación Biomédica, Universidad Diego Portales, Santiago, Chile
R. Nieuwenhuys, Netherlands Institute for Neuroscience, Amsterdam, The Netherlands
M.A. Raghanti, Department of Anthropology and School of Biomedical Sciences, Kent State University, Kent, OH, USA
G. Roth, Brain Research Institute, University of Bremen, Bremen, Germany
P.T. Schoenemann, Department of Anthropology, Indiana University, Bloomington, IN, USA
K. Semendeferi, Anthropology Department, University of California, San Diego, San Diego, CA, USA
C.C. Sherwood, Department of Anthropology, The George Washington University, Washington, DC, USA
K. Teffer, Anthropology Department, University of California, San Diego, San Diego, CA, USA
E.J. Vallender, New England Primate Research Center, Harvard Medical School, Southborough, MA, USA
C.P.E. Zollikofer, Anthropological Institute and Museum, University of Zurich, Zurich, Switzerland

Preface

Evolutionary neuroscience is undergoing vast changes that are facilitated by new methods for studying developmental neurobiology, evolutionary genetics, comparative neuroanatomy (including neurochemistry, cytoarchitecture, neuronal connectivity), and paleoneurology. The goal of this volume is to provide a synthetic source of information about the state-of-the-art research that has important implications for the evolution of the brain and cognition in primates, including humans. The contributors have been carefully selected, not only because of their particular areas of expertise but also because they are internationally renowned scientists who have demonstrated an ability to synthesize and interpret findings within a wider (big-picture) framework.

Chapters are organized into five sections devoted to genes and development, comparative neuroanatomy, human brain evolution, theories of neural organization, and cognition: from neurons to behavior. Primate brains did not evolve out of thin air, of course, and an introductory chapter by Francisco Aboitiz and Juan F. Montiel reminds us that human brains are not as exceptional as some might think (or hope) (Chapter 1). Their discussion focuses on two main events in brain evolution, the origin of mammals and that of primates. Within this context, comparative data regarding neurogenetics, developmental neurogenesis, cytoarchitecture, and neuroanatomy suggest that the brain design for primates is highly conserved and that it evolved before the first primate-like animal existed.

A thread regarding the conserved nature of brain evolution runs through all of the sections, as shown in the chapters by Eric J. Vallender; Christine J. Charvet and Barbara L. Finlay; Jon H. Kaas; Chet C. Sherwood et al.; Christopher Cherniak; Michel A. Hofman; and Gerhard Roth and Ursula Dicke. A related theme is illustrated by Kaas' discussion of important substrates that evolved long before the split between chimpanzee and hominin lineages, as evidenced by a characteristic pattern of areal organization found in the brains of all primates (Chapter 5). Thus, early primates acquired numerous cortical features that distinguish them from living ones, such as an array of new visual areas and an increased density of neurons in primary visual cortex. Posterior parietal cortex also expanded in association with reaching and grasping, and motor cortex became specialized for hand use. Despite the conserved substrates of primate brain evolution, Kaas' chapter reveals that large-brained primates evolved a greater number of cortical areas, some of which became highly specialized.

The tremendous shifts in the size, structure, and function of the brain during primate evolution are ultimately caused by changes at the genetic level. Understanding what these changes are and how they effect the phenotypic changes observed lies at the heart of understanding evolutionary change. Vallender in Chapter 2 focuses on understanding the genetic basis of primate brain evolution, considering the substrates and mechanisms through which genetic change occurs. He also discusses the implications that our current understandings and tools have for what we have already discovered and where our studies will head in the future. Zoltán Molnár and Gavin Clowry in Chapter 3 argue that the increased cortical neural populations afforded by the emergence and variation of the neuronal progenitor cells have led to the evolution of the primate neocortex and that the further diversification and compartmentalization of the germinal zone may have been the driving force behind increased cell numbers in larger brains.

Although genetic specification of cortical precursor neurons sets the framework for neurogenesis, the cerebral cortex establishes connections during development by responding to statistical regularities in external stimuli. Henry Kennedy and Colette Dehay in Chapter 16 hypothesize that this process depends largely on self-organization in the developing cortex. In the same spirit, Cherniak, in Chapter 17, applies combinatorial network optimization theory to questions about the evolution of connectivity in nervous systems and finds that similar wiring minimization governs invertebrate and vertebrate nervous systems, from placement of the entire brain in the body to subcellular organization. Thus, "save wire" may be a generative principle for nervous system organization.

The wiring of nervous systems of higher primates consists mostly of white matter. As described by Hofman in Chapter 18, tensions generated by white matter were probably a driving force in the development of cortical folding patterns, which along with local wiring permit the fitting of a relatively large sheetlike cortex into a compact skull and keep cortical connections short and efficient. At a more specific level, Kate Teffer and Katerina Semendeferi in Chapter 9 show that human brain evolution was characterized by distinct changes in the local circuitry and interconnectivity of the prefrontal cortex (in contrast to other areas), a region that is extremely important for higher cognitive functions. These chapters about connectivity are intriguing because the limit to any intelligent system lies in its ability to process and integrate large amounts of information and to compare signals with as many memory states as possible in a minimum amount of time.

Some chapters focus on other aspects of microstructure in primate brains. Based on a relatively new technique for quantifying the absolute number of neurons and nonneurons in brains, Suzana Herculano-Houzel in Chapter 15 argues that evolutionary changes resulted in more economical scaling in primate brains that permitted larger numbers of neurons relative to brain size compared to other mammals. Significantly, her cellular data suggest that body size might not be relevant for determining species' neurobehavioral performances. Instead, cognitive abilities might be a function of species' total number of neurons, an increasing fraction of which occurs in the cerebral cortex and cerebellum in larger brains. Sherwood et al. in Chapter 11 describe how deviations from scaling predictions are interpreted as strong evidence for evolutionary specializations, one example of which is the frequency of specific cell types in certain parts of the brains of higher primates (e.g., von Economo neurons). Daniel Buxhoeveden in Chapter 10 focuses on minicolumns, which may be the most fundamental functional units contained within the vertical columns that comprise the cerebral cortex. Buxhoeveden shows that the largest minicolumns in primates occur in apes and humans, especially in areas that are important for human cognition, such as the frontal cortex and left auditory association cortex.

Primates have relatively large brains for mammals, and humans have, by far, the largest brain of any primate. This book demonstrates that debates about the selective forces that operated during primate brain size evolution are alive and well. Louis Lefebvre in Chapter 19 employs a meta-analysis that incorporates a large number of investigations to illustrate the various ways of defining and quantifying encephalization in primates. Correlates of increased encephalization vary in the studies he examined and include measures of lifestyle (e.g., group living, diet), cognitive attributes (e.g., social learning, tactical deception), life history (lifespan), brain evolution (e.g., brain size, microcephaly genes), and evolutionary trade-offs (e.g., slower development, higher metabolisms).

In Chapter 4, Charvet and Finlay focus on the social brain hypothesis as emblematic of theories that postulate that brain and isocortex size selectively enlarged to confer a specific behavioral or cognitive trait, and make a strong argument that the mediating variable between brain size and behavioral complexity is the underlying developmental schedule during neurogenesis and brain maturation rather than selection for any particular behavior.

Although most chapters in this volume are about one part of the brain, namely, the cerebral cortex, several authors consider more specific regions. Carol E. MacLeod in Chapter 8 describes current thinking about the evolution of the cerebellum and its influence on higher cognitive functions. The amygdala's role in processing emotional states, especially fear, is the focus of Joseph E. LeDoux's Chapter 21, which also critiques the classic theory about the evolution of the limbic system. Rudolf Nieuwenhuys in Chapter 7 reviews the extensive literature on the insula in a historical perspective and discusses its large number of different functions including pain perception, speech production, and the processing of social emotions.

The contributions by Alexandra de Sousa and Eugénia Cunha; Dean Falk; and Christoph P.E. Zollikofer provide information about the fossil record of hominid brain evolution. De Sousa and Cunha in Chapter 14 note the importance of pathology for evolutionary studies and discuss paleogenetics, fossil analyses related to ontogenetic development (life histories), and the hypothetical brain morphology of different hominin taxa along with descriptions of their social/cognitive/material cultural associations. Falk's Chapter 12 focuses on hominin endocasts and discusses the coevolution of brain size, shape, and convolution (sulcal) patterns in human ancestors. In Chapter 13, Zollikofer offers clear definitions of relevant concepts (e.g., heterochrony, paedomorphism, neoteny) along with a discussion of how computer-assisted paleoanthropology (CAP) is used to noninvasively reconstruct hominin cranial ontogenies (and endocasts).

A number of chapters explore the evolution of neurological substrates that were critical for the emergence of higher cognitive capabilities in humans. Sherwood et al. observe that human cognition appears to be most unique in abilities related to "theory of mind" (mentalizing) and language and theorize that modification of particular cortical areas that are associated with these specializations (e.g., Broca's area and medial prefrontal cortex) provided the neural basis for the emergence of advanced human cognition. Several other chapters reinforce the idea that the emergence of language was pivotal for human cognitive evolution. In Chapter 6, Michael C. Corballis hypothesizes that the predominance of the left cerebral hemisphere for manual and linguistic functions played a special role during human evolution and proposes that language entails neurological circuits that were once specialized for manual grasping.

Noting that many of the cortical structures that are relevant for language increased disproportionately in size during human brain evolution, P. Thomas Schoenemann in Chapter 22 theorizes that the overall increase in brain size and its associated increase in numbers of specialized cortical areas paved the way for language. Roth and Dicke in Chapter 20 place the discussion within a comparative context by examining the various ways intelligence is measured in animal studies and comparing its forms and degrees in primates. They observe that, although the human brain has the highest information processing capacity and intelligence among animals, humans fit the general trends for other primates and mammals. The one exception they note is, again, syntactical language, which they regard as a potent "intelligence amplifier."

Together, these remarkable contributions impart a sense, not only of what is currently known about primate brain evolution but also of where the field is headed. The chapters reveal a discipline that is dynamic, integrative, and on the move. One cannot read this volume without realizing that the field is going to be in a completely different place in 10 years. Each contribution details an emerging foundation for this future transformation. Many readers will be left with a sense of awe about the realized and potential intricacies of neurological evolution in primates including *Homo sapiens*.

We would like to express our sincere gratitude to Dick F. Swaab, Series Editor of *Progress in Brain Research*, who brought up the idea to produce a *PBR* volume on brain evolution. We also very much appreciate the continuous support and excellent guidance of Ben G. Davie from Elsevier's Office in London.

Dean Falk
Michel A. Hofman

Contents

SECTION I

Introduction

M. A. Hofman and D. Falk (Eds.)
Progress in Brain Research, Vol. 195
ISSN: 0079-6123

CHAPTER 1

From tetrapods to primates: Conserved developmental mechanisms in diverging ecological adaptations

Francisco Aboitiz[†,*] and Juan F. Montiel[†,‡]

[†] *Departamento de Psiquiatría, Centro Interdisciplinario de Neurociencias Escuela de Medicina, Pontificia Universidad Católica de Chile, Santiago, Chile*
[‡] *Facultad de Medicina, Centro de Investigación Biomédica, Universidad Diego Portales, Santiago, Chile*

Abstract: Primates are endowed with a brain about twice the size that of a mammal with the same body size, and humans have the largest brain relative to body size of all animals. This increase in brain size may be related to the acquisition of higher cognitive skills that permitted more complex social interactions, the evolution of culture, and the eventual ability to manipulate the environment. Nevertheless, in its internal structure, the primate brain shares a very conserved design with other mammals, being covered by a six-layered neocortex that, although expands disproportionately to other brain components, it does so following relatively well-defined allometric trends. Thus, the most fundamental events generating the basic design of the primate and human brain took place before the appearance of the first primate-like animal. Presumably, the earliest mammals already displayed a brain morphology radically different from that of their ancestors and that of their sister group, the reptiles, being characterized by the presence of an incipient neocortex that underwent an explosive growth in subsequent mammal evolution. In this chapter, we propose an integrative hypothesis for the origin of the mammalian neocortex, by considering the developmental modifications, functional networks, and ecological adaptations involved in the generation of this structure during the cretaceous period. Subsequently, the expansion of the primate brain is proposed to have relied on the amplification of the same, or very similar, developmental mechanisms as those involved in its primary origins, even in different ecological settings.

Keywords: antihem; cortical hem; neocortex; dorsal pallium; olfaction; subventricular zone; ventral pallium; vision.

*Corresponding author.
Tel.: +56-2-354-3808; Fax: +56-2-665-1951
E-mail: faboitiz@uc.cl

DOI: 10.1016/B978-0-444-53860-4.00001-5

Introduction

"He who understands baboon would do more towards metaphysics than Locke."
–Charles *Darwin*, 1838: Notebook M.

Among the most conspicuous characters of primates—and humans—are their large brain and their advanced cognitive and social capacities, which are especially evident in the human species. In fact, their closeness to humans is perhaps the main reason why primates grasp our attention so significantly. As Darwin recalled, primates may provide us with fundamental clues about our own nature. Nonetheless, when we analyze in some detail the primate (and the human) brain, it is surprising to note that many of the fundamental elements of its architecture were established long before the origin of primates. As we will see, many key innovations in brain evolution took place in the previous history when mammals acquired their defining characteristics. Primates inherited from early mammals a highly conserved brain design, which includes a six-layered neocortex that is tightly and reciprocally connected with the dorsal thalamus directly, and indirectly with the basal ganglia and the cerebellum. It also has a strong projection beyond the brainstem that reaches the spinal cord (the corticospinal tract), and the hugest fiber tract found in vertebrates, connecting both cerebral hemispheres (the corpus callosum). Although there is a tremendous variability in the overall sizes of the mammalian brains, the growth of its main components is highly correlated across the different species so that when the brain grows, it tends to do so as a whole (even if some regions tend to grow disproportionately to others; Finlay et al., 1998). Nevertheless, the olfactory system displays some allometric independence of the rest of the brain; this is an important point as primates are notorious among mammals for their reduced olfactory system. In this way, even if primates have a brain about twice as large as other mammals of equal body size, and humans have a much larger brain than that expected for any other primate of its body size, the different brain components and their cellular organization tend to follow highly regular allometric trends in mammals, primates, and even humans. Notwithstanding these general constraints, there is also strong evidence of variability in the size and organization of specific sensory and motor regions in the neocortex of different species, according to particular ecological adaptations (Krubitzer and Kahn, 2003), and there are undoubtedly unique neural adaptations involved in the origin of human speech (Aboitiz et al., 2010).

However, although the mammalian brain can be described according to a specific and regular plan, when comparing mammals with other vertebrates the situation changes dramatically, particularly for the most expanding brain regions. Among other innovations, the origin of vertebrates is marked by the origin of complex brain evaginations like the cerebral hemispheres and the cerebellum, which have expanded independently in each lineage. Starting from an ancestral plan of olfactory-driven cerebral hemispheres and a very rudimentary cerebellum, each vertebrate group acquired a different brain configuration and subsequently tended to maintain it, only making it increasingly elaborate. Thus, each major vertebrate group is characterized by its own brain architecture, which most of the time is highly difficult to compare with other groups (Northcutt, 1981). In this way, what we observe in vertebrate brain evolution is a highly divergent morphological evolution between vertebrate classes, but a relatively conserved developmental process within each class, in which the main trend is (at least in mammals) to increase brain size following relatively well-defined allometric constraints.

What pressures drove this early divergence and the subsequent elaboration of evolutionary trends in each class, and what were the developmental processes underlying these events, are among the most fundamental questions of evolutionary neurobiology. In this context, the main goal of this chapter is to propose that a conserved developmental brain patterning process is shared, at least among amniotes (reptiles, birds, and mammals), and that variations or modulations of specific parameters within this framework may yield important morphological innovations. However, this is not to say that evolution is a developmentally guided process. The modulation of the mentioned developmental parameters is probably a result of natural selection, in which specific ecological circumstances favor distinct developmental solutions in each case. Thus, there is a complex interplay between ecological adaptations and embryological variation that has to be unveiled in order to get a comprehensive view of the evolutionary process. In this chapter, we will concentrate on two main events of brain evolution: the origin of mammals, when the architectural plan of the mammalian brain was established, and the origin of primates, a critical step in the evolution of our kind.

Nonmammalian brains and the problem of homology

The origin of vertebrates is one of the most discussed and transcendental processes in animal evolution. It involved a series of major genetic and developmental innovations like a double duplication of the genome, the differentiation of a neural crest that facilitated the development of complex sensory organs, and a branchial skeleton that allowed a more efficient respiration and new modes of feeding. In cephalochordates (Amphioxus, a basal group close to the vertebrate/invertebrate divergence), this event was preceded by other modifications like the inversion of the dorsoventral polarity, where the central nervous system became localized in the dorsal aspect of the animal as opposed to a ventral localization in invertebrates, and the process of neurulation, which results in a neural tube with a hollow cavity that, when expanding, generates distinct vesicles, as opposed to the basically ganglionar central nervous system that is observed in invertebrates. Despite these morphological innovations, the basic genetic mechanisms underlying brain regionalization and neuronal specification and differentiation are highly conserved in both vertebrates and invertebrates (Aboitiz and Montiel, 2007).

There are two main evaginations in the cerebrum of vertebrates: the cerebral hemispheres and the cerebellum. The paired cerebral hemispheres are already present in all vertebrates, but there is disagreement about whether a cerebellum proper can be distinguished in the more basal vertebrates (agnathans, in which the gill arches do not yet differentiate into jaws). Jawed vertebrates display both paired hemispheres and a well-differentiated cerebellum, which may have enlarged in association with more mobile habits and the opportunities to search for prey.

As mentioned, the morphology of the hemispheres is highly divergent in all vertebrate groups (see Butler and Hodos, 1996). Among agnathans, the cerebral hemispheres are dominated by extensive olfactory projections, but in jawed vertebrates, thalamic sensory inputs colonize large aspects of these structures. Concomitant with this, there is a tendency to increase in size of specific brain components in each group. In cartilaginous fishes, a central nucleus develops in the medioventral aspect of the hemispheres, while in bony fishes, the cavity of the hemispheres bulges to the outside, in a process called eversion that dramatically distorts the embryonic topography. Finally, among terrestrial vertebrates (tetrapods), amphibians (and the phylogenetically close lungfishes) have quite simplified cerebral vesicles, with little signs of expansion or neuronal migration. Amniotes, on the other hand, are characterized by a conspicuous process of brain expansion, resulting in two main patterns of cerebral organization (Medina and Abellán, 2009; Striedter, 2005) (see Fig. 1). On one hand, reptiles and birds (sauropsids) display a growing structure in the ventrolateral cerebral hemisphere (the dorsal ventricular ridge, DVR) that grows into the cerebral ventricle.

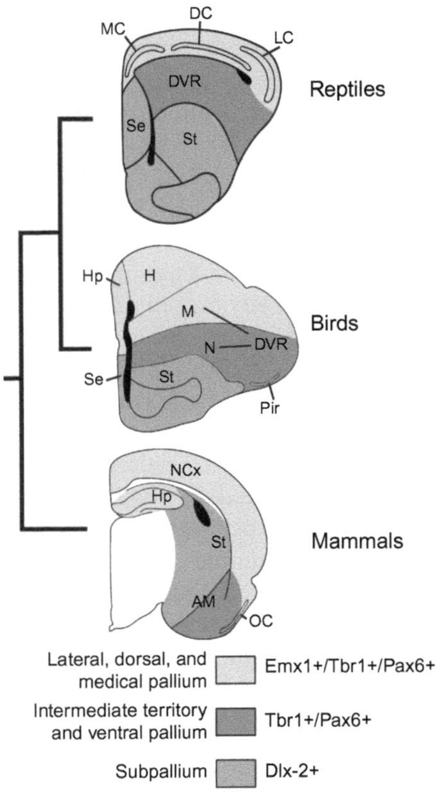

Fig. 1. Brain diversification in amniotes. The cerebral hemispheres can be divided into a pallium dorsally located (cyan and purple) and a ventral subpallium, corresponding to the basal ganglia and other nuclei (brown). The pallium is itself subdivided into a medial/dorsal/lateral pallium (cyan) and a ventral pallium (purple). The ventral pallium gives rise to the dorsal ventricular ridge of reptiles, the nidopallium of birds, and to parts of the amygdalar complex and related structures in mammals. The medial, dorsal, and lateral pallia express the markers Emx1, Tbr1, and Pax6, while the ventral pallium expresses Tbr1 and Pax6 but not Emx1. The subpallium expresses Dlx genes. DC, dorsal cortex (reptiles); DVR, dorsal ventricular ridge (reptiles); H, hyperpallium (birds); Hp, hippocampus; LC, lateral (olfactory) cortex (reptiles); M, mesopallium (birds); MC, medial cortex (hippocampus, reptiles); N, nidopallium (birds); NC, neocortex (mammals); OC, olfactory cortex (mammals); Se, septum; ST, striatum. Modified from Medina and Abellán (2009) with permission. (For interpretation of the references to color in this figure legend, the reader is referred to the Web version of this chapter.)

Sauropsids also develop a rudimentary three-layered cortex in the dorsal hemisphere, with limited radial migration. On the other hand, mammals show a distinct six-layered neocortex characterized by extensive radial and tangential neuronal migrations. In reptiles, and especially in birds, the lateroventral DVR acquires prominence and capitalizes much of the thalamic input, particularly the sensory pathways that relay in the mesencephalon before reaching the thalamus (collothalamic pathways). However, in mammals, the neocortex receives most thalamic sensory inputs and rapidly expands as a sheet that in species with large brains like primates becomes highly convoluted.

In these conditions, comparing brains of different vertebrate groups is not an easy task. From the earliest days of comparative neuroanatomy, the establishment of brain homologies across vertebrates has been a matter of intense debates. One particularly conflictive point has been to determine which structure is homologous to the mammalian neocortex in other species, especially in reptiles and birds. As said, in sauropsids, the brain structure that expands most and receives the main sensory input is the DVR (more specifically, its anterior component, the ADVR), which has been interpreted by different authors to correspond to parts of the mammalian striatum, amygdala, or neocortex (for reviews, see Aboitiz and Montiel, 2007; Striedter, 2005). While it is presently acknowledged by most present-day comparative neuroscientists that the ADVR is a pallial component (i.e., it does not correspond to the corpus striatum), there have been strong disagreements on whether it fits the amygdalar region or parts of the neocortex of mammals. Connectional evidence points to a strong similarity between the sensory inputs to the ADVR and those to the lateral neocortex (the auditory cortex and the extrastriate visual areas, both receiving collothalamic input; Butler and Hodos, 1996; Kartén, 1969), while developmental evidence has been unequivocal in establishing that the DVR arises from an embryonic region termed the ventral pallium, that in mammals gives rise to parts of the amygdala and neighboring structures (Aboitiz and Montiel, 2007; Medina and Abellán, 2009; Puelles et al., 1999; Smith Fernández et al., 1998; Striedter, 2005) (Fig. 1).

Attempting to escape from this apparently endless controversy, one of us recently proposed a hypothesis that may point to conciliate some aspects of these different views (Aboitiz, 2011). While developmental and genetic evidence strongly supports the concept that the neocortex originates in the dorsal aspect of the hemisphere (dorsal pallium) and the ADVR originates in the lateral hemisphere (ventral pallium), the genetic and cellular processes involved in the expansion of both structures may be partly comparable. However, before getting into more details on this proposal, it will be important to briefly review some basic aspects of mammalian neocortical development. We will first summarize the main events involved in neocortical neurogenesis and then will describe the patterning mechanisms by which the identity of different components of the cerebral hemispheres becomes specified.

Neocortical development: The basics

As mentioned, the neocortex of mammals is a laminated structure consisting of six layers (five cellular layers and a superficial marginal zone, bearing horizontally oriented axons, apical dendritic shafts, and scarce neurons). In the neocortex, there are two main types of neurons: excitatory pyramidal and spiny stellate cells on one hand and inhibitory interneurons on the other. Neurons in the neocortex originate in the deep ventricular surface (the ventricular zone, VZ), arising from asymmetric divisions of the radial glia, a cell type that works both as the primary progenitor of neurons and as a scaffolding for radial neuronal migration (i.e., from the ventricular surface to the pial surface). Excitatory neocortical neurons are produced in the VZ of the dorsal pallium and migrate largely following the radial glia processes (although there is an important proportion of tangential migrations), to make up the different cortical layers observed in the adult. However, neocortical inhibitory interneurons are generated in the ventral hemisphere (more specifically, the medial and caudal ganglionic eminences of the subpallium) and migrate tangentially (obliquely to the cerebral surface) to populate the developing neocortex. Tangential migration of subpallial interneurons into the pallium has been confirmed in sauropsids and amphibians (Métin et al., 2007) and was probably acquired in the first jawed vertebrates, as the lamprey was shown to lack a medial ganglionic eminence (MGE) and the genetic markers for migrating interneurons (Sugahara et al., 2011). Nonetheless, a recent study indicates that subpallial interneurons of reptiles and birds, when transplanted into the MGE of mouse embryos, are able to migrate to the pallium but do not penetrate the developing cortex as transplanted mammalian neurons do (Tanaka et al., 2011). This points to a new signaling mechanism that allows the incorporation of interneurons into the mammalian neocortex.

In the VZ of the cortical neuroepithelium, radial glia first produce excitatory neurons, many of which migrate radially to make up the embryonic preplate and the deepest cortical layers of the adult (Fig. 2). Later in development, divisions of the same radial glia produce cells called intermediate progenitors, which detach from the ventricular surface and aggregate in a zone overlying the VZ, the subventricular zone (SVZ). In the SVZ, cells undergo one to three more cell divisions and then migrate to make up the superficial layers of the adult neocortex. Thus, early generated neurons contribute to the deep cortical layers, while neurons generated in successively later moments are incorporated into progressively more superficial layers, generating the inside-out neurogenetic gradient that is characteristic of the neocortex. Note that although this

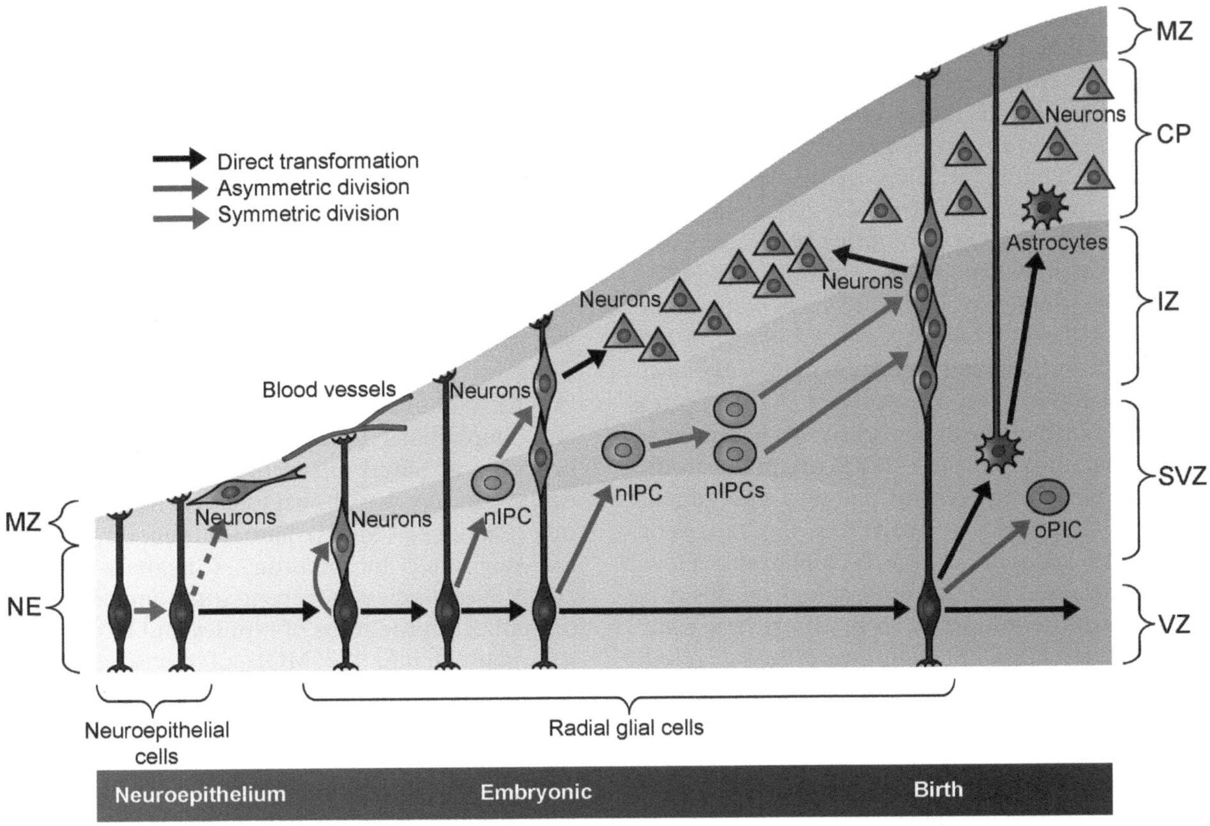

Fig. 2. Histogenesis of the cerebral cortex. After dividing symmetrically, neuroepithelial cells differentiate as radial glia cells that produce asymmetric divisions where one daughter cell is a neuron (in early stages), or a neuronal intermediate progenitor cell (nIPC), which preferentially locates in the subventricular zone (SVZ), keeps dividing, and then differentiates and migrates to make up the cortical plate (CP), which will become the adult neocortex. At later stages, radial glia may differentiate into astrocytes or generate an oligodendrocyte precursor cell (oIPC). MZ, marginal zone; NE, neuroepithelium; IZ, intermediate zone; VZ, ventricular zone. Modified from Kriegstein and Alvarez-Buylla (2009) with permission.

description serves as a general guide, there are observations of intermediate progenitors dividing in the VZ or in sites other than the VZ, contributing neurons that do not necessarily make up the superficial layers. According to recent models of neocortical growth, early tangential expansion of the neocortex is based primarily on the divisions of primary progenitors, which enlarge the surface of the VZ. However, late tangential growth and radial thickening (generation of superficial layers) of the neocortex depend mainly on the proliferation of intermediate progenitors (Pontious et al., 2008) and other glial-like neurons located in the SVZ (Reillo et al., 2010; Wang et al., 2011). In line with this interpretation, a recent hypothesis suggests that the SVZ results from a spatial constraint when there is a high rate of progenitor division and the VZ becomes unable to contain all the dividing progenitors, resulting in some of them migrating into the more superficial SVZ (Striedter and Charvet, 2009).

Interestingly, while a VZ has been described in all vertebrates that have been studied, the SVZ appears only in some species. Among mammals, the SVZ extends from the lateroventral aspect of the hemisphere to the dorsal pallium. Across species, the growth of the SVZ appears to correlate with the development of the superficial neocortical layers, being especially complex in primates and minimal in marsupials (see below) (Cheung et al., 2010). A SVZ has also been reported in the developing brains of crocodiles and birds (both comprising the taxon archosauria) but apparently is not present in other reptiles or in amphibians (Charvet et al., 2009; Cheung et al., 2007).

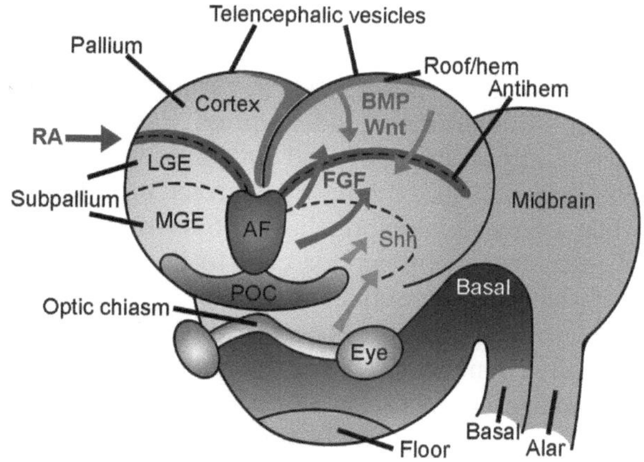

Fig. 3. Signaling centers in the embryonic brain. The telencephalic vesicles or cerebral hemispheres are patterned by the combined action of different signaling centers like the anterior forebrain (AF, violet) secreting FGFs, the dorsal hem (red), secreting Wnts and BMPs, and the antihem (green), which specifies the ventral pallium. Other signaling elements are retinoic acid (RA) laterally and sonic hedgehog (Shh) ventrally. LGE, lateral ganglionic eminence; MGE, medial ganglionic eminence; POC, commissural preoptic area. Modified from Medina and Abellán (2009) and Sur and Rubenstein (2005) with permission. (For interpretation of the references to color in this figure legend, the reader is referred to the Web version of this chapter.)

Cortical patterning

Underlying the neurogenetic process described above, there is a regionalization process in which the cortical neuroepithelium acquires its identity on the basis of the expression of regulatory genes that control the process of differentiation, yielding its characteristic adult phenotype. Molecular evidence indicates that the embryonic cerebral hemispheres are patterned according to several signaling centers from which morphogens are produced and expressed in gradients in different directions (Medina and Abellán, 2009; O'Leary and Sahara, 2008; Sur and Rubenstein, 2005) (Fig. 3). According to these studies, the combination of several of these gradients has been proposed to regulate the differentiation and cell proliferation in specific brain components. Thus, modulation of such gradients may yield important changes in brain development, expanding some regions and reducing others.

Two dorsal signaling centers are the cortical hem and the anterior telencephalon. The cortical hem is located in the dorsomedial hemisphere and expresses signaling molecules like Gli3, bone morphogenetic proteins, and Wnts, which are required for the differentiation of the medial and the dorsal pallium (hippocampus and neocortex, respectively) and are expressed in a posteromedial-high to anterolateral-low gradient (Subramanian and Tole, 2009). The anterior telencephalon is the source of several types of fibroblast growth factors (FGFs) that are present in a rostral-high to caudal-low gradient that overlaps the opposing gradients of other dorsal morphogens (Shimogori et al., 2004).

Of particular interest in this context is the gene *Pax6*, which is expressed in an anterolateral-high to posterodorsal-low gradient in the developing pallium that is complementary to the dorsal gradients described above (it is also expressed in lower levels in the lateral ganglionic eminence of the subpallium). Pax6 participates in the differentiation of both ventral pallial and dorsal pallial structures (Cocas et al., 2011; Stoykova et al., 2000). In the dorsal pallium, Pax6 patterns the lateral neocortex and is required for the generation of the superficial neocortical layers (but it also affects deep layers), by promoting the generation of intermediate progenitor cells that locate transiently in the SVZ of the cerebral vesicle and migrate superficially (O'Leary and Sahara, 2008; Tarabykin et al., 2001). Animals that underexpress Pax6 display early neurogenesis, reduced progenitor proliferation, and increased inhibitory neuronal production, while animals overexpressing Pax6 have an excess of intermediate progenitor cells to become excitatory neurons (Sansom et al., 2009).

As the superficial neocortical layers represent the phylogenetically newest component of the mammalian neocortex (Aboitiz and Montiel, 2007; Aboitiz et al., 2003), we originally proposed that the upregulation of Pax6 in early mammaliaforms was a key event in the generation of both the SVZ in the developing pallium and the superficial layers of the neocortex, a proposal that has been confirmed in several studies (Aboitiz et al., 2003; Cheung et al., 2010; Georgala et al., 2011; Tuoc et al., 2009). In the ventral pallium, Pax6 promotes the differentiation of the antihem, a signaling center producing secreted frizzled-related proteins that neutralize the action of dorsally derived signals like Wnts, and, together with other molecules like epidermal growth factors and FGFs, contributes to the differentiation of ventral pallial phenotypes (Assimacopoulos et al., 2003). Consistent with these descriptions, in the *Pax6−/−* mutant mouse, the ventral and dorsal pallial areas are highly dysgenic and express subpallial markers (Cocas et al., 2011; Quinn et al., 2007; Stoykova et al., 2000; Tole et al., 2005). The Pax6 mutant develops an accumulation of cells in the lateral hemisphere that protrudes into the ventricle, which to some authors is reminiscent of the DVR of reptiles (Molnár, 2011). However, it is not clear that this structure derives from the ventral pallium (Quinn et al., 2007).

Therefore, the dorsal hemisphere, and particularly the cerebral cortex, is patterned by at least three complementary signaling centers that generate overlapping but opposite gradients, contributing to regionalize the roof of the brain. There is an opposite action between Pax6 activity and the gene *Emx2*, a downstream target of signals emanating from the cortical hem, in the differentiation of neocortical regions. Mice deficient in Pax6 develop an expanded visual cortex posteriorly, concomitant with a reduction of frontal areas, while in *Emx2* mutants, the phenotype is opposite to this, with enlarged frontal areas and shrinkened visual regions (O'Leary and Sahara, 2008). On the other hand, FGF8, expressed by the anterior forebrain, is regulated by *Emx2* expression and is required for the differentiation of frontal cortical areas (Shimogori et al., 2004).

Toward a unifying hypothesis of amniote brain evolution

In light of the above evidence, the expansion of the dorsal pallium (and the origin of the neocortex) in mammals was proposed to result from the combined upregulation of the dorsal signaling factors (cortical hem and anterior forebrain) and *Pax6* expression from the lateral hemisphere (Aboitiz, 2011; Aboitiz and Montiel, 2007) (Fig. 4). Nevertheless, despite high levels of *Pax6* expression, which influence the development of both ventral pallial and dorsal pallial structures, the antihem (and the ventral pallium) remains limited in size in mammals. This is proposed to result from the increased expression of dorsal-derived factors from the cortical hem and the anterior forebrain, which antagonize lateral pallial signals and constrain the amplification of the antihem (Aboitiz, 2011). In this line, the ventral pallial neuroepithelium has been described to become highly compressed between the Emx-1-positive (dorsal pallium) and the Dlx-positive zones (subpallium) during late mammalian development (Smith Fernández et al., 1998).

Considering that *Pax6* has a similar expression pattern in all tetrapods, it was also suggested that this gene is a likely candidate for promoting the expansion of the DVR of sauropsids, by upregulating the antihem (Aboitiz, 2011). However, unlike in mammals, in reptiles the proposed amplification of Pax6 would not be accompanied

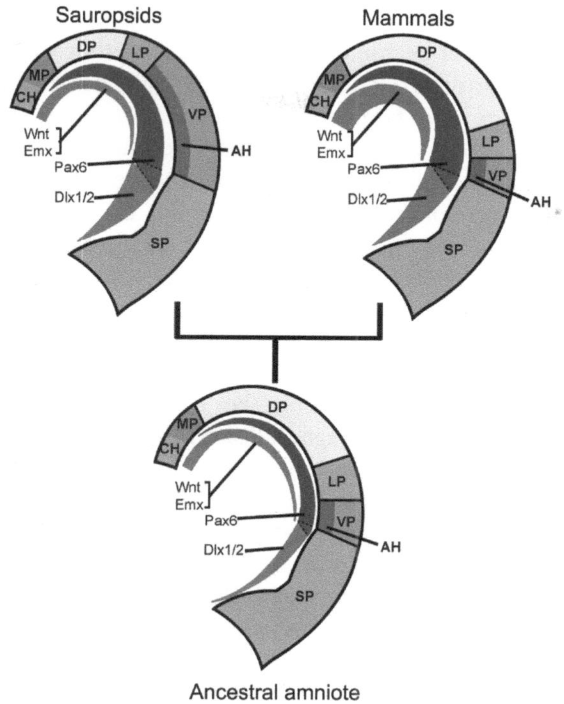

Fig. 4. Proposed scenario for pallial evolution in amniotes. In the ancestral amniote, there were complementary gradients expressing dorsal factors like Wnts and Emxs from the cortical hem (CH), and ventrolateral factors like Pax6, which not only specified the antihem (AH) and the ventral pallium (VP) but also affected to some degree the dorsal pallium (DP). In the origin of sauropsids, there was an upregulation of ventral factors (e.g., Pax6), which resulted in expansion of the antihem and the ventral pallium, while in mammals, there was a concomitant expansion of ventral signals and dorsal signals from the cortical hem. The latter prevented the expansion of the antihem but allowed Pax6 to contribute to late neurogenesis in the cortical neuroepithelium. MP, medial pallium; LP, lateral pallium; SP, subpallium. See also Aboitiz (2011).

by an upregulation of the dorsal cortical hem and its signaling molecules (see Abellán et al., 2010), thus liberating the ventral pallium from the restrictive effect of dorsal signals. Consistent with this view, there is a reduced progenitor pool in the septum and medial/dorsal pallium of birds compared to mammals (Charvet, 2010), suggesting a weaker influence of dorsal telencephalic signals in sauropsids. In these conditions, the dorsal hemisphere (medial and dorsal cortex) is relatively more simple in reptiles than in mammals, and conversely, the ventral pallium grows to a larger relative size in reptiles and birds than in mammals. This hypothesis also specifies that in both mammals and birds, there will be phenotypic similarities between neurons deriving from the ventral pallium and the dorsal pallium, as both structures depend importantly on Pax6 signaling for their development (Aboitiz, 2011).

The expansion of the dorsal pallium in mammals may have taken place either by a ventral shift of its lateral boundaries at the expense of ventral pallial territory, or by an expansion of its intrinsic proliferative activity, or perhaps more likely by the combination of both factors. In any case, the brain of ancestral amniotes was small and highly undifferentiated so that a pure boundary shift may not account for the expansion of the mammalian neocortex. This proposal is consistent with the tangential incorporation of transient embryonic neurons from different pallial regions into the developing neocortex (Puelles, 2011). In this line, transient ventral pallial neurons that reach the neocortex (Teissier et al., 2010) might have contributed to drag the collothalamic sensory afferents that in sauropsids end in the ventral pallium, into the lateral neocortex of mammals (Aboitiz, 2011).

Thus, the brain of mammals is characterized by a highly differentiated cortical hem, producing factors like Wnts that induce early neural proliferation and radial organization, and a concomitant and delayed expression of Pax6, which promotes the late division of neuronal precursors and contributes to the radial expansion of the neocortex. On the other hand, in sauropsids, the cortical hem remains more conservative while there is an expansion of the antihem, possibly due to an upregulation of *Pax6* expression resulting in the differentiation of a DVR in the ventral pallium and a rudimentary cortex in the dorsal pallium. The upregulation of Pax6 in reptiles is perhaps not as high as in mammals, as the dorsal pallium does not undergo significant radial growth;

however, in birds, and especially parrots, the hyperpallium (dorsal cortex) may become an important anatomic structure, perhaps due to an increasing influence of Pax6 and dorsal morphogens. Note that recent reports are suggestive of some radial and laminar organization in the bird's nidopallium (homologue to the reptilian ADVR) (Wang et al., 2010), although it clearly lacks the more obvious feature of the mammalian neocortex, which is the pyramidal neuron with radially oriented dendrites (however, exogenous reelin is capable to induce a strong radial organization in the avian DVR; Nomura et al., 2008). Further, it is not clear whether these characters are ancestral to amniotes as they have not been described in reptiles.

The ecological context and the elaboration of cortical networks

As mentioned above, evolutionary explanations have two sides that must be integrated: one are the intrinsic mechanisms involved in the morphological transformations and the other are the ecological circumstances that generated the selection of these traits. In this perspective, together with a developmental or genetic hypothesis, one ideally has to provide an explanation of the behavioral context in which these structures were acquired.

Amniotes are characterized by an amniotic cavity covering the egg, which makes it possible to lay the egg on ground instead of depending on water for reproduction. This innovation resulted in a successful colonization of the land by two lineages that split very early in history. One were the stem reptiles (ancestors to all modern reptiles and birds), and the other were the synapsid reptiles that eventually gave rise to mammals. Cynodonts, a group of late synapsids, were small animals with a mammal-like body that gave rise to true mammals, characterized mainly by the presence of the ear ossicles detached from the articulation of the lower jaw. However, this diagnosis is biased in the sense that bones are the only body part that is well preserved in fossils, but mammals acquired many other characters not observed in other vertebrate groups, like hair, mammary glands, a diaphragm, and homeothermy (the latter is shared with birds). Other important skeletal changes included a different locomotion, the development of a secondary palate that separates the oral and the nasal cavities (this character has also developed independently in crocodiles and some lizards), and the appearance of turbinate bones in the nasal cavity. The secondary palate and the turbinate bones, together with a more efficient respiration produced by the new gait and the origin of the diaphragm, allowed the maintenance of moist air in the nose and the possibility of exploiting the sense of olfaction, which was reduced in the ancestral terrestrial vertebrates (Kielan-Jaworowska et al., 2004).

The standard picture of early mammal evolution is that mammals were small, nocturnal animals that lived hiding from the large dominant reptiles. Although the evidence is consistent with a predominantly nocturnal way of life (Heesy and Hall, 2010), recent data indicate that mesozoic mammals were a diversified group with different ecological specializations, that were immersed in a complex ecosystem including early birds, the emerging flowering (scenting) plants, and insects (Zhou et al., 2003). Within this context, senses like olfaction and hearing, which were very well developed in early mammals, may have become critical, for orienting and recognition, and for detecting small prey like insects, predators, or conspecifics.

According to a recent study, brain expansion is associated to increased dispersal and invasive behavior of new territories in all terrestrial vertebrates (Amiel et al., 2011), which suggests that this character was an important element in colonizing different habitats. Fossil evidence indicates that, in the lineage leading to mammals, brain expansion was a late event, more or less coincident with the acquisition of modern mammalian characters. The early mammal-like reptile *Probainognathus* had slender, elongated cerebral hemispheres bearing a small dorsal slope that has been interpreted to represent a forerunner of the neocortex (Quiroga, 1980). However, hemispheric expansion is observed much later in fossils like *Triconodon*, who has a well-differentiated middle ear (see Kielan-Jaworowska et al., 2004). More recent fossil evidence indicates that in early mammals, olfactory enhancement was strongly associated to successive increases in brain size, from *Morganucodon* to *Hadrocodium* to crown mammalia, with the development of olfactory turbinals in the latter (Rowe et al., 2011).

In amphibians and reptiles, there are limited olfactory projections into the hippocampus (Lynch, 1986), which possibly participates in learning (particularly spatial learning). On the other hand, the brain of so-called primitive mammals is strongly dominated by olfaction, which is apparently an essential component of behavior in these animals. In rodents, the hippocampus contains a map receiving olfactory input together with inputs from other sensory modalities that contributes to generate a multimodal representation of space for episodic and spatial memory (Ergorul and Eichenbaum, 2004). In addition, the olfactory cortex provides an important projection into the dorsomedial nucleus of the thalamus, which is connected to the frontal cortex and participates in planning and decision making (Staubli et al., 1987).

Others and we have previously proposed that these olfactory circuits, including the hippocampus and the dorsal pallium (the nascent neocortex), were put to use by the first mammals to make predominantly olfactory-based maps of space, in which odors not only served as labels to routes and places but were also used to detect predators and conspecifics (Aboitiz and Montiel, 2007; Aboitiz et al., 2003; Lynch, 1986). Within this framework, sensory projections that in reptiles are directed to the DVR (ventral pallium) became included into this associative network by virtue of the growth of the dorsal pallium. Further, these incoming projections may have also contributed to pallial expansion as well.

The mammalian olfactory cortex and hippocampus, and all cortical structures of reptiles differ from the neocortex in displaying a tangential organization of their inputs, which are superficially arranged in the neuron-free layer I and contact in series several apical dendrites in their path. In mammals and reptiles, incoming axons into these structures follow a route specified by early, tangentially migrating, pioneer neurons that locate superficially (Aboitiz and Montiel, 2007). In this way, olfaction imprinted its signature into the organization of neocortical circuits, by favoring the tangential organization of inputs into laminar structures where instead of a point-to point topography of connections, inputs are delivered into many distributed neurons that serve as coincidence detectors (Lynch, 1986; Shepherd, 2011). This arrangement enables the generation of multiple combinations of inputs in scattered populations of neurons. In the case of olfaction, this is particularly beneficial when it comes to distinguish a wide variety of odorants and combinations between them.

We have previously suggested that the inside-out neurogenetic gradient that is also characteristic of the neocortex (i.e., deep layers are generated first and more superficial layers are produced in successively later periods) was part of a developmental strategy for establishing synaptic contacts between the late-produced neurons and the superficial axons, as the cortex began to grow radially (in thickness) (Aboitiz and Montiel, 2007; Aboitiz et al., 2003) (Fig. 5). This is the case in the mammalian hippocampus, with a clear inside-out gradient and a superficial array of afferent projections. Perhaps due to the tangential expansion of the neocortex (in surface), axons developed a shorter, alternative route, traveling through the subcortical intermediate zone (future white matter) and entering the neocortex radially, as is observed in primates and other mammals with well-developed neocortices. In this process, pioneer neurons that guide the incoming axons began to migrate below the cortex, perhaps concomitantly with the elaboration of the cortical subplate (SPl, a transient

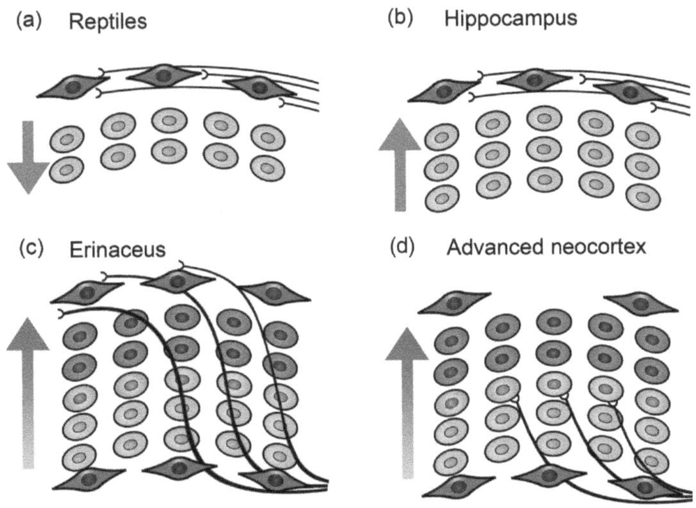

Fig. 5. Proposed model for the origin of cortical lamination. In reptiles (a), the neurogenetic gradient is outside-in (deep layers are generated after superficial layers), and afferents run in the superficial marginal zone, with many pioneer neurons (red, dark gray) running superficially. The hippocampal cortex (b) has an increased radial development and the afferents are still located superficially. In a primitive mammal (*Erinaceus*, c), afferents run below the cortex following pioneer neurons located in the subplate but penetrate obliquely to it and then run tangentially in the superficial layer for some distance. Finally, in species with a more differentiated neocortex (d), afferents enter radially and terminate in the mid-layers of the neocortex. Modified from Aboitiz and Montiel (2007). (For interpretation of the references to color in this figure legend, the reader is referred to the Web version of this chapter.)

embryonic layer located below the cortical plate, that largely disappears in late development; see below). An intermediate condition is found in mammals with small cortices like the hedgehog, where axons enter the neocortex from below but many of them travel obliquely to reach the superficial layer and grow tangentially there for some distance (Aboitiz et al., 2003).

Expansion of the neocortex in mammal evolution

Summarizing the above proposal, both the mammalian and the avian brains originated at least partly as a consequence of the differential modulation of conserved signaling centers in the early vertebrate telencephalon. The modulation of specific centers may differ in each case, but the underlying patterning mechanism is proposed to be conserved. Further, we suggest that there were adaptations to specific ecological circumstances that contributed to the laminar development of the mammalian neocortex. The layout of the primate brain (and that of other mammals) was established at this point; from then on, variations occurred within a constrained developmental program, consisting in a large part of size increases of the whole neocortex or parts of it.

Once the neocortex arose in early mammals, it maintained its fundamental six-layered architecture but expanded greatly in tangential size in some orders, leading to highly convoluted brains. While this expansion is concomitant with the growth of many other brain parts, it grows much more rapidly than them, dwarfing the sizes of other structures in large-brained mammals. This process of expansion was associated with the multiplication of new cortical areas, from a primary and secondary visual area, an auditory area, a primary and supplementary somatosensory area, a motor area, and perhaps a few others in the most likely common ancestor, to more than 50 in the macaque and possibly more in the human (Krubitzer and Kahn, 2003; Striedter, 2005). A large portion of these areas receives unimodal sensory inputs, but there are many of them receiving multimodal input as well.

Although the main process associated to cortical expansion has been the addition of new areas instead of the expansion of preexisting areas (Changizi and Shimojo, 2005), there are well-documented instances of increases or decreases in the size of specific sensory and motor areas in relation to behavioral specializations (Krubitzer and Kahn, 2003), like the representation of the "beak" of the platypus or that of the nose in the star-nosed mole. Further, the expansion of a given cortical area is likely concomitant with the segregation of heterogeneous inputs within each area, leading to functional microspecializations as those observed in the bat's auditory cortex. In this line, Krubitzer and Kahn (2003) have proposed that this process of segregation may eventually lead to the generation of new cortical areas. For example, while in primates there is a well-defined somatosensory area 2, separated from area 1, in the flying fox, the homologous to the former may be inserted as patches within area 1.

What are the developmental processes involved in the increase and eventual separation of cortical areas? It is likely that this results from the interaction of two factors, which were initially proposed as alternative mechanisms, but evidence has favored the coexistence of both of them. One is that the growth and the multiplication of new areas results from developmental modulations of the morphogenetic gradients generated by the signaling sources that pattern the neocortex (the cortical hem, the anterior forebrain, and the antihem). As mentioned above, a reduction of "anterior-dorsal" signals like FGF and Emx results in the expansion of frontal and somatosensory areas at the expense of visual and auditory areas, while a reduction in Pax6 signaling generates an opposite phenotype with an expansion of visual areas with a reduction of frontal regions (O'leary and Sahara, 2008). Further, if FGF8 (an anteriorizing signal) is artificially placed in the occipital cortex of a normal developing mouse, it develops a second somatosensory representation of the vibrissae, caudal to the original, oriented in a mirror-image manner with respect to the normal area (Shimogori et al., 2004). Thus, the generation of new signaling sources might contribute to the generation of new areas, and interestingly, to the generation of the mirror-image arrangement that is observed in several contiguous sensory areas of the neocortex.

The second possible developmental mechanism considers the role of neuronal inputs to different parts of the brain as determinants of the identity and size of specific cortical areas. For example, bilateral eye enucleation in rats during early development results in a profound disarrangement of the visual cortex, including a dramatic reduction in size that is taken over by somatosensory regions (Kahn and Krubitzer, 2002). In the same line, mice breeds that develop extra whiskers display an increased number of barrels (each barrel represents a whisker) in their somatosensory cortex (Welker and Van der Loos, 1986). The interaction with thalamic input is both ways, as early lesions in the neocortex affect the development of specific thalamic nuclei (Huffman et al., 1999) and mutants of *Emx2* (a gene not expressed in thalamic nuclei) display anomalous corticothalamic projections (Bishop et al., 2003).

However, genetic alterations in thalamic development seem not to distort significantly the patterning process of the neocortex. For example, mutants of *Gbx2* generate a profoundly distorted thalamus, but the expression and distribution of several genetic markers of neocortical regions are spared (Miyashita-Lin et al., 1999). Thus, it appears that genetic patterning processes establish a spatial blueprint or protomap in the developing neocortex, in which different kinds of afferences will prefer some regions over others. Subsequently, the normal competition between these inputs establishes the typical adult distribution of cortical areas. Altering either the early patterning process or the process of synaptic stabilization during development may result in changes in the adult cortical map.

New tracts in the mammalian neocortex

Associated with the increase in size of the neocortex, there is a massive increase in the numbers of axons projecting into and out of the neocortex from subcortical regions like the thalamus, corpus striatum, and brainstem, particularly the pontine nuclei that participate in a corticocerebellar loop. This extended connectivity resulted in the acquisition of two major tracts that are not observed in other vertebrates: one is the corticospinal tract and the other is the extensive corticocortical connectivity that makes up the subcortical white matter. Within the latter outstands the corpus callosum connecting both cerebral hemispheres.

The corticospinal tract connects different areas of the neocortex (especially the motor cortex but other areas too) with motor neurons in the spinal cord. The length of this tract along the spinal cord has been correlated with manual dexterity, being especially long in primates. However, the laminar penetration of corticospinal axons into the ventral horn of the medulla has no effect in dexterity (Iwaniuk et al., 1999). For example, the fossorial armadillo displays very little dexterity, with a particularly short corticospinal tract but a high laminar penetration, which suggests that the laminar penetration of the tract may be related to other functional abilities like strength, required for digging in this species.

The other major neocortical projection connects both cerebral hemispheres, a character that is also unique to mammals. In monotremes and marsupials, interhemispheric fibers use the anterior commissure to cross the midline. In small-brained marsupials, these fibers take a convoluted pathway around subcortical nuclei to reach the anterior commissure, but larger-brained species have developed a shortcut to the commissure via the fasciculus aberrans, that penetrates the internal capsule (Shang et al., 1997). The corpus callosum of placental mammals can be seen as a further step in this trend to minimize the axonal distance between hemispheres.

In early mammalian development, a glial wedge is formed over the hippocampal commissure, establishing a mechanical bridge between the hemispheres and allowing callosal axons to cross the midline. Presumably, at some stage, early placental mammals acquired this wedge, making it possible to develop a shortcut for these interhemispheric axons. However, the molecular mechanisms involved in midline crossing are based on highly conserved signaling processes, dependent on the genes *Robo* and *Slit*, which perform the same function in other decussations, in both vertebrates and invertebrates (Shu and Richards, 2001).

Others and we have speculated that an early function of interhemispheric connections was to establish continuity between the two sensory hemirepresentations in each hemisphere in early visual and somatosensory areas, each containing a map of the contralateral hemifield (Aboitiz and Montiel, 2003). Analyzing fiber composition in the posterior callosum across several mammal species, we found that the most common fiber diameter (representing the bulk of callosal fibers) was relatively constant despite large differences in brain size, indicating that in large brains, there is a cost in time for interhemispheric transmission. Nevertheless, in larger-brained species, the large-diameter fibers tend to increase their diameter in correlation with interhemispheric distance, skewing the distribution curve of axonal sizes to the right (Olivares et al., 2001). Interestingly, many large-diameter fibers tend to be located in callosal regions connecting sensory areas, suggesting that these participate primarily in transfer of sensory information. Subsequently, Caminiti et al. (2009) largely confirmed our findings in projections of the motor cortex of primates including man.

Primates arrive

Primates belong to a relatively ancient lineage of placental mammals, whose oldest fossil representative, *Purgatorius* (Plesiadapiformes), dates from the late Cretaceous or early Paleocene and was contemporaneous with the latest dinosaurs. These were arboreal animals similar to a tree shrew, in which many of the defining characters of primates were making their appearance. Anthropoids, or modern primates (old world and new world monkeys, apes, humans, and the nocturnal tarsier), are characterized by the presence of two fundamental adaptations: grasping extremities with opposable thumbs and nails instead of claws, and frontal vision, with a robust visual system and a large brain. Other typical, but not usually acknowledged, characteristics of primates are the reduced olfactory system, and the absence of a vomeronasal system in Old World Monkeys (features that are also observed in birds) (Williams et al., 2010).

One hypothesis to explain frontal vision in modern primates is that these were initially nocturnal animals who benefited from the convergence of the two visual fields by optimizing the perception of detail, a similar strategy as that used by owls. In addition, primates display an eye optically designed for high acuity vision. This hypothesis is partly supported by paleontological evidence suggesting that stem anthropoids were small, nocturnal insectivore–frugivore animals. However, a transition into diurnality may have taken place very early in anthropoid evolution (Williams et al., 2010).

When primates colonized diurnal niches, frontal vision became useful to measure depth and to move in the tridimensional canopy, and to manipulate objects or food with their hands. Trichromatic color vision was then reacquired in modern primates (monkeys) after early mammals had lost it. This was possibly an adaptation to take advantage of the spectral information provided by angiosperms, not only in their flowers but more importantly in the fruits many of them fed upon (Vorobyev, 2004). The acquisition of color vision occurred separately in Old and New World monkeys, as their respective chromatic capacities are based on different genetic modifications (Hunt et al., 1998).

Increase in brain size

As mentioned, the brain of anthropoids, particularly the neocortex, is largely dominated by their visual system, which is accompanied by a brain size that in average roughly doubles that of a similar sized nonprimate mammal. This may be partly a consequence of visual development, which is also related to the complex social life of these animals, in which face and emotion perception plays a fundamental role (Barton, 2004; Dobson and Sherwood, 2011; Shultz and Dunbar, 2010).

Most of this book will be discussing the different aspects of primate brain evolution, and we need not to delve into a deep discussion on this point. However, it may be appropriate to mention that processes similar as those involved the origin of the mammalian neocortex may have been at work in the evolution of the large primate brain. For example, the supragranular layers of the primate neocortex display a notorious expansion, which is paralleled by a concomitant development of the SVZ in the germinal neuroepithelium. More specifically, in the cortical germinal zone of primates and other mammals with a folded cerebral cortex, there is a massive outer subventricular zone (OSVZ) that contributes to the generation of supragranular neurons (Kennedy et al., 2007; see also Chapter 16). The OSVZ contains large numbers of intermediate progenitors and radial glia-like cells, with a long basal process toward the pial surface, but detached from the ventricular surface, which are actively dividing both symmetrically (self-renewing) and asymmetrically (generating neuronal progenitors that keep proliferating) (Fietz et al., 2010; Hansen et al., 2010). More recently, Reillo et al. (2010) characterized these glia-like cells as *Pax6*-expressing intermediate radial glia cells (IRGCs), located both in the SVZ and in the OSVZ, but being much more abundant in the latter. The basal process of these cells is obliquely located, favoring the tangential dispersal of radially migrating neurons, thus contributing to increase neocortical surface area. However, these IRGCs may not be exclusive of primates or mammals with large brains, as a small number of obliquely oriented radial glia have been observed in the outer VZ of rodents, perhaps representing an evolutionary forerunner of the former (Shitamukai et al., 2011; Wang et al., 2011).

Another structure that shows a differential expansion in primate brains is the SPl, an embryonic laminar compartment below the developing neocortex, that was mentioned above in the context of the development of the radial organization of neocortical inputs. Based on morphological and genetic criteria, some elements of the SPl can be considered ancestral in amniote phylogeny, but this structure dramatically increases its complexity in placental mammals, especially in large-brained species (Aboitiz et al., 2003; Kostovic and Rakic, 1990; Montiel et al., 2011). The SPl also shows an increasing diversity of connectional targets in carnivores and primates compared to rodents, sending robust projections through the corpus callosum and to the superior colliculus (Del Rio et al., 2000). Further, the primate SPl displays a continuous addition of neurons until advanced stages of corticogenesis (Lukaszewicz et al., 2005). The SPl is particularly well developed in humans, being divided into an upper and a lower compartment with different proportions of SPl neuronal loss during development (Kostovic and Rakic, 1990). Thus, the development of the SPl is very likely to provide a crucial support to the development of the primate and especially the human neocortex.

An additional innovation is that cortical interneurons have a dual origin in primates, not only from the ventral subpallium as in other mammals but also from the dorsal cortical VZ and SVZ, which express subpallial markers (Zecevic et al., 2011). This characteristic is likely associated with increased interneuronal diversity. Interestingly, in addition to its role in excitatory neurogenesis, Pax6 in the human has been found to play a role in the generation of both ventrally and dorsally generated interneurons (Mo and Zecevic, 2008), and human cortical radial glia have been found to be able to generate GABAergic interneurons (Yu and Zecevic, 2011).

Associated to this increase in brain size and histological complexity, in primates, there has been a proliferation of new cortical regions, especially visual (Barton, 2004; Dobson and Sherwood, 2011). One possibility is that the modulation of specific morphogenetic signaling systems like those mentioned previously has played a role in areal differentiation, in combination with posterior thalamic development and a concomitant increase in the complexity of sensory afferences. In this context, the visual area MT of primates, which emerges quite early in cortical development, has been proposed to serve as a signaling center triggering the proliferation of new areas and regulating the orientation of the visual maps (Bourne and Rosa, 2006). Another expanding brain region is the frontal cortex and its connected systems, like the dorsomedial thalamic nucleus and the anterior striatum, likely associated with the development of a complex social life.

Humans and language

Even if we do not remain self-centered, the human species has by far been the most successful of all primates and is the only animal that has developed such a complex sociality. Again, the large brain that we have is probably a significant factor in this development, even if most cultural advances are likely to have occurred after our brain acquired its present size (Aboitiz et al., 2006). However, beside brain size *per se*, there is another characteristic that has been considered to be fundamental in human evolution and is the ability to communicate using vocal language.

Communication depends on a highly complex and multimodal neural network, encompassing different brain systems involved in recognizing others actions and intentions, and controlling one own's behavior in order to maintain social contact (Arbib, 2005; Chapter 22). Many of these systems have been identified in the monkey and have been proposed to work as a scaffolding from which human language emerged. Nevertheless, the evolutionary beginnings of what is especially unique to us, vocal language, have been more elusive to establish. A crucial innovation in this sense has been claimed to be the direct control from motor or premotor areas over phonatory motor neurons in the brainstem (Jurgens, 2002). Interestingly, vocal learning birds have a direct connection from the arcopallium in the cerebral hemisphere to the motor neurons controlling the syrinx (Gahr, 2000), suggesting an important convergence between these species and humans. Further, we originally proposed that neural systems homologous to the human's "language networks" of the left hemisphere (i.e., Broca's and Wernicke's areas, the arcuate fasciculus, and other tracts connecting them) were present in a rudimentary form in the monkey brain, having expanded significantly during human evolution (Aboitiz and García, 1997), a proposal that was later confirmed by new findings (Petrides and Pandya, 2009; Rilling et al., 2008) (Fig. 6). However, only in early humans, this circuit developed as an auditory-vocal sensorimotor device that enabled to generate complex vocalizations and generating elaborate internal representations that were maintained in working memory (Aboitiz et al., 2010), thus marking the beginnings of phonology. This was a key innovation in human evolution, allowing the development of new modes of

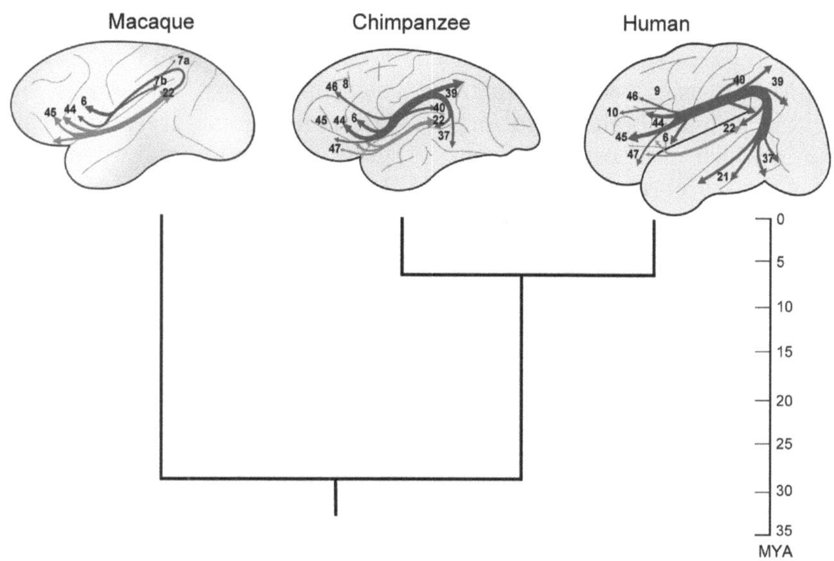

Fig. 6. Evolution of language-related circuitry in the primate brain. In the macaque, there are two main pathways connecting auditory areas (area 22) with the frontal cortex (areas 6, 44, and 45), one running via the temporal lobe (blue) and the other looping around the sylvian fissure (largely corresponding to the arcuate fasciculus and the inferior longitudinal fasciculus) (red). In the chimpanzee, and much further in the human brain, the dorsal route becomes the most conspicuous, connecting many other areas as well. Modified from Rilling et al. (2008) with permission. MYA, million years ago. (For interpretation of the references to color in this figure legend, the reader is referred to the Web version of this chapter.)

communication that substantially promoted cultural development and the eventual acquisition of the modern human mind.

Discussion

In this review, we have made a *tour de force* which, although definitely not exhaustive, has attempted to discuss a significant fraction of the events involved in the evolution of the mammalian and primate brains. Other subjects like neuronal diversity, mechanisms of cortical layering, and the development of specific neural connectivities have been less attended to especially considering that they will be amply covered in this volume. Nonetheless, a few general statements may emerge from this effort. First, at least in amniotes, there seem to be conserved developmental mechanisms involved in brain expansion, despite their highly divergent brain morphologies. These include similar strategies for increasing brain size via the late proliferation of neural progenitors in different regions of the embryonic cerebral vesicle, perhaps using similar genetic networks. Second, the primate brain has a long history behind one in which the most fundamental elements (i.e., the cerebral cortex) were acquired some time before the origin of primates, possibly in a highly different adaptive context. Notably, it is yet by no means clear whether the laminar design of the neocortex is in any sense more advanced than the apparently "nuclear" design of the similarly sized avian brains. Nevertheless, in both birds and mammals, their respective designs were maintained during the diversification within each group.

It must be noted that the telencephalon is not the only brain component that has amplified its cell populations in amniotes. The developing cerebellum displays an external granule layer in its pial surface, formed by the tangential migration of neuronal precursors from the rhombic lip. This is a transit amplification zone for neuronal precursors, in a way reminiscent of the SVZ, which makes up the granule cell layer in the adult cerebellar cortex, the most numerous neuronal type in the amniote brain. Interestingly, recent studies have shown absence of an external granule layer and its markers in either bony fish or cartilaginous fish, indicating that this is an amniote innovation (Chaplin et al., 2010). It is uncertain at this point whether the development of the external granular layer and the SVZ were related events in amniote brain evolution, but both may have contributed to the development of highly complex, large-scale networks controlling behavior.

Primates inherited a "dorsalized" brain design that was acquired in early mammals, which originated in a highly specific context, related to the preponderance of olfactory networks associated to their ecological adaptations. For some not well-known reason, the overall design of this brain has remained stable (perhaps due to developmental constraints or simply because it works well and permits the addition of extra levels of complexity). However, the primate brain is usually larger than that of other mammals, with the implications that some regions like the neocortex and some areas in the neocortex tend to grow disproportionally to the rest of the brain. In this scenario, more specialized areas appear and highly complex connectivity networks develop to an extent probably not observed in other mammals (birds remain to be more studied in this context). As mentioned above, early primates were characterized by their acute vision and more than once redeveloped the trichromatic color vision that was lost in early mammals. This character continues to be a salient feature of this group, together with their high dexterity, which developed as a consequence of their arboreal lifestyle. Such adaptations implied profound sensory and motor changes in the primate brain, including the development of visual areas and an increased cortical control of movement via a more robust corticospinal tract.

Finally, nonhuman primates are our most direct relatives, and as Darwin said, we can learn much of ourselves by studying them. However, perhaps the main point proposed in this chapter is that the fundamental characteristics of the primate brain were acquired in a crucial evolutionary moment, which is the origin of the first mammals. In our view, this event marked the architectural definition of the mammalian brain. Subsequently, there were many modifications, but all within the framework of a conserved design in which increases in size (and their direct consequences) were perhaps among the most important evolutionary changes.

Abbreviations

ADVR	anterior-dorsal ventricular ridge
AF	anterior forebrain
AH	antihem
BMP	bone morphogenetic protein
CH	cortical hem
CP	cortical plate
DC	dorsal cortex
DP	dorsal pallium
DVR	dorsal ventricular ridge
FGF	fibroblast growth factor
H	hyperpallium
Hp	hippocampus
IRGC	intermediate radial glia cell
IZ	intermediate zone
LC	lateral cortex
LGE	lateral ganglionic eminence
LP	lateral pallium
M	mesopallium
MC	medial cortex
MGE	medial ganglionic eminence
MP	medial pallium
MYA	million years ago
MZ	marginal zone
N	nidopallium
NC	neocortex
NE	neuroepithelium
nIPC	neuronal intermediate progenitor cell
OC	olfactory cortex
OSVZ	outer subventricular zone
POC	preoptical commissural area
RA	retinoic acid
Se	septum
SFRP	secreted frizzled-related protein
Shh	sonic hedgehog
SP	subpallium
SPl	subplate
ST	striatum
SVZ	subventricular zone
VP	ventral pallium
VZ	ventricular zone

References

Abellán, A., Menuet, A., Dehay, C., Medina, L., & Rétaux, S. (2010). Differential expression of LIM-homeodomain factors in Cajal-Retzius cells of primates, rodents, and birds. *Cerebral Cortex*, *20*, 1788–1798.

Aboitiz, F. (2011). Genetic and developmental homology in amniote brains. Towards conciliating radical views of brain evolution. *Brain Research Bulletin*, *84*, 125–136.

Aboitiz, F., Aboitiz, S., & García, R. (2010). The phonological loop: A key innovation in human evolution. *Current Anthropology*, *51*(S1), S55–S65.

Aboitiz, F., & García, R. (1997). The evolutionary origin of the language areas in the human brain. A neuroanatomical perspective. *Brain Research Reviews*, *25*, 381–396.

Aboitiz, F., García, R., Bosman, C., & Brunetti, E. (2006). Cortical memory mechanisms and language origins. *Brain and Language*, *98*, 40–56.

Aboitiz, F., & Montiel, J. (2003). One hundred million years of interhemispheric communication: The history of the corpus callosum. *Brazilian Journal of Medical and Biological Research*, *36*, 409–420.

Aboitiz, F., & Montiel, J. (2007). Origin and evolution of the vertebrate telencephalon, with special reference to the mammalian neocortex. *Advances in Anatomy, Embryology, and Cell Biology*, *193*, 1–111.

Aboitiz, F., Morales, D., & Montiel, J. (2003). The evolutionary origin of the mammalian isocortex: Towards an integrated developmental and functional approach. *The Behavioral and Brain Sciences*, *26*, 535–552.

Amiel, J. J., Tingley, R., & Shine, R. (2011). Smart moves: Effects of relative brain size on establishment success of invasive amphibians and reptiles. *PloS One*, *6*, e18277.

Arbib, M. (2005). From monkey-like action recognition to human language: An evolutionary framework for neurolinguistics. *The Behavioral and Brain Sciences*, *28*, 105–167.

Assimacopoulos, S., Grove, E. A., & Ragsdale, C. W. (2003). Identification of a Pax6-dependent epidermal growth factor family signaling source at the lateral edge of the embryonic cerebral cortex. *The Journal of Neuroscience, 23*, 6399–6403.

Barton, R. A. (2004). Binocularity and brain evolution in primates. *Proceedings of the National Academy of Sciences of the United States of America, 101*, 10113–10115.

Bishop, K. M., Garel, S., Nakagawa, Y., Rubenstein, J. L., & O'Leary, D. D. (2003). Emx1 and Emx2 cooperate to regulate cortical size, lamination, neuronal differentiation, development of cortical efferents, and thalamocortical pathfinding. *The Journal of Comparative Neurology, 457*, 345–360.

Bourne, J. A., & Rosa, M. G. (2006). Hierarchical development of the primate visual cortex, as revealed by neurofilament immunoreactivity: Early maturation of the middle temporal area (MT). *Cerebral Cortex, 16*, 405–414.

Butler, A. B., & Hodos, W. (1996). *Comparative vertebrate neuroanatomy. Evolution and adaptation.* New York: Wiley-Liss.

Caminiti, R., Ghaziri, H., Galuske, R., Hof, P. R., & Innocenti, G. M. (2009). Evolution amplified processing with temporally dispersed slow neuronal connectivity in primates. *Proceedings of the National Academy of Sciences of the United States of America, 106*, 19551–19556.

Changizi, M. A., & Shimojo, S. (2005). Parcellation and area-area connectivity as a function of neocortex size. *Brain, Behavior and Evolution, 66*, 88–98.

Chaplin, N., Tendeng, C., & Wingate, R. J. T. (2010). Absence of an external germinal layer in zebrafish and shark reveals a distinct, anamniote groundplan of cerebellum development. *The Journal of Neuroscience, 30*, 3048–3057.

Charvet, C. J. (2010). A reduced progenitor pool population accounts for the rudimentary appearance of the septum, medial pallium and dorsal pallium in birds. *Brain, Behavior and Evolution, 76*, 289–300.

Charvet, C. J., Owerkowicz, T., & Striedter, G. F. (2009). Phylogeny of the telencephalic sub-ventricular zone in sauropsids: Evidence for the sequential evolution of pallial and subpallial subventricular zones. *Brain, Behavior and Evolution, 73*, 285–294.

Cheung, A. F., Kondo, S., Abdel-Mannan, O., Chodroff, R. A., Sirey, T. M., Bluy, L. E., et al. (2010). The subventricular zone is the developmental milestone of a 6-layered neocortex: Comparisons in metatherian and eutherian mammals. *Cerebral Cortex, 20*, 1071–1081.

Cheung, A. F., Pollen, A. A., Tavare, A., Deproto, J., & Molnár, Z. (2007). Comparative aspects of cortical neurogenesis in vertebrates. *Journal of Anatomy, 211*, 164–176.

Cocas, L. A., Georgala, P. A., Mangin, J. M., Clegg, J. M., Kessaris, N., Haydar, T. F., et al. (2011). *Pax6* is required at the telencephalic pallial–subpallial boundary for the generation of neuronal diversity in the postnatal limbic system. *The Journal of Neuroscience, 31*, 5313–5324.

Del Rio, J. A., Martinez, A., Auladell, C., & Soriano, E. (2000). Developmental history of the subplate and developing white matter in the murine neocortex. Neuronal organization and relation-ship with the main afferent systems at embryonic and perinatal stages. *Cerebral Cortex, 10*, 784–801.

Dobson, S. D., & Sherwood, C. C. (2011). Correlated evolution of brain regions involved in producing and processing facial expressions in anthropoid primates. *Biological Letters, 7*, 86–88.

Ergorul, C., & Eichenbaum, H. (2004). The hippocampus and memory for "what," "where," and "when". *Learning & Memory, 11*, 397–405.

Fietz, S. A., Kelava, I., Vogt, J., Wilsch-Bräuninger, M., Stenzel, D., Fish, J. L., et al. (2010). OSVZ progenitors of human and ferret neocortex are epithelial-like and expand by integrin signaling. *Nature Neuroscience, 13*, 690–699.

Finlay, B. L., Darlington, R. B., & Nicastro, N. (1998). Developmental structure in brain evolution. *The Behavioral and Brain Sciences, 24*, 263–278.

Gahr, M. (2000). Neural song control system of hummingbirds: Comparison to swifts, vocal learning (songbirds) and nonlearning (suboscines) passerines, and vocal learning (budgerigars) and nonlearning (dove, owl, gull, quail, chicken) nonpasserines. *The Journal of Comparative Neurology, 426*, 182–196.

Georgala, P. A., Manuel, M., & Price, D. J. (2011). The generation of superficial cortical layers is regulated by levels of the transcription factor pax6. *Cerebral Cortex, 21*, 81–94.

Hansen, D. V., Lui, J. H., Parker, P. R., & Kriegstein, A. R. (2010). Neurogenic radial glia in the outer subventricular zone of human neocortex. *Nature, 464*, 554–561.

Heesy, C. P., & Hall, M. I. (2010). The nocturnal bottleneck and the evolution of mammalian vision. *Brain, Behavior and Evolution, 75*, 195–203.

Huffman, K. J., Molnár, Z., Van Dellen, A., Kahn, D. M., Blakemore, C., & Krubitzer, L. (1999). Formation of cortical fields on a reduced cortical sheet. *The Journal of Neuroscience, 19*, 9939–9952.

Hunt, D. M., Dulai, K. S., Cowing, J. A., Julliot, C., Mollon, J. D., Bowmaker, J. K., et al. (1998). Molecular evolution of trichromacy in primates. *Vision Research, 38*, 3299–3306.

Iwaniuk, A. N., Pellis, S. M., & Whishaw, I. Q. (1999). Is digital dexterity really related to corticospinal projections? A re-analysis of the Heffner and Masterton data set using modern comparative statistics. *Behavioral Brain Research, 101*, 173–187.

Jurgens, U. (2002). Neural pathways underlying vocal control. *Neuroscience and Biobehavioral Reviews, 26*, 235–258.

Kahn, D. M., & Krubitzer, L. (2002). Massive cross-modal cortical plasticity and the emergence of a new cortical area in developmentally blind mammals. *Proceedings of the National Academy of Science USA, 99*, 11429–11434.

Kartén, H. J. (1969). The organization of the avian telencephalon and some speculations on the phylogeny of the amniote telencephalon. *Annals of the New York Academy of Science, 167*, 164–179.

Kennedy, H., Douglas, R., Knoblauch, K., & Dehay, C. (2007). Self-organization and pattern formation in primate cortical networks. *Novartis Foundation Symposium*, *288*, 178–194.

Kielan-Jaworowska, Z., Cifelli, R. L., & Luo, Z. (2004). *Mammals from the age of dinosaurs.* New York: Columbia University Press.

Kostovic, I., & Rakic, P. (1990). Developmental history of the transient subplate zone in the visual and somatosensory cortex of the macaque monkey and human brain. *The Journal of Comparative Neurology*, *297*, 441–470.

Kriegstein, A., & Álvarez-Buylla, A. (2009). The glial nature of embryonic and adult neural stem cells. *Annual Review of Neuroscience*, *32*, 149–184.

Krubitzer, L., & Kahn, D. M. (2003). Nature versus nurture revisited: An old idea with a new twist. *Progress in Neurobiology*, *70*, 33–52.

Lukaszewicz, A., Savatier, P., Cortay, V., Giroud, P., Huissoud, C., Berland, M., et al. (2005). G1 phase regulation, area-specific cell cycle control, and cytoarchitectonics in the primate cortex. *Neuron*, *47*, 353–364.

Lynch, G. (1986). *Synapses, circuits, and the beginnings of memory.* Cambridge: MIT Press.

Medina, L., & Abellán, A. (2009). Development and evolution of the pallium. *Seminars in Cellular and Developmental Biology*, *20*, 698–711.

Métin, C., Álvarez, C., Moudoux, D., Vitalis, T., Pieau, C., & Molnár, Z. (2007). Conserved pattern of tangential neuronal migration during forebrain development. *Development*, *134*, 2815–2827.

Miyashita-Lin, E. M., Hevner, R., Wassarman, K. M., Martinez, S., & Rubenstein, J. L. (1999). Early neocortical regionalization in the absence of thalamic innervation. *Science*, *285*, 906–909.

Mo, Z., & Zecevic, N. (2008). Is Pax6 critical for neurogenesis in the human fetal brain? *Cerebral Cortex*, *18*, 1455–1465.

Molnár, Z. (2011). Evolution of cerebral cortical development. *Brain, Behavior and Evolution*, *78*, 94–107.

Montiel, J. F., Wang, W. Z., Oeschger, F. M., Hoerder-Suabedissen, A., Tung, W. L., García-Moreno, F., et al. (2011). Hypothesis on the dual origin of the mammalian subplate. *Frontiers in Neuroanatomy*, *5*, 25 doi:10.3389/fnana.2011.00025.

Nomura, T., Takahashi, M., Hara, Y., & Osumi, N. (2008). Patterns of neurogenesis and amplitude of Reelin expression are essential for making a mammalian-type cortex. *PloS One*, *3*, e1454.

Northcutt, R. G. (1981). Evolution of the telencephalon in nonmammals. *Annual Review of Neuroscience*, *4*, 301–350.

O'Leary, D. D., & Sahara, S. (2008). Genetic regulation of arealization of the neocortex. *Current Opinion in Neurobiology*, *18*, 90–100.

Olivares, R., Montiel, J., & Aboitiz, F. (2001). Species differences and similarities in the fine structure of the mammalian corpus callosum. *Brain, Behavior and Evolution*, *57*, 98–105.

Petrides, M., & Pandya, D. N. (2009). Distinct parietal and temporal pathways to the homologues of Broca's area in the monkey. *PLoS Biology*, *7*, e1000170.

Pontious, A., Kowalczyk, T., Englund, C., & Hevner, R. F. (2008). Role of intermediate progenitor cells in cerebral cortex development. *Developmental Neuroscience*, *30*, 24–32.

Puelles, L. (2011). Pallio-pallial tangential migrations and growth signaling: new scenario for cortical evolution? *Brain, Behavior and Evolution*, *78*, 108–127.

Puelles, L., Kuwana, E., Puelles, E., & Rubenstein, J. L. R. (1999). Comparison of the mammalian and avian telencephalon from the perspective of gene expression data. *European Journal of Morphology*, *37*, 139–150.

Quinn, J. C., Molinek, M., Martynoga, B. S., Zaki, P. A., Faedo, A., Bulfone, A., et al. (2007). Pax6 controls cerebral cortical cell number by regulating exit from the cell cycle and specifies cortical cell identity by a cell autonomous mechanism. *Developmental Biology*, *302*, 50–65.

Quiroga, J. (1980). The brain of the mammal-like reptile *Probainognathus jenseni* (*Therapsida*, *Cynodontia*), A correlative paleo-neurological approach to the neocortex at the reptile-mammal transition. *Journal für Hirnforschung*, *21*, 299–336.

Reillo, I., de Juan Romero, C., García-Cabezas, M. A., & Borrell, V. (2010). A role for intermediate radial glia in the tangential expansion of the mammalian cerebral cortex. *Cerebral Cortex*, *21*, 1674–1694.

Rilling, J. K., Glasser, M. F., Preuss, T. M., Ma, X., Zhao, T., Hu, X., et al. (2008). The evolution of the arcuate fasciculus revealed with comparative DTI. *Nature Neuroscience*, *11*, 426–428.

Rowe, T. B., Macrini, T. E., & Luo, Z. X. (2011). Fossil evidence on origin of the mammalian brain. *Science*, *332*, 955–957.

Sansom, S. N., Griffiths, D. S., Faedo, A., Kleinjan, D. J., Ruan, Y., Smith, J., et al. (2009). The level of the transcription factor Pax6 is essential for controlling the balance between neural stem cell self-renewal and neurogenesis. *PLoS Genetics*, *5*, e1000511.

Shang, F., Ashwell, K. W., Marotte, L. R., & Waite, P. M. (1997). Development of commissural neurons in the wallaby (Macropus eugenii). *The Journal of Comparative Neurology*, *387*, 507–523.

Shepherd, G. M. (2011). The microcircuit concept applied to cortical evolution: from three-layer to six-layer cortex. *Frontiers in Neuroanatomy*, *5*(30). doi:10.3389/fnana.2011.00030.

Shimogori, T., Banuchi, V., Ng, H. Y., Strauss, J. B., & Grove, E. A. (2004). Embryonic signaling centers expressing BMP WNT and FGF proteins interact to pattern the cerebral cortex. *Development*, *131*, 5639–5647.

Shitamukai, A., Konno, D., & Matsuzaki, F. (2011). Oblique radial glial divisions in the developing mouse neocortex induce self-renewing progenitors outside the germinal zone that resemble primate outer subventricular zone progenitors. *The Journal of Neuroscience*, *31*, 3683–3695.

Shu, T., & Richards, L. J. (2001). Cortical axon guidance by the glial wedge during the development of the corpus callosum. *The Journal of Neuroscience*, *21*, 2749–2758.

Shultz, S., & Dunbar, R. (2010). Encephalization is not a universal macroevolutionary phenomenon in mammals but is associated with sociality. *Proceedings of the National Academy of Science of the United States of America*, *107*, 21582–21586.

Smith Fernández, A., Pieau, C., Repérant, J., Boncinelli, E., & Wassef, M. (1998). Expression of the Emx-1 and Dlx-1 homeobox genes define three molecularly distinct domains in the telencephalon of mouse, chick, turtle and frog embryos: Implications for the evolution of telencephalic subdivisions in amniotes. *Development*, *125*, 2099–2111.

Staubli, U., Schottler, F., & Nejat-Bina, D. (1987). Role of dorsomedial thalamic nucleus and piriform cortex in processing olfactory information. *Behavioral Brain Research*, *25*, 117–129.

Stoykova, A., Treichel, D., Hallonet, M., & Gruss, P. (2000). Pax 6 modulates the dorsoventral patterning of the mammalian telencephalon. *The Journal of Neuroscience*, *20*, 8042–8050.

Striedter, G. F. (2005). *Principles of brain evolution.* Sunderland, MA: Sinauer Associates.

Striedter, G. F., & Charvet, C. J. (2009). Telencephalon enlargement by the convergent evolution of expanded subventricular zones. *Biological Letters*, *5*, 134–137.

Subramanian, L., & Tole, S. (2009). Mechanisms underlying the specification, positional regulation, and function of the cortical hem. *Cerebral Cortex*, *19*(Suppl. 1), i90–i95.

Sugahara, F., Aota, S., Kuraku, S., Murakami, Y., Takio-Ogawa, Y., Hirano, S., et al. (2011). Involvement of Hedgehog and FGF signalling in the lamprey telencephalon: Evolution of regionalization and dorsoventral patterning of the vertebrate forebrain. *Development*, *138*, 1217–1226.

Sur, M., & Rubenstein, J. L. (2005). Patterning and plasticity of the cerebral cortex. *Science*, *310*, 805–810.

Tanaka, D. H., Oiwa, R., Sasaki, E., & Nakajima, K. (2011). Changes in cortical interneuron migration contribute to the evolution of the neocortex. *Proceedings of the National Academy of Sciences of the United States of America*, *108*, 8015–8020.

Tarabykin, V., Stoykova, A., Usman, N., & Gruss, P. (2001). Cortical upper layer neurons derive from the subventricular zone as indicated by Svet1 gene expression. *Development*, *128*, 1983–1993.

Teissier, A., Griveau, A., Vigier, L., Piolot, T., Borello, U., & Pierani, A. (2010). A novel transient glutamatergic population migrating from the pallial-subpallial boundary contributes to neocortical development. *Journal of Neuroscience*, *30*, 10563–10574.

Tole, S., Remedios, R., Saha, B., & Stoykova, A. (2005). Selective requirement of Pax6, but not Emx2, in the specification and development of several nuclei of the amygdaloid complex. *The Journal of Neuroscience*, *25*, 2753–2760.

Tuoc, T. C., Radyushkin, K., Tonchev, A. B., Piñon, M. C., Ashery-Padan, R., Molnár, Z., et al. (2009). Selective cortical layering abnormalities and behavioral deficits in cortex-specific Pax6 knock-out mice. *The Journal of Neuroscience*, *29*, 8335–8349.

Vorobyev, M. (2004). Ecology and evolution of primate colour vision. *Clinical & Experimental Optometry*, *87*, 230–238.

Wang, Y., Brzozowska-Prechtl, A., & Karten, H. J. (2010). Laminar and columnar auditory cortex in avian brain. *Proceedings of the National Academy of Sciences of the United States of America*, *107*, 12676–12681.

Wang, X., Tsai, J.-W., LaMonica, B., & Kriegstein, A. R. (2011). A new subtype of progenitor cell in the mouse embryonic neocortex. *Nature Neuroscience*, *5*, 555–561.

Welker, E., & Van der Loos, H. (1986). Quantitative correlation between barrel-field size and the sensory innervation of the whiskerpad: A comparative study in six strains of mice bred for different patterns of mystacial vibrissae. *The Journal of Neuroscience*, *6*, 3355–3373.

Williams, B. A., Kay, R. F., & Kirk, E. C. (2010). New perspectives on anthropoid origins. *Proceedings of the National Academy of Sciences of the United States of America*, *107*, 4797–4804.

Yu, X., & Zecevic, N. (2011). Dorsal radial glial cells have the potential to generate cortical interneurons in human but not in mouse brain. *The Journal of Neuroscience*, *31*, 2413–2420.

Zecevic, N., Hu, F., & Jakovcevski, I. (2011). Interneurons in the developing human neocortex. *Developmental Neurobiology*, *71*, 18–33.

Zhou, Z., Barrett, P. M., & Hilton, J. (2003). An exceptionally preserved lower Cretaceous ecosystem. *Nature*, *421*, 807–814.

SECTION II

Genes and development

M. A. Hofman and D. Falk (Eds.)
Progress in Brain Research, Vol. 195
ISSN: 0079-6123

CHAPTER 2

Genetic correlates of the evolving primate brain

Eric J. Vallender*

New England Primate Research Center, Harvard Medical School, Southborough, MA, USA

Abstract: The tremendous shifts in the size, structure, and function of the brain during primate evolution are ultimately caused by changes at the genetic level. Understanding what these changes are and how they effect the phenotypic changes observed lies at the heart of understanding evolutionary change. This chapter focuses on understanding the genetic basis of primate brain evolution, considering the substrates and mechanisms through which genetic change occurs. It also discusses the implications that our current understandings and tools have for what we have already discovered and where our studies will head in the future. While genetic and genomic studies have identified many regions undergoing positive selection during primate evolution, the findings are certainly not exhaustive and functional relevance remains to be confirmed. Nevertheless, a strong foundation has been built upon which future studies will emerge.

Keywords: genetic evolution; molecular evolution; catarrhine; hominoid; hominin; FOXP2; microcephaly; opsin; olfaction; pleiotropy; gene regulation; divergence; polymorphism.

Introduction

Evolution at the most basic level occurs in the genome. In the simplest formulation, mutations occur that give rise to phenotypes upon which selection acts. While overly simplistic, this can form a useful framework to begin our understanding of the genetics of primate brain evolution. If we begin with the premise that the primate brain generally, and the human brain specifically, systematically differs from that of other species, then it follows naturally that this must result from a genetic heritage. Understanding the genetic changes that have led to the phenotypic changes that we observe in the primate brain may ultimately lead to a better understanding of the brain itself, the differences between species, and the neuropathologies with which we struggle.

Many studies exist comparing genetic similarities between species for specific classes of genes. In particular, it has been shown that genes expressed in the brain show fewer differences between species than genes expressed in other

*Corresponding author.
Tel.: 508-624-8194; Fax: 508-624-8166
E-mail: eric_vallender@hms.harvard.edu

DOI: 10.1016/B978-0-444-53860-4.00002-7

tissues or more ubiquitously (Duret and Mouchiroud, 2000; Miyata et al., 1994). Unsurprising, perhaps, given the enormous complexity of the brain and the hugely deleterious effects of even the most subtle change. And yet, even to an untrained eye, the brain of a mouse, marmoset, chimpanzee, and human are immediately and obviously different. Phenotypic change has obviously occurred and this can only mean a genetic underpinning. While there are many contributing factors to this paradox, a significant understanding can be reached simply by noting the different domains in which these observations exist.

When we speak of brain evolution, we are conflating at least two disparate concepts. The first is a basic change of brain function, how neurons communicate with one another, the neurotransmitter mechanisms by which the brain operates. The second is a change in structure and development, neurons continue to function similarly, but now there are more of them or they are in different places. It is important to appreciate these differences in order to make sense of the evolutionary studies that focus on brain genetics.

When we compare very divergent classes of organisms, say insects to mammals, the genes and proteins of basic brain function may change dramatically. Mammalian usage of catecholamines, norepinephrine and epinephrine, compared to insect usage of phenolamines, octopamine and tyramine, is an excellent example (Caveney et al., 2006; Vincent et al., 1998). And yet, with few exceptions, the basic physiological functionality of the brain is remarkably similar between more closely related species such as mammals generally or primates specifically. Nevertheless, the differences that do emerge tend to be correlated with significant change. In primates, change at this basic functional level is felt most strongly in those systems associated with a shift from olfactory perception to visual perception.

Canonical gene evolution studies in primate perception

One of the most salient and extreme examples of genetic evolution in primates is the wholesale loss of olfactory receptors (Gilad et al., 2003a,b; Young et al., 2002). Mammalian olfactory receptors, a gene superfamily consisting of more than 1000 genes, form a significant portion of the mammalian genome. This extensive diversity is likely the result of olfactory receptors specific binding to odorant molecules. But this specificity that leads to such variety overall also leads to significant losses when a given organism is not exposed to the odorant. In rodents and dogs, only 20% of the olfactory receptor genes are nonfunctional and yet in humans fully 60% of olfactory receptors have undergone pseudogenization (Dong et al., 2009). While initially focused on the human genome, this finding has also held up across other nonhuman primate species, correlating well with the relative roles of visual and olfactory perception.

Few studies are as dramatic as the evolution of the primate olfactory subgenome, yet we can observe similar findings in other sensory domains. Perhaps unexpectedly, given their similarity to olfactory receptors, taste receptors have also repeatedly shown selective signatures across species including primates (Soranzo et al., 2005; Wooding et al., 2004, 2006). Divergence between species and polymorphism within species has been widely observed for both the bitter taste receptor gene family and the sweet taste receptor genes. Hypothesized to reflect changes in diet and perhaps the ability to distinguish between nutritious and harmful foods, the evolution of taste receptor genes parallels that of the olfactory genes.

The evolution of trichromatic vision in primates is also a prominent example of molecular evolution (Shyue et al., 1995; Zhou and Li, 1996). Humans, apes, and old-world monkeys are trichromatic because they have three opsin genes each of which detects light at a different wavelength. The blue (short-wave) opsin is shared across primates and is located autosomally, while the green (middle-wave) and red (long-wave) opsins are located on the X chromosome. While bichromatic platyrrhines only possess one X-chromosome opsin gene, trichromatic catarrhines have two. Starting from this observation, it has been possible to reconstruct the molecular evolution of the X-chromosome opsins, beginning with a gene duplication event approximately 35–40 million years ago, after the divergence of catarrhines from new-world monkeys, and the subsequent functional divergence of the two genes allowing them to separately recognize the red and green wavelengths (Fig. 1). Interestingly, new-world monkeys have multiple alleles at the X-chromosome opsin gene, allowing female homozygotes to effectively have trichromatic vision, while all males are obligatorily color-blind. The exception here is in howler monkeys where a duplication, akin to what is observed in catarrhines, has fixed the two platyrrhine alleles (Boissinot et al., 1998).

These examples, though, are the exception rather than the rule; changes are large and obvious. In the olfactory system, hundreds of genes are inactivated, a not-so-subtle change at a level that cannot be overlooked. The visual system represents an extreme phenotypic change and a strong single-gene effect. Neither of these situations generalize. The evolution of most of the genes is subtle, with changes that affect

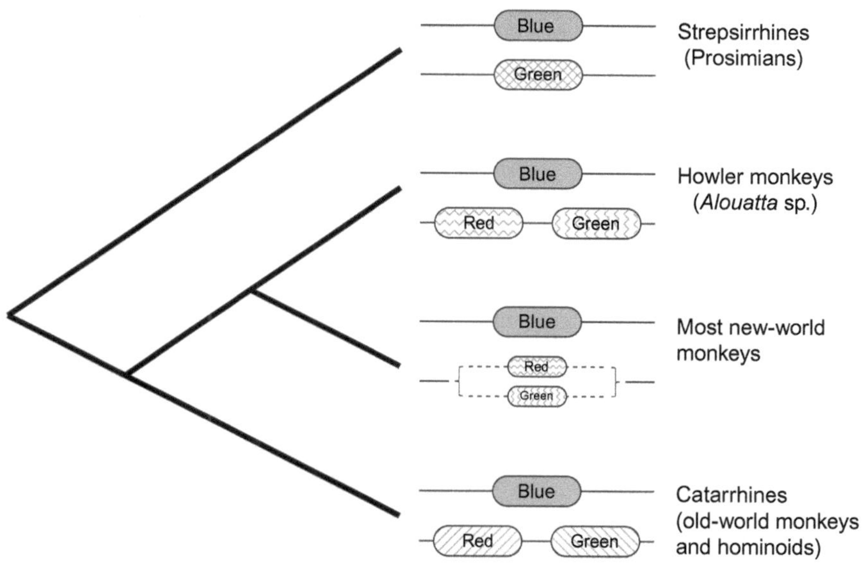

Fig. 1. Molecular evolution of opsin genes in the primates. The blue opsin, on autosomes has remained conserved across primate species. In the ancestral primate, and in extant strepsirrhines, only a green opsin is present on the X chromosome. In catarrhines, this gene has been duplicated and been functionally altered to detect both red and green. In most new-world monkeys, and seemingly the ancestral platyrrhine, only one opsin exists on the X chromosome though it has multiple alleles and female heterozygotes may have trichromatic vision. A separate duplication event in howler monkeys has fixed red and green opsins on the X chromosome as well. (For interpretation of the references to color in this figure legend, the reader is referred to the Web version of this chapter.)

function hiding among large numbers of obfuscatory neutral differences. Most phenotypes, especially when dealing with the brain, are the result of many interacting genes and proteins, and associating specific genetic changes with observable phenotypic differences is extremely difficult. It is to this, however, that we set our mind when we seek to understand the genetic differences underlying primate brain evolution.

Gene gain and loss

Before applying genetic and molecular evolution to questions of the brain, it is useful to discuss it in a more general sense. Despite inferences elsewhere, there is no reason to believe that the molecular evolution of the primate, or human, brain is inherently any different than the genetic evolution of any other character or trait. The brain, as a complex system, may be inclined toward certain categories of change, but it is governed by the same factors that are seen elsewhere.

As demonstrated in the emergence of primate trichromatic vision, gene gain through duplication can be a major source of evolutionary novelty (Ohno, 1970; Zhang, 2003b). Following the initial duplication event, selective pressures are temporarily relaxed as genetic redundancy can hide evolutionary missteps. By far, the most common fate following a duplication event is the most evolutionarily uninteresting, the pseudogenization of one copy and the perseverance, in a largely unaltered form, of the second. However, in rare instances both copies of the duplicated gene can persist either through neofunctionalization (Hughes, 1994), where one copy evolves a new functionality separate from the original, or by subfunctionalization (Force et al., 1999), where the two copies divide among each other the functional characteristics of the original. In this latter case, multiple functions of a single gene can thus be disjoined allowing for the separate evolution of each.

Yet, while the most evolutionarily satisfying from a theoretical point of view, it is unclear the degree to which novel genes actually contribute to short timescale evolutionary events, such as primate evolution. The number of gene gains found along any primate lineage is fairly small, despite the relative simplicity in identifying them. This combination of their paucity and their potential to generate major evolutionary change has made gene gains a major focus for evolutionary inquiry. Yet despite this fact, there are relatively few examples of gene gains with functional effects in primates and even fewer that involve brain development or function.

Among the more interesting cases are those of the DUF1220 domain family of genes (Popesco et al., 2006). This gene family has expanded greatly in primates and seems to be uniquely and specifically expressed in neurons, but the overall function of the gene remains unclear. Other gene families, such as the *morpheus* genes (Johnson et al., 2001) and the *KRAB-ZNF* genes (Nowick et al., 2010), have also showed rapid evolution in primates, though again their functional relevance is unclear.

The duplication of glutamate dehydrogenase (*GLUD*), on the other hand, represents a more clear functional picture (Burki and Kaessmann, 2004). In old-world monkeys, like most mammals, there exists only a single *GLUD* gene. It encodes for a protein that recycles the excitatory neurotransmitter glutamate. In species with only one *GLUD* gene, this protein is ubiquitously expressed. In apes, however, a duplication event has led to the emergence of a second glutamate dehydrogenase, *GLUD2*. This gene is expressed uniquely in the nerve tissues and testis (Shashidharan et al., 1994) and appears to have undergone positive selection to optimize its function in environments, like the brain, with high GTP concentrations (Plaitakis et al., 2003). The difficulty here, however, is that while the specific protein function is understood, a phenotypic consequence is less clear.

The other side of the coin to gene gain is gene loss. Like gene gain, gene loss is relatively simple to identify and the change in function is easy to interpret; the functional gene is either there or it is not. Typically, gene loss is envisioned as a loss of a trait. The pseudogenization of the olfactory system is a good example. In species with particular intact olfactory receptors, individuals have the capacity to recognize certain odors; individuals in species where these receptors have been lost cannot. The gene loss corresponds to a trait loss. Of course, genes and their products operate in complex systems and their loss can have far-reaching effects that may ultimately result in a phenotypic trait gain.

One of the most interesting examples of gene loss and the role it may play in primate brain evolution also demonstrates this circuity. MYH16 is a protein from the gene family that makes up the myosin strands in skeletal muscle. It was identified in 2004 as having been lost in humans (Stedman et al., 2004). Further investigation found that *MYH16* was expressed uniquely in masticatory muscles, those muscles that attach the jaw to the skull and allow for a greater bite strength and chewing ability. The loss of this gene in *Homo sapiens* also seemed to correlate with the dietary transition to cooked meat and the loss of a need for a heavy masticatory apparatus. This also released structural constraints for attaching these muscles to the skull and allowed for an enlargement of the human brain. Naturally, much of this latter is conjecture based upon circumstantial evidence that has subsequently been challenged (Perry et al., 2005). As we will see repeatedly, the basic functional change is undebatable, the gene and protein were there in chimpanzees but lost in humans, but the question of how this change fits into an evolutionary framework is more difficult.

So far our discussions have focused on the most obvious, yet rare, molecular evolutionary events. By far, the most common of evolutionary changes are much more subtle; the functional element is still there as before but it is changed slightly. This can occur either in protein-coding regions or in regulatory regions and by many different kinds of genetic changes (Fig. 2). In protein-coding regions, we often think of point mutations that can change the encoded amino acid. If the amino acid is important and/or the change is drastic enough, the proteins function can be changed. In regulatory regions, point mutations can have similar effects, but so too can insertions or deletions, or other mutational events that would have the singular effect in protein-coding regions of pseudogenization. Because regulatory regions seem to be much more labile in their structural and functional organization, they can tolerate changes that coding regions cannot.

These latter changes occur much more commonly than gene gain or gene loss. Small changes to an individual's genome occur in every generation. Most times, these changes are nearly neutral, that is to say, they have little to no effect on the overall selective fitness of an organism. Changes that do have a noticeable effect are almost always deleterious, again think of genetic diseases. It is upon this background of changes that evolution operates. Identifying changes that have impacted the observable differences between species thus becomes a major challenge to understanding the genetic basis behind evolutionary change.

Detecting adaptive genetic change

When we compare two species at the genetic level, we find many fixed differences. Of course, the more evolutionarily distant the two species the greater the genetic difference, but even closely related species show millions of differences. Indeed, even populations within a single species will show fixed differences relative to each other. The issue becomes separating out changes that matter, functionally relevant changes, from those that do not. Ideally, this would be accomplished via direct functional assay: If a receptor is exactly the same except for this specific change, does it bind its ligand differently? If this promoter region is exactly the same except for this mutation, is the expression

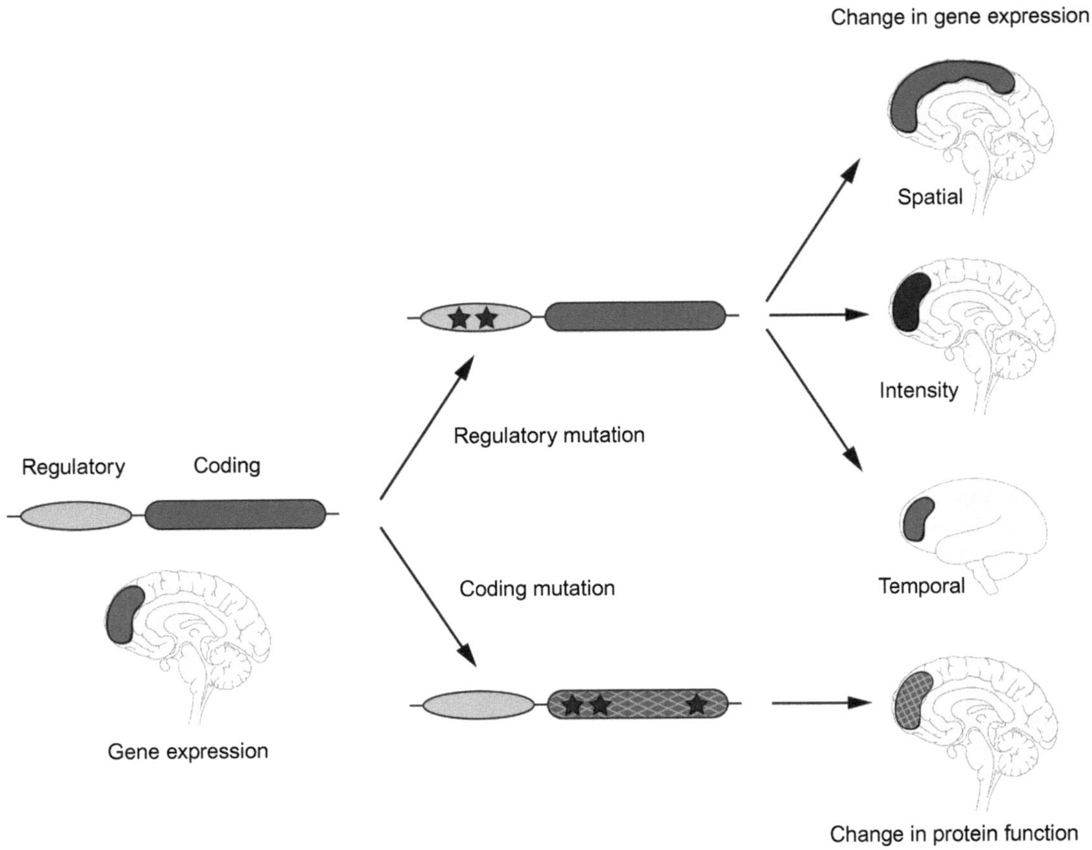

Fig. 2. Schematic demonstrating the various effects of regulatory and coding mutations on genes.

pattern of its gene changed? This is the gold standard for an unequivocal statement of functional change, yet it is obviously not tenable on large scales. Further, it can be difficult, if not impossible, to find the relevant assay for detecting a functional effect. Because of this, we have developed computational and bioinformatic approaches to identify likely candidates.

One of the most common and widespread ways to detect functionally important differences between species is to look for mutations that have fixed at a faster rate than they would have if they were behaving neutrally. This positive selection of the mutations argues that they were functionally important. Many algorithms have been developed around this basic premise (Goldman and Yang, 1994; Li et al., 1985; Nei and Gojobori, 1986; Yang and Nielsen, 2000). In proteins, this approach measures the rate of fixation of non-synonymous or replacement changes, those that do change an amino acid, to the rate of fixation of synonymous changes, those that do not change an amino acid. The synonymous rate is alternatively called K_S or d_S and is assumed to represent the neutral mutation rate, while the non-synonymous rate is called K_A or d_N and is reflective of protein change. The ratio of these rates (K_A/K_S, d_N/d_S, or ω) can be used to infer the selective pressure on the gene. A neutrally evolving gene, one under no selective pressure, would have a K_A/K_S rate equal to one. Most genes are under negative selection, the protein has an evolved function that it cannot stray from, and have K_A/K_S rates less than 1. (The average across genes in primates hovers around 0.2 (Gibbs et al., 2007; The Chimpanzee Sequencing and Analysis Consortium, 2005).) Positive selection, genes whose function has changed during evolution, shows K_A/K_S rates above 1.

One of the findings of the early genomic studies was that genes expressed in the brain tended to have lower K_A/K_S rates than genes expressed in other tissues (Duret and Mouchiroud, 2000; Miyata et al., 1994). These results applied across multiple brain regions (Tuller et al., 2008) and across multiple species comparisons, including primates where selection on the brain was anticipated (Khaitovich et al., 2005). In some ways, this is unsurprising, the brain is extremely complicated and changes that disturb this delicate balance are likely to have deleterious effects. On the other hand, this conservation should not be taken as a lack of selection either. Rather, it highlights the complexities of detecting positive selection in this manner.

This method for detecting positive selection is very sensitive to dilution, either spatial or temporal. The first is the result of parts of genes being under differential selective pressures. In the context of the brain, a G-protein-coupled receptor is likely to be under strong negative selection in its transmembrane domains, while the intra- or extracellular domains may be under less strong selection or even positive selection (Fig. 3). If the entire gene is used as the unit of selection, then the positive selection in, say, the ligand-binding site may be obscured by the negative selection in the transmembrane domains. Time can also dilute the effects of selection. If the selective event was specific and short-lived, then K_A/K_S levels may be higher during those periods but lost overall. Pairwise species comparisons necessarily entail the entirety of the evolutionary history separating the species, so if the selective event only happened for a short period, then it can be obscured. This is particularly important when

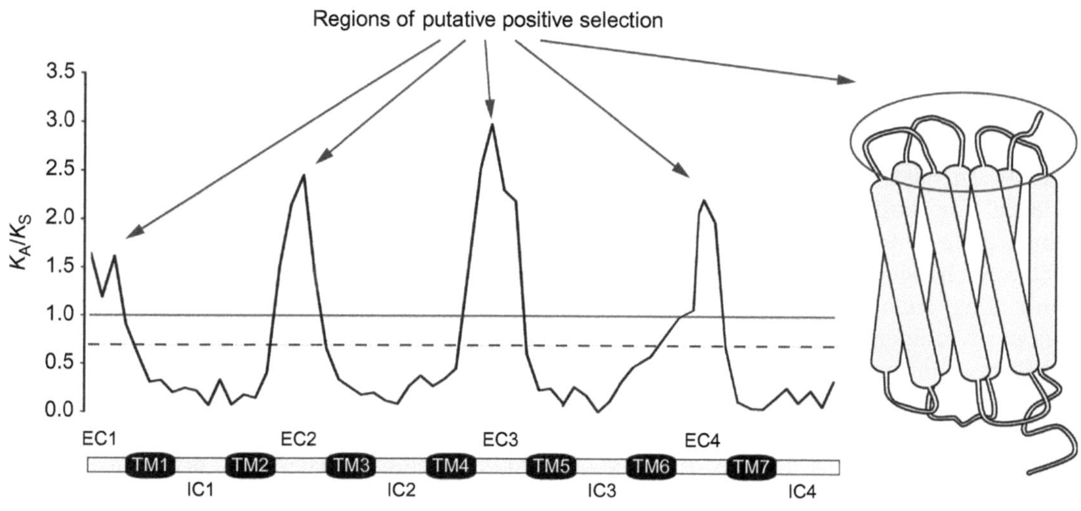

Fig. 3. Hypothetical sliding window analysis of a seven-transmembrane domain G-protein-coupled receptor (after Choi and Lahn, 2003). K_A/K_S ratios in transmembrane (TM) and intracellular (IC) domains are low indicative of negative selection, while K_A/K_S ratios for extracellular (EC) domains are above 1, indicative of positive selection. The dashed line indicates the gene average K_A/K_S, under 1 and evidence of overall negative selection.

considering brain evolution in primates because encephalization events are generally short and discreet rather than long and pervasive. Methodologies have been developed that can take these factors into account (Yang, 2007), but the smaller the physical region or time frame, the smaller the K_S, the greater the noise in the system.

There are many examples of genes putatively involved in the evolution of the primate brain that show signatures of positive selection. Some of the most extensive evidence is found in genes associated with primary microcephaly. Primary microcephaly is a disease in which the brain is reduced in size, but without any other gross abnormalities in structure (Dobyns, 2002). This malformation has been of interest in the evolutionary community because the condition seemingly recapitulates ancestral primate brain characters, the suggestion being that the same genes that cause the disease may also have been those genes under selection to produce modern brain phenotypes (Kaindl et al., 2010; Woods et al., 2005). Over the years, K_A/K_S-based methods have been used to identify positive selection on multiple genes implicated in the disorder: *ASPM* (Evans et al., 2004b; Kouprina et al., 2004; Zhang, 2003a), *MCPH1* (Evans et al., 2004a; Wang and Su, 2004), *CDK5RAP2* and *CENPJ* (Evans et al., 2006), and *CEP152* (Guernsey et al., 2010). In each of these cases, however, the evidence for positive selection has come from comparative genomics and specific attributable functional change remains elusive.

Protein-coding changes can be more or less straightforward to identify bioinformatically but difficult to ascertain function. Noncoding changes are the inverse, with functionality yielding easily, but identification much more slowly. The difficulty in regulatory regions lies in identifying which changes are functionally relevant and which are not. Initial studies used either protein-coding synonymous sites (Wong and Nielsen, 2004) or surrounding intronic regions (Haygood et al., 2007) to calibrate for neutral mutations. While there was some limited success with these methods, functional sites in the test regions are likely to be swamped out by neutral sites, significantly reducing power. As our understanding of gene regulation improves, these test regions can be better defined and better tools developed. Indeed, more recent attempts incorporate transcription factor-binding motifs to define regions (Hoffman and Birney, 2010).

The specific successes in regulatory regions, especially regarding the brain, are few, the gene encoding the opioid neuropeptide precursor, *PDYN*, being the significant exception (Rockman et al., 2005). For this gene, an identified *cis*-regulatory element was shown to have accumulated an exceptionally high number of mutations since the divergence of humans from chimpanzees is consistent with a hypothesis of positive selection. More importantly, however, functional assays employed showed that the humanized regulatory element drove much higher levels of expression compared to the chimpanzee element *in vitro*. Interestingly, this finding has recently been extended to other members of the opioidergic system (Cruz-Gordillo et al., 2010), though it remains unclear if the fact that both of these findings arise in a single system is biologically meaningful or it simply represents a relative paucity of study elsewhere.

Implications of genetic change

Our ability to detect genetic change has significantly impacted the way that we pursue questions of genetic evolution. It has long been appreciated that regulatory change is likely to play a major role in brain evolution. Building on the work of others (Ohno, 1972), in 1975, King and Wilson (1975) noted that "The organismal differences between chimpanzees and humans would then result chiefly from genetic changes in a few regulatory systems, while amino acid substitutions in general would rarely be a key factor in major adaptive shifts." Little has occurred in the intervening years that would force us to reassess this basic belief. Rather, it has been buttressed by subsequent work and the emergence of the field of evolutionary developmental biology (Carroll, 2008). Yet the preponderance of work to date has focused on protein change rather than regulatory change, largely because of the tools at our disposal.

Proteins, seemingly especially brain proteins, have been optimized for their specific roles; there simply is not a lot of room for them to change without causing major effects. This is especially true because of genetic pleiotropy. In essence, proteins serve many functions simultaneously and any change that might be made affecting one of these functions necessarily affects all of them. While it is true that not all proteins are pleiotropic, it seems likely that many of the developmental proteins that would most likely account for the evolutionary differences that we see in the primate brain are. As we have seen, one way in which these pleiotropic effects can be overcome is through gene duplication, yet this is rare. One major finding that has come following the publication of the nonhuman primate genomes is that there simply are not a tremendous number of species-unique genes. The gene complement of each of the primate species is largely the same enough, so the exceptions warrant significant attention even in the absence of any functional understanding. Again, while a theoretically satisfying method to generate novelty, gene duplication does not seem to be a major mechanism for adaptation in primates.

Because of what is currently feasible, the focus so far has been predominantly on protein change. While it may only represent a portion of the genetic underpinnings of primate brain evolution, it has been tractable and significant changes have been identified. As we move forward, however, it will be increasingly important to expand our studies of regulatory regions. Many attempts are being made to address this disparity with varying degrees of success, yet it is clear that the next major step forward in evolutionary genetics will follow our ability to solve this problem.

Phenotypic change in the primate brain

When speaking of the evolution of the primate brain, we are, by and large, talking about developmental differences. The most obvious of these is simply an overall expansion in size. The volume of the brain in humans is roughly eight times the size of that of large new-world monkeys, six times that of old-world monkeys, and three to four times that of apes (Jerison, 1973). Generally speaking, an enlargement of the brain in its entirety, or at least the whole brain excluding the olfactory regions, seems to be a common mechanism upon which selection acts in mammals generally including primates (Finlay and Darlington, 1995). It has also been observed, however, that neocortex increases in the brains of apes and humans are particularly pronounced (Kaas, 2005; Semendeferi et al., 2002). Of course, brain size, or neocortex size, is not the direct phenotypic output upon which selection is acting. Rather we are assuming that behavioral complexity or intelligence is somehow correlating with the increase in size. It is this behavioral output that selection is ultimately acting upon, introducing another obfuscatory layer to evolutionary analysis.

In primate evolution, we see several major bouts of brain growth, encephalization events. In fact, each major divergence event, Haplorrhini and Strepsirrihini, Catarrhini and Platyrrhini, Cercopithecoidea and Hominoidea, seems to correlate with a brain size expansion. In more recent human evolution, we also see brain size expansions, from *Australopithecus* to *Homo*, in the emergence of *H. erectus*, and finally in the emergence of *H. sapiens*. Each of these encephalization events has presumably left marks in the genome. It is not known, however, if each of these events was the result of succeeding changes to the same genes or if each event utilized novel genes.

It is also important not to overlook other sources of phenotypic differences. While large-scale structural differences can be the most obvious to observe, connectivity changes can have significant impacts as can more subtle microstructural changes. We also have numerous examples from human and rodent literature of variation associated with neurotransmission having significant effects on behavior and, in humans at least, this variation seemingly under variable selective regimes.

Implications of phenotypic change for genetic evolution

The context of brain evolution has several important implications for genetic studies: what we look for and when we look for it. In general, when we are considering primate brain evolution, we are focusing on developmental changes: changes in growth rates, changes in developmental timing, changes in patterns of expression. As previously mentioned, developmental change can seemingly be more easily effected through changes in gene regulation than by changes in protein sequence and function (Carroll, 2008). This may help explain why studies of primate evolution generally can have power to detect selection in other systems, but not in the brain despite its relative importance to adaptation.

When we do undertake studies of protein evolution, it is informative to consider where we find our best signals of selection, at least relative to the brain. We have already mentioned the positive selection detected in the genes associated with primary microcephaly. Although undoubtedly functionally complex, it is interesting to note that these genes, when mutated in humans, have a decidedly nonpleiotropic effect. Indeed, their initial selection for consideration as relevant substrates for selection was precisely because the disease that they associated with resulted in a decrease in brain size without any other significant pathologies. In other words, the focus, even in proteins, has been on genes most likely to have singular functions.

The other effect that this emphasis on developmental change has had on primate brain evolution can be seen in the genes identified. In several of the early pseudogenomic studies, an accelerated rate of evolution in brain genes was dominated by genes of predominantly developmental function (Dorus et al., 2004; Khaitovich et al., 2005). Other studies have also broadly identified "transcription factors" as a category of genes rapidly evolving in primates relative to rodents (Gibbs et al., 2007). In both cases, it is primarily by gathering together genes of like function that these patterns emerge. If this primate phenotypic evolution is occurring primarily by many changes of very small effect, then this is the sort of pattern that we would anticipate.

One last gene set worthy of mention is the class of genes involved in the apoptotic pathways, particularly the extrinsic apoptosis pathways. These genes were first identified as undergoing positive selection in a large-scale scan comparing humans and chimpanzees (Nielsen et al., 2005). Focused follow-up studies confirmed and extended the findings (da Fonseca et al., 2010; Vallender and Lahn, 2006). Apoptosis, programmed cell death, is important in many developmental processes including brain growth. Knockouts of many of these genes in mice result in dramatic brain phenotypes (Putcha et al., 2002). Again, however, the question of pleiotropy is significant. The role of apoptosis in neuronal development is undisputed, yet so too is the role of apoptosis in numerous other processes including immune response, a known major driver of selection.

Another important confound that this introduces revolves around the specific encephalization event implicitly under study. Many studies of recent evolution, almost always in humans, focus on polymorphism levels and have an effective resolution of only the last several million years. In humans, this means at best the changes in the brain since the Australopithecines. In reality, demographic effects, including bottlenecks and the "Out-of-Africa" migration of *H. sapiens* (and possibly *H. erectus*), likely create an even shorter timeframe of effective resolution.

Studies that focus on the differences between chimpanzees and humans, with or without an outgroup to stratify changes, encompass all the successive encephalization events in the terminal *Homo* lineage. The primary difficulty with these studies is that stochastic noise in the neutral rate of evolution is great, often dominating the selective signal. This is a pervasive problem with short lineages and likely extends across all ape pairwise comparisons. Some power will be gained when the gorilla and orangutan genomes become widely available and all four primary great ape lineages can be integrated, though it seems likely that stochastic effects will continue to dominate.

When old-world monkeys are included in the analyses, this short lineage effect is largely ameliorated. The cost of this, however, is that the time period that dominates the selective effects seen corresponds to the lineage separating the apes from the last common catarrhine ancestor. In effect, this means that many studies focusing on the human–old-world monkey (rhesus macaque) comparison may actually be more generally studies of the evolution of the ape brain rather than the human brain.

To date few studies have included new-world monkeys or other more basal primate lineages, though this is certain to change in the future. These studies will be informative, but particularly for understanding brain evolution during the Catarrhini–Platyrrhini or Haplorrhini–Strepsirrihini events. Perhaps, the most important of the genomes yet to be sequenced for understanding brain evolution, particularly in the neocortex, is the gibbon. This lineage separated from the apes approximately 10 million years after the old-world monkeys and may help isolate and elucidate the changes involved in the emergence of the ape brain.

Surveys of genetic evolution

There have been several large-scale surveys of genetic evolution that have emerged over the past decade. Yet while there has been significant comment arising from these studies, it has been difficult to translate them into biological understanding. This is in part because of the large differences in their findings. The initial focus of many of these surveys was on divergence between human and chimpanzee (Arbiza et al., 2006; Bustamante et al., 2005; Clark et al., 2003; Nielsen et al., 2005; The Chimpanzee Sequencing and Analysis Consortium, 2005), and indeed, much of the confusion lies in the nature of the human–chimpanzee comparison itself. The short lineage effects can make small differences in methodologies, including the species used as outgroup, alignment method, divergence algorithm, and universe of genes under study, have large impact.

At the same time, there have been studies focusing on polymorphism within humans as a means to detect more recent positive selection (Carlson et al., 2005; International HapMap Consortium, 2005; Kelley et al., 2006; Sabeti et al., 2007; Voight et al., 2006; Williamson et al., 2007). Again, there have been few consensus findings. Like the human–chimpanzee comparisons, variability in methodologies may have a large effect on these polymorphism-based studies. Many of these approaches are only recently being pioneered as data developed from large-scale human diversity studies become available. In addition to methodological uncertainty, however, polymorphism studies suffer from biological difficulties. Demographic effects complicate any study and range of resolution can have major effects on results. Recent selection study designs may capture only the most recent *H. erectus* to *H. sapiens* speciation or may extend further. This general issue is a common one when comparing genome-wide surveys; different study designs focus on different time frames. Perhaps the most noteworthy finding that can be drawn from this is that the same genes do not seem to be underlying each adaptive event.

Across genomic studies, genes involved in any aspect of brain or nervous system biology regularly fail to appear. Several possible interpretations exist for this. In human–chimp and human–old-world monkey studies, the focus is almost exclusively, and necessarily so, on protein-coding sequence. A lack of brain genes in these studies seems to confirm a primacy of regulatory substrates for evolution. Polymorphism studies, however, are ambivalent to substrate and still do not offer significant insight into brain evolution. An explanation for this could be that we simply do not understand enough about gene regulation and our annotation of positively selected regions is deficient, but perhaps more likely is that selective pressures are not sufficient to reach significance levels required for genome-wide detection. It is for this reason that candidate gene studies continue to be the most useful.

Candidate gene studies

Candidate gene studies are often driven first by phenotype. This can mean that either functional studies, often in rodents, or diseases in humans have implicated a gene in brain development. Indeed, it is often the neurodevelopmental research that drives the evolutionary studies rather than vice versa. This has not only allowed for a better focus on genes with phenotypic effects of interest, but it also lets us focus more carefully on the evolutionary questions. Significance thresholds for candidate gene studies need not account for multiple testing to the same degree as whole genome studies. Also the increased scrutiny lends itself to more subtle effects and nuanced interpretations. Researchers on candidate gene studies also, in general, have the greater personal investment required to follow up and understand the functional implications of the evolutionary changes. When so often the functionally relevant changes are hidden among selectively and functionally neutral changes, this increased motivation is crucial.

To this end, it is useful to compare several recent candidate gene studies (Fig. 4), beginning by reconsidering the situation of the genes implicated in primary microcephaly and their evolutionary history. The first two primary microcephaly genes identified and for which evolutionary studies were undertaken were *ASPM* and *microcephalin* (*MCPH1*). For each of these genes, the first studies, conducted independently in multiple labs, focused on interspecies comparisons. For *microcephalin*, the findings indicated positive selection in the lineages leading from old-world monkeys to the great apes (Evans et al., 2004a; Wang and Su, 2004). While elevated rates of protein evolution were found elsewhere, notably in the lineages following the divergence between gorillas and the hominins, they are not significant. In *ASPM*, on the other hand, significant positive selection is seen both in the human terminal lineage and in the lineage separating the great apes from the lesser apes (Evans et al., 2004b; Kouprina et al., 2004; Zhang, 2003a).

Several early studies of polymorphism in humans on these same genes also found evidence of nonneutral evolution (Evans et al., 2005; Mekel-Bobrov et al., 2005). This implication led to an interpretation that these genes were also under more recent selection. Later findings, however, suggested that the patterns of polymorphism seen in *ASPM* and *microcephalin*, while not expected under neutrality, were not uncommon in the *H. sapiens* genome (Currat et al., 2006; Yu et al., 2007). This suggested that demographic effects, rather than selection effects, were primarily responsible for the observed patterns of variation. Additionally, studies were unable to identify any phenotypic effect for these polymorphisms (Dobson-Stone et al., 2007; Woods et al., 2006).

The difficulty in conducting functional studies of these genes has played a significant role in interpreting evolutionary genetic findings. Like most genes implicated in developmental processes, *in vitro* studies of function are difficult to do. While *ex vivo* or *in vivo* models can be used to understand their roles and functions, it is difficult to assess the impact of specific mutations in a shared context or environment. This ultimately

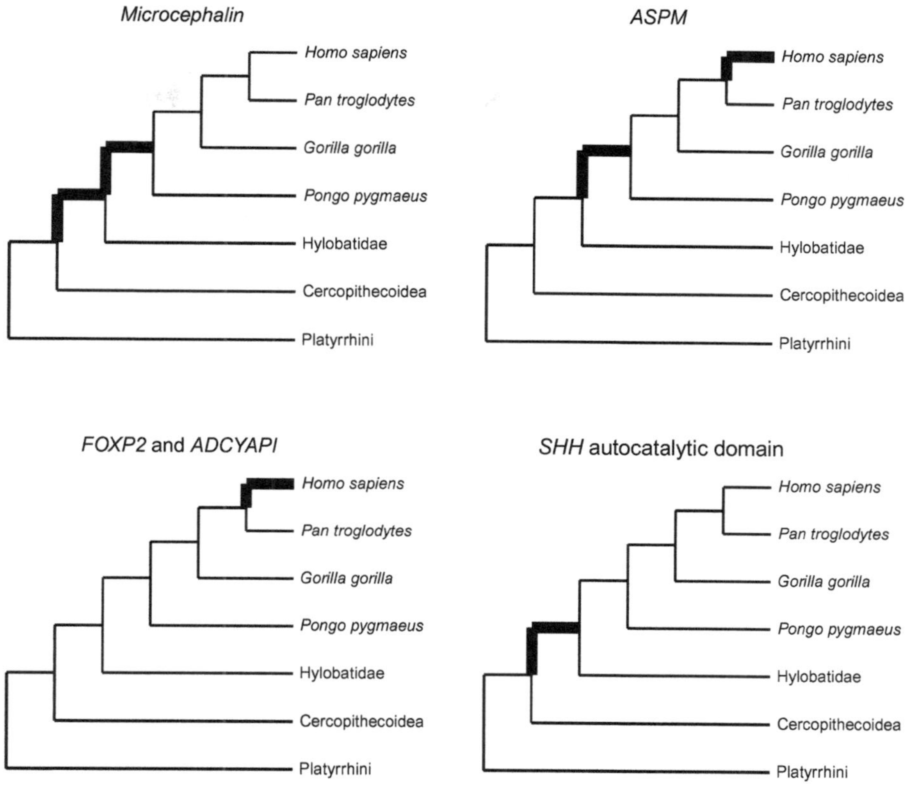

Fig. 4. Cladograms representing consensus findings on evolutionary histories of specific genes. Lineages in bold have been identified as undergoing positive selection for the gene noted.

has led to a difficulty in attributing definitive and unambiguous meanings to the explicit and unequivocal patterns in the genomes.

Another candidate gene study worthy of mention is *FOXP2*. Like the genes implicated in primary microcephaly, *FOXP2* initially garnered interest because of its association with a pathological disorder, in this case, a speech and language deficit (Lai et al., 2001). Evolutionary studies followed that showed a remarkable conservation of the gene across mammals, but significant multiple amino acid changes in the human lineage (Enard et al., 2002; Zhang et al., 2002). Since these studies, several additional findings have driven interest in the gene. The first was evidence that *FOXP2* plays an important role in song-learning in birds (Haesler et al., 2004). The second used "humanized" mice, transgenic mice with the human version of *FOXP2*, to demonstrate "qualitatively different ultrasonic vocalizations" (Enard et al., 2009). *In vitro* and *in vivo* studies were then developed that showed that human and chimpanzee *FOXP2* differentially regulated downstream gene expression (Konopka et al., 2009). Most recently, a study has shown that differentially expressed genes in the developing cerebral cortex, particularly in the Broca and Wernicke areas, show accelerated evolution in humans and are enriched for transcriptional targets of *FOXP2* (Lambert et al., 2011).

Perhaps most notable about *FOXP2* is how completely its evolutionary importance is driven first by functional concerns. Of all the genes considered here, and indeed perhaps among all the genes implicated in primate brain evolution, the case for *FOXP2* is perhaps the strongest yet the evidence at the genetic level is the weakest. Contrast *FOXP2* with *ADCYAP1*. *ADCYAP1* encodes a protein that has been shown to play a part in the transition from proliferative to differentiated states during neurogenesis (Dicicco-Bloom et al., 1998). It is one of the most divergent genes between humans and chimpanzees and has strong genetic evidence for positive selection in humans (Wang et al., 2005). In essence, the pattern of evolution for these two genes is the same, yet for one the functional evidence dominates, while for the other, the genetic evidence dominates. It is informative, and yet not surprising, where the balance of confidence lies. It also provides an instructive guideline for evolutionary studies moving forward.

Nontraditional substrates of evolution

In the past, evolutionary studies have focused primarily on point mutations and primary DNA sequences. As our understanding of biological function grows, so too have the possibilities for evolutionary action. Perhaps not surprisingly most of these revolve around gene regulation. Some of these novel substrates are natural extensions of existing concepts, such as the emergence of noncoding RNA genes. Though they operate in a manner unlike proteins or *cis*-regulatory sequences, positive selection on these regulatory nucleic acids can be identified in a similar manner. Indeed, already a human noncoding RNA has been identified that shows signatures of positive selection (Pollard et al., 2006). While the precise function of this gene is unknown, it is found in the neurons of the developing neocortex. Again, functional studies lag behind genetic implications, but the prospect of the finding is there.

Three other substrates for evolution recently under exploration similarly represent outgrowths from existing studies. Selection on regulatory regions may occur through selection on epigenetic markers (Enard et al., 2004), and early studies have identified differences in methylation patterns between humans and nonhuman primates in the brain (Farcas et al., 2009). Whole genes may not be emerging or vanishing as rapidly as we once thought, but the intriguing possibility is arising that alternative splice forms may be more variable between species than previously thought. Again, one early piece of evidence for this was a comparison between the human and nonhuman primate brain (Lin et al., 2010). Finally, plasticity in posttranslational modifications on proteins may represent another source of regulatory evolution.

The autocatalytic domain of the sonic hedgehog (*SHH*) gene was shown to be under positive selection in the lineage separating old-world monkeys from apes (Dorus et al., 2006). This was peculiar as *SHH* is one of the most preeminent developmental regulator genes with many functions across many tissues and developmental stages. A potential explanation may be found in the nature of those changes, a statistically nonrandom excess of serines and threonines, residues often the targets of posttranslational modifications. This raises the intriguing possibility that it is these regulatory effects that underlie the observed phenomena.

The future of primate brain evolution genetics

When viewing these studies from afar, one cannot help but notice the proliferation coinciding with the emergence and publication of the human, chimp, and rhesus genomes as well as the major human polymorphism surveys. The amount of data generated by these studies provided tremendous fodder for evolutionary geneticists eager to expand our understanding of brain evolution. Yet too often, these studies were "one-and-done"; genetic evidence implicating many genes in brain evolution has gone without follow-up.

This will be the next major hurdle faced by those seeking to understand primate brain evolution.

The mantra of evolutionary genetics, particularly regarding the primate brain, going forward will be function, function, and function. Descriptive genetic studies have demonstrated both their utility and their drawbacks. Leads have emerged by the dozens, if not hundreds, but until an unambiguous functionality can be associated with the putative selective mutations little will advance. This recognition is setting in and is now driving studies. Increasingly, we are seeing more studies focused on gene expression, a combined result of its likely importance in developmental regulation and the practicality of assaying these changes. This is not to say that there are not other approaches still to be taken or substrates to be found, simply that research has transitioned to the next lowest hanging fruit.

Perhaps the most important thing to emerge during this decade of explosive growth in primate evolutionary genetics is the increasing appreciation for the interconnectedness between fields. Researchers studying neurodevelopment can more easily and with more confidence interpret their findings through a comparative evolutionary framework. Evolutionary geneticists have an increasing appreciation and focus on the basic functional work of developmental neurobiologists. As we come to view primate brain evolution within a more complete and gestalt framework, we are increasingly likely to understand both its genetic and phenotypic roots and its implications for human health, biology, and our understanding of ourselves.

Acknowledgment

This work was supported by grants from the National Institutes of Health: AA019688 (E. J. V.) and RR000168.

References

Arbiza, L., Dopazo, J., & Dopazo, H. (2006). Positive selection, relaxation, and acceleration in the evolution of the human and chimp genome. *PLoS Computational Biology*, *2*, e38.

Boissinot, S., Tan, Y., Shyue, S. K., Schneider, H., Sampaio, I., Neiswanger, K., et al. (1998). Origins and antiquity of X-linked triallelic color vision systems in New World monkeys. *Proceedings of the National Academy of Sciences of the United States of America*, *95*, 13749–13754.

Burki, F., & Kaessmann, H. (2004). Birth and adaptive evolution of a hominoid gene that supports high neurotransmitter flux. *Nature Genetics*, *36*, 1061–1063.

Bustamante, C. D., Fledel-Alon, A., Williamson, S., Nielsen, R., Hubisz, M. T., Glanowski, S., et al. (2005). Natural selection on protein-coding genes in the human genome. *Nature*, *437*, 1153–1157.

Carlson, C. S., Thomas, D. J., Eberle, M. A., Swanson, J. E., Livingston, R. J., Rieder, M. J., et al. (2005). Genomic regions exhibiting positive selection identified from dense genotype data. *Genome Research*, *15*, 1553–1565.

Carroll, S. B. (2008). Evo-devo and an expanding evolutionary synthesis: A genetic theory of morphological evolution. *Cell*, *134*, 25–36.

Caveney, S., Cladman, W., Verellen, L., & Donly, C. (2006). Ancestry of neuronal monoamine transporters in the Metazoa. *Journal of Experimental Biology*, *209*, 4858–4868.

Choi, S. S., & Lahn, B. T. (2003). Adaptive evolution of MRG, a neuron-specific gene family implicated in nociception. *Genome Research*, *13*, 2252–2259.

Clark, A. G., Glanowski, S., Nielsen, R., Thomas, P. D., Kejariwal, A., Todd, M. A., et al. (2003). Inferring nonneutral evolution from human-chimp-mouse orthologous gene trios. *Science*, *302*, 1960–1963.

Cruz-Gordillo, P., Fedrigo, O., Wray, G. A., & Babbitt, C. C. (2010). Extensive changes in the expression of the opioid genes between humans and chimpanzees. *Brain, Behavior and Evolution*, *76*, 154–162.

Currat, M., Excoffier, L., Maddison, W., Otto, S. P., Ray, N., Whitlock, M. C., et al. (2006). Comment on "Ongoing adaptive evolution of ASPM, a brain size determinant in Homo sapiens" and "Microcephalin, a gene regulating brain size, continues to evolve adaptively in humans". *Science*, *313*, 172.

Da Fonseca, R. R., Kosiol, C., Vinar, T., Siepel, A., & Nielsen, R. (2010). Positive selection on apoptosis related genes. *FEBS Letters*, *584*, 469–476.

Dicicco-Bloom, E., Lu, N., Pintar, J. E., & Zhang, J. (1998). The PACAP ligand/receptor system regulates cerebral cortical neurogenesis. *Annals of the New York Academy of Sciences*, *865*, 274–289.

Dobson-Stone, C., Gatt, J. M., Kuan, S. A., Grieve, S. M., Gordon, E., Williams, L. M., et al. (2007). Investigation of MCPH1 G37995C and ASPM A44871G polymorphisms and brain size in a healthy cohort. *NeuroImage*, *37*, 394–400.

Dobyns, W. B. (2002). Primary microcephaly: New approaches for an old disorder. *American Journal of Medical Genetics*, *112*, 315–317.

Dong, D., He, G., Zhang, S., & Zhang, Z. (2009). Evolution of olfactory receptor genes in primates dominated by birth-and-death process. *Genome Biology and Evolution*, *1*, 258–264.

Dorus, S., Anderson, J. R., Vallender, E. J., Gilbert, S. L., Zhang, L., Chemnick, L. G., et al. (2006). Sonic Hedgehog, a key development gene, experienced intensified molecular evolution in primates. *Human Molecular Genetics*, *15*, 2031–2037.

Dorus, S., Vallender, E. J., Evans, P. D., Anderson, J. R., Gilbert, S. L., Mahowald, M., et al. (2004). Accelerated evolution of nervous system genes in the origin of Homo sapiens. *Cell*, *119*, 1027–1040.

Duret, L., & Mouchiroud, D. (2000). Determinants of substitution rates in mammalian genes: Expression pattern affects selection intensity but not mutation rate. *Molecular Biology and Evolution*, *17*, 68–74.

Enard, W., Fassbender, A., Model, F., Adorjan, P., Paabo, S., & Olek, A. (2004). Differences in DNA methylation patterns between humans and chimpanzees. *Current Biology*, *14*, R148–R149.

Enard, W., Gehre, S., Hammerschmidt, K., Holter, S. M., Blass, T., Somel, M., et al. (2009). A humanized version of Foxp2 affects cortico-basal ganglia circuits in mice. *Cell*, *137*, 961–971.

Enard, W., Przeworski, M., Fisher, S. E., Lai, C. S., Wiebe, V., Kitano, T., et al. (2002). Molecular evolution of FOXP2, a gene involved in speech and language. *Nature*, *418*, 869–872.

Evans, P. D., Anderson, J. R., Vallender, E. J., Choi, S. S., & Lahn, B. T. (2004a). Reconstructing the evolutionary history of microcephalin, a gene controlling human brain size. *Human Molecular Genetics*, *13*, 1139–1145.

Evans, P. D., Anderson, J. R., Vallender, E. J., Gilbert, S. L., Malcom, C. M., Dorus, S., et al. (2004b). Adaptive evolution of ASPM, a major determinant of cerebral cortical size in humans. *Human Molecular Genetics*, *13*, 489–494.

Evans, P. D., Gilbert, S. L., Mekel-Bobrov, N., Vallender, E. J., Anderson, J. R., Vaez-Azizi, L. M., et al. (2005). Microcephalin, a gene regulating brain size, continues to evolve adaptively in humans. *Science*, *309*, 1717–1720.

Evans, P. D., Vallender, E. J., & Lahn, B. T. (2006). Molecular evolution of the brain size regulator genes CDK5RAP2 and CENPJ. *Gene*, *375*, 75–79.

Farcas, R., Schneider, E., Frauenknecht, K., Kondova, I., Bontrop, R., Bohl, J., et al. (2009). Differences in DNA methylation patterns and expression of the CCRK gene in human and nonhuman primate cortices. *Molecular Biology and Evolution*, *26*, 1379–1389.

Finlay, B. L., & Darlington, R. B. (1995). Linked regularities in the development and evolution of mammalian brains. *Science*, *268*, 1578–1584.

Force, A., Lynch, M., Pickett, F. B., Amores, A., Yan, Y. L., & Postlethwait, J. (1999). Preservation of duplicate genes by complementary, degenerative mutations. *Genetics*, *151*, 1531–1545.

Gibbs, R. A., Rogers, J., Katze, M. G., Bumgarner, R., Weinstock, G. M., Mardis, E. R., et al. (2007). Evolutionary and biomedical insights from the rhesus macaque genome. *Science*, *316*, 222–234.

Gilad, Y., Bustamante, C. D., Lancet, D., & Paabo, S. (2003a). Natural selection on the olfactory receptor gene family in humans and chimpanzees. *American Journal of Human Genetics*, *73*, 489–501.

Gilad, Y., Man, O., Paabo, S., & Lancet, D. (2003b). Human specific loss of olfactory receptor genes. *Proceedings of the National Academy of Sciences of the United States of America*, *100*, 3324–3327.

Goldman, N., & Yang, Z. (1994). A codon-based model of nucleotide substitution for protein-coding DNA sequences. *Molecular Biology and Evolution*, *11*, 725–736.

Guernsey, D. L., Jiang, H., Hussin, J., Arnold, M., Bouyakdan, K., Perry, S., et al. (2010). Mutations in centrosomal protein CEP152 in primary microcephaly families linked to MCPH4. *American Journal of Human Genetics*, *87*, 40–51.

Haesler, S., Wada, K., Nshdejan, A., Morrisey, E. E., Lints, T., Jarvis, E. D., et al. (2004). FoxP2 expression in avian vocal learners and non-learners. *Journal of Neuroscience*, *24*, 3164–3175.

Haygood, R., Fedrigo, O., Hanson, B., Yokoyama, K. D., & Wray, G. A. (2007). Promoter regions of many neural- and nutrition-related genes have experienced positive selection during human evolution. *Nature Genetics*, *39*, 1140–1144.

Hoffman, M. M., & Birney, E. (2010). An effective model for natural selection in promoters. *Genome Research*, *20*, 685–692.

Hughes, A. L. (1994). The evolution of functionally novel proteins after gene duplication. *Proceedings. Biological Sciences*, *256*, 119–124.

International Hapmap Consortium. (2005). A haplotype map of the human genome. *Nature*, *437*, 1299–1320.

Jerison, J. H. (1973). *Evolution of the brain and intelligence*. New York: Academic Press.

Johnson, M. E., Viggiano, L., Bailey, J. A., Abdul-Rauf, M., Goodwin, G., Rocchi, M., et al. (2001). Positive selection of a gene family during the emergence of humans and African apes. *Nature*, *413*, 514–519.

Kaas, J. H. (2005). From mice to men: The evolution of the large, complex human brain. *Journal of Biosciences*, *30*, 155–165.

Kaindl, A. M., Passemard, S., Kumar, P., Kraemer, N., Issa, L., Zwirner, A., et al. (2010). Many roads lead to primary autosomal recessive microcephaly. *Progress in Neurobiology*, *90*, 363–383.

Kelley, J. L., Madeoy, J., Calhoun, J. C., Swanson, W., & Akey, J. M. (2006). Genomic signatures of positive selection in humans and the limits of outlier approaches. *Genome Research*, *16*, 980–989.

Khaitovich, P., Hellmann, I., Enard, W., Nowick, K., Leinweber, M., Franz, H., et al. (2005). Parallel patterns of evolution in the genomes and transcriptomes of humans and chimpanzees. *Science*, *309*, 1850–1854.

King, M. C., & Wilson, A. C. (1975). Evolution at two levels in humans and chimpanzees. *Science*, *188*, 107–116.

Konopka, G., Bomar, J. M., Winden, K., Coppola, G., Jonsson, Z. O., Gao, F., et al. (2009). Human-specific transcriptional regulation of CNS development genes by FOXP2. *Nature*, *462*, 213–217.

Kouprina, N., Pavlicek, A., Mochida, G. H., Solomon, G., Gersch, W., Yoon, Y. H., et al. (2004). Accelerated evolution of the ASPM gene controlling brain size begins prior to human brain expansion. *PLoS Biology*, *2*, E126.

Lai, C. S., Fisher, S. E., Hurst, J. A., Vargha-Khadem, F., & Monaco, A. P. (2001). A forkhead-domain gene is mutated in a severe speech and language disorder. *Nature*, *413*, 519–523.

Lambert, N., Lambot, M. A., Bilheu, A., Albert, V., Englert, Y., Libert, F., et al. (2011). Genes expressed in specific areas of the human fetal cerebral cortex display distinct patterns of evolution. *PloS One*, *6*, e17753.

Li, W. H., Wu, C. I., & Luo, C. C. (1985). A new method for estimating synonymous and nonsynonymous rates of nucleotide substitution considering the relative likelihood of nucleotide and codon changes. *Molecular Biology and Evolution*, *2*, 150–174.

Lin, L., Shen, S., Jiang, P., Sato, S., Davidson, B. L., & Xing, Y. (2010). Evolution of alternative splicing in primate brain transcriptomes. *Human Molecular Genetics*, *19*, 2958–2973.

Mekel-Bobrov, N., Gilbert, S. L., Evans, P. D., Vallender, E. J., Anderson, J. R., Hudson, R. R., et al. (2005). Ongoing adaptive evolution of ASPM, a brain size determinant in Homo sapiens. *Science*, *309*, 1720–1722.

Miyata, T., Kuma, K., Iwabe, N., & Nikoh, N. (1994). A possible link between molecular evolution and tissue evolution demonstrated by tissue specific genes. *Japanese Journal of Genetics*, *69*, 473–480.

Nei, M., & Gojobori, T. (1986). Simple methods for estimating the numbers of synonymous and nonsynonymous nucleotide substitutions. *Molecular Biology and Evolution*, *3*, 418–426.

Nielsen, R., Bustamante, C., Clark, A. G., Glanowski, S., Sackton, T. B., Hubisz, M. J., et al. (2005). A scan for positively selected genes in the genomes of humans and chimpanzees. *PLoS Biology*, *3*, e170.

Nowick, K., Hamilton, A. T., Zhang, H. M., & Stubbs, L. (2010). Rapid sequence and expression divergence suggest selection for novel function in primate-specific KRAB-ZNF genes. *Molecular Biology and Evolution*, *27*, 2606–2617.

Ohno, S. (1970). *Evolution by gene duplication.* New York: Springer-Verlag.

Ohno, S. (1972). Argument for genetic simplicity of man and other mammals. *Journal of Human Evolution*, *1*, 651–662.

Perry, G. H., Verrelli, B. C., & Stone, A. C. (2005). Comparative analyses reveal a complex history of molecular evolution for human MYH16. *Molecular Biology and Evolution*, *22*, 379–382.

Plaitakis, A., Spanaki, C., Mastorodemos, V., & Zaganas, I. (2003). Study of structure-function relationships in human glutamate dehydrogenases reveals novel molecular mechanisms for the regulation of the nerve tissue-specific (GLUD2) isoenzyme. *Neurochemistry International*, *43*, 401–410.

Pollard, K. S., Salama, S. R., Lambert, N., Lambot, M. A., Coppens, S., Pedersen, J. S., et al. (2006). An RNA gene expressed during cortical development evolved rapidly in humans. *Nature*, *443*, 167–172.

Popesco, M. C., Maclaren, E. J., Hopkins, J., Dumas, L., Cox, M., Meltesen, L., et al. (2006). Human lineage-specific amplification, selection, and neuronal expression of DUF1220 domains. *Science*, *313*, 1304–1307.

Putcha, G. V., Harris, C. A., Moulder, K. L., Easton, R. M., Thompson, C. B., & Johnson, E. M. Jr. (2002). Intrinsic and extrinsic pathway signaling during neuronal apoptosis: Lessons from the analysis of mutant mice. *The Journal of Cell Biology*, *157*, 441–453.

Rockman, M. V., Hahn, M. W., Soranzo, N., Zimprich, F., Goldstein, D. B., & Wray, G. A. (2005). Ancient and recent positive selection transformed opioid cis-regulation in humans. *PLoS Biology*, *3*, e387.

Sabeti, P. C., Varilly, P., Fry, B., Lohmueller, J., Hostetter, E., Cotsapas, C., et al. (2007). Genome-wide detection and characterization of positive selection in human populations. *Nature*, *449*, 913–918.

Semendeferi, K., Lu, A., Schenker, N., & Damasio, H. (2002). Humans and great apes share a large frontal cortex. *Nature Neuroscience*, *5*, 272–276.

Shashidharan, P., Michaelidis, T. M., Robakis, N. K., Kresovali, A., Papamatheakis, J., & Plaitakis, A. (1994). Novel human glutamate dehydrogenase expressed in neural and testicular tissues and encoded by an X-linked intronless gene. *Journal of Biological Chemistry*, *269*, 16971–16976.

Shyue, S. K., Hewett-Emmett, D., Sperling, H. G., Hunt, D. M., Bowmaker, J. K., Mollon, J. D., et al. (1995). Adaptive evolution of color vision genes in higher primates. *Science*, *269*, 1265–1267.

Soranzo, N., Bufe, B., Sabeti, P. C., Wilson, J. F., Weale, M. E., Marguerie, R., et al. (2005). Positive selection on a high-sensitivity allele of the human bitter-taste receptor TAS2R16. *Current Biology*, *15*, 1257–1265.

Stedman, H. H., Kozyak, B. W., Nelson, A., Thesier, D. M., Su, L. T., Low, D. W., et al. (2004). Myosin gene mutation correlates with anatomical changes in the human lineage. *Nature*, *428*, 415–418.

The Chimpanzee Sequencing and Analysis Consortium. (2005). Initial sequence of the chimpanzee genome and comparison with the human genome. *Nature*, *437*, 69–87.

Tuller, T., Kupiec, M., & Ruppin, E. (2008). Evolutionary rate and gene expression across different brain regions. *Genome Biology*, *9*, R142.

Vallender, E. J., & Lahn, B. T. (2006). A primate-specific acceleration in the evolution of the caspase-dependent apoptosis pathway. *Human Molecular Genetics*, *15*, 3034–3040.

Vincent, J. D., Cardinaud, B., & Vernier, P. (1998). Evolution of monoamine receptors and the origin of motivational and emotional systems in vertebrates. *Bulletin of the National Academy of Medicine*, *182*, 1505–1514 discussion 1515–1516.

Voight, B. F., Kudaravalli, S., Wen, X., & Pritchard, J. K. (2006). A map of recent positive selection in the human genome. *PLoS Biology*, *4*, e72.

Wang, Y. Q., Qian, Y. P., Yang, S., Shi, H., Liao, C. H., Zheng, H. K., et al. (2005). Accelerated evolution of the pituitary adenylate cyclase-activating polypeptide precursor gene during human origin. *Genetics*, *170*, 801–806.

Wang, Y. Q., & Su, B. (2004). Molecular evolution of microcephalin, a gene determining human brain size. *Human Molecular Genetics*, *13*, 1131–1137.

Williamson, S. H., Hubisz, M. J., Clark, A. G., Payseur, B. A., Bustamante, C. D., & Nielsen, R. (2007). Localizing recent adaptive evolution in the human genome. *PLoS Genetics*, *3*, e90.

Wong, W. S., & Nielsen, R. (2004). Detecting selection in non-coding regions of nucleotide sequences. *Genetics*, *167*, 949–958.

Wooding, S., Bufe, B., Grassi, C., Howard, M. T., Stone, A. C., Vazquez, M., et al. (2006). Independent evolution of bitter-taste sensitivity in humans and chimpanzees. *Nature*, *440*, 930–934.

Wooding, S., Kim, U. K., Bamshad, M. J., Larsen, J., Jorde, L. B., & Drayna, D. (2004). Natural selection and molecular evolution in PTC, a bitter-taste receptor gene. *American Journal of Human Genetics*, *74*, 637–646.

Woods, C. G., Bond, J., & Enard, W. (2005). Autosomal recessive primary microcephaly (MCPH): A review of clinical, molecular, and evolutionary findings. *American Journal of Human Genetics*, *76*, 717–728.

Woods, R. P., Freimer, N. B., De Young, J. A., Sicotte, N. L., Service, S. K., et al. (2006). Normal variants of Microcephalin and ASPM do not account for brain size variability. *Human Molecular Genetics*, *15*, 2025–2029.

Yang, Z. (2007). PAML 4: Phylogenetic analysis by maximum likelihood. *Molecular Biology and Evolution*, *24*, 1586–1591.

Yang, Z., & Nielsen, R. (2000). Estimating synonymous and nonsynonymous substitution rates under realistic evolutionary models. *Molecular Biology and Evolution*, *17*, 32–43.

Young, J. M., Friedman, C., Williams, E. M., Ross, J. A., Tonnes-Priddy, L., & Trask, B. J. (2002). Different evolutionary processes shaped the mouse and human olfactory receptor gene families. *Human Molecular Genetics*, *11*, 535–546.

Yu, F., Hill, R. S., Schaffner, S. F., Sabeti, P. C., Wang, E. T., Mignault, A. A., et al. (2007). Comment on "Ongoing adaptive evolution of ASPM, a brain size determinant in Homo sapiens." *Science*, *316*, 370.

Zhang, J. (2003a). Evolution of the human ASPM gene, a major determinant of brain size. *Genetics*, *165*, 2063–2070.

Zhang, J. Z. (2003b). Evolution by gene duplication: An update. *Trends in Ecology & Evolution*, *18*, 292–298.

Zhang, J., Webb, D. M., & Podlaha, O. (2002). Accelerated protein evolution and origins of human-specific features: Foxp2 as an example. *Genetics*, *162*, 1825–1835.

Zhou, Y. H., & Li, W. H. (1996). Gene conversion and natural selection in the evolution of X-linked color vision genes in higher primates. *Molecular Biology and Evolution*, *13*, 780–783.

M. A. Hofman and D. Falk (Eds.)
Progress in Brain Research, Vol. 195
ISSN: 0079-6123

CHAPTER 3

Cerebral cortical development in rodents and primates

Zoltán Molnár[†,*] and Gavin Clowry[‡]

[†] *Department of Physiology, Anatomy and Genetics, University of Oxford, Oxford, United Kingdom*
[‡] *Institute of Neuroscience, Newcastle University, Newcastle upon Tyne, United Kingdom*

Abstract: Rodents and primates both show considerable variation in the overall size, the radial and tangential dimensions, folding and subdivisions into distinct areas of their cerebral cortex. Our current understanding of brain development is based on a handful of model systems. A detailed comparative analysis of the cellular and molecular mechanisms that regulate neural progenitor production, cell migration, and circuit assembly can provide much needed insights into the working of neocortical evolution. From the limited comparative data currently available, it is apparent that the emergence and variation of the neuronal progenitor cells have led to the production of increased neuronal populations and the evolution of the cortex. Further diversification and compartmentalization of the germinal zone together with changing proportions of radial glia in the ventricular zone and various intermediate progenitors in the subventricular zone may have been the driving force behind increased cell numbers in larger brains both in rodents and primates. Radial and tangential migratory patterns are both present in rodents and primates, but in different proportions. There are apparent differences between mouse and human in the generation and elaboration of the interneuronal subtypes and also in gene expression patterns associated with the appearance of distinct cortical areas. The increased cortical dimensions and the formation of a more elaborate cortical architecture in primates require a larger and more compartmentalized transient subplate zone during development. More comparative analysis in rodent and primate species with large, small, and smooth and folded brains is needed to reveal the biological significance of the alterations in these cortical developmental programs.

Keywords: cerebral cortex; neurogenesis; neural progenitors; neural migration; ventricular zone; subventricular zone; subplate zone; thalamocortical projections.

Introduction

The cerebral cortex has been historically considered as part of the brain in which resides the vastly increased cognitive capacity that

*Corresponding author.
Tel.: +44-865-272-169; Fax: +44-1865-272488
E-mail: zoltan.molnar@dpag.ox.ac.uk

DOI: 10.1016/B978-0-444-53860-4.00003-9

distinguishes us from other species (Willis, 1664; see Molnár, 2004). Understanding the evolution and development of this complex structure is therefore central to our understanding of human intelligence and creativity, and also of disorders of cognitive functions. Evolutionary expansion in size and complexity of the human cerebral cortex is probably a result of changes in both the molecular mechanisms of both cell proliferation and phenotypic differentiation. Although the basic principles are similar in all mammalian species, the modulation of developmental mechanisms leads to the emergence of new neuronal subclasses and the addition of more specialized cortical areas in human and nonhuman primates. Cortical expansion in primates is not just quantitative; there are some unique types of neurons and novel cytoarchitectonic areas identified by their gene expression, connectivity, and functions that do not exist in rodents (Clowry et al., 2010). Many specifically human psychiatric and neurological conditions have developmental origins, they involve the cortex among other structures, but the causes and remedies are not known. The prevalence of developmental disorders in the population is high [schizophrenia (1:140, Saha et al., 2005); autism spectrum disorders (1:85, Baird et al., 2006); attention deficit hyperactivity disorder (1:20, Polanczyk et al., 2007); childhood epilepsy (1:120 Oka et al., 2006)]. In spite of recent progress (Bystron et al., 2008; Meyer, 2007; Rakic, 2009), we are only beginning to understand basic neural developmental mechanisms and their involvement in the pathomechanisms of several debilitating diseases. The human cerebral cortex has some unique genetic, molecular, cellular, and anatomical features that are not always replicated in the usual animal models. Rodents are extremely valuable for the investigation of brain development but cannot provide insight into aspects that are specifically primate or even human. Therefore, there is a need to investigate primate cortical development to link our knowledge in simpler rodent model systems to that of humans. This chapter aims to compare some key cerebral cortical developmental steps in rodents and primates such as the formation of the first postmitotic cell layers of the preplate, the compartmentalization of the germinal zone, the formation of the earliest connections, the development of the cortical plate and subplate, and the development of functionally distinct cortical areas and of functional specializations in one or the other hemisphere. We shall emphasize the lack of systematic and quantitative comparative work in this area that is essential to validate our currently used disease models.

Rodent and primate cortices demonstrate much heterogeneity in their radial and tangential dimensions and folding patterns

All mammalian cerebral neocortices have a uniform laminar structure that has been historically divided into six layers (Brodmann, 1908; Economo and Koskinas, 1925, 2008). The layering is apparent even in Nissl-stained sections because each layer contains different cell types with distinct concentrations and distributions of RNA and their somata populating the cerebral cortex at different depths (Jones, 2000; Lorente de No, 1949; Peters and Yilmaz, 1993; Ramón y Cajal, 1909). This layering is reflected in the differences in gene expression patterns in the adult (Belgard et al., 2011). The deep layers of the cortex, V and VI, form longer distance projections to subcortical targets (including thalamus, striatum, basal pons, tectum, and spinal cord) and to the opposite hemisphere. Some of the layer V pyramidal neurons are among the largest cells of the brain with the longest connections, but even this single layer can contain several subtypes (Molnár and Cheung, 2006; Molyneaux et al., 2007). The upper layers (I–IV) contain smaller pyramidal neurons that tend to form more shorter range intracortical connections, although interhemispheric connections from these layers do occur, and are believed to process information more locally (Gilbert and Wiesel, 1979; Toyama et al., 1974).

Our current understanding of adult brain structure and brain development is based on a handful of animal models. The common house mouse is usually used as a rodent model, while macaque and human brains serve as traditional primate models. However, the order Rodentia consists of 2277 recognized species (Wilson and Reeder, 2005) making up 42% of mammalian diversity, while the primate order contains 376 species. Thus, generalizations of any nature must be made with caution, as both anatomical and morphological variations are present in both of these orders. Rodent and primate species show much variation in cortical cytoarchitecture, in number of functional areas, in composition of cortical layers (Krubitzer and Kaas, 2005). This emphasizes the need for comparative studies of several other species. Figure 1 (adopted from Cheung et al., 2007) accentuates that there are examples of a considerable amount of variation of brain and cortex size between rodent species and also between primate species; moreover, the complexity of the cortical folding in the representative species can also differ. There are examples for lissencephalic primates (Potto) and gyrencephalic rodents (Capybara); therefore, generalizations from a handful of species can be dangerous (Cheung et al., 2007; Clowry et al., 2010; Molnár et al., 2006; Pillay and Manger, 2007).

The quantitative aspects of cortical differences between rodents and primates are dealt with in the chapters of Charvet and Finlay and Herculano-Houzel in more detail. These chapters emphasize that both radial and tangential parameters show great deal of variations in these orders. It has been widely believed that in spite of

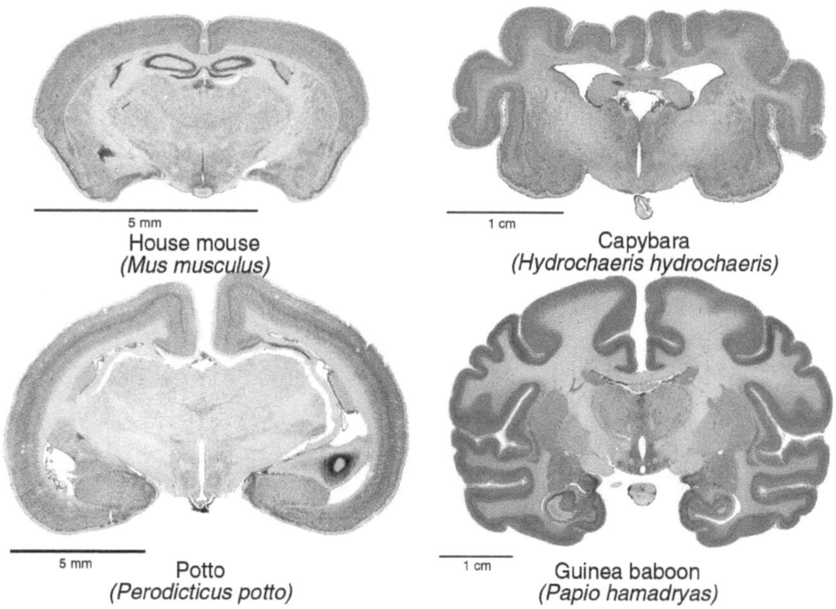

Fig. 1. In rodents and primates, the overall size, the radial and tangential dimensions, and the folding of the cerebral cortex vary considerably. The examples demonstrate that the shape and size of mammalian brains are different in spite of the basic uniformity of the six-layered mammalian neocortex. Examples for lissencephalic (left column) and gyrencephalic (right column) brains from rodents (upper row) and from primates (lower row). Images were taken from the University of Wisconsin–Madison Brain Collection (http://www.brainmuseum.org/). Reprinted with permission from Blackwell-Wiley Journal of Anatomy 211, 164-176 Cheung et al. (2007).

the variations in cortical thickness and in the relative proportions of the layers across various species, the number of neurons in a cortical column is largely constant in different species (Rockel et al., 1980). This concept derived from the original studies of Powell and colleagues who quantified neuronal cell bodies in a 30-μm wide radial strip through the depth of the neocortex in different functional areas (motor, somatic sensory, area 17, frontal, parietal, and temporal) in mouse, rat, cat, monkey, and human (Rockel et al., 1980). These studies emphasized that the same absolute number of neurons (110 per radial strip) was found in all areas of all species with only one exception: the binocular part of area 17 of primates, which had approximately 2.5 times more neurons. While it is still valid that the tangential extent of the cortex shows much more variation than the radial, the rigidly conserved cortical cell number doctrine has not been supported by recent quantitative work (see e.g., Chapter 10). The issue of neuronal numbers has been examined with novel methodologies, and great variations have been revealed (Herculano-Houzel et al., 2008; Chapter 15 of this volume). Even using traditional methods (counting in Nissl-stained sections in an arbitrary radial strip of 100 μm) in mouse and macaque suggests that the variation of cell counts in different functional regions is greater than previously reported (Fig. 2; Cheung et al., 2007, 2010). Although these more recent counts still emphasize the increased cell number in the primary visual cortex in macaque compared to other cortical areas in macaque or mouse, it is also apparent that the neurons in mouse S1 outnumber the macaque S1 considerably in an arbitrarily selected 100-μm radial strip. The neuron/glia ratio also shows great areal and species-specific variation (Cheung et al., 2007; Fig. 2b).

Further confirmation of the substantial differences in cortical cell numbers has been provided in studies that have been extending the quantification to marsupials. Marsupials have a six-layered dorsal cerebral cortex that appears very similar to that of other mammals. However, comparisons of cortical neuron numbers in a unit column of adults revealed that adult South American gray short-tailed opossums and tammar wallabies possess just half of the cerebral cortical neurons in an arbitrary unit column in the primary somatosensory cortex compared with the mouse (Cheung et al., 2010). These differences were seen in both infragranular

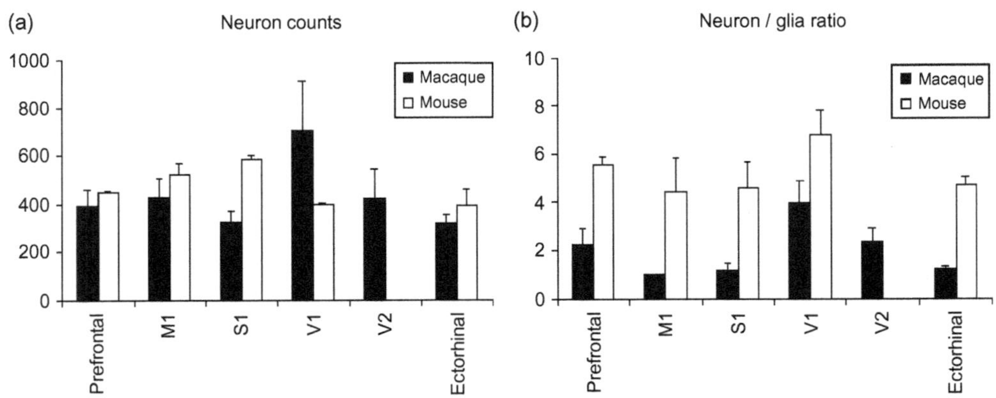

Fig. 2. Average (±SEM) (a) neuronal cell count and (b) neuron/glia ratio in mouse and macaque. Data were obtained from three radial strips in each functional region (except M1 of macaque, $n=2$). The variation of cell count in different functional regions is greater than previously reported, and one-way ANOVA analysis indicates that there are significant differences between regions in the neuronal cell count in mouse ($P<0.005$). Reprinted by permission of Blackwell-Wiley Journal of Anatomy 211, 164-176 Cheung et al. (2007).

(lower) and supragranular (upper) layers equally. Interestingly, the glial numbers were similar to the ones observed in corresponding areas in mouse. These observations on the quantitative differences between different mammalian species and cortical areas emphasize the need for further quantitative comparisons.

General developmental pattern of the mammalian cerebral cortex

The six layers of the neocortex develop from the thin-walled telencephalic vesicles over the course of days in mice and months in humans (Rakic, 2009). Cell division primarily takes place in the germinal zone surrounding the lateral ventricle—named the ventricular zone (VZ)—and cells have to migrate to their final position in the cortex (Bystron et al., 2008). Pyramidal neurons and interneurons are generated in different parts of the VZ. Pyramidal neurons arise from the VZ directly underneath the telencephalic wall and migrate radially toward the pial surface to form the cortical plate, whereas in rodents at least, interneurons are generated in the lateral and medial ganglionic eminences (future striatum or basal ganglia) and migrate laterally into the cortex (de Carlos et al., 1996; Marin and Rubenstein, 2003; Parnavelas, 2000). But irrespective of their site of origin, cortical neurons eventually settle in the cortex in an inside-out manner, such that late-born cells (even the tangentially migrating ones) settle in progressively more outward layers.

The first postmitotic neurons form the preplate (also referred to as primordial plexiform layer (PPL)). The preplate is on the outside of the cerebral wall and is populated by neurons in an outside-in sequence, such that later born cells settle underneath the earlier born cells and thereby "push" the older ones outward. The preplate and VZ are separated by the intermediate zone (IZ), which consists of tangentially oriented fibers and will eventually develop into the white matter underneath the cortex. Some of the earliest cells in the preplate derive from the subpallium and arrive in the preplate by tangential migration (Bystron et al., 2006; Meyer, 2007).

Subsequently, putative pyramidal cells migrate into the preplate and split it into two parts: the presubplate at the inner edge and the marginal zone at the outside. Although initially derived from the same cell population, and indistinguishable from each other, the cells located in the marginal zone develop different neurochemical properties from those of the subplate. The marginal zone contains the reelin expressing Cajal–Retzius cells (Meyer, 2007).

Distinctions in the preplate stage between primate and rodent

One would expect that these earliest steps of development are almost identical between studied rodents (mouse) and primates (human and nonhuman primates). However, this does not seem to be the case. Bystron and colleagues demonstrated that the first postmitotic neurons, the predecessor neurons in the PPL of the human dorsal cerebral cortex at Carnegie Stage 13 (4–4.5 gestational weeks (GWs)), were not seen in mouse (Bystron et al., 2006). Using markers for newly generated neurons, it was demonstrated that predecessor cells and their tangentially oriented processes populate the PPL in the telencephalon, with a clear inner to outer density gradient prior to local cortical neurogenesis (Bystron et al., 2006; Carney et al., 2007). These predecessor neurons are not immunoreactive for reelin and are different from Cajal–Retzius neurons. At later stages of development, comparing the preplate in mouse and human, Meyer and her colleagues demonstrated that the earliest preplate has a more complex structure in humans than in rodents (Meyer, 2007). Human preplate includes the Cajal–Retzius cells, important in guiding cell migration, and pioneer neurons that form the first axons to leave the cortex (Suárez-Solá et al., 2009). The role of both of these early-generated cell populations are more complicated in human than in mouse. In addition, layer I in the fetal cerebral cortex of human and nonhuman primates contains a large subpial granular layer that does not exist in rodents (Brun, 1965; Gadisseux et al., 1992; Zecevic and Rakic, 2001).

Cortical germinal zones in rodents and primates

Evolution must have acted on the existing basic developmental apparatus in the telencephalon in order to produce the observed diversity of adult forms. Comparative studies can show where mechanisms are shared and where they are distinct and contribute toward the comprehension of major cellular and molecular mechanisms of increased cortical neurogenesis in mammals. Therefore, there is increasing interest in understanding the alterations in neurogenesis in various mammals that are related to the variations in cell numbers, folding, and overall brain size (Kriegstein et al., 2006; Lui et al., 2011; Rakic et al., 2009). These studies have revealed that the VZ is present in the inner lining of the neural tube in all vertebrates; its cortical section lies at the most anterior portion of the telencephalon and is a proliferative layer of radially oriented neuroepithelial progenitor cells (historically called radial "glia"), whose nuclei go through a series of stereotyped movements (interkinetic nuclear migration) during mitotic cycles and which ultimately give rise to several types of cells within the cerebral cortex (Bystron et al., 2008). In early vertebrate embryonic corticogenesis, neuroepithelial cells in the VZ divide symmetrically to produce two identical cells, thus increasing the pool of neuroprogenitors. In mammals, a subset of neuroepithelial cells will differentiate into radial glial cells (RGCs) that extend a process to the cortical plate and guide migrating neurons to their ultimate destination, which follows an inside-out gradient (Rakic, 2009). Although radial "glia" progenitor cells are present in all vertebrates, the rate at which they produce progenitors and the length of time over which neurogenesis occurs can vary (Dehay and Kennedy, 2007; Rakic, 1995) and these differences can account for some of the tremendous variability observed in the size of the cortical sheet in different species.

Intermediate progenitors amplify the output of the cortical germinal zone

Due to the development of techniques imaging and following cell division and migration, additional neurogenic divisions were discovered outside the VZ in mouse cortex in 2004. Three separate studies demonstrated that radial glia can divide asymmetrically to produce a replacement RGC and either a mature cortical neuron that travels to its appropriate layer or an intermediate progenitor cell (IPC) that translocates as an entire cell to the zone adjacent to the VZ (subventricular zone, SVZ) (Haubensak et al., 2004; Miyata et al., 2004; Noctor et al., 2004). These progenitors do not demonstrate interkinetic nuclear migration during their mitotic cycles. Where this distinct embryonic mammalian progenitor compartment is larger, the number of neurons generated is amplified by increasing the rate and duration of neurogenesis. Instead of producing one neuron, RGCs can undergo asymmetric division to produce another RGC progenitor and one IPC (or basal progenitor cell) that subsequently migrates to the SVZ. Once in the SVZ, the IPCs undergo symmetric divisions to produce either two identical neurons destined for the same cortical layer or two daughter IPCs that propagate the cycle. IPCs can undergo 1–3 symmetric divisions to considerably amplify neuron production. Relative to the single neuron output by asymmetric division of RGCs, this two-step pattern of neurogenesis increases the output of neurons while maintaining a pool of neuroprogenitors (Martínez-Cerdeño et al., 2006; Pontious et al., 2008). The regulation of these progenitors is likely to be important in determining brain size.

We examined the pattern of dividing cells and the orientation of mitotic spindles in the VZ of opossums and wallabies where cortical neuronal numbers in an arbitrary unit column are considerably lower than in the mouse (Cheung et al., 2010). We found that the basic patterns of the cell division and the presence of an organized SVZ are conserved (Cheung et al., 2010). However, the emergence of a distinctive band of dividing cells in the SVZ occurs later in the opossum [postnatal day 14 (P14)] and the tammar wallaby (P40) than in mouse and rat (E14-15) (Carney et al., 2007; Cheung et al., 2010). This late appearance of the SVZ divisions raises a question about the existence of the neurogenic intermediate progenitors in the SVZ zone in marsupial cortex (Abdel-Mannan et al., 2008; Puzzolo and Mallamaci, 2010; Reynolds et al., 1985) or suggests that there is only a rudimentary small population. Interestingly, the comparisons of selected genes with known cortical VZ and SVZ expression in the mouse and opossum revealed very similar expression patterns (Cheung et al., 2010; Chodroff et al., 2010; Puzzolo and Mallamaci, 2010). Even the SVZ-specific vascular patterning was apparent in opossum (Cheung et al., 2010). These observations suggest that the reductions in the number of cortical neurons in an arbitrary unit column in opossum compared to mouse are associated with the reduction and late appearance of the SVZ in marsupials. It also suggests that, in spite of these variations in timing and proportions of SVZ progenitors of the entire proliferative pool, the cortical SVZ and the intermediate SVZ progenitors have been conserved across all studied mammals.

Several transcription factors are expressed at specific times and in specific regions during cerebral cortex formation in mouse (Britanova et al., 2008; Guillemot et al., 2006; Tarabykin et al., 2001). Differential expression of these genes in a specific spatiotemporal pattern has been hypothesized to promote neuronal diversity (Hevner et al., 2006; Molnár and Cheung, 2006; Molyneaux et al., 2007). The transcription factors Tbr2, Cux1, Cux2, Svet1, Satb2, and NeuroD specifically label the mammalian pallial SVZ during cortical neurogenesis and therefore might serve a critical role in its formation and pattern of neurogenesis (Britanova et al., 2008; Guillemot et al., 2006; Hevner, 2006). This spatiotemporal pattern of transcription factor expression and neurogenesis may be coupled to changes in the transient vasculature in the germinal zone. The dense and elaborate vasculature in the cortical germinal zone shows signs of dynamic rearrangements during cortical neurogenesis and dividing neuronal profiles are commonly observed close to the vessels (Javaherian and Kriegstein, 2009; Nie et al., 2010; Stubbs et al., 2009).

Multiple progenitor subtypes in the cortical germinal zone in mouse and human

Smart et al. (2002) described cytoarchitectonic distinctions within the germinal zone in macaque and human cortices (Fig. 3; see also Chapter 16). This consisted of a compartmentalized SVZ, subdivided into an inner subventricular zone (ISVZ) and outer subventricular zone (OSVZ) separated by a thin layer of fibers, the inner fiber layer (IFL) (Fig. 3a). Subsequent studies revealed characteristic gene expression patterns of these zones in carnivores and primates (Bayatti et al., 2008a; Fietz et al., 2010; Fish et al., 2008; Hansen et al., 2010). The progenitors in the OSVZ have a radial arrangement, reaching to the pial surface, but they have no endfeet and their nuclei stop short of the ventricular surface. The nuclei of these progenitors line up within the OSVZ, and they demonstrate interkinetic nuclear migration in the direction opposite to that of the VZ radial glia progenitors (Hansen et al., 2010). They move toward the pia (but they do not leave the germinal zone) and subsequently move away, back toward the germinal zone. These OSVZ progenitors were initially associated exclusively with developing primate cortex and were considered specific characteristics of the primate lineage (Hansen et al., 2010; Smart et al., 2002), but similar cells have recently been described in ferrets and more recently in mice (Fietz et al., 2010; Wang et al., 2011, respectively).

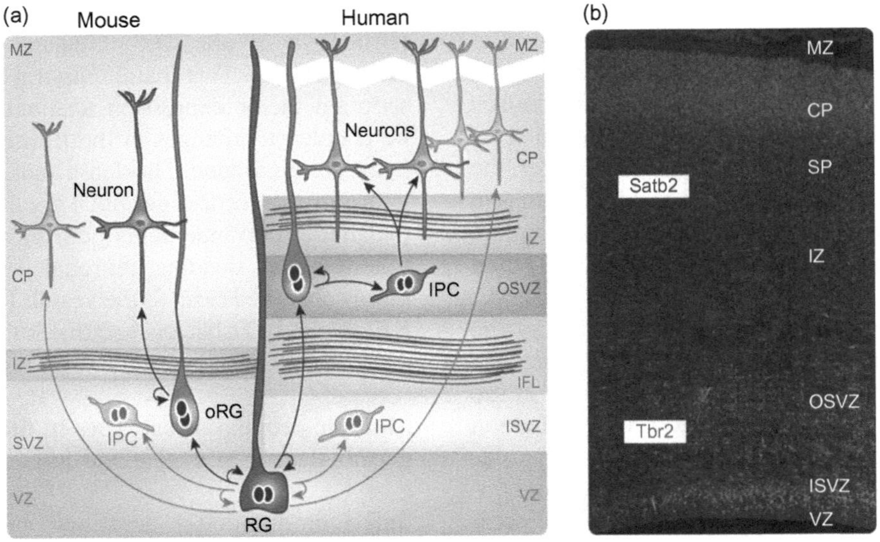

Fig. 3. (a) Comparisons between mouse and human cortices during neurogenesis. Similar progenitor cell types reside in the mouse (left) and human (right) developing cortex. These include the radial glial cells (RGCs; brown), which divide at the ventricular surface to generate an IPC (orange), an oRG progenitor cell (purple), or a neuron (blue) while also renewing themselves (circular arrows). The mouse and human OSVZ or oRG progenitor cells both lack contact with the ventricular (apical) surface (bottom) but maintain their pial (basal) processes to the pial surface. The "apical" and "basal" terms are used on cell biological, rather than anatomical, grounds and reflect the relationship with the basal membrane. The human cortex differs from that of the mouse in two respects. First, human but not mouse cortex has a cytoarchitectonically distinct extra germinal zone, the OSVZ (dark green) that is separated from the ISVZ by an inner fiber layer. Second, oRG cells in mouse generate neurons directly through self-renewing asymmetric division, whereas human oRG cells generate neurons through transit-amplifying cells (i.e., IPCs). Reprinted by permission from Macmillan Publishers Ltd: Nature Neuroscience 14, 1-3 Molnár et al. (2011). (b) In human, in addition to ISVZ progenitors, there are many more Tbr2+ IPCs in the OSVZ. Therefore, the difference in these Tbr2+ IPCs could be a key factor in the production of cortical size and evolution. Both of these differences could have evolved as mechanisms for increased production of neurons (reproduced with permission from Clowry et al., 2010). MZ, marginal zone; CP, cortical plate; SP, subplate; IZ, intermediate zone; OSVZ, outer subventricular zone; IFL, inner fiber layer; ISVZ, inner subventricular zone; VZ, ventricular zone; oRG, outer radial glia; IPC, intermediate progenitor cell; RGC, radial glial cell; Satb2, AT-rich DNA-binding protein. (For interpretation of the references to color in this figure legend, the reader is referred to the Web version of this chapter.)

Studies in human and ferret have revealed the distinct morphology and somal translocation behavior of OSVZ radial glia (oRGs; Fietz et al., 2010; Hansen et al., 2010). Besides having a process directed to the pia and lacking any contact with the ventricular surface, these cells also express progenitor-specific markers such as Sox2 and Pax6 and are negative for the IPC marker Tbr2 in human and ferret. These studies prompted the search for similar cells in the mouse and revealed that proliferative cell divisions occur within the upper SVZ in mice during embryonic development (Wang et al., 2011). In fact, mice possess similar oRGs, but these are much less abundant than in human or in ferret. Moreover, oRGs in mouse do not reside in a cytoarchitectonically separate zone (Fig. 3a). The Kriegstein laboratory exploited methods of real-time recording to study labeled progenitor neurons in slice cultures and combined these with additional labeling methods (Hansen et al., 2010; Wang et al., 2011). This approach revealed all three progenitors—oRG, radial glia, and IPC—with their distinct patterns of cell division in mice and in human.

What is the cell lineage in rodents and primates?

Based on their birthdating studies in the macaque visual cortex, Smart and his colleagues suggested that the SVZ is a major compartment for the generation of the upper cortical layers (Lukaszewicz et al., 2006; Smart et al., 2002). These observations were further supported in the mouse by evidence for similarities of the gene expression patterns between the VZ and lower layers (Er81, Otx1, etc.) together with the similarities between the SVZ and the upper cortical layers (Satb2, NeuroD2) (Tarabykin et al., 2001). The hypothesis postulating that the intermediate progenitors of the SVZ contribute to the upper supragranular layers has been called "the upper layer hypothesis" (Pontious et al., 2008) and it applies to both primates and rodents. Indeed, an analysis of NEX-positive intermediate progenitors in mouse demonstrated the upper layer lineage for this particular subtype of intermediate progenitors (Wu et al., 2005).

However, there are several observations that do not fit the upper layer hypothesis. Hevner and colleagues demonstrated that some intermediate progenitors contribute to all layers throughout all stages (early, middle, and late) of cortical neurogenesis in mice (Kowalczyk et al., 2009; Pontious et al., 2008). These authors also analyzed the phenotype of several transcription factor knock-out mice. Our own investigations in the Tbr2 conditional knock-out mice revealed that the SVZ mitotic profiles are reduced together with the expression of several SVZ genes (including Svet1, Satb2, NeuroD2, Brn2, Cux2, NeuroD6, Tbr2) (Arnold et al., 2008). The resulting reduction of cortical thickness is accompanied by reduced gene expression specific for the upper (Svet1, Satb2), but not the lower, layers (Otx1, Tbr1, Er81, Foxp2). This is suggestive of an upper layer differentiation defect (Arnold et al., 2008). However, a detailed analysis of neuronal and glial cell numbers with methods similar to the one used for Fig. 2 in Tbr2 cKO suggests that while the upper layers suffer gene expression defects, the reduction of cell numbers extends to both upper and lower layers (Arnold et al., 2008; A. Cheung and Z. Molnár, unpublished observations). Moreover, in the Tbr2 GFP mouse cortex, both lower and upper cortical plate neurons express GFP. The persisting GFP expression in all cortical layers suggests that most of these neurons originated through Tbr2-positive intermediate progenitors in mice (Stubbs et al., 2009). In the human cortex, TBR2 expression is observed in the SVZ from the earliest stages of cortical plate formation when Layer VI and V neurons are being born (Bayatti et al., 2008a,b), and thus, it is likely that IPCs contribute to production of cells in this layer. We have very little information on the overlap between the Tbr2, Brn2, Cux2, NeuroD6, NeuroD2, and Nex populations in the SVZ in rodents or primates. To resolve these issues, an analysis of cell lineage is required with selective labeling of single intermediate progenitor clones.

Compartmentalization of the germinal zone is not primate specific

There is current debate about whether this new set of oRG progenitors are really specific to primates or whether they are associated with all gyrencephalic mammalian brains. Recent work, including our own, suggests that distinct OSVZ progenitors are found in the gyrencephalic cortex of ferrets (Fietz et al., 2010), in developing pig brains (M. Perenyei, N. Vashistha, F. García-Moreno, I. Holms, and Z. Molnár, unpublished observations) and in the large brains of Amazonian rodents, such as the agouti (García-Moreno et al., 2011). We have also been studying marmoset brains; these small primates possess a brain with a lissencephalic cortex, and therefore the study of their germinal zones could contribute to the resolution of the above debate (García-Moreno et al., 2011). Interestingly, marmoset contains just as numerous oRG progenitors as gyrencephalic human or ferret; moreover, there is a clear cytoarchitectonic distinction with IFL in their SVZs. The cortical development of more species with gyrencephalic and lissencephalic brains should be systematically analyzed to establish correlations between the presence or the absence of cytoarchitectonic distinctions between ISVZ and OSVZ in brains with sulci and gyri in both primates and rodents. This should include species with different (long–short) gestational periods and with either large or small brains. Systematic studies on selected species can reveal whether the OSVZ radial progenitors are present in all mammals but in different proportions and whether the OSVZ might only appear as a cytoarchitectonically distinct layer in the cortical germinal zone in species where the increased mitotic output is required, that is, in larger, more complex, and relatively more rapidly developing brains.

The origin and role of the IFL that separates the ISVZ from the OSVZ is unknown. These fibers can be revealed with immunohistochemical methods (see examples in Fig. 4) and with diffusion tensor MRI imaging (Vasung et al., 2011). One possibility is that the IFL houses the fibers from the newly generated projection neurons. Time-lapse recordings of newly born neurons in the mouse cortex revealed that they extend projections into the SVZ and VZ during early cortical development (Lickiss, 2011; Lickiss et al., 2011; Noctor et al., 2004). If this is the case, then the direct or indirect influence of these projections might contribute to the regulation of neuronal production, perhaps through transmitter release or another effect. Another possibility is that the early thalamocortical projections constitute the outer fiber layer and some might contribute to the IFL. Afferents from the thalamus to cortex have been shown to provide a major influence on areal identity and differentiation during embryonic development in the fetal monkey (Dehay et al., 1996; Rakic et al., 1991). Smart et al. (2002) considered acetylcholinesterase as an early marker of the geniculocortical projections. If any of the fibers in IFL or OFL indeed originate from the thalamus, then thalamic fibers in species with cytoarchitectonic distinctions between OSVZ and ISVZ would be much more closely associated with the germinal zone than in species without such distinctions with the possibility that the ascending pathways influence rates of proliferation in the cortex, ultimately contribute to setting up distinct proliferative programs in the germinal zone, and possibly determine cortical cytoarchitecture (Carney et al., 2007; Dehay et al., 2001; Lukaszewicz et al., 2005). While the growth and fasciculation patterns appear similar in primates and rodents, the early macaque monkey thalamic projections exhibit interactions with the germinal zone through short descending fibers (Carney et al., 2002). It has been demonstrated *in vitro* in the mouse cortex that thalamic explants can alter cell-cycle parameters in embryonic progenitor cells (Dehay et al., 2001; Dehay and Kennedy, 2007) and accelerate neuronal migration in cortical explants *in vitro* (Edgar and Price, 2001). The further study of proliferative compartments together with the IFL and OFL after experimental manipulations of the thalamic input will be important. Such manipulations have been performed indirectly by removing the retina at embryonic stages in macaque that leads to cell death in the thalamus and cause the reduction and "respecification" of the primary visual cortex in the adult (Dehay et al., 1996; Rakic, 1988). Cortical progenitor proliferation was altered after such manipulations in ferret (Reillo et al., 2011).

Is the generation of inhibitory neurons different in rodent and primate?

The origin of inhibitory interneurons in the neocortex in rodents and primates has been controversial. While it is generally accepted that in mice the interneurons are born almost entirely outside the dorsal telencephalon, in the subpallium, from which they migrate tangentially into the cortex (de Carlos et al., 1996; reviewed

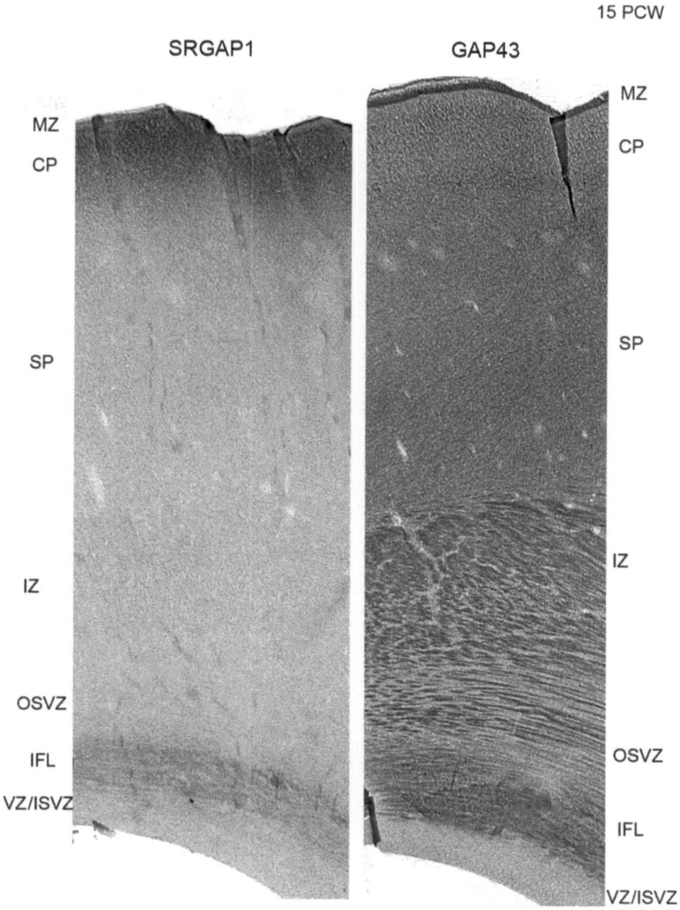

Fig. 4. Immunohistochemistry for SRGAP1 (ROBO1 receptor-associated protein) and GAP43 reveals the periventricular fibers in the developing cortex in 15PCW human. The SRGAP1 immunoreactivity is mainly associated with the internal fiber layer (IFL), whereas GAP43 is more widespread and involves IFL, outer subventricular zone (OSVZ), and intermediates zone (IZ). Abbreviations: MZ, marginal zone; CP, cortical plate; SP, subplate; VZ, ventricular zone; ISVZ, internal subventricular zone. Scale bar=500μm. We thank Dr. B.K. Ip for providing the images for this figure.

in Marín and Rubenstein, 2003; Parnavelas, 2000), in human, some local neurogenesis has been suggested. Rakic and colleagues used immunohistochemistry and retroviral labeling of slice preparations to show that at later stages of human fetal development in prospective visual cortex, the majority of interneurons are generated within the cortical progenitor zones (Letinic et al., 2002). This interpretation was supported by the use of double-label immunohistochemistry (Mo and Zecevic, 2008; Zecevic et al., 2005) and in the analysis of malformations that involve deletion of the ganglionic eminence (Fertuzinhos et al., 2009). Further studies in the macaque monkey have confirmed similar observations (Petanjek et al., 2009). However, the most recent study of the neurons birthdated in cultured in 15GW human cortical cultures did not reveal postmitotic cells double labeled for GABA and BrdU immunoreactivity in the SVZ (Hansen et al., 2010), whereas they were detected histologically in the 20 GW fetus (Zecevic et al., 2011).

This discrepancy may be explained by differences in expression of GABA markers in different regions of the cortex during early development. In the rodent, parvalbumin- and somatostatin-positive inhibitory neurons arise from the medial ganglionic eminence and mostly populate the rostral two-thirds of the cortex, whereas calretinin (CR)-positive inhibitory neurons arise from the caudal ganglionic eminence (CGE) and mostly populate the caudal pole (Xu et al., 2004), although they do find their way to all parts of the cortex (Polleux et al., 2002). However, Bayatti et al. (2008a) found that in human neocortex, the numbers of CR-positive neurons increased from about 12 postconceptional week (PCW), particularly in the SVZ, confirming the previous observations of Meyer et al. (2002). Further, at this stage of development, the rostral cortex was far more densely populated with CR neurons than the caudal pole, suggesting that they were unlikely to arise by migration from the CGE. CR interneurons are more prevalent in the adult primate than in the rodent (Conde et al., 1994; Gabbott et al., 1997; Wonders and Anderson, 2005), and it has been previously suggested that these extra numbers of inhibitory neurons might be generated in the cortex itself (Wonders and Anderson, 2005). More specifically, at least initially, such interneurons may be generated at the rostral pole of the neocortex. This hypothesis is supported by our mRNA expression analysis that finds a number of markers for GABAergic neurons including GAD 1 and 2, transcription factors including those of the DLX family. GABA receptor subunits and CR are more highly expressed rostrally than caudally in the human neocortex between 8 and 12 PCWs (Al-Jaberi et al., 2011).

Thus, intracortical generation of GABAergic interneurons in the primate may occur at different times and in different locations, explaining Hansen et al.'s (2010) results that showed no generation of GABAergic neurons in slices from 15 GW brain, especially if these slices originated from occipital cortex. Our preliminary investigations suggest distinct waves of generation from the rostral pole of the cortex at 7 and 12 PCWs, but with the possibility that this might include, or even consist entirely of, migration into the cortex from the subpallium at this location (Al-Jaberi et al., 2011). Other evidence strongly suggests intracortical interneuron generation at mid-gestational stages (Letinic et al., 2002; Zecevic et al., 2011). Inhibitory interneurons play a crucial role in cognitive processing, fine-tuning the oscillations in neural activity in distributed networks that underlie learning and memory (Klausberger and Somogyi, 2008; Whittington et al., 2011). The hypothesis that the expanded repertoire of inhibitory interneuron subtypes present in primates may in part underlie their greater cognitive powers has been influential since the time of Ramón y Cajal (1909). Failures in proliferation and migration of the specific classes of inhibitory interneurons have been implicated in diverse conditions including autism, epilepsy, and schizophrenia, and we must question whether our mouse models are appropriate for the study of these diseases (Hannan et al., 1999; Jones, 2009; Levitt, 2005; Lewis and Hashimoto, 2007).

Thalamocortical development and the subplate in rodents and primates

The development of thalamocortical projections has been extensively examined in rodents, but we have less detailed data in human and nonhuman primates. While the general pattern of axon outgrowth and fasciculation is similar, there are some interesting differences (Molnár, 2000; Fig. 5). Carney and colleagues compared the early development of the thalamocortical axons in the macaque monkey (Carney et al., 2002) with carbocyanine dye tracing with mouse and rat. At E45, the thalamocortical fibers exhibit fasciculation changes upon entering the internal capsule and at the corticostriatal boundary as individual fibers turn into the intermediate zone to the dorsal

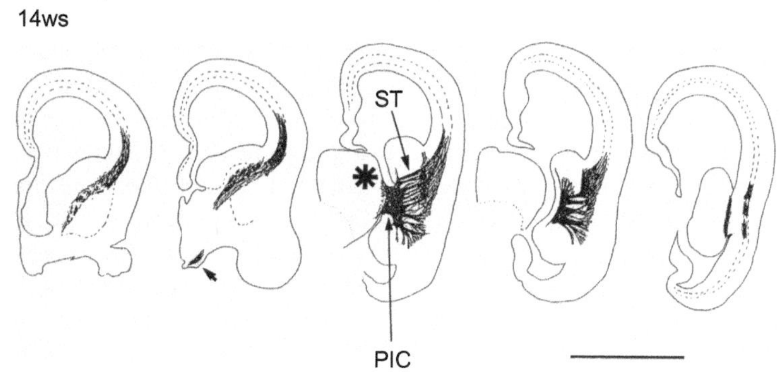

Fig. 5. Thalamocortical projections labeled with carbocyanine dye DiI in a 14-postconceptional-week (PCW) human embryo. A single DiI crystal was placed to the dorsolateral aspect of the thalamus (depicted with *). This labeled the thalamocortical projections as they enter through the early internal capsule (PIC) and then form fascicles through the striatum (ST). Some of the label was taken up by the retinal axons toward the chiasm as they can be seen on ventral aspect of the thalamus in the more anterior section (arrow). With kind permission from Springer Science+Business Media: Development of Thalamocortical Connections (1998) Molnar, Z.

telencephalon in macaque. These are similar to the patterns observed in human (Molnár, 1998; Fig. 5) and in rat and mouse (Molnár et al., 1998a,b). In the macaque, labeled thalamocortical axons can be observed below the cortical plate as early as E40 (Carney et al., 2002). At E71, the majority of the thalamocortical fibers are accumulating in the subplate, while a subset of thalamocortical fibers traverses the cortical plate to reach the marginal zone (Carney et al., 2002), as has been observed in the cat (Shatz and Luskin, 1986) and in rat and mouse (Molnár et al., 1998a,b).

One of the most striking differences in brain development between rodents (mouse) and primates (macaque and human) is the presence of the large subplate compartment and the more protracted accumulation of thalamocortical projections in the subplate before the ultimate thalamocortical connections with layer 4 develop (Rakic, 1977; Fig. 4). The subplate was first described in the human cortex, then in the fetal macaque (Rakic, 1977), rat (Rickmann et al., 1977), and then in carnivores (Luskin and Shatz, 1985). It was defined as a transient zone below the cortical plate and above the intermediate zone in the developing cortex (Bystron et al., 2008; Rakic, 1977). The developing subplate zone contains residential subplate cells, and numerous other migrating cells and fibers extending through the region. Subplate cells initially contribute to the preplate that is then split into the subplate and marginal zone by the subsequent arrival of cortical plate cells (Marin-Padilla, 1971). The subplate has received renewed attention because of its functional relevance in cerebral cortex development and because new methods emerged to study gene expression patterns or circuit properties (Ayoub and Kostovic, 2009; Kanold and Luhmann, 2010). During the past two decades, knowledge about the subplate has been extended to include functional and molecular properties pointing to a structure with heterogeneous cell populations and a highly dynamic ontogeny (Antonini and Shatz, 1990; Hoerder-Suabedissen et al., 2009; Oeschger et al., 2011; Piñon et al., 2009).

The cortical subplate shows important differences between humans and rodents (reviewed by Ayoub and Kostovic, 2009; Kanold and Luhmann, 2010; Montiel et al., 2011; Wang et al., 2010). This transient, developmental structure is the first site of synapse formation in the cortex and the first region to receive inputs from the thalamus and other regions; its functioning is crucial to establishing the correct wiring and functional maturation of the cerebral cortex

(Allendoerfer and Shatz, 1994). In humans, its extra size and complexity may reflect its role in also receiving inputs from many diverse cortical regions as part of its role in establishing corticocortical connectivity (Vasung et al., 2010). The subplate zone reaches its largest size and survives longest in the regions subjacent to the association cortices that contain transient corticocortical and callosal connections before they enter the cortical plate (Table 1). Recent research comparing MRI data with histological sections has demonstrated how vastly larger and more elaborate the human subplate is compared with that of other species (Kostovic and Vasung, 2009; Kostovic et al., 2002; Molnár et al., 2011; Perkins et al., 2008; Wang et al., 2010, 2011; Figs. 6 and 7). The vulnerability of the subplate to hypoxic ischemia in premature babies, when it achieves its greatest extent and complexity, has been demonstrated with *in vivo* imaging (Ferriero and Miller, 2010; McQuillen and Ferriero, 2005). Our group presented data on the differences in gene expression and complexity of the subplate between species (Bayatti et al., 2008a; Hoerder-Suabedissen et al., 2009; Wang et al., 2009, 2010).

It has been recently proposed that an embryonic subplate was present in an ancestral mammal and that additional populations evolved as cortical development and connectivity became more complex. This hypothesis was initially articulated by Aboitiz and colleagues (Aboitiz et al., 2005; Chapter 1 of this volume), and more recently further strengthened based on comparative gene expression data (Molnár et al., 2011; Wang et al., 2010). These studies demonstrated that subplate cellular components are not exclusive to mammals; some cell populations expressing the same cohort of genes are also present in sauropsids. In the human cortex, subplate subpopulations show increased compartmentalization which segregate into sublayers (Bayatti et al., 2008a; Wang et al., 2010). However, to further test these ideas, we shall need to be able to distinguish the ancestral and new populations of subplate cells by birthdating, and by differential gene expression or connectivity. Evolution of the mammalian cortex required the modification of developmental programs; some of these started to rely on novel populations of subplate neurons possibly characterized by different targets of connectivity.

Functional specification of the neocortex

The division of the neocortex into functional areas (the cortical map) differs little between individuals in any given species (Brodmann, 1908; Rakic et al., 2009). Previous work on rodent

Table 1. Table summarizing course of development of the human subplate

Human subplate development		Presubplate 10–13PCW	Early subplate 13–18PCW	Maximal subplate 18–32PCW
Events		Synapse formation, GABAergic terminals	Condensation of lower cortical plate with presubplate	Accumulation of thalamic afferents in upper SP followed by invasion of CP from 24PCW
New innervation		Catecholaminergic	Thalamic Basal forebrain	Corticocortical
Neuronal markers	Upper SP	TBR1 calretinin	NURR1 NPY NADPH-d	CTGF
	Lower SP		KCC2	
Size (% of cerebral wall)		10%	35%	50%

Taken from Bayatti et al. (2008a), Judaš et al. (2010), Vasung et al. (2010), and Wang et al. (2010).

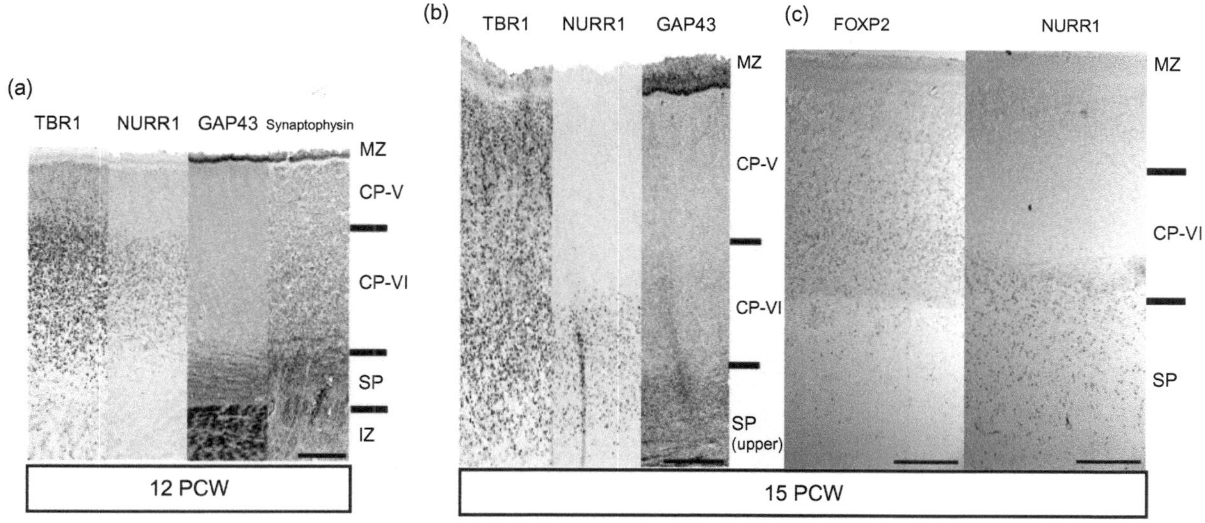

Fig. 6. Comparisons of TBR1, NURR1, GAP43, synaptophysin, and FOXP2 expression patterns on serial paraffin sections from the dorsolateral part of the human frontal cortex at (a) 12 and (b and c) 15 postconceptional weeks (PCW). (a) A distinct subplate (SP) at 12PCW is moderately immunopositive for GAP43 and strongly immunopositive for synaptophysin, whereas the cortical plate (CP) shows no expression of these markers. At 12PCW, TBR1 is expressed throughout putative layer 6 and also in the SP. NURR1 is confined to the lower portion of the CP corresponding to putative layer 6. Scale bar: 200μm. (b) TBR1, NURR1, and GAP43 expression at 15PCW. By this stage, NURR1-immunoreactive cells are more numerous in the SP zone. Subplate zone is labeled with GAP43, whereas the lower part of the CP is GAP43-negative. TBR1 is expressed throughout the SP and CP. The bulk of the NURR1 expression starts in layer 6 of the CP and is increasing in the upper SP at later developmental stages. Scale bar: 200μm. (c) NURR1 and FOXP2 expression in the human cerebral cortex at 15PCW. At 15PCW, immunohistochemistry for FOXP2 and NURR1 on adjacent sections demonstrated partial overlap in the expression pattern. FOXP2, a marker associated with layer 6 cortical neurons (in the adult? Here, it looks more widespread), is expressed more superficially than NURR1, but there is a clear overlap between the two zones. NURR1 is expressed in both the lower layer 6 and the upper SP. Scale bar: 200μ m. CP-VI, cortical plate-layer VI; CP-V, cortical plate-layer V; FOXP2, forkhead box P2; GAP43, growth-associated protein 43 kDa; IZ, intermediate zone; MZ, marginal zone; NURR1, nuclear receptor-related 1; TBR1, T-box brain gene 1. Reprinted by permission from Blackwell-Wiley, Journal of Anatomy 214, 368-381 Wang et al. (2010).

development has identified certain transcription factors (e.g., Pax6, Sp8, Emx2, Coup TF1) expressed in gradients across the neocortex that appear to control regional expression of cell adhesion molecules and organization of area-specific thalamocortical afferent projections (López-Bendito and Molnár, 2003; Muzio et al., 2002; O'Leary et al., 2007; Rakic et al., 2009). Although there may be common mechanisms, the human neocortex is composed of different and more complex local areas. Studies of mRNA expression in early fetal neocortex have confirmed that many genes show differences in expression across the anterior–posterior axis including transcription factors implicated in the emergence of distinct cortical areas such as EMX2 and COUP TFI, which specify posterior regions (Bayatti et al., 2008b; Ip et al., 2010), although "rostral" markers such as PAX6 failed to show significant gradients of expression except at the very earliest stages of cortical plate formation (Table 2 and Fig. 8). Strikingly, EMX2 expression, which is confined to the proliferative zones in rodents, was found to be expressed in postmitotic neurons in the cortical plate of humans (Bayatti et al., 2008b; Fig. 9). The FGF receptor 3, which shows marked posterior expression in rodents, was also found to be highly expressed posteriorly in humans (Ip et al.,

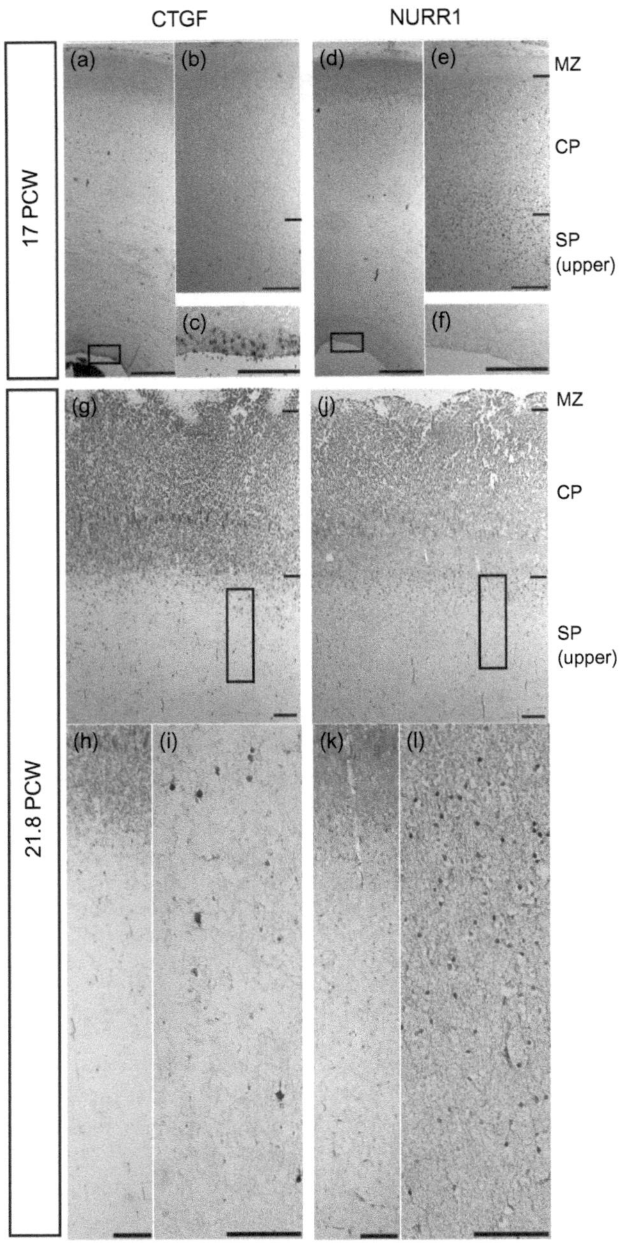

Fig. 7. Connective tissue growth factor (CTGF) and nuclear receptor-related 1 (NURR1) expression in the 17 and 21.8PCW human cortex revealed with *in situ* hybridization and immunoreactivity. (a–c) At 17PCW, *in situ* hybridization of CTGF expression showed no CTGF-positive cells in the subplate (SP) (a and b), but the presence of numerous CTGF-positive cells in the ventricular zone (c). (d–f) At 17PCW, NURR1 immunoreactivity was located around the border between layer 6 and the SP zone (e), but there was none in the ventricular zone (f). Scale bars: 1mm (a and d) and 200μm (b, c, e, and f). *In situ* hybridization for CTGF and immunohistochemistry for NURR1 revealed SP populations with a slightly different distribution at 21.8PCW in the human parietal cerebral cortex. (g–i) CTGF-positive cells were sparsely scattered throughout the SP zone. (j–l) NURR1-immunoreactive cells were located around the border between the layer 6 and the SP zone, with the majority of cells present in a superficial subcompartment of the SP. Scale bars: 200μm (g, h, j, and k); 100μm (i and l). CP, cortical plate; MZ, marginal zone. Reprinted by permission of Blackwell-Wiley Journal of Anatomy 217, 368-381 Wang et al. (2010).

Table 2. At 8PCW, *EMX2* and *PAX6* are localized within the proliferative zones of the developing cortex in opposing rostro-lateral/caudo-medial gradients

	8 PCW Rostral/lateral Caudal/medial	9 PCW Rostral/lateral Caudal/medial	10 PCW Rostral/lateral Caudal/medial	12 PCW Rostral/lateral Caudal/medial
SVZ +VZ	↑Pax6 Pax6↓ Emx2↓ Emx2↑	↑Pax6 ↑Pax6 Emx2↓ Emx2↓	↑Pax6 ↑Pax6 Emx2↓ Emx2↓	↑Pax6 ↑Pax6 Emx2↓ Emx2↓
SVZ only	↑Tbr2 Tbr2↓ ↑NeuroD NeuroD↓	↑Tbr2 Tbr2↓	↑Tbr2 ↑Tbr2	↑Tbr2 ↑Tbr2
SVZ + IZ +CP	↑Tbr1 Tbr1↓	↑NeuroD NeuroD↓ ↑Tbr1 Tbr1↓	↑NeuroD NeuroD↓ ↑Tbr1 Tbr1↓	↑NeuroD ↑NeuroD ↑Tbr1 ↑Tbr1
CP only		Emx2↓ Emx2↑	Emx2↓ Emx2↑	Emx2↓ Emx2↑

From 9–12PCW onward, the majority of *EMX2* expression is found in the cortical plate, where a similar gradient exists to that observed in the SVZ/VZ. By this time, the PAX6 gradient has disappeared and expression is widespread throughout the SVZ/VZ only. However, the mRNA of transcription factors downstream of PAX6 in neurogenesis form PAX6-like gradients in different compartments during this time period. *TBR2* exhibits a *PAX6*-like gradient within the SVZ until 10PCW. *NEUROD*, which initially is not expressed in the CP, also forms a gradient within the SVZ at 8 PCW, which extends to the SP/IZ and CP from 9PCW onwards. The *TBR1* gradient is also observed from 8PCW and encompasses all compartments outside of the VZ, most prominently within the CP, until 12PCW. Thus as *EMX2*-expressing cells migrate from the VZ to the CP, they pass through compartments expressing genes downstream of *PAX6* which exhibit *PAX6*-like gradients. Reprinted by permission of Blackwell-Wiley European Journal of Neuroscience 28, 1449-1456 Bayatti et al. (2008).

2010), and our recent unpublished observations suggest it is expressed in the cortical plate in humans, rather than being confined to proliferative layers, as it is in rodents (Zhang et al., 2006). FGFR3 signaling is believed to control expression of EMX2 and other areal transcription factors in rodent progenitor cells (Sansom and Livesey, 2009), but our initial data suggest a more protracted role for FGF/transcription factor interaction in human development that extends to postmitotic neurons and may be necessary for determining the more complex and elaborate cortical areas of the human cortex.

One set of genes found to be more highly expressed at the anterior pole of the human cortex compared to the caudal pole included those characteristic of corticofugal projection neurons: *CTIP2*, *ROBO1*, *SRGAP1*, *ER81*, *S100A10*, and *Protocadherin 17* (Ip et al., 2010, 2011). Coexpression of CTIP2, ROBO1, and SRGAP1, in particular, in the anterior neocortex between 8 and 15PCW, may mark the early location of the human motor cortex, including its corticospinal projection neurons which express ROBO1 and SRGAP1 in their axons in the medullary pyramids (Ip et al., 2011). Development of the prefrontal cortex appears to be limited at this stage. Prefrontal cortex is of far greater size and complexity in primates and particularly humans in comparison to rodents in the mature cortex, and its later development may include expression of a set of genes not expressed in rodents (Johnson et al., 2009).

Lateralization in cortical representation

Humans show stereotyped lateralization of hemispheric function. Approximately 90% of humans are right-handed implying that fine motor control is predominately controlled from the left

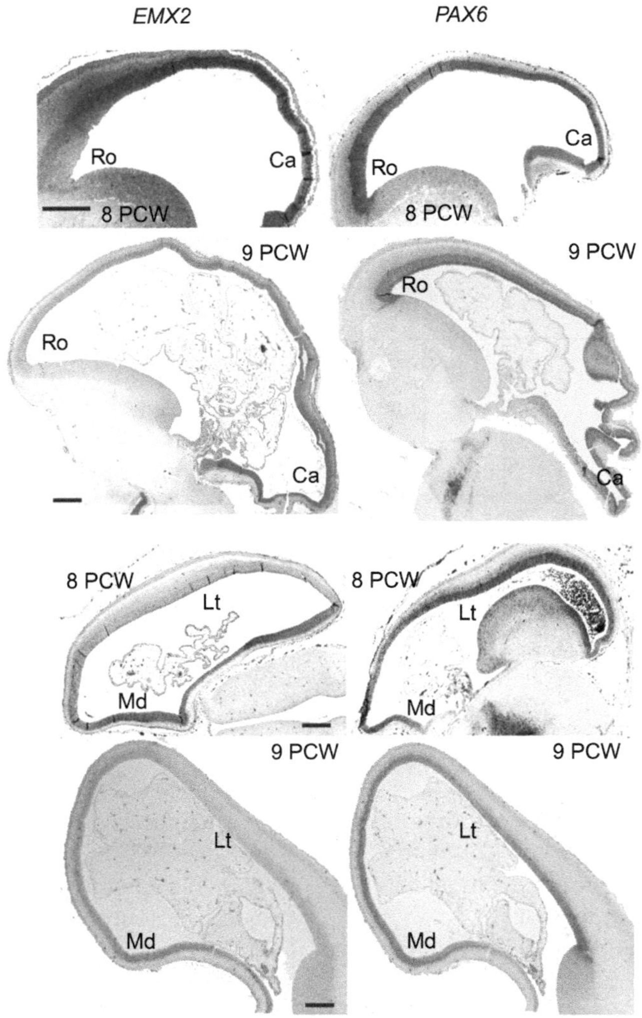

Fig. 8. *In situ* hybridization analysis in human at 8PCW revealed that *EMX2* and *PAX6* mRNA were expressed in reciprocal rostro-caudal and medio-lateral gradients. In sections at 9PCW, *EMX2* mRNA maintained a similar caudal-rostral and medial-lateral gradient to that seen at 8PCW. The previously observed *PAX6* expression gradients, however, disappeared. Ro, rostral; Ca, caudal; Md, medial; Lt, lateral. Scale bars: 500μm. Reprinted by permission of Blackwell-Wiley from European Journal of Anatomy 28 1449-1456 Bayatti et al. (2008).

hemisphere (Provins et al., 1982). Individual rodents (Tang and Verstynen, 2002) and other primates (McGrew and Marchant, 1997) may show a preference for use of one limb or the other for manual tasks, but overall, there is a 50/50 split suggesting this occurs by chance. However, in chimpanzees, there is a bias toward predominant use of the right hand for communication by gesticulation (Meguerditchian et al., 2010). In human, language function is lateralized to an even higher degree with language perception and production areas found predominantly in the left hemisphere (Hickok and Poeppel, 2007). Asymmetry of function exists alongside macro- and microanatomical differences; in humans,

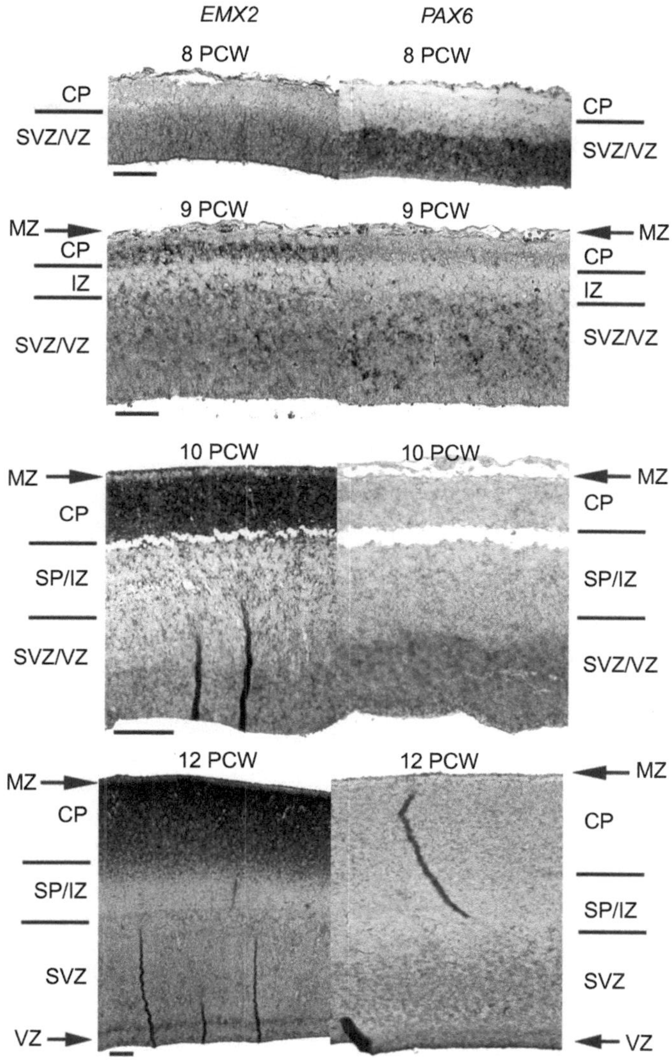

Fig. 9. *In situ* hybridization revealed a changing laminar distribution of *EMX2* mRNA between 8 and 12PCW in human cerebral cortex. Expression was observed only in the subventricular and ventricular zones (SVZ/VZ) at 8PCW. However, by 9PCW, expression was also noted in the cortical plate (CP). Subsequently, expression of *EMX2* intensified in the CP at 10 and 12PCW, while still present in the subventricular (SVZ) zone and VZ at lower intensities. At 12PCW, the highest level of *EMX2* mRNA expression was found in the cortical plate most proximal to the marginal zone (MZ). *PAX6* was observed predominantly in the proliferative zones (SVZ/VZ) of the developing cortex. *PAX*6 mRNA was observed most intensely at 8PCW in the VZ, after which a decrease was observed at 9PCW. Staining intensity of *PAX6* decreased at 10 and 12PCW due to the decrease in the relative thickness of the VZ. Sections for *EMX2* and *PAX6* are taken from the caudal and rostral poles, respectively. Scale bars: 100μm (a, c), 200μm (e, g). SP, subplate; IZ, intermediate zone. Reprinted by kind permission of Blackwell-Wiley from European Journal of Anatomy 28, 1449-1456 Bayatti et al. (2008).

it has been shown that the right frontal lobe and left occipito-parietal lobe are larger than in the opposite hemisphere (Mackay et al., 2003; Weinberger et al., 1982). Whether similar asymmetries are observed in other primates is controversial (Buxhoeveden et al., 2001; Cantalupo et al., 2009; Zilles et al., 1996). The ability to use language to convey and understand abstract ideas is uniquely human, and it has been suggested that this highest of cognitive functions is dependent upon the division of processing between the two hemispheres (Cook, 2002). In addition, defects in the process of lateralization may underlie neurodevelopmental psychiatric disorders such as schizophrenia (Crow, 2008).

The study of development of cerebral asymmetry is only just beginning. *In utero* MRI has revealed that asymmetry in the human perisylvian region emerges as early as 23 PCW (Habas et al., 2011). This suggests that the origins of asymmetry lie in the earliest gene-directed events of cortical development. A recent profile of gene expression in the developing human cerebral cortex has revealed a human-specific pattern, particularly in Broca and Wernicke's areas related to language (Johnson et al., 2009). Transcriptome analyses of these rapidly evolving brain regions have found evidence of human-specific regulation of gene expression in these regions by rapidly evolving microRNAs and by the language-associated transcription factor FOXP2 (Konopka et al., 2009; Pollard et al., 2006). It has also been proposed that sex differences in cerebral asymmetry and language could be explained by a gene pair on the X/Y chromosomes that has only very recently been established in evolution, with protocadherin 11X/Y the prime candidate due to the potential roles of this molecule in neural development (Priddle and Crow, 2009).

Conclusions

There is surprisingly little work on primate cortical development in contrast to rodent model systems. However, it is also clear that certain aspects of brain development (preplate formation, compartmentalization of cortical germinal zone, generation of interneurons, cell migration, thalamocortical development, subplate zone, functional arealization, hemispheric lateralization) have specific properties in human and nonhuman primates. The systematic studies of these differences might lead the way to the understanding of human neuropathologies of some neurological and psychiatric diseases.

Acknowledgments

The work in ZM's laboratory is supported by MRC, BBSRC, and the Wellcome Trust. GC and colleagues have received support from the Wellcome Trust and the Anatomical Society of Great Britain and Ireland. This review contains the overview of projects that have been developed in various original publications and reviews with members of the ZM and GC laboratories over the years. We thank Dr. Bui Kar for providing the images for Fig. 4. We are very grateful to Ray Guillery, Tony Rockel, Navneet Vasistha, and Miklos Perenyei for their comments on an earlier version of this chapter.

References

Abdel-Mannan, O., Cheung, A. F., & Molnár, Z. (2008). Evolution of cortical neurogenesis. *Brain Research Bulletin*, *75*, 398–404.

Aboitiz, F., Montiel, J., & García, R. R. (2005). Ancestry of the mammalian preplate and its derivatives: Evolutionary relicts or embryonic adaptations? *Reviews in the Neurosciences*, *16*(4), 359–376.

Al-Jaberi, N. A. A., Bayatti, N., Lindsay, S., & Clowry, G. J. (2011). The early development of human neocortical GABAergic interneurons. *IBRO World Congress Abstracts*, *D024*.

Allendoerfer, K. L., & Shatz, C. J. (1994). The subplate, a transient neocortical structure: Its role in the development of connections between thalamus and cortex. *Annual Review of Neuroscience*, *17*, 185–218.

Antonini, A., & Shatz, C. J. (1990). Relation between putative transmitter phenotypes and connectivity of subplate neurons during cerebral cortical development. *The European Journal of Neuroscience*, *2*, 744–761.

Arnold, S. J., Huang, G. J., Cheung, A. F., Era, T., Nishikawa, S., Bikoff, E. K., et al. (2008). The T-box transcription factor Eomes/Tbr2 regulates neurogenesis in the cortical subventricular zone. *Genes & Development*, *22*, 2479–2484.

Ayoub, A. E., & Kostovic, I. (2009). New horizons for the subplate zone and its pioneering neurons. *Cerebral Cortex*, *19*, 1705–1707.

Baird, G., Simonoff, E., Pickels, A., Chandler, S., Loucas, T., Meldrum, D., et al. (2006). Prevalence of disorders of the autism spectrum in a population cohort of children in South Thames: The Special Needs and Autism Project (SNAP). *Lancet*, *368*, 210–215.

Bayatti, N., Moss, J. A., Sun, L., Ambrose, P., Ward, J. F., Lindsay, S., et al. (2008). A molecular neuroanatomical study of the developing human neocortex from 8 to 17 postconceptional weeks revealing the early differentiation of the subplate and subventricular zone. *Cerebral Cortex*, *18*, 1536–1548.

Bayatti, N., Sarma, S., Shaw, C., Eyre, J. A., Vouyiouklis, D. A., Lindsay, S., et al. (2008). Progressive loss of PAX6, TBR2, NEUROD and TBR1 mRNA gradients correlates with translocation of EMX2 to the cortical plate during human cortical development. *The European Journal of Neuroscience*, *28*, 1449–1456.

Belgard, T. G., Marques, A. C., Oliver, P. L., Abaan, H. O., Sirey, T. M., Hoerder-Suabedissen, A., et al. (2011). A transcriptomic atlas of mouse neocortical layers. *Neuron*, *71*, 605–616.

Britanova, O., de Juan Romero, C., Cheung, A., Kwan, K. Y., Schwark, M., Gyorgy, A., et al. (2008). Satb2 is a postmitotic determinant for upper-layer neuron specification in the neocortex. *Neuron*, *57*, 378–392.

Brodmann, K. (1908). Beitraege zur histologischen Lokalisation der Grosshirnrinde. VI. Mitteilung: Die Cortexgliederung des Menschen. *Journal für Psychologie und Neurologie*, *10*, 231–246.

Brun, A. (1965). The subpial granular layer of the foetal cerebral cortex in man. Its ontogeny and significance in congenital cortical malformations. *Acta Pathologica et Microbiologica Scandinavica*, *179*(Suppl.), 3–98.

Buxhoeveden, D. P., Switala, A. E., Litaker, M., Roy, E., & Casanova, M. F. (2001). Lateralisation of minicolumns in human planum temporale is absent in nonhuman primate cortex. *Brain, Behavior and Evolution*, *57*, 349–358.

Bystron, I., Blakemore, C., & Rakic, P. (2008). Development of the human cerebral cortex: Boulder Committee revisited. *Nature Reviews. Neuroscience*, *9*, 110–122.

Bystron, I., Rakic, P., Molnár, Z., & Blakemore, C. (2006). The first neurons of the human cerebral cortex. *Nature Neuroscience*, *9*, 880–886.

Cantalupo, C., Oliver, J., Smith, J., Nir, T., Taglialatela, J. P., & Hopkins, W. D. (2009). The Chimpanzee brain shows human-like perisylvian like asymmetries in white matter. *The European Journal of Neuroscience*, *30*, 431–438.

Carney, R. S., Bystron, I., López-Bendito, G., & Molnár, Z. (2007). Comparative analysis of extra-ventricular mitoses at early stages of cortical development in rat and human. *Brain Structure & Function*, *212*, 37–54.

Carney, R., Molnár, Z., Giroud, P., Cortay, V., Berland, M., Kennedy, H., et al. (2002). Thalamocortical projections in the developing primate cortex. *FENS Abstract*, *1*, A007.6.

Cheung, A. F., Kondo, S., Abdel-Mannan, O., Chodroff, R. A., Sirey, T. M., Bluy, L. E., et al. (2010). The subventricular zone is the developmental milestone of a 6-layered neocortex: Comparisons in metatherian and eutherian mammals. *Cerebral Cortex*, *20*, 1071–1081.

Cheung, A. F., Pollen, A. A., Tavare, A., DeProto, J., & Molnár, Z. (2007). Comparative aspects of cortical neurogenesis in vertebrates. *Journal of Anatomy*, *211*, 164–176.

Chodroff, R. A., Goodstadt, L., Sirey, T. M., Oliver, P. L., Davies, K. E., Green, E. D., et al. (2010). Long noncoding RNA genes: Conservation of sequence and brain expression among diverse amniotes. *Genome Biology*, *11*, R72.

Clowry, G., Molnár, Z., & Rakic, P. (2010). Renewed focus on the developing human neocortex. *Journal of Anatomy*, *217*, 276–288.

Conde, F., Lund, J. S., Jacobowitz, D. M., Baimbridge, K. G., & Lewis, D. A. (1994). Local circuit neurons immunoreactive for calretinin, calbindin D-28k or parvalbumin in monkey prefrontal cortex: Distribution and morphology. *The Journal of Comparative Neurology*, *341*, 95–116.

Cook, N. D. (2002). Bihemispheric language: How the two hemispheres collaborate in the processing of language. In T. J. Crow (Ed.), *The speciation of modern Homo sapiens* (pp. 169–194). Oxford: Oxford University Press.

Crow, T. J. (2008). The 'big bang' theory of the origin of psychosis and the faculty of language. *Schizophrenia Research*, *102*, 31–52.

de Carlos, J. A., López-Mascaraque, L., & Valverde, F. (1996). Dynamics of cell migration from the lateral ganglionic eminence in the rat. *The Journal of Neuroscience*, *16*, 6146–6156.

Dehay, C., Giroud, P., Berland, M., Killackey, H., & Kennedy, H. (1996). Contribution of thalamic input to the specification of cytoarchitectonic cortical fields in the primate: Effects of bilateral enucleation in the fetal monkey on the boundaries, dimensions, and gyrification of striate and extrastriate cortex. *The Journal of Comparative Neurology*, *367*, 70–89.

Dehay, C., & Kennedy, H. (2007). Cell-cycle control and cortical development. *Nature Reviews Neuroscience*, *8*, 438–450.

Dehay, C., Savatier, P., Cortay, V., & Kennedy, H. (2001). Cell-cycle kinetics of neocortical precursors are influenced by embryonic thalamic axons. *The Journal of Neuroscience*, *21*, 201–214.

Economo, C., & Koskinas, G. N. (1925). *Die Cytoarchitektonik der Hirnrinde des erwachsenen Menschen.* Wien: Springer Verlag.

Edgar, J. M., & Price, D. J. (2001). Radial migration in the cerebral cortex is enhanced by signals from thalamus. *The European Journal of Neuroscience*, *13*, 1745–1754.

Ferriero, D. M., & Miller, S. P. (2010). Imaging selective vulnerability in the developing nervous system. *Journal of Anatomy*, *217*, 429–435.

Fertuzinhos, S., Krsnik, Z., Kawasawa, Y. I., Rasin, M. R., Kwan, K. Y., Chen, J. G., et al. (2009). Selective depletion of molecularly defined cortical interneurons in human holoprosencephaly with severe striatal hypoplasia. *Cerebral Cortex*, *19*, 2196–2207.

Fietz, S. A., Kelava, I., Vogt, J., Wilsch-Bräuninger, M., Stenzel, D., Fish, J. L., et al. (2010). OSVZ progenitors of human and ferret neocortex are epithelial-like and expand by integrin signaling. *Nature Neuroscience*, *13*, 690–699.

Fish, J. L., Dehay, C., Kennedy, H., & Huttner, W. B. (2008). Making bigger brains—The evolution of neural-progenitor-cell division. *Journal of Cell Science*, *121*, 2783–2793.

Gabbott, P. L., Jays, P. R., & Bacon, S. J. (1997). Calretinin neurons in human medial prefrontal cortex (areas 24a, b, c, 32', and 25). *The Journal of Comparative Neurology*, *381*, 389–410.

Gadisseux, J. F., Goffinet, A. M., Lyon, G., & Evrard, P. (1992). The human transient subpial granular layer: An optical, immunohistochemical, and ultrastructural analysis. *The Journal of Comparative Neurology*, *324*, 94–114.

García-Moreno, F., Vasistha, N.A., Trevia, N., Bourne, J. A., & Molnár, Z. (2011). Compartmentalization of cerebral cortical Germinal zones in a lissencephalic primate and gyrencephalic rodent. *Cerebral Cortex*, [Epub ahead of print]. doi: 10.1093/cercor/bhr312.

Gilbert, C. D., & Wiesel, T. N. (1979). Morphology and intracortical projections of functionally characterised neurones in the cat visual cortex. *Nature*, *280*, 120–125.

Guillemot, F., Molnár, Z., Tarabykin, V., & Stoykova, A. (2006). Molecular mechanisms of cortical differentiation. *The European Journal of Neuroscience*, *23*, 857–868.

Habas, P. A., Scott, J. A., Roosta, A., Rajagopalan, V., Kim, K., Rousseau, F., et al. (2011). Early folding patterns and asymmetries of the normal human brain detected from in utero MRI. *Cerebral Cortex,* [Epub ahead of print]. doi:10.1093/cercor/bhr053.

Hannan, A. J., Servotte, S., Katsnelson, A., Sisodiya, S., Blakemore, C., Squier, M., et al. (1999). Characterization of nodular neuronal heterotopia in children. *Brain*, *122*(2), 219–238.

Hansen, D. V., Lui, J. H., Parker, P. R., & Kriegstein, A. R. (2010). Neurogenic radial glia in the outer subventricular zone of human neocortex. *Nature*, *464*, 554–561.

Haubensak, W., Attardo, A., Denk, W., & Huttner, W. B. (2004). Neurons arise in the basal neuroepithelium of the early mammalian telencephalon: A major site of neurogenesis. *Proceedings of the National Academy of Sciences of the United States of America*, *101*, 3196–3201.

Herculano-Houzel, S., Collins, C. E., Wong, P., Kaas, J. H., & Lent, R. (2008). The basic uniformity of the cerebral cortex. *Proceedings of the National Academy of Sciences of the United States of America*, *105*, 12593–12598.

Hevner, R. F. (2006). From radial glia to pyramidal-projection neuron: Transcription factor cascades in cerebral cortex development. *Molecular Neurobiology*, *33*, 33–50.

Hevner, R. F., Hodge, R. D., Daza, R. A., & Englund, C. (2006). Transcription factors in glutamatergic neurogenesis: Conserved programs in neocortex, cerebellum, and adult hippocampus. *Neuroscience Research*, *55*, 223–233.

Hickok, G., & Poeppel, D. (2007). The cortical organization of speech processing. *Nature Reviews. Neuroscience*, *8*, 393–402.

Hoerder-Suabedissen, A., Wang, W. Z., Lee, S., Davies, K. E., Goffinet, A. M., Rakić, S., et al. (2009). Novel markers reveal subpopulations of subplate neurons in the murine cerebral cortex. *Cerebral Cortex*, *19*, 1738–1750.

Ip, B. K., Bayatti, N., Howard, N. J., Lindsay, S., & Clowry, G. J. (2011). The corticofugal neuron associated genes ROBO1, SRGAP1 and CTIP2 exhibit an anterior to posterior gradient of expression in early fetal human neocortex development. *Cerebral Cortex*, *21*, 1395–1407.

Ip, B. K., Wappler, I., Peters, H., Lindsay, S., Clowry, G. J., & Bayatti, N. (2010). Investigating gradients of gene expression involved in early human cortical development. *Journal of Anatomy*, *217*, 300–312.

Javaherian, A., & Kriegstein, A. (2009). A stem cell niche for intermediate progenitor cells of the embryonic cortex. *Cerebral Cortex*, *19*(Suppl. 1), i70–i77.

Johnson, M. B., Kawasawa, Y. I., Mason, C. E., Krsnik, Z., Coppola, G., Bogdanović, D., et al. (2009). Functional and evolutionary insights into human brain development through global transcriptome analysis. *Neuron*, *62*, 494–509.

Jones, E. G. (2000). Microcolumns in the cerebral cortex. *Proceedings of the National Academy of Sciences of the United States of America*, *97*, 5019–5021.

Jones, E. G. (2009). The origins of cortical interneurons: Mouse versus monkey and human. *Cerebral Cortex*, *19*(9), 1953–1956.

Judaš, M., Sedmak, G., Pletikos, M., & Jovanov-Milosevic, N. (2010). Populations of subplate and interstitial neurons in fetal and adult human telencephalon. *Journal of Anatomy*, *217*, 381–399.

Kanold, P. O., & Luhmann, H. J. (2010). The subplate and early cortical circuits. *Annual Review of Neuroscience*, *33*, 23–48.

Klausberger, T., & Somogyi, P. (2008). Neuronal diversity and temporal dynamics: The unity of hippocampal circuit operations. *Science*, *321*(5885), 53–57.

Konopka, G., Bomar, J. M., Winden, K., Coppola, G., Jonsson, Z. O., Gao, F., et al. (2009). Human-specific transcriptional regulation of CNS development genes by FOXP2. *Nature*, *462*, 213–217.

Kostovic, I., Judas, M., Rados, M., & Hrabac, P. (2002). Laminar organization of the human fetal cerebrum revealed by histochemical markers and magnetic resonance imaging. *Cerebral Cortex, 12*, 536–544.

Kostovic, I., & Vasung, L. (2009). Insights from in vitro fetal magnetic resonance imaging of cerebral development. *Seminars in Perinatology, 33*, 220–233.

Kowalczyk, T., Pontious, A., Englund, C., Daza, R. A., Bedogni, F., Hodge, R., et al. (2009). Intermediate neuronal progenitors (basal progenitors) produce pyramidal-projection neurons for all layers of cerebral cortex. *Cerebral Cortex, 19*, 2439–2450.

Kriegstein, A., Noctor, S., & Martínez-Cerdeño, V. (2006). Patterns of neural stem and progenitor cell division may underlie evolutionary cortical expansion. *Nature Reviews. Neuroscience, 7*, 883–890.

Krubitzer, L., & Kaas, J. (2005). The evolution of the neocortex in mammals: How is phenotypic diversity generated? *Current Opinion in Neurobiology, 15*, 444–453.

Letinic, K., Zoncu, R., & Rakic, P. (2002). Origin of GABAergic neurons in the human neocortex. *Nature, 417*, 645–649.

Levitt, P. (2005). Disruption of interneuron development. *Epilepsia, 46*(Suppl 7), 22–28.

Lewis, D. A., & Hashimoto, T. (2007). Deciphering the disease process of schizophrenia: The contribution of cortical GABAergic neurons. *International Reviews of Neurobiology, 78*, 109–131.

Lickiss, T. (2011). Development of directed early axon growth in the cerebral neocortex. In D.Phil thesis, University of Oxford, Oxford.

Lickiss, T., Cheung, A., Hutchinson, C., Taylor, J., & Molnár, Z. (2011). Examining the relationship between early axon growth and transcription factor expression in the developing cerebral cortex. *Journal of Anatomy* (in revision).

López-Bendito, G., & Molnár, Z. (2003). Thalamocortical development: How are we going to get there? *Nature Reviews. Neuroscience, 4*, 276–289.

Lorente de No, R. (1949). Cerebral cortex: Architecture, intracortical connections, motor projections. In J. F. Fulton (Ed.), *Physiology of the nervous system* (pp. 288–330). (3rd ed.). Oxford: Oxford University Press.

Lui, J. H., Hansen, D. V., & Kriegstein, A. R. (2011). Development and evolution of the human neocortex. *Cell, 146* (1), 18–36.

Lukaszewicz, A., Cortay, V., Giroud, P., Berland, M., Smart, I., Kennedy, H., et al. (2006). The concerted modulation of proliferation and migration contributes to the specification of the cytoarchitecture and dimensions of cortical areas. *Cerebral Cortex, 16*(Suppl. 1), i26–i34.

Lukaszewicz, A., Savatier, P., Cortay, V., Giroud, P., Huissoud, C., Berland, M., et al. (2005). G1 phase regulation, area-specific cell cycle control, and cytoarchitectonics in the primate cortex. *Neuron, 47*, 353–364.

Luskin, M. B., & Shatz, C. J. (1985). Studies of the earliest generated cells of the cat's visual cortex: Cogeneration of subplate and marginal zones. *The Journal of Neuroscience, 5*, 1062–1075.

Mackay, C. E., Barrick, T. R., Roberts, N., DrLisi, L. E., Maes, F., Vandermeulen, D., et al. (2003). Application of a new image analysis technique to study brain asymmetry in schizophrenia. *Psychiatry Research: Neuroimaging, 124*, 25–35.

Marin-Padilla, M. (1971). Early prenatal ontogenesis of the cerebral cortex (neocortex) of the cat (Felis domestica). A Golgi study. I. The primordial neocortical organization. *Z Anat Entwicklungsgesch, 134*(2), 117–145.

Marín, O., & Rubenstein, J. L. (2001). A long, remarkable journey: tangential migration in the telencephalon. *Nature Reviews Neuroscience, 2*(11), 780–790.

Marín, O., & Rubenstein, J. L. (2003). Cell migration in the forebrain. *Annual Review of Neuroscience, 26*, 441–483.

Martínez-Cerdeñ, V., Noctor, S. C., & Kriegstein, A. R. (2006). The role of intermediate progenitor cells in the evolutionary expansion of the cerebral cortex. *Cerebral Cortex, 16*(Suppl. 1), i152–i161.

McGrew, W. C., & Marchant, L. F. (1997). On the other hand: Current issues in and meta-analysis of the behavioural laterality of hand function in non-human primates. *American Journal of Physical Anthropology, 104*, 201–232.

McQuillen, P. S., & Ferriero, D. M. (2005). Perinatal subplate neuron injury: Implications for cortical development and plasticity. *Brain Pathology, 15*(3), 250–260.

Meguerditchian, A., Vauclair, J., & Hopkins, W. D. (2010). Captive chimpanzees use their right hand to communicate with each other: Implications for the origin of the cerebral substrate for language. *Cortex, 46*, 40–48.

Meyer, G. (2007). Genetic control of neuronal migrations in human cortical development. *Advances in Anatomy, Embryology, and Cell Biology, 189*, 1–111.

Meyer, G., Perez-Garcia, C. G., & Gleeson, J. G. (2002). Selective expression of doublecortin and LIS1 in developing human cortex suggests unique modes of neuronal movement. *Cerebral Cortex, 12*, 1225–1236.

Miyata, T., Kawaguchi, A., Saito, K., Kawano, M., Muto, T., & Ogawa, M. (2004). Asymmetric production of surface-dividing and non-surface-dividing cortical progenitor cells. *Development, 131*, 3133–3145.

Mo, Z., & Zecevic, N. (2008). Is Pax6 critical for neurogenesis in the human fetal brain? *Cerebral Cortex, 18*, 1455–1465.

Molnár, Z. (1998). *Development of thalamocortical connections.* Heidelberg: Springer (264p).

Molnár, Z. (2004). Thomas Willis (1621–1675), the founder of clinical neuroscience. *Nature Reviews. Neuroscience, 5*, 329–335.

Molnár, Z., Adams, R., & Blakemore, C. (1998a). Mechanisms underlying the early establishment of thalamocortical connections in the rat. *The Journal of Neuroscience, 18*, 5723–5745.

Molnár, Z., Adams, R., Goffinet, A. M., & Blakemore, C. (1998b). The role of the first postmitotic cortical cells in the development of thalamocortical innervation in the reeler mouse. *The Journal of Neuroscience, 18*(15), 5746–5765.

Molnár, Z. (2000). Development and evolution of thalamocortical interactions. *European Journal of Morphology, 38*(5), 313–320.

Molnár, Z., & Cheung, A. F. (2006). Towards the classification of subpopulations of layer V pyramidal projection neurons. *Neuroscience Research, 55*, 105–115.

Molnár, Z., Métin, C., Stoykova, A., Tarabykin, V., Price, D. J., Francis, F., et al. (2006). Comparative aspects of cerebral cortical development. *European Journal of Neuroscience, 23*, 921–934.

Molnár, Z., Vasistha, N. A., & Garcia-Moreno, F. (2011). Hanging by the tail: Progenitor populations proliferate. *Nature Neuroscience, 14*, 1–3.

Molyneaux, B. J., Arlotta, P., Menezes, J. R., & Macklis, J. D. (2007). Neuronal subtype specification in the cerebral cortex. *Nature Reviews. Neuroscience, 8*, 427–437.

Montiel, J. F., Wang, W. Z., Oeschger, F. M., Hoerder-Suabedissen, A., Tung, W. L., García-Moreno, F., et al. (2011). Hypothesis on the dual origin of the Mammalian subplate. *Frontiers in Neuroanatomy, 5*, 25.

Muzio, L., DiBenedetto, B., Stoykova, A., Boncinelli, E., Gruss, P., & Mallamaci, A. (2002). Conversion of cerebral cortex into basal ganglia in Emx2(−/−) Pax6(Sey/Sey) double-mutant mice. *Nature Neuroscience, 5*(8), 737–745.

Nie, K., Molnár, Z., & Szele, F. G. (2010). Proliferation but not migration is associated with blood vessels during development of the rostral migratory stream. *Developmental Neuroscience, 32*, 163–172.

Noctor, S. C., Martínez-Cerdeñ, V., Ivic, L., & Kriegstein, A. R. (2004). Cortical neurons arise in symmetric and asymmetric division zones and migrate through specific phases. *Nature Neuroscience, 7*, 136–144.

Oeschger, F. M., Wang, W. Z., Lee, S., García-Moreno, F., Goffinet, A. M., Arbonés, M. L., et al. (2011). Gene expression analysis of the embryonic subplate. *Cerebral Cortex*, [Epub ahead of print]. doi: 10.1093/cercor/bhr197.

Oka, E., Ohtsuka, Y., Yoshinaga, H., Murakami, N., Kobayashi, K., & Ogino, T. (2006). Prevalence of childhood epilepsy and distribution of epileptic syndromes: A population-based survey in Okayama, Japan. *Epilepsia, 47*, 626–630.

O'Leary, D. D., Chou, S. J., & Sahara, S. (2007). Area patterning of the mammalian cortex. *Neuron, 56*, 252–269.

Parnavelas, J. G. (2000). The origin and migration of cortical neurones: New vistas. *Trends in Neurosciences, 23*, 126–131.

Perkins, L., Hughes, E., Srinivasan, L., Allsop, J., Glover, A., Kumar, S., et al. (2008). Exploring cortical subplate evolution using magnetic resonance imaging of the fetal brain. *Developmental Neuroscience, 30*(1–3), 211–220.

Petanjek, Z., Berger, B., & Esclapez, M. (2009). Origins of GABAergic neurons in the cynomolgous monkey. *Cerebral Cortex, 19*, 249–262.

Peters, A., & Yilmaz, E. (1993). Neuronal organization in area 17 of cat visual cortex. *Cerebral Cortex, 3*, 49–68.

Pillay, P., & Manger, P. R. (2007). Order-specific quantitative patterns of cortical gyrification. *The European Journal of Neuroscience, 25*(9), 2705–2712.

Piñon, M. C., Jethwa, A., Jacobs, E., Campagnoni, A., & Molnár, Z. (2009). Dynamic integration of subplate neurons into the cortical barrel field circuitry during postnatal development in the Golli-tau-eGFP (GTE) mouse. *The Journal of Physiology, 587*(Pt. 9), 1903–1915.

Polanczyk, G., Silva de Lima, M., Horta, B. L., Biederman, J., & Rohde, L. A. (2007). The worldwide prevalence of ADHD: A systematic review and metaregression analysis. *The American Journal of Psychiatry, 164*, 942–948.

Pollard, K. S., Salama, S. R., Lambert, N., Lambot, M. A., Coppens, S., Pedersen, J. S., et al. (2006). An RNA gene expressed during cortical development evolved rapidly in humans. *Nature, 443*(7108), 167–172.

Polleux, F., Whitford, K. L., Dijkhuizen, P. A., Vitalis, T., & Ghosh, A. (2002). Control of cortical interneuron migration by neurotrophins and PI3-kinase signaling. *Development, 129*(13), 3147–3160.

Pontious, A., Kowalczyk, T., Englund, C., & Hevner, R. F. (2008). Role of intermediate progenitor cells in cerebral cortex development. *Developmental Neuroscience, 30*(1–3), 24–32.

Priddle, T. H., & Crow, T. J. (2009). The protocadherin 11X/Y gene pair as a putative determinant of cerebral dominance in Homo sapiens. *Future Neurology, 4*, 4.

Provins, K. A., Milner, A. D., & Kerr, P. (1982). Asymmetry of manual preference and performance. *Perceptual and Motor Skills, 54*, 179–194.

Puzzolo, E., & Mallamaci, A. (2010). Cortico-cerebral histogenesis in the opossum Monodelphis domestica: Generation of a hexalaminar neocortex in the absence of a basal proliferative compartment. *Neural Development, 19*, 8.

Rakic, P. (1977). Prenatal development of the visual system in rhesus monkey. *Philosophical Transactions of the Royal Society of London. Series B, Biological Sciences, 278*(961), 245–260.

Rakic, P. (1988). Specification of cerebral cortical areas. *Science, 241*(4862), 170–176.

Rakic, P. (1995). A small step for the cell, a giant leap for mankind: A hypothesis of neocortical expansion during evolution. *Trends in Neurosciences, 18*, 383–388.

Rakic, P. (2009). Evolution of the neocortex: A perspective from developmental biology. *Nature Reviews. Neuroscience, 10*, 724–735.

Rakic, P., Ayoub, A. E., Breunig, J. J., & Dominguez, M. H. (2009). Decision by division: Making cortical maps. *Trends in Neurosciences*, *32*, 291–301.

Rakic, P., Suñer, I., & Williams, R. W. (1991). A novel cytoarchitectonic area induced experimentally within the primate visual cortex. *Proceedings of the National Academy of Sciences of the United States of America*, *88*(6), 2083–2087.

Ramón y Cajal, S. (1909). *Histologie du Systeme Nerveux de l'Homme et des Vertébrés (translated by L. Azoulay)*. Madrid: Instituto Ramón y Cajal de CSIC (1952–1955).

Reillo, I., de Juan Romero, C., García-Cabezas, M. A., & Borrell, V. (2011). A role for intermediate radial glia in the tangential expansion of the mammalian cerebral cortex. *Cerebral Cortex*, *21*, 1674–1694.

Reynolds, M. L., Cavanagh, M. E., Dziegielewska, K. M., Hinds, L. A., Saunders, N. R., & Tyndale-Biscoe, C. H. (1985). Postnatal development of the telencephalon of the tammar wallaby (Macropus eugenii): An accessible model of neocortical differentiation. *Anatomy and Embryology*, *173*, 81–94.

Rickmann, M., Chronwall, B. M., & Wolff, J. R. (1977). On the development of non-pyramidal neurons and axons outside the cortical plate: The early marginal zone as a pallial anlage. *Anatomy and Embryology*, *151*(3), 285–307.

Rockel, A. J., Hiorns, R. W., & Powell, T. P. (1980). The basic uniformity in structure of the neocortex. *Brain*, *103*, 221–244.

Saha, S., Chant, D., Welham, J., & McGrath, J. (2005). A systematic review of the prevalence of schizophrenia. *PLoS Medicine*, *2*, e141 Analysis of 188 studies published between 1965 to 2002 yields no difference between men and women and suggests that current text-book estimates for lifetime morbid risk are too high.

Sansom, S. N., & Livesey, F. J. (2009). Gradients in the brain: The control of the development of form and function in the cerebral cortex. *Cold Spring Harbor Perspectives in Biology*, *1*, a002519.

Shatz, C. J., & Luskin, M. B. (1986). The relationship between the geniculocortical afferents and their cortical target cells during development of the cat's primary visual cortex. *The Journal of Neuroscience*, *6*, 3655–3668.

Smart, I. H., Dehay, C., Giroud, P., Berland, M., & Kennedy, H. (2002). Unique morphological features of the proliferative zones and postmitotic compartments of the neural epithelium giving rise to striate and extrastriate cortex in the monkey. *Cerebral Cortex*, *12*, 37–53.

Stubbs, D., DeProto, J., Nie, K., Englund, C., Mahmud, I., Hevner, R., et al. (2009). Neurovascular congruence during cerebral cortical development. *Cerebral Cortex*, *19*(Suppl. 1), i32–i41.

Suárez-Solá, M. L., González-Delgado, F. J., Pueyo-Morlans, M., Medina-Bolívar, O. C., Hernández-Acosta, N. C., González-Gómez, M., et al. (2009). Neurons in the white matter of the adult human neocortex. *Frontiers in Neuroanatomy*, *3*, 7.

Tang, A. C., & Verstynen, T. (2002). Early life modulates 'handedness' in rats. *Behavioural Brain Research*, *131*, 1–7.

Tarabykin, V., Stoykova, A., Usman, N., & Gruss, P. (2001). Cortical upper layer neurons derive from the subventricular zone as indicated by Svet1 gene expression. *Development*, *128*, 1983–1993.

Toyama, K., Matsunami, K., Ono, T., & Tokashiki, S. (1974). An intracellular study of neuronal organization in the visual cortex. *Experimental Brain Research*, *21*, 45–66.

Vasung, L., Huang, H., Jovanov-Milošević, N., Pletikos, M., Mori, S., & Kostović, I. (2010). Development of axonal pathways in the human fetal fronto-limbic brain: Histochemical characterization and diffusion tensor imaging. *Journal of Anatomy*, *217*, 400–417.

Vasung, L., Jovanov-Milošević, N., Pletikos, M., Mori, S., Judaš, M., & Kostović, I. (2011). Prominent periventricular fiber system related to ganglionic eminence and striatum in the human fetal cerebrum. *Brain Structure & Function*, *215*, 237–253.

Von Economo, C., & Koskinas, G. N. (2008). *Atlas of the cytoarchitectonics of the adult human cerebral cortex (translated, revised and edited by L.C. Triarhou)*. Basel: Karger (p. 946).

Wang, W. Z., Hoerder-Suabedissen, A., Oeschger, F. M., Bayatti, N., Ip, B. K., Supramaniam, V., et al. (2010). Subplate in the developing cortex of mouse and human. *Journal of Anatomy*, *217*, 368–381.

Wang, W. Z., Oeschger, F. M., Lee, S., & Molnár, Z. (2009). High quality RNA from multiple brain regions simultaneously acquired by laser capture microdissection. *BMC Molecular Biology*, *10*, 69.

Wang, W. Z., Oeschger, F. M., Montiel, J. F., García-Moreno, F., Hoerder-Suabedissen, A., Krubitz- er, L., et al. (2011). Comparative aspects of subplate zone studied with gene expression in sauropsids and mammals. *Cerebral Cortex*, *21*, 2187–2203.

Weinberger, D. R., Luchins, D. J., Morihisa, J., & Wyatt, R. J. (1982). Assymmetric volumes of the right frontal and left occipital regions of the human brain. *Annals of Neurology*, *11*, 97–100.

Whittington, M. A., Cunningham, M. O., LeBeau, F. E. N., Racca, C., & Traub, R. D. (2011). Multiple origins of the cortical gamma rhythm. *Developmental Neurobiology*, *71*, 92–106.

Willis, T. (1664). *Cerebri Anatome: Cui accessit nervorum descriptio et usus.* London.

Wilson, D. E., & Reeder, D. A. (2005). *Mammal species of the world: A taxonomic and geographic reference.* Baltimore, Maryland, USA: The Johns Hopkins University Press (Vol. 1, pp. 1–1945).

Wonders, C., & Anderson, S. A. (2005). Cortical interneurons and their origins. *The Neuroscientist*, *11*(3), 199–205.

Wu, S. X., Goebbels, S., Nakamura, K., Nakamura, K., Kometani, K., Minato, N., et al. (2005). Pyramidal neurons of upper cortical layers generated by NEX-positive progenitor cells in the subventricular zone. *Proceedings of the National Academy of Sciences of the United States of America*, *102*(47), 17172–17177.

Xu, Q., Cobos, I., De La Cruz, E., Rubenstein, J. L., & Anderson, S. A. (2004). Origins of cortical interneuron subtypes. *The Journal of Neuroscience*, *24*(11), 2612–2622.

Zecevic, N., Chen, Y., & Filipovic, R. (2005). Contributions of cortical subventricular zone to the development of the human cerebral cortex. *The Journal of Comparative Neurology*, *491*(2), 109–122.

Zecevic, N., Hu, F., & Jakovcevski, I. (2011). Interneurons in the developing human neocortex. *Developmental Neurobiology*, *71*, 18–33.

Zecevic, N., & Rakic, P. (2001). Development of layer I neurons in the primate cerebral cortex. *The Journal of Neuroscience*, *21*(15), 5607–5619.

Zhang, X., Ibrahimi, O. A., et al. (2006). Receptor specificity of the fibroblast growth factor family. *The Journal of Biological Chemistry*, *281*(23), 15694–15700.

Zilles, K., Dabringhaus, A., Geyer, S., Amunts, K., Qü, M., Schleicher, A., et al. (1996). Structural asymmetries in the human forebrain and the forebrain of non-human primates and rats. *Neuroscience and Biobehavioral Reviews*, *20*, 593–605.

M. A. Hofman and D. Falk (Eds.)
Progress in Brain Research, Vol. 195
ISSN: 0079-6123

CHAPTER 4

Embracing covariation in brain evolution: Large brains, extended development, and flexible primate social systems

Christine J. Charvet* and Barbara L. Finlay

Behavioral and Evolutionary Neuroscience Group, Department of Psychology, Cornell University, Ithaca NY, USA

Abstract: Brain size, body size, developmental length, life span, costs of raising offspring, behavioral complexity, and social structures are correlated in mammals due to intrinsic life-history requirements. Dissecting variation and direction of causation in this web of relationships often draw attention away from the factors that correlate with basic life parameters. We consider the "social brain hypothesis," which postulates that overall brain and the isocortex are selectively enlarged to confer social abilities in primates, as an example of this enterprise and pitfalls. We consider patterns of brain scaling, modularity, flexibility of brain organization, the "leverage," and direction of selection on proposed dimensions. We conclude that the evidence supporting *selective* changes in isocortex or brain size for the *isolated* ability to manage social relationships is poor. Strong covariation in size and developmental duration coupled with flexible brains allow organisms to adapt in variable social and ecological environments across the life span and in evolution.

Keywords: evolution; primate; cortex; social; variation.

Introduction

Beware Procrustes bearing Occam's Razor.[1]

*Corresponding author.
Tel.: +1-607-255-3996; Fax: (607) 255-8433
E-mail: charvetcj@gmail.com

Numerous attempts have been made to account for the high intelligence and impressively large isocortex of primates, and especially humans. Theories about the evolution of large brains can be broadly divided into two classes, organized by direction of causality. The first class focusses on the energetic requirements of large brain production, looking for an innovation allowing greater or more efficient energy acquisition or

[1]Attributed to Lise Menn, Department of Linguistics, University of Colorado. Procrustes is the Greek innkeeper of mythology who either sawed off the legs of his guests or stretched them on a rack to make them fit his beds. Occam's razor, of course, is the scientific principle of parsimony, to prefer the simplest theory from a set of contenders to explain some phenomenon.

DOI: 10.1016/B978-0-444-53860-4.00004-0

utilization, assuming the benefits of a large brain to be manifest (e.g., the expensive-tissue hypothesis, Aiello and Wheeler (1995); the radiator hypothesis, Falk (1990) and Falk and Gage (1998); the introduction of cooking, Carmody and Wrangham (2009); the conscription of alloparental care, Isler and Van Schaik (2009)). These theories have intrinsic bidirectional, "racheting," or exaptive aspects explored by their authors. In the case of the "expensive-tissue" hypothesis, reallocation of metabolic costs from gut to brain permits a larger brain to be supported, enhancing memory for the location and seasonality of high-quality foods, which then allow further increases in brain size, and so forth. In the case of fire use for food preparation, selection for the cognitive competency to produce fire not only allows more nutrients to be extracted from foodstuffs but also ritualizes eating as an intrinsically social activity. Biparental care for young allows more numerous and/or larger offspring but also immediately provides more opportunities for learning from conspecifics, typically over a longer period. We argue here that the multiple types and levels of causality embedded in most of these scenarios should be acknowledged at the outset.

The second class of theories about primate brain size and intelligence focusses on the selection for specific behavioral or cognitive competencies as the essential change in brain organization (e.g., the social brain hypothesis, Dunbar (1992); symbol-making, Deacon (1990); recursion in the language facility, Hauser et al. (2002); and many more, Finlay (2007) and Sherwood et al. (2008)). We examine a particular hypothesis, the social brain hypothesis, as emblematic of this approach, not as a theory *per se* deserving of special criticism! We will argue that a focus on brain size and a specific behavioral adaptation neglects to consider coordinated variations in developmental schedules, body, brain, and brain region size. This focus is so much an obligatory feature of the basic discriminative methods of analysis anthropologists and comparative anatomists have been using, that it rarely receives any scrutiny. These tactics simultaneously focus attention not only away from the wide variations in social systems within species but also away from the necessary coordination and covariance of changes in size (i.e., body, brain, brain region) and developmental duration across species. We conclude that brain or isocortex size is unlikely to have *selectively* expanded to manage social systems as many energetic, ecological, and behavioral factors together coordinate changes in development and brain structure. We argue that it is essential to consider the coordinated nature of variations in size and time between and within species to understand human brain evolution.

The social brain hypothesis

The social structure of primates is highly diverse. Some primates are monogamous, while other primates are polygamous or polyandrous (Smuts et al., 1986). Some primates live in groups forming simple or complex social hierarchies, while other primates are mainly solitary (Smuts et al., 1986). The finding that brain size and specifically the relative size of the isocortex positively correlate with group size has been used to argue that bigger groups require large brains and larger isocortices to manage social relationships (Dunbar, 1992, 1993, 2009; Dunbar and Shultz, 2007a,b; Lehmann and Dunbar, 2009; Shultz and Dunbar, 2007). In support of this hypothesis is the finding that the residuals of a phylogenetically controlled linear regression between brain and body size were found to positively correlate with aspects of social structure and behavior related to group size in primates (Pérez-Barbería et al., 2007). Moreover, the size of the isocortex, in particular, relative to the rest of the brain, was found to positively correlate with group size in primates (Fig. 1; Dunbar, 1992). Taken together, these correlative analyses have been used to argue that larger brains and proportionately enlarged isocortices endow

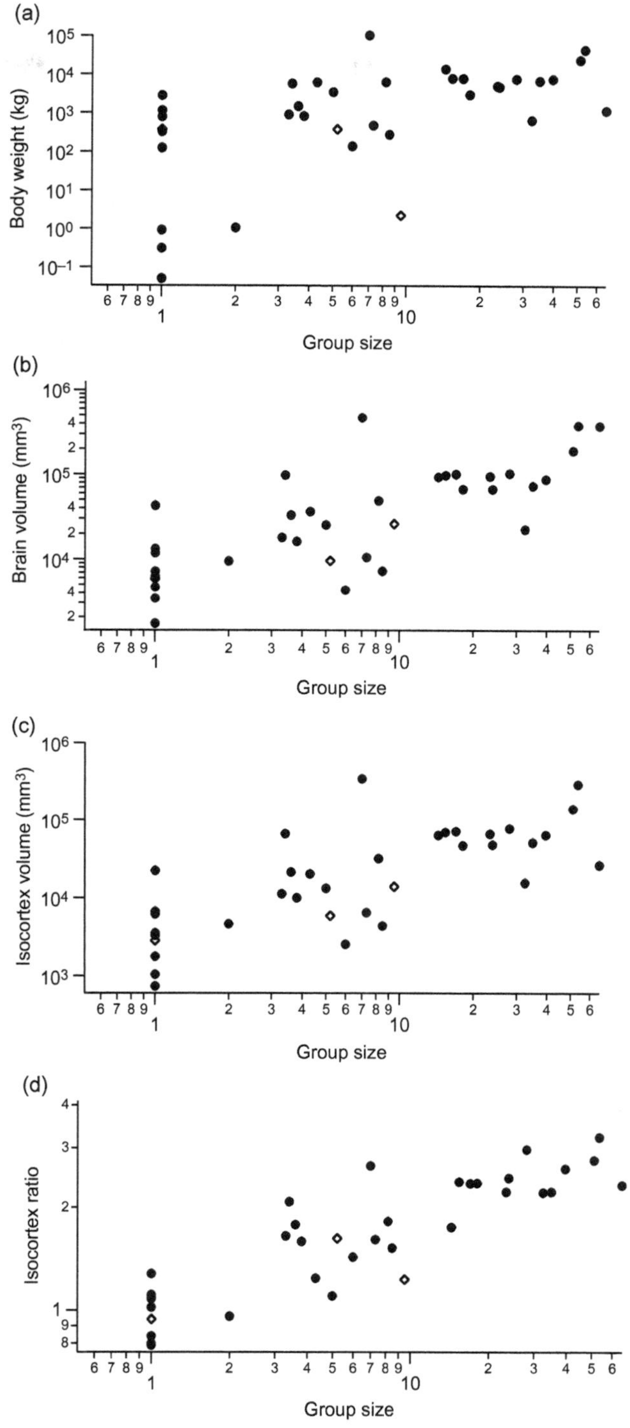

Fig. 1. Body weight (a), brain volume (b), isocortex volume (c), and the size of the isocortex relative to the rest of the brain (d) are regressed against group size in primates. Open diamonds are averages of species. Data are from Dunbar (1992); isocortex volume is gray matter with a white matter component as detailed in Stephan et al. (1981).

primates with increased cognitive "computational power" to manage larger number of possible social interactions and relationships (Dunbar, 1992, 1993, 2009). Although these analyses attempt to correct for possible confounds in cross-species variations in overall brain and body size, the analyses used to support the social brain hypothesis neglect to consider the behavioral setting of allometric and developmentally coordinated changes in brain region size: to consider the brain's intrinsic plasticity and its fundamental role in learning and adapting to both physical and social environments. In addition, the magnitude of the residualized change in brain size attributable to discriminable social factors or identified behavioral capacities of any kind in most cases is extremely small compared to the range over which brains vary. This notable and persistent discrepancy has caused speculation that there might be "two kinds" of size: shared and unshared (residualized) variation in brain size might be physically distinct in some way (Aboitiz, 1996). This is an interesting idea, but one that still has no empirical support.

Coordinated changes in time and size

Overall body size varies widely between primate species ranging from 56g in gray mouse lemur (*Microcebus murinus)* to 105kg in gorillas (*Gorilla gorilla*). The brain occupies 3% of the overall body volume in the gray mouse lemurs compared with 0.4% of the overall body in gorillas (Stephan et al., 1981). Well-known comparative analyses of brain and body size across mammalian species show that the overall brain size scales with a negative allometry when regressed against the body (Armstrong, 1982; Jerison, 1973; Jerison, 1979). That is, as bodies expand, brains get proportionately smaller.

The size of the isocortex also varies widely between primate species ranging from 0.74 cm^3 in the gray mouse lemur to approximately 1000 cm^3 in humans (Stephan et al., 1981). The size of the isocortex occupies only 44% of the overall brain in the gray mouse lemur compared to 80% of their entire brain in humans (*Homo sapiens*; Stephan et al., 1981). It might therefore seem reasonable to assume that the proportionally enlarged isocortex of humans sets us apart from other primates. However, it has long been established that the size of the brain of humans, and the isocortex particularly, is an allometrically scaled-up version of its close relatives (Hofman, 1989, See Chapter 18).

Mammalian and nonmammalian vertebrates exhibit a conserved pattern of brain scaling (Figs. 2 and 3; Darlington et al., 1999; Finlay et al., 2011; Finlay and Darlington, 1995; Reep et al., 2007; Yopak et al., 2010). Some brain regions (e.g., medulla) scale with a negative allometry when regressed against the rest of the brain. Other brain structures such as the isocortex scale with a positive allometry when regressed against the rest of the brain (Finlay and Darlington, 1995) so that as mammalian brain sizes expand, the more they come to be dominated by the volume of the isocortex (Fig. 4).

Isocortical subdivisions vary widely between primate species. For instance, the frontal cortex occupies 19% of the overall brain in a monk sakai (*Pithecia monachus*), whereas the frontal cortex occupies approximately 42% of the overall brain in humans (Smaers et al., 2010, 2011). Isocortical regions, such as "primary visual cortex" or "frontal cortex" (variously defined), also exhibit distinct allometric scaling with the rest of the isocortex (Fig. 3; Bush and Allman, 2004; Kaskan et al., 2005; Smaers et al., 2010, 2011). Other isocortical subdivisions (e.g., primary somatosensory isocortex) scale with a negative allometry when regressed against the rest of the isocortex (Fig. 3; Kaskan et al., 2005). Taken together, these findings demonstrate that as primate brains get bigger, the isocortex and, in particular, the frontal and visual cortices become disproportionately enlarged relative to the rest of the brain.

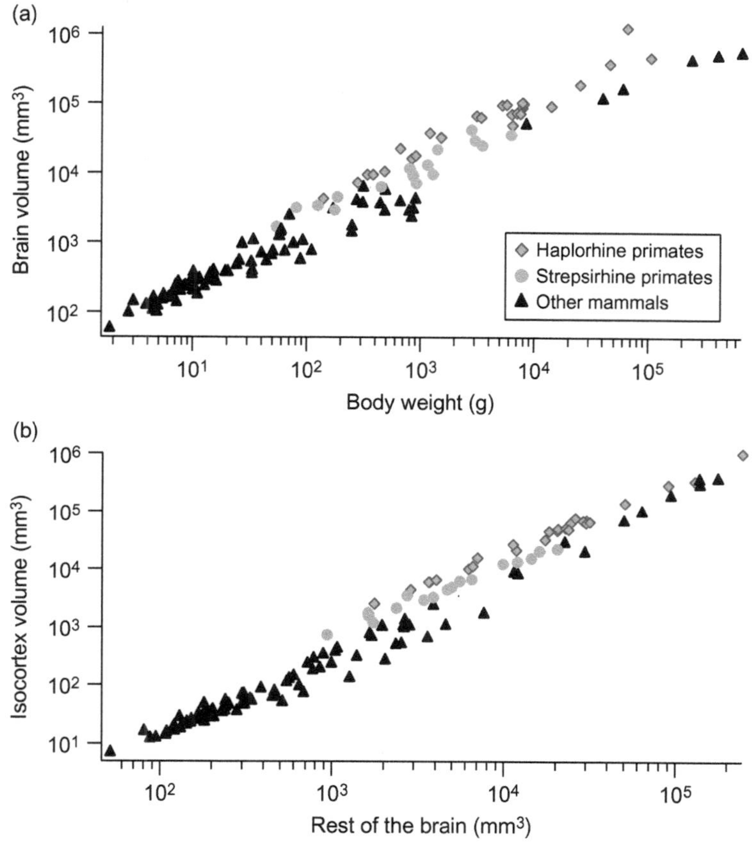

Fig. 2. Brain volume as a function of body weight (a) and isocortex volume as a function of the rest of the brain (b) in primates and other mammals. These data show that the overall brain and isocortex size is disproportionately expanded in primates relative to many other mammals, but that the size of the brain and isocortex strongly covaries within taxonomic groups. Data are from Stephan et al. (1981).

How might these coordinated changes in body, brain, and brain region size arise in development? Finlay and Darlington (1995) have argued that the conserved pattern of brain scaling is mirrored in a conserved pattern in the sequence of developmental events. Allometric variations in adult brain size arise because of the exponential increase in the progenitor pool population when developmental schedules lengthen. That is, progenitor cells that exit the cell cycle late in development benefit from an exponential multiplication of cells relative to progenitor cells that exit the cell cycle early in development (Finlay et al., 2001). The effect of stretching developmental schedules is that structures that are born late in development (e.g., isocortex) become proportionately enlarged relative to the overall brain.

We should emphasize here that we are not using this description of development as an argument for overwhelming "developmental constraint," brain or behavioral uniformity, but rather to lay out the reasonably simple basis for the nonlinear behavior of the vertebrate brain "Bauplan" as it enlarges. Within the context of this Bauplan, new phenotypes emerge, arising from such diverse developmental mechanisms as

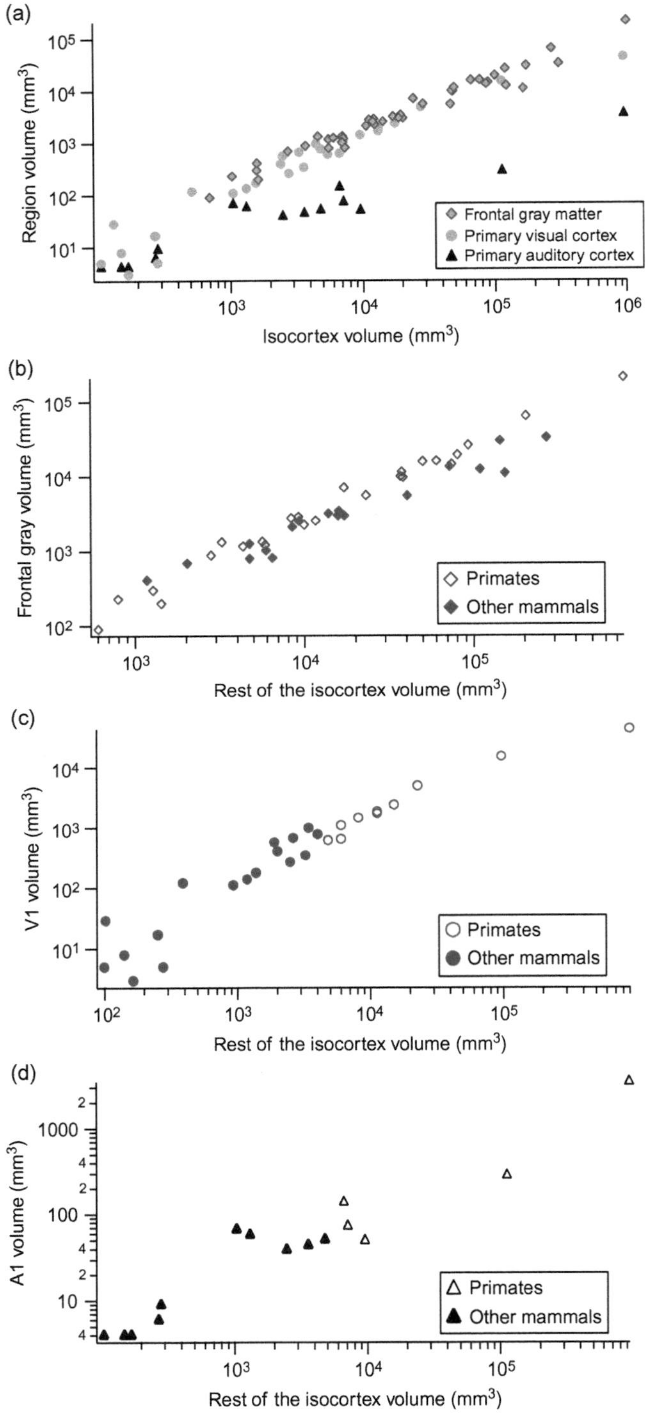

Fig. 3. Isocortical regions are plotted against the overall isocortex volume (a) or the rest of the isocortex volume (b–d) in several mammalian species. The frontal gray matter and the primary visual cortex (V1) expand faster than the size of the primary auditory cortex (A1) as overall brain size increases. These observations suggest that various isocortical regions expand with a distinct allometry. Data are from Kaskan et al. (2005) and Smaers et al. (2010).

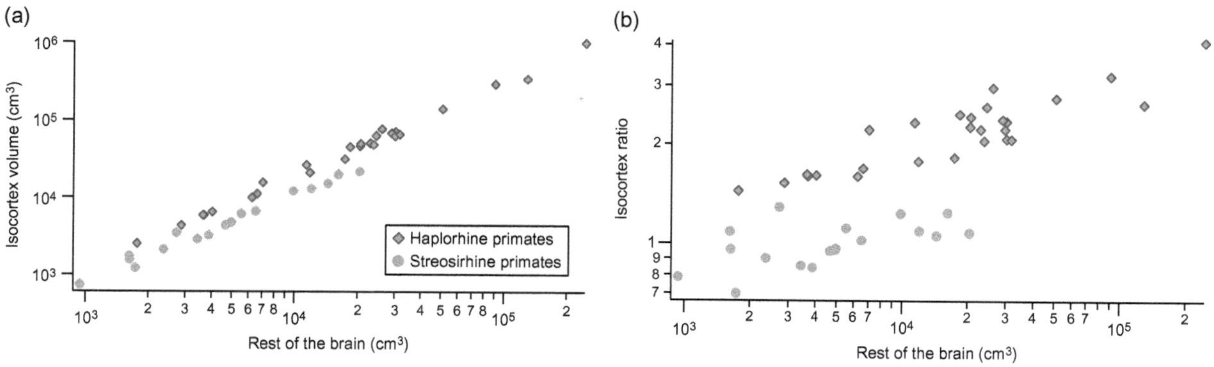

Fig. 4. Isocortex volume (a) and isocortex volume relative to the rest of the brain (b) are plotted against the rest of the brain (i.e., brain–isocortex) in haplorhine and strepsirhine primates. As brains expand, the isocortex becomes disproportionately enlarged in strepsirhine and haplorhine primates. This is evident from the observation that the relative size of the isocortex increases as the rest of the brain expands. Data are from Stephan et al. (1981).

changing rates of cell division (Charvet et al., 2011), early alterations in gene expression patterns that specify brain regions (McGowan et al., 2010; Menuet et al., 2007; Sylvester et al., 2010), heterochronic shifts of "clocks" of cell specification with respect to the establishment of precursor pools to change neuron proportions (Finlay, 2008; Dyer et al., 2009), and extended or limited receptor distributions altering motivational strengths (Young and Wang, 2004). Rather than ignoring "common variance" in brain size by examining residuals, we argue that the stability of this fundamental plan is so strong that it demands a functional explanation of its own.

Variation in size and time

Selective changes in brain region size between taxonomic groups are called grade shifts (Barton and Harvey, 2000). Primate suborders exhibit a number of grade shifts in brain and brain region size. Haplorhine primates (i.e., new-world and old-world monkeys and apes) exhibit a disproportionately enlarged brain and a disproportionately enlarged isocortex relative to that of strepsirhine primates (i.e., lemurs, lorises, galagos; Figs. 2 and 4; Barton and Harvey, 2000; Finlay et al., 1998; Stephan et al., 1981). Both haplorhine and strepsirhine primates exhibit a disproportionately enlarged isocortex relative to many other mammals. However, within each primate suborder, the size of the isocortex is extremely predictable when regressed against the rest of the brain (Fig. 2).

Comparative developmental studies in primates and other mammals show that the grade shifts just described in isocortex size may arise due to selective alterations in the timing of developmental schedules. Comparative analyses of isocortex generation and development showed that haplorhine primates (i.e., rhesus monkeys, humans) selectively delay isocortical neurogenesis compared to rodents (i.e., rats, mice, hamsters, spiny mice, guinea pigs; Clancy et al., 2000, 2001, 2007; Finlay et al., 1998). Relative delay in isocortical neurogenesis entails that the isocortical progenitor pool population will multiply exponentially relative to other nondelayed structures and the isocortex will expand in neuron number and size. Note that we have described at this point two separate, but logarithmically additive, ways of increasing relative cortex size. Increase in duration of development alone to produce a larger brain, with rate of cell production unchanged, automatically increases the relative proportion of the isocortex. The primate isocortex expands disproportionately still more, adding to the fundamental nonlinearity of allometric scaling an increase in the dedicated precursor cell pool for the cortex, by delaying isocortical stem-cell cessation with respect to the schedule established in rodents and insectivores, and thereby producing a "grade shift."

Some intrinsic difficulties on the use of residuals and ratios in allometric studies

In a series of studies beginning in the early 1990s, Dunbar and his colleagues showed that the relative size of the isocortex or brain size and residuals derived from a linear regression of brain and body size positively correlate with group size or related measures of group size in primates (Dunbar, 1992, 2009; Dunbar and Shultz, 2007a,b; Pérez-Barbería et al., 2007). In parallel with statistical practices in the field, the first studies looked at basic regressions between two variables. The next set of studies used more elaborate multiple regression techniques and phylogenetic contrasts to eliminate the statistical problem of non-independence of taxonomic relationships. Recent studies attempted to determine aspects of temporal emergence of the correlated features examined using discretized variables in conjunction with extensive phylogenetic analyses (Pérez-Barbería et al., 2007). As statistical analyses flourish, it is rare to see any representation of primary data, and the basic "visual" sense of the strength of association, magnitude of results, or amount of variation has tended to fade. In this chapter, following the historical progression of the analyses described, we will plot the basic data relating brain, body, and group size in primates, then add in taxonomic variability, and finally consider the range of variation and a few of the measurement issues in group size, but will go back and plot the basic data on which these claims are established. We should emphasize that we do not contest that there is a relationship between relative brain size, and (possibly) relative isocortex volume and social complexity, generally speaking. The sophistication of the statistical analyses is undoubted. What we do contest are the basic assumptions of the techniques, the causal relationships implied, and the claim that the relationship between social competence and relative brain size, compared to any of a number of other measures of behavioral complexity, is unique.

First, we describe problems with statistical comparisons between groups involving basic allometric relationships between brain parts, to set out very basic issues, which antedate the social brain hypothesis. A number of studies examining the potential mechanisms underlying species-specific adaptations or developmental disorders have focused on the relative sizes of parts of the brain. The initial problem (not a problem of the social brain studies) is "relative to what"? For example, suppose it is shown that the relative size of the frontal cortex is greater in autistic individuals relative to healthy individuals (Carper and Courchesne, 2005; Courchesne et al., 2011), even correcting for a somewhat greater brain size in the autistic group by taking a ratio of frontal cortex to brain volume overall. If individual variation in humans follows primate brain allometry, increases in brain size will produce an even greater increase in the proportion of cortex, and frontal cortex will be a greater proportion still (Fig. 4). "Correcting" for brain size by taking a simple ratio of frontal cortex to brain size between two groups with differing brain sizes will invariably demonstrate relatively more frontal cortex in the group with the larger mean brain size, but this is simply a predictable outcome of the underlying allometry and no indication of any unusual hypertrophy or pathology of the frontal cortex. Although this problem plagues a number of comparative studies in which two species are compared, or brains with a developmental disorder that are compared to normal brains, fortunately, for studies of primate brain evolution, we have ample information to be able to predict the different allometries of various brain divisions.

The use of residuals derived from allometric equations relating the size of two structures is a common method to compare brain region size across species. In the case of the social brain hypothesis, the finding that residuals derived from the linear regression between brain and body size correlates with group size must account for grade shifts in brain size between haplorhine and strepsirhine primates, and when that is done, a significant statistical relationship remains (Pérez-Barbería et al., 2007). The brains of haplorhine primates are disproportionately enlarged relative to those of strepsirhine primates. However, brain and body size strongly covary within haplorhine and strepsirhine primates. Fitting a linear regression through brain and body size in primates (haplorhines and strepsirhine primates) would fit a linear regression with a different slope and intercept than those obtained by fitting two separate linear regressions through the brain and body of haplorhine and strepsirhine primates (Fig. 5). Returning to the basic data, we look at the amount of association between relative brain size and social group size in these two taxonomic groups. In the case of the social brain hypothesis, the brain versus body residual values obtained for both haplorhine and strepsirhine primates correlate more strongly with group size than does the brain versus body residual values derived from separate linear regressions for haplorhine versus strepsirhine primates (Pérez-Barbería et al., 2007). However, the correlation coefficients derived from the residuals of brain to body size in both scenarios are surprisingly low, and it would appear that something about the grade shift in relative brain and cortex size between these two taxa is accounting for most of the effect. Further analyses from the same laboratory group and others considering phylogenetic contrasts, other behavioral measures, and more elaborate statistics generally

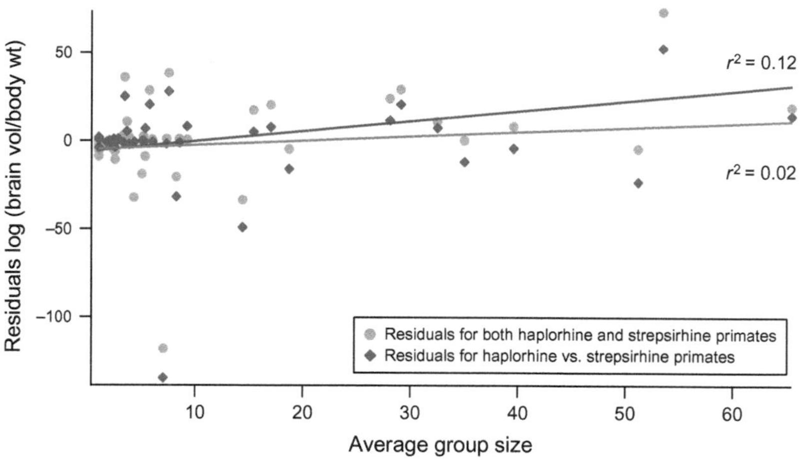

Fig. 5. Residuals derived from a linear regression through the brain size and body size are correlated against group size in primates. In one scenario, residuals are derived from a linear regression through the brain and body size of both haplorhine and strepsirhine primates (red line). In another scenario, residuals are derived from two separate linear regressions derived for haplorhine and strepsirhine primates (blue line). The correlation coefficient between group size and brain versus body size residual values derived from both haplorhine and strepsirhine primates is higher than the correlation coefficient between group size and brain versus body size residuals obtained for strepsirhine and haplorhine primates. Data are from Pérez-Barbería et al. (2007). Although the authors examined the geometric means rather than the arithmetic means of primate group size, the low correlation coefficients between group size and brain versus body size residuals in primates do not support the claim that group size correlates with brain size in primates. (For interpretation of the references to color in this figure legend, the reader is referred to the Web version of this chapter.)

demonstrate a statistically significant but clearly very small effect in residual change in brain size (Barton, 1993; Dunbar, 1993).

We return to the basic data in Fig. 1 of the regressions of primate body, brain isocortex volume, and "isocortex ratio" with group size (taken from Dunbar, 1992). The social brain hypothesis posits that isocortex (either volume or ratio) and group size are positively correlated (isocortex volume residualized with respect to both body size and brain volume). Of interest here is the relationship of body size to social group size, which opens the possibility of many other causal routes between brain size and group size than the ones mentioned here—for example, niche, range size, and type of food consumed. Further, consider other behaviors demonstrated to vary with brain size: "innovation" in the wild, successful invasion of new territories, residual mortality when corrected for body size, and laboratory measures of learning ability. These all correlate with each other and with relative brain size (González-Lagos et al., 2010; Lefebvre and Sol, 2008; Lefebvre et al., 2002; Reader et al., 2011). The strength of the association between isocortex size and group size that would be left after partialling out capability to innovate, or general learning ability, seems unlikely to be significant. Put another way, we would suggest that virtually any reasonable measure of cognitive or behavioral complexity—working memory grammatical sequence learning, innovation and so on—would show the same relationship to relative brain size.

Variation in social structure within a species

The size of social systems and social structures varies widely within a species. Humans form groups that are variable in size (Dunbar, 1993; Zhou et al., 2005). Some humans are polygamous or polyandrous, whereas others are monogamous. Some humans pair-bond with a single individual for life, while others pair-bond for short bouts. Given the variation in social structure within humans, it seems difficult to assign a specific sociality index or group size to humans. Indeed, Dunbar has refrained from estimating group size in humans (Dunbar, 1993). Instead, estimation of group size for humans is based on analyses of group size and brain size of nonhuman primates.

Strepsirhine primates are considered to be solitary or form small groups. In contrast, haplorhine primates are thought to aggregate in large groups. However, there is considerable variation within species in each of these taxonomic groups (Fig. 6; Smuts et al., 1986). Among strepsirhine primates, lemurs such as the white sifaka (*Propithecus verreauxi*) have been observed to be solitary but they may also form groups of up to 13 individuals (Smuts et al., 1986). Among haplorhine primates, rhesus monkeys form groups that range from nine to well over 100 individuals (Berman et al., 1997; Smuts et al., 1986). Gorillas are considered to form large and complex social systems but some members of these species are actually solitary (Smuts et al., 1986). Collectively, these observations suggest that the size of social system varies widely within a species and that estimates of mean group size neglect to consider the wide variation in the size of social systems within each primate species.

Food resource distribution contributes systematically to social organization in vertebrates generally, and mammals and primates specifically (Chapman, 1990a). Resource availability in part determines the decision of birds to contribute alloparental care to relatives, rather than to seek independent reproduction (Emlen, 1974). For instance, in primates, it has been reported that spider monkeys (*Ateles geoffroyi*) from Costa Rica vary from 1 to 35 individuals and 50% of the variance in mean subgroup size can be predicted from the size, density, and distribution of food patches (Chapman, 1990b). These observations suggest that primates are actually highly flexible in modifying the size of their social system in response to resource availability. In the case of humans (Betzig, 2009), it has been persuasively argued that, in cases where resources are physically

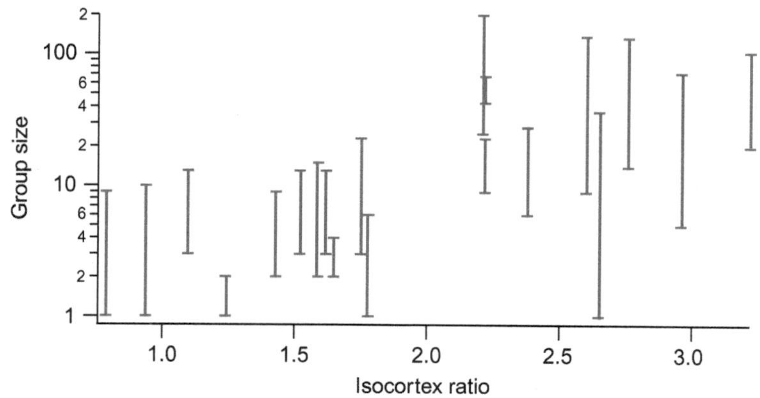

Fig. 6. The range of reported group size within primates is plotted against the relative size of the isocortex. These data show that group size varies extensively within primate species. Data on group size ranges are from Izawa (1976), Smuts et al. (1986), Koenig (1995), and Higham et al. (2009).

stationary, and can be monopolized, as in traditional agrarian societies, the resulting social structure can be characterized as "eusocial," with reproductive activity limited to a few individuals, who can control large harems, with explicitly designated nonreproductive castes.

The social brain hypothesis, quite obviously, does not argue for a complete absence of contribution of other factors to group size, but only for a constraint on maximal tolerable group size related to brain size. When the actual range of natural variation in primate societies is considered, however, the conceptualization of how any individual might be selected on to cope with a particular group size becomes suspect, and the kind of explanations offered seem more like general capacity arguments, rather than a numerical limit on the number of individuals to be remembered.

Constancy in size and time within a species: An unusual example from human pygmies

Developmental schedules in primates, and mammals generally, subsuming brain growth, body growth, maturational milestones, and life span are very highly intercorrelated. Moreover, the initial production of brain tissue is exceptionally predictable as a constant function per unit time (Passingham, 1985), and brain volume, both relative and absolute, is highly correlated with life span. For example, several studies have shown that variations in body size are associated with variations in postnatal growth, life-history schedules, and life expectancy within humans and across species (Charnov, 1991; Migliano et al., 2007; Nettle, 2010). One recent study showed that human pygmies from two different continents reach adult stature and sexual maturity and die earlier than taller individuals (Fig. 7; Migliano et al., 2007). These findings suggest that size, developmental schedules, as well as the overall life span length covary within a species. It is not clear what factors might have caused changes in size, developmental schedules, and the overall life span in pygmies, but there is no evidence that we know of to suggest that a simple social structure or a reduced social system in pygmies caused them to be smaller than taller individuals, or the reverse.

Constancy in brain architecture fosters variation in brain function

The plasticity of the isocortex described in present neuroscience work, and in functional imaging,

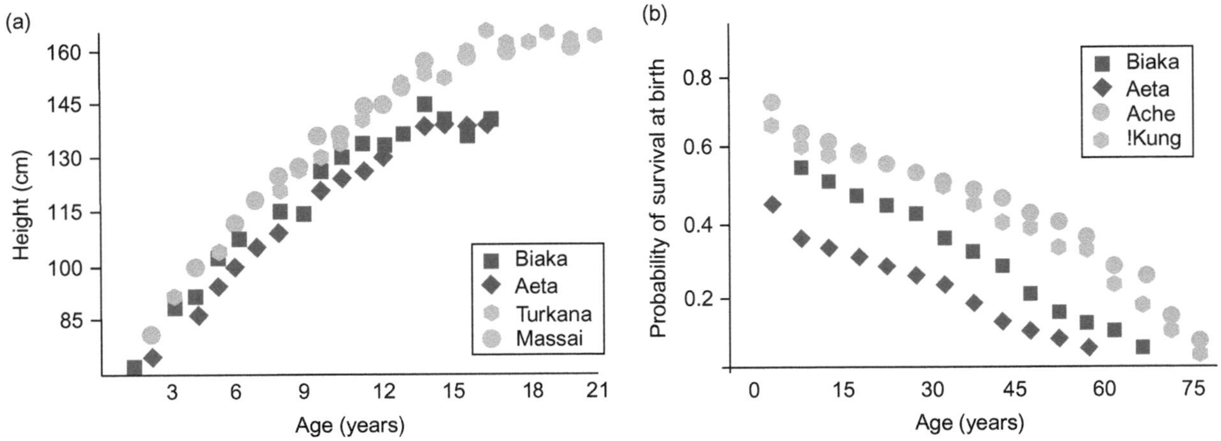

Fig. 7. The Biaka and Aeta are taller than Turkana, Massai, Ache, and !Kung. (a) Biaka and Aeta reach adult height later than Turkana and Massai. (b) Biaka and Aeta have a reduced survival probability at birth than Biaka and Aeta do. Data are from Migliano et al. (2007).

is increasingly at odds with the specificity and modularity of brain function that is often presumed in studies relating brain region size to specific cognitive abilities—this enterprise termed "neuroecology." There is tremendous plasticity in what sensory modalities a brain region may process. It is well known that the visual cortex or auditory cortex may reallocate function to process information from other modalities in the absence of visual input or auditory input. Experimental studies have found that early removal of large regions of the visual cortex, superior colliculus, and the brachium of the inferior colliculus of developing ferrets gives rise to a novel and functional visual pathway that projects through the medial geniculate nucleus to the primary auditory cortex (Roe et al., 1993; Sur et al., 1988). In naturally blind humans, functional magnetic resonance imaging studies show increased blood oxygen level-dependent (BOLD) signals in the visual cortex during tactile discrimination (Sadato et al., 1996; Sathian and Stilla, 2010), auditory-spatial tasks (Collignon et al., 2011), and sentence comprehension tasks (Bedny et al., 2011). Reallocation of function is not restricted to long-term deprivations of one sensory modality. Short-term loss of vision also leads to increased BOLD signal in the occipital cortex during tactile discrimination tasks (Kauffman et al., 2002; Merabet et al., 2008). Collectively, these observations suggest that there is tremendous flexibility in processing information from various modalities. We argue that this flexibility allows organisms to respond to changing social and environmental situations throughout the life span.

At the core of the social brain hypothesis is the assumption that primates with larger isocortices can manage social groups because they have more cognitive capacity for social information than primates that form smaller groups. Does this claim about specific intelligence track well onto specific abilities subserved by cortical areas in humans? Unfortunately, there is little connection between these literatures. In particular, it has been suggested that the size and activation of frontal and temporal isocortices correlate with basic measures of intelligence in humans. In support of this argument is the finding that the size of the isocortex and activation of the frontal and temporal cortices correlate with measures of intelligence and social cognitive performance in humans (Jung and Haier, 2007; Powell et al., 2010). However, the isocortex obviously mediates more behaviors involved in social situations, and social behaviors are mediated by a multitude of brain regions (Anderson, 2010). Within the temporal cortex, Broca's area is a neural substrate for the perception and production of speech. However, Broca's area is also involved in other tasks such as imitation (Makuuchi, 2005), anticipation of movement, and imagery of motion (Binkofski et al. 2000). Therefore, the finding that activation and size of the isocortex is correlated with measures of intelligence or social cognitive performance does not, in and of itself, show that the expanded isocortex confers specific increased cognitive abilities to manage social situations.

Predictable relationships in brain architecture and brain size

Coordinated variations in developmental length, life span, and brain size appear to have evolved in a number of mammalian taxa. Instances of convergent evolution suggest that this covariation may be intrinsically linked. Similar to haplorhine primates, elephants and cetaceans exhibit an extended period of development, an extended life span, and a large brain (Armstrong, 1982; Haug, 1987; Hofman, 1983, 1993). The observation that elephants and cetaceans differ from their sister groups and most other mammals in being large and having prolonged developmental periods and extended life spans suggests that changes in time and size have evolved together multiple times.

Elephants and cetaceans are considered to be among the longest-lived mammals. Among cetaceans, whales such as the bowhead whale (*Balaena mysticetus*) have been estimated to live over 100 years (George et al., 1999). The Asian elephant (*Elephas maximus*) exhibits some of the longest recorded life spans of land animals, with an estimated maximum life span of approximately 65–86years (Weigl, 2005; Wiese and Willis, 2004). Elephants and whales not only exhibit extended life spans, but they also exhibit extended periods of postnatal development. For instance, the Asian elephant reaches adult stature at around 17 years of age and it is estimated that the bowhead whales reach sexual maturity after 22 years of age (George et al., 1999). Bowhead whales are among the largest animals weighing 100,000kg and adult Asian elephants are among the largest land animals weighing approximately 3000kg. The finding that some elephants and cetaceans are among the largest and most long-lived animals suggests that these taxa expanded and prolonged the duration of developmental length and life span. Taken together, these observations suggest that time and size vary together. The coordinated variation in size and time may entail predictable consequences for behavior.

Some elephants, cetaceans, and primates are well known for their cooperation. We argue that an extended duration of postnatal development entails an extended period of postnatal parental care, which may foster affiliate behavior and cooperation directed toward juveniles or adult group members. In support of this argument is the observation that some primates and elephants not only receive parental care, but they also receive allomaternal or alloparental care (Lee, 1997; Rapaport and Haight, 1987; Riedman, 1982). Some elephants and primates display evidence of life-long affiliate and cooperative behavior toward kin (Langergraber et al., 2007) and nonkin (de Waal et al., 2008; Langergraber et al., 2007; Plotnik et al., 2011). Evidence that long-life histories covary with animal cooperation is also found among nonmammalian vertebrates such as corvids (Møller, 2006; Seed et al., 2008). Taken together, these findings suggest that species that exhibit long-life histories also exhibit evidence of animal cooperation and affiliate behavior. However, it is not clear if sociality may foster coordinated changes in the organism's overall size, brain size, or cortex size. It is possible that coordinated changes in developmental schedules foster changes in social behavior, and the record of that learning process is the mature isocortex.

Causal scenarios, which depend on covariation, give development a central role

We now return to the bidirectional forms of brain change described in the introductory sections and develop some causal scenarios, which may link the morphological and general behavioral capacities under consideration here. Particularly, we are interested in those behaviors, which allow the brain to play a causal role in its own construction. In primates, big bodies and big brains take longer to make and require more resources. Those larger primate infants are necessarily going to have a longer developmental period to learn in, and because their larger brain size is likely to be associated with biparental or possibly alloparental care, more individuals populate its extended developmental period, both at any moment and over time. This extended learning period may enable these large infants to better develop the categorization skills to differentiate individuals and their motivations, learn elaborate methods of food processing, or learn the unique characteristics and affordances of foraging sites, depending on what the social and natural ecology presents. By its essential, covarying nature, however, a big-brained mammal has an extended developmental period, populated by at least one and often many caretakers, which will make social complexity in large-brained creatures a high probability. While we are uncertain whether group size *per se* is a good measure of a broader notion of social complexity, which must certainly be a multivariate entity, we suggest that the real mediating variable between brain size and behavioral complexity might be developmental time, and not simply the number of neurons available to discriminate individuals.

This view of brain evolution is quite distinct from the one that emerges from the differentiating, residual variance view of brain part evolution. In the view of brain evolution set out by Dunbar and colleagues, additional brain goes directly to improve capacity limitation on the ability to either remember or orchestrate the interactions of a set number of individuals, a direct mapping of a social problem defined numerically to a volume of committed tissue. In our view, the extended developmental schedule required to make a large brain and the size of the brain itself must be considered as one variable. The extreme conservation of this relationship across mammals suggests that there have been few advantages in attempts to decouple rate of production and size of the end product. The motivations each developing organism brings to the environment in combination with its plastic brain allow the information represented in the physical and social environment to construct the mature organism on which natural selection will act.

Acknowledgments

This work was supported by an NSF grant #0849612 to B. L. F. and by the Eunice Kennedy Shriver National Institute of Child Health and Human Development fellowship #F32HD067011 to C. J. C. The content is solely the responsibility of the authors and does not necessarily represent the official views of the Eunice Kennedy Shriver National Institute of Child Health and Human Development or the National Institutes of Health.

References

Aboitiz, F. (1996). Does bigger mean better? Evolutionary determinants of brain size and structure. *Brain, Behavior and Evolution*, *47*, 225–245.

Aiello, L. C., & Wheeler, P. (1995). The expensive-tissue hypothesis: The brain and the digestive system in human and primate evolution. *Current Anthropology*, *36*, 199–221.

Anderson, M. L. (2010). Neural reuse: A fundamental organizational principle of the brain. *The Behavioral and Brain Sciences*, *33*, 245–266 discussion 266–313.

Armstrong, E. (1982). A look at relative brain size in mammals. *Neuroscience Letters*, *34*, 101–104.

Barton, R. A. (1993). Independent contrasts analysis of neocortical size and socioecology in primates. *The Behavioral and Brain Sciences*, *16*, 694–695.

Barton, R. A., & Harvey, P. H. (2000). Mosaic evolution of brain structure in mammals. *Nature*, *405*, 1055–1058.

Bedny, M., Pascual-Leone, A., Dodell-Feder, D., Fedorenko, E., & Saxe, R. (2011). Language processing in the occipital cortex of congenitally blind adults. *Proceedings of the National Academy of Sciences of the United States of America*, *108*, 4429–4434.

Berman, C. M., Rasmussen, K. L. R., & Suomi, S. J. (1997). Group size, infant development and social network in free-ranging rhesus monkey. *Animal Behavior*, *53*, 405–421.

Betzig, L. (2009). But what is government itself but the greatest of all reflections on human nature? *Politics and the Life Sciences*, *28*, 102–105.

Binkofski, F., Amunts, K., Stephan, K. M., Posse, S., Schormann, T., Freund, H. J., et al. (2000). Broca's region subserves imagery of motion: a combined cytoarchitectonic and fMRI study. *Human Brain Mapping*, *11*, 273–285.

Bush, E. C., & Allman, J. M. (2004). The scaling of frontal cortex in primates and carnivores. *Proceedings of the National Academy of Sciences of the United States of America*, *101*, 3962–3966.

Carmody, R. N., & Wrangham, R. W. (2009). The energetic significance of cooking. *Journal of Human Evolution*, *57*, 379–391.

Carper, R. A., & Courchesne, E. (2005). Localized enlargement of the frontal cortex in early autism. *Biological Psychiatry*, *57*, 126–133.

Chapman, C. A. (1990a). Ecological constraints on group size in three species of neotropical primates. *Folia Primatologica: International Journal of Primatology*, *55*, 1–9.

Chapman, C. A. (1990b). Association patterns of spider monkeys: The influence of ecology and sex on social organization. *Behavioral Ecology and Sociobiology*, *26*, 409–414.

Charnov, E. L. (1991). Evolution of life history variation among female mammals. *Proceedings of the National Academy of Sciences of the United States of America*, *88*, 1134–1137.

Charvet, C. J., Striedter, G. F., & Finlay, B. L. (2011). Evo-devo and brain scaling: Candidate developmental mechanisms for variation and constancy in vertebrate brain evolution. *Brain, Behavior and Evolution*, *78*, 248–257.

Clancy, B., Darlington, R. B., & Finlay, B. L. (2000). The course of human events: Predicting the timing of primate neural development. *Developmental Sciences*, *3*, 57–66.

Clancy, B., Darlington, R. B., & Finlay, B. L. (2001). Translating developmental time across mammalian species. *Neuroscience*, *105*, 7–17.

Clancy, B., Kersh, B., Hyde, J., Anand, K. J. S., Darlington, R. B., & Finlay, B. L. (2007). Web-based method for translating neurodevelopment from laboratory species to humans. *Neuroinformatics*, *5*, 79–94.

Collignon, O., Vandewalle, G., Voss, P., Albouy, G., Charbonneau, G., Lassonde, M., et al. (2011). Functional specialization for auditory-spatial processing in the occipital cortex of congenitally blind humans. *Proceedings of the National Academy of Sciences of the United States of America*, *108*, 4435–4440.

Courchesne, E., Campbell, K., & Solso, S. (2011). Brain growth across the life span in autism: Age-specific changes in anatomical pathology. *Brain Research*, *22*, 138–145.

Darlington, R. B., Dunlop, S. A., & Finlay, B. L. (1999). Neural development in metatherian and eutherian mammals: Variation and constraint. *The Journal of Comparative Neurology*, *411*, 359–368.

de Waal, F. B., Leimgruber, K., & Greenberg, A. R. (2008). Giving is self-rewarding for monkeys. *Proceedings of the National Academy of Sciences of the United States of America*, *105*, 13685–13689.

Deacon, T. W. (1990). Rethinking mammalian brain evolution. *American Zoologist*, *30*, 629–705.

Dunbar, R. I. M. (1992). Neocortex size as a constraint on group size in primates. *Journal of Human Evolution*, *22*, 469–493.

Dunbar, R. I. M. (1993). Coevolution of neocortical size, group size, and language in humans. *The Behavioral and Brain Sciences*, *16*, 681–735.

Dunbar, R. I. (2009). The social brain hypothesis and its implications for social evolution. *Annals of Human Biology*, *36*, 562–572.

Dunbar, R. I. M., & Shultz, S. (2007a). Understanding primate brain evolution. *Philosophical Transactions of the Royal Society of London. Series B, Biological Sciences*, *362*, 649–658.

Dunbar, R. I., & Shultz, S. (2007b). Evolution in the social brain. *Science*, *317*, 1344–1347.

Dyer, M. A., Martins, R., da Silva Filho, M., Muniz, J. A., Silveira, L. C., Cepko, C. L., et al. (2009). Developmental sources of conservation and variation in the evolution of the primate eye. *Proceedings of the National Academy of Sciences of the United States of America*, *106*, 8963–8968.

Emlen, J. T. (1974). An urban bird community in Tucson, Arizona: Derivation, structure, regulation. *The Condor*, *76*, 184–197.

Falk, D. (1990). Brain evolution in Homo: The "radiator" theory. *The Behavioral and Brain Sciences*, *13*, 333–381.

Falk, D., & Gage, T. B. (1998). Radiators are cool: A response to Braga & Boesch's published paper and reply. *Journal of Human Evolution*, *35*, 307–312.

Finlay, B. L. (2007). E pluribus unum: Too many unique human capacities and too many theories. In S. Gangestad & J. Simpson (Eds.), *The evolution of mind: Fundamental questions and controversies* (pp. 294–304). New York: Guilford Press.

Finlay, B. L. (2008). The developing and evolving retina: Using time to organize form. *Brain Research*, *1192*, 5–16.

Finlay, B. L., & Darlington, R. B. (1995). Linked regularities in the development and evolution of mammalian brains. *Science*, *268*, 1578–1584.

Finlay, B. L., Darlington, R. B., & Nicastro, N. (2001). Developmental structure in brain evolution. *The Behavioral and Brain Sciences*, *24*, 263–278.

Finlay, B. L., Hersman, M. N., & Darlington, R. B. (1998). Patterns of vertebrate neurogenesis and the paths of vertebrate evolution. *Brain, Behavior and Evolution*, *52*, 232–242.

Finlay, B. L., Hinz, F., & Darlington, R. B. (2011). Mapping behavioural evolution onto brain evolution: The strategic roles of conserved organization in individuals and species. *Philosophical Transactions of the Royal Society of London. Series B, Biological Sciences*, *366*(1574), 2111–2123.

George, J. C., Bada, J., Zeh, J., Scott, L., Brown, S. E., O'Hara, T., et al. (1999). Age and growth estimates of bowhead whales (Balaena mysticetus) via aspartic acid racemization. *Canadian Journal of Zoology*, *77*, 571–580.

González-Lagos, C., Sol, D., & Reader, S. M. (2010). Large-brained mammals live longer. *Journal of Evolutionary Biology*, *23*, 1064–1074.

Haug, H. (1987). Brain sizes, surfaces, and neuronal sizes of the cortex cerebri: A stereological investigation of man and his variability and a comparison with some mammals (primates, whales, marsupials, insectivores, and one elephant). *American Journal of Anatomy*, *180*, 126–142.

Hauser, M. D., Chomsky, N., & Fitch, W. T. (2002). The faculty of language: What is it, who has it, and how did it evolve? *Science*, *298*, 1569–1579.

Higham, J. P., Warren, Y., Adanu, J., Umaru, B. N., MacLarnon, A. M., Sommer, V., et al. (2009). Living on the edge: Life-history of olive baboons at Gashaka-Gumti National Park, Nigeria. *American Journal of Primatology*, *71*, 293–304.

Hofman, M. A. (1983). Energy metabolism, brain size and longevity in mammals. *The Quarterly Review of Biology*, *58*, 495–512.

Hofman, M. A. (1989). On the evolution and geometry of the brain in mammals. *Progress in Neurobiology*, *32*, 137–158.

Hofman, M. A. (1993). Encephalization and the evolution of longevity in mammals. *Journal of Evolutionary Biology*, *6*, 209–227.

Isler, K., & Van Schaik, C. P. (2009). Why are there so few smart mammals (but so many smart birds)? *Biology Letters*, *5*, 125–129.

Izawa, K. (1976). Group sizes and composition of monkeys in the upper Amazon basin. *Primates*, *17*, 367–399.

Jerison, H. J. (1973). *Evolution of the brain and intelligence.* New York: Academic Press.

Jerison, H. J. (1979). The evolution of diversity in brain size. In M. E. Hahnet al. (Ed.)*Development and evolution of brain size* (pp. 29–57). New York: Academic Press.

Jung, R. E., & Haier, R. J. (2007). The Parieto-Frontal Integration Theory (P-FIT) of intelligence: Converging neuroimaging evidence. *The Behavioral and Brain Sciences*, *30*, 135–154 discussion 154–187.

Kaskan, P. M., Franco, E. C., Yamada, E. S., Silveira, L. C., Darlington, R. B., Darlington, L. C., et al. (2005). Peripheral variability and central constancy in mammalian visual system evolution. *Proceedings of the Royal Society B: Biological Sciences*, *272*, 91–100.

Kauffman, T., Théoret, H., & Pascual-Leone, A. (2002). Braille character discrimination in blindfolded human subjects. *Neuroreport*, *13*, 571–574.

Koenig, A. (1995). Group size, composition, and reproductive success in wild common marmosets (*Callithrix jacchus*). *American Journal of Primatology*, *35*, 311–317.

Langergraber, K. E., Mitani, J. C., & Vigilant, L. (2007). The limited impact of kinship on cooperation in wild chimpanzees. *Proceedings of the National Academy of Sciences of the United States of America*, *104*, 7786–7790.

Lee, P. C. (1997). Allomothering among African elephants. *Animal Behavior*, *35*, 278–291.

Lefebvre, L., Nicolakakis, N., & Boire, D. (2002). Tools and brains in birds. *Behaviour*, *139*, 939–973.

Lefebvre, L., & Sol, D. (2008). Brains, lifestyles and cognition: Are there general trends? *Brain, Behavior and Evolution*, *72*, 135–144.

Lehmann, J., & Dunbar, R. I. (2009). Network cohesion, group size and neocortex size in female-bonded Old World primates. *Proceedings of the Royal Society B: Biological Sciences*, *276*, 4417–4422.

Makuuchi, M. (2005). Is Broca's area crucial for imitation? *Cerebral Cortex*, *15*, 563–570.

McGowan, L., Kuo, E., Martin, A., Monuki, E. S., & Striedter, G. (2010). Species differences in early patterning of the avian brain. *Evolution*, *65*, 907–911.

Menuet, A., Alunni, A., Joly, J. S., Jeffrey, W. R., & Rétaux, S. (2007). Expanded expression of Sonic Hedgehog in *Astyanax* cavefish: Multiple consequences on forebrain development and evolution. *Development*, *134*, 845–855.

Merabet, L. B., Hamilton, R., Schlaug, G., Swisher, J. D., Kiriakopoulos, E. T., Pitskel, N. B., et al. (2008). Rapid and reversible recruitment of early visual cortex for touch. *PLoS One*, *3*, e3046.

Migliano, A. B., Vinicius, L., & Lahr, M. M. (2007). Life history trade-offs explain the evolution of human pygmies. *Proceedings of the National Academy of Sciences of the United States of America*, *104*, 20216–20219.

Møller, A. P. (2006). Sociality, age at first reproduction and senescence: Comparative analyses of birds. *Journal of Evolutionary Biology*, *19*, 682–689.

Nettle, D. (2010). Dying young and living fast: Variation in life history across English neighborhoods. *Behavioral Ecology*, *21*, 387–395.

Passingham, R. E. (1985). Rates of brain development in mammals including man. *Brain, Behavior and Evolution*, *26*, 167–175.

Pérez-Barbería, F. J., Shultz, S., & Dunbar, R. I. (2007). Evidence for coevolution of sociality and relative brain size in three orders of mammals. *Evolution*, *61*, 2811–2821.

Plotnik, J. M., Lair, R., Suphachoksahakun, W., & de Waal, F. B. (2011). Elephants know when they need a helping trunk in a cooperative task. *Proceedings of the National Academy of Sciences of the United States of America*, *108*, 5116–5121.

Powell, J. L., Lewis, P. A., Dunbar, R. I., García-Fiñana, M., & Roberts, N. (2010). Orbital prefrontal cortex volume correlates with social cognitive competence. *Neuropsychologia*, *48*, 3554–3562.

Rapaport, L., & Haight, J. (1987). Some observations regarding allomaternal caretaking among captive Asian elephants (*Elephas maximus*). *Journal of Mammalogy*, *68*, 438–442.

Reader, S. M., Hager, Y., & Laland, K. N. (2011). The evolution of primate general and cultural intelligence. *Philosophical Transactions of the Royal Society of London. Series B, Biological Sciences*, *366*, 1017–1027.

Reep, R. L., Finlay, B. L., & Darlington, R. B. (2007). The limbic system in Mammalian brain evolution. *Brain, Behavior and Evolution*, *70*, 57–70.

Riedman, M. L. (1982). The evolution of alloparental care and adoption in mammals and birds. *The Quarterly Review of Biology*, *57*, 405–435.

Roe, A. W., Garraghty, P. E., Esguerra, M., & Sur, M. (1993). Experimentally induced visual projections to the auditory thalamus in ferrets: Evidence for a W cell pathway. *The Journal of Comparative Neurology*, *334*, 263–280.

Sadato, N., Pascual-Leone, A., Grafman, J., Ibanez, V., Deiber, M.-P., Dold, G., et al. (1996). Activation of the primary visual cortex by Braille reading in blind subjects. *Nature*, *380*, 526–528.

Sathian, K., & Stilla, R. (2010). Cross-modal plasticity of tactile perception in blindness. *Restorative Neurology and Neuroscience*, *28*, 271–281.

Seed, A. M., Clayton, N. S., & Emery, N. J. (2008). Cooperative problem solving in rooks (*Corvus frugilegus*). *Proceedings of the Royal Society B: Biological Sciences.*, *275*, 1421–1429.

Sherwood, C. C., Subiaul, F., & Zawidzki, T. W. (2008). A natural history of the human mind: Tracing evolutionary changes in brain and cognition. *Journal of Anatomy*, *212*, 426–454.

Shultz, S., & Dunbar, R. I. (2007). The evolution of the social brain: Anthropoid primates contrast with other vertebrates. *Proceedings of the Royal Society B: Biological Sciences*, *274*, 2429–2436.

Smaers, J. B., Schleicher, A., Zilles, K., & Vinicius, L. (2010). Frontal white matter volume is associated with brain enlargement and higher structural connectivity in anthropoid primates. *PLoS One*, *5*, e9123.

Smaers, J. B., Steele, J., Case, C. R., Cowper, A., Amunts, K., & Zilles, K. (2011). Cortex evolution: Human brains are the extreme of a lateralized ape trend. *Brain, Behavior and Evolution*, *77*, 67–78.

Smuts, B., Cheney, D., Seyfarth, R., Wrangham, R., & Struhsaker, T. (1986). *Primate societies.* Chicago: University of Chicago Press.

Stephan, H., Frahm, H., & Baron, G. (1981). New and revised data on volumes of brain structures on insectivores and primates. *Folia Primatologica: International Journal of Primatology*, *35*, 1–29.

Sur, M., Garraghty, P. E., & Roe, A. W. (1988). Experimentally induced visual projections into auditory thalamus and cortex. *Science*, *242*, 1437–1441.

Sylvester, J. B., Rich, C. A., Loh, Y. H., van Staaden, M. J., Fraser, G. J., & Streelman, J. T. (2010). Brain diversity evolves via differences in patterning. *Proceedings of the National Academy of Sciences of the United States of America*, *107*, 9718–9723.

Weigl, R. (2005). *Longevity of mammals in captivity; from the living collections of the world.* Stuttgart: Kleine Senckenberg-Reihe.

Wiese, R. J., & Willis, K. (2004). Calculation of longevity and life expectancy in captive elephants. *Zoo Biology*, *23*, 365–373.

Yopak, K. E., Lisney, T. J., Darlington, R. B., Collin, S. P., Montgomery, J. C., & Finlay, B. L. (2010). A conserved pattern of brain scaling from sharks to primates. *Proceedings of the National Academy of Sciences of the United States of America*, *107*, 12946–12951.

Young, L. J., & Wang, Z. (2004). The neurobiology of pair bonding. *Nature Neuroscience*, *7*, 1048–1054.

Zhou, W. X., Sornette, D., Hill, R. A., & Dunbar, R. I. (2005). Discrete hierarchical organization of social group sizes. *Proceedings of the Royal Society B: Biological Sciences*, *272*, 439–444.

SECTION III

Comparative neuroanatomy

M. A. Hofman and D. Falk (Eds.)
Progress in Brain Research, Vol. 195
ISSN: 0079-6123

CHAPTER 5

The evolution of neocortex in primates

Jon H. Kaas*

Department of Psychology, Vanderbilt University, Nashville, TN, USA

Abstract: We can learn about the evolution of neocortex in primates through comparative studies of cortical organization in primates and those mammals that are the closest living relatives of primates, in conjunction with brain features revealed by the skull endocasts of fossil archaic primates. Such studies suggest that early primates had acquired a number of features of neocortex that now distinguish modern primates. Most notably, early primates had an array of new visual areas, and those visual areas widely shared with other mammals had been modified. Posterior parietal cortex was greatly expanded with sensorimotor modules for reaching, grasping, and personal defense. Motor cortex had become more specialized for hand use, and the functions of primary motor cortex were enhanced by the addition and development of premotor and cingulate motor areas. Cortical architecture became more varied, and cortical neuron populations became denser overall than in nonprimate ancestors. Primary visual cortex had the densest population of neurons, and this became more pronounced in the anthropoid radiation. Within the primate clade, considerable variability in cortical size, numbers of areas, and architecture evolved.

Keywords: prosimians; tarsiers; anthropoids; sensory cortex; motor cortex.

Introduction

This review focuses on the areal and structural organization of neocortex in primates. This is an especially challenging topic to review, as primates constitute a highly diversified taxon consisting of 14 families and at least 350 extant species. Primates emerged as a distinct line of evolution around 80 million years ago (Martin, 2004) and diversified within three early branches leading to present-day prosimians, tarsiers, and anthropoids (monkeys, apes, and humans). Primates have adapted to a wide range of environments, allowing them to vary in size 5000-fold from the mouse lemur at 40g to the male gorilla, sometimes over 200kg. Neocortex is a major part of the brain of all primates, especially so in humans where cerebral cortex occupies 80% of the brain mass and contains 16 billion neurons (Avzevedo et al., 2009). Despite a huge variability in absolute brain

*Corresponding author.
Tel.: +1-615-3223029; Fax: +1-615-3438449
E-mail: jon.h.kaas@vanderbilt.edu

DOI: 10.1016/B978-0-444-53860-4.00005-2

size, and the size of neocortex, a characteristic pattern of areal organization has been found in all studied primates. Numbers of cortical areas, connection patterns, and structural and functional organization of areas all vary. Yet, a set of cortical areas exists in all primates, and some of these areas appear to be unique with primates. Here, we first consider the organization of neocortex in a prosimian primate that has been extensively studied, the African galago, compare the organization of neocortex in primates with that of the close relatives of primates to suggest how the primate pattern might have evolved, and briefly considering some of the major variations in cortical organization and structure that we see across the primate radiation.

Cortical organization in prosimian galagos: Comparisons with other primates

Prosimian primates have been of special value to those interested in the evolution of primates because they resemble the early ancestors of primates more closely than members of the tarsier and anthropoid radiations. However, early primates were small, had smaller brains relative to body size, and depended less on vision and more on olfaction than modern prosimians (Martin, 2004; Ni et al., 2003). While the prosimian radiation of galagos, lorises, and lemurs includes a range of adaptations, it is presently unclear how much prosimians vary in cortical organization, as only galagos have been well studied.

In galagos, neocortex is subdivided into a number of areas that are also recognized in other primates. These areas are shown on a flattened surface view of the neocortex so that the relationship of proposed cortical areas can be seen, including areas not apparent on a dorsolateral view of the brain as they are hidden on the ventral and medial surfaces of the brain, and in the few fissures found in galago cortex (Fig. 1). Here, we describe some of the major functional divisions of neocortex in galagos and discuss those divisions in relation to what is known in other primates and nonprimate relatives of primates.

Visual cortex

The full extent of visual cortex and the total number of visual areas is not known in galagos, as it can be difficult to define cortical areas, and even in the more extensively studied macaques, there are uncertainties. However, a collection of visual areas has been defined (Fig. 1), and these areas also exist in other primates. The primary and secondary visual areas, V1 and V2, exist in nearly all mammals, but these areas have specialized features in primates. Thus, galagos (Rosa et al., 1997) and other primates have greatly expanded representations of central vision in V1 and other visual areas. In addition, V1 has sublayers of layer 4 that are different from those in other taxa. The inner half of layer 4 receives inputs from the parvocellular layers of the lateral geniculate nucleus and these inputs are especially important in detailed object vision, while the outer half of layer 4 receives inputs from the magnocellular layers, and these inputs are important in detecting motion and change (Casagrande and Kaas, 1994). The superficial layers of cortex express a dot-like distribution of functional modules, the so-called blobs, which may be important in color vision (Casagrande and Kaas, 1994; Preuss and Kaas, 1996). Because these features are not found in the nonprimate members of the Euarchontoglire clade (rabbits, rodents, flying lemurs, tree shrews, and primates), they must have evolved in the line leading to early primates (Kaas, 2005). However, both primates and tree shrews have orderly arrangements of orientation-selective neurons in V1, while these neurons are randomly distributed in rodents (Van Hooser et al., 2006). Thus, columns of orientation-selective neurons in V1 likely evolved in the common ancestors of the tree shrews and primates.

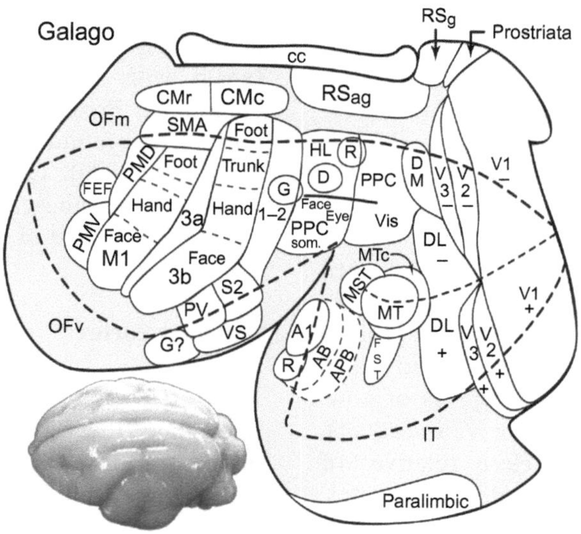

Fig. 1. Areal subdivisions of neocortex in a prosimian primate, galago (*Otolemur garnetti*). For orientation, a dorsolateral view of the brain is on the lower left. The larger figure is of the neocortex after it has been removed from the rest of the brain, fissures opened, and flattened so that all of the cortical surface can be seen, and cortical areas can be depicted relative to each other. Visual areas include primate visual cortex (V1), the second visual area (V2), the third visual area (V3), the dorsomedial visual area (DM), the dorsolateral visual area (DL, also known as V4), the middle temporal visual area (MT), the MT crescent (MTc), the middle superior temporal area (MST), and the fundal area of the superior temporal sulcus (FST) which has dorsal and ventral subdivisions. Inferior temporal cortex (IT) contains several visual areas, but they have not been well defined. Auditory cortex includes a core of two primary areas, primary auditory cortex (A1) and the rostral auditory area (R), as well as a surrounding belt of as many as eight secondary areas (AB) and an adjoining auditory parabelt (APB) with two major divisions. Somatosensory cortex includes a primary area (3b or S1), a proprioceptive area (3a), a secondary area caudal to S1 (areas 1–2), a second area (S2), a parietal ventral area (PV), a ventral somatosensory area (VS), possibly a gustatory (taste) area (G?), and other less well-defined areas in insular cortex. Motor cortex includes a primary area (M1), a ventral premotor area (PMV), a dorsal premotor area (PMD), a supplementary motor area (SMA), a frontal eye field (FEF), and rostral (CMr) and caudal (CMc) cingulate motor areas. Posterior parietal cortex (PPC) has a large caudal division with visual inputs (Vis), and a large rostral division with somatosensory inputs (Som.) and a somatotopic organization from hindlimb (HL) to face and eye. Territories within PPC are indicated where reaching (R) or defensive (D) movements can be evoked with electrical stimulation. A territory for grasping (G) movements is marked in areas 1–2. Foot (F) and hand (H) regions in M1 are marked and representations of upper (+) and lower (−) visual field representation are indicated for some visual areas. Retrosplenial agranular (RSag) and retrosplenial granular (RSg) areas are marked. The thicker dotted line outlines the extent of neocortex visible on a lateral view of the intact brain, the dotted line through visual areas indicates the location of the representation of the zero horizontal meridian, and dashed lines in M1 and 3b delimit the representation segments for major body parts. Medial (OFm) and ventral (OFv) orbital–frontal regions are marked. Corpus callosum (CC) is marked.

In almost all studied mammals, V1 is bordered along most of its perimeter by the second visual area, V2 (Rosa and Krubitzer, 1999). In anthropoid primates, V2 is uniquely subdivided into repeating sets of three types of band-like modules that cross the width of the V2 belt. These bands can be revealed by stains for cytochrome oxidase (CO) or myelin and have been characterized as CO-dense thick or thin bands, or CO-light (pale) bands. Each type of band has different inputs from modules and layers in V1 and outputs to other cortical areas, and neurons of different functional properties (Casagrande and Kaas, 1994). Studies of connection patterns in prosimian primates suggest that V2 is subdivided into the same three types of modules in these primates as well, but the V2 modules are only weakly apparent in CO or myelin preparations (Collins et al., 2001; Preuss et al., 1993). Rodents and tree shrews do not have this type of modular organization in V2. Thus, the V2 bands evolved with the first primates and became histologically more distinct in anthropoid primates.

V3 is another visual area common to all primates. This conclusion was in question until recently when connection patterns with V1 were used to clearly identify V3 in prosimian galagos (Lyon and Kaas, 2002a), and new and old world monkeys (Lyon and Kaas, 2002b,c). As of now, there is no compelling evidence for V3 in rodents, rabbits, or tree shrews, the close relatives of primates, so V3 may have emerged with primates. If so, the V3 described in cats and other carnivores evolved independently.

In all primates studied, V1 projects to a densely myelinated area in the middle of the upper temporal lobe, the middle temporal (MT) visual area, where neurons are sensitive to stimulus orientation and direction of motion (Kaskan et al., 2010). Because of its histological distinctiveness (Allman and Kaas, 1971), MT has been histologically identified in humans (Tootell and Taylor, 1995) well before it was possible by imaging (fMRI), and in tarsiers (Wong et al., 2010), which are unavailable for experimental study. An area highly similar to MT has not been identified in tree shrews, rodents, or rabbits, raising the likelihood that MT is new with primates, or that MT is an area that has been so modified in primates that homologues of MT are not recognizable as MT in other members of the Euarchontoglire clade (Kaas and Preuss, 1993).

Finally, all primates appear to have visual areas, termed here the dorsolateral visual area, DL (Allman and Kaas, 1974), and the dorsomedial visual area, DM (Allman and Kaas, 1975). These areas have been less well defined than V1, V2, and MT, and their boundaries have been adjusted by various investigators. DL is also termed V4 and DM also termed V3a. Together with MST, FST, and MTc, areas associated with MT (Kaas and Morel, 1993), DL, and DM are likely components of visual cortex in all primates (again with no certain homologues in other mammals). Prostriata (Fig. 1) is a limbic visual area that is found in most mammals (Rosa and Krubitzer, 1999). Other visual areas have been proposed, but these areas have not been well defined in a range of primate taxa.

Auditory cortex

Auditory cortex has been less extensively studied in primates than visual cortex, and little is known about auditory cortex organization in galagos. The standard model for the early stages of processing in auditory cortex of primates now includes a core of three primary or primary-like areas, A1 (the primary area), R (the rostral area), and RT (the rostrotemporal area), surrounded by a belt of eight secondary areas, bordered laterally by a parabelt, a third level of cortical processing of two divisions (Kaas and Hackett, 2000). This model appears to apply to old world macaque monkeys and new world monkeys (Kaas, 2011), and recent imaging and histological evidence from humans is consistent with the model (Sweet et al., 2005; Woods et al., 2010). In a review, Brugge (1982) provided evidence for two core auditory areas in galagos, R and A1, and one of the most distinctive of the belt areas (CM), so it is reasonable to conclude that all primates share at least two core areas and some belt areas of auditory cortex. Areas A1, R, some or all of the belt, and perhaps the parabelt are likely common to all primates (Fig. 1). Since all or nearly all mammals have a primary auditory area or areas, as well as secondary fields (Kaas, 2011), some of the areas proposed for primates probably were retained from early nonprimate ancestors. However, the common existence of more than one primary area in mammals makes the identification of homologous primary areas difficult.

In macaque monkeys, it is clear that larger regions of cortex are involved in processing auditory signals (Poremba and Mishkin, 2007), and this is the case for humans as well where specializations for language occur (Scott and Johnsrude, 2003). Thus, it seems likely that major differences exist in the number and organization of higher auditory areas across primate taxa.

Somatosensory cortex

The somatosensory cortex of galagos includes a primary area, 3b, as in other primates, a narrow area 3a with proprioceptive inputs along the rostral border of area 3b, and a band of cortex along the caudal border of area 3b with inputs from area 3b that resembles area 1 or area 1 plus area 2 of anthropoid primates (Fig. 1). Area 3b is homologous to S1 as defined in most nonprimate mammals (Kaas, 1983), and area 3a, with inputs from a proprioceptive nucleus in the thalamus, the ventroposterior superior nucleus in primates, is homologous to a dysgranular representation of muscle and joint receptors along the rostral border of S1 that has been described in rats, cats, raccoons, and other mammals (Kaas, 2007). The band of somatosensory cortex just caudal to area 3b of galagos does not respond well to somatosensory stimuli in anesthetized galagos but is in the position of area 1, or area 1 plus area 2, of anthropoid primates. In all primates, area 3b contains a systematic representation of cutaneous receptors of the contralateral half of the body, in addition to representations of the ipsilateral teeth and tongue (Kaas et al., 2006). Thus, both the contralateral and the ipsilateral tongue and teeth are represented in area 3b of each hemisphere. Area 3b consistently represents the contralateral body surface from foot to face in a mediolateral sequence across cortex (Fig. 1). In primates, a large portion of the representation is devoted to the glabrous hand, and a larger portion is devoted to the face, teeth, and tongue. In galagos and most monkeys, little cortex is used to represent the tail, but the new world monkeys that have evolved a prehensile tail with a ventral pad of glabrous sensory skin have also evolved a large representation of the tail in medial area 3b (Felleman et al., 1983). In anthropoid primates, area 1, the strip of cortex immediately caudal to area 3b, does respond well to tactile stimulation, and a systematic representation of the contralateral body surface in area 1 forms a mirror image of the representation in area 3b. In addition, a third representation of at least much of the contralateral body surface, including face, hand, and forearm, exists in area 2 in anthropoid primates (Pons et al., 1985). The areas 1–2 region in galagos are similar in location, architecture, and connections to a band of cortex along the caudal border of S1 in tree shrews, rats, and other studied mammals and thus, as for areas 3a and 3b (S1), appear to be an early feature of cortical organization in the evolution of mammals. We have tentatively termed this strip of cortex in galagos area 1–2 (Fig. 1), as this area has features of both areas 1 and 2 of anthropoid primates. However, there is no evidence yet that area 1–2 of galagos includes two parallel strips that are homologous to areas 1 and 2 in anthropoid primates. Our current hypothesis, based on present evidence, is that a single area in most mammals, and in prosimian primates, subdivided to become two distinct areas with the evolution of anthropoid primates.

Galagos have additional somatosensory areas in the cortex of the upper bank of the lateral sulcus and the insula in the depths of the sulcus (Wu and Kaas, 2003). Two of these areas, the second somatosensory area, S2, and the parietal ventral area, PV, have been described in a number of nonprimate mammals, as well as in several species of monkeys (Coq et al., 2004; Disbrow et al., 2003; Krubitzer and Kaas, 1990) and humans (Eickhoff et al., 2007). Both S2 and PV represent the contralateral body surface from the head along the 3b border to foot deep in the sulcus. The two representations mirror each other and depend on inputs from area 3b for activation. As S2 and PV have been identified in a number of mammals, including opossums, these two areas have been retained in primates from early mammals (Kaas, 2007). Galagos have caudal and rostral divisions of the ventral somatosensory area (VS in Fig. 1) that are also found in monkeys and humans but have not been reported for non-primate mammals, raising the possibility that VSc and VSr are areas that evolved in early primates. Other areas in the cortex of the lateral sulcus have been defined in galagos by patterns of connections with S2 and PV, and these include a parietal rostral area and a retroinsular area (Wu and Kaas, 2003). The parietal rostral area is in the expected location of gustatory cortex (G? in the unfolded lateral sulcus in Fig. 1), a region that responds to taste stimuli, but has not been well defined in primates (Kaas et al., 2006). Parts of the insula are involved in nociception, and in humans, empathy for the pain of others (Keysers et al., 2010). The insula is greatly expanded in humans and is involved in mediating empathy, pleasant touch, and other socially relevant functions. Likely, the insula is one of the most changed regions of the human brain.

Motor cortex

Motor cortex in galagos (Fig. 1) includes a number of areas that are also found in other primates (Wu et al., 2000). A primary motor area, M1, just rostral to areas 3a and 3b of somatosensory cortex, has been identified in all studied placental mammals, and thus has an ancient origin. In primates, M1 has a large region devoted to hand movements, and in some primates, M1 has distinct rostral and caudal divisions (Preuss et al., 1997). M1 is more specialized for movements of individual digits in macaque monkeys than it is in galagos, as M1 organization reflects types of use of the hand. Galagos and other primates also have dorsal and ventral premotor areas, PMD and PMV, a supplementary motor area (SMA), and caudal and rostral cingulate motor areas (CMr and CMc). A dorsal premotor area has been found in rodents and tree shrews (Remple et al., 2007) that may correspond to either SMA or PMD of primates. PMV may have emerged with the first primates. Both PMD and PMV have functionally distinct subdivisions in macaque monkeys (Fujii et al., 2000; Geyer et al., 2000; Luppino et al., 1999), suggesting further evolution of premotor cortex. An anterior part of ventral premotor cortex that is involved with orofacial movements has been suggested to be a homologue of the much more extensive Broca's area of the left cerebral hemisphere in humans (Petrides et al., 2005). Finally, the frontal eye field (FEF) where electrical stimulation produces eye movements is an area that has been identified in galagos, new and old world monkeys, and humans. The eye movements are evoked via connections with the superior colliculus and brainstem motor centers, but galagos appear to differ from monkeys in having very few projections from the FEF to the superior colliculus. As cortex rostral to M1 has few if any projections to the superior colliculus in most non-primate mammals, the sparseness or lack of such a projection in galagos may reflect the primitive condition.

Posterior parietal sensorimotor cortex

The organization of posterior parietal cortex is variable across primate taxa. Compared to rodents and tree shrews with very little cortex that can be considered posterior parietal cortex, all primates have a large posterior parietal region that is especially enlarged in humans. In galagos, PPC can be divided into two large regions, a posterior division with inputs from a collection of higher visual areas, and a rostral division with inputs from higher order somatosensory areas (Fig. 1). The caudal division of PPC gets inputs mainly from visual areas MT, MST, MTc, and DM, areas considered to be components of the dorsal stream of visual processing that mediate visuomotor guidance via connections with PPC (Goodale and Milner, 1992; Ungerleider and

Mishkin, 1982). Visual inputs to rostral PPC largely depend on projections from caudal PPC to rostral PPC. Electrical stimulation of rostral, but not caudal, PPC evokes complex movements that depend on connections with motor and premotor cortex (Kaas et al., 2011). The types of movements evoked by electrical stimulation of rostral PPC depend on the location of the stimulating electrode so that hindlimb movements are evoked from the most medial sites, forelimb movements from more lateral sites, and face and eye movements from the most lateral sites. Even from anesthetized galagos, functionally meaningful complex behaviors can be evoked when electrical stimulation continues for the duration of the behavior (about 0.5s). Thus, reaching, defensive movements of the arm to protect the face and grasping movements can be evoked from a caudorostral sequence of locations in or near PPC (locations R, D, and G in Fig. 1). Other locations produce face, eye, and other movements. Note especially that grasping movements were evoked from a location that was largely in areas 1–2, providing evidence that part of areas 1–2 is functionally related to PPC. A similar arrangement of reach, defense, and grasp zones exist in new world monkeys and in macaque monkeys (Gharbawie et al., 2011; Kaas et al., 2011), with the difference that these zones have a more rostrocaudal arrangement, especially in macaques, and these zones have more direct visual inputs from dorsal stream visual areas in macaques. Overall, the differences suggest that most of PPC is occupied by an expanded rostral division of PPC in anthropoid primates and that visual inputs to this expanded rostral division have become more direct and more important.

As the grasp zone in macaque monkeys is largely in area 2 and the grasp zone in new world monkeys is in cortex immediately caudal to area 1, it appears that new world monkeys have an area 2 that is separated from area 1, something that has been questioned. However, the location of the grasp zone in areas 1–2 in galagos suggests that galagos do not have separate areas 1 and 2.

PPC in humans is a greatly expanded part of neocortex (Hill et al., 2010) and it contains subdivisions that may be homologous to those in monkeys, as well as those that may have been elaborated or developed in the ancestors of modern humans, allowing new abilities such as the extensive use of tools and the use of gestures for communication (Frey, 2007). A number of human abilities likely depend on PPC.

Prefrontal cortex

Prefrontal cortex of primates is a large division of neocortex that is thought to be especially important in mediating cognitive and social aspects of human behavior. Most mammals have an orbital region with lateral and medial components, as indicated for galagos in Fig. 1. There is also a granular frontal region rostral to the FEF that is less expansive in galagos than in anthropoid primates (Preuss and Goldman-Rakic, 1991), and possibly absent in mammals other than primates (Preuss, 1995). Many of the higher cognitive and social abilities of humans are attributed to frontal cortex, especially granular frontal cortex, and frontal cortex is larger in humans than in other primates. However, frontal cortex as a whole is not proportionately larger in humans than expected for a primate brain (Semendeferi et al., 2002). The functional organization of prefrontal cortex likely differs considerably across primate taxa, with evolved elaborations and multiple subdivisions in the human brain.

The evolution of structural and cellular differences in cortical areas in primates

Primate brains of all sizes differ from rodent brains, and likely all other mammals, in having more densely packed and overall smaller neurons, and this largely reflects the smaller neurons of neocortex in primates (Herculano-Houzel et al., 2007). In addition, neuron sizes and densities vary across cortical areas within primate species, and across primate taxa for homologous areas (Collins et al., 2010). When the densities of neurons were compared across the cortical sheet in galagos, new world monkeys, and old world macaques and baboons, neuron densities across cortical areas varied the least in galagos and the most in the larger brains of macaques and baboons. Even greater variability might be expected in human brains. Overall, the highest densities were observed in primary visual cortex across these primates, while secondary visual cortex and primary somatosensory cortex also had higher density values in the old world anthropoids. Lower cell densities were found in association areas with larger neurons. High densities of small neurons are useful in areas that need to segregate many inputs during processing for an analysis of details, while areas with large neurons of low densities are more useful for summing inputs for a more global analysis.

As galagos appear to have the fewest cortical areas, with macaques having more, and humans likely having many more, primates with large brains and many cortical areas have the advantage of being able to specialize some areas for analyzing local detail and others for integrative processing. Primates with fewer cortical areas need to preserve the more general functions of areas, and thus areas are less specialized and have less variability in neuron densities across areas. In large-brained primates with more cortical areas, we can expect some of these areas to become highly specialized, because other areas can be retained for a broad range of functions. But, such specializations come at a greater cost in small-brained mammals with few cortical areas, since this limits options.

As a clear exception, present-day tarsiers are such highly specialized visual predators that they eat only small invertebrates and vertebrates, and no plants. The ability to be an effective, nocturnal visual predator depends on a highly specialized visual system (Collins et al., 2005; Wong and Kaas, 2010). In tarsiers, a single visual area, V1, occupies over 20% of neocortex, more than in any other primate, and V1 is more distinctly divided into layers and sublayers than in any other primate (Fig. 2). The large V1 with densely packed small neurons preserves the detail of

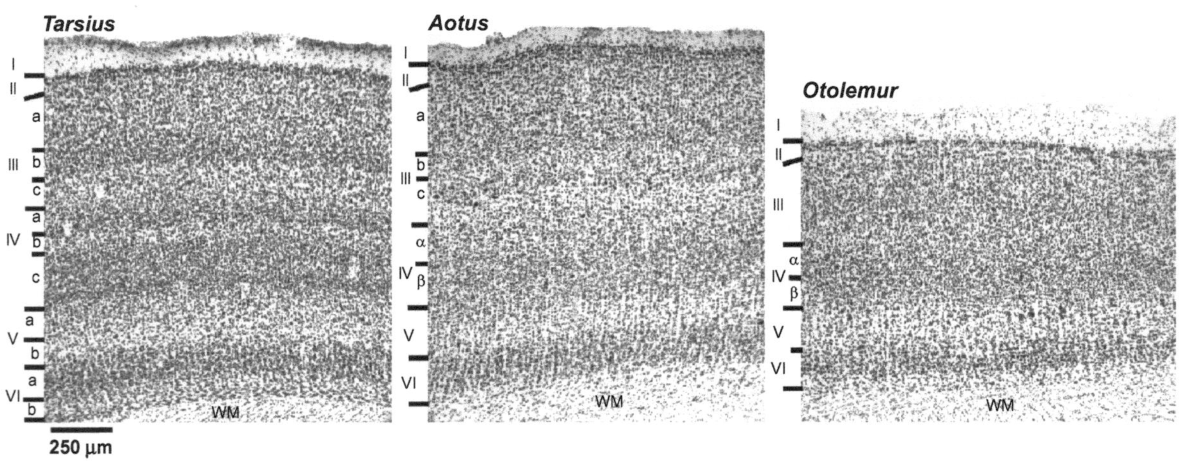

Fig. 2. Nissl-stained sections from primary visual cortex (V1) of *Tarsius* and *Otolemur* (galago) and *Aotus* (a nocturnal new world monkey) for comparison. Note the more obvious layering and sublayering of V1 in *Tarsius*, less in *Aotus*, and the least in *Otolemur*. WM, white matter. Roman numbers mark the six cortical layers, while letters are used for sublayers. See Collins et al. (2005) for a full description of the tarsier visual system.

central vision, so that tarsiers can detect small cryptic prey. The distinctive lamination of V1, reminiscent of the laminated optic tectum of birds of prey, reflects the specializations of cellular morphologies of different input and output layers and sublayers for different functional roles in tarsiers.

Epilogue

Given the limitations of space and present understandings, this review only touched on the huge topic of the evolution of neocortex in primates. Here, we tried to reconstruct from comparative and fossil evidence the organization of the neocortex of the common primate ancestor of all extant primates. We know that primates emerged over 80 million years ago as a branch of the Euarchontoglire superclade (Murphy et al., 2004). This branch included several lines of archaic primates that became extinct, and the stem euprimates that led to the present-day galagos, lorises, tarsiers, and the greatly varied anthropoid monkeys, apes, and hominids (humans and extinct species more closely related to us than chimpanzees). The closest living relatives of primates are the Scandentia (tree shrews) and Dermoptera (flying lemurs) of the Archontan branch of Euarchontoglires. The more distant Glires branch includes rodents and lagomorphs. The extinct archaic primates had smaller brains with less neocortex than modern primates, and it is difficult to deduce how their cortex was organized. Perhaps the most we can do is compare the cortical organization of extant primates with the most primitive cortical features to the cortical features of tree shrews and rodents, as flying lemurs are not available for experimental study. When we do that, we can surmise that the proportionately smaller neocortex of archaic primates had less posterior parietal, frontal, and temporal cortex, and that the ventral visual stream for object vision via temporal cortex and the dorsal visual stream for visually guided motor behavior were less developed than in any extant primates. Additionally, the premotor areas of frontal cortex were likely less developed. Thus, motor abilities and motor flexibility would have been less pronounced in archaic primates. Finally, a less expansive prefrontal cortex would suggest that archaic primate behavior was more dependent on ongoing sensory events, and less on social experience and previous environmental events.

In contrast to these uncertain possibilities, comparative studies of cortical organization in members of extant primate taxa are now extensive enough to extract many shared features of cortical organization that likely reflect those retained from a common stem euprimate ancestor. It is likely that all of the cortical areas illustrated for extant galagos (Fig. 1) were present in that common ancestor, as these areas and subdivisions of areas can be identified in other primates. However, this proposal is incomplete, perhaps in need of correction, and it should be evaluated further in ongoing studies of cortical organization in prosimians and other branches of the primate radiation. Cortical organization in galagos needs further study, especially in temporal, posterior parietal, prefrontal, and medial wall regions. Nevertheless, the evidence to date suggests that prosimian galagos present a very good model of what the neocortex of early euprimates was like, not only in terms of areal organization but also in terms of structural and connectional organization.

What is more difficult to deduce and fully describe is how neocortex became modified in the many branches of the primate radiation. Here, we barely touched on this important topic. The challenge is great given the many primate species, the difficulty or impossibility of conducting experimental studies on many of these species, and the major gaps in the radiation produced by extinctions. While modern humans and chimpanzees are separated from a common ancestor by only a few million years, our brains are three times larger, with most of this increase over the past 2 million years of hominin evolution. Only relatively recently, within thousands of years, we have become the only surviving species within the formerly varied hominin branch. This loss, plus the limited ways in which the brains of apes and humans can be studied, makes it very difficult to reconstruct the evolution of the human brain, although information on how neocortex is organized in humans is being rapidly acquired. As fMRI, optical imaging and other recent technical advances have greatly magnified what can be learned about brain organization, a much better understanding of the evolution of cortical organization in primates can be expected in the near future.

References

Allman, J. M., & Kaas, J. H. (1971). A representation of a visual field in the caudal third of the middle temporal gyrus of the owl monkey (*Aotus trivirgatus*). *Brain Research*, *31*, 85–105.

Allman, J. M., & Kaas, J. H. (1974). A crescent-shaped cortical visual area surrounding the middle temporal area (MT) in the owl monkey (*Aotus trivirgatus*). *Brain Research*, *81*, 199–213.

Allman, J. M., & Kaas, J. H. (1975). The dorsomedial cortical visual area: A third tier area in the occipital lobe of the owl monkey (Aotus trivirgatus). *Brain Research*, *100*, 473–487.

Avzevedo, F. A. C., Carvalho, L. R. B., Grinberg, L. T., Farfel, J. M., Ferretti, R. E. J., Leite, R. E. P., et al. (2009). Equal numbers of neuronal and nonneuronal cells make the human brain an isometrically scaled-up primate brain. *The Journal of Comparative Neurology*, *513*, 532–541.

Brugge, J. F. (1982). Auditory areas in primates. In C. N. Woolsey (Ed.), *Cortical sensory organization* (pp. 59–70). Clifton, NJ: Humana Press.

Casagrande, V. A., & Kaas, J. H. (1994). The afferent, intrinsic, and efferent connections of primary visual cortex in primates. In A. Peters & K. Rockland (Eds.), *Cerebral cortex*, (Vol. 10), (pp. 201–259). New York: Plenum Press.

Collins, C. E., Airey, D. C., Young, N. A., Leitch, D. B., & Kaas, J. H. (2010). Neuron densities vary across and within cortical areas in primates. *Proceedings of the National Academy of Sciences USA*, *107*(36), 15927–15932.

Collins, C. E., Hendrickson, A., & Kaas, J. H. (2005). Overview of the visual system of Tarsius. *The Anatomical Record Part A*, *287*(1), 1013–1025.

Collins, C. E., Stepniewska, I., & Kaas, J. H. (2001). Topographic patterns of V2 cortical connections in a prosimian primate (*Galago garnetti*). *The Journal of Comparative Neurology*, *431*, 155–167.

Coq, J. O., Qi, H.-X., Collins, C. E., & Kaas, J. H. (2004). Anatomical and functional organization of somatosensory areas of the lateral fissure of the New World titi monkey (*Callicebus moloch*). *The Journal of Comparative Neurology*, *476*(4), 363–387.

Disbrow, E., Litinas, E., Recanzone, G. H., Padberg, J., & Krubitzer, L. (2003). Cortical connections of the second somatosensory area and the parietal ventral area in macaque monkeys. *The Journal of Comparative Neurology*, *462*, 382–399.

Eickhoff, S. B., Grefkes, C., Zilles, K., & Fink, G. R. (2007). The somatotopic organization of cytoarchitectonic areas on the human parietal operculum. *Cerebral Cortex*, *17*, 1800–1811.

Felleman, D. J., Nelson, R. J., Sur, M., & Kaas, J. H. (1983). Representations of the body surface in areas 3b and 1 of postcentral parietal cortex of Cebus monkeys. *Brain Research*, *268*(1), 15–26.

Frey, S. H. (2007). Neurological specializations for processing faces and objects. In J. H. Kaas & T. M. Preuss (Eds.), *Evolution of nervous systems*, (Vol. 4), (pp. 395–406). Oxford: Academic Press.

Fujii, N., Mushiake, H., & Tanji, J. (2000). Rostrocaudal distinction of the dorsal premotor area based on oculomotor involvement. *Journal of Neurophysiology*, *83*, 1764–1769, Rapid Communication.

Geyer, S., Zilles, K., Luppino, G., & Matelli, M. (2000). Neurofilament protein distribution in the macaque monkey dorsolateral premotor cortex. *European Journal of Neuroscience*, *12*, 1554–1566.

Gharbawie, O. A., Stepniewska, I., Qi, H. X., & Kaas, J. H. (2011). Multiple parietal-frontal networks mediate grasping in macaque monkeys. *The Journal of Neuroscience*, *31*, 11660–11677.

Goodale, M. A., & Milner, A. D. (1992). Separate visual pathways for perception and action. *Trends in Neuroscience*, *15*, 20–25.

Herculano-Houzel, S., Collins, C. E., Wong, P., & Kaas, J. H. (2007). Cellular scaling rules for primate brains. *Proceedings of the National Academy of Sciences of the United States of America*, *104*, 3562–3567.

Hill, J., Inder, T., Neil, J., Dierker, D. L., Harwell, J., & Van Essen, D. (2010). Similar patterns of cortical expansion during human development and evolution. *Cerebral Cortex*, *8* (3), 13135–13140.

Kaas, J. H. (1983). What, if anything, is SI? The organization of "first somatosensory area" of cortex. *Physiological Review*, *63*, 206–231.

Kaas, J. H. (2005). The evolution of visual cortex in primates. In J. Kremers (Ed.), *The structure, function and evolution of the primate visual system* (pp. 267–283). Chichester, UK: John Wiley and Sons.

Kaas, J. H. (2007). The evolution of sensory and motor systems in primates. In J. H. Kaas & T. M. Preuss (Eds.), *Evolution of nervous systems*, (Vol. 4), (pp. 35–57). London: Academic Press.

Kaas, J. H. (2011). The evolution of auditory cortex: The core areas. In J. A. Winer & C. E. Schreiner (Eds.), *The auditory cortex* (pp. 407–427). New York: Springer-Verlag.

Kaas, J. H., Gharbawie, O. A., & Stepniewska, I. (2011). The organization and evolution of dorsal stream multisensory motor pathways in primates. *Frontiers in Neuroanatomy, 5* (34), 1–7.

Kaas, J. H., & Hackett, T. A. (2000). Subdivisions of auditory cortex and processing streams in primates. *Proceedings of the National Academy of Sciences of the United States of America, 97*, 11793–11799.

Kaas, J. H., & Morel, A. (1993). Connections of visual areas of the upper temporal lobe of owl monkeys: The MT crescent and dorsal and ventral subdivisions of FST. *The Journal of Neuroscience, 3*, 534–546.

Kaas, J. H., & Preuss, T. M. (1993). Archontan affinities as reflected in the visual system. In F. Szalay, M. Novacek & M. McKenna (Eds.), *Mammalian phylogeny* (pp. 115–128). New York: Springer-Verlag.

Kaas, J. H., Qi, H.-X., & Iyengar, S. (2006). Cortical network for representing the teeth and tongue in primates. *The Anatomical Record, 288A*, 182–190.

Kaskan, P. M., Dillenburger, B. C., Lu, H. D., Roe, A. W., & Kaas, J. H. (2010). Orientation and direction-of-motion response in the middle temporal visual area (MT) of New World Owl Monkeys as revealed by intrinsic-signal optical imaging. *Frontiers in Neuroanatomy, 4*, 23.

Keysers, C., Kaas, J. H., & Gazzola, V. (2010). Somatosensation in social perception. *Nature Reviews. Neuroscience, 11*(6), 417–428.

Krubitzer, L. A., & Kaas, J. H. (1990). The organization and connections of somatosensory cortex in marmosets. *The Journal of Neuroscience, 10*(3), 952–974.

Luppino, G., Muvata, A., Govoni, P., & Matelli, M. (1999). Largely segregated parietofrontal connections linking rostral intraparietal cortex (areas AIP and VIP) and the ventral premotor cortex (areas F5 and F4). *Experimental Brain Research, 128*, 181–187.

Lyon, D. C., & Kaas, J. H. (2002a). Connectional evidence for dorsal and ventral V3, and other extrastriate areas in the prosimian primate, *Galago garnetti. Brain, Behavior and Evolution, 59*, 114–129.

Lyon, D. C., & Kaas, J. H. (2002b). Evidence for a modified V3 with dorsal and ventral halves in macaque monkeys. *Neuron, 33*, 453–461.

Lyon, D. C., & Kaas, J. H. (2002c). Evidence from V1 connections for both dorsal and ventral subdivisions of V3 in three species of New World monkeys. *The Journal of Comparative Neurology, 449*, 281–297.

Martin, R. D. (2004). Palaeontology: Chinese lantern for early primates. *Nature, 427*, 22–23.

Murphy, W. J., Pevzner, P. A., & O'Brien, J. O. (2004). Mammalian phylogenomics comes of age. *Trends in Genetics, 20* (12), 631–639.

Ni, X., Wang, Y., Hu, Y., & Li, C. (2003). A euprimate skull from the early Eocene of China. *Nature, 427*, 65–68.

Petrides, M., Cadoret, G., & Mackey, S. (2005). Orofacial somatomotor responses in the macaque monkey homologue of Broca's area. *Nature, 435*, 1235–1238.

Pons, T. P., Garraghty, P. E., Cusick, C. G., & Kaas, J. H. (1985). The somatotopic organization of area 2 in macaque monkeys. *The Journal of Comparative Neurology, 241*, 445–466.

Poremba, A., & Mishkin, M. (2007). Exploring the extent and function of higher-order auditory cortex in rhesus monkeys. *Hearing Research, 229*, 14–23.

Preuss, T. M. (1995). Do rats have prefrontal cortex? The Rose-Woolsey-Akert program reconsidered. *Journal of Cognitive Neuroscience, 7*, 1–24.

Preuss, T. M., Beck, P. D., & Kaas, J. H. (1993). Areal, modular, and connectional organization of visual cortex in a prosimian primate, the slow loris (*Nycticebus coucang*). *Brain, Behavior and Evolution, 42*(6), 321–335.

Preuss, T. M., & Goldman-Rakic, P. S. (1991). Architectonics of the parietal and temporal association cortex in the strepsirhine primate *Galago* compared to the anthropoid primate *Macaca. The Journal of Comparative Neurology, 310*, 475–506.

Preuss, T. M., & Kaas, J. H. (1996). Cytochrome oxidase 'blobs' and other characteristics of primary visual cortex in a lemuroid primate, *Cheirogaleus medius. Brain, Behavior and Evolution, 47*(2), 103–112.

Preuss, T. M., Stepniewska, I., Jain, N., & Kaas, J. H. (1997). Multiple divisions of macaque precentral motor cortex identified with neurofilament antibody SMI-32. *Brain Research, 767*, 148–153.

Remple, M. S., Reed, J. L., Stepniewska, I., Lyon, D. C., & Kaas, J. H. (2007). The organization of frontoparietal cortex in the tree shrew (*Tupaia belangeri*): II Connectional evidence for a frontal-posterior parietal network. *The Journal of Comparative Neurology, 501*, 121–149.

Rosa, M. G. P., Casagrande, V. A., Preuss, T. M., & Kaas, J. H. (1997). Visual field representation in striate and prostriate cortices of a prosimian primate (*Galago garnetti*). *Journal of Neurophysiology, 77*, 3193–3217.

Rosa, M. G. P., & Krubitzer, L. A. (1999). The evolution of visual cortex: Where is V2? *Trends in Neuroscience, 22*, 242–248.

Scott, S. K., & Johnsrude, I. S. (2003). The neuroanatomical and functional organization of speech perception. *Trends in Neuroscience, 26*, 100–107.

Semendeferi, K., Lu, A., Schenker, N., & Damasio, H. (2002). Humans and great apes share a large frontal cortex. *Nature, 5*, 272–276. On-line (February, 2002).

Sweet, R. A., Dorph-Petersen, K. A., & Lewis, D. A. (2005). Mapping auditory core, lateral belt, and parabelt cortices in the human superior temporal gyrus. *The Journal of Comparative Neurology*, *491*, 270–289.

Tootell, R. B., & Taylor, J. B. (1995). Anatomical evidence for MT and additional cortical visual areas in humans. *Cerebral Cortex*, *5*(1), 39–55.

Ungerleider, L. G., & Mishkin, M. (1982). Two cortical visual systems. In D. G. Ingle, M. A. Goodale & R. J. Q. Mansfield (Eds.), *Analysis of visual behavior* (pp. 549–586). Cambridge: MIT Press.

Van Hooser, S. D., Heimel, J. A., Chung, S., & Nelson, S. B. (2006). Lack of patchy horizontal connectivity in primary visual cortex of a mammal without orientation maps. *Journal of Neuroscience*, *26*, 7680–7692.

Wong, P., Collins, C. E., & Kaas, J. H. (2010). Overview of sensory systems of Tarsius. *International Journal of Primatology*, *31*, 1002–1031.

Wong, P., & Kaas, J. H. (2010). Architectonic subdivisions of neocortex in the Galago (*Otolemur garnetti*). *Anatomical Record*, *293*(6), 1033–1069.

Woods, D. L., Herron, T. J., Cate, A. D., Yund, E. W., Stecker, G. C., Rinne, T., et al. (2010). Functional properties of human auditory cortical fields. *Frontiers in Systems Neuroscience*, *4*, 1–13.

Wu, C. W.-H., Bichot, N. P., & Kaas, J. H. (2000). Converging evidence from microstimulation, architecture, and connections for multiple motor areas in the frontal and cingulate cortex of prosimian primates. *The Journal of Comparative Neurology*, *423*, 140–177.

Wu, C. W., & Kaas, J. H. (2003). Somatosensory cortex of prosimian galagos: Physiological recording, cytoarchitecture, and corticocortical connections of anterior parietal cortex and cortex of the lateral sulcus. *The Journal of Comparative Neurology*, *457*(3), 263–292.

M. A. Hofman and D. Falk (Eds.)
Progress in Brain Research, Vol. 195
ISSN: 0079-6123

CHAPTER 6

Lateralization of the human brain

Michael C. Corballis*

Department of Psychology, University of Auckland, Auckland, New Zealand

Abstract: It is often suggested that cerebral asymmetry, when a consistent direction of asymmetry prevails, is unique to humans. We now know that many other species exhibit directionally consistent cerebral and behavioral asymmetries. Nevertheless, the predominance of left-cerebral dominance for language and manual functions may have played a special role in human evolution, even though precursors may be found in other animals—and especially in the great apes. I argue that the common cerebral asymmetry for these functions derives from the origins of language in manual gestures. These, in turn, may originate in specialized circuits for grasping that have been identified in primates, and lateralization may have been progressively introduced as praxic and linguistic functions became more complex.

Keywords: evolution; genetics; gesture; handedness; language; mirror neurons; praxis; signed languages; speech.

Introduction

A major issue in the understanding of human cerebral asymmetry is whether it is unique to humans and perhaps even defines our species. This idea was expressed, albeit with some apprehension, in my own book, *The Lopsided Ape* (Corballis, 1991), but has been declared more recently and adamantly by Crow (2004, 2005a,b, 2008), who argues that a late rearrangement of the X and Y chromosomes gave rise to human speciation through the emergence of cerebral asymmetry, language, theory of mind, and a disposition to schizophrenia. He refers to this as the "Broca–Annett" axiom, in recognition of Broca's (1861) pioneering discovery of cerebral asymmetry for speech and Annett's (2002) genetic theory of handedness and cerebral asymmetry. Annett herself regarded right-handedness as a uniquely human characteristic, involving "some small change ... in the genome that gave a slight weighting in favor of right-handedness" (Annett, 1985, p. 400). McManus (2002), who developed a similar genetic model, reached the same conclusion: "... right-handedness, and the D gene that causes it, are specifically human characteristics" (p. 233). Annett and McManus both

*Corresponding author.
Tel.: +649 3737555; Fax: +649 3737450
E-mail: m.corballis@auckland.ac.nz

DOI: 10.1016/B978-0-444-53860-4.00006-4

suppose that the same gene was also responsible for the left-cerebral dominance for language.

The idea that cerebral asymmetry is unique to humans, though, has been challenged by the now abundant evidence for both cerebral and behavioral asymmetries in a wide range of vertebrate species, from birds and fish to mammals and primates, much of it summarized in a volume edited by Rogers and Andrew (2002). The division of opinion is nowhere more evident than in the exchanges between Crow and Rogers, first in the journal *Laterality* (Crow, 2004; Rogers, 2004) and then in the *Behavioral and Brain Sciences* (Crow, 2005b; Vallortigara and Rogers, 2005).

A possible, if partial, reconciliation may lie in growing evidence that cerebral asymmetry is multidimensional so that different dimensions may have different evolutionary trajectories. Evidence from unilateral brain injury (Bryden et al., 1983) and functional transcranial Doppler imaging (Whitehouse and Bishop, 2009) suggests that left-cerebral language dominance and right-hemisphere dominance for spatial attention are independent, although a study using functional magnetic resonance imaging (fMRI)-based indices of functional asymmetry did show a small but significant correlation between the two, but zero correlation between the attentional bias and handedness (Badzakova-Trajkov et al., 2010) see Fig. 1 for more detail. More generally, a factor analysis of asymmetries of intrinsic activity in the human brain suggests at least four independent asymmetry factors, corresponding to brain areas involved in vision, internal thought (the so-called default network), attention, and language (Liu et al., 2009).

In this chapter, I focus on language and manual activities, where the case for human distinctiveness, if not uniqueness, may be the strongest. The asymmetries for language and handedness are correlated, suggesting a common evolutionary trajectory (e.g., Knecht et al., 2000). Language is itself widely understood to be uniquely human, which implies *a fortiori* that its predominantly left-hemispheric representation is also unique to our species—although we shall see that precursors to language in other species may also be lateralized. Moreover, humans are especially skilled at manual tasks, if not uniquely so, suggesting that the predominance of right-handedness is a distinctive human trait—hence the term *dexterity*—although again there is growing evidence for species-level right-handedness in chimpanzees.

I suggest that the commonality between linguistic and manual asymmetries arises because language itself emerged from manual gestures, and that the neural system involved in both became lateralized in the course of evolution, especially in the later stages when both gesture and language—or its precursors—became increasingly complex and independent of external input. This derives from a general point that bilateral symmetry is adaptive primarily in the context of interactions with the natural environment, where there is no systematic distinction between events to the right and left of the organism. This is why most animals are bilaterally symmetrical and indeed belong to the order known as the Bilateria, which dates from well before the Cambrian (Chen et al., 2004). Given the absence of systematic left–right bias in the natural world, animals have evolved a body plan that is for the most part bilaterally symmetrical, with symmetrically placed limbs and sense organs (Corballis and Beale, 1976; Gardner, 1967). To the extent that the brain is concerned with perception and movement, it too will tend to bilateral symmetry. But with the evolution of functions that are generated within the brain, and that are less dependent on the spatial environment, bilateral symmetry may become an impediment, imposing unnecessary constraints on both structure and function. This is perhaps why the internal organs of the body, including the heart and stomach, are asymmetrical, and why brain regions associated with language and skilled movement are lateralized. In this view, cerebral lateralization for these functions probably emerged late in primate evolution but became most prominent in our own species.

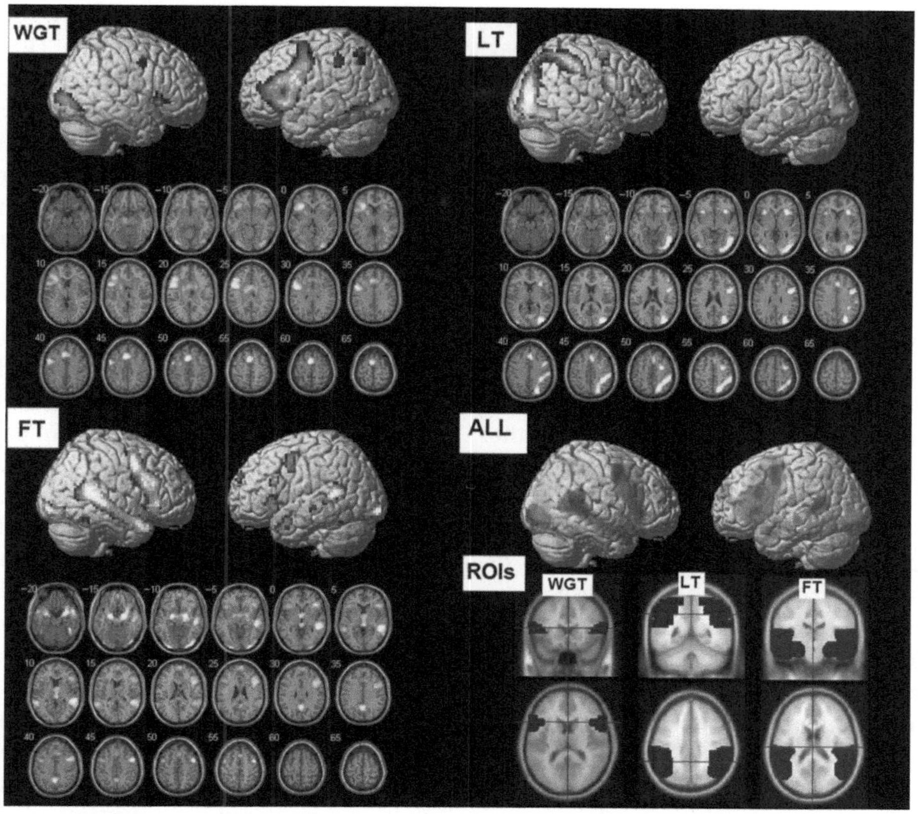

Fig. 1. Asymmetrical activation from fMRI scans elicited from three tasks: (a, upper left) word generation task (WGT) in which participants silently generated words beginning with designated letters; (b, upper right) line bisection task (LT) in which participants judged whether a short vertical line was or was not in the center of a horizontal line; (c, lower left) faces task (FT) in which participants watched short videos of faces expressing emotions; (d, middle right) superimposed scans from all three tasks; (e, lower right) regions of interest (ROIs) over which laterality indices were calculated. Asymmetries shown were highly significant despite nearly half (48 of 107) participants being left-handed, indicating that handedness has a relatively weak influence on cerebral asymmetries. Reproduced from Badzakova-Trajkov et al. (2010), which provides further details. (For color version of this figure, the reader is referred to the Web version of this chapter.)

To set the stage for the discussion of laterality, I begin with the argument for the gestural origins of language.

The gestural theory of language origins

The idea that language evolved from manual gestures has a long if intermittent history, dating from at least the seventeenth century (e.g., Bulwer, 1974/1644), and with occasional support over the ensuing centuries (e.g., Condillac, 1971; Critchley, 1939, 1975; Rousseau, 1781; Vico, 1744; Wundt, 1900). Even Darwin pointed to it, if only obliquely:

> I cannot doubt that language owes its origins to the imitation and modification of various natural sounds, and man's own distinctive cries, *aided by signs and gestures*
> Darwin (1896, p. 87; *emphasis added*)

In more recent times, the gestural theory gained considerable impetus from a seminal paper by the anthropologist Hewes (1973) and has been developed from different perspectives in a number of books published in the past 20 years (e.g., Armstrong, 1999; Armstrong et al., 1995; Corballis, 1991, 2002; Donald, 1991; Givòn, 1995; Rizzolatti and Sinigaglia, 2006; Tomasello, 2008).

Although some have supposed that language emerged only within the past 100,000 years or so (e.g., Chomsky, 2010), a gestural perspective suggests a much earlier and more gradual evolution, going even before the emergence of the hominins some 6 or 7 million years ago. Our primate ancestors, like most nonhuman primates today, were arboreal, with limbs adapted to climbing and manipulation. The movements of the forelimbs in primates are under precise cortical control and can therefore, in principle at least, serve as devices for intentional expression, whereas vocalization is under limbic control, with at best limited intentionality. Jürgens (2002) summarizes evidence that electrical stimulation of two areas of the mediofrontal cortex can produce vocalizations, the anterior cingulate gyrus (part of the limbic system), and the supplementary motor area (SMA). Stimulation of the SMA produces vocalization only in humans, while stimulation of the anterior cingulate gyrus produces vocalization only in nonhuman mammals, including rhesus monkey, squirrel monkey, cat, and bat. Jürgens writes:

> As the vocalizations of monkeys, cats, and bats are almost completely genetically determined in their acoustic structure, while the vast majority of human vocalizations are more or less completely learned, the difference in cortical representation might reflect the different role motor learning plays in vocal behavior of these species (p. 246).

Jürgens also writes that stimulation of the face area of the motor cortex not only elicits movements of the lips, tongue, and jaw in mammals but also elicits movements of the vocal folds only in primates, including humans. However, direct connections between motor cortex and the nucleus ambiguus, which supplies nerves to both larynx and pharynx, seem to exist only in humans (see also Ploog, 2002) and may hold the key to why only humans among the primates appear capable of cortically induced fine motor control of vocalization—in short, why only humans can speak.

Even chimpanzees, our closest nonhuman relatives, seem to have difficulty with the voluntary production of vocal sounds. Jane Goodall, who spent many years observing chimpanzees in the wild, once wrote that "(t)he production of sound in the absence of the appropriate emotional state seems to be an almost impossible task for a chimpanzee" (Goodall, 1986, p. 125). David Premack, who has worked extensively with chimpanzees in captivity, notes similarly that chimpanzees "lack voluntary control of their voice" (Premack, 2007, p. 13, 866).

These observations perhaps need some qualification, since chimpanzees do show some variety in their vocalizations, and can extract social information from them. For example, they can extract information about the status of another chimpanzee based on its vocalization, and they can distinguish the screams of a victim from that of an aggressor (Slocombe et al., 2010). Hopkins et al. (2007b) report that captive chimpanzees produce atypical sounds, a "raspberry" and an "extended grunt," to attract human attention and suggest that these sounds are intentional. Pant hoot calls among chimpanzees also display some cultural variations (Arcadi et al., 1998). Even so, such variations pale beside the flexibility of manual action in chimpanzees and other primates. Attempts to teach great apes anything resembling human speech have been conspicuously unsuccessful (e.g., Hayes, 1952), whereas moderate success in teaching them a form of language has been achieved through manual actions, either by having them point to symbols on a keyboard (Savage-Rumbaugh et al., 1998) or by using a simplified form of sign language (Gardner and Gardner, 1969; Patterson, 1978).

Great apes display flexible manual action in many other ways too. Chimpanzees groom each other, and it has even been suggested that their grooming behavior is the precursor to language (Dunbar, 1998). They make and use tools, including sticks for fishing termites out of holes (Bogart and Pruetz, 2008), and spears for jabbing into the hollow trunks of trees to extract bushbabies, which they then eat (Pruetz and Bertolani, 2007). Chimpanzees in the Loango National Park in Gabon use tool sets of up to five different stick and bark tools to extract honey from hives (Boesch et al., 2009). Pollick and de Waal (2007) compared manual gestures directly with orofacial movements and vocalizations in the natural communications of captive chimpanzees and bonobos and found manual gestures to be much less tied to context and more variable between groups, implying intentionality.

Manual gestures therefore provide a much better platform for the emergence of an intentional form of communication than do vocal calls. Indeed, it seems reasonable to suppose that language emerged from communicative acts such as pointing, and eventually from more complex actions such as pantomime. Donald (1991) has proposed that pantomime emerged in the genus *Homo* from some 2 million years ago. *Homo* also marks the beginning of the increase in brain size and the emergence of obligate bipedalism, both of which may have been driven in part by the adaptive advantages of a form of communication capable of transmitting information about complex events—and perhaps especially about events displaced from the present in both space and time (Corballis, 2009). Through a process of conventionalization (Burling, 1999), pantomime would have given way to a more arbitrary mapping of gestures onto meaning, whether in abstract hand signals or subsequently in spoken words.

Elements of pantomime are still present in dance, ballet, and mime. They also persist in signed languages. For example, in Italian Sign Language, some 50% of the hand signs and 67% of the bodily locations of signs stem from iconic representations, in which there is a degree of spatiotemporal mapping between the sign and its meaning (Pietrandrea, 2002). According to Emmorey (2002), American Sign Language includes some signs that are purely arbitrary, but many more are iconic. Further, we all resort to pantomime when trying to communicate with people who speak a different language. But, of course, the dominant mode of present-day language is speech, although movements of the hand and face do play a role in normal conversation (Goldin-Meadow and McNeill, 1999; McNeill, 1985, 1992). We may surmise that vocal components were gradually blended with manual and facial gesture and eventually assumed dominance (Corballis, 2002).

Speech as gesture

One objection to the gestural theory has been expressed by the linguist Robbins Burling:

> [T]he gestural theory has one nearly fatal flaw. Its sticking point has always been the switch that would have been needed to move from a visual language to an audible one
> Burling (2005, p. 123)

One answer to this is that speech itself is better conceived as a gestural system than as a sound-based one. Traditionally, speech has been regarded as made up of discrete elements of sound, called phonemes, despite the fact that phonemes do not exist as discrete units in the acoustic signal (Joos, 1948) and are not discretely discernible in mechanical recordings of sound, such as a sound spectrograph (Liberman et al., 1967). One reason for this is that the acoustic signals corresponding to individual phonemes vary widely, depending on the contexts in which they are embedded. So long as speech is considered in auditory terms, then, it must be assumed that the acoustic signal undergoes complex transformation for individual phonemes to be perceived as such. Yet we can perceive speech at remarkably high rates, up to at least 10–15 phonemes per second, which seems at odds with the idea that some complex, context-dependent transformation is necessary (Studdert-Kennedy, 1998).

These problems led to the so-called motor theory of speech perception, whereby speech is perceived in terms of how it is produced, rather than how it sounds (Liberman et al., 1967; see also Galantucci et al., 2006). This, in turn, led to the concept of articulatory phonology (Browman and Goldstein, 1995), in which speech is described in terms of the movements, or gestures, of the six articulatory organs—namely, the lips, the velum, the larynx, and the blade, body, and root of the tongue. Each of these organs is controlled separately so that individual speech units comprise different combinations of movements. The distribution of action over these articulators means that the elements overlap in time, which makes possible the high rates of production and perception. Although phonemes are not evident in mechanical recordings, such as the sound spectrograph, speech gestures are. Studdert-Kennedy (2005) writes that "as a unit of phonetic action the gesture can be directly observed by a variety of recording techniques, including X-ray, magnetic resonance imaging, and palatography (p. 57)." In other words, we seem to be wired to perceive speech as a series of gestures rather than as a sequence of sounds.

The mirror system

The gestural theory of language origins was boosted by the discovery of mirror neurons in area F5 in the premotor cortex of the primate brain (Rizzolatti and Arbib, 1998; Rizzolatti et al., 1988). The critical property of these neurons is that they fire not only when the animal executes a hand movement to grasp an object but also when the animal observes another individual making the same movement. The neurons, thus, represent a mapping of perception onto action—as also implied by the motor theory of speech perception. Mirror neurons were later understood to be part of a more general mirror system, incorporating the superior temporal sulcus and area PF in the inferior parietal lobule (Rizzolatti et al., 2001). More recent work, though, suggests an even wider distribution of mirror neurons in the brain, perhaps even confirming James's (1890) prescient remark that "Every representation of a movement awakens in some degree the actual movement which is its object" (p. 526). Work on the macaque now suggests that mirror neurons are located in many parts of the frontal cortex, including motor cortex, ventral and dorsal premotor cortex, as well as in inferior and superior parietal cortex, and in primary and supplementary somatosensory areas (see Kemmerer and Gonzalez-Castillo, 2010, for summary and references).

Although mirror neurons have been widely inferred from brain imaging in humans (e.g., Buccino et al., 2004; Gazzola and Keysers, 2009; Grézes et al., 2003), Mukamel et al. (2010) have now provided direct evidence from single-cell recordings in patients with intractable epilepsy. Extracellular activity was recorded while the patient executed and observed grasping movements of the hands as well as facial movements. Cells with mirror properties were found in a variety of areas. As expected, these included not only the medial frontal and temporal areas but also the SMA and the hippocampus and its environs.

Notwithstanding the evidence that the mirror system now appears to be ubiquitous, perhaps supporting the general notion of "embodied cognition" (e.g., Barsalou, 2008) rather than one specific to particular actions, language itself may have emerged from the components of the mirror system linked specifically to manual grasping in the primate. Rizzolatti and Sinigaglia (2010), while recognizing the ubiquity of mirror neurons, suggest that the circuit in the macaque brain that includes F5, parietal area PFG, and the anterior intraparietal area is "special" in that it permits the understanding of goal-directed actions "from the inside"; that is, it "gives the observer a first-person understanding of the motor goals and intentions of the motor goals and intentions of other individuals" (p. 264)—see Fig. 2. Although Rizzolatti and Sinigaglia do not specifically mention language, understanding of others' intentions is itself a communicative act, and the sharing of intentions was probably critical to the evolution of language in humans (e.g., Tomasello, 2008).

Brain-imaging studies provide more direct evidence that the same frontoparietal network is activated by both mimed actions and words describing those actions (Kemmerer and Gonzalez-Castillo, 2010; Peran et al., 2010; Xu et al., 2009). Discussing their findings, Xu et al. (2009) suggest that the areas identified since the mid-nineteenth century as language areas are part of a system more broadly concerned with linking meaning to symbols, "whether these are words, gestures, images, sounds, or objects" (p. 20,664). As suggested above, though, the true nature of this system may not be symbolic so much as "embodied," grounded in movement. We may surmise that, in the course of evolution, the system initially specialized for grasping provided the basis for the subsequent emergence of an intentional communication system based on manual gestures but eventually conventionalized so that other forms of representation, including spoken and written words, as well as more abstract manual gestures, could suffice to carry the message.

The mirror system in primates, though, appears to be responsive to manual actions but not to vocalization. This is not to say that it is unresponsive to sounds. Kohler et al. (2002) showed that mirror neurons responded not only to the sight of manual actions but also to the sound that they elicit—such as the cracking of nuts or the tearing of paper. What is important, though, is that these neurons did not respond to vocalizations made by their conspecifics. This shows that the incorporation of vocalization into the mirror system must

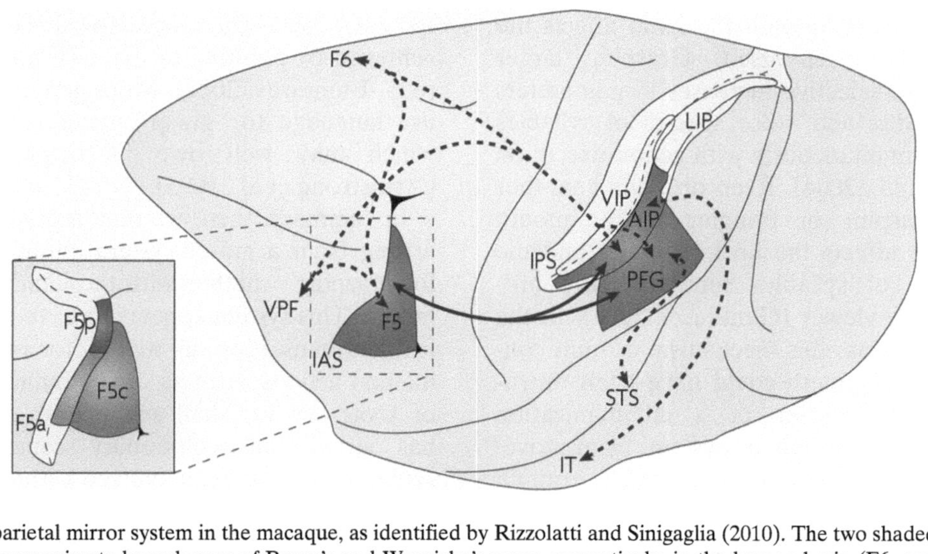

Fig. 2. Frontoparietal mirror system in the macaque, as identified by Rizzolatti and Sinigaglia (2010). The two shaded areas (F5 and AIP/PFG) are approximate homologues of Broca's and Wernicke's areas, respectively, in the human brain (F6, presupplementary motor area; IAS, inferior limb of arcuate sulcus; IPS, intraparietal sulcus; IT, inferior temporal lobe; LIP, lateral intraparietal sulcus; PFG, area between parietal areas PF and PG; STS, superior temporal sulcus; VIP, ventral intraparietal area; VPF, ventral prefrontal area). Reproduced with permission from Rizzolatti and Sinigaglia (2010). (For color version of this figure, the reader is referred to the Web version of this chapter.)

have been a later evolutionary development. Nevertheless, the mirror system does provide some insight into a possible transition from manual gesture to speech. Some neurons in area F5 in the monkey fire when the animal makes movements to grasp an object with either the hand or the mouth (Rizzolatti et al., 1988). Petrides et al. (2005) have identified an area in the monkey brain just rostral to premotor area 6 that is involved in control of the orofacial musculature. This area is also considered a homologue of part of Broca's area. The close neural associations between hand and mouth may be related to eating rather than communication, but later exapted for gestural and finally, vocal language.

The connection between hand and mouth can also be demonstrated behaviorally in humans. In one study, people were instructed to open their mouths while grasping objects, and the size of the mouth opening increased with the size of the grasped object; conversely, when they opened their hands while grasping objects with their mouths, the size of the hand opening also increased with the size of the object (Gentilucci et al., 2001). Grasping with the hand affects the kinematics of speech itself. Grasping larger objects induces selective increases in parameters of lip kinematics and voice spectra of syllables pronounced simultaneously with action execution (Gentilucci et al., 2004). Even observing another individual grasping or bringing to the mouth larger objects affects the lip kinematics and the voice spectra of syllables simultaneously pronounced by the viewer (Gentilucci, 2003). In the course of evolution, this mechanism of joint control of hand and mouth could have been instrumental in the transfer of a communication system, based on the mirror system, from movements of the hand to movements of the mouth (Gentilucci and Corballis, 2006).

To be involved in language, whether spoken or signed, the mirror system required a further property not evident in primates, that of intransitivity. In the primate, mirror neurons are *transitive* in that they respond only to actions involving the grasping of an object, even if that object is hidden behind a screen but the animal knows it is there (Umiltà et al., 2001). In humans, evidence from neuroimaging suggests that the mirror system responds also to *intransitive* actions, where a movement occurs in the absence of an object to be grasped (Rizzolatti and Sinigaglia, 2006). This suggests that the human mirror system incorporated actions that were symbolic rather than object related. This may have arisen not from symbolic representation *per se* but from the capacity of language to refer to events not present in the here-and-now. Indeed, language may have evolved precisely because our hominin ancestors evolved a capacity for "mental time travel" (Suddendorf and Corballis, 2007), and as suggested earlier, language emerged at least in part to enable us to refer to objects and actions that occurred, or will occur, at other times and places (Corballis, 2009). This is almost tangible in American Sign Language, where in the course of conversation absent objects are notionally "placed" at different locations in front of the speaker, and reference to those objects is achieved by pointing to the appropriate location (e.g., Emmorey, 2002). More generally, one may use language to "grasp" meaning, a metaphor which may well owe its origin to gesture (Armstrong et al., 1995).

In summary, then, we may regard language as arising from a motor system initially specialized for grasping, whether with the hands or with the mouth. This system appears to have been adapted for communication, as well as for more complex manual actions, such as the manufacture and use of tools—as we shall see below. Arbib (2005) has outlined an evolutionary scenario, whereby syntax may have been evolved within this system, leading ultimately to the complex language systems, whether spoken or signed, that characterizes our species.

I now consider cerebral asymmetry in the context of this scenario.

Handedness and language lateralization

If language originated in manual gesture, we may surmise that lateralization was induced as both communication and manipulative skills became more complex, and less reliant on direct input from the environment. This may explain both the predominance of a right-hand preference, implying left-hemispheric dominance, and the left-cerebral specialization for language, whether spoken or signed. These two asymmetries are correlated, suggesting a common source, but the correlation is not perfect. This was recognized early on from cases of so-called crossed aphasia (e.g., Kennedy, 1916), in which language disturbances arise from lesions ipsilateral to the dominant hand. For example, Penfield and Roberts (1959) report on 53 cases of right-handers with aphasia following right-hemisphere injury and 66 left-handed aphasics with left-hemisphere injury. Although some 95–99% of right-handers are left-cerebrally dominant for language, we now know that this is also true of some 70–80% of left-handers (Badzakova-Trajkov et al., 2010; Knecht et al., 2000; Pujol et al., 1999; Rasmussen and Milner, 1977; Warrington and Pratt, 1973). Left-handers, then, are not simply reversed right-handers.

Despite the imperfect correlation, there have been attempts to account for variations in both handedness and language lateralization in terms of a single hypothetical gene. The most successful models postulate two alleles, one biasing handedness to the right and language lateralization to the left, and the other leaving both asymmetries to chance. In McManus's (2002) version, these two alleles are labeled D (for dextral) and C (for chance), while in Annett's (2002) version, the gene is termed the "right-shift" (RS) gene and the alleles labeled RS^+ and RS^-, respectively. In both cases, the assumption is that the same gene influences cerebral asymmetry, but the "chance" allele operates independently with respect to the two asymmetries. In order to provide more precise estimates of the relations between handedness and cerebral asymmetry for language, Annett makes the further assumption that the RS gene operates additively on handedness, but in dominant-recessive fashion on cerebral asymmetry. This is shown in Fig. 3.

Despite the moderate success of such models in accounting for inheritance patterns, handedness and cerebral asymmetry for language have different neural correlates. The principal brain region involved in the determination of handedness appears to be the primary motor cortex (M1). In one study, activation in the contralateral motor cortex of both right- and left-handed subjects was consistently greater during simple movements of the dominant than during those of the nondominant hand, and the degree of cortical asymmetry was correlated with degree to which one hand was preferred over the other as measured by the Edinburgh Handedness Inventory (Dassonville et al., 1997). That is, the

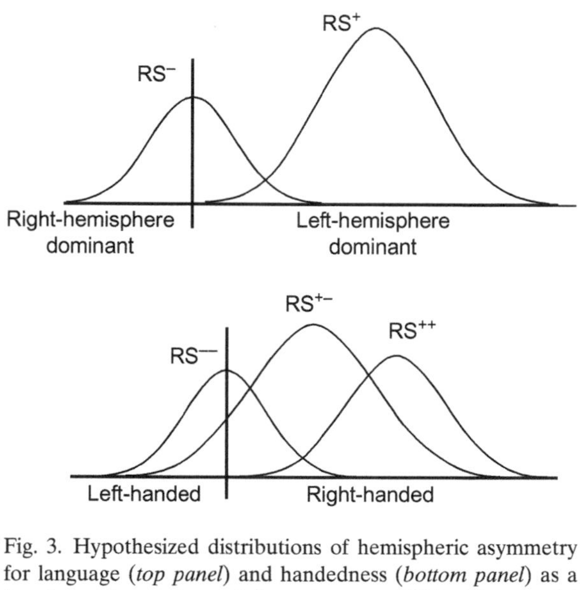

Fig. 3. Hypothesized distributions of hemispheric asymmetry for language (*top panel*) and handedness (*bottom panel*) as a function of genotype (after Annett, 2002). The model assumes normal distributions of relative hemisphere dominance and relative handedness, respectively. The RS gene is assumed to operate in dominant-recessive fashion on hemispheric asymmetry for language, and in additive fashion on handedness.

asymmetry in M1 reflected both the direction and strength of handedness. These relationships were not evident in activity in the premotor cortex or the SMA. The effects for left-handers were essentially opposite to those for right-handers, but an earlier study suggested that the asymmetry was less marked for left- than for right-handers (Kim et al., 1993).

These differences are accompanied by anatomical differences. In right-handers, the left central sulcus is deeper than the right, and the opposite is true for left-handers, and these differences were also evident in an increased amount of neuropil in the region of hand representation in Brodman's area 4 opposite the preferred hand (Amunts et al., 1996). Hammond (2002) has suggested that the origins of handedness lie in the more diffuse intrahemispheric connections in the primary motor cortex contralateral to the dominant hand, and that these diffuse connections provide a substrate for learning. Voxel-based morphometry shows that the volume of gray matter in M1 can be substantially altered as a consequence of activity in the contralateral hand, increasing with increased activity and decreasing with inactivity (Granert et al., 2011).

In contrast, cerebral asymmetry for language does not appear to involve M1 but applies over wide regions of the left hemisphere, including the classic Broca's and Wernicke's areas (Dick et al., 2001). A closer neural correspondence comes not from handedness *per se* but from cerebral asymmetry in the control of skilled manual actions or praxis.

Language and praxis

Praxis includes actions like using tools such as a comb or a toothbrush, as well as pantomime. Correspondingly, the term "apraxia" refers to the loss of ability to perform such actions, usually as a consequence of brain injury. Cerebral asymmetry for praxis is somewhat independent of handedness itself and indeed appears to be more closely related to cerebral asymmetry for language. In a study of 90 epileptic patients undergoing the intracarotid amobarbital procedure (IAP) for unilateral cerebral evaluation prior to surgery, Meador et al. (1999) found that those with left-cerebral language dominance made more praxic errors following left IAP (inactivating the left hemisphere) than did those with right-cerebral or bilateral language. Right IAP reversed this pattern. These effects were independent of handedness, and patients with atypical language dominance showed more bilateral representation of both language and praxic representation.

There are also recorded instances of apraxia following right-hemisphere lesions in both left-handers (e.g., Valenstein and Heilman, 1979) and right-handers (Raymer et al., 1999), again suggesting that the asymmetry underlying praxis is not the same as that underlying handedness itself but may lie closer to that underlying language. Lausberg et al. (1999) describe a left-handed patient with callosal section who showed severe apraxia of the left hand but was well able to execute actions with the right hand, whether to verbal command or through imitation of meaningless actions. He was also able to learn a new visuomotor skill only with the right hand. Finally, Johnson-Frey et al. (2005) found that two patients with full callosotomy were more accurate in pantomiming actions based on pictured stimuli when the stimuli were presented to the isolated left hemisphere and the corresponding pantomimes were made with the right hand, despite the fact that the patients were of opposite handedness.

Earlier, I noted the work of Xu et al. (2009) showing that mimed actions activated brain areas overlapping activated by speech. These areas are predominantly left-hemispheric (see Fig. 4). The mimed actions included those that might be broadly described as praxis, including pantomime (e.g., mimicking the threading of a needle) and the production of emblems (e.g., finger to lip to indicate "be quiet"). The same system that

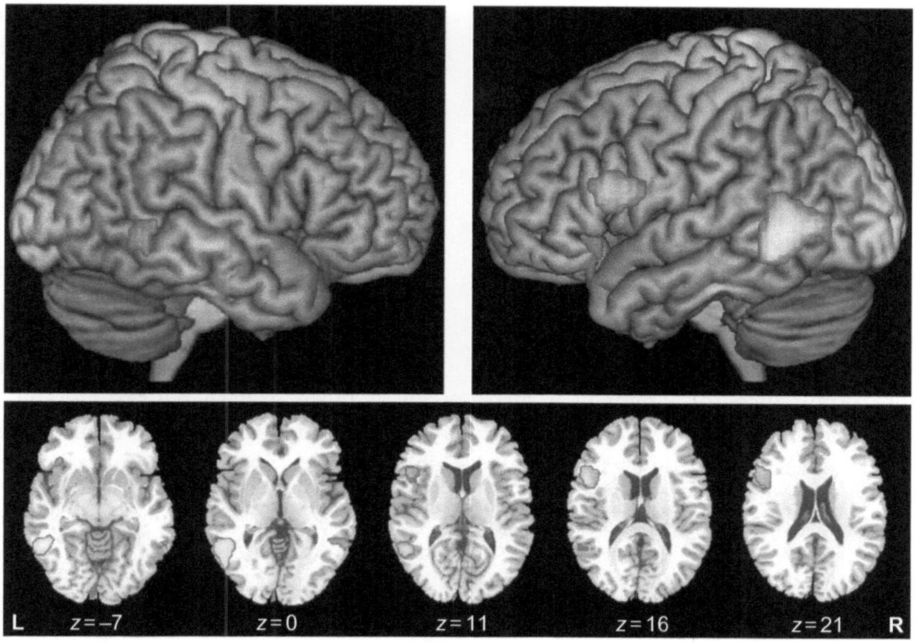

Fig. 4. Common areas of activation in processing symbolic gestures and spoken speech, showing marked left-hemispheric bias. This figure is reproduced from Xu et al. (2009), which provides further details. (For color version of this figure, the reader is referred to the Web version of this chapter.)

underlies pantomime is also activated by the retrieval and planning of learned motor actions (Frey, 2008), indicating that the symbolic system identified by Xu et al. is part of a more general left-cerebral system for praxis. Roy and Arbib (2005) refer to this system as the "syntactic motor system," with application beyond language itself to skilled movements.

The left-hemispheric representation of both praxis and language is further supported by neuroimaging studies showing activation of Broca's area when people make meaningful arm gestures (Buccino et al., 2001; Decety et al., 1997; Gallagher and Frith, 2004; Grèzes et al., 1998), or even imagine them (Gerardin et al., 2000; Grafton et al., 1996; Hanakawa et al., 2003; Kuhtz-Buschbeck et al., 2003). Moreover, signed languages, which have their origins in pantomime, are also represented predominantly in the left-cerebral hemisphere (Corina, 2002; Hickok and Bellugi, 2000), and Broca's area, in particular, is activated in signers while signing and in speakers while speaking (Horwitz et al., 2003).

Left-cerebral dominance for language, and presumably praxis as well, is evident in anatomical as well as functional asymmetries. One such asymmetry is the Yakovlevian torque, in which right frontal lobe is wider and protrudes forward relative to that on the left, and the left hemisphere is wider and protrudes rearward relative to that on the left. Crow (2005a,b) has suggested that this asymmetry is critical to functional asymmetries in the human brain. In the majority of human brains, the left temporal planum, which forms part of Wernicke's area, is larger on the left than on the right (Geschwind and Levitsky, 1968), an asymmetry also apparent in infant brains (Wada et al., 1975). This asymmetry is correlated with functional asymmetry for language in adults (Foundas et al., 1996). Broca's area is also larger on the left than on the right in most individuals (Amunts et al., 1999; Falzi et al., 1982). The volumes and functional anisotropy (FA) of tracts between Broca's and Wernicke's areas have also been shown to be greater in the left hemisphere, and these asymmetries are correlated with left-hemisphere dominance for language processing as measured from fMRI (Powell et al., 2006). Glasser and Rilling (2008) have similarly used diffusion tensor imaging to demonstrate a leftward bias in the arcuate fasciculus, which connects temporal regions with the inferior frontal cortex. Left-hemisphere dominance for language is also associated with lowered FA in the corpus callosum (Häberling et al., 2011), supporting Gazzaniga's (2000) conjecture that cerebral asymmetry may be accomplished through callosal pruning.

The puzzle of handedness

The fact that cerebral asymmetry for praxis is more closely related to that for language than is handedness itself raises the question of why the strong preference for one hand, usually the right, exists in the human population. It seems to have relatively little to do with praxis. The neural substrates are different; as we have seen, handedness appears to depend on asymmetries in area M1, whereas cerebral asymmetry for praxis and language is evident in a broader network involving prefrontal and temporoparietal regions. Yet there is a correlation, albeit imperfect, between handedness and praxic asymmetries.

Perhaps, the mechanisms responsible for praxic asymmetry only loosely influence asymmetry in the adjacent areas of M1. Perhaps, handedness is simply more malleable, subject to environmental pressures leading to a reduction in right-handedness. Indeed, the strong random component to handedness, suggested by Annett (2002) and McManus (2002), might lead us to expect some regression toward bilateral symmetry, which in biological terms is the "default" condition (Palmer, 2004). Yet another possibility, suggested earlier by Levy and Nagylaki (1972), is that two separate genes control handedness. One controls cerebral asymmetry itself and is directly manifest in lateralization of language and perhaps of praxis, and the other determines whether cerebral control over the dominant hand is ipsilateral or contralateral. Thus, for example, in those who are left-handed but left-cerebrally dominant for praxis, control of the dominant hand may be ipsilateral.

To resolve the puzzle of handedness, we must await more precise neural, genetic, and correlational evidence on the relations between the handedness and the cerebral asymmetries for language and praxis.

Comparative perspectives

I have focused on cerebral asymmetries for language and manual action, since these are especially prominent in our own species and can be linked through a common neural network for praxis. This network derives from the mirror system, dedicated initially to grasping, which in primates appears to be bilaterally organized (Rizzolatti and Sinigaglia, 2006). This raises the question of when the system became more complex and began to lateralize.

Although handedness appears to be only loosely related to asymmetry of the praxic system in humans, chimpanzees and other great apes do appear to show species-level right-handedness. Captive chimpanzees show an overall preference for the right hand, at least for certain activities such as extracting peanut butter from a glass tube (Hopkins, 1996), using an anvil (Hopkins et al., 2007a), gestural communication (Hopkins and Leavens, 1998; Meguerditchian et al., 2010), and throwing (Hopkins et al., 2005). Although it has been suggested that this bias is restricted to chimpanzees in captivity (McGrew et al., 2007), some studies of chimpanzees in the wild have shown a right-handed bias for a number of activities, such as leaf folding for drinking water and nut cracking (Biro et al., 2006), termite fishing (Lonsdorf and Hopkins, 2005), nut cracking and wedge-dipping for drinking water (Boesch, 1991), and ant dipping and extraction of palm heart (Humle and Matsuzawa, 2009). The incidence of right-handedness in chimpanzees in these studies is typically around 65%, considerably lower than the incidence of right-handedness in humans, which is around 88–90%.

At least some of these activities revealing right-handedness might be better described as praxic rather than simply based on preference. They include communicative gesture, again supporting the gestural origins of language. We saw earlier that the temporal planum is generally larger on the left in humans, and this appears to be true also of chimpanzees (Gannon et al., 1998; Hopkins et al., 1998), but not of rhesus monkeys or baboons (Wada et al., 1975). The leftward asymmetry in the temporal planum in the chimpanzee is correlated with a right-handed bias in gestural communication (Hopkins and Nir, 2010), and a leftward asymmetry in the cytoarchitecture of the areas homologous to Wernicke's area is also interconnected with a Broca's area homologue in the chimpanzee (Spocter et al., 2010).

Curiously, a left-hemisphere advantage for vocalization appears to exist in species much more distant than the primates, including frogs (Bauer, 1993), mice (Ehert, 1987), and canaries (Halle et al., 2004)—as well as in rhesus monkeys (Hauser and Anderson, 1994; Poremba et al., 2004) and Japanese macaques (Heffner and Heffner, 1984). This raises the possibility that cerebral asymmetry in the praxic system was triggered by a preexisting asymmetry in vocalization (Corballis, 2003). Vocalization is internally generated and relatively independent of the spatial environment, and although it is not under cortical control in most species, including primates, its gradual incorporation into the praxic system, and consequent cortical representation, may have biased the system to the left.

Conclusions

Human right-handedness has been noted and remarked upon since Biblical times, and left-cerebral dominance for language has been consistently observed since Broca's pioneering work 150 years ago. Not surprisingly, these two asymmetries are commonly linked and regarded as characteristic of our species—perhaps even a defining characteristic, as indeed I once suggested in my book *The Lopsided Ape* (Corballis, 1991). Subsequent developments, though, suggest a more complex evolutionary scenario.

In this chapter, I have developed the hypothesis that the link between manual action and language derives from the gestural origins of language itself. That is, language emerged from the primate mirror system—a system initially specialized for the perception and production of manual grasping. At some point in hominin evolution, and perhaps in the evolution of the great apes, this system was expanded to encompass communication. At first, communication was primarily manual, but gestures expanded to include facial movements, ultimately including vocalization. The switch from manual to vocal gestures was probably gradual, and even today manual gestures typically accompany speech—although speech is sufficiently dominant to enable us to communicate with little loss of intelligibility by phone or the radio.

Since present-day signed languages have all of the linguistic sophistication of spoken ones, the reasons for the switch were presumably practical rather than linguistic. Speech allows communication at night or when visual contact is blocked, and vocalization is less physically demanding than is manual gesture. Perhaps, more importantly, speech frees the hands for manual activities, such as carrying things or using tools, and allows the speaker to explain manual operations while demonstrating them (Corballis, 2002, 2003). The switch to vocalization might also be regarded as a partitioning of resources, with the hands evolving the capacity for complex nonverbal skills, from the construction of tools to the playing of musical instruments, while the mouth and vocal tract carry the main burden of language. Restricting language primarily to the mouth and its internal movable parts might also be considered an early instance of miniaturization.

Although the mirror system in primates appears to be bilateral, the system may have become lateralized, generally favoring the left hemisphere, as it developed increased programming complexity, and less reliance on spatial input from the natural environment. This complexity applies not only to language but also to learned manual actions. Whatever the precise mechanisms controlling handedness and cerebral asymmetry for praxis and language, the degree of lateralization does appear to be more pronounced in humans than in our closest relatives, the great apes, but probably did not emerge as a "big bang" in our own species, as suggested by Crow (2008). It is more consistent with Darwinian principles to suppose that lateralization itself emerged incrementally and perhaps derives in part from asymmetries that have quite ancient origins. One possibility is that vocal asymmetry played a role, perhaps as vocalization was introduced into the system, culminating in speech. As language grew more complex, the asymmetry itself may have intensified, perhaps to avoid the restrictions on processing created by interhemispheric transfer. Conversely, the more pronounced asymmetry in humans may have occurred because the flowering of the mirror system gave rise to activities that are themselves distinctively human, including language and our extraordinary manual skill. It may not be altogether inappropriate to regard cerebral asymmetry as a characteristic that is especially exploited by humans, even if not unique to us.

This should not be taken to imply that lateralization constitutes some end state that evolution drives toward. Even in humans, lateralization must be balanced against the advantages of bilateral symmetry. A minority of people fail to show the typical asymmetries of right-handedness and left-cerebral representation of language, and as we have seen, these divergences are best explained in terms of chance rather than the reversal of a genetic disposition to asymmetry. That is, left-handedness, and perhaps right-hemispheric representation of language, results from the lack of the genetic bias, and some individuals are better characterized as ambidextrous or as having language represented bilaterally. Across the population, then, greater adaptiveness might be maintained by preserving a balance between asymmetry, with advantages in language and programmed action, and symmetry, with advantages in spatial awareness and perhaps hunting (or sporting) activity. This balance might be maintained by a heterozygotic advantage, maintaining both the lateralizing influence (RS^+ or D allele) and the chance influence (RS^- or C allele) in the population (Annett, 2002).

To conclude, cerebral asymmetry in manual control and language is not a defining characteristic of our species. There are precursors in our primate forebears, and a not insubstantial minority of humans does not show the asymmetries typical of the majority. Rather, language itself and the complexities of human manual skills are the true markers of our species. Nevertheless, lateralization is of considerable importance, for two reasons. First, the correlation between asymmetry for language and that for manual action adds support to the theory that language evolved from manual gestures. Second, and more generally, lateralization has important implications for the understanding and evolution of function. Given that bilateral symmetry is the default condition, there must be special circumstances that favor bilateral asymmetry, causing the symmetry of the body plan to be selectively overruled. Language and praxis provide one such set of circumstances.

Acknowledgments

Some of the work described in this chapter was supported by a grant from the Marsden Fund, administered by the Royal Society of New Zealand. I thank Gjurgjica Badzakova-Trajkov, Isabelle Häberling for their laboratory support, and Michael Arbib, Tim Crow, Maurizio Gentilucci, Chris McManus, Giacomo Rizzolatti, and Leslie Rogers for discussion over the years.

References

Amunts, K., Schlaug, G., Schleicher, A., Steinmetz, H., Dabringhaus, A., Roland, P. E., et al. (1996). Asymmetry in the human motor cortex and handedness. *NeuroImage, 4*, 216–222.

Amunts, K., Schleicher, A., Burgel, U., Mohlberg, H., Uylings, H. B. M., & Zilles, K. (1999). Broca's region revisited: Cytoarchitecture and intersubject variability. *The Journal of Comparative Neurology, 412*, 319–341.

Annett, M. (1985). *Left, right, hand and brain: The right shift theory.* Hillsdale, NJ: Lawrence Erlbaum Associates.

Annett, M. (2002). *Handedness and brain asymmetry: The right shift theory.* Hove, UK: Psychology Press.

Arbib, M. A. (2005). From monkey-like action recognition to human language: An evolutionary framework for neurolinguistics. *The Behavioral and Brain Sciences, 28*, 105–168.

Arcadi, A. C., Robert, D., & Boesch, C. (1998). Buttress drumming by wild chimpanzees: Temporal patterning, phase integration into loud calls, and preliminary evidence for individual differences. *Primates, 39*, 505–518.

Armstrong, D. F. (1999). *Original signs: Gesture, sign, and the source of language.* Washington: Gallaudet University Press.

Armstrong, D. F., Stokoe, W. C., & Wilcox, S. E. (1995). *Gesture and the nature of language.* Cambridge, MA: Cambridge University Press.

Badzakova-Trajkov, G., Häberling, I. S., Roberts, R. P., & Corballis, M. C. (2010). Cerebral asymmetries: Complementary and independent processes. *PloS One, 5*(3), e9682.

Barsalou, L. W. (2008). Grounded cognition. *Annual Review of Psychology, 59*, 617–645.

Bauer, R. H. (1993). Lateralization of neural control for vocalization by the frog (*Rana pipiens*). *Psychobiology, 21*, 243–248.

Biro, D., Sousa, C., & Matsuzawa, T. (2006). Ontogeny and cultural propagation of tool use by wild chimpanzees at Bossou, Guinea: Case studies in nut cracking and leaf folding. In T. Matsuzawa, M. Tomonaga & M. Tanaka (Eds.), *Cognitive development in chimpanzees* (pp. 476–508). Tokyo, Japan: Springer.

Boesch, C. (1991). Handedness in wild chimpanzees. *International Journal of Primatology, 12*, 541–558.

Boesch, C., Head, J., & Robbins, M. M. (2009). Complex tool sets for honey extraction among chimpanzees in Loango National Park, Gabon. *Journal of Human Evolution, 56*, 560–569.

Bogart, S. L., & Pruetz, J. D. (2008). Ecological context of savanna chimpanzee (*Pan troglodyte verus*) termite fishing at Fongoli, Senegal. *American Journal of Primatology, 70*, 605–612.

Broca, P. (1861). Remarques sur la siège de la faculté du langage articulé, suivies d'une observation d'aphémie. *Bulletin de la Société Anatomique de Paris, 2*, 330–357.

Browman, C. P., & Goldstein, L. F. (1995). Dynamics and articulatory phonology. In T. van Gelder & R. F. Port (Eds.), *Mind as motion*. Cambridge, MA: MIT Press.

Bryden, M. P., Hécaen, M., & DeAgostini, M. (1983). Patterns of cerebral organization. *Brain and Language, 20*, 249–262.

Buccino, G., Binkofski, F., Fink, G. R., Fadiga, L., Fogassi, L., Gallese, V., et al. (2001). Action observation activates premotor and parietal areas in a somatotopic manner: An fMRI study. *The European Journal of Neuroscience, 13*, 400–404.

Buccino, G., Lui, F., Canessa, N., Paterri, I., Lagravinese, G., Benuzzi, F., et al. (2004). Neural circuits underlying imitation learning of hand actions: An event-related fMRI study. *Neuron, 42*, 323–334.

Bulwer, J. (1974). *Chirologia: Or the natural language of the hand.* London: Harper (Reprinted, with commentary and notes by J. W. Cleary, Carbondale and Edwardsville, IL: Southern Illinois University Press. Original work published 1644).

Burling, R. (1999). Motivation, conventionalization, and arbitrariness in the origin of language. In B. J. King (Ed.), *The origins of language: What nonhuman primates can tell us* (pp. 307–350). Washington, DC: School of American Research Press.

Burling, R. (2005). *The talking ape.* New York: Oxford University Press.

Chen, J.-Y., Bottjer, D. J., Oliveri, P., Dornbos, S. Q., Gao, F., Ruffins, S., et al. (2004). Small bilaterian fossils from 40 to 55 million years before the Cambrian. *Science, 305*, 218–222.

Chomsky, N. (2010). Some simple evo devo theses: How true might they be for language? In R. K. Larson, V. Déprez & H. Yamakido (Eds.), *The evolution of human language* (pp. 45–62). Cambridge, UK: Cambridge University Press.

Condillac, E (1971). *An essay on the origin of human knowledge: Being a supplement to Mr. Locke's Essay on the Human Understanding.* A facsimile reproduction of the 1756 translation by T. Nugent of Condillac's 1747 essay. Gainesville, FL: Scholars' Facsimiles and Reprints.

Corballis, M. C. (1991). *The lopsided ape.* New York: Oxford University Press.

Corballis, M. C. (2002). *From hand to mouth: The origins of language.* Princeton, NJ: Princeton University Press.

Corballis, M. C. (2003). From mouth to hand: Gesture, speech, and the evolution of right-handedness. *The Behavioral and Brain Sciences, 26*, 198–208.

Corballis, M. C. (2009). The evolution of language. *Annals of the New York Academy of Sciences*, *1156*, 19–43.

Corballis, M. C., & Beale, I. L. (1976). *The psychology of left and right.* Hillsdale, N.J.: Lawrence Erlbaum Associates.

Corina, D. P. (2002). Aphasia in users of signed languages. In P. Coppens, Y. Lebrun & A. Basso (Eds.), *Aphasia in atypical populations* (pp. 261–310). Mahwah, NJ: Erlbaum.

Critchley, M. (1939). *The language of gesture.* London: Arnold.

Critchley, M. (1975). *Silent language.* London: Arnold.

Crow, T. J. (2004). Directional asymmetry is the key to the origin of modern *Homo sapiens* (the Broca-Annett axiom): A reply to Rogers' review of *The Speciation of Modern Homo Sapiens*. *Laterality*, *9*, 233–242.

Crow, T. J. (2005a). The cerebral torque and directional asymmetry for hand use are correlates of the capacity for language in *Homo sapiens*. *The Behavioral and Brain Sciences*, *28*, 595–596.

Crow, T. J. (2005b). Who forgot Paul Broca? The origin of language as a test case for speciation theory. *Journal of Linguistics*, *41*, 133–156.

Crow, T. J. (2008). The 'big bang' theory of the origin of psychosis and the faculty of language. *Schizophrenia Research*, *102*, 31–52.

Darwin, C. (1896). *The descent of man and selection in relation to sex* (2nd ed.). New York: Appleton (1st ed. published 1871).

Dassonville, P., Zhu, X.-H., Uĝurbil, K., Kim, S.-G., & Ashe, J. (1997). Functional activation in motor cortex reflects the direction and the degree of handedness. *Proceedings of the National Academy of Sciences of the United States of America*, *94*, 14015–14018.

Decety, J., Grezes, J., Costes, N., Perani, D., Jeannerod, M., Procyk, E., et al. (1997). Brain activity during observation of actions. Influence of action content and subject's strategy. *Brain*, *120*, 1763–1777.

Dick, F., Bates, E., Wulfeck, B., Utman, J. A., Dronkers, N. F., & Gernsbacher, M. A. (2001). Language deficits, localization, and grammar: Evidence for a distributed model of language breakdown in aphasic patients and neurologically intact individuals. *Psychological Review*, *108*, 759–788.

Donald, M. (1991). *Origins of the modern mind.* Cambridge, MA: Harvard University Press.

Dunbar, R. I. M. (1998). *Grooming, gossip, and the evolution of language.* Cambridge, MA: Harvard University Press.

Ehert, G. (1987). Left hemisphere advantage in the mouse brain for recognizing ultrasonic communication calls. *Nature*, *325*, 249–251.

Emmorey, K. (2002). *Language, cognition, and brain: Insights from sign language research.* Hillsdale, NJ: Erlbaum.

Falzi, G., Perrone, P., & Vignolo, L. (1982). Right–left asymmetry in anterior speech region. *Archives of Neurology*, *39*, 239–240.

Foundas, A. L., Leonard, C. M., Gilmore, R. L., Fennell, E. B., & Heilman, K. M. (1996). Pars triangularis asymmetry and language dominance. *Proceedings of the National Academy of Sciences of the United States of America*, *93*, 719–722.

Frey, S. H. (2008). Tool use, communicative gesture and cerebral asymmetries in the modern human brain. *Philosophical Transactions of the Royal Society B*, *363*, 1951–1957.

Galantucci, B., Fowler, C. A., & Turvey, M. T. (2006). The motor theory of speech perception reviewed. *Psychonomic Bulletin and Review*, *13*, 361–377.

Gallagher, H. L., & Frith, C. D. (2004). Dissociable neural pathways for the perception and recognition of expressive and instrumental gestures. *Neuropsychologia*, *42*, 1725–1736.

Gannon, P. J., Holloway, R. L., Broadfield, D. C., & Braun, A. R. (1998). Asymmetry of chimpanzee planum temporale: Human-like brain pattern of Wernicke's area homolog. *Science*, *279*, 220–221.

Gardner, M. (1967). *The ambidextrous universe.* London: Penguin.

Gardner, R. A., & Gardner, B. T. (1969). Teaching sign language to a chimpanzee. *Science*, *165*, 664–672.

Gazzaniga, M. S. (2000). Cerebral specialization and interhemispheric communication: Does the corpus callosum enable the human condition? *Brain*, *123*, 1293–1326.

Gazzola, V., & Keysers, C. (2009). The observation and execution of actions share motor and somatosensory voxels in all tested subjects: Single-subject analyses of unsmoothed fMRI data. *Cerebral Cortex*, *19*, 1239–1255.

Gentilucci, M. (2003). Grasp observation influences speech production. *The European Journal of Neuroscience*, *17*, 179–184.

Gentilucci, M., Benuzzi, F., Gangitano, M., & Grimaldi, S. (2001). Grasp with hand and mouth: A kinematic study on healthy subjects. *Journal of Neurophysiology*, *86*, 1685–1699.

Gentilucci, M., & Corballis, M. C. (2006). From manual gesture to speech: A gradual transition. *Neuroscience and Biobehavioral Reviews*, *30*, 949–960.

Gentilucci, M., Santunione, P., Roy, A. C., & Stefanini, S. (2004). Execution and observation of bringing a fruit to the mouth affect syllable pronunciation. *The European Journal of Neuroscience*, *19*, 190–202.

Gerardin, E., Sirigu, A., Lehericy, S., Poline, J. B., Gaymard, B., Marsault, C., et al. (2000). Partially overlapping neural networks for real and imagined hand movements. *Cerebral Cortex*, *10*, 1093–1104.

Geschwind, N., & Levitsky, W. (1968). Human brain: Left–right asymmetries in temporal speech region. *Science*, *161*, 186–187.

Givòn, T. (1995). *Functionalism and grammar.* Philadelphia, PA: John Benjamins.

Glasser, M. F., & Rilling, J. K. (2008). DTI tractography of the human brain's language pathways. *Cerebral Cortex*, *18*, 2471–2482.

Goldin-Meadow, S., & McNeill, D. (1999). The role of gesture and mimetic representation in making language the province of speech. In M. C. Corballis & S. E. G. Lea (Eds.), *The descent of mind* (pp. 155–172). Oxford, UK: Oxford University Press.

Goodall, J. (1986). *The chimpanzees of Gombe: Patterns of behavior.* Cambridge, MA: Harvard University Press.

Grafton, S. T., Arbib, M. A., Fadiga, L., & Rizzolatti, G. (1996). Localization of grasp representations in humans by positron emission tomography. 2. Observation compared with imagination. *Experimental Brain Research*, *112*, 103–111.

Granert, O., Peller, M., Gaser, C., Groppa, S., Hallett, M., Knutzen, A., et al. (2011). Manual activity shapes structure and function in contralateral human motor hand area. *NeuroImage*, *54*, 32–41.

Grézes, J., Armony, J. L., Rowe, J., & Passingham, R. E. (2003). Activations related to "mirror" and "canonical" neurones in the human brain: An fMRI study. *NeuroImage*, *18*, 928–937.

Grèzes, J., Costes, N., & Decety, J. (1998). Top-down effect of strategy on the perception of human biological motion: A PET investigation. *Cognitive Neuropsychology*, *15*, 553–582.

Häberling, I. S., Badzakova-Trajkov, G., & Corballis, M. C. (2011). Callosal tracts and patterns of hemispheric dominance: A combined fMRI and DTI study. *NeuroImage*, *54*, 779–786.

Halle, F., Gahr, M., & Kreutzer, M. (2004). Effects of unilateral lesions of HVC on song patterns of male domestic canaries. *Journal of Neurobiology*, *56*, 303–314.

Hammond, G. (2002). Correlates of human handedness in primary motor cortex: A review and hypothesis. *Neuroscience and Biobehavioral Reviews*, *26*, 285–292.

Hanakawa, T., Immisch, I., Toma, K., Dimyan, M. A., Van Gelderen, P., & Hallett, M. (2003). Functional properties of brain areas associated with motor execution and imagery. *Journal of Neurophysiology*, *89*, 989–1002.

Hauser, M. D., & Anderson, K. (1994). Functional lateralization for auditory temporal processing in adult, but not infant rhesus monkeys: Field experiments. *Proceedings of the National Academy of Sciences of the United States of America*, *91*, 3946–3948.

Hayes, C. (1952). *The ape in our house.* London: Gollancz.

Heffner, H. E., & Heffner, R. S. (1984). Temporal lobe lesions and perception of species-specific vocalizations by Japanese macaques. *Science*, *226*, 75–76.

Hewes, G. W. (1973). Primate communication and the gestural origins of language. *Current Anthropology*, *14*, 5–24.

Hickok, G., & Bellugi, U. (2000). The signs of aphasia. In F. Boller & J. Grafman (Eds.), *Handbook of neuropsychology* (2nd ed.). Amsterdam: Elsevier.

Hopkins, W. D. (1996). Chimpanzee handedness revisited: 55 years since Finch (1941). *Psychonomic Bulletin & Review*, *3*, 449–457.

Hopkins, W. D., & Leavens, D. A. (1998). Hand use and gestural communication in chimpanzees (*Pan troglodytes*). *Journal of Comparative Psychology*, *112*, 95–99.

Hopkins, W. D., Marino, L., Rilling, J. K., & MacGregor, L. A. (1998). Planum temporale asymmetries in great apes as revealed by magnetic resonance imaging (MRI). *NeuroReport*, *9*, 2913–2918.

Hopkins, W. D., & Nir, T. M. (2010). Planum temporale surface area and grey matter asymmetries in chimpanzees (*Pan troglodytes*): The effect of handedness and comparison with findings in humans. *Behavioral Brain Research*, *208*, 436–443.

Hopkins, W. D., Russell, J. L., & Cantalupo, C. (2007a). Neuroanatomical correlates for tool use in chimpanzees: Implications for theories on the evolution of language. *Psychological Science*, *18*, 971–977.

Hopkins, D., Russell, J. L., Cantalupo, C., Freeman, H., & Schapiro, S. J. (2005). Factors influencing the prevalence and handedness for throwing in captive chimpanzees (*Pan troglodytes*). *Journal of Comparative Psychology*, *119*, 363–370.

Hopkins, W. D., Taglialatela, J. P., & Leavens, D. A. (2007b). Chimpanzees differentially produce novel vocalizations to capture the attention of a human. *Animal Behaviour*, *73*, 281–286.

Horwitz, B., Amunts, K., Bhattacharyya, R., Patkin, D., Jeffries, K., Zilles, K., et al. (2003). Activation of Broca's area during the production of spoken and signed language: A combined cytoarchitectonic mapping and PET analysis. *Neuropsychologia*, *41*, 1868–1876.

Humle, T., & Matsuzawa, T. (2009). Laterality in hand use across four tool-use behaviors among the wild chimpanzees of Bossou, Guinea, West Africa. *American Journal of Primatology*, *71*, 40–48.

James, W. (1890). *Principles of psychology*, (Vol. II). London: MacMillan.

Johnson-Frey, S. H., Funnell, M. G., Gerry, V. E., & Gazzaniga, M. S. (2005). A dissociation between tool use skills and hand dominance: Insights from left- and right-handed callosotomy patients. *Journal of Cognitive Neuroscience*, *17*, 262–272.

Joos, M. (1948). *Acoustic phonetics.* Baltimore, MD: Linguistic Society of America. Language Monograph No. 23.

Jürgens, U. (2002). Neural pathways underlying vocal control. *Neuroscience and Biobehavioral Reviews*, *26*, 235–258.

Kemmerer, D., & Gonzalez-Castillo, J. (2010). The two-level theory of verb meaning: An approach to integrating the semantics of action with the mirror neuron system. *Brain and Language*, *112*, 54–76.

Kennedy, F. (1916). Stock-brainedness, the causative factor in the so-called 'crossed aphasias'. *The American Journal of the Medical Sciences*, *152*, 849–859.

Kim, S.-G., Ashe, J., Hendrich, K., Ellermann, J. M., Merkle, H., Uğurbil, K., et al. (1993). Functional magnetic resonance imaging of motor cortex: Hemispheric asymmetry and handedness. *Science*, *261*, 615–617.

Knecht, S., Dräger, B., Deppe, M., Bobe, L., Lohmann, H., Flöel, A., et al. (2000). Handedness and hemispheric language dominance in healthy humans. *Brain*, *123*, 2512–2518.

Kohler, E., Keysers, C., Umilta, M. A., Fogassi, L., Gallese, V., & Rizzolatti, G. (2002). Hearing sounds, understanding actions: Action representation in mirror neurons. *Science*, *297*, 846–848.

Kuhtz-Buschbeck, J. P., Mahnkopf, C., Holzknecht, C., Siebner, H., Ulmer, S., & Jansen, O. (2003). Effector-independent representations of simple and complex imagined finger movements: A combined fMRI and TMS study. *The European Journal of Neuroscience*, *18*, 3375–3387.

Lausberg, H., Gottert, R., Munssinger, U., Boegner, F., & Marx, P. (1999). Callosal disconnection syndrome in a left-handed patient due to infarction of the total length of the corpus callosum. *Neuropsychologia*, *37*, 253–265.

Levy, J., & Nagylaki, T. (1972). A model for the genetics of handedness. *Genetics*, *72*, 117–128.

Liberman, A. M., Cooper, F. S., Shankweiler, D. P., & Studdert-Kennedy, M. (1967). Perception of the speech code. *Psychological Review*, *74*, 431–461.

Liu, H., Stufflebeam, S. M., Sepulcrea, J., Heddena, T., & Buckner, R. L. (2009). Evidence from intrinsic activity that asymmetry of the human brain is controlled by multiple factors. *Proceedings of the National Academy of Sciences of the United States of America*, *106*, 20499–20503.

Lonsdorf, E. V., & Hopkins, W. D. (2005). Wild chimpanzees show population-level handedness for tool use. *Proceedings of the National Academy of Sciences of the United States of America*, *102*, 12634–12638.

McGrew, W. C., Marchant, L. F., & Hunt, K. D. (2007). Ethoarchaeology of manual laterality: Well digging by wild chimpanzees. *Folia Primatologica*, *78*, 240–244.

McManus, C. (2002). *Right hand, left hand.* London: Weidenfeld and Nicolson.

McNeill, D. (1985). So you think gestures are nonverbal? *Psychological Review*, *92*, 350–371.

McNeill, D. (1992). *Hand and mind: What gestures reveal about thought.* Chicago, IL: University of Chicago Press.

Meador, K. J., Loring, D. W., Lee, K., Hughes, M., Lee, G., Nichols, M., et al. (1999). Cerebral lateralization: Relationship of language and ideomotor praxis. *Neurology*, *53*, 2028–2031.

Meguerditchian, A., Vauclair, J., & Hopkins, W. D. (2010). Captive chimpanzees use their right hand to communicate with each other: Implications for the origin of the cerebral substrate for language. *Cortex*, *46*, 40–48.

Mukamel, R., Ekstrom, A. D., Kaplan, J., Iacoboni, M., & Fried, I. (2010). Single-neuron responses in humans during execution and observation of actions. *Current Biology*, *20*, 750–756.

Palmer, A. R. (2004). Symmetry breaking and the evolution of development. *Science*, *306*, 828–833.

Patterson, F. (1978). Conversations with a gorilla. *National Geographic*, *154*, 438–465.

Penfield, W., & Roberts, L. (1959). *Speech and brain mechanisms.* Princeton, NJ: Princeton University Press.

Peran, P., Demonet, J.-F., Cherubini, A., Carbebat, D., Caltagirone, C., & Sabatini, U. (2010). Mental representations of action: The neural correlates of the verbal and motor components. *Brain Research*, *1328*, 89–103.

Petrides, M., Cadoret, G., & Mackey, S. (2005). Orofacial somatomotor responses in the macaque monkey homologue of Broca's area. *Nature*, *435*, 1325–1328.

Pietrandrea, P. (2002). Iconicity and arbitrariness in Italian Sign Language. *Sign Language Studies*, *2*, 296–321.

Ploog, D. (2002). Is the neural basis of vocalisation different in non-human primates and *Homo sapiens*? In T. J. Crow (Ed.), *The speciation of modern* Homo sapiens (pp. 121–135). Oxford, UK: Oxford University Press.

Pollick, A. S., & de Waal, F. B. M. (2007). Apes gestures and language evolution. *Proceedings of the National Academy of Sciences of the United States of America*, *104*, 8184–8189.

Poremba, A., Mallow, M., Saunders, R. C., Carson, R. E., Herscovitch, P., & Mishkin, M. (2004). Species-specific calls evoke asymmetrical activity in the monkey's temporal poles. *Nature*, *427*, 448–451.

Powell, H. W. R., Parker, G. J. M., Alexander, D. C., Symms, M. R., Boulby, P. A., Wheeler-Kingshott, C. A. M., et al. (2006). Hemispheric asymmetries in language-related pathways: A combined functional MRI and tractography study. *NeuroImage*, *32*, 388–399.

Premack, D. (2007). Human and animal cognition: Continuity and discontinuity. *Proceedings of the National Academy of Sciences of the United States of America*, *104*, 13861–13867.

Pruetz, J. D., & Bertolani, P. (2007). Savanna chimpanzees, *Pan troglodytes verus*, hunt with tools. *Current Biology*, *17*, 412–417.

Pujol, J., Deus, J., Losilla, J. M., & Capdevila, A. (1999). Cerebral lateralization of language in normal left-handed people studied by functional MRI. *Neurology*, *52*, 1038–1043.

Rasmussen, T., & Milner, B. (1977). The role of early left-brain injury in determining lateralization of cerebral speech functions. *Annals of the New York Academy of Sciences*, *299*, 355–369.

Raymer, A. M., Merians, A. S., Adair, J. C., Schwartz, R. L., Williamson, D. J., Rothi, L. J., et al. (1999). Crossed apraxia: Implications for handedness. *Cortex*, *35*, 183–199.

Rizzolatti, G., & Arbib, M. A. (1998). Language within our grasp. *Trends in Neurosciences*, *21*, 188–194.

Rizzolatti, G., Camardi, R., Fogassi, L., Gentilucci, M., Luppino, G., & Matelli, M. (1988). Functional organization of inferior area 6 in the macaque monkey. II. Area F5 and the control of distal movements. *Experimental Brain Research*, *71*, 491–507.

Rizzolatti, G., Fogassi, L., & Gallese, V. (2001). Neurophysiological mechanisms underlying the understanding and imitation of action. *Nature Reviews*, *2*, 661–670.

Rizzolatti, G., & Sinigaglia, C. (2006). *Mirrors in the brain: How our minds share actions and emotions.* Oxford: Oxford University Press.

Rizzolatti, G., & Sinigaglia, C. (2010). The functional role of the parieto-frontal mirror circuit: Interpretations and misinterpretations. *Nature Reviews Neuroscience, 11*, 264–274.

Rogers, L. J. (2004). A review of *The Speciation of Modern Homo Sapiens* by Tim Crow. *Laterality, 9*, 225–232.

Rogers, L. J., & Andrew, R. J. (2002). *Comparative vertebrate lateralization.* Cambridge, UK: Cambridge University Press.

Rousseau, J.-J. (1781). *Essai sur l'Origine des Langues.* Paris: A. Belin.

Roy, A. C., & Arbib, M. A. (2005). The syntactic motor system. *Gesture, 5*, 7–37.

Savage-Rumbaugh, S., Shanker, S. G., & Taylor, T. J. (1998). *Apes, language, and the human mind.* Oxford: Oxford University Press.

Slocombe, K. E., Kaller, T., Call, J., & Zuberbühler, K. (2010). Chimpanzees extract social information from agonistic screams. *PloS One, 5*(7), e11473.

Spocter, M. A., Hopkins, W. D., Garrison, A. R., Bauernfeind, A. L., Stimpson, C. D., Hof, P. R., et al. (2010). Wernicke's area homologue in chimpanzees (*Pan troglodytes*) and its relation to the appearance of modern human language. *Proceedings of the Royal Society B: Biological Sciences, 277*, 2165–2174.

Studdert-Kennedy, M. (1998). The particulate origins of language generativity: From syllable to gesture. In J. R. Hurford, M. Studdert-Kennedy & C. Knight (Eds.), *Approaches to the evolution of language* (pp. 169–176). Cambridge, UK: Cambridge University Press.

Studdert-Kennedy, M. (2005). How did language go discrete? In M. Tallerman (Ed.), *Language origins: Perspectives on evolution* (pp. 48–67). Oxford, UK: Oxford University Press.

Suddendorf, T., & Corballis, M. C. (2007). The evolution of foresight: What is mental time travel, and is it unique to humans? *The Behavioral and Brain Sciences, 30*, 299–351.

Tomasello, M. (2008). *The origins of human communication.* Cambridge, MA: MIT Press.

Umiltà, M. A., Kohler, E., Gallese, V., Fogassi, L., Fadiga, L., Keysers, C., et al. (2001). I know what you are doing. A neurophysiological study. *Neuron, 31*, 155–165.

Valenstein, E., & Heilman, K. M. (1979). Apraxic agraphia with neglect-induced paragraphia. *Archives of Neurology, 36*, 506–508.

Vallortigara, G., & Rogers, L. J. (2005). Survival with an asymmetrical brain: Advantages and disadvantages of cerebral lateralization. *The Behavioral and Brain Sciences, 28*, 575–633.

Vico, G. (1744). *La scienza nuova.* Naples: Stamperia Muziana (1954 edition by F. Nicolini, Giambattista Vico – Opere. Verona: Stamperia Valdonega).

Wada, J. A., Clarke, R., & Hamm, A. (1975). Cerebral hemispheric asymmetry in humans: Cortical speech zones in 100 adult and 100 infant brains. *Archives of Neurology, 32*, 239–246.

Warrington, E. K., & Pratt, R. T. C. (1973). Language laterality in left handers assessed by unilateral ECT. *Neuropsychologia, 11*, 423–428.

Whitehouse, A. J. O., & Bishop, D. V. M. (2009). Hemispheric division of function is the result of independent probabilistic biases. *Neuropsychologia, 47*, 1938–1943.

Wundt, W. (1900). *Die Sprache.* Leipzig: Engelmann (2 vols).

Xu, J., Gannon, P. J., Emmorey, K., Smith, J. F., & Braun, A. R. (2009). Symbolic gestures and spoken language are processed by a common neural system. *Proceedings of the National Academy of Sciences of the United States of America, 106*, 20665–20669.

M. A. Hofman and D. Falk (Eds.)
Progress in Brain Research, Vol. 195
ISSN: 0079-6123

CHAPTER 7

The insular cortex: A review

Rudolf Nieuwenhuys*

Netherlands Institute for Neuroscience, Amsterdam, The Netherlands

Abstract: The human insular cortex forms a distinct, but entirely hidden lobe, situated in the depth of the Sylvian fissure. Here, we first review the recent literature on the connectivity and the functions of this structure. It appears that this small lobe, taking up less than 2% of the total cortical surface area, receives afferents from some sensory thalamic nuclei, is (mostly reciprocally) connected with the amygdala and with many limbic and association cortical areas, and is implicated in an astonishingly large number of widely different functions, ranging from pain perception and speech production to the processing of social emotions. Next, we embark on a long, adventurous journey through the voluminous literature on the structural organization of the insular cortex. This journey yielded the following take-home messages: (1) The meticulous, but mostly neglected publications of Rose (1928) and Brockhaus (1940) are still invaluable for our understanding of the architecture of the mammalian insular cortex. (2) The relation of the insular cortex to the adjacent claustrum is neither ontogenetical nor functional, but purely topographical. (3) The insular cortex has passed through a spectacular progressive differentiation during hominoid evolution, but the assumption of Craig (2009) that the human anterior insula has no homologue in the rhesus monkey is untenable. (4) The concept of Mesulam and Mufson (1985), that the primate insula is essentially composed of three concentrically arranged zones, agranular, dysgranular, and granular, is presumably correct, but there is at present much confusion concerning the more detailed architecture of the anterior insular cortex. (5) The large spindle-shaped cells in the fifth layer of the insular cortex, currently known as von Economo neurons (VENs), are not only confined to large-brained mammals, such as whales, elephants, apes, and humans, but also occur in monkeys and prosimians, as well as in the pygmy hippopotamus, the Atlantic walrus, and Florida manatee. Finally, we point out that the human insula presents a unique opportunity for performing an in-depth comparative analysis of the relations between structure and function in a typical sensory and a typical cognitive cortical domain.

Keywords: agranular zone; awareness; cognitive functions; cytoarchitecture; dysgranular zone; granular zone; insula; insular cortex; spindle cells; von Economo neurons.

*Corresponding author.
Tel.: +31-20-5665500
E-mail: RudolfN@planet.nl

DOI: 10.1016/B978-0-444-53860-4.00007-6

Introduction

The human insular cortex forms a distinct, but entirely hidden cerebral lobe, situated in the depth of the Sylvian fissure. The primordial insula is initially located on the free lateral surface of the cerebral hemisphere, but during further development, it lags behind and becomes gradually overgrown by adjacent regions of the hemispheres. As regards the structure of the insular cortex, Mesulam and Mufson (1985), Mufson et al. (1997), and Bonthius et al. (2005) recognized three concentrically arranged zones, a rostroventral agranular zone, a caudodorsal granular zone, and a wide, intermediate dysgranular (Ig) zone. The terms agranular and granular refer to the absence or presence of an internal granular layer (IV). The intermediate zone is termed dysgranular because granule cells are rather scarce in layer IV and do not display complete laminar differentiation. A special feature of the anterior insular cortex (AIC) is that its fifth layer contains, in addition to pyramidal neurons, numerous large spindle-shaped cells. Similar cells have been observed in the anterior cingulate cortex (ACC; von Economo, 1926). These distinctive elements, which have been termed: "von Economo neurons" (VENs; Allman et al., 2005; Ngowyang, 1932), have attracted much attention in the recent literature. It was initially believed that these cells are unique to humans and great apes (Allman et al., 2005; Nimchinsky et al., 1999), but it has been reported meanwhile that they also occur in whales (Hof and van der Gucht, 2007) and elephants (Hakeem et al., 2009). VENs are hypothesized to form part of circuits underlying decision making, complex social cognition, and self-awareness (Allman et al., 2005, 2010, 2011a,b).

Neuroanatomical studies

Experimental neuroanatomical studies in monkeys, mostly rhesus macaques, summarized by Mesulam and Mufson (1985), Mufson et al. (1997), and Augustine (1996), have shown that the insular cortex receives afferents from the dorsal thalamus and from several sensory cortical areas, is reciprocally connected with the amygdala as well as with several limbic and association cortical areas, and projects to the premotor cortex and the ventral striatum. Moreover, there is an abundance of local intrainsular connections (Friedman et al., 1986).

The thalamic nuclei which project to the insula include the ventral posterior superior (VPS) nucleus and the ventral posterior inferior (VPI) nucleus, the ventromedial posterior (VMPo) nucleus, and the parvocellular part of the ventral posteromedial (VPMpc) nucleus.

The VPS and VPI nuclei form both part of the shell surrounding the ventral posterior complex. They receive afferents from the vestibular nuclear complex and project to several cortical areas, including an area in the posterosuperior part of the insula and adjacent operculum known as the *parietoinsular vestibular cortex* (*PIVC*; Büttner-Ennever and Gerrits, 2004).

The VMPo nucleus is in receipt of nociceptive and thermoreceptive spinothalamic lamina I neurons (Craig et al., 1994) and projects to the posterosuperior part of the insular cortex (Brooks et al., 2005). This projection area may be designated as the *insular nociceptive and thermoreceptive cortex* (*INTC*; Fig. 1a). The parvocellular part of the VPMpc nucleus has different medial and lateral sectors, with different fiber connections. The medial sector (VPMpc, med) receives gustatory projections from the most rostral part of the nucleus of the solitary tract and sends its efferents to the granular, anterosuperior part of the insula, and the adjacent portion of the frontal operculum. This area, which represents the *primary gustatory cortex* (*GI*), projects in its turn to a more basally situated, dysgranular insular area, which hence, may be designated as the *secondary gustatory cortex* (*GII*; Pritchard and Norgren, 2004; Fig. 1a). The lateral sector of the parvocellular part of VPM (VPMpc, lat) receives general visceral information from the caudal part of the nucleus of the

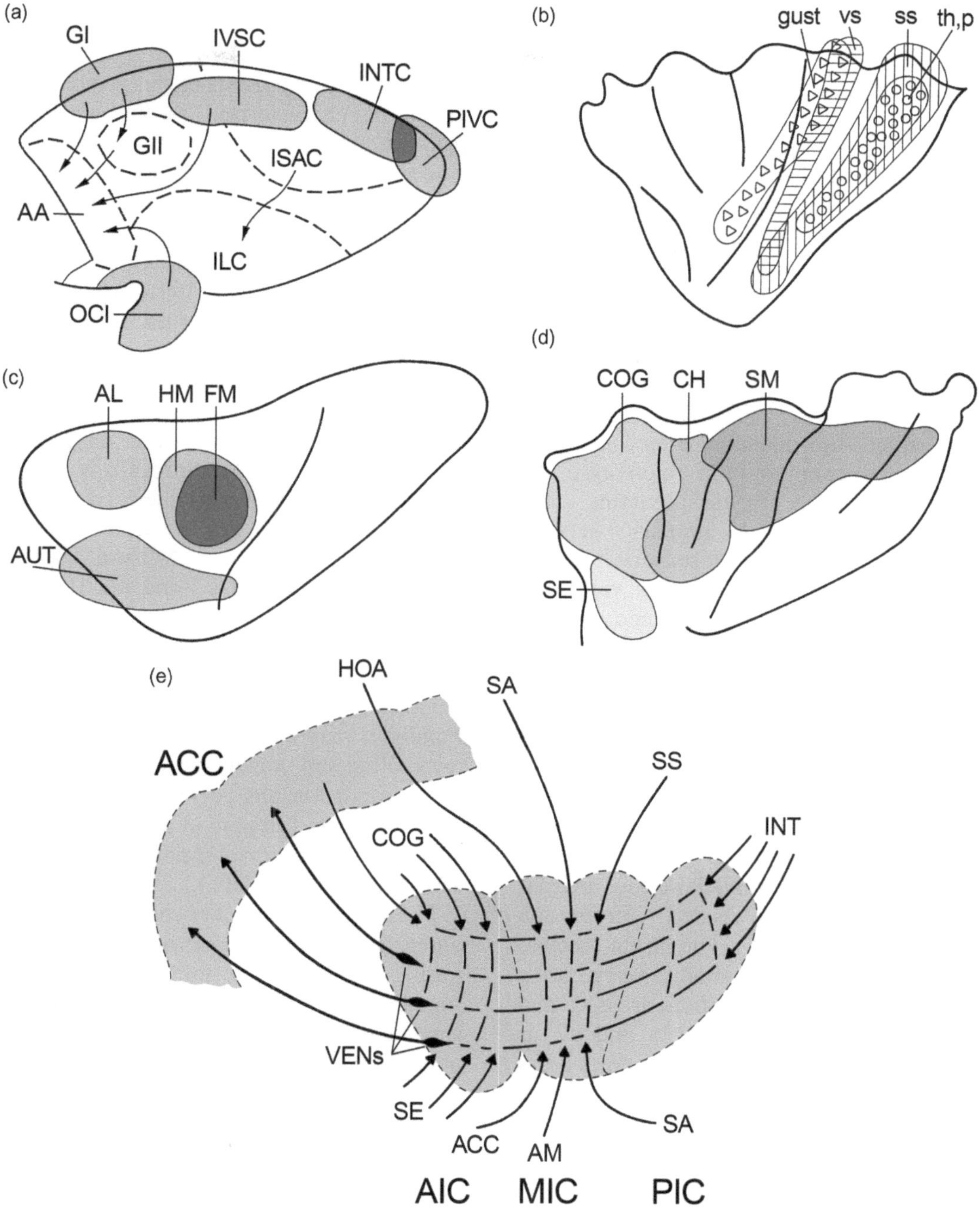

Fig. 1. Functional subdivision of the insular cortex. (a) The approximate position of functional areas and the course of some fiber connections in the insula of the rhesus monkey. Primary sensory areas are shaded (Nieuwenhuys et al., 2008). AA, agranular anterior zone; GI, GII, primary and secondary gustatory areas; ILC, insular limbic cortex; INTC, insular nociceptive and thermoreceptive cortex; ISAC, insular somatic association cortex; IVSC, insular viscerosensory cortex; OCI, primary olfactory cortex. (b) Sensory areas detected in the human insula by electrocortical stimulation (Stephani et al., 2011). Gust, gustation; th, thermosensation; p, pain; ss, somatosensation; vs, viscerosensation. (c) Organization of the human anterior insular cortex, as revealed by a meta-analysis of neuroimaging studies (Mutschler et al., 2009). Localization of responses related to auditory and

solitary tract, which attains the nucleus via a synaptic relay in the external medial parabrachial nucleus. VPMpc, lat projects to an insular area located directly posterior to the gustatory areas. This area, which is known as the *insular viscerosensory cortex* (*IVSC*; Fig. 1a), shows an organotopical ordering. Physiological mapping experiments revealed that neurons responding to gastrointestinal sensations are located in its anterior part, whereas neurons responding to cardiovascular and respiratory afferents are located more posteriorly (Saper, 2002). If we survey the data concerning the thalamic projections just discussed, it appears that in monkeys, the superior tier of the insula contains four anteroposteriorly arranged primary sensory areas; gustatory, general viscerosensory, somatosensory (pain and temperature), and vestibular (Fig. 1a). The insula is, as mentioned, also in receipt of sensory projections from other cortical areas. Fibers originating from the olfactory prepiriform cortex project to an agranular anterior (AA) zone of the insula. This zone also receives afferents from the primary and secondary gustatory cortices and from the primary viscerosensory insular cortex (Fig. 1a). It participates with the caudal orbitofrontal cortex in the formation of an "orbital network," involved in the analysis and integration of food-related information (Carmichael and Price, 1996). The insula receives afferents from the primary somatosensory cortex SI, the somatosensory association areas 5 and 7b, the primary vestibular areas 3a and 2v, as well as from the auditory association areas surrounding the primary auditory cortex. All of these sensory cortical areas project to the posterosuperior portion of the insula, which may hence be characterized as the *insular somatic association cortex* (*ISAC*). Some high-order association areas, including the anterior orbitofrontal cortex (area 11), the prefrontal cortex (areas 45, 46), and the polymodal association cortex occupying the banks of the superior temporal sulcus (Seltzer and Pandya, 1991), are also known to be connected with the insula. Limbic structures, including the entorhinal (area 28), perirhinal (areas 35, 36), posterior orbitofrontal (areas 13, 14), temporopolar (area 38), and cingulate (areas 23, 24) cortices, as well as the amygdaloid complex, are strongly and reciprocally connected with the anterobasal sector of the insula. I designate this sector as the *insular limbic cortex* (*ILC*; Fig.1a). Mesulam and Mufson (1985) have suggested that the ISAC and the ILC represent way stations in a somatolimbic projection (Fig. 1a), and that this projection may provide a means for interrelating events in the extrapersonal world with relevant motivational states. Finally, it may be mentioned that the insula projects to the supplementary and presupplementary motor areas forming part of the medial premotor cortex (area 6; Luppino et al., 1993), as well as to the ventral and dorsal striatum (Chikama et al., 1997; Fudge et al., 2005).

Before leaving the fiber connections, and before entering into a discussion of the functions of the insula, it should be mentioned that complex functions at cortical levels are never related to single centers or circuits, but rather to distributed neural networks. The spots or patches of cortical activity, visualized in neuroimaging studies, figure as nodes or humps in these networks.

language processing (AL), hand and foot movements (HM, FM), and peripheral autonomic changes related to emotional processes. (d) Functionally distinct regions in the human insula, delineated on the basis of another meta-analysis of functional neuroimaging experiments (Kurth et al., 2010b). Separate chemical sensory (CH), cognitive (COG), social-emotional (SE), and sensorimotor (SM) domains could be distinguished. N.B: The figure actually shows the results obtained in the right insula but has been reversed in order to facilitate comparison with (b) and (c). (e) Craigs' (2009) concept of the processing of information in the insular cortex. The posterior (PIC), middle (MIC), and anterior insular cortex (AIC) represent three consecutive stages or levels in an anteriorly directed processing stream. AIC is strongly and reciprocally connected with the anterior cingulate cortex (ACC). AM, amygdala; COG, cognitive networks; HOA, high-order association cortices; INT, interoceptive thalamic afferents; SE, social-emotional networks; SS, somatosensory cortices; SA, sensory association cortices; VENs, von Economo neurons.

Noninvading techniques, such as diffusion tensor imaging and resting-state functional magnetic resonance imaging (fMRI), render it possible to study these networks *in vivo*. It is important to note that one and the same locus may be activated during different tasks, and that different functional networks may share one or several humps. For a specification of the term network, see Cole and Schneider (2007, p. 356).

Functional studies

The insular cortex was long regarded simply as a viscerosensory and visceromotor region, based on the studies of Penfield and Rasmussen (1950) and Penfield and Faulk (1955). These authors performed electrical stimulation of the insula during presurgical evaluations of a large number of temporal lobe epilepsy patients. During the past decades of the twentieth century, our insights into the functional significance of the insula rapidly progressed, enabling Augustine (1996) to list in an authoritative review no less than 10 different insular functions. These included, apart from viscerosensory and visceromotor functions, vestibular, somatosensory, somatomotor, motor association, limbic integration, and language functions. Recent clinicopathological studies (reviewed in Flynn, 1999; Ibanez et al., 2010), intraoperative electrical stimulation studies in epilepsy patients (Afif et al., 2010; Mazzola et al., 2006; Ostrowsky et al., 2002; Stephani et al., 2011), and an avalanche of fMRI and positron emission tomography (PET) studies (summarized by Craig, 2009; Kurth et al., 2010b; Mutschler et al., 2009) have greatly increased our insights into the functional significance of the insular cortex. Recently, much of our knowledge on the structure, function, and pathology of the insula was compiled in a special issue of Brain Structure and Function (Craig, 2010a). It is no exaggeration to say that the insula is a focus of attention and even a "hot spot" of current neuroscience. The following summary of the functions and dysfunctions of the insula is an update and extension of a previous survey (Nieuwenhuys et al., 2008).

1. *Auditory functions*: Connectivity studies (see above) and single-cell recordings in monkeys (Bieser, 1998) indicate that the posterior insula is involved in auditory processing. Further, Bamiou et al. (2003) reviewed clinical data showing that bilateral damage of the insula may lead to auditory agnosia, and functional imaging studies demonstrating that the insular cortex participates in several key auditory processes, such as tuning into novel auditory stimuli and allocating auditory attention.
2. *Vestibular functions*: Neurons in the PIVC of the rhesus monkey respond to vestibular stimuli (Grusser et al., 1990a). However, most of these elements also responded to somatosensory and visual information and were hence, classified as polymodal vestibular units (Grusser et al., 1990b). Caloric stimulation combined with fMRI revealed in humans a vestibular cortical network with right hemispheric dominance, comprising, among many other cortical areas, the posterior as well as the anterior portions of the insula (Fasold et al., 2002).
3. *Somatosensory functions*: Physiological studies in rhesus monkeys (Robinson and Burton, 1980a,b; Schneider et al., 1993) have shown that neurons in the granular part of the insula respond to innocuous cutaneous stimuli. An early PET study (Burton et al., 1993) revealed that the human insula can be activated by vibrotactile stimulation. Further, in humans, somatosensory activation of the posterior insula was shown by direct electrical stimulation in patients undergoing surgery for temporal lobe epilepsy (Ostrowsky et al., 2002; Stephani et al., 2011).
4. *Pain and temperature perception*: The involvement of the insula in protopathic sensibility is documented by clinical evidence. Thus,

Biemond (1956) described a patient in whom a lesion involving the insula and SII was associated with a dramatic loss of pain perception, and Birklein et al. (2005) reported that isolated insular infarction may lead to contralateral elimination of cold, cold pain, and pinprick perception. Neuroimaging studies and preoperative electrical stimulation studies in epilepsy patients have also demonstrated the responsiveness of the insular cortex to painful stimulation, but there is no unanimity with regard to the exact localization of the pain-related activity. Thus, some authors, including Ostrowsky et al. (2002) and Alkire et al. (2004), reported a focus of pain-related activity in the dorsal posterior insula, whereas others, including Brooks et al. (2005) and Afif et al. (2010), found it rather in the dorsal part of the mid-insula. Isnard et al. (2011) analyzed brain activity in a patient during spontaneously painful epileptic seizures. They found that the attacks originated from a very limited locus in the posterior third of the right insula and propagated from there to other pain-related cortical areas, such as the parietal operculum and the anterior cingulate gyrus. Finally, Hua et al. (2005), using fMRI, demonstrated that in humans, innocuous thermal stimuli activated the dorsal posterior insular cortex, and that the foci were arranged in an anteroposterior somatotopic pattern.

5. *Viscerosensation*: We have seen that the posterior insular cortex receives a projection from the viscera, which is relayed in the nucleus of the solitary tract, the parabrachial complex, and the thalamus. The classical experiments of Penfield and Faulk (1955) have shown that electrical stimulation of the insula in humans produces nausea and a variety of gastric and abdominal sensations. Stephani et al. (2011) who, just like Penfield and Faulk, performed preoperative electrical stimulation of the insula in epilepsy patients, found that viscerosensory phenomena could be exclusively elicited from an elongated zone, situated directly posteriorly to the central insular sulcus (Fig. 1b).
6. *Taste*: The results of experimental neuroanatomical studies (see above), and of stimulation and ablation experiments in monkeys (Bagshaw and Pribram, 1953; Sudakov et al., 1971), point to the presence of a gustatory center in the anterior insula (Fig. 1b). The detection of taste-sensitive neurons in the same area further documents the presence of this center (Yaxley et al., 1990). Recent evidence from neuroimaging studies (Small, 2010), and from intracerebral electric stimulations in epilepsy patients (Stephani et al., 2011), indicate the presence of a comparable gustatory center in the human insula. However, it appeared that this center is located in the central parts of the insula, that is, considerably further caudally than in the monkey (Fig. 1a and b). This remarkable interspecies difference in the localization of the primary gustatory cortex will be further discussed below. Verhagen et al. (2004) reported that, apart from numerous taste-sensitive neurons, the primary gustatory cortex in the rhesus monkey also contains elements responding to nontaste properties of oral stimuli, related to the texture (viscosity, grittiness) or temperature of food. Neurons responding to combinations of these inputs also appeared to be present. These findings further substantiate the concept that the anterior insula and the adjacent caudal orbitofrontal cortex are involved in the analysis and integration of food-related information.
7. *Olfaction*: Electrical stimulation of the anterior insula in humans may lead to olfactory sensations (Penfield and Faulk, 1955), and a recent meta-analysis of neuroimaging data (Kurth et al., 2010b) revealed that olfactory stimulation consistently activates an area in the posterocentral part of the right anterior insula.

8. *Visceromotor control*: Electrical stimulation of the insular region in humans may elicit a variety of visceromotor phenomena, including vomiting and other alterations of the gastrointestinal tract, respiratory arrest, as well as changes in cardiac rate and rhythm control, and blood pressure (Oppenheimer, 1994, 2006; Penfield and Faulk, 1955). Ischemic strokes involving the insula are frequently accompanied by cardiac arrhythmias and other electrocardiographic abnormalities (Tatschl et al., 2006). These findings indicate that the insula plays a role in autonomic regulation. It is remarkable that these heart dysfunctions occur much more frequently after right-sided strokes than after left-sided ones (Colivicchi et al., 2004; Meyer et al., 2004; Oppenheimer, 2006). These findings suggest that the autonomic control of cardiac activity is lateralized and is mediated by the right-sided insular cortex.
9. *Somatomotor control*: A meta-analysis of functional neuroimaging data (Mutschler et al., 2009) showed that hand and foot motor tasks activate the posterosuperior part of the insular cortex, just in front of the central insular sulcus, which divides the insula in anterior and posterior lobules (Fig. 1c). In another recent meta-analysis of fMRI data (Kurth et al., 2010b), the area activated by motor tasks was situated more posteriorly, and overlapped with the representation of somatosensory stimuli (Fig. 1d). Afif et al. (2010), who explored the human insular cortex, using intracerebral electrical stimulation, also reported an overlap of the areas from which somatosensory and somatomotor responses could be elicited. According to their mappings, this "sensorimotor" area is situated in the anterosuperior part of the posterior insular lobule.
10. *Motor plasticity*: There is clinical evidence that the insular cortex has a role in poststroke recovery of motor function. Weiller et al. (1992) investigated patients with complete recovery of upper limb function from striatocapsular motor strokes. PET imaging revealed that during the performance of motor tasks by the recovered hand, there was significantly greater activation than in control subjects in both anterior insulae, and in several other structures involved in motor control, among which the prefrontal and cingulate cortices on both sides, the ipsilateral premotor cortex and basal ganglia, and the contralateral cerebellum.
11. *Speech production*: The sensory speech area of Wernicke and the motor speech area of Broca are interconnected by a large fiber system, known as the arcuate fasciculus. Some of the fibers of this fasciculus reach Broca's area, by way of the capsula extrema, passing directly beneath the insular cortex. On account of these anatomical relations, it has long been thought that speech disturbances following ischaemic strokes in the insular region are due to interruption of the arcuate fasciculus, rather than to damage of the insular cortex *per se* (Damasio and Damasio, 1980). However, a study involving computerized lesion reconstruction in a series of aphasia patients (Dronkers, 1996) revealed that in patients with a disorder in coordinating the movements for speech articulation, an area situated in the anterosuperior insula was specifically damaged. Functional imaging studies on the cerebral correlates of language production (Bohland and Guenther, 2006) also suggest that the anterior insula forms part of the brain network of speech motor control. Moreover, Afif et al. (2010) evoked speech disturbances, such as speech arrest and episodes of reduced voice intensity, by electrical stimulation of the posterior third of the anterior insular lobule. Nevertheless, several recent reviewers (Ackermann and Riecker, 2010; Jones et al., 2010) maintain that the specific role of the insula in speech production still requires further clarification.

12. *Cognitive control*: Cole and Schneider (2007) review functional neuromaging studies, showing that the AIC is involved in a large variety of cognitive control tasks. They adduce evidence, indicating that this insular region forms, together with other cortical regions, among which the ACC, the pre-supplementary motor area, the dorsal premotor cortex, the dorsolateral prefrontal cortex, and the posterior parietal cortex, a highly interconnected cognitive control network.
13. *Bodily awareness*: Some patients with hemiplegia after right-sided strokes deny their paralysis (anosognosia for hemiplegia) and are convinced that their paralyzed limbs function normally. They may experience their limb(s) as not belonging to them and may even attribute them to other persons. Lesion analyses with modern imaging techniques revealed that in such patients, the disturbances are consistently associated with damage to the right insula. These findings suggest a prominent role of the right insula in our sense of limb ownership as well as in our self-awareness of actions. Hence, Karnath and Baier (2010), from whom the data just cited are derived, hypothesized that the right insular cortex forms a central node of the network involved in human body scheme representation.
14. *Self-recognition*: Craig (2009) cites neuroimaging studies, showing that the act of seeing and recognizing one's own image produces activity in the AIC.
15. *Individual emotions*: Neuroimaging studies summarized by Phan et al. (2002) strongly suggest that the anterior insula is preferentially involved in the evaluative, experiential, and expressive aspects of specific individual emotions, such as happiness, sadness, fear, and disgust.
16. *Social emotions*: Social species are so characterized because they form organizations that extend beyond the individual. Social neuroscience is a new and rapidly expanding field of research, aimed at investigating the neural structures and processes underlying social interactions and behavior (Cacioppo and Decety, 2011). Recent investigations in the field of social neuroscience, reviewed by Lamm and Singer (2010), indicate that the anterior insula is consistently involved in social emotions, including empathy and compassion, as well as in interpersonal phenomena such as fairness and cooperation.
17. *Schizophrenia*: Studies using magnetic resonance morphometry (Crespo-Facorro et al., 2000; Takahashi et al., 2004, 2005) revealed morphological abnormalities of reduced cortical surface area and gray matter volume in the insula of schizophrenic patients.
18. *Conduct disorder*: Conduct disorder (CD) is characterized by repetitive and chronic aggressive and antisocial behavior in which the basic rights of others or major age-appropriate norms or rules of society are violated and that has a variety of implications such as school refusal, social communication problems, and legal involvement (American Psychiatric Association, 1994). Sterzer et al. (2007), using an fMRI-based morphometry protocol, demonstrated that gray matter volume in bilateral AIC and left amygdala was significantly reduced in CD patients, compared to healthy control subjects. According to the authors mentioned, these findings suggest a critical role for the AIC in regulating social behavior.
19. *Frontotemporal dementia*: Frontotemporal dementia (FTD) refers to a family of clinical syndromes caused by frontotemporal lobar degeneration (Seeley, 2008). Clinical subtypes of this form of dementia include progressive nonfluent aphasia (PNFA) and a behavioral variant (bvFTD; Seeley, 2010). PNFA is associated with effortful, nonfluent, often agramatical speech, whereas bvFTD is characterized by dramatic changes in social-emotional processing. Seeley (2010) reported

that in both of these disorders, the degeneration clearly involves the anterior insula. In cases of bvFTD, the VENs appeared to be specifically affected.

20. *Drug addiction*: Functional imaging studies, summarized by Naqvi and Bechara (2009, 2010), have demonstrated that activity in the anterior insula is correlated with conscious urges to take drugs.

This list, which is by no means exhaustive, may suffice to show that the human insula, that is, a small and entirely hidden patch of cortex, taking up less than 2% of the total cortical surface area, is implicated in an astoundingly large number of different functions and dysfunctions.

Current research is not only concerned with the individual functions and dysfunctions of the insular cortex but also focused on the regional functional organization within the insula, and on the putative functional significance of the insula as a whole. As for regional functional organization, the results of the recent studies of Stephani et al. (2011), Mutschler et al. (2009), and Kurth et al. (2010b) should be especially mentioned.

Stephani et al. (2011) subjected epilepsy patients to an invasive presurgical evaluation of the insula with implanted depth electrodes. They found somatosensory symptoms to be restricted to the posterior insula and a subgroup of painful or warmth sensations in the dorsal posterior insula. General viscerosensory symptoms were elicited by more anterior electrode contacts, whereas gustatory symptoms could be evoked by stimulation of a still more anteriorly situated zone, which ventrally extended into the anterior insular lobule (Fig. 1b). Mutschler et al. (2009) explored the functional organization of the human AIC, on the basis of a meta-analysis of previously published neuroimaging studies reporting insula effects. They found auditory and language tasks to preferentially activate an area in the dorsal part of the anterior insula. Motor tasks involving both the upper and the lower extremity appeared to activate a posterior-anterior insula region, adjacent to the sulcus centralis insulae. The ventral part of the anterior insula was activated by peripheral physiological changes such as measured by cardiovascular activity or sympathetic skin responses. The amygdaloid complex was significantly coactivated during these changes (Fig. 1c). Kurth et al. (2010b) also performed a meta-analysis of fMRI experiments. Their analysis revealed four functionally distinct regions in the human insula, which they designated as sensorimotor, chemical sensory (olfaction and taste), cognitive, and social-emotional (Fig. 1d). Sensorimotor tasks activated the mid-posterior insular region, and olfacto-gustatory stimuli activated the central insula. Cognitive tasks, related to attentional processes, language processing, active speech, and memory retrieval, activated the anterodorsal insular region, whereas the anteroventral insular region was activated by personal emotional and social-emotional tasks.

As regards the putative functional significance of the insular cortex as a whole, Craig (2009, 2010b, 2011) has recently made an attempt at synthesizing the enormous amount of experimental neuroanatomical, physiological, clinical, and neuroimaging data available on the insula, into a coherent picture of human insular functioning (Fig. 1e). His concept may be summarized as follows: (1) The posterior, middle, and anterior sectors of the insula represent three different stages or levels of integration in a posteriorly-to-anteriorly directed processing stream. (2) Afferents from the solitary nucleus, and from the phylogenetically new pathway from lamina I spinal neurons, converge upon the posterior insula, where they provide a primary interoceptive representation of the physiological condition of the body. (3) The mid-insula is to be considered as a polymodal integrative zone, where the interoceptive information from the posterior insula is rerepresented and associated with inputs from multiple other sources. Prominent among these are higher-order sensory cortices, providing emotionally salient information from the external world, and the cingulate cortex and amygdala, providing homeostatic information related to the current motivational state. (4) The information thus gathered is conveyed to the anterior insula and there once again integrated with novel types of information from many sources. Imaging studies have shown that the anterior insula is involved in an astonishing number of activities, associating this sector with attention, vocalization, and music, cognitive control, perceptual decision making, self-recognition, time perception, and emotional awareness. The unique convergence of so much information, so many activities, and so many tasks leads to the proposition that the AIC contains the anatomical substrate for the evolved capacity of humans to be aware of themselves, others, and the environment. In brief: *The AIC fulfills the requirements to represent the neural correlate of awareness.* (5) Neuroimaging studies have shown that the AIC and the ACC are often jointly activated in subjects experiencing emotional feelings. Because the ACC underpins motivations and behavior, it seems likely that the AIC and ACC serve as complementary limbic sensory and motor systems that work together in a way similar to the somatosensory and somatomotor cortices. (6) The AIC and ACC, including the VONs they contain, are salient components of a core control network that guides all mental activity and behavior in adult humans. (7) The human AIC has no clear homologue in the rhesus monkey.

Evidence in support of Craigs' concept has been recently published by Cole and Schneider (2007), Menon and Uddin (2010), Medford, and Critchley (2010), and Jones et al. (2010).

During the first half of the twentieth century, a number of important studies on the insula have appeared. Cajal (1900, 1911), Brodmann (1909), von Economo and Koskinas (1925), Rose (1935), and Bailey and von Bonin (1951) included descriptions of the insular region in their general treatises on the structure of the cerebral cortex. Moreover, Rose (1928) and Brockhaus (1940), both working in the institute of Cécile and Oskar Vogt, produced very detailed descriptions of the cytoarchitecture (and in Brockhaus' case also of the myeloarchitecture) of the human insular cortex. Two of the authors mentioned, viz. Brodmann (1909) and Rose (1928), not only confined themselves to the human insula but also included a number of other mammalian species in their analyses. The primary aim of the present chapter is to highlight the principal results of the classical studies just listed and to specify their bearing on our current insights in the structural and functional organization of the insular cortex. The plan of the remainder of this chapter is as follows: Following a brief description of the gross morphology of the human insula, there are three sections dealing consecutively with: the cytoarchitecture of the human insula, the comparative anatomy of the insular cortex, and the occurrence of special neurons in the insular cortex. In each of these sections, the presentation of data is followed by a shorter or longer commentary. The chapter concludes with, a synopsis of our current state of knowledge of the architecture of the human insula.

Gross morphology of the human insula

The human insula or island of Reil forms a distinct cerebral lobe, which is buried in the depth of the sulcus lateralis. It is covered by adjacent parts of the frontal, parietal, and temporal lobes, known as the orbitofrontal, frontoparietal, and temporal opercula (Fig. 2a). The insula is shaped like a triangle, the apex of which is directed anterobasally. A deep sulcus circularis separates the insular lobe from the surrounding opercula. Owing to the triangular shape of the insula, separate anterior, superior, and inferior parts can be distinguished on the sulcus circularis (Fig. 2a). A distinct sulcus centralis insulae, passing obliquely from posterosuperior to anterobasal, divide the insula into a larger lobulus anterior and a smaller lobulus posterior. Several authors, including Retzius (1896, 1902), Brodmann (1909), von Economo and Koskinas (1925), and Afif and Mertens (2010), have indicated that the sulcus centralis insulae represent a continuation of the sulcus centralis Rolando on the dorsolateral convexity of the cerebral hemisphere. The lobulus

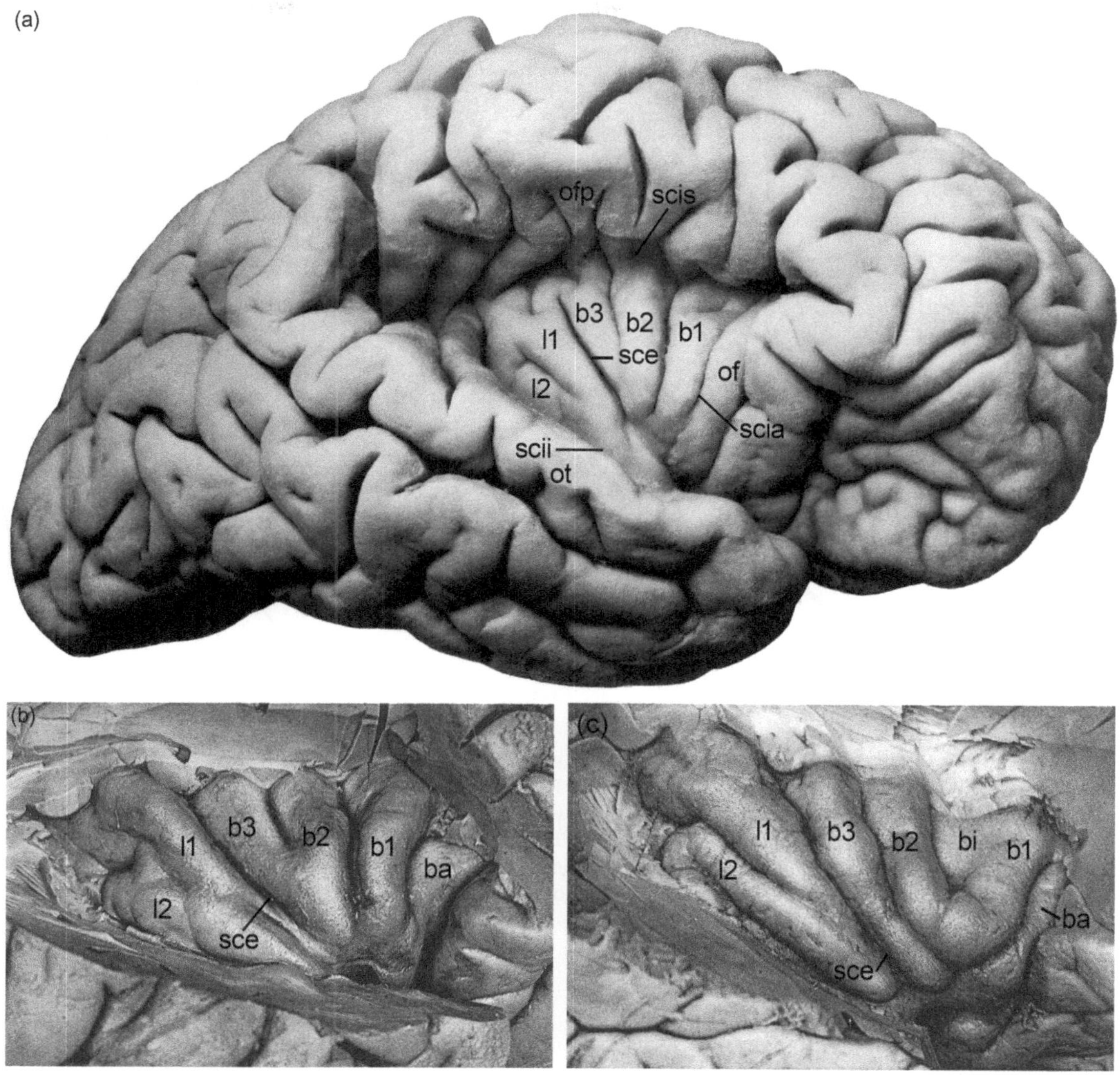

Fig. 2. The human insula. In (a), the insula is exposed, by pulling apart the opercula; in (b) and (c), the opercula are resected. (a) is reproduced from Retzius (1896); (b) and (c) are reproduced from Retzius (1902). ba, gyrus brevis accessorius; bi, gyrus brevis intermedius; b1, b2, b3, gyrus brevis primus -secundus, -tertius; l, gyrus longus; I1, I2, gyrus longus primus, -secundus; of, operculum orbitofrontale; ofp, operculum frontoparietale; ot, operculum temporale; sce, sulcus centralis insulae; scia, scii, scis, sulcus circularis insulae, pars anterior, -pars inferior, -pars superior.

posterior is in some cases represented by a single gyrus, the gyrus longus (Fig. 3b), but it is generally incompletely divided into two gyri, known as the gyrus longus primus and the gyrus longus secundus (Figs. 2 and 3a and c). The lobulus anterior is commonly composed of three short gyri, the gyrus brevis primus, -secundus, and -tertius, which converge toward the apex (Fig. 2a and b). Occasionally, a smaller or larger gyrus brevis intermedius is interposed between the first and second short gyrus

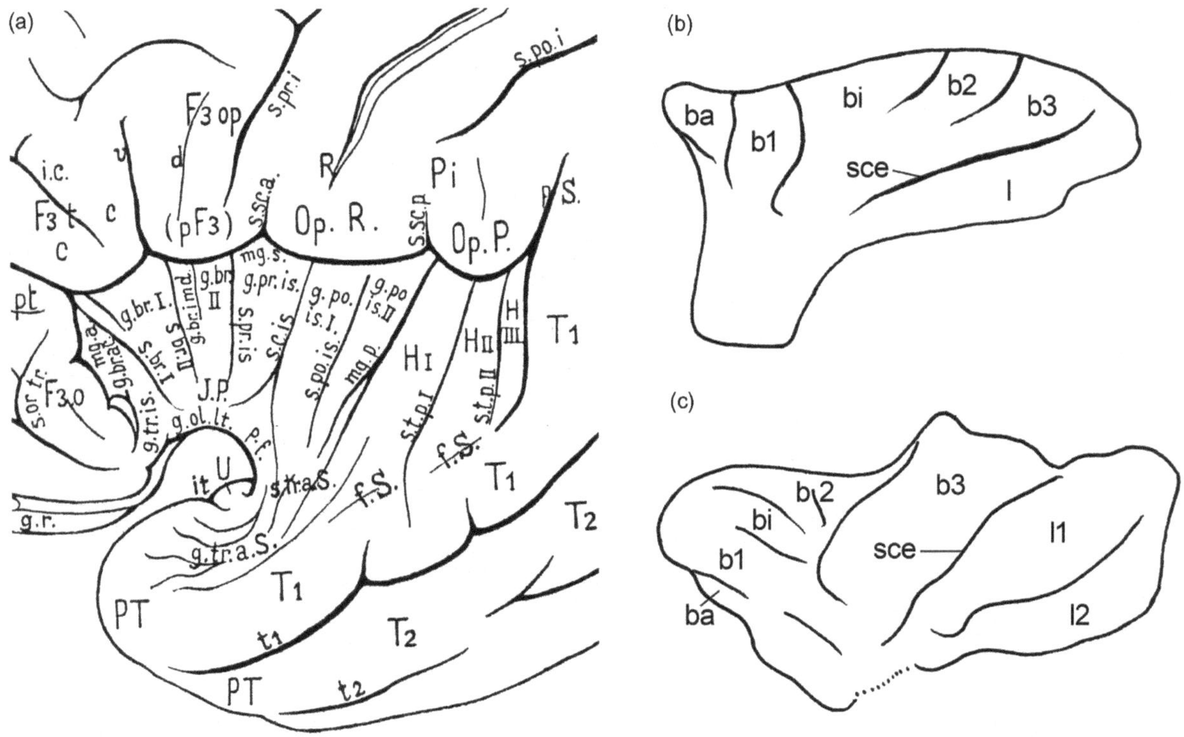

Fig. 3. Outlines of the human insular region, according to (a), von Economo and Koskinas (1925), (b) Rose (1928, 1935), and (c) Brockhaus (1940). In (a) the original labeling is maintained. Relevant abbreviations include: F3.o, gyrus frontalis tertius, pars orbitalis; F3op, gyrus frontalis tertius, pars opercularis; F3t, gyrus frontalis tertius, pars triangularis; f. S., fissura Sylvii; g. pr. is., gyrus praecentralis insulae (I1); g. po. is. I, II, gyrus postcentralis insulae primus et secundus; HI, II, III, Gyrus Heschl primus, secundus et tertius; IP, pole of insula; mg.a, -p, -s, margo anterior, -posterior, -superior sulci circularis insulae; Op.P, operculum parietale; Op.R, operculum Rolando; Pi, lobus parietalis inferior; R, sulcus rolando; for meaning of abbreviations in (b) and (c), see legend of Fig. 2.

(Figs. 2c and 3c). A gyrus brevis accessorius of varying size occupies a transitory position between the anterior insular lobule and the deep part of the orbitofrontal operculum (Figs. 2c and 3). Small gyrus transversus insulae connect the basal part of the gyrus brevis accessorius with the mediobasal part of the frontal lobe (Fig. 3a). Because of their relationship with the sulcus centralis insulae, the gyrus brevis tertius and the gyrus longus (primus) are often alternatively designated as gyrus praecentralis- and gyrus postcentralis insulae, respectively. The tapered anterobasal portion of the insula is known as the limen insulae. The synopsis of the macroscopic anatomy of the human insula just presented, and the nomenclature used therein are principally based on the studies of Ture et al. (1999) and Naidich et al. (2004), to which the reader is referred for details.

Tramo et al. (1995) measured the surface area of the cerebral cortex and its gross morphological subdivisions in 10 pairs of monozygotic twins *in vivo*, using magnetic resonance imaging. They found that the total cortical surface area is, on the average, 1906 cm^2, and that the surface areas of the left and right insulae are, on the average, 16.7 and 17.0 cm^2, respectively. This means that the surface of the insulae takes about 1.8% of the total cortical surface area.

Cytoarchitecture of the human insula

Presentation of data

Brodmann (1909) distinguished two cytoarchitectonic zones in the human insula, a granular posterior zone (J. post) and an AA zone (J. Ant). (The designation *agranular* indicates that an internal granular layer (IV) is entirely lacking.) The border between these two zones was found to correspond roughly, but not exactly to the central sulcus of the insula (Fig. 4a). Brodmann (1909, p. 146) states that it is difficult to divide the two insular zones into individual fields, and he adds that such a parcellation must await further investigation.

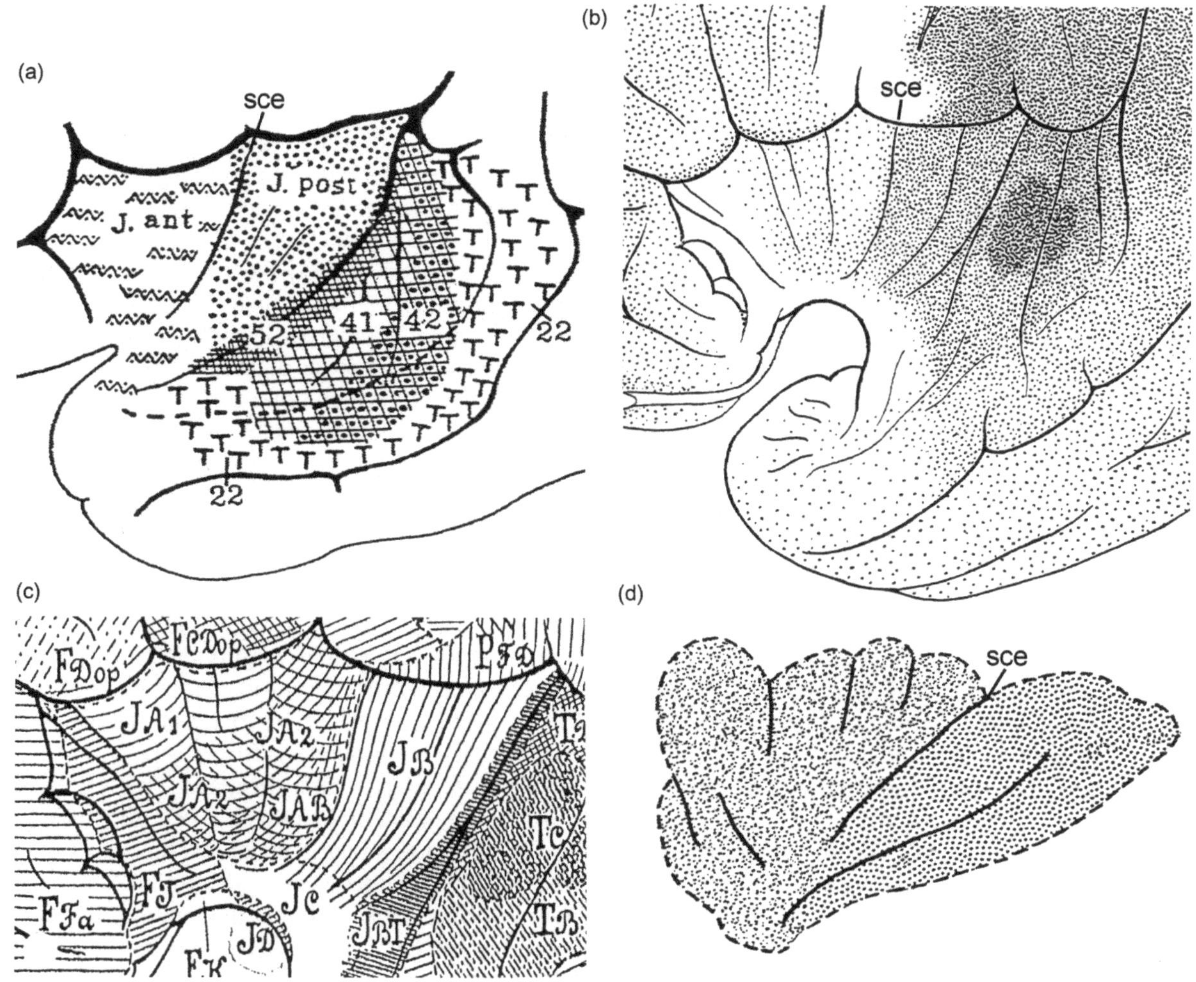

Fig. 4. Cytoarchitecture of the human insula. (a) Brodmann's (1909) map of the insular region and the exposed superior aspect of the superior temporal gyrus. J. ant, agranular anterior insular zone; J. post, granular posterior insular zone; sce, sulcus centralis insulae. (b) Semidiagrammatic representation of the density of granule cells in the insula and surrounding cortical areas. Reproduced from von Economo and Koskinas (1925). (c) The insular region of the map of von Economo and Koskinas (1925). For explanation, see text. (d) Map of the insula, according to Bailey and von Bonin (1951).

The claustrum is a thin sheet of gray matter, embedded in the white matter of the cerebral hemispheres and largely situated between the putamen and the insular cortex. It is separated from the latter by a thin layer of fibers, known as the capsula extrema. It is important to note that Brodmann considered the claustrum as a cellular sublayer, derived from the innermost layer (VI) of the insular cortex.

von Economo and Koskinas (1925) divided the insular region into three cytoarchitectonic formations, which they designated as area frontoinsularis (FJ), area praecentralis insulae (JA), and area postcentralis insulae (JB). Area FJ is a typical agranular formation, which covers the gyrus transversus insulae, the gyrus brevis accessorius, as well as adjacent strips of the orbitofrontal operculum and the gyrus brevis I (Figs. 3a and 4c). Its fifth layer contains the remarkable spindle cells, which will be discussed in a later section. Areas JA and JB correspond to the anterior and posterior zones of Brodmann. von Economo and Koskinas emphasized, however, that JA is not entirely devoid of granule cells (Fig. 4b). The authors mentioned pointed out that in areas JA and JB, the cells in the superior half of the fifth layer are exceptionally densely arranged and present themselves collectively in Nissl preparations as a highly characteristic dark band (Inselgürtel), halfway the thickness of the cortex. On account of differences in the density of granule cells in layer IV, JA was subdivided into an anterosuperior field JA1 and a posteroinferior field JA2 (Fig. 4b and c). The authors designated a transitional zone between JA and JB as JAB (Fig. 4c).

von Economo and Koskinas took issue with the view of Brodmann that the claustrum represents a split-off layer of the insular cortex. They emphasized that neither during development nor in the adult brain, there is any structural connection between these two entities.

The meticulous cytoarchitectonic analysis of the human insular cortex, carried out by Rose (1928, 1935), was based on transverse serial sections of a single specimen designated as A 40. In the reconstruction, which Rose prepared from this series, the insular region makes the impression of being dorsoventrally compressed with, as a consequence, an unusual, almost horizontal course of the sulcus centralis (Figs. 3b and 5a–c). Rose divided the insular cortex in no less than 31 different areas, all of which were described, layer by layer, in great detail, and all of which received full Latin names, such as area insularis agranularis posterior dorsalis (ai7) or area insularis eugranularis gyri centralis anterioris caudalis ventralis (i21). The positions of most of these areas are indicated in Fig. 5b. The presence and disposition of small granule cells, particularly in the fourth (IV) cortical layer, formed an important criterium in Rose's parcellation. Thus, he distinguished three principal regions, a regio insularis agranularis (Iag), in which a lamina granularis (IV) is entirely lacking; a regio insularis propeagranularis (Ipag), in which a continuous and distinct lamina granularis (IV) is also absent, but in which the adjacent zones of layers III and V contain scattered granule elements; and a regio insularis granularis, which is characterized by the presence of a continuous granular (IV) layer (Fig. 5a). The agranular and propeagranular insular regions are relatively small and form together the rostrobasal zone of the insula (Fig. 5a). Rose emphasized that the propeagranular region forms a zone of transition between the agranular and the granular insular regions. The very extensive granular region, which occupies the entire dorsal insular zone, was subdivided by Rose into four subregions, termed subregio insularis eugranularis frontalis (Ief), subregio insularis tenuigranularis frontalis (Itf), subregio insularis eugranularis caudalis (Iec), and subregio insularis tenuigranalaris caudalis (Itc; Fig. 5a). In the eugranular subregions, a wide and distinct granular layer was present, but in the tenuigranular subregions, this layer was narrow and its constituent elements were rather loosely arranged. On the basis of additional differences in the disposition of the granule cells, and a variety of other criteria, all of the regions and subregions discussed were divided into a number of separate areas (Fig. 5b). The number of areas per (sub)region varied from 2 to 13, and the total number of areas amounted, as already mentioned, to 31. It will not be attempted here to epitomize Rose's (1928) description of all of these areas, which run to almost 60 pages in his treatise. However, because the presence and the density of granule cells plays a very prominent role in the literature on the insular cortex, I have used the data provided by Rose on this matter, for the preparation of a semidiagrammatic pictorial representation of this particular feature (Fig. 5c). It will be seen that the density of the elements, forming the granular layer (IV), shows considerable differences among the various areas forming the large caudal eugranular subregion. To give some idea, I add here the characterization of the lamina granularis (IV) in some of the eugranular fields, in Rose's own words: i1, "hebt sich deutlich ab"; i2, "ebenso gut ausgeprägt wie in i1"; i18, "sehr gut ausgebildet"; i19, "viel dichter als in i18"; i20, "schmal aber gut ausgeprägt"; i21, "lockerer und breiter als in i20"; i23, "sehr gut ausgeprägt." All in all, Rose distinguished 31 cytoarchitectonic areas in the human insular cortex, 3 agranular, 6 propeagranular, 8 tenuigranular, and 14 eugranular.

Brockhaus (1940), working, like Rose, in the institute of Cécile and Oskar Vogt, produced 12 years after the appearance of Rose's study, another very detailed account on the human insula, in which he did not confine himself to the cytoarchitecture but also took the myeloarchitecture into consideration. His study was based on six series, designated as A 18, A 39, A 40, A 61, A 65, and A 66.

Brockhaus regarded the presence of the claustrum as the defining structural feature of the insula, hence he designated the cortex covering this region as claustrocortex. However, unlike Brodmann (1909), he did not consider the claustrum as a product of the insular cortex, but rather as a separate subcortical formation. Brockhaus distinguished three principal regions within the insula, which he designated as allocortex claustralis (Acl), mesocortex claustralis (Mcl), and isocortex claustralis (Icl; Fig. 5d). The small Acl, which is characterized by poor lamination and absence of granule cells, occupies a basal position. The very extensive, dorsally situated Icl shows, just like the remainder of the isocortex, a very pronounced laminar pattern, within which a lamina granularis (IV) is always distinguishable. The Mcl forms, topographically and structurally, a transitional zone between Acl and Icl. The lamination is less distinct in Mcl than in Icl. A granular layer is only feebly developed and in places hard to distinguish. Brockhaus divided these three principal regions into a varying number of areas and subareas (Fig. 5e), all of which received, just as in Rose's study, full Latin names. For some of these (sub)areas he used abbreviations, identical to those of Rose, without intending to indicate homologies. It is not feasible to specify all of the cytoarchitectonic (and myeloarchitectonic) features on which Brockhaus's parcellation was based. However, here too I have used the data on presence and density of granule cells in the various areas, provided by that author, for the preparation of a semidiagrammatic "granularity map" (Fig. 5f).

The Acl, also designated as subregio allocorticalis insularis (ai), was divided into an area allocorticalis insularis oralis (aio) and an area allocorticalis insularis caudalis (aic).

Within the Mcl, two formations, the formatio mesocorticalis insularis oralis (mio) and -insularis caudalis (mic), were distinguished. The mio was divided into separate anterior and posterior areas (mioa, miop), and both of these appeared to be subdivisible into dorsal and ventral subareas: mioad, mioav, miopd, and miopv. Within the mic, separate dorsal, intermediate, and ventral areas were delineated: micd, mici, and micv.

The anterobasal part of the insula is occupied by a field, which Brockhaus designated as area insularis biregionalis (ibr). Although this field clearly forms a structural unit, it appeared to be divisible into two subareas, one belonging to the isocortical region, ibri, and the other to the mesocortical region, ibrm (Fig. 5e).

The very large isocortical insular region (Icl) occupies the entire dorsal portion of the insula (Fig. 5d). This region is, according to Brockhaus,

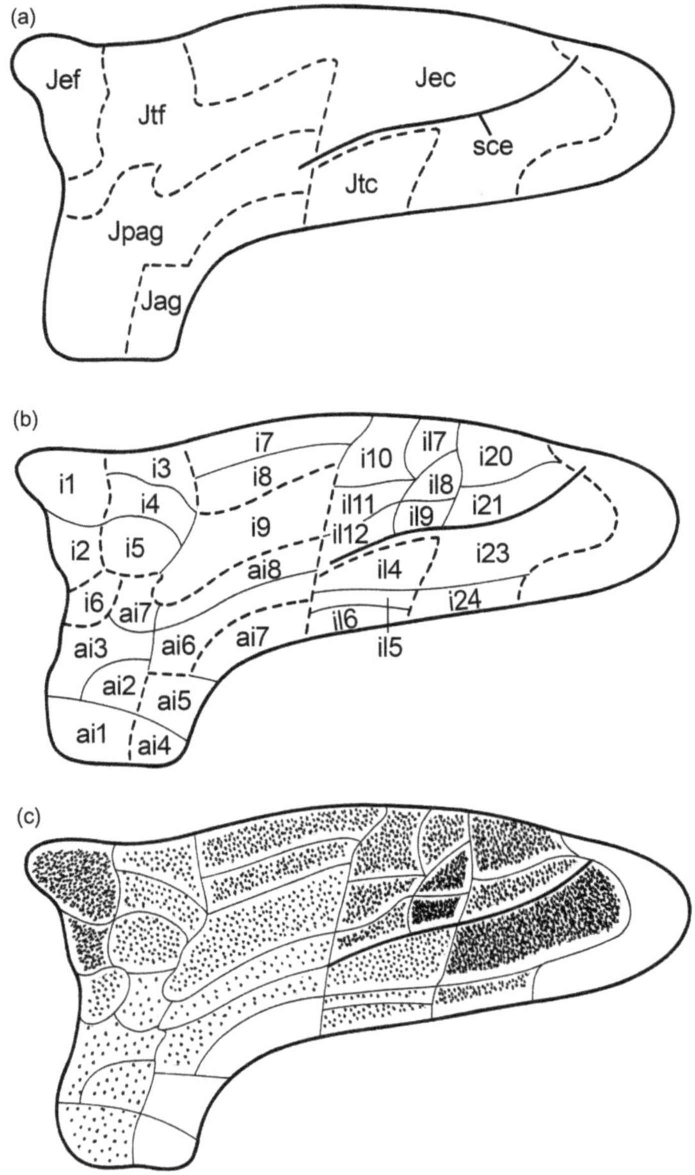
(a)
Jef
Jtf
Jec
sce
Jtc
Jpag
Jag
(b)
i7
i3
i8
i1
i4
i10
il7
i20
il8
il11
il9
i21
i9
il12
i2
i5
ai8
il4
i23
i6
ai7
i24
il6
ai3
ai6
ai7
il5
ai2
ai5
ai1
ai4
(c)

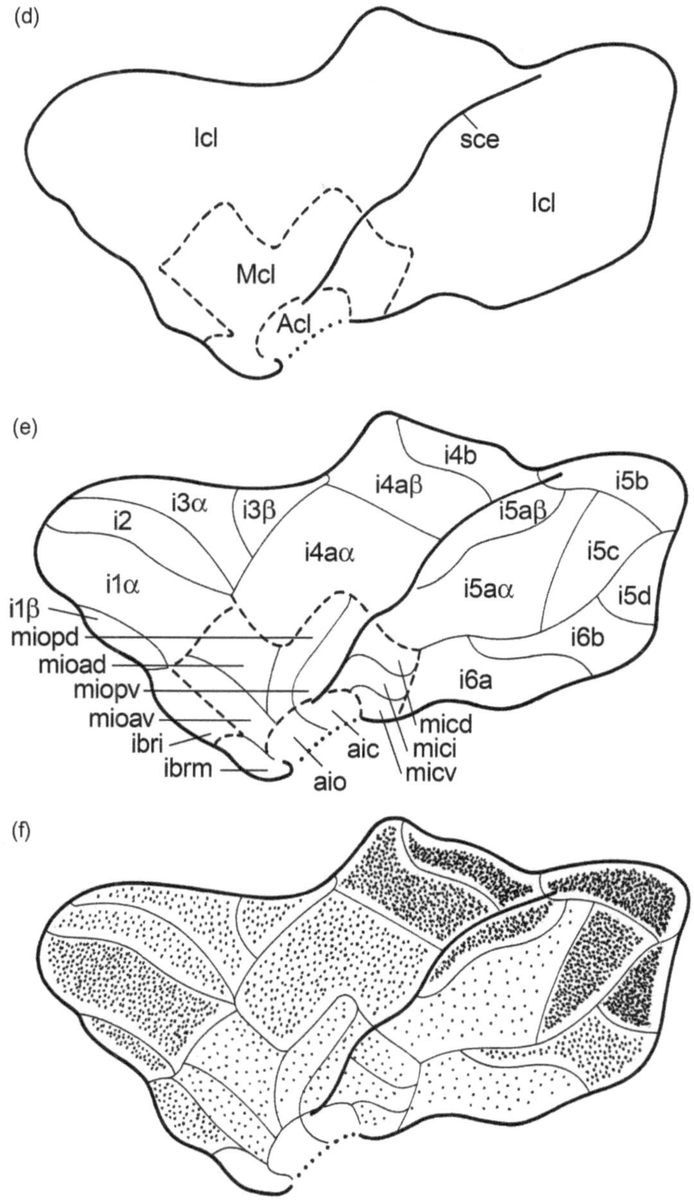

Fig. 5. Cytoarchitecture of the human insula. Comparison of the subdivisions of Rose (1928, 1935): left panel and Brockhaus (1940): right panel: (a) and (d) principal regions; (b) and (e) cytoarchitectonic areas; (c) and (f) semidiagrammatic representation of the "granularity" of the various areas. See text for further explanation.

divisible into six districts (Bezirke), four of which, i1–i4, are situated in front of the sulcus centralis insulae, the other two, i5 and i6, behind this sulcus. The districts i1–i3 are identical with the cytoarchitectonic areas i1–i3, the districts i4 and i6 each comprise two areas: i4a, i4b, i6a, and i6b, whereas district i5 is composed of four areas: i5a–d. Moreover, four of the isocortical insular areas, i1, i3, i4a, and i5a, appeared to be divisible into two subareas: i1α, i1β, i3α, i3β, i4aα, i4aβ, i5aα, and i5aβ (Fig. 5e). Figure 5f shows that, with regard to "granularity," the various (sub)areas contained within the Icl show considerable differences.

In summary, the insular cortex contains, according to Brockhaus' parcellation, 26 (sub) areas: 2 allocortical, 8 mesocortical, and 16 isocortical. We have seen that Rose (1928, 1935), working earlier in the same institute as Brockhaus (1940), had delineated 31 cytoarchitectonic areas in the human insula. At the end of his paper, Brockhaus presents a detailed comparison between the results of Rose and himself in tabular form (p. 377, Tabelle 2). Brockhaus indicates that this comparison was based on a postanalysis (Nachuntersuchung) of the histological material of brain A 40, that is, the brain studied by Rose. In the following examples of Brockhaus's results, the (sets of) areas as distinguished by Rose are placed in brackets (Fig. 5c and f): i1α (i1, i5); i1β (i2); i2 (i4); i4aβ (i9); i5aβ (i10, i17); i5c (i18+i19, i21); i6a (i13+i14); miopd+miopv (ai6); aic (ai5). Brockhaus (1940, p. 336) mentions specifically that lack of time had prevented him from preparing a reconstruction of the insula of brain A 40. Finally, it is of note that Brockhaus's parcellation of the human insular cortex was not only based on a cytoarchitectonic analysis but also based on an, equally detailed, myeloarchitectonic analysis.

The last parcellation of the insular cortex to be discussed here is the one Bailey and von Bonin (1951) presented in their monograph: "The Isocortex of Man." Unfortunately, these authors were not entirely consistent in their interpretation. They indicate in two of their figures (Frontispiece and Fig. 115c) that the sulcus centralis insulae mark the boundary between two different regions, which in the text (p. 229) are specified as "anterior agranular" and "posterior eulaminate" (Fig. 4d). However, they mention in their description of a series of horizontal sections (pp. 183–185) that in the dorsal zone of the insula, the eulaminate cortex extends anteriorly to the central sulcus, covering the posterior short gyrus. The authors indicate that in cortices of the eulaminate type, the external and the internal granular layers II and IV are well developed.

Commentary

The basic subdivision of the human insular cortex

From the data reviewed above, it appears that at the middle of the twentieth century, with regard to the subdivision of the insular cortex, two different views existed. One group of authors (Brodmann, 1909; Bailey and von Bonin, 1951; von Economo and Koskinas, 1925; Fig. 4) divided the insular cortex into two cytoarchitectonic areas, an anterior agranular and a posterior granular area, and indicated that the sulcus centralis insulae marks the (approximate) boundary between these two areas. The other group (Brockhaus, 1940; Rose, 1928, 1935; Fig. 5) distinguished a much larger number of areas and did not consider the sulcus centralis as a fundamental landmark. The first view may be designated for brevity as the "anterior–posterior" concept. For an optimal evaluation and characterization of the second view, acquaintance with the work of two more recent students of the insula, viz. Mesulam and Mufson (1985) is indispensable. These authors advocated a subdivision of the insular cortex in three zones, which are concentrically arranged around the anterobasally situated apex of the insula. This view may be denoted as the "concentric" concept. In what follows, the "anterior–posterior" concept will be discussed first, then the "concentric" concept will be considered, next, we return to the classic studies of Rose

(1928, 1935) and Brockhaus (1940), and finally, we will comment on some recent publications on the human insula.

The "anterior–posterior" concept

This concept goes in fact back to Betz (1874), who claimed that the human cerebral cortex is divided by the plane of the central sulcus (fissure of Rolando) into an anterior domain in which pyramidal cells predominate, and a posterior domain where granule cells prevail. Brodmann and von Economo and Koskinas indicated that the sulcus centralis insulae represent a continuation of the sulcus centralis on the dorsolateral convexity of the cerebral hemisphere and (roughly) marks the boundary between agranular and granular fields (Fig. 4a and b). The central sulcus of Rolando represents not only a morphological but also a functional boundary, separating the precentral primary somatomotor cortex from the postcentral primary somatosensory cortex. Because the noninsular and insular central sulci are directly continuous, the question arises whether the latter also separates motor from sensory domains. Electrical stimulation studies in epilepsy patients (Mazzola et al., 2006; Ostrowsky et al., 2000, 2002; Stephani et al., 2011) have shown that the posterior insular lobule, that is, the region posterior to the central sulcus, is occupied by four partially overlapping zones, sensitive to general somatosensory, thermoreceptive+nociceptive, general viscerosensory, and gustatory stimuli, respectively (Fig. 1b). The zones sensitive to general somatosensory, temperature, and painful stimuli show a distinct somatotopic organization (Brooks et al., 2005; Hua et al., 2005; Ostrowsky et al., 2002). Stephani et al. (2011) found that the basal part of the area sensitive to gustatory stimuli extends rostral to the central sulcus into the anterior insular lobule (Fig. 1b). The extensive meta-analysis of fMRI-data, performed by Kurth et al. (2010b), revealed the presence of two gustatory areas in the insula, a larger bilateral area in the mid-dorsal insula (roughly corresponding to the area marked "CH" in Fig. 1d) and a smaller area in the right anterior insula. As regards the localization of motor functions in the human insula, Mutschler et al. (2009) reported, on the basis of a meta-analysis of neuroimaging data, that motor tasks involving both the upper and the lower extremity activate a posterior region in the anterior insular lobule, situated directly in front of the sulcus centralis insulae (HM+FM in Fig. 1c). This region roughly corresponds to the anterobasal part of the senorimotor domain (SM in Fig. 1d), delineated by Kurth et al. (2010b). However, the data reviewed in the section "Functional studies" of this chapter have sufficiently shown that the human anterior insular lobule is primarily concerned with higher cognitive and emotional tasks, rather than with simple motor activities. Concluding this discussion of the "anterior–posterior" concept, it may be stated that (1) because the insular region situated posterior to the central sulcus, subserves somatosensory and viscerosensory functions, it may be characterized as "sensory"; (2) although the insular region, situated anterior to the central sulcus, harbors a somatomotor area, the well-known fact that it subserves a multitude of other non-somatomotor functions renders it incorrect to characterize the entire region as "motor."

The "concentric" concept

It has already been indicated that this concept was first enunciated by Mesulam and Mufson (1985). In a very influential paper, entitled "The Insula of Reil in Man and Monkey," these authors presented a detailed description of the structure of the insula and surrounding regions of the rhesus monkey. As regards the general structure of the insula, they state that "Contrary to widespread opinion, the fundamental vector of organization is not directed anteroposteriorly since the most anterior part is not necessarily its least well-developed component. Instead, the principal vector has a radial orientation emanating from the allocortical focus provided by the piriform olfactory cortex" (l. c. p. 183). They then pass on with a description of the three insular zones or belts, already mentioned in the introductory section this chapter, that is, (1) an innermost (periallocortical) agranular belt (Ia-p), so named because it lacks identifiable zones of small, granule cells; (2) a wide intermediate (proisocortical) belt (Idg), which is characterized as dysgranular, because the granule cells in layers II and IV do not display complete laminar differentiation; and (3) an outer granular belt (Ig), showing a typical isocortical structure with fully demarcated granule cell layers II and IV (Fig. 11b). Mesulam and Mufson emphasized that the belt-like organization of the insular cortex fits into a general olfactocentric plan of cortical organization. Following the detailed and well-illustrated description of the insular cortex of the monkey, Mesulam and Mufson devote a single paragraph to the human insula, in which they state that "Despite the considerably greater size, the general architectonic plan of the human insula is very similar to that observed in the macaque brain," (l. c. p. 192), and also "The concentric organization … can also be identified in the human brain." (l. c. p. 192). In a later paper (Mufson et al., 1997), it is indicated that the general plan of the human insula is not only similar to that in the macaque monkey but also resembles that in the baboon. It was possible to identify a distinct agranular Ia-p sector, coextensive with the piriform cortex, in all of these three species. With regard to the structure of the human insula, the authors mentioned that the transition of Ia-p to Idg is marked by a gradual increase in the density of layer II and layer IV granule cells, and that the visualization of fully defined granule cell laminae in both layer II and layer IV indicates the beginning of Ig. In both papers, the very succinct descriptions of the human insular cortex are accompanied by a few photomicrographs, but in none a cytoarchitectonic map of the human insula is presented. In spite of this serious limitation, practically all of the later students of the human insula have adopted the interpretation and the nomenclature introduced by Mesulam and Mufson (1985). The authors of one review article (Shelley and Trimble, 2004) went even so far as to label a simplified version of Mesulam and Mufsons' map of the insula of the rhesus monkey, as showing the architectonic organization of the human insula.

Once again, the studies of Rose and Brockhaus

The analyses of the structure of the human insula, performed by these two neuroanatomists, reviewed above, show several points of resemblance: both were carried out in the famous "Vogt-Vogt" institute; both were very voluminous (175 vs. 99 pages) and extremely detailed; in both, the insular cortex is parceled into a large number of cytoarchitectonic areas (31 vs. 26); and both are largely neglected in the current literature. A comparison concerning content of these two studies yields the following results:

(1) The reconstructions of the human insula, prepared by these two authors, differ considerably in overall shape and sulcal pattern Fig. 3b and c. (It should be mentioned that in Fig. 3c, the labeling of the small structure, situated anterobasally to the gyrus brevis primus, as gyrus brevis accessorius is mine.) Special features of the two reconstructions are that in Rose's reconstruction, the gyrus brevis intermedius is very strongly developed, whereas the (undivided) posterior insular lobule is very small (Fig. 3b), and that in the reconstruction of Brockhaus, the gyrus brevis tertius, as well as the posterior insular lobule, are both exceptionally large (Fig. 3c).
(2) In both reconstructions, all ventricular sulci coincide with cytoarchitectonic boundaries (Figs. 3b,c and 5). This is remarkable because in most other cortical regions, such a strict matching between sulci and cytoarchitectonic borders does not obtain.
(3) The sulcus centralis insulae are no exception to the rule just formulated, but Figs. 5c and 5f show clearly that, neither in Rose's map nor in Brockhaus' map, this sulcus separates an anterosuperior agranular domain, from a

posteroinferior granular domain, as it does in the maps of Brodmann (Fig. 4a), von Economo and Koskinas (Fig. 4b and c), and Bailey and von Bonin (Fig. 4d). Rather, the two reconstructions show a concentric organization of the principal insular regions. This is obvious in Brockhaus' reconstruction, where the consecutive allocortical (Acl), mesocortical (Mcl), and iscortical (Icl) zones are evidently concentrically arranged (Fig. 5d). In Rose's reconstruction, such an arrangement is less obvious; yet, this author clearly indicates that the propeagranular region (Ipag) forms a zone of transition between the agranular (Iag) and the granular (Ief+Itf+Itc+Iec) regions (Fig. 5a). Thus, it may be concluded *a posteriori*, that the subdivisions of the human insula, presented by Rose and Brockhaus, both principally answer to the Mesulam and Mufson's "concentric" concept.

(4) In any more detailed comparison of the results of Rose's and Brockhaus' cytoarchitectonic analyses of the human insular cortex, the considerable differences in overall shape of their respective maps, already alluded to, should be taken into consideration. If we do so, it seems obvious that Rose's Iag and Ipag regions correspond to Brockhaus' Acl and Mcl zones, respectively. The same holds true for some strongly granular sectors in both maps, to wit, Iaf of Rose and i1 of Brockhaus as well as Iec of Rose and the complex formed by the areas i4aβ, i4b, i5aβ, and i5b–d of Brockhaus. These correspondences tally, by and large, with those presented by Brockhaus at the end of his paper. However, several problems remain. Thus, it is unclear why Brockhaus compared the large tenuigranular area i9 of Rose with his eugranular area i4aβ. Moreover, Brockhaus homologized several areas, situated above the sulcus centralis insulae in the map of Rose, with areas located below that sulcus in his map. In the following examples, the "suprasulcal" areas/complexes in Rose's map are mentioned firstly, and their alleged "infrasulcal" equivalents in Brockhaus' map follow in brackets: i11+i12 (i5aα); i10+i17 (i5aα); i20 (i5b); i18+i19+i21 (i5c). Finally, if we study the distribution of the homologous entities, as proposed by Brockhaus, over the various insular gyri, as delineated by the two authors (the reader should compare here Fig. 3b and c with Fig. 5), the degree of correspondence appears to be surprisingly small.

The studies of Bonthius et al. *(2005) and Kurth* et al. *(2010a)*

The cytoarchitecture of the human insula was recently studied by Bonthius et al. (2005). These authors were interested in the pathology of the insular cortex in Alzheimer's disease. The reason for their neuroanatomical excursion was that they wanted to find out whether the density of Alzheimer disease-related pathology within the insula depends on cytoarchitecture. The results of their mapping, which was based on five control human brains stained with thionin or the Gallyas cell stain, is shown in Fig. 6a. It was found that the human insular cortex, like that of the monkey, as described by Mesulam and Mufson (1985; Fig. 11b), is composed of three concentrically arranged architectonic regions, a small anterobasal agranular region (Iag), a somewhat larger posterosuperior granular region (Ig), and a wide Ig region. Still more recently, Kurth et al. (2010a) subjected the human posterior insular cortex to a thorough probabilistic cytoarchitectonic analysis, based on 10 postmortem brains and using an observer-independent approach. In the introductory part of their paper, Kurth et al. mentioned that neuroimaging experiments have shown that the posterior insula is involved in the cortical processing of a wide variety of different stimuli. They then pose the question how this astonishing functional diversity maps to the architectonic structure of the insula and state that the classic maps of that structure offer no reliable basis for tackling this question.

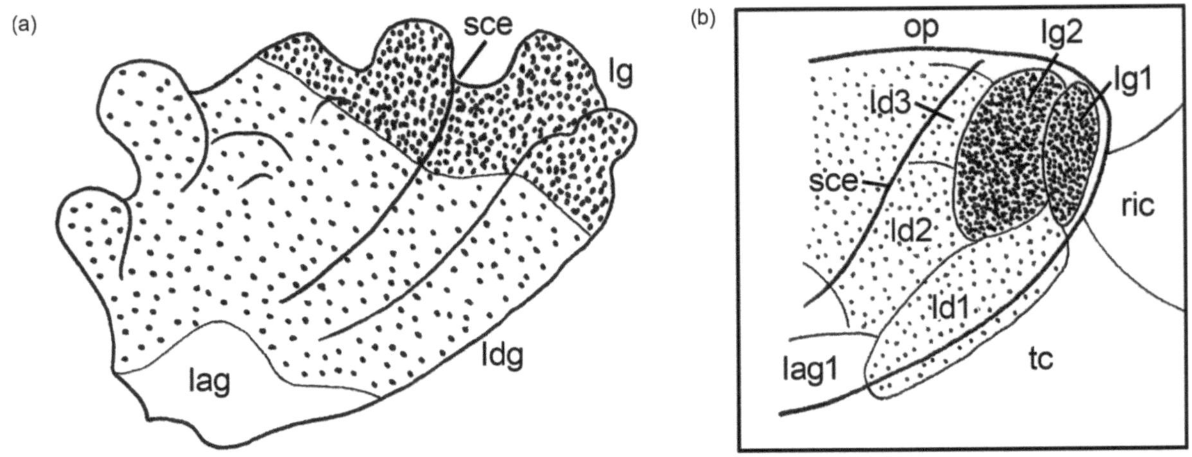

Fig. 6. Cytoarchitecture of the human insula. (a) Subdivision according to Bonthius et al. (2005). Iag, agranular cortex; Idg, dysgranular cortex; Ig, granular cortex. (b) Subdivision of the posterior insular region according to Kurth et al. (2010a). Ig1, Ig2, granular areas; Id1–3, dysgranular areas; Iag1, agranular area; op, operculum parietale; ric, retroinsular cortex; sce, sulcus centralis insulae; tc, temporal cortex.

Their specification of the shortcomings of these maps may be quoted in full (l. c. p. 1449):

"First of all, it has to be emphasized that that all classic architectonic maps purely relied on visual inspection of histological sections, using subjective nonstandardized criteria and very small sample sizes. This presumably led to the differences in the available maps, which evidently question the generalizability of such maps for anatomical reference. Second, classical cytoarchitectonic maps are only available as schematic drawings and hence cannot be matched with the 3D data sets resulting from neuroimaging experiments. Further, these atlases also lack information about intersubject variability in the size and extent of cortical areas provoking the misleading view that a particular cortical location is necessary and in all subjects corresponding to a particular area. Finally, it has to be emphasized that most influential anatomical data in the insular cortex are derived from studies in nonhuman primates (Mesulam and Mufson, 1985). Though a basic comparability between the insular cortex of humans and nonhuman primates can be assumed, straightforward extrapolation must be discussed with caution."

Kurth et al. (2010a) delineated three areas in the posterior insula, which they designated as Ig1, 1g2, and Id1 (Fig. 6b). Ig1 and Ig2 are both typical granular areas, characterized by a well developed and clearly demarcated layer IV. Area Id1 shows a considerably less distinctly developed layer IV characterizing it as dysgranular. The authors also partially delineated three other areas, two dysgranular (Id2, Id3) and one agranular (Iag1; Fig. 6b).

If we survey the results of the studies of Rose (1928), Brockhaus (1940), Bonthius et al. (2005), and Kurth et al. (2010a), it appears that all of these authors distinguished a number of distinct architectonic areas in the human insular cortex, and that the maps produced by all of them are of the "concentric" type, which means that a ventrally situated agranular zone is separated from a dorsocaudally situated granular zone, by a more or less extensive dysgranular zone (Figs. 5 and 6). The number of areas distinguished by these authors differs considerably, however, ranging from 3 to 31.

As regards methodology, the present author agrees largely with the criticism, leveled by Kurth et al. (2010a), against the approach of the classic architectonists. It is true that these authors purely relied on visual inspection of their histological material, used subjective, nonstandardized criteria and small sample sizes, and paid little attention to the range variability between different individuals. Yet, with regard to Rose (1928) and Brockhaus (1940), the authors of the two exhaustive studies on the architecture of the human insula reviewed above, it may be stated that (i) both were very careful observers; (ii) both recorded their observations in great detail; (iii) both documented their observations with numerous photomicrographs; and (iv) both adduced additional evidence for their cytoarchitectonic parceling of the human insular cortex. Rose buttressed his observations with an extensive study of nonhuman species (see next section), whereas Brockhaus performed, in parallel to his cytoarchitectonic analysis, an equally detailed myeloarchitectonic analysis. It is telling that all of the areas, which Kurth et al. (2010a; Fig. 6b) delineated in the human posterior insula, on the basis of an observer-independent probilistic analysis, can be readily equated with areas in the "classic" maps of Brockhaus (1940; Fig. 5d–f) and, to a somewhat lesser extent, also with those in the maps of Rose (1928; Fig. 5a–c; Table 1).

The comparative anatomy of the insula

Presentation of data

As the title of his classical work "Vergleichende Lokalisationslehre der Groszhirnrinde" indicates,

Table 1. Comparison of areas in the human posterior insular cortex

Kurth et al. (2010a)	Brockhaus (1940)	Rose (1928)
Ig1	i5d	i24
Ig2	i5c	i23
Id1	i6a+b	i15+i16
Id2	i5aα	i14
Id3	i5aβ	–
Iag1	Acl	ai7

α, alfa; β, beta.

Brodmann (1909) did not only confine himself to the cytoarchitecture of the human cortex but also included a number of other mammals in his analysis. The list of species studied by him includes a marsupial, the wallaby *Macropus dorsalis*; an insectivore, the hedgehog *Erinaceus europaeus*; a bat, the flying fox *Pteropus rufus*; a lagomorph, the rabbit *Oryctolagus cuniculus*; a rodent, the ground squirrel *Spermophilus citellus*; a carnivore, the kinkajou *Potos flavus*; a prosimian, the black lemur *Lemur macaco*; and two monkeys, the marmoset *Callithrix jacchus* and the guenon *Cercocebus fulginosus*. (The species names are derived from a list, which Garey (2006) added to his excellent English translation of Brodmanns' book.) Brodmann states that the insular cortex could be easily delimited in all species studied, thanks to the presence of the claustrum, which he considered as a sublayer, VIc, of the deepest cortical layer (Fig. 7c and d). He qualifies the insula as the most constant, and as a unit, most conspicuous, homogeneous zone of the mammalian cortex. In an earlier study, Brodmann (1905) distinguished four different cytoarchitectonic areas in the insula of the monkey, which he designated as areas 13–16. Of these, areas 13 and 14 are large and occupy a superior position, whereas areas 15 and 16 are small and occupy a basal position in the insula. Area 14, which is typically agranular, is situated in front of area 13, which possesses a distinct inner granular layer. The boundary between these two large areas appeared to be formed by a line marking a prolongation of the central sulcus of Rolando. The small, basal areas 15 and 16 were also agranular and of a "ganz rudimentär gebauter Typus" (l. c. p. 146). Their position corresponds approximately to that of the regions Acl+Mcl in Brockhaus' map of the human insula (Fig. 5d). Brodmann made no attempt to delineate all of the four areas, just discussed in the insula of other species. Rather, he designated the insular cortex there, as a whole, as areas 13–16 (Fig. 7a and b). However, he emphatically and repeatedly stressed that in all mammals, the insular cortex is differentiated into an agranular anterior

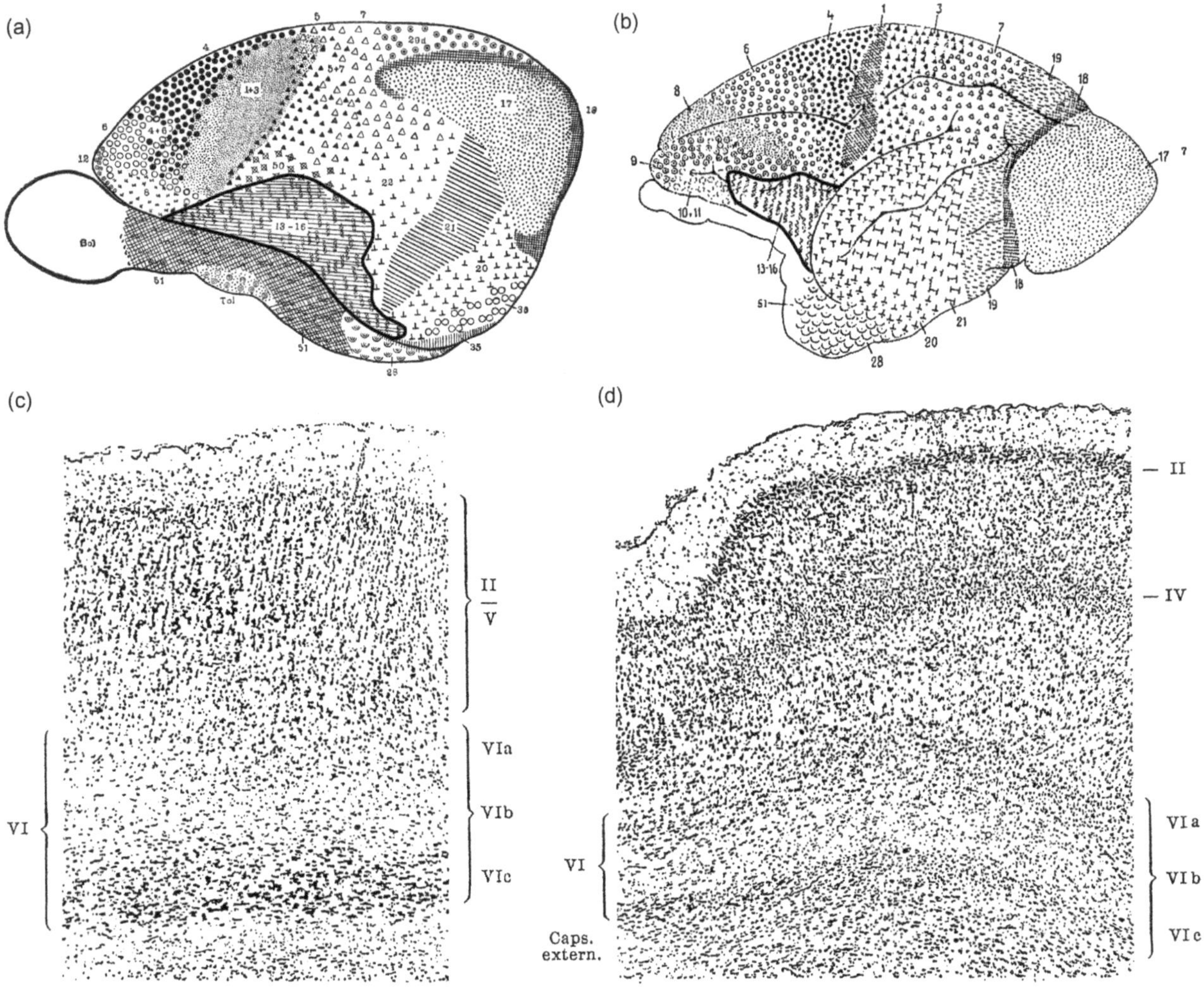

Fig. 7. Cytoarchitecture of the cerebral cortex according to Brodmann (1909): cytoarchitectonic maps of the cortex of the rabbit *Oryctolagus cuniculus* (a) and the black lemur *Lemur macaco* (b). The insular cortex is marked by a thick line. In the rabbit, this cortex is entirely situated on the free lateral surface of the cerebral hemisphere. In the lemur, this cortex is partly "operculized," hence only its anterior portion is visible. (c, d) Sections through the insular cortex of the wallaby *Macropus dorsalis*, stained according to Nissl; (c) agranular insular cortex; the internal granular layer (IV) is completely absent, and the layers II–V are fused so that the lamination is severely regressed; (d) granular insular cortex; layers II and IV are well differentiated. It is of note that, according to current views, the laminae VIa, VIb, and VIc represent the entire lamina VI, the capsula extrema, and the claustrum, respectively.

and a granular posterior half (Fig. 7c and d). As regards its overall position, Brodmann indicated that the insular cortex in "simple" mammals, such as the hedgehog and the rabbit, is situated entirely on the free lateral surface of the hemispheres (Fig. 7a), that it is partly covered by adjacent cortical formations in carnivores and prosimians (Fig. 7b), and that it is totally "operculated" in primates. It may be added that the latter is also the case in cetaceans, in which the cortex as a whole, as well as the insula, are both very well developed (Fig. 8).

To my knowledge, the only comprehensive, comparative cytoarchitectonic study, specifically devoted to the insula, is that of Rose (1928). In that study, which he dedicated to the memory of Korbinian Brodmann, Rose presented detailed descriptions of the insulae of the bat *Vespertilio murinus*, the hedgehog *E. europaeus*, the mouse *Mus musculus*, the squirrel *Sciurus vulgaris*, the rabbit *Oryctolagus cuniculus*, the lemur *Lemur catta*, the hamadryas baboon *Cynocephalus hamadryas*, and the human. Rose observed a clear progressive differentiation of the insula in the series of mammals studied, the number of delimitable cytoarchitectonic areas ranging from 1 in the bat, via 2 in the hedgehog, 4 in the mouse,

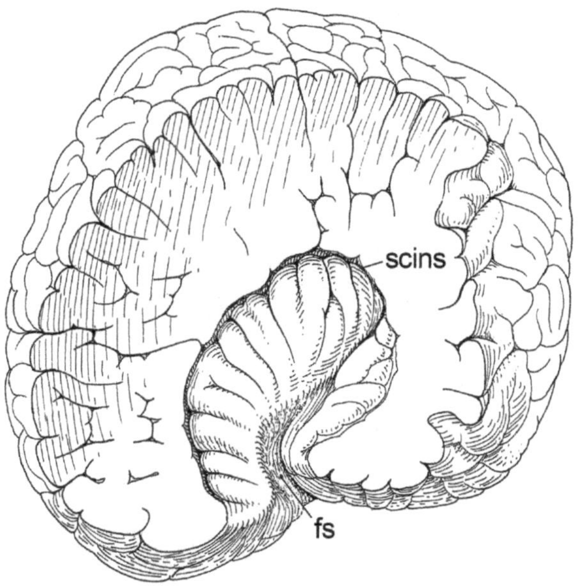

Fig. 8. Lateral view of the left cerebral hemisphere of the bottle-nosed dolphin *Tursiops truncatus*. The opercula have been removed to reveal the insula. fS, fissura lateralis Sylvii; scins, sulcus circularis insulae. Reproduced with permission from Nieuwenhuys et al. (1998).

5 in the squirrel, 6 in the rabbit, and 8 in the lemur, to 22 in the baboon. In all species studied, the insular cortex bordered basally on the olfactory prepiriform cortex. Just as in the analysis of the human insula, three basic types of cortex, agranular, propeagranular, and granular, were distinguished. The insular cortex of the bat, which comprises a single agranular field, ai (Fig. 9a), was considered by Rose as an extremely primitive structure. In the hedgehog, two insular areas, ai1 and ai2, were found, which were both of the agranular type. In the mouse, the squirrel (Fig. 9b), the rabbit, and the lemur (Fig. 9c and e), the insular cortex could be divided into a dorsal granular zone, and a ventral agranular zone, the latter comprising two fields in the mouse, three in the squirrel and the rabbit, and four in the lemur. Rose designated all of these agranular fields with the letters ai, followed by a number (Fig. 9). In the four animals mentioned, the granular zone could be divided into an anterior sector, in which the lamina granularis (IV) was poorly developed, and a posterior sector in which the granular layer appeared to be better developed. The anterior granular zone was found to consist of a single field, i1; the posterior granular zone comprised a single field, i2, in the mouse and the squirrel; and two fields, i2 and i3, in the rabbit and the lemur. Rose designated the ventral zone of the insula of the lemur, which comprises the fields ai1–4, as regio Iag, the anterodorsal sector, which comprises the field i1, as regio Ipag, and the posterodorsal sector, which comprises the fields i2 and i3, as regio insularis granularis (Ig; Fig. 10a). The insula of the baboon is largely "operculated" and contains a shallow groove, the sulcus centralis insulae, which divide the insula into lobulus insulae anterior and lobulus insulae posterior (Fig. 9d). Rose divided the insular cortex of the baboon, just like that of the lemur, into three principal regions, regio Iag, -propeagranularis (Ipag), and -granularis (Ig). He indicated that the lamina granularis (IV) is well delineable throughout the granular region, but that the degree of development of this layer rendered it possible to subdivide the granular

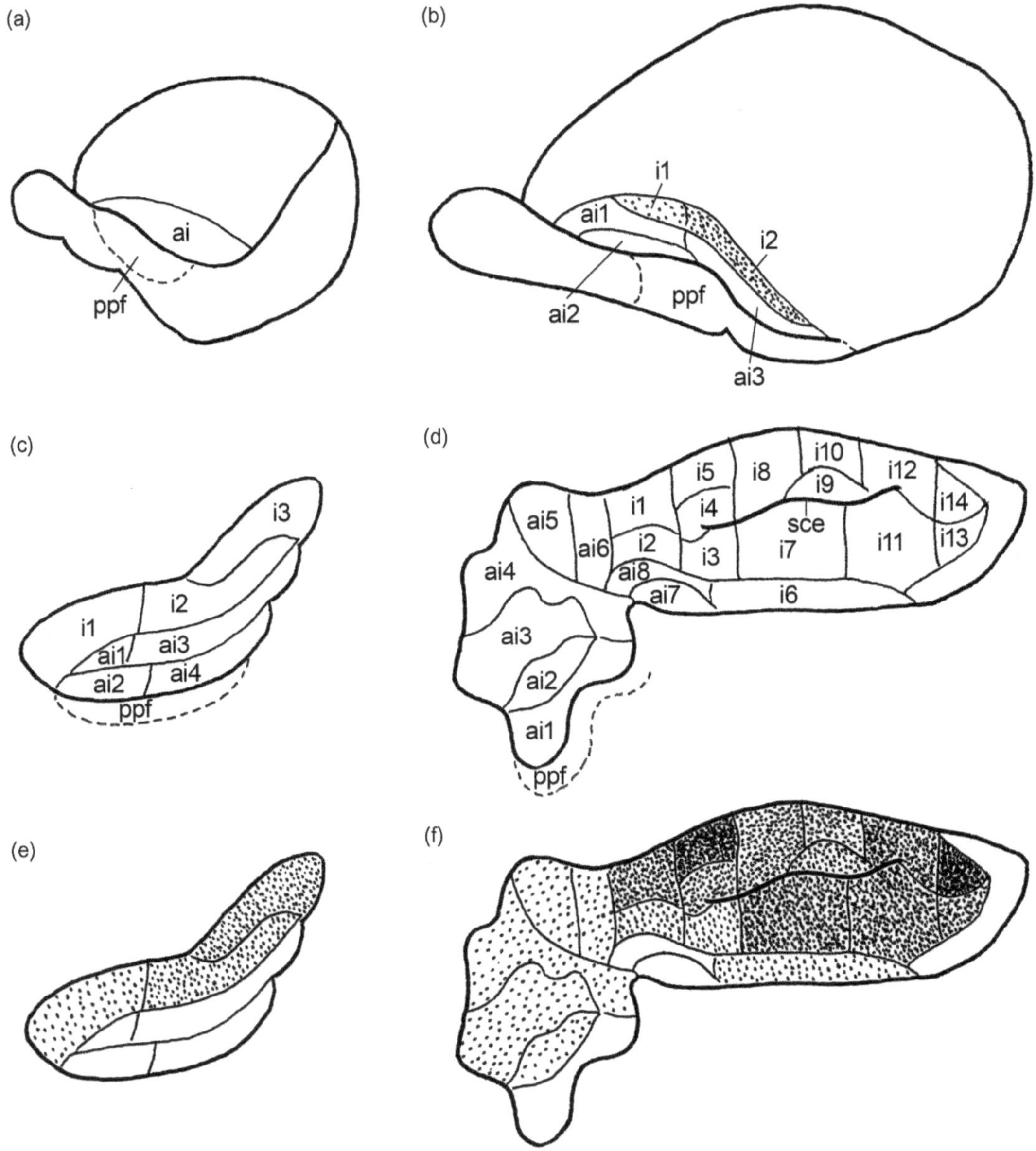

Fig. 9. The cytoarchitecture of the insular cortex in various mammals, according to Rose (1928). (a) The bat *Vespertilio murinus*; (b) the squirrel *Sciurus vulgaris*; (c, e) the ring-tailed lemur *Lemur catta*; (d, f) the baboon *Papio hamadryas*. In (a) and (b), the entire cerebral hemisphere, but in (c–f), only the insular region is shown. In (b), (e), and (f), the "granularity" of the various insular areas is indicated semidiagrammatically. ai, ai1, ai2, etc., agranular insular fields; i1, i2, etc., insular granular fields; ppf, prepiriform cortex; sce, sulcus centralis insulae.

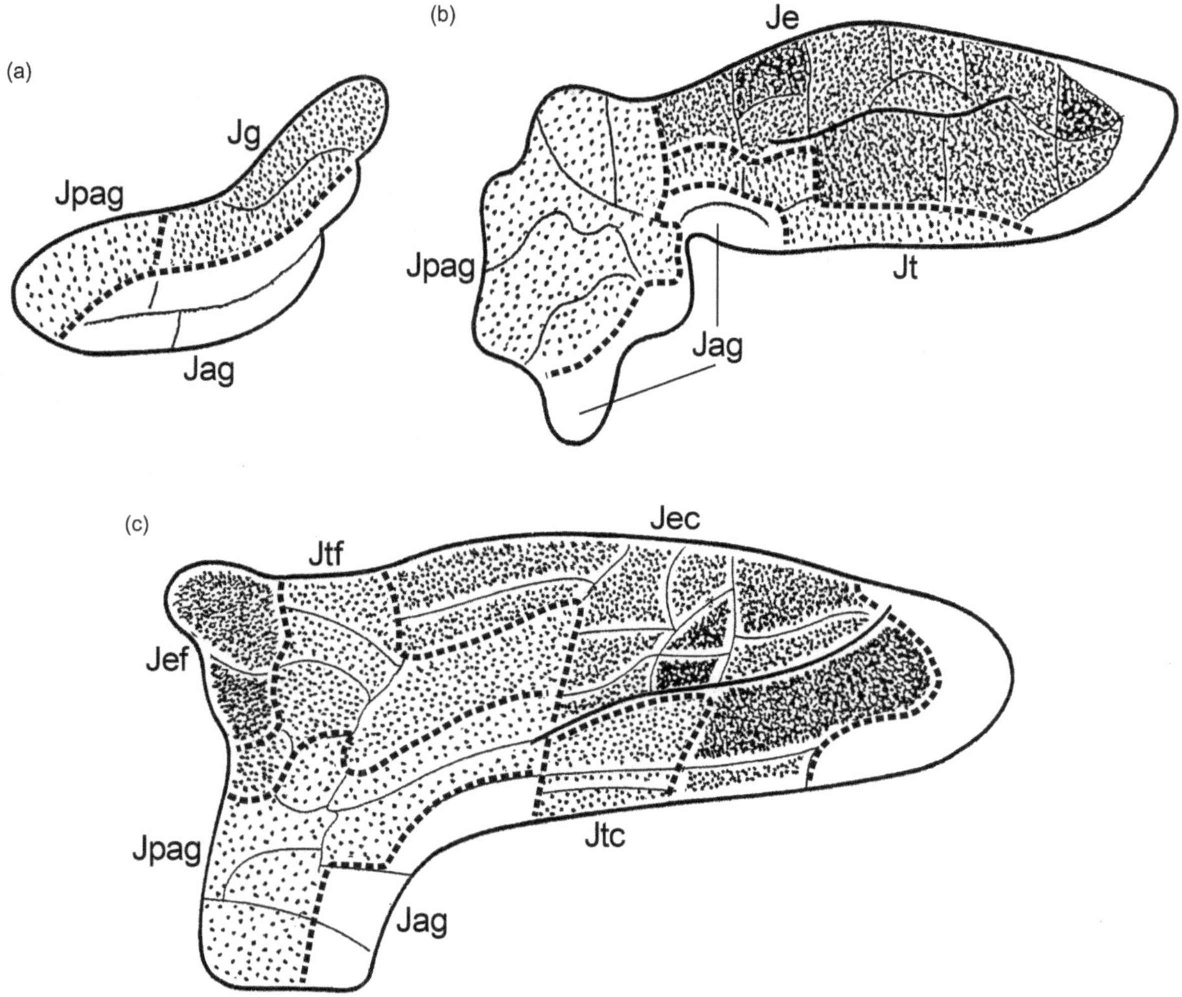

Fig. 10. The cytoarchitectonic differentiation within the principal subdivisions of the insular cortex in (a) the ring-tailed lemur *Lemur catta*, (b) the baboon *Papio hamaddryas,* and (c) the human, according to Rose (1928). Jag, regio insularis agranularis; Je, subregio insularis eugranularis; Jec, Jef, subregio insularis eugranularis caudalis, -frontalis; Jg, regio insularis granularis; Jpag, regio insularis propeagranularis; Jt, subregio insularis tenuigranularis; Jtc, Jtf, subregio insularis tenuigranularis caudalis, -frontalis.

region into two subregions, a subregio It, in which lamina IV is only feebly developed, and a subregio insularis eugranularis (Ie), with stronger development of lamina IV (Fig. 10b). Rose distinguished several separate cytoarchitectonic areas within each of the (sub)regions mentioned. Thus he divided the regio Iag into three areas, ai1, ai7, and ai8; the regio Ipag into five areas, ai2–6; the subregio It into three areas, i2, i3, and i6; and the subregio insularis eugranularis (Ie) into no less than 11 areas, i1, i4, i5, i7–14 (Figs. 9d,f and 10b). Rose noticed that the "granularity" of the various areas located within the eugranular subregion shows considerable differences. Thus, he characterized lamina granularis (IV) in i1 as "gut ausgeprägt," in i5 as "sehr gut ausgeprägt," in i9 as "gut ausgeprägt, jedoch weniger dicht als in i9," and in i12 as "schön ausgebildet" (Fig. 9f).

In the general discussion section of his megapaper, Rose (1928) emphasized the following points: (1) An agranular insular region is present in all mammals investigated. (2) In some (bat, hedgehog), the entire insular cortex is of the agranular type. (3) The granular insular region shows a progressive development, with regard to both size and complexity of organization. As for size, in those "lower" mammals, which do possess a granular insular region (mouse, squirrel, rabbit), this region is smaller than the granular one; in the lemur, these relations are reversed and in the baboon and the human the granular insular region exceeds the agranular region by far in size. As for complexity of organization, the number of cytoarchitectonic areas delineated within the granular insular region, increased in the series of mammals investigated from 2 in the mouse, the squirrel' and the lemur, to 14 in the baboon, and to 22 in the human. (4) A separate propeagranular insular region could be distinguished in the lemur, the baboon, and the human. In this region, the granular layer (IV) is not distinguishable as a separate layer. Groups of small granule cells are present, however, in the deeper zone of layer III and the superficial zone of layer V. The propeagranular region forms always a transitional zone between the agranular and the granular insular cortex. The propeagranular region comprises a single cytoarchitectonic area in the lemur, five areas in the baboon, and six areas in the human. (5) Differences in the width and the cell density of the granular layer (IV), made it possible to distinguish in the baboon and the human, separate tenuigranular and eugranular subregions within the granular insular region. The tenuigranular subregion comprised three cytoarchitctontc areas in the baboon and seven in the human. (6) In the human, the tenuigranular and eugranular insular subregions can both be subdivided into separate frontal and caudal districts. The frontal eugranular district, Ief, is a special feature of the human insula.

Commentary

The relation between the insular cortex and the claustrum

Brodmann (1909) and Rose (1928) were both of the opinion that the insular cortex and the claustrum develop from one and the same anlage. However, de Vries (1910) and Landau (1919) adduced embryological evidence, suggesting that these two structures develop from different matrix zones, and this view has been fully confirmed by numerous recent gene expression studies, reviewed in Nieuwenhuys et al. (2008). It may be categorically stated that the relations between the insular cortex and the claustrum, as observed in adult mammals, are neither embryological nor functional, but purely topographical.

Differences between the results of the comparative studies of Brodmann and Rose

One of the most striking general results of Brodmann's comparative studies was that the number of delimitable cytoarchitectonic areas within the cerebral cortex varies considerably among mammals. Thus, he delineated 12 of such areas in the hedgehog, 14 in the lemur, 26 in the marmoset monkey, and 44 in the human. It might be expected that this difference in arealization would also be reflected in the insular cortex, but Brodmann does not mention such differences. Rather, he labeled the insular cortex in all species investigated collectively as areas 13–16, without specifying the extent of these four areas (Fig. 7a and b). Yet, he emphasized that in all species studied, the insula could be readily divided into an anterior agranular and a posterior granular half, and that further parceling of the insular cortex was fraught with difficulties. We have seen that these two points played no role, whatsoever, in the work of Rose. In none of the cytoarchitectonic maps of the insula he produced, a vertical borderline, separating an agranular from a granular region can be discerned, and in all of these maps, the insular cortex is parcellated into a number of cytoarchitectonic areas, ranging from 1 to 31 (Figs. 5b,c, 9, and 10).

Are agranular insular cortices primitive?

Rose held that an entirely agranular insular cortex, as observed in the bat (Fig. 9a) and the hedgehog, represents a simple and primitive condition. This view is untenable in light of the results of a recent study of Butti and Hof (2010). These authors studied the cytoarchitecture of the insular cortex in a considerable number of "less common species." They found, *inter alia*, that the insular cortex is completely agranular in the pigmy hippopotamus, *Hexaprotodon liberiensis*; the Western bongo, *Tragelaphus eurycerus*; the Florida manatee, *Trichechus manatus*; and the minke whale, *Balaenoptera acutorostrata*.

Comparison of the results of Rose's cytoarchitectonic analysis of the insular cortex of the baboon, with those of similar studies on the insula of the macaque

The cytoarchitectonic organization of the insular cortex of the rhesus monkey has been studied by von Bonin and Bailey (1947), Roberts and Akert (1963), Mesulam and Mufson (1985), Friedman et al. (1986), Mufson et al. (1997), and Gallay et al. (2011). In their monograph "The isocortex of *Macaca mulatta*," von Bonin and Bailey (1947) did not present a separate map of the insular cortex of that species. They mention, however, that this cortex can be divided into a smaller anterior area and a larger posterior area. Using the nomenclature of von Economo and Koskinas (1925; Fig. 4c), they designated the anterior area as IA and the posterior area as IB. They indicated that in IA, the fourth cortical layer is just hinted at by the presence of a few scattered granules, but that in IB, the fourth layer is fairly thick and contains small granules with very few slightly larger cells.

Roberts and Akert (1963; Fig. 11a), Mesulam and Mufson (1985; Fig. 11b), Friedman et al. (1986), Mufson et al. (1997), and Gallay et al. (2011; Fig. 11c) all indicated that the insular cortex of the rhesus monkey can de divided into three architectonic fields, an anteroventral agranular field, a dysgranular intermediate field, and a dorsocaudal granular field. All of these authors emphasized that the changes in "granularity" occur gradually over the length of the insula, and that, hence, the precise boundaries between the fields are sometimes difficult to identify. The recent study of Gallay et al. (2011) was specified by the authors as a multiarchitectonic analysis because it was not only based on Nissl material but also based on the use of various immunohistochemical stainings. Changes in density and laminar distributions of the neurochemical markers rendered it possible to divide each of the three insular areas into two or three subareas, and to add a small, dorsocaudally situated "hypergranular" area. Thus, the total number of concentrically arranged architectonic units, distinguished by these authors in the insula of the rhesus monkey, amounted to eight (Fig. 11c). Interestingly, Cerliani et al. (2011) recently reported on the basis of a preliminary probabilistic tractography study that in the insula of the rhesus monkey, there is a continuous variation in connectivity along a rostroventral-to-dorsocaudal axis, and that the gradients in the shift of this variation are oriented more or less perpendicular to this axis (Fig. 11d). Comparison of Fig. 11d with Fig. 11a–c shows that the gradients in the shift of connectivity just mentioned run strikingly parallel to the architectonic borders, discussed above. It is also of note that Mufson et al. (1997) did not only study the insula of the rhesus monkey but also that of the baboon *Papio papio*. They found that the insulae of these two species closely resemble each other with regard to architecture but show a salient difference in gross structure. This difference is that the surface of the insula of the macaque is essentially smooth, whereas that of the baboon is

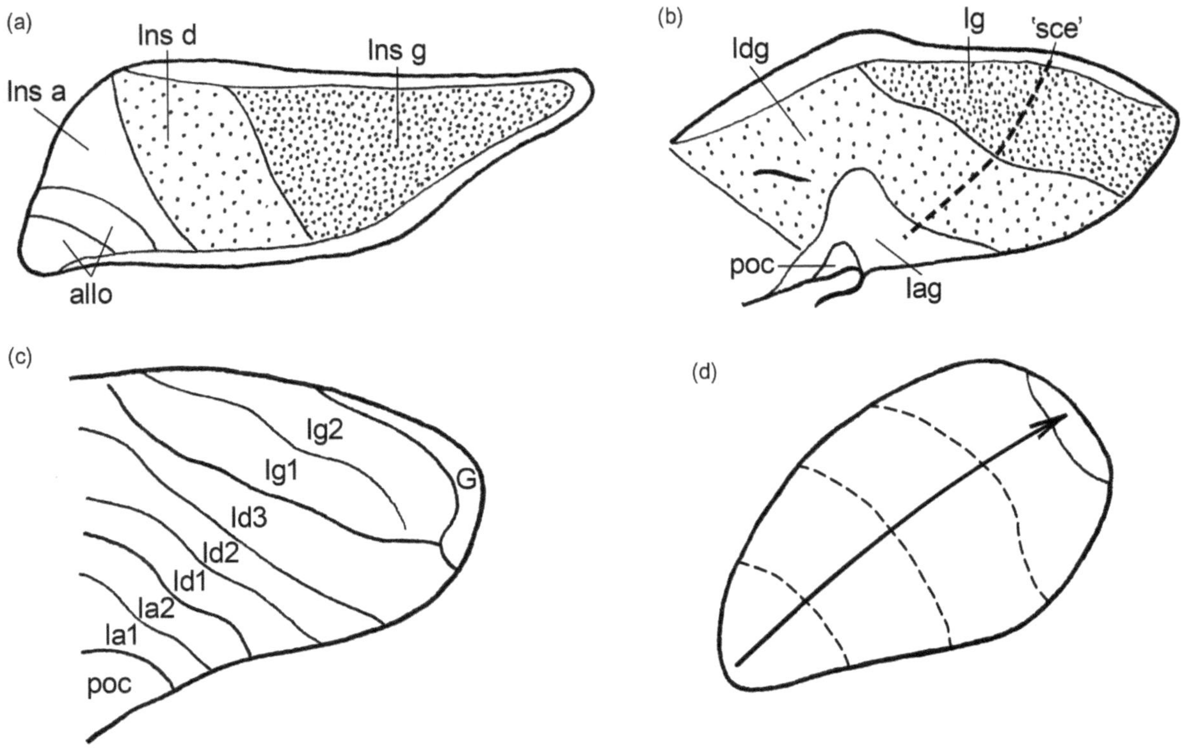

Fig. 11. Subdivisions of the insular cortex of the rhesus monkey. (a) Cytoarchitectonic organization, according to Roberts and Akert (1963). (b) Idem, according to Mesulam and Mufson (1985). In these two parcellations, the insular cortex is divided into agranular (Iag, Ins a), dysgranular (Idg, Ins d), and granular (Ig, Ins g) regions; "sce," virtual sulcus centralis insulae, explained in the text. (c) Multiarchitectonic organization according to Gallay et al. (2011), based on the use of multiple immunohistochemical techniques, in addition to Nissl and myelin staining. The agranular (Ia), dysgranular (Id), and granular (Ig) domains are all subdivided into two or three subdomains. Moreover, a small dorsocaudally situated "hypergranular" domain (g), which extends into the adjacent opercula, could be distinguished. (d) Preliminary results of a probabilistic tractography analysis performed in a single-macaque specimen by Cerliani et al. (2011). There appeared to be a continuous variation in connectivity along a rostroventral-to-dorsocaudal axis (arrow). The thin, dashed curves indicate the orientation of gradients, rather than the borders of connectivity domains.

sculptured by a distinct sulcus centralis insulae. The authors mentioned published an architectonic map of the insula of the macaque (their Fig. 6, reproduced here in a simplified form in Fig. 11b) and a photograph of the lateral aspect of the insula of the baboon (their Fig. 5b). Comparison of these two figures revealed that the shape of the insula is very similar in these two species. If we now transfer the course of the sulcus centralis insulae, as observed in the picture of the insula of the baboon, to the map of the insula of the rhesus monkey (dashed line in Fig. 11b), it becomes abundantly clear that this sulcus does not coincide with any of the concentrically arranged cytoarchitectonic borders but is rather oriented perpendicular to them.

The data on the architecture of the insular cortex of the rhesus monkey, just reviewed, indicate that the organization of this cortex complies with the "concentric" concept, rather than with the "anterior–posterior" concept. If we return now to Rose's map of the insular cortex of the hamadryas baboon (Fig. 9d and f), it is at first sight hard to decide to which concept this insula answers. However, if we (1) equate the propeagranular (Ipag) and tenuigranular (It) regions of the baboon (Fig. 10b), with the dysgranular domain of the macaque (Fig. 11a and b), and (2) take into consideration that the sulcus centralis insulae in the baboon do not separate an agranular domain from a granular domain, it becomes clear that the organization of the insula in the latter species also meets the "concentric" concept.

Comparison of the insular cortex of the rhesus monkey, with that of the human

A careful comparison of the architectonic organization of the insula of the rhesus monkey with that of the human is of great interest for several reasons. One of these is that practically all of our current knowledge of the fiber connections of the insula is based on experimental hodological studies on the rhesus monkey (see section "Neuroanatomical studies"), and that the results of this comparison are indispensable for a sound transfer of this body of hodological knowledge to the human insula. The data assembled in the present and the previous section of this chapter show that the "concentric" concept of Mesulam and Mufson (1985) offers a solid basis for this comparison, but that for the introduction of a sound parceling, going beyond the classic tripartition into agranular, dysgranular, and granular zones, the following additional research is required: (a) We have seen that a recent multiarchitectonic analysis (Gallay et al., 2011; Fig. 11c) has shown that the insular cortex of the rhesus monkey is divisible into no less than eight concentric subzones. Comparable studies, aimed at answering the question whether such subzones are also present in the human insula, are highly wanted. (b) We have seen that, according to the classic studies of Rose (1928; Fig. 5a–c) and Brockhaus (1940; Fig. 5d–f), the human insula can be divided into a large number of cytoarchitectonic areas, and that Rose (1928; Figs. 9d–f and 10b) produced a comparably detailed parceling of the insular cortex of a nonhuman primate. We have also seen that Kurth et al. (2010a), so far as the human posterior insula is concerned, were able to confirm the results of Rose and Brockhaus, on the basis of a modern observer-independent probabilistic analysis. It will be clear that a definitive assessment of the pending homologies requires that Kurth and collaborators extend their mapping study over the human AIC and perform a similar analysis of the insular cortex of the rhesus monkey.

Before closing this section, attention should be paid to the recent presumption of Craig (2009, 2010b, 2011) that the human AIC has no clear homologue in the rhesus monkey. Craig adduces the following pieces of evidence for this presumption:

(1) The insular cortex is disproportionately (about 30%) enlarged in the human relative to the macaque monkey. He bases this statement reportedly on a quantitative neuroimaging study of Semendeferi and Damasio (2000).
(2) The primary gustatory cortex extends to the anterior limit of the insula in macaques but only to the middle of the insula in humans. As regards the first point, Craig misquotes here the paper of Semendeferi and Damasio (2000). These authors studied the volumes of the insula (and other brainparts) in hominoids, *and not* in the rhesus monkey. Semendeferi and Damasio (2000, p. 324) actually stated: "The range of values of the human insula overlaps with the range of all other hominoids, most of which are around the lower end of the human variation." As regards the second point, this difference in position of the primary gustatory cortex is undisputable, but I should like to present an alternative explanation for this difference (and the implicit nuclear shift). In general, expansion of the cerebral cortex is not the result of the addition of *novel* territories or areas. Brodmann (1909) delineated three

different cytoarchitectonic fields in the human posterior parietal region, which he designated with the numbers 7, 39, and 40. In the corresponding region of the cortex of two different monkeys (the guenon and the marmoset), he could only distinguish a single area, which he numbered 7. However, Brodmann did not conclude that the areas 39 and 40 are novel human acquisitions, but rather that area 7 in monkeys represents a still undifferentiated precursor zone for all posterior parietal areas in the human. Recently, Petrides and Pandya (2002) established that areas 44 and 45 in the human cortex, which together constitute Broca's speech region, have small, but distinct homologues in the rhesus monkey. I believe that the gradual increase in complexity of cognitive functions (such as vocal communication) during hominoid evolution has been correlated with a gradual expansion of numerous cortical regions, including Broca's speech region and the AIC. Hence, I consider it likely that the caudal displacement of the primary gustatory cortex (and of other insular sensory areas as well, see Fig. 1a and b) is the result of the considerable increase in size of the insular cognitive domain (Fig. 1c and d). The following comparative neuroanatomical observations argue in favor of this concept. (1) The architectonic analyses of Rose (1928; Fig. 10) have shown that the dysgranular anterior insular region (Jpag plus Jtf) increases considerably in size, in the series lemur→baboon→human. (2) Cerliani et al. (2011) recently analyzed the overall connectivity of the human insula, using a probilistic tractography procedure, and it has already been mentioned that the same study also contains preliminary data on the rhesus monkey. It appeared that in both species there is a continuous variation in connectivity along a rostroventral-to-dorsocaudal axis, and that the gradients in the shift of connectivity are oriented more or less perpendicularly to this axis. If we now compare the dashed lines indicating the course of these gradients in the monkey (Fig. 11d) with those in the human (Fig. 14b), it appears that in the latter, the dorsal and most ventral parts of these lines are caudally deflected. I consider it likely that the dorsal deflection is caused by the expansion of the cognitive domain. The ventral deflection might well be evoked by an analogous expansion of the social-emotional domain (Fig. 1d).

Special neurons in the insular cortex

Presentation of data

In a publication, mainly devoted to the acoustic cortex, Cajal (1900) pays brief attention to the insula. His description is based on Golgi-preparations of the insular cortex (no further specification) of a child of 1 month. This description is accompanied by an illustration, which is reproduced here in Fig. 11a. The capitals in bold used in what follows correspond to those in the figure. Cajal mentions that the fifth layer of the insular cortex contains, apart from numerous ordinary pyramids (A, B), aberrant pyramidal elements of two types, which he designates as triangular, bifid, or trifid cells (E, F), and as fusiform cells (C, D).

The triangular, bifid, or trifid cells have two, three, or more ascending dendritic branches which extend into the first (plexiform) layer of the cortex, and a single-basal dendrite forming, at a certain distance from the soma, a tuft of diverging dendrites. The axon arises from the lower end of the basal dendritic stem.

The fusiform cells are characterized by a radially oriented dendritic shaft, in which the soma forms a simple, spindle-shaped thickening. The apical dendrite ascends to the plexiform layer. The basal dendritic shaft is generally devoid of branches and forms, at a varying distance from the soma, a tuft of descending dendrites. The soma and the initial portion of the apical dendrite give rise to some delicate horizontal dendrites.

The axon arises, also in the elements of this type, from the end of the basal dendritic stem and descends to the white matter. Before leaving the cortex, it gives rise to some collateral branches.

von Economo and Koskinas (1925) mention that their typically agranular area FJ (Fig. 4c), which forms a transitional zone between the frontal operculum and the insula proper, contains numerous slender, elongated, fusiform cells, which to their knowledge had not been described before (Figs. 12b,c and 13c). They designate these elements as "Stäbchen-oder Korkzieherzellen,"

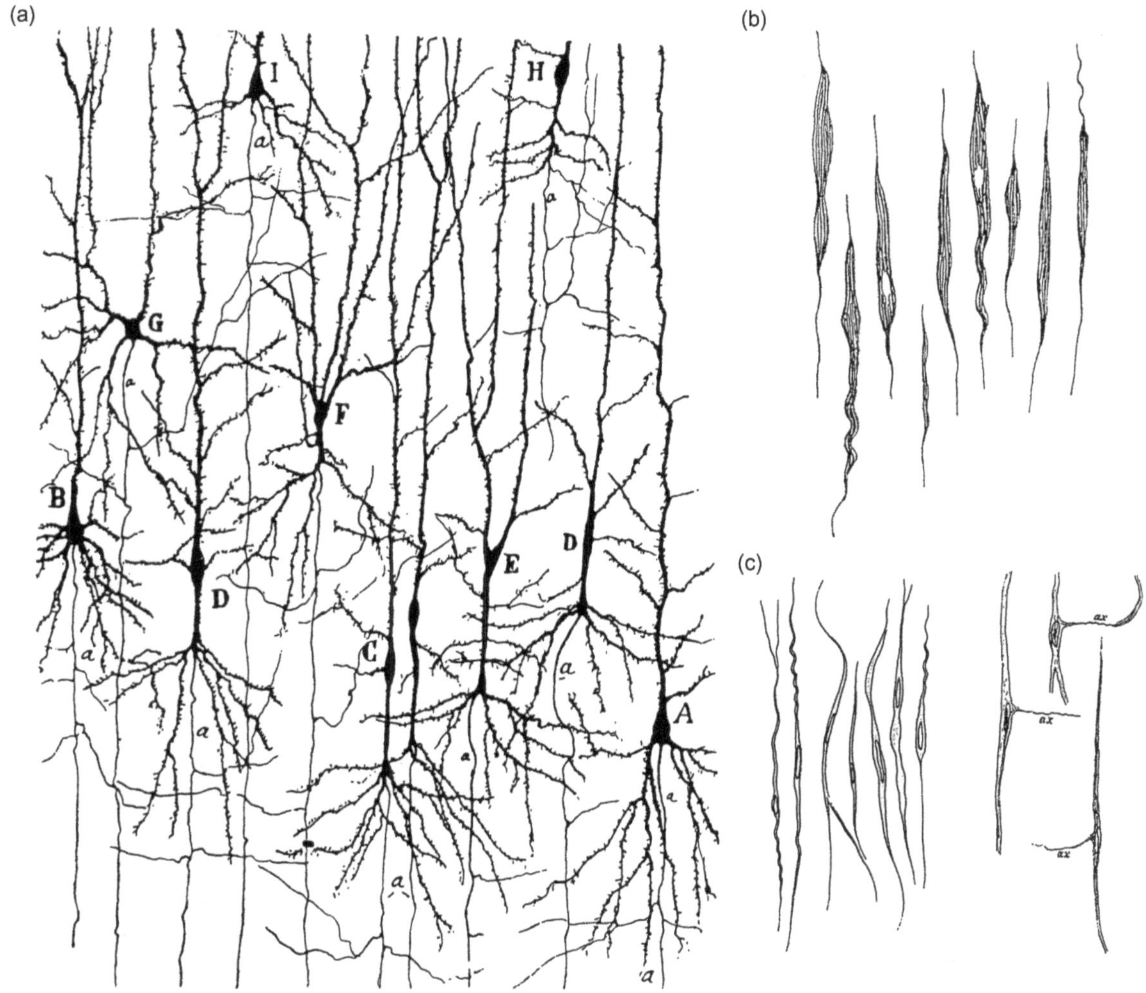

Fig. 12. Special cells in the human insular cortex. (a) Section through the insular cortex of a child of 1-month old. Golgi technique. The fifth layer, which is shown here, contains, in addition to ordinary large pyramids (A, B), fusiform cells with descending dendritic tufts (C, D), and cells provided with two or more ascending dendritic shafts prolonged up to the first layer (E, F). Reproduced from Cajal (1900). (b) "Rod-like" and "corkscrew" cells in the gyrus transversus insulae, reproduced from von Economo and Koskinas (1925). (c) Special cells in the deep layers of the gyrus cinguli and the gyrus transversus insulae. The elements in the left half of the figure are drawn from preparations stained with thionin; those in the right half are drawn from preparations impregnated according to Bielschowsky. Reproduced from von Economo (1926). ax, axon.

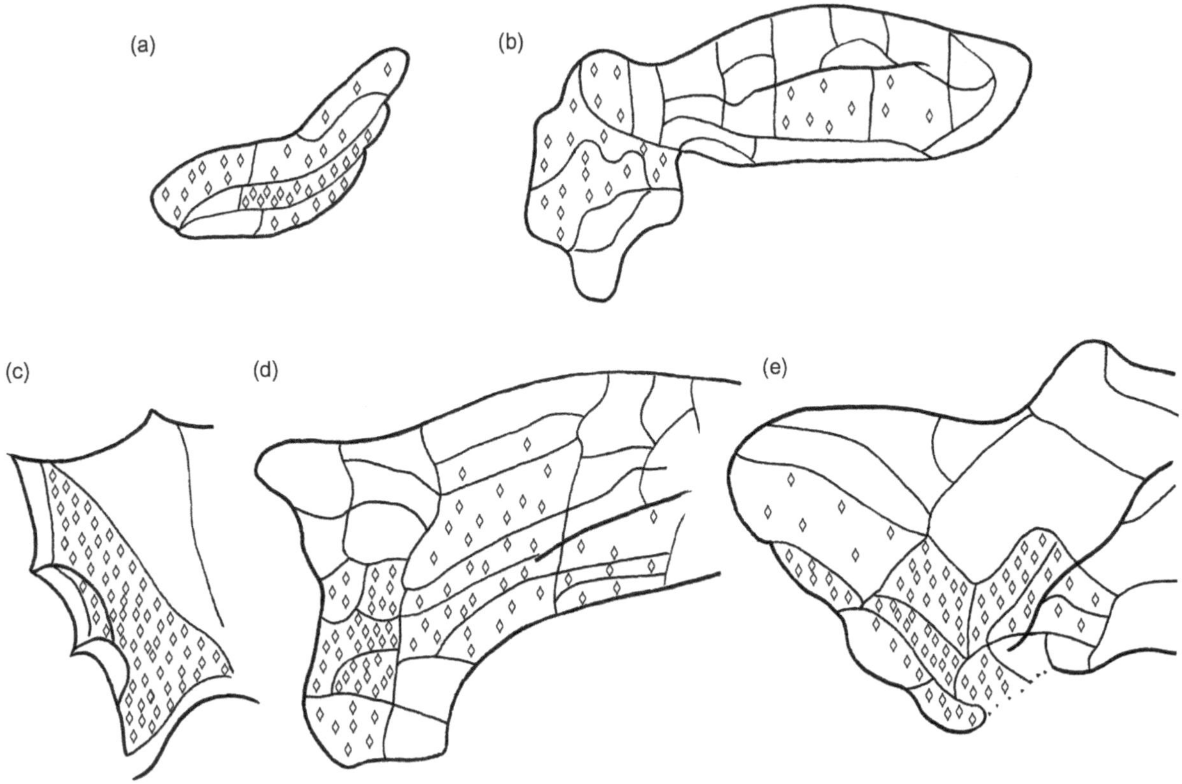

Fig. 13. The distribution of spindle cells over the insular cortex in (a) the ring-tailed lemur (Rose, 1928); (b) the baboon (Rose, 1928); (c), (d), and (e) the human, according to von Economo and Koskinas (1925), Rose (1928), and Brockhaus (1940), respectively.

the latter term referring to the spiral-shaped, corkscrew-like appearance of their processes. These elements are not only confined to the insula but also occur in the ACC. They are abundant in the fifth layer of the cortical areas mentioned, but scattered elements of the same type are also found in the third layer. von Economo (1926) emphasizes in a subsequent publication once again that these elements represent a new type of special cells. Moreover, he reports to have observed in silver-impregnated preparations that the axons of these cells originate from the soma and pass horizontally over some distance (Fig. 12c).

Rose (1928) reports the presence of typical spindle cells in the insula of the lemur, the baboon, and the human, but not in the insula of any of the "lower" mammalian species he examined. "Auf das Auftreten von Spindelzellen in der V. Schicht der *Regio insularis agranularis* wird deshalb hingewiesen, weil diese Elemente bei niederen Säugetieren nicht, oder wenigstens nicht in ihrer typischen Form nachweisbar sind" (l. c. p. 607). In the lemur, these elements were found in the posterior agranular, propeagranular, and granular regions (Figs. 10a and 13a); in the baboon, they were scattered over the propeagranular region and over the granular areas i7 and i11 (Figs. 9d,e and 13b); and in the human, they were concentrated in the anterior propeagranular cortex and occurred more scattered in the posterior propeagranular and tenuigranular regions (Figs. 10c and 13d).

Ngowyang (1932) confirms in a short paper the observations of Rose (1928) on the distribution of the spindle cells in the human insular cortex. He mentions that cells of this type also occur in the insula of the chimpanzee, but not in that of the cat. Moreover, he observed that the most basal part of the human insular cortex contains, in addition to typical spindle cells, elements in which the upper part of the soma passes over into two dendritic trunks. These, so-called fork cells (Gabelzellen), obviously correspond to the bifid cells described by Cajal (Fig. 12a and e). The paper of Ngowyang (1932) is also of historical interest because that author appears to have been the first who designated the spindle cells in the insular cortex as "the special cells of von Economo" (von Economosche Spezialzellen).

The observations of Brockhaus (1940) on the distribution of spindle cells in the human insula appear to correspond largely to those of Rose (1928), particularly so, if allowance is made to the differences in overall shape of the reconstructions of the insula, prepared by these two authors (Fig. 13d and e).

Commentary

Conclusions from the data reviewed

It can be concluded that the fifth layer of the anteroventral human insular cortex contains numerous spindle-shaped neurons, which most probably represent aberrant pyramidal cells. Similar cells have been observed in the insular cortex of the chimpanzee, the baboon, and the lemur, but neither in the insula of the cat nor in that of any "lower" mammal. There is a difference of opinion concerning the origin and course of the axons of the spindle cells. Cajal (1900; Fig. 12a) indicates that these axons originate from the end of the basal dendritic trunk and descend to the subcortical white matter, but von Economo (1926; Fig. 12c) observed that they arise from the lateral aspect of the soma and course horizontally. This observation has got some support from a recent study of Brune et al. (2010). These authors presented a photomicrograph of a large, cresyl-violet-stained VEN with an elongated soma, from which an axon tapers laterally. Nimchinsky et al. (1995) demonstrated that a lipophilic dye, injected into the cingulum bundle, backfills spindle cells in the ACC, thus indicating that they project their axons into the white matter. The destination of the axons of the spindle cells in both the cingulate gyrus and the anterior insula is still unknown.

The current focus on the insular spindle cells

The large spindle cells in the anterior insula, and in the ACC, were rediscovered by Nimchinsky et al. (1995, 1999) and Allman et al. (2005). In order to avoid confusion with other uses of the term spindle cell, they opted to call them, after its alleged discoverer, VENs. Nimchinsky et al. (1999) reported, on the basis of an extensive comparative study, that these cells only occur in great apes and humans, and not in other primates nor in any other of the numerous mammalian species investigated. Hence, these elements were considered as a phylogenetically recent specialization in hominoid evolution. Their large size and the fact that they send their axon out of the cerebral cortex (see above) have led to the hypothesis that these cells are specialized for rapid transmission of information over long distances (Nimchinsky et al., 1995). Allman et al. (2005) have suggested that the VENs are part of the neural circuitry involved in social awareness and may participate in fast, intuitive decisions in complex and rapidly changing social situations. As such, they could be part of the circuitry supporting human social networks. It has been established that the VENs are strongly and selectively affected in a particular variant of FTD, that is characterized by severe deficits in the patient's ability to realize the emotional impact of their demeanor on others (Seeley, 2008, 2010). We have seen already that according to Craig (2009), the AIC and the ACC, including the VENs they contain, represent salient components of a core control network that guides all mental activity and behavior in adult humans (Fig. 1e).

In the mean time, it has been established that VENs are not only confined to humans and apes but also occur in the anterior insula and the anterior cingulate gyrus of whales (Hof and van der Gucht, 2007) and elephants (Hakeem et al., 2009). In light of these findings, the concept that VENs are typical for hominoid evolution had to be abandoned. It was now proposed that these neurons have emerged independently several times in large-brained animals with wide and complex social networks, and that they may reflect a specialization for the rapid transmission of crucial social information in very large brains (Hakeem et al., 2009). Craig (2009) pointed out that all animals possessing VENs have the exclusive capacity to pass the mirror test for self-recognition, indicating the presence of self-awareness. The fully neglected fact that Rose, as early as1928, established that the spindle-shaped neurons, now known as VENs, are also present in the insula of the, not particularly large-brained, hamadryas baboon, and even in the lemur, will necessitate further adaptations of the theory concerning the specific functions of these remarkable elements. Butti and Hof (2010) recently reported that VENs also occur in the insular cortex of the pygmy hippopotamus, the Atlantic walrus, as well as the Florida manatee. It will be quite a job to subject all of these animals to the mirror test for self-recognition.

The architecture of the human insular cortex: synopsis and perspective

All studies on the cytoarchitectecture of the human insular cortex carried out so far agree that this cortex contains a rostroventral agranular zone and a dorsocaudal granular zone. The agranular zone and its immediate surroundings are connected with limbic structures, such as the amygdala and the posterior orbitofrontal and anterior cingulate cortices. They are involved in the processing of autonomic and food-related information, and neuroimaging studies have shown that they are strongly activated during personal emotional and social-emotional tasks. The anteroventral insular zone contains a concentration of VENs (Fig. 13c–e), and it is of note that this zone coincides with one of Flechsig's (1920) late-myelinating association areas (Fig. 14a).

The granular zone is a constant feature in all maps of the human insula, but there appear to be considerable differences with regard to its rostral extent. In the maps of Brodmann (1909; Fig. 4a), von Economo and Koskinas (1925; Fig. 4c), and Bailey and von Bonin (1951; Fig. 4d), the rostral border of the granular zone coincides (approximately) with the central insular sulcus; in the recently produced map of Kurth et al. (2010a; Fig. 6b), this zone is confined to the dorsocaudal part of the posterior insular lobule, whereas in the maps of Rose (1928; Fig. 5c), Brockhaus (1940; Fig. 5f), and Bonthius et al. (2005; Fig. 6a), it clearly extends into the dorsal part of the anterior insular lobule. Moreover, Rose (1928) and Brockhaus (1940) found some clearly granular areas in the morphologically most rostral parts of the insula, which so far have not been observed by other investigators. These discrepancies, with regard to the rostral extent of the granular zone in the human insula, need to be clarified. However, there can be no doubt that the dorsocaudal core region of the granular zone subserves sensory functions. This core region is, according to the mapping studies of Brockhaus (1940) and Kurth et al. (2010a), differentiated into several cytoarchitectonic areas, and the electrical stimulation experiments of Stephani et al. (2011; Fig. 1b) have revealed the presence of spatially separated zones within this region, from which gustatory, general viscerosensory, general somatosensory, and pain sensations could be elicited. These functional findings are in full harmony with the results of experimental hodological studies in monkeys (see Section "Introduction"). Combination of the data reviewed warrants the conclusion that the structural units in the posterior insular core region are at the same time functional units. They answer to the "Rindenfelder" of Vogt

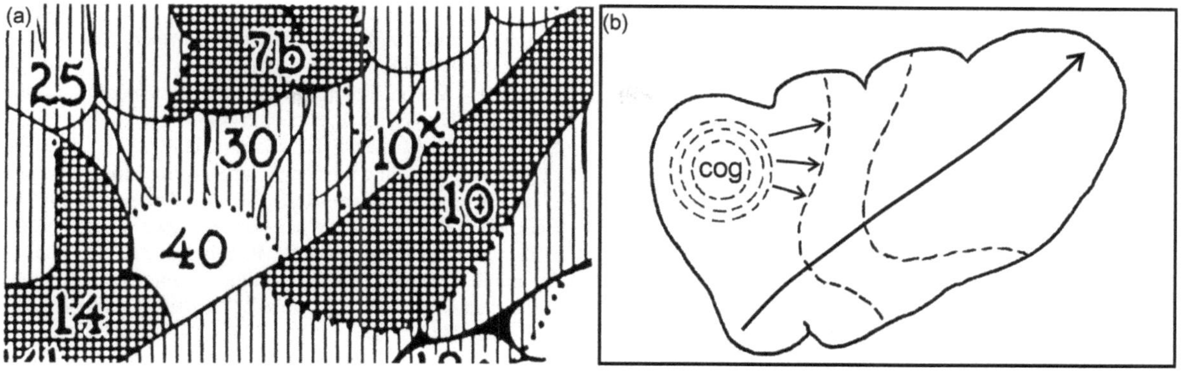

Fig. 14. The human insula. (a) A small part of Flechsig's (1920) myelogenetic map, showing that one of the late-myelinating association areas (numbered 40), distinguished by that author, is situated in the ventral insula. (b) Diagrammatic representation of the results of an *in vivo* probabilistic tractography study on the human insula, by Cerliani et al. (2011). It was found that the overall connectivity pattern of the insula changes gradually along a rostroventral-to-dorsocaudal axis (large arrow), and that the gradients in the shift of connectivity are oriented more or less perpendicular to this axis (dashed curves). The present author considers it likely that the course of the "gradient curves" has been influenced by the evolutionary expansion (small arrows) of a cognitive domain (cog) in the rostrodorsal insula.

and Vogt (1954), the "functional fields" of Roland and Zilles (1998), and the "discrete structural and functional modules" of Ongur et al. (2003).

In the maps of Brodmann (1909; Fig. 4a), von Economo and Koskinas (1925; Fig. 4c), and Bailey and von Bonin (1951; Fig. 4d), the agranular and granular insular zones border directly to each other, but in the map of Bonthius et al. (2005; Fig. 6a), these two zones are separated by a dysgranular zone, and this situation concurs with that seen in several maps of the insular cortex of the macaque (Fig. 11a and b). Neuroimaging studies, summarized by Mutschler et al. (2009; Fig. 1c) and Kurth et al. (2010b; Fig. 1d), have shown that the anterodorsal part of the insula, that is, an area falling within the confines of Bonthius' dysgranular zone, contains a domain which is critically involved in numerous cognitive functions including attentional processes, memory retrieval, and language production. In the section on the comparative anatomy of the insula of the present chapter, we adduce comparative anatomical and general connectional evidence, suggesting that this domain has expanded considerably during hominoid evolution (Fig. 14b). Neuroimaging studies have shown that complex functions or tasks never lead to the isolated activation of a single cortical locus but always induce synchronous activity in multiple cortical loci. Thus, it has been established that six different cortical areas, the anterodorsal insula, the dorsolateral prefrontal cortex, the ACC, the dorsal premotor cortex, the so-called inferior frontal junction, and the posterior parietal cortex, are consistently involved in a large variety of cognitive control tasks. Using fMRI techniques, Cole and Schneider (2007) demonstrated that these coactive regions form a functionally connected cognitive control network. So, the anterodorsal insular domain under consideration represents a functional node in a cognitive network. It would be highly interesting to know how the structural and functional organization of this typically cognitive insular domain relate to the corresponding features of the typically sensory insular domain discussed above.

Acknowledgments

The author thanks Dr. Michel Hofman for asking him to contribute to this volume, and for *not asking* him to reduce the chapter to the limits originally set for it, Dr. Leonardo Cerliani for many, most stimulating discussions on "our" insula, Mr. Ton Put and Mr. Wil Maas for help with the illustrations, and to Suzanne Bakker M.Sc. for moral support and reference management. Finally, the author wants to acknowledge especially the invaluable and continuous assistance of Dr. Jenneke Kruisbrink, the librarian of our institute, with literature retrieval. Without her help and without access to the network of libraries and the collection of the Deutsche Zentralbibliothek fuer Medizin (ZBMed) in Cologne, this chapter would not have been possible.

References

Ackermann, H., & Riecker, A. (2010). The contribution(s) of the insula to speech production: A review of the clinical and functional imaging literature. *Brain Structure and Function*, *214*, 419–433.

Afif, A., & Mertens, P. (2010). Description of sulcal organization of the insular cortex. *Surgical and Radiologic Anatomy*, *32*, 491–498.

Afif, A., Minotti, L., Kahane, P., & Hoffmann, D. (2010). Anatomofunctional organization of the insular cortex: A study using intracerebral electrical stimulation in epileptic patients. *Epilepsia*, *51*, 2305–2315.

Alkire, M. T., White, N. S., Hsieh, R., & Haier, R. J. (2004). Dissociable brain activation responses to 5-Hz electrical pain stimulation: A high-field functional magnetic resonance imaging study. *Anesthesiology*, *100*, 939–946.

Allman, J. M., Tetreault, N. A., Hakeem, A. Y., Manaye, K. F., Semendeferi, K., Erwin, J. M., et al. (2010). The von Economo neurons in frontoinsular and anterior cingulate cortex in great apes and humans. *Brain Structure and Function*, *214*, 495–517.

Allman, J. M., Tetreault, N. A., Hakeem, A. Y., Manaye, K. F., Semendeferi, K., Erwin, J. M., et al. (2011a). The von Economo neurons in the frontoinsular and anterior cingulate cortex. *Annals of the New York Academy of Sciences*, *1225*, 59–71.

Allman, J. M., Tetreault, N. A., Hakeem, A. Y., & Park, S. (2011b). The von economo neurons in apes and humans. *American Journal of Human Biology*, *23*, 5–21.

Allman, J. M., Watson, K. K., Tetreault, N. A., & Hakeem, A. Y. (2005). Intuition and autism: A possible role for Von Economo neurons. *Trends in Cognitive Sciences*, *9*, 367–373.

Augustine, J. R. (1996). Circuitry and functional aspects of the insular lobe in primates including humans. *Brain Research Reviews*, *22*, 229–244.

Bagshaw, M. H., & Pribram, K. H. (1953). Cortical organization in gustation (*Macaca mulatta*). *Journal of Neurophysiology*, *16*, 499–508.

Bailey, P., & von Bonin, G. (1951). *The isocortex of man.* Urbana: University of Illinois Press.

Bamiou, D. E., Musiek, F. E., & Luxon, L. M. (2003). The insula (Island of Reil) and its role in auditory processing. Literature review. *Brain Research. Brain Research Reviews*, *42*, 143–154.

Betz, W. (1874). Anatomischer Nachweis zweier Gehirnzentra. *Zentralblatt fuer medizinische Wissenschaft*, *12*, 578–580.

Biemond, A. (1956). The conduction of pain above the level of the thalamus opticus. *Archives of Neurology and Psychiatry*, *75*, 231–244.

Bieser, A. (1998). Processing of twitter-call fundamental frequencies in insula and auditory cortex of squirrel monkeys. *Experimental Brain Research*, *122*, 139–148.

Birklein, F., Rolke, R., & Muller-Forell, W. (2005). Isolated insular infarction eliminates contralateral cold, cold pain, and pinprick perception. *Neurology*, *65*, 1381.

Bohland, J. W., & Guenther, F. H. (2006). An fMRI investigation of syllable sequence production. *NeuroImage*, *32*, 821–841.

Bonthius, D. J., Solodkin, A., & Van Hoesen, G. W. (2005). Pathology of the insular cortex in Alzheimer disease depends on cortical architecture. *Journal of Neuropathology and Experimental Neurology*, *64*, 910–922.

Brockhaus, H. (1940). Die Cyto- und Myeloarchitektonik des Cortex claustralis und des Claustrum beim Menschen. *Journal fuer Psychologie und Neurologie*, *49*, 249–348.

Brodmann, K. (1905). Beiträge zur histologischen Lokalisation der Grosshirnrinde. Dritte Mitteilung: Die Rindenfelder der niederen Affen. *Journal fuer Psychologie und Neurologie*, *4*, 177–226.

Brodmann, K. (1909). *Vergleichende Lokalisationslehre der Grosshirnrinde in ihren Prinzipien dargestellt und Grund des Zellenbaues.* Leipzig: J.A. Barth.

Brooks, J. C., Zambreanu, L., Godinez, A., Craig, A. D., & Tracey, I. (2005). Somatotopic organisation of the human insula to painful heat studied with high resolution functional imaging. *NeuroImage*, *27*, 201–209.

Brune, M., Schobel, A., Karau, R., Benali, A., Faustmann, P. M., Juckel, G., et al. (2010). Von Economo neuron density in the anterior cingulate cortex is reduced in early onset schizophrenia. *Acta Neuropathologica*, *119*, 771–778.

Burton, H., Videen, T. O., & Raichle, M. E. (1993). Tactile-vibration-activated foci in insular and parietal-opercular cortex studied with positron emission tomography: Mapping the second somatosensory area in humans. *Somatosensory and Motor Research*, *10*, 297–308.

Butti, C., & Hof, P. R. (2010). The insular cortex: A comparative perspective. *Brain Structure and Function*, *214*, 477–493.

Büttner-Ennever, J. A., & Gerrits, N. M. (2004). Vestibular system. In G. Paxinos & J. K. Mai (Eds.), *The human nervous system* (pp. 1212–1240). Elsevier: Amsterdam.

Cacioppo, J. T., & Decety, J. (2011). Social neuroscience: Challenges and opportunities in the study of complex behavior. *Annals of the New York Academy of Sciences, 1224*, 162–173.

Cajal, S. R. (1900). Estudios sobre la corteza cerebral humana III: Corteza acústica. *Revista trimesta Micrográfica, 5*, 129–183.

Cajal, S. R. (1911). *Histologie du système nerveux de l'homme & des vertébrés.* Madrid: Instituto Ramon y Cajal.

Carmichael, S. T., & Price, J. L. (1996). Connectional networks within the orbital and medial prefrontal cortex of macaque monkeys. *The Journal of Comparative Neurology, 371*, 179–207.

Cerliani, L., Thomas, R. M., Jbabdi, S., Siero, J. C. W., Nanetti, L., Crippa, A., et al. (2011). Probabilistic tractography recovers a rostrocaudal trajectory of connectivity variability in the human insular cortex. *Human Brain Mapping*, (Published on line July 14, 2011) DOI: 10.1002/hbm.21338.

Chikama, M., McFarland, N. R., Amaral, D. G., & Haber, S. N. (1997). Insular cortical projections to functional regions of the striatum correlate with cortical cytoarchitectonic organization in the primate. *Journal of Neuroscience, 17*, 9686–9705.

Cole, M. W., & Schneider, W. (2007). The cognitive control network: Integrated cortical regions with dissociable functions. *NeuroImage, 37*, 343–360.

Colivicchi, F., Bassi, A., Santini, M., & Caltagirone, C. (2004). Cardiac autonomic derangement and arrhythmias in right-sided stroke with insular involvement. *Stroke, 35*, 2094–2098.

Craig, A. D. (2009). How do you feel—Now? The anterior insula and human awareness. *Nature Reviews Neuroscience, 10*, 59–70.

Craig, A. D. (Ed.), (2010a). Special issue on the Insula. *Brain Structure and Function, 214*, 395–680.

Craig, A. D. (2010b). The sentient self. *Brain Structure and Function, 214*, 563–577.

Craig, A. D., Bushnell, M. C., Zhang, E. T., & Blomqvist, A. (1994). A thalamic nucleus specific for pain and temperature sensation. *Nature, 372*, 770–773.

Crespo-Facorro, B., Kim, J., Andreasen, N. C., O'Leary, D. S., Bockholt, H. J., & Magnotta, V. (2000). Insular cortex abnormalities in schizophrenia: A structural magnetic resonance imaging study of first-episode patients. *Schizophrenia Research, 46*, 35–43.

Damasio, H., & Damasio, A. R. (1980). The anatomical basis of conduction aphasia. *Brain, 103*, 337–350.

de Vries, E. (1910). Bemerkungen zur Ontogenie und vergleichenden Anatomie des Claustrums. *Folia Neurobiologica, 4*, 481–513.

Dronkers, N. F. (1996). A new brain region for coordinating speech articulation. *Nature, 384*, 159–161.

Fasold, O., von Brevern, M., Kuhlberg, M., Ploner, C. J., Villringer, A., Lempert, T., et al. (2002). Human vestibular cortex as identified with caloric stimulation in functional magnetic resonance imaging. *NeuroImage, 17*, 1384–1393.

Flechsig, P. (1920). *Anatomie des menschlichen Gehirns und Rückenmarks auf myelogenetischer Grundlage.* Leipzig: Thieme.

Flynn, F. G. (1999). Anatomy of the insula functional and clinical correlates. *Aphasiology, 13*, 55–78.

Friedman, D. P., Murray, E. A., O'Neill, J. B., & Mishkin, M. (1986). Cortical connections of the somatosensory fields of the lateral sulcus of macaques: Evidence for a corticolimbic pathway for touch. *The Journal of Comparative Neurology, 252*, 323–347.

Fudge, J. L., Breitbart, M. A., Danish, M., & Pannoni, V. (2005). Insular and gustatory inputs to the caudal ventral striatum in primates. *The Journal of Comparative Neurology, 490*, 101–118.

Gallay, D. S., Gallay, M. N., Jeanmonod, D., Rouiller, E. M., & Morel, A. (2011). The insula of Reil revisited: Multiarchitectonic organization in macaque monkeys. *Cerebral Cortex*, (Published on line May 25, 2011) DOI: 101093/cercor/bhr104.

Garey, L. (2006). *Brodmann's localisation in the cerebral cortex: The principles of comparative localisation in the cerebral cortex based on the cytoarchitectonics.* Heidelberg: Springer.

Grusser, O. J., Pause, M., & Schreiter, U. (1990a). Localization and responses of neurones in the parieto-insular vestibular cortex of awake monkeys (*Macaca fascicularis*). *The Journal of Physiology, 430*, 537–557.

Grusser, O. J., Pause, M., & Schreiter, U. (1990b). Vestibular neurones in the parieto-insular cortex of monkeys (*Macaca fascicularis*): Visual and neck receptor responses. *The Journal of Physiology, 430*, 559–583.

Hakeem, A. Y., Sherwood, C. C., Bonar, C. J., Butti, C., Hof, P. R., & Allman, J. M. (2009). von Economo neurons in the elephant brain. *Anatomical Record, 292*, 242–248.

Hof, P. R., & Van der Gucht, E. (2007). Structure of the cerebral cortex of the humpback whale, Megaptera novaeangliae (Cetacea, Mysticeti, Balaenopteridae). *Anatomical Record, 290*, 1–31.

Hua, L. H., Strigo, I. A., Baxter, L. C., Johnson, S. C., & Craig, A. D. (2005). Anteroposterior somatotopy of innocuous cooling activation focus in human dorsal posterior insular cortex. *American Journal of Physiology. Regulatory, Integrative and Comparative Physiology, 289*, R319–R325.

Ibanez, A., Gleichgerrcht, E., & Manes, F. (2010). Clinical effects of insular damage in humans. *Brain Structure and Function, 214*, 397–410.

Isnard, J., Magnin, M., Jung, J., Mauguiere, F., & Garcia-Larrea, L. (2011). Does the insula tell our brain that we are in pain? *Pain, 152*, 946–951.

Jones, C. L., Ward, J., & Critchley, H. D. (2010). The neuropsychological impact of insular cortex lesions. *Journal of Neurology, Neurosurgery, and Psychiatry, 81*, 611–618.

Karnath, H. O., & Baier, B. (2010). Right insula for our sense of limb ownership and self-awareness of actions. *Brain Structure and Function*, *214*, 411–417.

Kurth, F., Eickhoff, S. B., Schleicher, A., Hoemke, L., Zilles, K., & Amunts, K. (2010a). Cytoarchitecture and probabilistic maps of the human posterior insular cortex. *Cerebral Cortex*, *20*, 1448–1461.

Kurth, F., Zilles, K., Fox, P. T., Laird, A. R., & Eickhoff, S. B. (2010b). A link between the systems: Functional differentiation and integration within the human insula revealed by meta-analysis. *Brain Structure and Function*, *214*, 519–534.

Lamm, C., & Singer, T. (2010). The role of anterior insular cortex in social emotions. *Brain Structure and Function*, *214*, 579–591.

Landau, E. (1919). The comparative anatomy of the nucleus amygdalae, the claustrum and the insular cortex. *Journal of Anatomy*, *53*, 351–360.

Luppino, G., Matelli, M., Camarda, R., & Rizzolatti, G. (1993). Corticocortical connections of area F3 (SMA-proper) and area F6 (pre-SMA) in the macaque monkey. *The Journal of Comparative Neurology*, *338*, 114–140.

Mazzola, L., Isnard, J., & Mauguiere, F. (2006). Somatosensory and pain responses to stimulation of the second somatosensory area (SII) in humans. A comparison with SI and insular responses. *Cerebral Cortex*, *16*, 960–968.

Medford, N., & Critchley, H. D. (2010). Conjoint activity of anterior insular and anterior cingulate cortex: Awareness and response. *Brain Structure and Function*, *214*, 535–549.

Menon, V., & Uddin, L. Q. (2010). Saliency, switching, attention and control: A network model of insula function. *Brain Structure and Function*, *214*, 655–667.

Mesulam, M. M., & Mufson, E. J. (1985). The insula of Reil in man and monkey. In A. Peters & E. G. Jones (Eds.), *Association and auditory cortices* (pp. 179–226). New York: Plenum.

Meyer, S., Strittmatter, M., Fischer, C., Georg, T., & Schmitz, B. (2004). Lateralization in autonomic dysfunction in ischemic stroke involving the insular cortex. *NeuroReport*, *15*, 357–361.

Mufson, E. J., Sobreviela, T., & Kordower, J. H. (1997). Chemical neuroanatomy of the primate insula cortex: Relationship to cytoarchitectonics connectivity function and neurodegeneration. In F. E. Bloom, et al. (Ed.), *Handbook of chemical neuroanatomy* (Vol. 13, pp. 377–454). Amsterdam: Elsevier.

Mutschler, I., Wieckhorst, B., Kowalevski, S., Derix, J., Wentlandt, J., Schulze-Bonhage, A., et al. (2009). Functional organization of the human anterior insular cortex. *Neuroscience Letters*, *457*, 66–70.

Naidich, T. P., Kang, E., Fatterpekar, G. M., Delman, B. N., Gultekin, S. H., Wolfe, D., et al. (2004). The insula: Anatomic study and MR imaging display at 1.5 T. *American Journal of Neuroradiology*, *25*, 222–232.

Naqvi, N. H., & Bechara, A. (2009). The hidden island of addiction: The insula. *Trends in Neurosciences*, *32*, 56–67.

Naqvi, N. H., & Bechara, A. (2010). The insula and drug addiction: An interoceptive view of pleasure, urges, and decision-making. *Brain Structure and Function*, *214*, 435–450.

Ngowyang, G. (1932). Beschreibung einer Art von Spezialzellen in der Inselrinde. *Journal für Psychologie und Neurologie*, *44*, 671–674.

Nieuwenhuys, R., ten Donkelaar, H. J., & Nicholson, C. (1998). *The central nervous system of vertebrates.* Heidelberg: Springer.

Nieuwenhuys, R., Voogd, J., & van Huijzen, C. (2008). *The human central nervous system.* Heidelberg: Springer.

Nimchinsky, E. A., Gilissen, E., Allman, J. M., Perl, D. P., Erwin, J. M., & Hof, P. R. (1999). A neuronal morphologic type unique to humans and great apes. *Proceedings of the National Academy of Sciences of the United States of America*, *96*, 5268–5273.

Nimchinsky, E. A., Vogt, B. A., Morrison, J. H., & Hof, P. R. (1995). Spindle neurons of the human anterior cingulate cortex. *The Journal of Comparative Neurology*, *355*, 27–37.

Ongur, D., Ferry, A. T., & Price, J. L. (2003). Architectonic subdivision of the human orbital and medial prefrontal cortex. *The Journal of Comparative Neurology*, *460*, 425–449.

Oppenheimer, S. (1994). Neurogenic cardiac effects of cerebrovascular disease. *Current Opinion in Neurology*, *7*, 20–24.

Oppenheimer, S. (2006). Cerebrogenic cardiac arrhythmias: Cortical lateralization and clinical significance. *Clinical Autonomic Research*, *16*, 6–11.

Ostrowsky, K., Isnard, J., Ryvlin, P., Guenot, M., Fischer, C., & Mauguiere, F. (2000). Functional mapping of the insular cortex: Clinical implication in temporal lobe epilepsy. *Epilepsia*, *41*, 681–686.

Ostrowsky, K., Magnin, M., Ryvlin, P., Isnard, J., Guenot, M., & Mauguiere, F. (2002). Representation of pain and somatic sensation in the human insula: A study of responses to direct electrical cortical stimulation. *Cerebral Ccortex*, *12*, 376.

Penfield, W., & Faulk, M. E. Jr. (1955). The insula; further observations on its function. *Brain*, *78*, 445–470.

Penfield, W., & Rasmussen, T. (1950). *The cerebral cortex of man.* New York: MacMillan.

Petrides, M., & Pandya, D. N. (2002). Comparative cytoarchitectonic analysis of the human and the macaque ventrolateral prefrontal cortex and corticocortical connection patterns in the monkey. *European Journal of Neuroscience*, *16*, 291–310.

Phan, K. L., Wager, T., Taylor, S. F., & Liberzon, I. (2002). Functional neuroanatomy of emotion: A meta-analysis of emotion activation studies in PET and fMRI. *NeuroImage*, *16*, 331–348.

Pritchard, T. C., & Norgren, R. (2004). Gustatory system. In G. Paxinos & J. K. Mai (Eds.), *The human nervous system* (pp. 1171–1196). Amsterdam: Elsevier.

Retzius, G. (1896). *Das Menschenhirn.* Stockholm: Königliche Buchdruckerei P.A. Norstedt und Söner.

Retzius, G. (1902). Zur Morphologie der Insula Reili. *Biologische Untersuchungen, Neue Folge X*, 14–20.

Roberts, T. S., & Akert, K. (1963). Insular and opercular cortex and its thalamic projection in *Macaca mulatta. Schweizer Archiv für Neurologie Neurochirurgie und Psychiatrie, 92*, 1–43.

Robinson, C. J., & Burton, H. (1980a). Organization of somatosensory receptive fields in cortical areas 7b, retroinsula, postauditory and granular insula of M. fascicularis. *The Journal of Comparative Neurology, 192*, 69–92.

Robinson, C. J., & Burton, H. (1980b). Somatic submodality distribution within the second somatosensory (SII), 7b, retroinsular, postauditory, and granular insular cortical areas of M. fascicularis. *The Journal of Comparative Neurology, 192*, 93–108.

Roland, P. E., & Zilles, K. (1998). Structural divisions and functional fields in the human cerebral cortex. *Brain Research. Brain Research Reviews, 26*, 87–105.

Rose, M. (1928). Die Inselrinde des Menschen und der Tiere. *Journal fuer Psychologie und Neurologie, 37*, 467–624.

Rose, M. (1935). Cytoarchtektonik und Myeloarchitektonik der Groszhirnrinde. In O. Bumke & O. Foerster (Eds.), *Handbuch der Neurologie* (Vol. I, pp. 589–778). Berlin: Springer.

Saper, C. B. (2002). The central autonomic nervous system: Conscious visceral perception and autonomic pattern generation. *Annual Review of Neuroscience, 25*, 433–469.

Schneider, R. J., Friedman, D. P., & Mishkin, M. (1993). A modality-specific somatosensory area within the insula of the rhesus monkey. *Brain Research, 621*, 116–120.

Seeley, W. W. (2008). Selective functional, regional, and neuronal vulnerability in frontotemporal dementia. *Current Opinion in Neurology, 21*, 701–707.

Seeley, W. W. (2010). Anterior insula degeneration in frontotemporal dementia. *Brain Structure and Function, 214*, 465–475.

Seltzer, B., & Pandya, D. N. (1991). Post-rolandic cortical projections of the superior temporal sulcus in the rhesus monkey. *The Journal of Comparative Neurology, 312*, 625–640.

Semendeferi, K., & Damasio, H. (2000). The brain and its main anatomical subdivisions in living hominoids using magnetic resonance imaging. *Journal of Human Evolution, 38*, 317–332.

Shelley, B. P., & Trimble, M. R. (2004). The insular lobe of Reil—its anatamico-functional, behavioral and neuropsychiatric attributes in humans—A review. *The World Journal of Biological Psychiatry, 5*, 176–200.

Small, D. M. (2010). Taste representation in the human insula. *Brain Structure & Function, 214*, 551–561.

Stephani, C., Fernandez-Baca Vaca, G., Maciunas, R., Koubeissi, M., & Luders, H. O. (2011). Functional neuroanatomy of the insular lobe. *Brain Structure and Function, 216*, 137–149.

Sterzer, P., Stadler, C., Poustka, F., & Kleinschmidt, A. (2007). A structural neural deficit in adolescents with conduct disorder and its association with lack of empathy. *NeuroImage, 37*, 335–342.

Sudakov, K., MacLean, P. D., Reeves, A., & Marino, R. (1971). Unit study of exteroceptive inputs to claustrocortex in awake, sitting, squirrel monkey. *Brain Research, 28*, 19–34.

Takahashi, T., Suzuki, M., Hagino, H., Zhou, S. Y., Kawasaki, Y., Nohara, S., et al. (2004). Bilateral volume reduction of the insular cortex in patients with schizophrenia: A volumetric MRI study. *Psychiatry Research, 131*, 185–194.

Takahashi, T., Suzuki, M., Zhou, S. Y., Hagino, H., Tanino, R., Kawasaki, Y., et al. (2005). Volumetric MRI study of the short and long insular cortices in schizophrenia spectrum disorders. *Psychiatry Research, 138*, 209–220.

Tatschl, C., Stollberger, C., Matz, K., Yilmaz, N., Eckhardt, R., Nowotny, M., et al. (2006). Insular involvement is associated with QT prolongation: ECG abnormalities in patients with acute stroke. *Cerebrovascular Diseases, 21*, 47–53.

Tramo, M. J., Loftus, W. C., Thomas, C. E., Green, R. L., Mott, L. A., & Gazzaniga, M. S. (1995). Surface area of human cerebral cortex and its gross morphological subdivisions: In vivo measurements in monozygotic twins suggest differential hemisphere effects of genetic factors. *Journal of Cognitive Neuroscience, 7*, 292–302.

Ture, U., Yasargil, D. C., Al-Mefty, O., & Yasargil, M. G. (1999). Topographic anatomy of the insular region. *Journal of Neurosurgery, 90*, 720–733.

Verhagen, J. V., Kadohisa, M., & Rolls, E. T. (2004). Primate insular/opercular taste cortex: Neuronal representations of the viscosity, fat texture, grittiness, temperature, and taste of foods. *Journal of Neurophysiology, 92*, 1685–1699.

Vogt, C., & Vogt, O. (1954). Gestaltung der topistischen Hirkforschung und ihre Förderung durch den Hirnbau und seine Anomalien. *Journal für Hirnforschung, 1*, 1–46.

von Bonin, G., & Bailey, P. (1947). *The neocortex of* Macaca mulatta. Urbana, Illinois: University of Illinois Press.

von Economo, C. (1926). Eine neue Art Spezialzellen des Lobus cinguli und Lobus insulae. *Zeitschrift fuer die gesamte Neurologie und Psychiatrie, 100*, 706–712.

von Economo, C. F., & Koskinas, G. N. (1925). *Die Cytoarchitektonik der Hirnrinde des Erwachsenen Menschen. Textband und Atlas mit 112 Mikrophotographischen Tafeln.* Wien-Berlin: Julius Springer.

Weiller, C., Chollet, F., Friston, K. J., Wise, R. J., & Frackowiak, R. S. (1992). Functional reorganization of the brain in recovery from striatocapsular infarction in man. *Annals of Neurology, 31*, 463–472.

Yaxley, S., Rolls, E. T., & Sienkiewicz, Z. J. (1990). Gustatory responses of single neurons in the insula of the macaque monkey. *Journal of Neurophysiology, 63*, 689–700.

M. A. Hofman and D. Falk (Eds.)
Progress in Brain Research, Vol. 195
ISSN: 0079-6123

CHAPTER 8

The missing link: Evolution of the primate cerebellum

Carol MacLeod*

Department of Anthropology, Langara College, Vancouver, BC, Canada

Abstract: The cerebellum has too often been seen as the "little brain," subservient to the "big brain," the cerebrum. That is changing, as neuroimaging uncovers the cerebellum as the "missing link" in the neurological underpinnings of many cognitive domains. Connections between the neocortex and the cerebellum are now more precisely defined, with functionally localized areas of cerebellar cortex understood for cognitive tasks in humans. Comparative volumetric studies of the primate cerebellum have isolated some elements of circuitry, and our field is moving toward a better integration with the neurosciences in a systematic comparative framework. The next decade may show great advances, as relatively noninvasive techniques of neuroimaging have the potential to build a comparative model of the evolution of primate neurocircuitry.

Keywords: neocerebellum; dentate nucleus; inferior olivary nucleus; hominoid; neuroanatomy; primates.

Introduction: The cerebellum and cognition

Just as the early Neanderthal remains discovered in the nineteenth century were misunderstood as pathological moderns, or the Taung child of Raymond Dart seen as an extinct baboon (Tobias, 1971), it took many decades of subsequent fossil discoveries before these finds were understood as "missing links" in the complex expression of hominin prehistory. Fossil remains are not the only vestiges of hominin evolution, however, as evolutionary forms can be "unearthed" in primate anatomy by comparing structures in several extant primate taxa. As brain structures are highly allometric, any unexpected and significant break in patterning can be indicative of selection for an augmented neurological function. Such differential expansion can be seen in the hominoid cerebellum, especially the lateral hemispheres (MacLeod et al., 2003). Increasingly, neuroscientists are including the cerebellum in their models of functional circuitry, opening new vistas of collaboration between anthropology and the neurosciences.

*Corresponding author.
Tel.: +1-604-323-5722; Fax: +1-604-323-5555
E-mail: cmacleod@langara.bc.ca

DOI: 10.1016/B978-0-444-53860-4.00008-8

Twenty-five years ago, the groundbreaking work of Leiner et al. (1986) seeded researchers' minds to the possibility that the cerebellum participated in cognitive tasks in humans. Subsequent advances in neuroimaging provided much of the evidence. A reevaluation of the cerebellum was inevitable, but it would never have come so quickly if not for these three visionaries. In the last decade, research on cerebellar participation in cognitive domains has increased exponentially. The following paragraphs only dust the surface of this research.

Germinal studies in the 1990s like those of Raichle et al. (1994) explored lateral cerebellar participation in language tasks such as generating verbs from nouns, for example, "cut" for scissors, "chew" for bone. The connection between the right lateral cerebellum and phonemic fluency (Schweizer et al., 2010), temporal organization of a prearticulatory verbal code (inner speech) (Ackermann et al., 2007), and syntax or the temporal organization of spatial relations (Kotz and Schwartze, 2010), are now well accepted. Language processes of the left frontal and temporal lobes appear to function in concert with the right lateral cerebellum (Marien et al., 2001), consistent with neuroanatomical connections, although there is also evidence for left cerebellar hemisphere participation as well (see Murdoch, 2010 for review). The lateral cerebellum is part of a circuit with the superior temporal gyrus, the frontal operculum, and BA38 that underlies the perception of song, melody, and speech (Brown et al., 2004), with implications for the evolution of language.

In observing cerebellar-lesioned patients, Fiez et al. (1992) noticed that, unlike control subjects, cerebellar patients did not improve over time in a linguistic task of generating verbs from nouns. When controls were imaged in this task (Raichle et al., 1994), the cerebellum was very active in the first few trials, but no longer participated as the task became routinized. Numerous studies have shown that blood flow to the cerebellum decreases markedly after a task has been learned, implying that the cerebellum is important in the acquisition of new behavioral patterns (procedural learning), be they spatial, temporal, linguistic, or operational. Doyon (1997) postulates that the learned, automatic movements are then stored in cerebral cortical circuitry, contrary to the opposite "take over" view of stereotyped cerebellar patterning (see Sakai et al., 2004 discussed below).

Our understanding of cerebellar participation in working memory is still incomplete, and it may be more modality-specific than we realize. There is strong evidence that the cerebellum is active in verbal working memory, but inconsistent evidence that it is active in visual working memory (Ben-Yehuda et al., 2007), although an imaging study by Hautzel et al. (2009) shows strong bilateral activation of the cerebellum in both verbal and abstract working memory tasks. Verbal working memory is a complex process (Baddeley, 1992), recruiting several functions, including verbal rehearsal in which the cerebellum is involved. However, studies of cerebellar-lesioned patients do not support such an applied view of innate speech, and some researchers present evidence that the cerebellum is involved at a deeper level involving timing and error correction (Ben-Yehuda et al., 2007).

Both procedural learning (Lalonde, 1997) and working memory (Molinari and Leggio, 2007) can be associated with the cerebellum in rats and monkeys. Area 46 of the dorsolateral prefrontal cortex is active in both monkey and human working memory (Luebke et al., 2010), although it is extremely complicated to reliably localize common functions in such wildly divergent cognitions. Nonetheless, area 46 in *Cebus* is a multimodal associative area that has been directly linked to the cerebellum (Kelly and Strick, 2003), thereby showing a neuroanatomical basis of cerebellar participation in cognition in nonhuman primates.

Cerebellar participation in visuospatial tasks with a cognitive component (Kim et al., 1994) is well established, but the cerebellar role in the shifting of attention between visual and auditory stimuli (Courchesne et al., 1994) has been questioned. This may be because a common thread running through cerebellar activity is the ability to correct errors and organize activities into new patterns. Rather than the cerebellum being active in stereotyped tasks, it is most important in the converting of complex tasks to simpler patterns, as shown by work in divided and selective attention (Hahn et al., 2008). In this fMRI study, researchers tested whether there were brain structures specific to divided attention, in which two stimulus dimensions must be processed simultaneously, that could be distinguished from those used in selective attention, with only one stimulus dimension. The cerebellum was not active in simple selective attention tasks, but the left lateral cerebellum was the only brain structure that showed activity specific to the more complex task.

Imaging these activities, largely through fMRI, has led to connectivity models that link functionally localized areas of cerebral cortex with subcortical structures, but what the cerebellum actually does in these circuits underlying cognitive functions is still not understood. The cerebellum operates on very different principles than the neocortex and contributes at a deeper, almost abstract level of process. Marien et al. (2001), for example, do not think that there is any intrinsic linguistic generation in the lateral cerebellum, but that aphasia can be caused by the interruption of cerebellar excitement to the left hemisphere language areas. It is important to understand whether cerebellar participation in cognition is entirely dependent on neocortical connections, or whether it carries out more fundamental processing common to sensory, motor, and cognitive activities. Partial answers to this question may be found in cerebellar anatomy and connectivity.

Cerebellar structure and connectivity

The cerebellum has a compartmental structure organized in a series of parasagittal zones. Each major zone, the vermis (medial), paravermis, and neocerebellum (lateral), has its own output nucleus, the fastigial, interposed (globus and emboliform), and dentate nuclei, respectively (Altman and Bayer, 1997). The hemisphere folia, finely folded layers of cerebellar cortex, run perpendicular to these zones and form a series of 10 lobules divided into the anterior and posterior lobes. The vermis is the oldest part of the cerebellum, with the folia running orthogonal to the sagittal plane, while the hemisphere folia change direction in a medial to lateral direction, enabling their parcellation in fixed or scanned brains (MacLeod et al., 2003) (Fig. 1). The anterior cerebellum is primarily vermis, while the posterior cerebellum is primarily hemisphere. It is in the lateral part of the cerebellum that cognitive activity is most apparent in neuroimaging studies.

A somatotopic map is known for the medial cerebellum, but there is now accumulating evidence for localization of function within the cerebellar lobules connected to neocortical association areas. Stoodley and Schmahmann (2009) conducted a meta-analysis of cerebellar function as shown by fMRI and PET scans and followed up with a study on one individual to control for averaged peak activations (Stoodley et al., 2010). The two studies grouped comparable data into

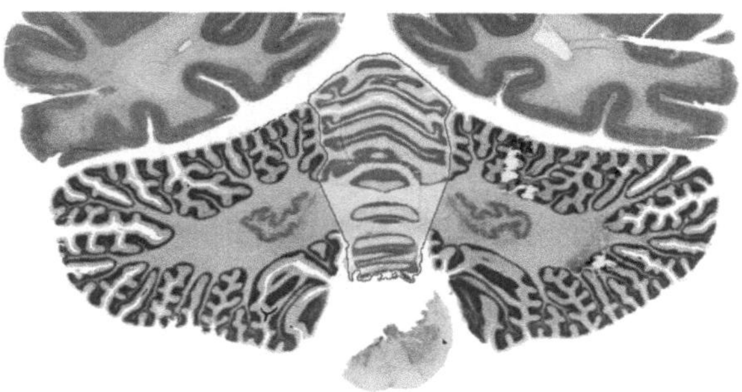

Fig. 1. Section of a gibbon cerebellum showing the parcellation of the vermis from the hemisphere, as done interactively on computer by author. Change in orientation of the folia from vermis to hemisphere defines the boundary. Dentate nucleus is visible in the white matter. Specimen is from the Zilles Collection.

seven domains: motor, somatosensory, language (cognitive), verbal working memory, executive, spatial, and limbic/emotional processing. Sensorimotor processing is concentrated in lobule V of the anterior cerebellum but is also found in posterior lobules VI and VIII (Fig. 2). Motor and somatosensory representations overlap and are right-lateralized. Language, working memory, executive, and spatial tasks are strictly confined to the posterior lateral cerebellum. The executive functions show a mosaic pattern of localization depending on the types of information being integrated. As the dentate nucleus is connected to the contralateral cerebral hemisphere, it is not surprising that linguistic functions are more prominent in the right cerebellar hemisphere, while visuospatial functions are more dominant in the left. Both involve Crus I and II, and VIIB, although they are not mirror images. The left–right dichotomy is not absolute, and most of the domains overlap somewhat, especially verbal working memory and language, but the degree of functional localization for cognitive tasks is well demonstrated in this metanalysis.

Functional circuitry magnetic resonance imaging (fcMRI) is a noninvasive technique that correlates resting signal fluctuations in the MR signal from brain regions with known anatomic connections (coherence). Allen et al.'s study (2005) correlates signal fluctuations in the dentate nucleus with temporal fluctuations in MR signals from all other brain voxels. There are three

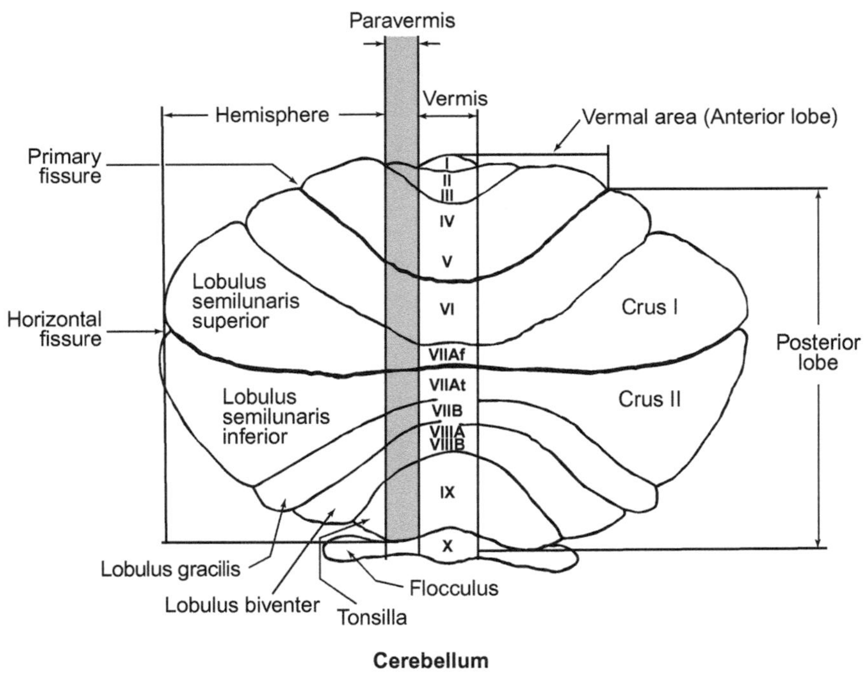

Fig. 2. Schematized drawing of cerebellar surface showing divisions of vermis, paravermis, and hemisphere. Primary fissure marks division between anterior and posterior cerebellum. Some conventional (clinical tradition) names are given on left side, while Schmahmann et al.'s terminology (1999, 2000) is given on midline and right side, following Larsell's Roman numeral designation of vermal and hemisphere lobules (e.g., Larsell and Jansen, 1970). Schmahmann et al. drop the distinction between hemisphere and vermal lobules in their descriptions, except for Crus I and Crus II, leading to some confusion. However, Schmahmann et al.'s terminology is now widely adopted in neuroimaging studies because the atlas gives standardized coordinates that allow accurate interstudy comparisons.

clusters of coherence, first within the cerebellum, then with subcortical structures such as the basal ganglia and the thalamus, and finally with neocortical areas. The left dentate shows strong coherence with the right inferior parietal (BA40), extending into the supramarginal and postcentral gyri as well as the angular gyrus. The right dentate shows strongest coherence with the left middle occipital (BA19) extending into BA31, 18, and 39 in the left hemisphere. Thus, the coherence is consistent with the Stoodley and Schmahmann (2009) study, that is, visuospatial in right parietal and language in left occipital, but not exactly as would be predicted, especially for language.

The left dentate has bilateral coherence with BA46 (working memory) in dorsolateral prefrontal cortex, whereas the right dentate shows only contralateral coherence with area 46. Allen et al. (2005) suggest that left dentate coherence may be partly explained by a bilateral projection of the dentate nucleus, but 97% of the dentate fibers reaching area 46 are contralateral in the Middleton and Strick (2001) tracer study in *Cebus*; only the interposed and fastigial nuclei are more bilaterally projecting. The bilateral signals in the cerebral cortex might be attributed to corticocortical signals, or even bilateral signals from the basal ganglia, as Middleton and Strick (1994) have established its participation in cerebellar circuits. Equally intriguing, dentate projections in humans may be more divergent than in *Cebus*.

Krienen and Buckner (2009) seeded frontal and cerebellar regions in an fcMRI study in 40 humans. Prefrontal regions active in cognition show coherence with posterior cerebellar hemispheres, especially Crus I and II, which when seeded show reciprocal coherence with these same prefrontal regions. Unilateral cortical seeds show bilateral cerebellar activity in the raw correlation maps, more than could be accounted by Brodal's finding (1979) that 10% of the projections from the pontine nuclei to the posterior lobe are ipsilateral. In Stoodley et al. (2010), verbal working memory is also bilateral in the cerebellum. Thus, connectivity between the neocortex and the cerebellum can be unambiguously demonstrated in the above studies, but bilateral activity, especially that associated with BA46, needs further investigation.

An fcMRI study by Habas et al. (2009) establishes coherence networks linking the neocortex and cerebellum with nonoverlapping localization. Both the left and right executive control networks show contralateral connectivity between Crus II and dorsolateral prefrontal cortex as well as the superior parietal lobe. Thus, there is no exclusive connectivity between Crus II and dorsolateral prefrontal cortex, in keeping with recent neuroimaging studies which find strong inferior parietal connections (O'Reilly et al., 2010) that, nonetheless, overlap with other cortical input in Crus II. Early tracer studies showed strong parietal–Crus II connections in the cat, but not in the monkey (Brodal, 1979). If significant differences in connectivity exist between cat and monkey, it is reasonable to assume that species-specific differences exist to some degree within the primate order. These neuroanatomical studies would be even more informative in a comparative primate sample.

Kelly and Strick (2003) examined cerebellar connectivity with M1 and Walker's area 46 in *Cebus*, using tracers that could pass through three or more synaptic junctures. Anterograde tracers from neocortex to pons to granule cells in cerebellar cortex show the same precise localization in both neocortex and cerebellum as the retrograde tracers connecting Purkinje cells to the dentate, then thalamus to M1 or area 46 (Fig. 3). Vermal lobules V–VI and hemisphere lobules VIIB–VIII both receive from and project to M1, while Crus II and vermal lobules VII and IX communicate with contralateral area 46. Dum and Strick (2003) isolated localized, nonoverlapping areas in the dentate nucleus projecting to M1 and area 46.

The localization in matching areas of cerebellar cortex, dentate nucleus, and thalamus from anterograde and retrograde tracers originating in small, precise neocortical areas has led Strick and colleagues to a model of closed, reciprocal

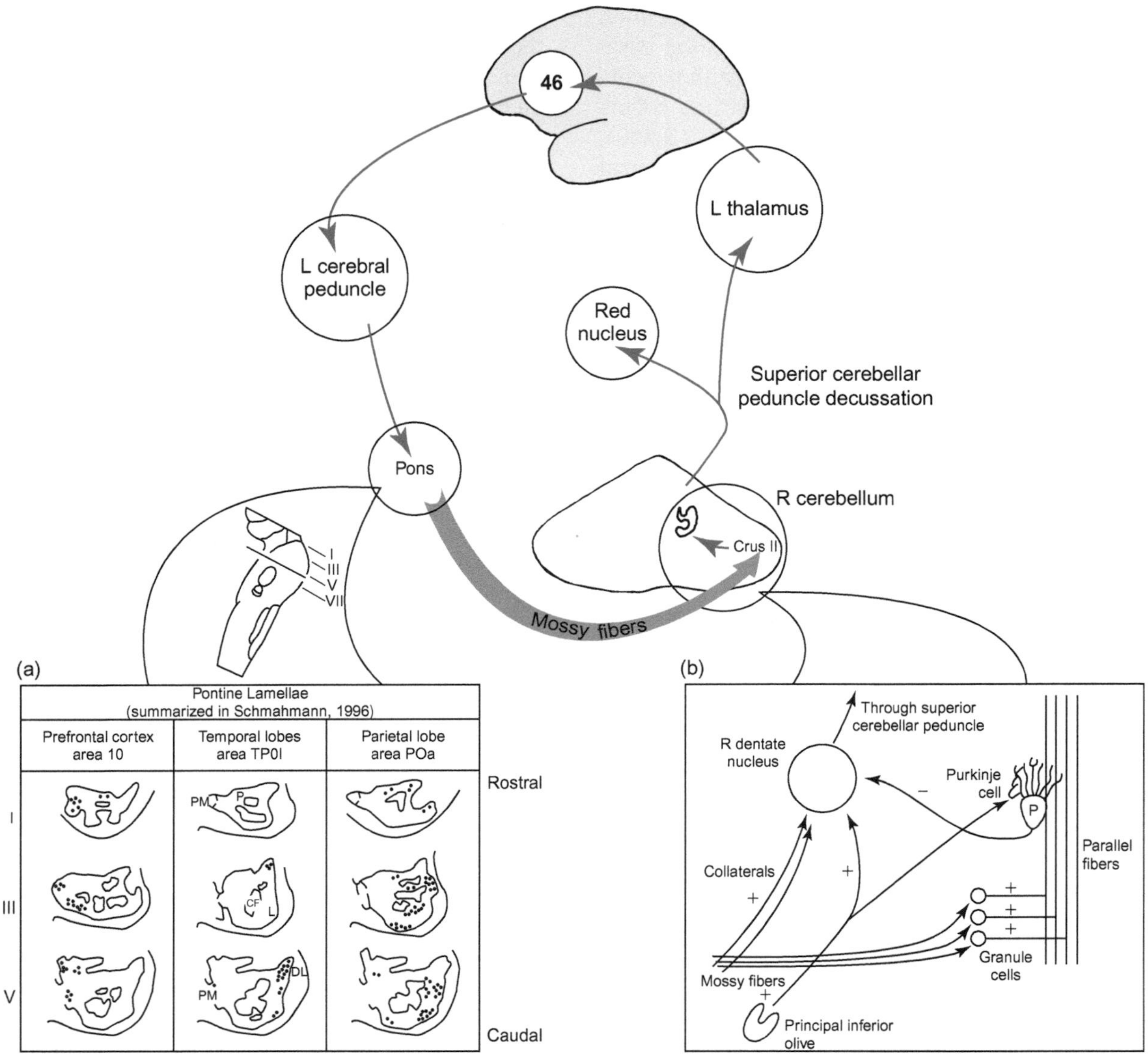

Fig. 3. Drawing showing connections between Walker's area 46 and Crus II in *Cebus* determined by Kelly and Strick (2003) and Dum and Strick (2003), a closed, reciprocal cortico-cerebello-cortical loop. Left highlight box (a) shows distribution of diverse neocortical projections to pontine nuclei in macaque from Schmahmann and Pandya summary (1997) of many years of research. Notice the varied, rostral–caudal pattern of interdigitating cortical projections on each lamella, allowing for the possibility of widely divergent cortical areas to influence one another by proximity. Anterograde tracers show projections from area 10 in first column, projections from the rostral upper bank of the superior temporal sulcus in the second column, and projections from the lower bank of the intraparietal sulcus in the third column. Figure 4 shows the nonoverlapping but interdigitating nature of pontine lamellae very clearly. Right highlight box (b) shows mossy and climbing fiber collaterals to dentate nucleus, which actually make up the excitatory pathway to area 46 that is modified by Purkinje cell inhibition. (Figure 3a is redrawn and modified, with permission, from Schmahmann, 1996).

loops for cerebellar–cerebrum interactions. Kelly and Strick (2003) conclude that separate areas of projection in cerebellar cortex for M1 and area 46 disprove Glickstein's hypothesis (2000) that the cerebellum is the primary route by which signals from cortical visual areas guide reaching movements. This hypothesis emerges from the fact that monkeys can still execute rapid and accurate arm movements when the fasciculus between the parietal and frontal lobe is cut, implying that subcortical links may be crucial to the dorsal visual stream between parietal and motor areas. When Stein et al. (1987) cooled the cerebellum, the ability of a monkey to track a moving target was severely perturbed. The putative path in this model is posterior neocortex—cerebellum—motor cortex, in contrast to Kelly and Strick's reciprocal, closed loops. Although Kelly and Strick argue that synthesis of information from diverse sources cannot be a major feature of cerebrocerebellar processing, they do see convergence operating on a smaller functional scale within localized areas of cerebellar cortex.

The possibility of cerebellar processing of recontextualized information has come from tracer studies examining neocortical projections to pontine nuclei (Brodal, 1979; Brodal and Bjaalie, 1997). Projections are localized in each section (lamella) and distributed in a rostrocaudal manner, with each section showing its own proportion of projections from different cortical areas. There is both "convergence" by which diverse areas of cortex project to defined areas of the pontine nucleus (Brodal and Bjaalie, 1997) and "divergence" by which mossy fibers from the pontine nuclei project to diverse areas of cerebellar cortex within large localized regions. Areas on each lamella are interdigitating, but not overlapping, as verified by studies of Schmahmann and Pandya (1997, summary) that traced projections from multimodal cortical areas in the macaque to the pons (Figs. 3a and 4). The recontextualization of cortical areas on the lamellae could be a mechanism for diverse areas of neocortex to influence one another (convergence). "Thus, we concluded that one particular cortical region has access to widely separated parts of the cerebellum, and that one particular part of the cerebellar cortex would receive convergent inputs from different parts of the cortex" (Brodal and Bjaalie, 1997, p. 240). Further, in the mossy fiber projections to the granule cells, pontine nuclei organization could partly account for the fractured somatotopy that Welker (1987) and others discovered on the cerebellar cortex of the cat, whereby the cerebral somatotopy of the whiskers was not duplicated in the cerebellar cortex but was rather broken up into smaller areas in new contexts (fractured somatotopy). Thus, in both the pons and the cerebellar cortex itself, there is a reconfiguration of cortical somatotopy, although the nature of this recontextualization is not really understood nor is it clear that Welker's findings could apply to more cognitive cerebellar functions (Schmahmann and Pandya, 1997). Strick et al. (2009) point out that the dentate output channels are overlapping for some areas, showing a functional rather than spatial correlation, since these areas of overlap do not follow neocortical somatotopy. If this functional reorganization does not take place in the cerebellum, then where does it take place?

Thus, if cerebellar connectivity is examined more closely, especially within the pons, there are opportunities in these loops for cerebellar processing, such that a direct connection with the neocortex is not the only explanation for cerebellar participation in cognition. There is evidence for overlap of localized areas on cerebellar cortex (Lu et al., 2007), just as the neuroimaging studies discussed above demonstrate overlap of functional areas. This, as with Welker's fractured somatotopy, would be another possible platform for cerebellar integration of cerebral input.

Cerebellar microcircuitry

The finer level of microcircuitry within the cerebellum reveals how distinctive cerebellar

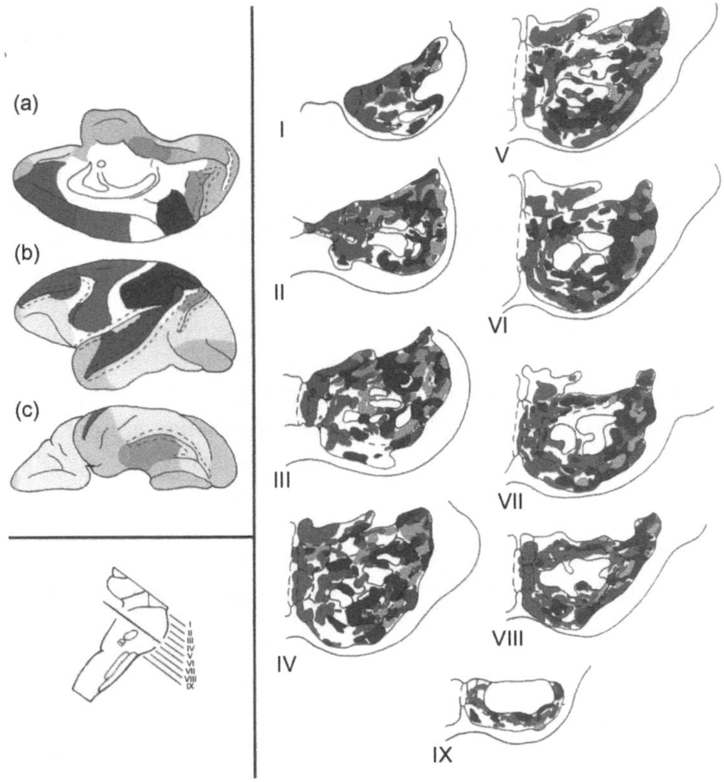

Fig. 4. Composite color-coded summary diagram illustrating distribution within the basilar pons of the rhesus monkey of projections derived from associative cortices in prefrontal (purple), posterior parietal (blue), temporal (red), and parastriate and parahippocampal regions (orange), and from motor, premotor, and supplementary motor areas (green). Medial (a), lateral (b), and ventral (c) surfaces of the cerebral hemisphere are shown at upper left. The plane of section through the basilar pons is at lower left, and rostrocaudal levels of pons I–IX are shown at right. Cerebral areas that have been shown to project to the pons by other investigators using either anterograde or retrograde tracers are depicted in white; those areas studied with both anterograde and retrograde studies and found to have no pontine projections are shown on the hemispheres in yellow; and those with no pontine projections according to retrograde studies by other investigators are shaded in gray. Dashed lines in hemisphere diagrams represent sulcal cortices. In the pons diagrams, dashed lines represent pontine nuclei, and solid lines depict the traversing corticofugal fibers. Pontine projections are presented as a whole, and this diagram does not illustrate the finding that each architectonic area has its own unique pattern of pontine terminations. Associative corticopontine projections are substantial and are not overshadowed by the motor cortico-pontine system. It is apparent that there is a complex mosaic of terminations in the pons. Each cerebralcortical region has preferential sites of pontine terminations. There is considerable interdigitation of terminations from some different cortical sites, but almost no overlap. This pattern is reminiscent of the fractured somatotopy shown in the sensory projections to the cerebellum. This figure was derived from a review of 80 *cases* previously reported in Schmahmann and Pandya (1989, 1991, 1993, 1995) and in abstract-form in Schmahmann and Pandya (1995). All *cases* were studied using the same experimental technique. Pontine terminations were mapped manually onto a standard outline of the pons. Inherent inaccuracies in this method are readily acknowledged, largely on the basis of between-case comparison. There are also unavoidable inaccuracies in the attempted precise transformation of the data from an actual transverse section of the pons to an idealized version. Open areas in the pons are likely to represent sites of termination of projections from cortices not studied by these investigators. (Reproduced with permission from Schmahmann, 1996). (For interpretation of the references to color in this figure legend, the reader is referred to the Web version of this chapter.)

processing is from that in the neocortex. Cytoarchitecture shows that functional localization in the neocortex and the cerebellar cortex are not the same. The neocortex has six layers with discernable variations in neuron size and distribution that define elaborately interconnected functional areas (Brodmann, 1912), but the cytoarchitecture of the cerebellum is uniform throughout (Eccles et al., 1967). Most of the cerebellar input is from mossy fibers carrying information from spinal tracts, brain stem nuclei, and the pons. They synapse in the cerebellar cortex with granule cells whose axons, the parallel fibers, synapse with interneurons and Purkinje cells, forming an on/off pattern of regulated excitement that, with enough synapses, eventually results in the Purkinje cell firing (Eccles et al., 1967). It takes a large number of parallel fiber synapses before the Purkinje cell will fire. The Purkinje cell is the only neuron to project to the rest of the brain via its own nucleus within its zone, but it is not regulated by mossy fibers alone. Each of the three major divisions of the inferior olive, a precerebellar nucleus in the medulla, projects to Purkinje cells in even finer zones in the cerebellum (Voogd and Ruigrok, 1997). The inferior olivary axon envelops the Purkinje cell in multiple synaptic nodes such that the firing of the olivary neuron elicits the firing of the Purkinje cell.

Both mossy fibers and inferior olivary projections send collateral fibers to the corresponding cerebellar nucleus. However, Purkinje cells are inhibitory. It is the tonic excitement in the nucleus from mossy fiber and inferior olive collaterals that actually completes the cortico-cerebello-cortical loop, leaving the most highly synapsed information in the cerebellum eloquent in its silence. It should be borne in mind that tracer studies show potential pathways through intracellular transport, and that a Purkinje cell that fires affects the *pattern* of cerebellar projection through inhibition of a dentate neuron but cannot directly participate in the return loop to the neocortex. It is challenging to understand how the inhibition of important information plays out in the neocortex.

Although the closed-loop model of Strick and colleagues is an important landmark in the connectivity of the neocortex and cerebellum, examination of the microcircuitry of the cerebellum, especially in the mossy fiber/climbing fiber interplay, leads to a more fundamental, almost abstract understanding of what the cerebellum might be doing in both motor and cognitive domains. These perspectives concern learning, timing, and patterning. Climbing fiber synapses on the Purkinje cell produce complex spikes when recorded by EEG, in contrast to the simple spikes of parallel-fiber synapses. Complex spikes are found when a learned motor pattern is perturbed, or when a new pattern needs to be learned. When learning has taken place, complex spikes are no longer present (Devor, 2002). The climbing fibers may alert the Purkinje cells to new mossy fiber patterns (Marr, 1969), or erase a Purkinje cell response to a defective pattern (Albus, 1971; Ito and Kano, 1982). The facilitation of learning by the inferior olive can be demonstrated with the nictitating response in rabbits, or with interruption and subsequent adjustment of walking on treadmills by cats, but it is not yet known whether it underlies more cognitively sophisticated forms of learning.

The inferior olive is in the unique position of receiving information from peripheral receptors before it reaches the cerebrum and so may be regulating timing at the most fundamental level (Welsh and Llinás, 1997). Indeed, the cerebellum receives collaterals from the spinal column tracts on their way to sensory cortex. As almost 95% of descending projections from the neocortex terminate in the pontine nuclei (Tomasch, 1969), it is reasonable to see the cerebellum as a sensory–motor interface that enables the motor system to make fine adjustments, increasing and smoothing processing efficiency (Bower, 1997), or as Thach proposes (1996), contextualizing the motor in the sensory. Motor timing and temporal perception have been specifically linked to the lateral cerebellum (Ivry et al., 1988).

Complex movement is not just a spatial sequence, but a temporal one. Sakai et al. (2004)

argue that any complex movement needs the structuring principle of rhythm. Your own signature, for example, has a particular rhythm to it, and it cannot be reproduced without it. How these rhythms emerge and change is attributed by Sakai and colleagues to the cerebellum, specifically the lateral cerebellum (posterior lobe) and the premotor cortex. These interlinked structures organize the motor performance into larger and larger units until the movement forms a "musical phrase," with the rhythm actually structuring the movement. The anterior cerebellum in concert with the premotor area maintains this structured rhythm. This explains the activity in the dentate nucleus/lateral cerebellum, when a task is being learned, and its extinction, when the task becomes automatic (Doyon, 1997).

Although cerebellar participation in cognitive activity is a burgeoning field of research, it is misleading to separate sensory and motor activities from the cognitive, since "there is not movement without cognition, and there is not cognition without movement" (Bloedel and Bracha, 1997). This is especially true for students of evolution searching for routes to abstract thought in feeding strategies and locomotory patterns. As neuroscience begins to understand how the cerebellum may be active in learning circuits and complex activities, integrating data from cognitive psychology and neuroimaging with cerebellar microcircuitry and physiology, neuroanthropologists follow with measures of the cerebellum, its components and connections, placed in the larger context of the comparative neuroanatomy of the primate order. The linking of structure and function to socioecology is another missing link in the evolution of the cerebellum.

Comparative cerebellar anatomy in primates

The first comparative studies of the primate cerebellum (Larsell and Jansen, 1970; Tilney and Riley, 1928) were incisive and descriptive, but not quantitative. It was not until Heinz Stephan and colleagues (Stephan and Andy, 1969; Stephan et al., 1970, 1981) undertook the task of compiling a database of fixed and mounted primate brains that comparative primate brain anatomy could be explored quantitatively. The Stephan database includes volumes of 42 brain structures in 21 prosimians and 27 anthropoids from fixed, sectioned brains stained with Nissl and myelin. The sample is widely representative of the primate order, although lacking orangutan and bonobo specimens. The Stephan database has been the single, most important influence on the field because of its scope and rigor. A drawback is that the species' means are based on one or two specimens in most cases, precluding an understanding of within-species biological variability. A study of the cerebellum and its associated nuclei by Shozo Matano (Matano, 1992; Matano and Hirasaki, 1997; Matano et al., 1985a, b) emerged from the Stephan database. There is no question that this study was ahead of its time because of its attempt to create a complex model of the evolution of cerebellar connectivity based on comparative volumes of the cerebellar-related structures such as the ventral pons, the inferior olivary nuclei, and the deep cerebellar nuclei.

With the advent of neuroimaging, the ambitious task of scanning primate brains *in vivo* was undertaken by the Yerkes Regional Primate Research Center. This provided 47 specimens, including two ape species missing from the Stephan data set (orangutans and bonobos), but without its broader species' representation. MRI may never have the resolution to permit the detailed examination of cytology and nuclei done with fixed brains, but it does allow exploration of macrocircuitry through diffusion tensor imaging (DTI) and fcMRI. The Yerkes project was the first step in a new age.

Volumetric analysis

The larger sample of ape brains in the Yerkes data set enabled Rilling and Insel (1998) to discern a differential increase in ape and human (hominoid) cerebellar volume over the rest of the anthropoids when regressed against brain volume. The slopes of the ape and monkey regression lines are identical, but the intercepts differ. Such a pattern has been discerned by Martin (1980) as a grade shift, a comprehensive change in neurological proportions in a large taxonomic group. The ape cerebellum is significantly larger, but the human cerebellum is below the regression line for apes, attributed by Rilling and Insel (1998) to the differential increase in the neocortex in modern humans, that is, an increase in the denominator. Semendeferi and Damasio (2000), in contrast, interpret the smaller human cerebellum to mean that the cerebellum did not expand as much as the cerebrum during hominin evolution, that is, a decrease in the numerator that would be reflected in a lower exponent of increase or slope, when regressed against the rest of the brain.

Exploring the evolution of the hominin cerebellum is much more difficult than measuring soft tissue in extant primates, but some paleoneurologists have applied their ingenuity to the task. Weaver (2005) used the Yerkes sample to estimate cerebellar volume from posterior fossa volume and found a predictable linear relationship between the two. She applied her formula to a large sample of hominin skulls and included extensive data from extant human measures of whole brain and cerebellum. She concluded that cerebellar and neocortex volume (derived from brain volume) underwent a mosaic expansion in the last 2 million years, with the human cerebellum only reaching its present size in absolute and relative terms in recent humans.

White (2005) also documented trends in the evolution of hominin cerebellar shape, applying multivariate and morphometric analyses to landmarks on macaque, ape, and fossil hominin endocasts. He concluded that hominin ancestors could be distinguished from apes and macaques through cerebellar shape, and that shape could be influenced by both cranial and functional pressures. For the moment, we must wait, patiently, for more fossils to see whether the differences in relative cerebellar size and shape found by Weaver and White are true indicators of significant species-specific and grade-specific differences or are simply artifacts of biological variation in a limited sample size.

As primate brains are still relatively scarce, small sample sizes hamper our ability to see significant interspecific differences in cerebellar volume in extant apes. Hopkins et al. (2009) measured eight bonobo and eight chimpanzee cerebella in subjects matched for age and sex and found the ratio of cerebellar volume to the rest of the brain significantly larger in chimpanzee than bonobo. Neither ratios nor multiple regression in my data set (described below) indicates such a distinction between 14 chimpanzees and 6 bonobos (MacLeod, 2000).

Sherwood et al. (2004) compared three mountain gorilla with three Western lowland gorilla brains (one female and two males each), and scanned postmortem with MRI. Mountain gorilla brains are extremely rare, and any information on them is highly prized. Aware of the vagaries of small sample size, the authors placed these two species in the context of a larger database of 23 chimpanzees and 6 orangutans and conducted a principal components analysis. Their data show no significant interspecific differences in brain structure volumes except for the hippocampus, striatum, and cerebellum. Chimpanzees are separated from the rest of the sample by a larger hippocampus. The mountain gorilla has a smaller cerebellum but larger striatum in comparison to the Western gorilla, but the small sample size hampers a definitive conclusion.

Connecting variations in size of neurological structures with socioecology is essential to the evolutionary perspective, but without sufficiently large databases, our conclusions remain tentative. However, the problem can be better defined if cerebellar divisions are parcellated into finer structures that exhibit functional localization shown by recent neuroimaging studies.

Anterior–posterior torque

Cantalupo et al. (2008) measured the anterior–posterior and right–left cerebellum for volumetric asymmetry in 53 chimpanzee MR scans. The authors used the same methodology as an earlier study of human cerebellar torque by Snyder et al. (1995), in which the cerebellum was divided into two equal divisions, anterior and posterior, with the vermal midline marking right and left halves. This does not correspond to the traditional division in which anterior and posterior lobes have an equal number of lobules, but not volume (Jansen and Brodal, 1954). In hominoids, the anterior lobe is much smaller than the posterior. If Cantalupo and Hopkin's methodology divides the cerebellum into two equal parts, then their findings cannot as easily translate to the work on cerebellar functional localization in anterior and posterior lobes.

Nonetheless, Cantalupo et al. (2008) offer insight into cerebellar asymmetry. In a study of 15 dextral and 8 nondextral human subjects, Snyder et al. (1995) found that the right anterior cerebellum was larger than the left, while the left posterior cerebellum was larger than the right in humans (anterior: L<R; posterior: L>R). Only right-handed individuals had a significant torque. Cantalupo et al. (2008) found evidence of cerebellar torque related to hand preference in the chimpanzee sample. The experimenters gave the chimpanzees three tasks: simple reaching for raisins, extracting peanut butter from a tube, and extracting honey or applesauce from a narrow tube using a probe (simulated ant fishing task). Only tasks that involved a tool attained significance for cerebellar asymmetry. In the posterior cerebellum, right-handed and ambiguously handed chimpanzees showed a leftward bias in comparison to the left-handed chimpanzees, and vice versa. However, only the left-handed tool users had a significant torque (i.e., anterior: L>R; posterior: L<R). This pattern exactly mirrors that found by Snyder et al. (1995) in right-handed humans.

In 2010, Cantalupo and Hopkins expanded their study to include throwing as an experimental condition. Those chimpanzees that threw objects consistently (n=32) had larger posterior cerebella than those that did not throw objects (n=30), but the nonthrowers had larger anterior cerebella than the throwers. Within the throwers, the 12 left-handers showed the same torqued pattern found in the 2008 study.

The directionality of the torque in both the human and chimpanzee studies makes some sense in light of the Stoodley and Schmahmann meta-review (2009), in which sensorimotor information is localized to the anterior cerbellum (IV, V, and posterior VI) largely on the right side, and visuospatial information is localized in the left posterior cerebellum. Sensorimotor information from the anterior cerebellum would aid the hand doing the throwing or probing, while visuospatial information in the posterior hemisphere would aid in the conceptual aspects of tool use. The correlation of torque with handedness could be a key to understanding comparative differences in the organization of hominoid brains, especially with regard to the emergence of lateralization in primate evolution. However, absolute differences in quadrant size are small, and the asymmetries may not be related to connectivity at all, but rather to developmental timing, as Catherine Best (1988) postulated for the cerebral petalia.

Cerebellar hemispheres

In order to discern the evolution of cerebellar cognitive processes, I undertook a volumetric study that parcellated the cerebral hemispheres from the vermis (medial cerebellum) and measured two nuclei closely associated with the neocerebellum, the dentate and the principal inferior olive. Our database (MacLeod et al., 2003) combines the Yerkes MR scans and the fixed-brain collection of the Institute für Hirnforschung in Dusseldorf, Germany (Zilles Collection) and measures brain and brain structure volumes for

97 anthropoids, including 33 great apes, 9 gibbons, 14 humans, and 41 New and Old World monkeys. It represents the largest *comprehensive* sample of ape brains yet collected. The composition of the sample allows the reliable separation of larger taxonomic units, or grades, such as hominoids (apes, humans) or monkeys, in a method of multiple regression that tests for significance and slope of the additional *x*-value of grade (1 or 0).

The cerebellar hemispheres, when regressed against the vermis, show a spectacular expansion in hominoids over monkeys (Fig. 5). The hemispheres are 2.7 times larger than would be expected in a monkey with a vermis of the same size. Humans fit perfectly on the hominoid regression line whether it is determined with or without them, but as with the cerebellum, are below predicted values when the hemispheres are regressed against the rest of the brain. In humans, the hemispheres represent 91.5% of the cerebellum in humans, 88.6% in great apes, and 82.2% in gibbons, thus accounting for the bulk of the cerebellum (MacLeod, 2000). If the hemispheres had stopped expanding at the same rate relative to the vermis in hominin evolution, the slopes of the two regression lines in Fig. 5 would not be parallel. This result would not support any decrease in the rate of cerebellar expansion in hominin evolution, but only an increase in neopallium size. The magnitude of the lateral hemisphere expansion dwarfs any species-specific differences but would not preclude the eventual uncovering of hominoid and fossil hominin hemisphere and vermis variation correlated with socioecology in much larger data sets.

Seidler et al. (1997) observed a wider space between the posterior fossae for the cerebellar

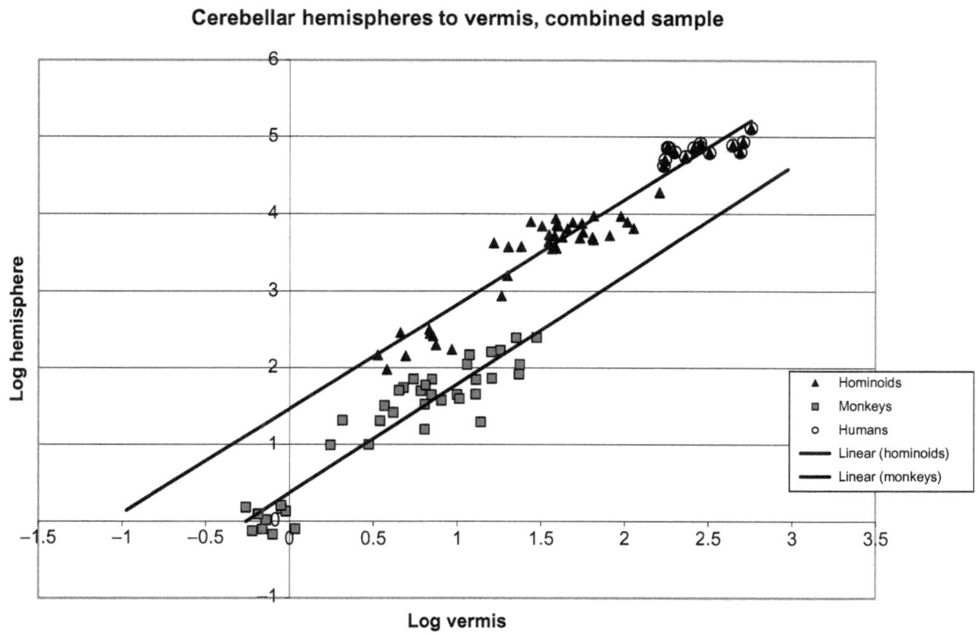

Fig. 5. Logged cerebellar hemisphere to vermis volumes for the combined Yerkes and Hirnforschung samples (MacLeod et al., 2003). SE is 0.268, and r^2 value is 0.968. Regression formula for monkeys is $y' = 0.367 + 1.4588x$, and for hominoids is $y' = 1.465 + 1.365x$. Slopes are statistically parallel, but the intercepts are significantly different, thus showing a grade shift in hemisphere to vermis proportions.

hemispheres in the Broken Hill and Petralona skulls and attributed this to a larger vermis than modern humans. A significant difference in hemisphere to vermis ratio in these skulls does not fit with the extant primate data, however. Instead, a wider vermal space could be a design feature related to the cranial base or simply a derived feature that also distinguishes robust from gracile australopiths (White and Falk, 1999).

Dentate and principal inferior olivary nuclei

In our study, the nuclei associated with the cerebellar hemispheres do not fall into their predicted pattern. Only the Hirnforschung database ($N=48$) could be used to measure the nuclei, since MR scans do not give high enough resolution. The dentate nucleus should expand in concert with the neocerebellum, being its only output nucleus. Instead, the dentate nucleus does not expand differentially in anthropoid evolution but increases only with overall brain volume (Fig. 6). In fact, the human dentate nucleus has the smallest volume relative to the cerebellar hemispheres of the primate sample (MacLeod, 2000). In contrast, the principal inferior olivary nucleus does show dramatic expansion along with the cerebellar hemispheres (MacLeod, 2000; MacLeod et al., 2001a,b).

A conservative rate of expansion for the dentate nucleus is not consistent with Matano's data (1985b). This is likely because of the techniques employed in our larger sample size, with several specimens per hominoid species. The higher magnification level, number of sections traced, and computer-aided measuring techniques all contributed to more robust results, but results which

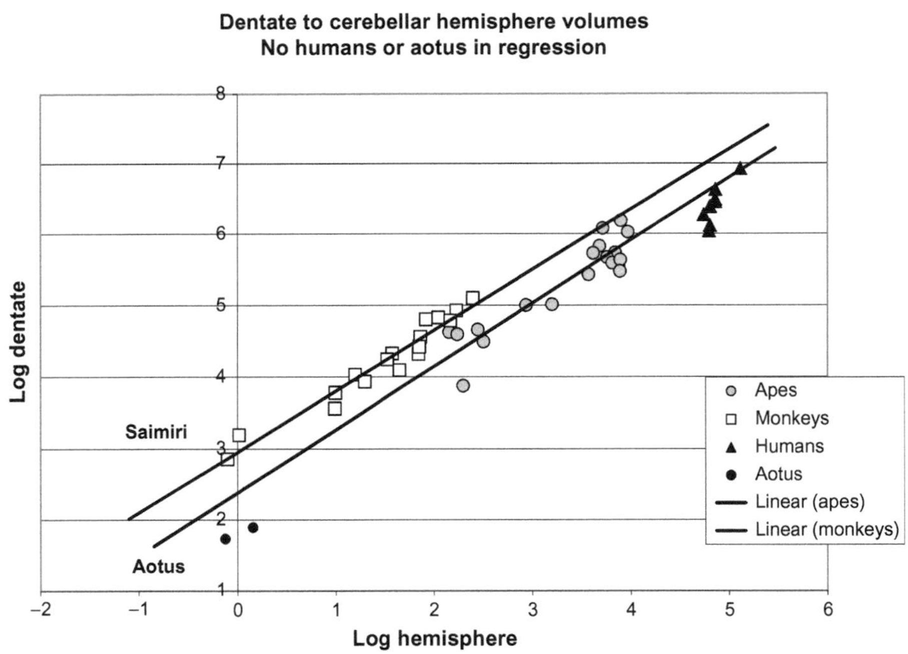

Fig. 6. Regression of dentate to cerebellar hemisphere (logged volumes). *Aotus* is an outlier and is dropped from the regression. Humans have the smallest dentate to hemisphere relation of all the other anthropoids and pull the hominoid regression line from parallel. They have been removed from the ape regression line to show the grade shift between monkeys and apes. Regression formulae: y' (monkey)=2.948+0.851x; y'(ape)=2.378+0.883x (from author's data).

overturn the received wisdom that the dentate nucleus, especially the neodentate, has shown a dramatic expansion in humans (Leiner et al., 1986). The dentate nucleus can be divided into the *paleodentate*, the microgyric region (dorsomedial) that is phylogenetically older, and the *neodentate*, a macrogyric but parvocellular region. The Kelly and Strick (2003) tracer study demonstrates that primary motor areas of the neocortex are connected with the paleodentate, while Area 46, involved in working memory, is connected to the neodentate. Dow (1942) argues that this parvocellular region is most advanced in humans, but his evidence is based on an early study by Demole (1927) that does not examine the dentate in a systematic comparative context. Matano (2001) measured the neodentate in two humans, two gorillas, and one chimpanzee from the Stephan collection and found a significantly higher ratio of neo-to-paleodentate in humans over the gorilla and chimpanzee. The criteria for separating macro- from microgyri are not explained; however, compounding the problem of a small sample size restricted to three species. The crux of the issue lies in the measurement protocols. There is no clear division between the magno- and parvocellular regions, and as Chan-Palay found when examining the dentate nucleus of the rat and the macaque (1977), elliptical neurons may appear large or small in brain sections depending on the angle at which the brain has been sectioned. An expanded neodentate is not consistent with our study's findings of a smaller than expected dentate, although the hominoid dentate does increase in concert with the rest of the brain, dominated of course by the neocortex.

The differential expansion of the lateral hemisphere and principal inferior olive allows an increase in neocerebellar processing power, but the lack of differential expansion of the dentate nucleus implies an adjustment in cerebellar processing itself. In our comparative sample, both the principal inferior olive and the dentate show significant shape changes in the hominoids, becoming much more convoluted, with more surface area than the rest of the anthropoids. Yet in spite of this common shape change, only one nucleus shows a differential volumetric increase.

Neocortical connections with the inferior olive are via the *zona incerta* (Schmahmann and Pandya, 1997) and the red nucleus (Altman and Bayer, 1997) but are known rather sketchily. Strong neocortical input may not be necessary to olivary participation in cognition if the olive is working at the level of timing and pattern. The exclusive synaptic connection of one Purkinje cell with an inferior olivary neuron could also explain why an increase in Purkinje cells in the neocerebellum would necessitate a corresponding increase in numbers of olivary neurons, although climbing fibers branch to synapse several Purkinje cells (Altman and Bayer, 1997). The dentate nucleus, on the other hand, could possibly be receiving a higher concentration of Purkinje cell axons in hominoids than monkeys, thus implying convergence. At this point, the surprising finding of changed proportions in these three critical elements cannot be explained without further exploration of inferior olivary and dentate nuclear function, as well as comparative measures of surface shape of the two nuclei as it relates to function.

Lobules

In keeping with the past decade's work in determining cerebellar functional localization and connectivity, Balsters et al. (2010) measured cerebellar lobules V through VIIIa in five females and five males in each sample of human, chimpanzee, and *Cebus* monkey MR scans. Of the three species, human Crus I and Crus II are significantly larger, while human lobules V and VI are smallest relative to overall cerebellar volume and to the rest of the lobules measured. While it is clear from the ratios that there has been a change in cerebellar proportions in humans over chimpanzees and capuchins, a larger representative primate sample would permit multiple regression to determine whether Crus I and II have a higher exponent of increase (steeper slope) and hence occupy a higher percentage of a larger cerebellum, and/or whether there has been a differential expansion of Crus I and II in humans (grade shift). Gross anatomy shows that Crus I and II occupy a huge part of the lateral cerebellar hemispheres in hominoids, but the grade shift of the lateral cerebellum relative to the vermis appears in hominoids, not humans alone.

Balsters et al. (2010) suggest that the expansion of Crus I and II is related to larger human prefrontal lobes because of their exclusive connections. The Habas et al. (2009) and the O'Reilly et al. (2010) studies remind us that Crus I and II are also part of networks strongly connected with Area 7 of the parietal lobes. Further, there is a great deal of controversy surrounding frontal and prefrontal lobe measures, with Semendeferi et al. (2002) arguing that humans do not show differential expansion of the prefrontal lobes, and Schoenemann et al. (2005) arguing the opposite. The disagreement is largely a methodological one based on the determination of boundaries of the prefrontal lobes and on the number of primates in the sample (Sherwood et al., 2005).

Smaers et al. (2010) use cytoarchitectural boundaries to parcellate frontal lobes in a large primate sample and distinguish white from gray matter. They show that both frontal white and gray matter hyperscale against the rest of the brain, whereas only the white matter of the non-frontal neopallium hyperscales. Frontal white matter is the most important factor in the relative enlargement of the anthropoid neopallium. There is also a significant correlation between frontal white matter volume and the basal ganglia, implying that "frontal white matter is at the heart of increased structural connectivity associated to brain enlargement and higher cognitive capacities" (Smaers et al., 2010, p. e9123). Parcellating the prefrontal lobes would clarify their analysis even more, although relative volume is not the only measure of evolutionary change in human prefrontal lobes. Increased minicolumn width (Semendeferi et al., 2011) and rapid metabolic evolution (Fu et al., 2011) are just two examples that could explain greater prefrontal input to the larger Crus I and II lobules in Balster et al.'s analysis.

Measuring prefrontal input

Using DTI, Ramnani et al. (2006) imaged the cerebral peduncles in macaque and human brains in order to determine the relative contribution of neocortical areas to the pontine nuclei and hence the cerebellum. They subdivided cerebral cortex into prefrontal, premotor, primary motor, somatosensory, parietal, visual cortices, and temporal lobes. Fibers from visual cortex and the temporal lobes are negligible in both humans and macaques. Primary motor areas dominate the cerebral efferents in macaques, but not in humans. Humans have much more representation from somatosensory cortex than macaques, although the implications of this are not discussed. The most striking finding of their work is the extent to which the human prefrontal lobes account for the lion's share of fibers in the cerebral peduncles in contrast to the macaques. Ramnani and colleagues demonstrate that cortical afferents may vary quantitatively within primates while still conserving basic connectivity. This study opens a new dimension of comparative primate anatomy.

Statistical treatment of volumes

The statistical analysis of large amounts of volumetric data can give us the big picture of the evolution of the primate cerebellum, while the collection of more numerous and specific brain structure volumes can refine these models. Clark et al. (2001) plotted the Stephan et al. (1981) insectivore and primate brain structures as ratios of total brain size. The cerebellum ratio is constant, at 13% of brain volume (±2%). The authors conclude that it cannot be yoked to the neocortex functionally, since the neocortex ratio is not constant across taxa but increases dramatically in hominoids. These conclusions do not ring true for our data set. The percentage of whole brain occupied by cerebellum by categories of humans, great apes, lesser apes, OW, and NW monkeys varies from 9.1% to 13.8%. These percentages are significantly different from one another, except between great and small apes. The variation is even greater when the lateral hemispheres are compared among taxonomic categories, from 6.2% to 12.2%. When the primate order is examined in greater detail, there are indeed important differences in cerebellar ratios that are not trivial.

In a complex multivariate analysis, de Winter and Oxnard (2001) demonstrate socioecological correlations with brain structure ratios for large mammalian groupings, including primates. This is convincing evidence that natural selection does indeed intervene in brain structure evolution. Their analysis shows distinctions in primate groupings based on diet and associated locomotory and cognitive adaptations, and not phylogenetic categories, with the exception of the strepsirrhine/haplorhine subgroups.

Barton and Harvey (2000) have proposed that primate brain evolution proceeds according to inherent functional connectivity. They demonstrate this by using CAIC (comparative analysis by independent contrasts) on six systems from the Stephan database. Whiting and Barton (2003), using the same statistical method applied to the Stephan database, also find evidence for correlated evolution between the neocortex and the cerebellum. The Barton and Harvey (2000) hypothesis that integrated neurological units evolve as a whole is especially relevant to the last decade's advances in understanding corticocerebellar connectivity, in which the cerebellum participates in a number of distinct circuits. As these circuits become more evident in human neuroimaging studies, they can be explored in the comparative primate databases to allow better modeling of primate cerebellar evolution.

It is logical that if structures are functionally connected, they must evolve in association with each other. However, our data show that this model must be more nuanced. If the cerebellar hemispheres are regressed against the rest of the brain, there is an obvious grade shift (MacLeod et al., 2003). If the neopallium and the neocerebellum were coevolving in step, the shift would not be apparent. Our data show that the lateral hemispheres expanded differentially with the hominoids, not humans alone. Within the cerebellar complex, our data demonstrate a reorganization of cerebellar circuitry in a selectively mosaic fashion (MacLeod et al., 2001b), since the principal inferior olive expands differentially with the lateral hemispheres, but the dentate nucleus volume expands only with the rest of the brain.

Our study reveals a high degree of biological variability in all of the structures measured. This makes the comparison of neural structures at the interspecific level much more problematic. Most comparative models use the Stephan database (1981), where there is no indication of variability and no possibility of applying statistical tests of significance. Methods such as CAIC may be too sensitive to the significance of interspecific means at the branch ends (noise) (Purvis and Rambaut, 1995). The fact that specimens, and not means, used in our study should make it more difficult to find high r^2 values, yet those regressions related to cerebellar circuitry all have values over 0.9, with some approaching 1.

A recent study of Smaers et al. (2011) examines a number of variables in corticocerebellar circuitry. Two major patterns of correlation emerge from their data set: a frontal lobe-basal ganglia-neocerebellum-dentate nucleus- and thalamus circuit and a posterior neocortex-pons-neocerebellum-dentate nucleus–thalamus circuit. Their results support the Barton and Harvey (2000) model of functionally correlated evolutionary change. However, the pattern of correlation does not always make sense anatomically. For example, there is a strong correlation between the dentate nucleus and the frontal lobes, as would be expected, and also a weaker but significant correlation between the fastigial nucleus and the frontal lobes. The fastigial nucleus has only sparse projections to the thalamus (Altman and Bayer, 1997), yet the pons, so closely connected with the frontal lobes in Ramnani et al.'s study (2006), shows no statistical correlation whatsoever with the frontal lobes. The total statistical picture is enigmatic.

Interpreting lateral cerebellar expansion

There will always be differences of opinion as to choice of statistical treatment of the data, but the method used in our study clearly shows a massive hominoid grade shift of the cerebellar hemispheres. If the lateral cerebellum had expanded at different times in different hominoid lineages in a mosaic fashion, one would expect a rather haphazard collection of volumes that would not yield parallel slopes between monkeys and hominoids (Fig. 5). Instead, the allometric patterning is explicable if this change in cerebellar proportions took place in the last common ancestor to extant hominoids, just as Jerison (1973) postulated significant changes in peripheral and central sensory systems of olfaction and vision in mammals as corresponding to periods of radiation into new niches by founding ancestors.

Since cognitive functions are primarily localized to the cerebellar hemispheres, it is likely that there was selection for these functions through hemisphere expansion. As the medial and anterior cerebellum are the locus of the execution of motor patterns, and the lateral cerebellum for the planning of these movements (Thach, 1996), the increase in the ratio of lateral to medial cerebellum would imply that the early hominoids had an augmented capacity for complexity of movement, and for the cognitive aspects of movement in structuring their niche. If cerebellar expansion were linked to motor coordination alone, there would have been a comparable increase in the vermis, associated with balance and coordination, and the execution of movement (Thach, 1996).

The possible ancestors of extant hominoids were not suspensory primates but quadrupedal branch walkers (Benefit, 1999). A variety of locomotory styles appeared in the early Miocene, leading Fleagle (1999) to describe early hominoid skeletal anatomy as versatile, comparable to the living spider monkeys or chimpanzees (1999). The expansion of the lateral cerebellum was not coincident with suspensory locomotion, but rather suspensory feeding in the hominoid frugivorous niche, perhaps because it offered more movement patterns in a three-dimensional space, with a greater capacity for visuospatial problem solving. In this scenario, both the locomotor and the cerebellar adaptations anticipate the later development of suspensory locomotion.

Early Miocene hominoids and cercopithecoids were largely frugivorous, as inferred from habitat and dentition (Andrews and Martin, 1992). Frugivory is associated with relatively large home ranges compared to folivorous primates (Clutton-Brock and Harvey, 1979), even in the case of the gibbon where body and group size are reduced. Larger ranges require enhanced memory capacity to return to fruiting trees that are patchily dispersed in space and time (Milton, 1981). Mapping the fruiting trees would recruit areas of the right inferior parietal lobes, areas that are associated with visuospatial cognition and map making that have extensive afferents to the lateral cerebellum. The learning of paths and important points on this map would be facilitated by another function of the lateral cerebellum, procedural learning (Doyon, 1997). A finer analysis of cerebellar expansion lobule by lobule (Balsters et al., 2010) in a representative primate sample size would greatly refine the evolutionary explanation for lateral hemisphere selection in the hominoids.

Summary

The vision of Leiner et al., first explored in 1986, has guided understanding of the emerging evidence of cerebellar participation in cognition. Connections between the neocortex, especially the prefrontal lobes, with functionally localized areas of the cerebellum, demonstrate the anatomical basis of this cognitive activity. Anthropologists are turning to models of functional connectivity to further explore the Barton and Harvey (2000) hypothesis of functionally integrated circuits underlying changes in primate brain evolution.

Corticocerebellar connectivity has been rigorously examined in some nonhuman primates using tracer studies. As the animals must be sacrificed in order to examine the circuits, very little comparative data exist. With the technology of DTI and fcMRI, circuits in a wide variety of primates can be studied noninvasively. It has been widely assumed that tracer studies in monkeys are conserved in apes and humans and that there is little prefrontal input to the cerebellum and, hence, no strong anatomical basis for cerebellar participation in cognition (Brodal and Bjaalie, 1997). The Ramnani et al. study (2006) shows that although the circuits remain conserved, the proportion of input via the pontine nuclei is radically different for humans. The human–monkey contrast does not show evolutionary history with only two or three specimens, however. The application of neuroimaging techniques to a large sample of primates could trace the evolution of functional circuits through the comparative method and give us a new dimension of primate brain evolution.

Neuroimaging studies do not precisely verify the presence of closed, reciprocal loops demonstrated by Strick and colleagues in *Cebus,* since Allen et al. (2005) and Krienen and Buckner (2009) show some localization overlap, and also bilateral activation from the left dentate nucleus. The reason may be that MR technology is not sufficiently precise enough to demonstrate closed loops, but there may be a difference in circuitry between *Cebus* and humans, just as Ramnani could demonstrate a distinctly human pattern in connectivity proportions. Application of nonhuman primate tracer models to humans may ignore important differences that must be understood in light of the comparative method. Anthropology has always argued that we cannot understand human functional anatomy without reference to our closest relatives. In this sense, neuroanthropology is a missing link for the neurosciences.

The hominoid shift in the functional model of hemisphere, dentate, and principal inferior olive underlines the fact that the principal inferior olive must not be ignored in models of corticocerebellar connectivity. The mossy fiber-climbing fiber dialectic is on a different plane than macrocircuitry, but understanding of the cerebellum can only come through integration of its many levels of operation, just as the pontine nuclei need to be explored in greater detail. Why would the principal inferior olivary nucleus expand differentially with the hemispheres, while the dentate nucleus expands with the rest of the brain? Does this indicate an augmented participation of the inferior olive in timing, patterning, and learning? Does the more conservative expansion of the dentate imply more complex cerebellar cortical processing in hominoids and even greater convergence upon the dentate nucleus? These are possibilities that should and will be explored by neuroscientists working at an entirely different level from comparative volumes. The exciting part for our field and theirs is that we can both begin to pose questions of mutual interest.

The title of this volume reflects our common goal of understanding the evolution of cognition through the comparative study of neuroanatomy. Perhaps it is still too early to reliably trace evolution "from neuron to behavior," but as our discipline integrates itself more and more with the neurosciences, the understanding of our past will not only be less obscure, but will in turn, shed light on neuroscientists' understanding of the brain.

Acknowledgments

Many thanks to the editors for their invitation to participate in this volume, and to Dr. Peter Strick for his clarification of some important points. Thanks to Langara College and especially the

Langara Research Committee, and to fellow Langarans, George Kropinski, for his expertise in preparing Figs. 2 and 3, and my colleague, Ariadne Bursewicz, for her insight. As always, thanks to Dr. Axel Schleicher and Professor Karl Zilles of the Institut für Hirnforschung for their inspiration and continuing support.

References

Ackermann, H., Mathiak, K., & Riecher, A. (2007). The contribution of the cerebellum to speech production and speech perception: Clinical and functional imaging data. *The Cerebellum, 6*, 202–213.

Albus, J. S. (1971). A theory of cerebellar function. *Mathematical Biosciences, 10*, 25–61.

Allen, G., McColl, R., Barnard, H., Ringe, W. K., Fleckenstein, J., & Cullum, C. M. (2005). Magnetic resonance imaging of cerebellar-prefrontal and cerebellar-parietal functional connectivity. *NeuroImage, 28*, 39–48.

Altman, J., & Bayer, S. A. (1997). *Development of the cerebellar system in relation to its evolution, structure, and functions.* Boca Raton: CRC Press.

Andrews, P., & Martin, L. (1992). Hominoid dietary evolution. In A. Whiten & E. M. Widdowson (Eds.), *Foraging strategies and natural diet of monkeys, apes and humans* (pp. 39–49). Oxford: Clarendon Press.

Baddeley, A. (1992). Working memory. *Science, 255*, 556–559.

Balsters, J. H., Cussans, E., Diedrichsen, J., Phillips, K. A., Preuss, T. M., Rilling, J. K., et al. (2010). Evolution of the cerebellar cortex: The selective expansion of prefrontal-projecting cerebellar lobules. *NeuroImage, 49*, 2045–2052.

Barton, R. A., & Harvey, P. H. (2000). Mosaic evolution of brain structure in mammals. *Nature, 405*, 1055–1058.

Ben-Yehuda, G., Guediche, S., & Fiez, J. A. (2007). Cerebellar contributions to verbal working memory: Beyond cognitive theory. *The Cerebellum, 6*, 193–201. doi:10.1080/14734220701286195.

Benefit, B. R. (1999). Victoriapithecus: The key to Old World monkey and Catarrhine origins. *Evolutionary Anthropology, 7*, 155–174.

Best, C. T. (1988). The emergence of cerebral asymmetries in early human development: A literature review and a neuroembryological model. In D. L. Molfese & S. J. Segalowitz (Eds.), *Brain lateralization in children: Developmental implications* (pp. 5–34). New York: The Guilford Press.

Bloedel, J. R., & Bracha, V. (1997). Duality of cerebellar motor and cognitive functions. In J. D. Schmahmann (Ed.), *The cerebellum and cognition, International Review of Neurobiology, 41*, (pp. 613–634). San Diego: Academic Press.

Bower, J. M. (1997). Is the cerebellum sensory for motors sake, or motor for sensorys sake: The view from the whiskers of a rat. In C. E. de Zeeuw, P. Strata & J. Voogd (Eds.), *The cerebellum: From structure to control* (pp. 463–498). Amsterdam: Elsevier.

Brodal, P. (1979). The pontocerebellar projection in the rhesus monkey: An experimental study with retrograde axonal transport of horseradish peroxidase. *Neuroscience, 4*, 193–208.

Brodal, P., & Bjaalie, J. G. (1997). Salient anatomic features of the cortico-ponto-cerebellar pathway. In C. E. de Zeeuw, P. Strata & J. Voogd (Eds.), *The cerebellum: From structure to control. Progress in brain research*, (Vol. 114), (pp. 227–250). Amsterdam: Elsevier.

Brodmann, K. (1912). Neue Ergenbnisse über die vergleichende histologische Lokalisation der Grosshirnrinde mit besonderer Berücksicktigung des Stirnhirns. *Anatomischer Anzeiger, 41*, 157–216.

Brown, S., Martinez, M. J., Hodges, D. A., Fox, P. T., & Parsons, L. M. (2004). The song system of the human brain. *Cognitive Brain Research, 20*, 363–375.

Cantalupo, C., Freeman, H., Rodes, W., & Hopkins, W. D. (2008). Handedness for tool use correlates with cerebellar asymmetries in chimpanzees (*Pan troglodytes*). *Behavioral Neuroscience, 122*, 191–198. doi:10.1037/0735-7044.122.1.191.

Cantalupo, C., & Hopkins, W. (2010). The cerebellum and its contribution to complex tasks in primates: A comparative perspective. *Cortex, 46*, 821–830. doi:10.1016/j.cortex.2009.10.004.

Chan-Palay, V. (1977). *Cerebellar dentate nucleus: Organization, cytology and transmitters.* Berlin: Springer-Verlag.

Clark, D. A., Mitra, P. P., & Wang, S. S.-H. (2001). Scalable architecture in mammalian brains. *Nature, 411*, 189–193.

Clutton-Brock, T. H., & Harvey, P. H. (1979). Home range size, population density and phylogeny in primates. In I. S. Bernstein & E. O. Smith (Eds.), *Primate ecology and human origins* (pp. 201–214). New York: Garland Press.

Courchesne, E., Townsend, J., Akshoomoff, N. A., Saitoh, O., Yeung-Courchesne, R., Lincoln, A. J., et al. (1994). Impairment in shifting attention in autistic and cerebellar patients. *Behavioral Neuroscience, 108*(5), 848–865.

de Winter, W., & Oxnard, C. E. (2001). Evolutionary radiations and convergences in the structural organization of mammalian brains. *Nature, 409*, 710–713.

Demole, V. (1927). Structure et connexions des noyaux denteles du cervelet. *Schweizer Archiv für Neurologie und Psychiatrie, 17*, 271–294.

Devor, A. (2002). The great gate: Control of sensory information flow to the cerebellum. *The Cerebellum, 1*, 27–34.

Dow, R. S. (1942). The evolution and anatomy of the cerebellum. *Biological Reviews of the Cambridge Philosophical Society, 17*, 179–220.

Doyon, J. (1997). Skill learning. *International Review of Neurobiology, 41*, 273–294.

Dum, R. P., & Strick, P. L. (2003). An unfolded map of the cerebellar dentate nucleus and its projections to the cerebral cortex. *Journal of Neurophysiology, 89*, 634–639.

Eccles, J. C., Ito, M., & Szentágothai, J. (1967). *The cerebellum as a neuronal machine.* New York: Springer-Verlag New York, Inc.

Fiez, J. A., Petersen, S. E., Cheney, M. K., & Raichle, M. E. (1992). Impaired non-motor learning and error detection associated with cerebellar damage. *Brain, 115*, 155–178.

Fleagle, J. C. (1999). *Primate adaptation and evolution* (2nd ed.). Toronto: Academic Press.

Fu, X., Giavalisco, P., Liu, X., Catchpole, G., Fu, N., Ning, Z.-B., et al. (2011). Rapid metabolic evolution in human prefrontal cortex. *Proceedings of the National Academy of Sciences of the United States of America, 108*(15), 6181–6186. doi:10.1073/pnas.1019164108.

Glickstein, M. (2000). How are visual areas of the brain connected to motor areas for the sensory guidance of movement? *Trends in Neurosciences, 23*, 613–617.

Habas, C., Kamdar, N., Nguten, D., Prater, K., Beckmann, C. F., Menon, V., et al. (2009). Distinct cerebellar contributions to intrinsic connectivity networks. *The Journal of Neuroscience, 29*(26), 8586–8594. doi:10.1523/JNEUROSCI.1868-2009.

Hahn, B., Wolkenberg, F. A., Ross, T. J., Myers, C. S., Heishman, S. J., Stein, D. J., et al. (2008). Divided versus selective attention: Evidence for common processing mechanisms. *Brain Research, 1215*, 137–146. doi:10.1016/j.brainres.2008.03.058.

Hautzel, H., Mottaghy, F. M., Specht, K., Müller, H.-W., & Krause, B. J. (2009). Evidence of a modality-dependent role of the cerebellum in working memory? An fMRI study comparing verbal and abstract n-back tasks. *NeuroImage, 47*, 2073–2082. doi:10.1016/j.neuroimage.2009.06.005.

Hopkins, W. D., Lyn, H., & Cantalupo, C. (2009). Volumetric and lateralized differences in selected brain regions of chimpanzees (*Pan troglodytes*) and bonobos (*Pan paniscus*). *American Journal of Primatology, 71*(12), 988–997. doi:10.1002/ajp. 20741.

Ito, M., & Kano, M. (1982). Long lasting depression of parallel fiber-Purkinje cell transmission induced by conjunctive stimulation of parallel fibers and climbing fibers in the cerebellar cortex. *Neuroscience Letters, 33*, 253–258.

Ivry, R. B., Keele, S. W., & Diener, H. C. (1988). Dissociation of the lateral and medial cerebellum in movement timing and movement execution. *Experimental Brain Research, 73*, 167–180.

Jansen, J., & Brodal, A. (1954). *Aspects of cerebellar anatomy.* Oslo: Johan Grundt Tanum Forlag.

Jerison, H. J. (1973). *Evolution of the brain and intelligence.* NY: Academic Press.

Kelly, R. M., & Strick, P. L. (2003). Cerebellar loops with motor cortex and prefrontal cortex in a nonhuman primate. *The Journal of Neuroscience, 23*, 8432–8444.

Kim, S.-G., Ugurbil, K., & Strick, P. L. (1994). Activation of a cerebellar output nucleus during cognitive processing. *Science, 265*, 949–951.

Kotz, S. A., & Schwartze, M. (2010). Cortical speech processing unplugged: A timely subcortico-cortical framework. *Trends in Cognitive Sciences, 14*(9), 392–399. doi:10.1016/j.tics.2010.06.005.

Krienen, F. M., & Buckner, R. L. (2009). Segregated fronto-cerebellar circuits revealed by intrinsic functional connectivity. *Cerebral Cortex, 19*, 2485–2497.

Lalonde, R. (1997). Visuospatial abilities. *International Review of Neurobiology, 41*, 191–215.

Larsell, O., & Jansen, J. (1970). *The comparative anatomy and histology of the cerebellum from monotremes through apes, Volume II.* Minneapolis: University of Minnesota Press.

Leiner, H. C., Leiner, A. L., & Dow, R. S. (1986). Does the cerebellum contribute to mental skills? *Behavioral Neuroscience, 100*(4), 443–454.

Lu, X., Miyachi, S., Ito, Y., Nambu, A., & Takada, M. (2007). Topographic distribution of output neurons in cerebellar nuclei and cortex to somatotopic map of primary motor cortex. *The European Journal of Neuroscience, 25*, 2374–2382.

Luebke, J., Barbas, H., & Peters, A. (2010). Effects of normal aging on prefrontal area 46 in the rhesus monkey. *Brain Research Reviews, 62*, 212.

MacLeod, C. E. (2000). *The cerebellum and its part in the evolution of the hominoid brain.* Burnaby, BC: Simon Fraser University Ph.D. dissertation.

MacLeod, C. E., Schleicher, A., & Zilles, K. (2001a). Natural selection and fine neurological tuning. *American Journal of Physical Anthropology*, (Suppl. 32), 100.

MacLeod, C. E., Zilles, K., Schleicher, A., & Gibson, K. R. (2001b). The cerebellum: An asset to hominoid cognition. In B. M. F. Galdikas, N. E. Briggs, L. K. Sheeran, G. L. Shapiro & J. Goodall (Eds.), *All apes great and small. African apes*, (Vol. I), (pp. 35–53). New York: Kluwer Academic/Plenum Publishers.

MacLeod, C. E., Zilles, K., Schleicher, A., Rilling, J. K., & Gibson, K. R. (2003). Expansion of the neocerebellum in *Hominoidea. Journal of Human Evolution, 44*, 401–429.

Marien, P., Engelborghs, S., Fabro, F., & de Deyn, P. P. (2001). The lateralized linguistic cerebellum: A review and a new hypothesis. *Brain and Language, 79*, 580–600.

Marr, D. (1969). A theory of cerebellar cortex. *The Journal of Physiology, 202*, 437–470.

Martin, R. D. (1980). Adaptation and body size in primates. *Zeitschrift für Morphologie und Anthropologie, 71*, 115–124.

Matano, S. (1992). A comparative neuroprimatological study on the inferior olivary nuclei (from the Stephan's collection). *Journal of Anthropological Society of Nippon, 100* (1), 69–82.

Matano, S. (2001). Brief communication: Proportions of the ventral half of the cerebellar dentate nucleus in humans and great apes. *American Journal of Physical Anthropology, 114*, 163–165.

Matano, S., Baron, G., Stephan, H., & Frahm, H. D. (1985a). Volume comparisons in the cerebellar complex of primates. II. Cerebellar nuclei. *Folia Primatologica, 44*, 182–203.

Matano, S., & Hirasaki, E. (1997). Volumetric comparisons in the cerebellar complex of anthropoids, with special reference to locomotor types. *American Journal of Physical Anthropology*, *103*, 173–183.

Matano, S., Stephan, H., & Baron, G. (1985b). Volume comparisons in the cerebellar complex of primates. I. Ventral pons. *Folia Primatologica*, *44*, 171–181.

Middleton, F. A., & Strick, P. L. (1994). Anatomical evidence for cerebellar and basal ganglia involvement in higher cognitive function. *Science*, *266*, 458–461.

Middleton, F. A., & Strick, P. L. (2001). Cerebellar projections to the prefrontal cortex of the primate. *The Journal of Neuroscience*, *21*, 700–712.

Milton, K. (1981). Distribution patterns of tropical plant foods as an evolutionary stimulus to primate mental development. *American Anthropologist*, *83*, 534–548.

Molinari, M., & Leggio, M. G. (2007). Cerebellar information processing and visuospatial functions. *The Cerebellum*, *6*, 214–220.

Murdoch, B. E. (2010). The cerebellum and language: Historical perspective and review. *Cortex*, *46*, 858–868.

O'Reilly, J., Beckmann, C. F., Tomassini, V., Ramnani, N., & Johansen-Berg, N. (2010). Distinct and overlapping functional zones in the cerebellum defined by resting state functional connectivity. *Cerebral Cortex*, *20*, 953–965. doi:10.1093/cercor/bhp157.

Purvis, A., & Rambaut, A. (1995). Comparative analysis by independent contrasts (CAIC): An Apple Macintosh application for analysing comparative data. *Computer Applications in the Biosciences: CABIOS*, *11*, 247–251.

Raichle, M. E., Fiez, J., Videen, T. O., MacLeod, A.-M. K., Pardo, J. V., Fox, P. T., et al. (1994). Practice-related changes in human brain functional anatomy during nonmotor learning. *Cerebral Cortex*, *4*, 8–26.

Ramnani, N., Behrens, T., Johansen-Berg, H., Richeter, M. C., Pinsk, M. S., Andersson, J. L. R., et al. (2006). The evolution of prefrontal inputs to the cortico-pontine system: Diffusion imaging evidence from macaque monkeys and humans. *Cerebral Cortex*, *16*, 811–818. doi:10.1093/cercor/bhj024.

Rilling, J. K., & Insel, T. R. (1998). Evolution of the cerebellum in primates: Differences in relative volume among monkeys, apes and humans. *Brain, Behavior and Evolution*, *52*, 308–314.

Sakai, K., Hikosaka, O., & Nakamura, K. (2004). Emergence of rhythm during motor learning. *Trends in Cognitive Sciences*, *8*(12), 547–553. doi:10.1016/j.tics.2004.10.005.

Schmahmann, J. D. (1996). From movement to thought: Anatomic substrates of the cerebellar contribution to cognitive processing. *Human Brain Mapping*, *4*, 174–198.

Schmahmann, J. D., & Pandya, D. N. (1989). Anatomical investigation of projections to the basis pontis from posterior parietal association cortices in rhesus monkey. *Journal of Comparative Neurology*, *289*, 53–73.

Schmahmann, J. D., & Pandya, D. N. (1991). Projections to the basis pontis from the superior temporal sulcus and superior temporal region in the rhesus monkey. *Journal of Comparative Neurology*, *308*, 224–248.

Schmahmann, J. D., & Pandya, D. N. (1993). Prelunate, occipitotemporal, and parahippocampal projections to the basis pontis in rhesus monkey. *Journal of Comparative Neurology*, *337*, 94–112.

Schmahmann, J. D., & Pandya, D. N. (1995). Prefrontal cortex projections to the basilar pons in rhesus monkey: implications for the cerebellar contribution to higher function. *Neuroscience Letters*, *199*, 175–178.

Schmahmann, J. D., & Pandya, D. N. (1997). The cererocerebellar system. In J. D. Schmahmann (Ed.), *The cerebellum and cognition* (pp. 31–60). San Diego: Academic Press.

Schmahmann, J. D., Doyon, J., McDonald, D., Holmes, C., Lavoie, K., Hurwitz, A. S., et al. (1999). Three-dimensional MRI atlas of the human cerebellum in proportional stereotaxic space. *NeuroImage*, *10*, 233–260.

Schmahmann, J. D., Doyon, J., Toga, A. W., Petrides, M., & Evans, A. C. (2000). *MRI atlas of the human cerebellum*. San Diego: Academic Press.

Schoenemann, P. T., Sheehan, M. J., & Glotzer, L. D. (2005). Prefrontal white matter volume is disproportionately larger in humans than in other primates. *Nature Neuroscience*, *8*, 242–252.

Schweizer, T. A., Alexander, M. P., Gillingham, S., Cusimano, M., & Stuss, D. T. (2010). Lateralized cerebellar contributions to word generation: A phonemic and semantic fluency study. *Behavioural Neurology*, *23*, 31–37.

Seidler, H., Falk, D., Stringer, C., Wilfing, H., Müller, G. B., zur Nedden, D., et al. (1997). A comparative study of stereolithographically modelled skulls of Petralona and Broken Hill: implications for future studies of middle Pleistocene hominid evolution. *Journal of Human Evolution*, *33*, 691–703.

Semendeferi, K., & Damasio, H. (2000). The brain and its main anatomical subdivisions in living hominoids using magnetic resonance imaging. *Journal of Human Evolution*, *38*, 317–332.

Semendeferi, K., Lu, A., Schenker, N., & Damasio, H. (2002). Humans and great apes share a large frontal cortex. *Nature Neuroscience*, *5*(3), 272–276.

Semendeferi, K., Teffer, K., Buxhoeveden, D. P., Park, M. S., Bludau, S., Amunts, K., Travis, K., & Buckwater, J. (2011). Spatial organization of neurons in the frontal pole sets humans apart from Great Apes. *Cerebral Cortex*, *21*, 1485–1497.

Sherwood, C. C., Cranfield, M. R., Mehlman, P. T., Lilly, A. A., Garbe, J. L., Whittier, C. A., et al. (2004). Brain structure variation in Great Apes, with attention to the mountain gorilla (*Gorilla beringei beringei*). *American Journal of Primatology*, *63*, 149–164.

Sherwood, C. C., Holloway, R. L., Semendeferi, K., & Hof, P. R. (2005). Is prefrontal white matter enlargement a human evolutionary specialization? *Nature Neuroscience*, *8*, 537–538.

Smaers, J. B., Schleicher, A., Zilles, K., & Vinicius, L. (2010). Frontal white matter volume is associated with brain enlargement and higher structural connectivity in anthropoid primates. *PloS One*, *5*(2), e9123.

Smaers, J. B., Steele, J., & Zilles, K. (2011). Modeling the evolution of cortico-cerebellar systems in primates. *Annals of the New York Academy of Sciences*, *1225*, 176–190. doi:10.1111/j.1749-6632.2011.06003x.

Snyder, P. J., Bilder, R. M., Wu, H., Bogerts, B., & Lieberman, J. A. (1995). Cerebellar volume asymmetries are related to handedness: A quantitative MRI study. *Neuropsychologia*, *33*, 407–419.

Stein, J. F., Miall, R. C., & Weir, D. J. (1987). The role of the cerebellum in the visual guidance of movement. In M. Glickstein, C. Yeo & J. Stein (Eds.), *Cerebellum and neuronal plasticity* (pp. 175–192). New York: Plenum Press.

Stephan, H., & Andy, O. J. (1969). Quantitative comparative neuroanatomy of primates: An attempt at a phylogenetic interpretation. *Annals of the New York Academy of Sciences*, *167*, 370–387.

Stephan, H., Bauchot, R., & Andy, O. J. (1970). Data on size of the brain and of various brain parts in insectivores and primates. In C. R. Noback & W. Montagna (Eds.), *The primate brain; Advances in primatology*, (Vol. I), (pp. 289–298). New York: Appleton-Century-Crofts.

Stephan, H., Frahm, H., & Baron, G. (1981). New and revised data on volumes of brain structures in insectivores and primates. *Folia Primatologica*, *35*, 1–29.

Stoodley, C. J., & Schmahmann, J. D. (2009). Functional topography in the human cerebellum: A meta-analysis of neuroimaging studies. *NeuroImage*, *44*, 489–501. doi:10.1016/j.neuroimage.2008.08.039.

Stoodley, C. J., Valera, E. M., & Schmahmann, J. D. (2010). An fMRI study of intra-individual functional topography in the human cerebellum. *Behavioural Neurology*, *23*, 65–79. doi:10.3233/BEN-2010-0268.

Strick, P., Dum, R. P., & Fiez, J. A. (2009). Cerebellum and nonmotor function. *Annual Review of Neuroscience*, *32*, 413–434.

Thach, W. T. (1996). On the specific role of the cerebellum in motor learning and cognition: Clues from PET activation and lesion studies in man. *The Behavioral and Brain Sciences*, *19*, 411–431.

Tilney, F., & Riley, H. A. (1928). *The brain from ape to man: A contribution to the study of the evolution and development of the human brain, Volumes I & II.* New York: Paul B. Hoeber, Inc.

Tobias, P. V. (1971). *The brain in hominid evolution.* New York: Columbia University Press.

Tomasch, J. (1969). The numerical capacity of the human cortico-ponto-cerebellar system. *Brain Research*, *13*, 476–484.

Voogd, J., & Ruigrok, T. J. H. (1997). Transverse and longitudinal patterns in the mammalian cerebellum. In C. I. DeZeeuw, P. Strata & J. Voogd (Eds.), *The cerebellum: From structure to control. Progress in brain research*, (Vol. 114), (pp. 21–38). Amsterdam: Elsevier.

Weaver, A. H. (2005). Reciprocal evolution of the cerebellum and neocortex in fossil humans. *Proceedings of the National Academy of Sciences of the United States of America*, *102* (10), 3576–3580.

Welker, W. (1987). Spatial organization of somatosensory projections to granule cell cerebellar cortex: Functional and connectional implications of fractured somatotopy (summary of Wisconsin studies). In J. S. King (Ed.), *New concepts in cerebellar neurobiology* (pp. 239–280). New York: Alan R. Liss, Inc.

Welsh, J., & Llinás, R. (1997). Some organizing principles for the control of movement based on olivocerebellar physiology. In C. I. DeZeeuw, P. Strata & J. Voogd (Eds.), *The cerebellum: From structure to control. Progress in brain research*, (Vol. 114), (pp. 449–462). Amsterdam: Elsevier.

White, D. D. (2005). *Size and shape of the cerebellum in Catarrhine primates and Plio-Pleistocene fossil hominins: A paleoneurological analysis of endocranial casts.* PhD dissertation, University of Albany, SUNY, NY.

White, D. D., & Falk, D. (1999). A quantitative and qualitative reanalysis of the endocast from the juvenile *Paranthropus* specimen L388y-6 from Omo, Ethiopia. *American Journal of Physical Anthropology*, *110*, 399–406.

Whiting, B. A., & Barton, R. A. (2003). The evolution of the cortico-cerebellar complex in primates: Anatomical connections predict patterns of correlated evolution. *Journal of Human Evolution*, *44*, 3–12. doi:10.1016/S0047-2484(02)00162-8.

SECTION IV

Human brain evolution

M. A. Hofman and D. Falk (Eds.)
Progress in Brain Research, Vol. 195
ISSN: 0079-6123

CHAPTER 9

Human prefrontal cortex: Evolution, development, and pathology

Kate Teffer and Katerina Semendeferi*

Anthropology Department, University of California, San Diego, San Diego, CA, USA

Abstract: The prefrontal cortex is critical to many cognitive abilities that are considered particularly human, and forms a large part of a neural system crucial for normal socio-emotional and executive functioning in humans and other primates. In this chapter, we survey the literature regarding prefrontal development and pathology in humans as well as comparative studies of the region in humans and closely related primate species. The prefrontal cortex matures later in development than more caudal regions, and some of its neuronal subpopulations exhibit more complex dendritic arborizations. Comparative work suggests that the human prefrontal cortex differs from that of closely related primate species less in relative size than it does in organization. Specific reorganizational events in neural circuitry may have taken place either as a consequence of adjusting to increases in size or as adaptive responses to specific selection pressures. Living in complex environments has been recognized as a considerable factor in the evolution of primate cognition. Normal frontal lobe development and function are also compromised in several neurological and psychiatric disorders. A phylogenetically recent reorganization of frontal cortical circuitry may have been critical to the emergence of human-specific executive and social-emotional functions, and developmental pathology in these same systems underlies many psychiatric and neurological disorders, including autism and schizophrenia.

Keywords: primate; frontal lobe; autism.

Introduction

The frontal lobe and the portion of it occupied solely by association cortex, the prefrontal cortex (hereafter PFC), are eternally popular areas to research in human brain evolution due to their functional attributes. The PFC comprises several Brodmann areas (BAs) anterior to the primary motor and premotor cortex (Fig. 1). The PFC is involved in higher-level cognitive processes grouped under the term of "executive functions" in humans, including mostly dorsolateral areas, like BA 9, 10, and 46 (Baddeley, 1992; Fuster, 2000a; Jurado and Rosselli, 2007), as well as in

*Corresponding author
Tel.: 858-822-0750; Fax: 858-534-5946
E-mail: ksemende@ucsd.edu

DOI: 10.1016/B978-0-444-53860-4.00009-X

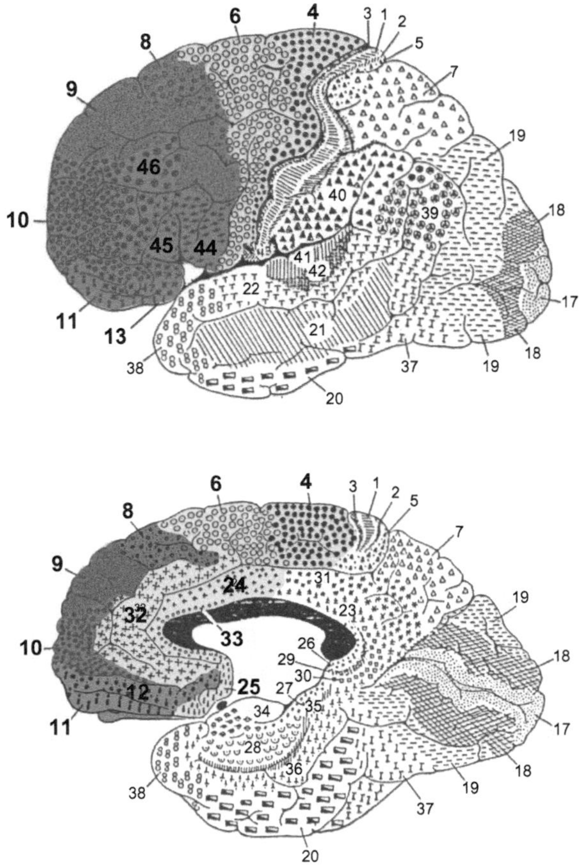

Fig. 1. Diagram of the human brain modified from Brodmann (1909) that illustrates the frontal lobe (all shaded regions rostral to the central sulcus), including the prefrontal cortex (only the darker gray shaded regions) and the anterior cingulate. According to Brodmann's classification scheme, "frontal region" included areas 8, 9, 10, 11, 12, 44, 45, 46, and 47 (of which 44, 45, and 47 he termed "subfrontal"); "precentral region" included areas 4 and 6; and anterior "cingulate region" included areas 24, 32, 33, and 25. Area 13 was at first identified by Brodmann only in nonhuman primates, not humans. It has subsequently been identified in humans as part of the orbital prefrontal. Interestingly, the term "prefrontal" was used by Brodmann only for orbitofrontal area 11 located in the rostroventral part of the frontal lobe. Contemporary use of the term prefrontal cortex refers usually either to all areas demarcated here as part of the frontal lobe, with the exception of the motor/premotor cortex (BA 4 and 6), BA 44, and BA 24, or to areas located only in the dorsolateral frontal lobe, mostly BA 9, 10, and 46. Another use of the term prefrontal is increasingly found in the imaging literature, where the term usually refers to areas "anterior to the genu of the corpus callosum."

language (mostly BA 44/45), emotional processing, and sociality (mostly BA 47, 10, 11, 13 in the orbitofrontal cortex; Beer et al., 2003; Fellows, 2007a,b; Habib et al., 1996; Stone et al., 1998). Executive functions include the organization of input from diverse sensory modalities, the maintenance of attention, the monitoring of information in working memory, and the coordination of goal-directed behaviors (Jurado and Rosselli, 2007; Miller, 2001; Miller and Cohen, 2001; Muller et al., 2002). Together, these abilities would have been necessary for navigating both the complex social groups and unpredictable, dangerous environments of our hominin ancestors. Thus, the capacities enabled by the PFC, while most are not exclusively human, are certainly a crucial aspect of what we think of as "human" in cognition. One of the most fundamental problems to be solved by any animal (Fuster, 2001a), human, or otherwise, living in a complex and ever-changing world, is how to make sense of this setting. There is variation in the environment, as well as in discernable patterns; navigating both the variation and the patterns are things at which humans excel and are activities largely subserved by the PFC.

Although the frontal lobe as a whole has not been differentially enlarged across human evolution (Semendeferi and Damasio, 2000; Semendeferi et al., 1997), there is increasing evidence for its reorganization, as some regions with known functional correlates are either bigger or smaller in the human brain than expected when compared with the same region in great apes. It is also increasingly important to look at microstructural differences in histology, given that humans do not stand out when gross measures such as whole frontal lobe volume are employed. In this chapter, we discuss comparative structural and microstructural work on the human PFC, including stereology, magnetic resonance imaging (MRI), minicolumn analysis, and diffusion tensor imaging (DTI), concentrating on the question of whether, and if so, in what ways the human PFC or any of its subdivisions differ from other primates, in particular, the apes.

A comparative exploration of PFC microstructure is all the more necessary given both that the PFC is one of the last regions of the brain to mature, based on most indices of development (Fuster, 2002), and that neurons in areas that develop later in life have more complex dendritic trees than those that mature earlier, such as primary somatosensory and primary motor cortex (Jacobs et al., 2001; Travis et al., 2005). Brain development is on the whole unusually prolonged in the human species, beginning *in utero* in the third gestational week and continuing well into adolescence. The evolutionary trade-off that arose between large brains and bipedality, and the ensuing difficulty with childbirth, led to the secondarily altricial state of the human newborn (Rosenberg and Trevathan, 1995) and thus an uncommonly lengthy period of brain maturation. Interestingly, the most markedly late-developing regions of the PFC, on the lateral aspect, are those involved in executive functions (Fuster, 2002).

The PFC is also affected in a number of conditions and disorders; its late maturation makes it particularly susceptible to disruption (Bradshaw and Sheppard, 2000; Dumontheil et al., 2008; Ghika, 2008). Some have also hypothesized that the brain regions that were most recently developed or altered in the course of human evolution, including prefrontal association cortex, are predominantly the site of disorders (Ghika, 2008). Here we discuss the impact of autism and schizophrenia on the PFC and frontal lobe in terms of histological and microscopic studies. Dorsolateral PFC, an especially late-developing region, exhibits abnormalities in both autism and schizophrenia, which is further characterized by abnormalities in medial frontal cortex.

Despite this chapter's focus on the anatomy of the PFC, we recognize that no region of the brain operates as a separate and monolithic entity; discussing the role of the PFC necessarily implies a role for other brain regions with which it shares extensive interconnections, including the basal ganglia, thalamus, brainstem, hippocampus, amygdala, and other neocortical regions (Ghashghaei and Barbas, 2002; Thorpe and Fabre-Thorpe, 2001). In addition to its intrinsic connections with other areas of the PFC, allowing access to emotional responses and other information, the lateral PFC is connected to occipital, temporal, and parietal cortices, and thus synthesizes visual, somatosensory, and auditory information at a high level of processing (Miller and Cohen, 2001). It receives input from other limbic structures by way of other prefrontal cortical regions. Further, even as we discuss the functional specialization of the major subdivisions, we appreciate the existence of extensive connections between these subdivisions (Barbas and Pandya, 1989; Wagner et al., 2001); while the orbital and medial regions of the PFC are thought to be involved in the processing and regulation of emotional behavior, and the lateral PFC is differentially implicated in language and the executive functions more traditionally associated with the PFC (Fuster, 2001b), it is well established that emotion plays an important role in many of the cognitive processes grouped under the term executive function, and vice versa (Bechara et al., 2000).

Development

The first brain structure to arise is the neural tube, which is formed in the third week of gestation from progenitor cells in the neural plate (Stiles and Jernigan, 2010). Neuron production begins in the sixth week. From gestational weeks 13 to 20, neuronal count increases exponentially in the neocortical part of the telencephalon (Dobbing and Sands, 1973), with 5.87×10^9 neurons at 20 weeks in the cortical plate and marginal zone (Samuelsen et al., 2003). Although it had traditionally been thought that all of neuronal proliferation and migration occurs by mid-gestation (Rakic, 1988; Sidman and Rakic, 1973), newer cell counts reported for mid-gestation brains are less than half the $19–23\times10^9$ neurons in the average adult human brain (Pakkenberg and Gundersen, 1997). Thus, it seems parsimonious to suggest that neurogenesis continues after mid-gestation (Shankle et al., 1999). Another study found that many regions of the cortex reached their peak cell density between 28 and 38 weeks (Rabinowicz et al., 1996). The future white matter in the intermediate zone also experiences a significant degree of growth from 13 to 20 weeks, with cell number increasing by 380% during this time period. Neurogenesis then slows to a linear rate of increase from gestational week 22 to birth, with total neuron number increasing to about 30×10^9 cells in full-term infants.

The chief external landmarks of the PFC, its primary sulci (superior frontal, inferior frontal, and precentral), develop during gestational weeks 25–26 (Stiles and Jernigan, 2010). In dorsolateral and lateral PFC, basic features of the dendritic arbors of pyramidal neurons emerge during gestational weeks 17–25 (Mrzljak et al., 1988). From weeks 26 to 34, layer III and V dendrites continue to mature, as spines develop, basal dendritic length increases, and interneurons differentiate in layer IV (Mrzljak et al., 1992). At birth, total brain weight is about 370g (Courchesne et al., 2000).

During early childhood, the brain quadruples in size and grows to roughly 90% of the adult volume by age 6 (Courchesne et al., 2000; Knickmeyer et al., 2008). The initial periods of rapid neurogenesis and synaptogenesis subsequently give way to a period of pruning and neuronal death to manage the overproduction of these components. Throughout childhood and adolescence, brain development is characterized by both growth and then decline in gray matter volume, and increases in white matter volume. During this period of brain development, dorsolateral and medial PFC expands nearly twice as much as some other regions, including medial occipital and insular cortex (Hill et al., 2010). Converging evidence from diverse methodologies has established that the frontal lobe, in particular, much of the PFC, matures late relative to much of the remainder of the cortex (Fig. 2). Regions of the temporal cortex, which, like the PFC, are higher-order association areas that also integrate diverse inputs from sensorimotor and other lower-order regions and develop late as well.

Cortical thickness is a useful gauge of overall maturity in developmental studies, since it is a composite measure that includes neurons, axons, dendrites, synapses, and glia. Prenatally, cortical thickness increases linearly throughout the entire brain as a function of age (Rabinowicz et al., 1996). If cortical thickness in children and adolescents is plotted as a function of age, the majority of PFC (including the lateral orbitofrontal, lateral prefrontal, medial and lateral frontal pole) follows a cubic trajectory or U-shape (Shaw et al., 2008). The development of the PFC is characterized by growth in early childhood, decrease in adolescence, and then a slight increase and stabilization in adulthood. This pattern is thought to be linked to the maturation of cortical circuits that underlie frontal lobe functioning, including language, decision-making, attention control, and working memory (Casey et al., 2005; Caviness et al., 1996; Giedd et al., 1996). Gray matter volume, also as measured by cortical thickness, reaches maximum volume in most of the frontal lobe between the ages of 11 and 12 (Giedd et al., 1999b). This contrasts with the cortex as a whole, where gray matter increases primarily from early childhood until the age of 6–9 (Courchesne et al., 2000). The dorsolateral PFC attains adult levels of cortical thickness particularly late, in early adolescence (Lenroot and Giedd, 2006).

From the ages of 5 to 11, regions in the PFC that correspond to Broca's area exhibit an increase in gray matter thickness relative to some neighboring regions (Sowell et al., 2004b), an occurrence that is thought to be related to the maturation of linguistic capacity. Developmentally normal and age-appropriate decreases in left dorsolateral PFC gray matter volume, likely reflecting

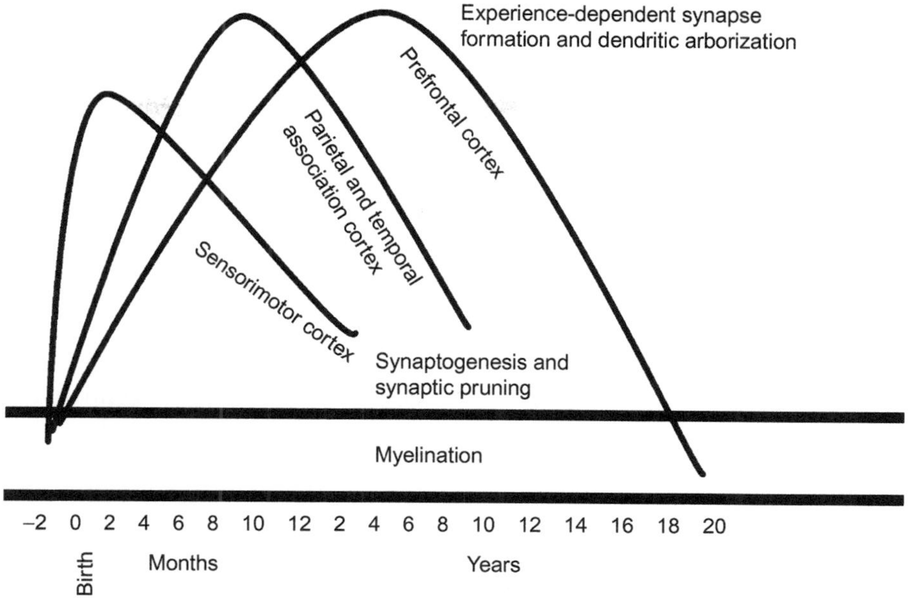

Fig. 2. A depiction of the time course of brain development in human prefrontal cortex, sensorimotor cortex, and parietal and temporal association cortex. Modified from Thompson and Nelson (2001).

neuronal pruning and increased myelination, were positively correlated with scores on vocabulary tests in one study (Sowell et al., 2004a). In the frontal lobe, this same phenomenon is positively correlated with scores on verbal memory tests (Sowell et al., 2001).

Unsurprisingly, the same U-shaped pattern reported by Shaw and colleagues is also described in the majority of neurodevelopmental studies examining gray matter in the PFC, regardless of what micro- or macrostructure is being examined; these gray matter components belong to the same neural circuits and have enduring reciprocal connections (Fields and Stevens-Graham, 2002). Gray and white matter both continue to experience macro- and microstructural changes throughout development, often even after adolescence, and these changes in structure parallel changes in functional organization that are in turn also reflected in behavior (Diamond, 2001).

As the brain increases in size throughout childhood and adolescence, many other microstructural changes occur as well, including dendritic and axonal growth and synaptogenesis. These microstructural changes are also heterochronous; most of these events occur earliest in sensorimotor cortex and other primary cortex, and latest in PFC and other higher-order association cortex that integrate and process information from primary cortex (Shankle et al., 1999). Neurogenesis begins during the sixth gestational week, and, at birth, neuronal density in the frontal cortex is similar to that in many other regions, including the visual cortex. This soon changes, however; while neuronal density in the visual cortex has decreased to adult values by only 5 months of age, neuron density does not peak until much later in childhood in the frontal lobe. Throughout the cortex as a whole, neuronal number increases 60–70% between 24 and 72 months postnatally (Shankle et al., 1999). From 1939 to 1967, J. L. Conel published a comprehensive histological data set in which he measured a number of microscopic features (including neuron packing density, total cortical and individual cortical layer thickness, myelinated large fiber density, large proximal dendrite density, and neuronal body size) in the developing human brain (Conel, 1939, 1941, 1947, 1951, 1955, 1959, 1963, 1967). His specimens ranged in age from birth to 5years. In all PFC areas included in a meta-analysis of J. L. Conel's data set (BA 8–11 and 44–46), neuronal number measurements increase at every age point postnatally (0–72 months), save for between 3 and 6months, where the neuron counts actually decrease. Neuron density is 55% higher in the frontal cortex of 2-year-olds than it is in adults, and 10% higher in 7-year-olds than in adults (Huttenlocher, 1990). Adult neuronal density in the frontal lobe is reached by 10 years of age (Huttenlocher, 1990).

Total gray matter volume is also greatest at the earlier stages of infancy, with sustained loss beginning around puberty. During infancy and childhood, gray matter volume in the frontal lobe is strongly and positively correlated with total brain volume, and a steep decline in gray matter proportional with volume occurs with age (Sowell et al., 2002). A longitudinal MRI study of gray matter development in juveniles from ages 4 to 21 discovered that the cortex matures in sequence from caudal to rostral (Gogtay et al., 2004) with an overall increase in gray matter from the ages of 4 to 12 and a decrease afterward (Pfefferbaum et al., 1994). An exception to this pattern is the frontal pole, where gray matter volume both peaks and begins to decrease earlier in childhood than it does in the rest of the PFC (Gogtay et al., 2004). However, cerebral energy metabolism studies have reported that lateral aspects of the PFC mature earlier in some ways than the most anterior regions, such as the frontal pole (Chugani and Phelps, 1991).

Within the frontal lobe, gray matter in the precentral gyrus develops the earliest, while more rostral regions, including the superior and inferior frontal gyri, mature later. The ventromedial areas of the PFC commonly reach maturity earlier than more lateral regions as well (Fuster, 2002). The dorsolateral PFC, a region involved in executive functioning, begins to lose gray matter only at the end of adolescence. Reduction in gray matter volume continues in the frontal lobe until adulthood and is most pronounced in adolescence and early adulthood (Sowell et al., 1999b). Although this decrease in gray matter volume in childhood is correlated with age, one study found that gray matter thinning in the frontal lobe is significantly and positively related to verbal memory abilities, independent of the age of the child (Sowell et al., 2002).

At the same time, as gray matter volume decreases throughout childhood and adolescence, white matter experiences a related enlargement in volume as fiber tracts grow and myelinate. Myelination begins in the 29th gestational week with the brain stem, and the development of white matter also typically follows a caudal to rostral progression (Flechsig, 1901, 1920). Humans (Giedd et al., 1999a; Gogtay et al., 2004; Klingberg et al., 1999; Levitt, 2003; Paus et al., 1999; Pfefferbaum et al., 1994; Watson et al., 2006) as well as our close relatives the chimpanzees (Watson et al., 2006) exhibit a nearly linear white matter volume increase and continued myelination until adolescence or early adulthood. Throughout the cortex as a whole, white matter volume increases 74% from infancy to mid-adolescence (Courchesne et al., 2000). The frontal lobe myelinates last, and in its most rostral regions myelination can continue well into the third decade of life (Sowell et al., 1999a). From ages 7 to 16, the frontal lobe experiences an increase in white matter volume that goes above and beyond what is expected from overall brain growth during these ages (Sowell et al., 2002). By 6 months of age, dendritic length is 5–10 times greater than it is at birth, yet dendritic length in the middle frontal gyrus is still only half of adult values at 2 years of age (Schade and Van Groenou, 1961). In contrast, adult dendritic length is reached in the visual cortex by the age of 1 year (Becker et al., 1984). The anteriomedial aspect of the frontal lobe, an area involved in attention and self-referential tasks (Zysset et al., 2003), is one of the last regions, along with the temporal lobe, to myelinate postnatally (Barkovich, 2005). Interestingly, in one study, white matter development in the dorsolateral regions of the frontal cortex appears to vary far more according to age than it does in ventral or subcortical regions (Sowell et al., 1999a).

As with the U-shaped pattern of gray matter maturation, the steady and nearly linear increase in white matter that is observed through early adulthood has been tied to age-appropriate changes in cognition and behavior. Unsurprisingly, strong effects have been shown for cognitive abilities in which the PFC is known to be involved, including working memory, inhibitory control, and language. Working memory capacity is positively correlated with the development of connectivity between superior frontal and parietal lobes (Nagy et al., 2004; Olesen et al., 2003). Maturing connections between the PFC and striatum are credited with the development of inhibitory control, as measured by a performance on a go-nogo task in children (Durston et al., 2002). "Inhibitory control" tasks rely on the more mature cognitive ability to suppress less relevant information and actions in favor of those more pertinent to the task at hand. Due to the immaturity of the PFC in the very young, children appear to differentially recruit these regions in inhibitory control tasks when compared to adults (Bunge et al., 2002; Durston et al., 2002).

Synaptogenesis begins *in utero* at around the 20th gestational week. Like many other neurodevelopmental processes, the formation and organization of synapses in the PFC increases after birth, reaches a peak, and is followed by pruning and decline. Also as with the other processes described in this chapter, synaptogenesis occurs later in the PFC than it does in other areas. The middle frontal gyrus of the PFC reaches peak synaptic density late in infancy at 3.5 years, while auditory and visual cortex attains peak density at 3 months (Huttenlocher and Dabholkar, 1997; Huttenlocher and De Courten, 1987). At the age of 3 months, synaptic density in the PFC is less than half of what it will eventually reach, and synapse elimination persists throughout adolescence. One early study suggested that synaptogenesis occurs at relatively the same time and rate in all parts of the neocortex in monkeys (Rakic et al., 1994).

The later maturation of more rostral regions of the cortex, particularly association cortex in the frontal lobe, is by now a well-established fact. Cerebral energy metabolism, a measure of regional activity, increases earlier in parietal, temporal, and occipital lobes (3 months) than it does in the PFC (8 months; Chugani and Phelps, 1986). Cytoarchitectonic asymmetries in PFC regions may also develop after they do in primary cortex (Amunts et al., 2003). Temporal association areas and regions with fronto-temporal connection also mature later. Diffusion tensor MRI (DTI) studies, which allow for the visualization of white matter tracts, have shown that a number of frontal connections mature more slowly than other white matter fibers. The uncinate fasciculus, which connects limbic system structures with the orbitofrontal cortex, a region involved with processing emotions and reward, and the head of the caudate nucleus, which possesses extensive interconnections with the PFC, do not fully mature until the third decade of life (Lebel et al., 2010). In infants, pyramidal neurons in regions that mature later, including the frontal lobe, have less complex dendritic trees than areas that mature early, such as primary sensorimotor cortices (Travis et al., 2005). Later in development, however, this trend is reversed; in adults, frontal association areas have the most complex dendritic trees (Jacobs et al., 1997, 2001).

During normal, healthy aging, gray matter volume and gray matter to white matter ratios decrease throughout the cortex. Whole brain volume decreases as well; in one study, individuals aged 71–80 years possess brain volumes that were close to those of healthy 2- and 3-year-old children, having decreased by about 26% (Courchesne et al., 2000). White matter volume in the cortex as a whole reaches a plateau around age 30 and decreases slightly but steadily in later years during normal aging.

Although, as described above, much of the cortex experiences age-related change, the PFC has long been reported to suffer the highest degree of change associated with aging (Jernigan et al., 2001; Salat et al., 1999a,b). Gray matter volume in PFC is disproportionately affected in healthy aging (Raz et al., 1997), particularly in comparison to sensorimotor cortex (Coffey, 1994; Cowell et al., 2007) but also in comparison to the temporal lobe (De Carli et al., 1994; Kemper, 1994; Raz, 1996). Many of the cognitive functions that are known to decline in senescence are those subserved by the PFC: working memory, behavioral inhibition, decision-making, and meta-memory (Salat et al., 1999a). The dorsolateral and orbitofrontal regions experienced a loss of gray matter volume at the rate of 4.9% per decade in one sample that comprised individuals ranging from 48 to 77 years (Raz et al., 1997). However, a later study by some of the same researchers reported a significant loss of volume from middle age in orbitofrontal gray matter and frontal white matter, but not in lateral frontal gray matter (Raz et al., 2010).

Another study examined total PFC volume, gray matter volume in PFC, and white matter volume in PFC in "young elderly" (mean age = 70 years) and "old elderly" (mean age = 90 years). These researchers found that from 70 to 90, there appeared to be a decrease in the gray matter to white matter ratio, and that in the very old, the decrease in white matter volume is greater than the loss in gray matter volume (Salat et al., 1999a). There were also significant negative correlations between age and total PFC volume and between age and white matter volume.

However, not all aspects of the PFC undergo change in healthy aging or change in the same way. PFC subregions appear to exhibit differential patterns of aging. One study partitioned PFC volume into many comparisons, including orbital versus dorsal regions, lateral versus medial regions, and right hemisphere versus left (Cowell et al., 2007). Age-related decreases in volume were significantly more prominent in medial than lateral PFC, particularly in male subjects. There was also significant age-related decline in dorsal medial PFC volume by age 70. In contrast to many reports on total gray and white matter volumes, synaptic density in layers III and V of the superior middle frontal gyrus (BA 9) remains largely constant from the years of 20 to 89 (Scheff et al., 2001). On the other hand, another study reported that synaptic density in the frontal lobe as a whole decreases with age (Masliah et al., 1993).

There is reduced activation in the aging PFC compared to younger subjects during visuospatial tasks (Solbakk et al., 2008) and working memory and attention tasks (Milham et al., 2002; Reuter-Lorenz, 2002). In the same vein, there also appear to be significant age-related declines in blood flow to the PFC (Waldemar, 1995). In contrast, some tasks, such as perceptual and episodic memory tasks, elicit greater response from the PFC of older individuals, which may denote a compensatory increase in recruitment of these regions (Cabeza et al., 2002; Grady, 2000).

Evolution

Gross anatomical cross-species comparisons: Frontal lobe

In many mammalian species, including primates, 9 out of 11 major brain regions (cerebellum, mesencephalon, diencephalon, olfactory cortex, parahippocampal cortex, hippocampus, neocortex, septum, and striatum) exhibit a robust covariance in size (Finlay et al., 2001). Deviations from allometry (Rilling, 2001) are of great interest to comparative neuroanatomists, as is the endeavor of accounting for structures whose growth does not scale with the rest of the brain. Yet, it has also been noted that an overemphasis on allometric relationships of large brain regions may obscure potentially important niche-specific adaptations of certain species (Holloway, 2001). In response to this, mosaic evolution argues that selection can act on specific brain regions, resulting in the enlargement or diminution of particular functional subsystems in response to environmental demands (Barton, 2001; Barton and Harvey, 2000). The fact that brain regions that are connected functionally or anatomically tend to evolve together, independent of other structures, supports the importance of mosaic evolution as a factor (Barton and Harvey, 2000). Individual brain regions are likely linked, but not tightly so, to the size of other brain regions; brain evolution is both limited by the rules of neural development and is possibly the site of species-specific adaptations (Striedter, 2005).

In absolute terms, the human frontal lobe is three times larger than that of our closest living relatives, the great apes, but the significance of this fact has been debated (Passingham, 2002) and is still being considered. Moreover, does thinking of a complex and heterogeneous neural system like the frontal lobe as a discrete entity obscure important differences? In the past two decades, an increasing number of studies have examined the human frontal lobe and PFC in comparison to other primates, utilizing a wide variety of methodologies including stereology, MRI, minicolumn analysis, and DTI. The comparative analysis of specific regions within the frontal cortex has the ability to inform debates on the evolution of the human frontal lobe. In this section, we discuss comparative structural and microstructural work on the human PFC, concentrating on the question of whether, and if so, in what ways, the human PFC or any of its subdivisions differ from other primates, in particular, the apes.

Due to the current impossibility of accurately identifying the PFC while relying solely on gross anatomical landmarks like sulci and gyri (Semendeferi et al., 2002), most imaging studies have examined the frontal lobe as a whole. Based on older studies, such as the classical work by Brodmann (1912) that employed samples of nonhuman primates that rarely included any great apes, or studies that only included one specimen per nonhuman species, it was long thought by many that human brain evolution was characterized by a disproportionate increase in the relative volume of the frontal lobes (Blinkov and Glezer, 1968; MacBride et al., 1999; Uylings and Van Eden, 1990). Although this finding was not without controversy at the time (Holloway, 1968; Passingham, 1973), a more recent comparison of the frontal cortex and its subdivisions in living specimens of humans and their closest living relatives using MRI revealed that this is not the case (Semendeferi and Damasio, 2000; Semendeferi et al., 1997, 2002). In the most recent of these studies (Semendeferi et al., 2002), the frontal cortex was parceled into two subdivisions that are accurately identifiable using gross anatomical markers: the cortex of the precentral gyrus and the remaining rostral frontal cortex on the dorsolateral, medial, and orbital surfaces of the frontal lobe. To date, this is the largest and most comprehensive attempt to examine the human frontal cortex in concert with that of the other living hominoids. The nonhuman sample comprises 6 chimpanzees, 3 bonobos, 2 gorillas, 4 orangutans, 4 gibbons, and 5 macaques; this group of 20 individual specimens was compared with 10 human specimens. Although the human frontal cortex is clearly bigger than that of the great apes in absolute terms, as a whole, it is not larger than expected for an ape brain of human size, based on both logarithmic and linear regressions (Semendeferi et al., 2002). Similarly, the proportion of the cortex occupied by the frontal cortex is not greater in humans than it is in great apes. The two partitions of the frontal cortex, the precentral gyrus and the rostral frontal cortex, were likewise no larger than expected in humans. These findings support previous work on the issue of absolute and relative frontal cortex size (Semendeferi and Damasio, 2000; Semendeferi et al., 1997).

When the relative volume of the dorsolateral, mesial, and orbital subdivisions of the frontal lobe is calculated in humans and great apes, also utilizing MRI scans, human values again do not differ from that expected for a human brain of great ape size (Semendeferi et al., 1997). Volumes were estimated using sulci that are homologous across great apes and humans. In all species, the orbital subdivision of the frontal lobe is the smallest, followed by the mesial and dorsolateral sectors. The relative volumes of all three subdivisions are very similar across all species, including humans. Cortical surface measurements, which were also performed for the entire frontal lobe and for the three subdivisions, likewise did not reveal any relative differences between humans and the other primates. Interestingly, an examination of the scaling of frontal cortex in mammalian species, including 25 primates, revealed that the order Primates is characterized by a hyperscaled frontal cortex that increases in size relative to both the isocortex and the whole brain with strongly positive allometry (Bush and Allman, 2004). The PFC is furthermore unique in primates in that it includes a small-celled granular layer absent in other mammals, although nonprimate mammals do possess analogues to the PFC (Uylings et al., 2003).

Although studying the frontal lobe as a whole is an important stage in our endeavor to determine the anatomical substrates underlying the uniquely human in human cognition, it is also merely a first step. The frontal lobe comprises numerous anatomical components and diverse functional areas, and, therefore, consideration of it as a discrete unit can only tell us so much. A number of recent studies have examined the relative size of gray and white matter in the frontal lobe or PFC, while others have examined the volume, neuron density, and columnar organization of functional subregions within the PFC.

When white matter is considered separately from gray matter, the human frontal lobe also remains undistinguished from apes in terms of overall relative volume (Schenker et al., 2005). White matter in the frontal lobes was divided into two sectors: the white matter immediately underlying the cortex, which was termed "gyral white matter," and the rest of white matter, or "core." The relative volume of gray matter and the two sectors of white matter in the frontal lobe was measured using MRIs of living humans and apes. The dorsolateral, mesial, and orbital subdivisions of the frontal lobe were outlined, and the relationship between cortex and gyral white matter within each subdivision was analyzed. In all three subdivisions of the frontal lobe, human values for core white matter volume were as large as expected. However, gyral white matter, which comprises white matter directly underneath the gyrus, was larger than anticipated in both the frontal and temporal lobes. Gyral white matter myelinates later in development than core white matter (Yakovlev and LeCours, 1967) and may connect neighboring cortical regions which lie on opposite sides of the gyrus (Van Essen, 1997). Thus, enlarged gyral white matter may indicate increased interconnectivity within and between adjacent cortical regions.

The question of whether the PFC, or its gray or white matter subdivisions, is differentially enlarged in humans is a matter that has received some attention as of late. As mentioned, it is nearly impossible to accurately identify the PFC, especially across a wide cross-section of species, based solely on gross anatomical features. One study measured gray matter, white matter, and total volumes for the PFC in humans, bonobos, chimpanzees, gorillas, orangutans, gibbons, and several monkey species (Schoenemann et al., 2005). In this study, the PFC was demarcated in MRIs, using the region of the frontal cortex anterior to the genu of the corpus callosum as a proxy definition for PFC. The authors concluded that PFC white matter is significantly larger in humans than in nonhuman primates, but that there is no difference between humans and other primates regarding PFC gray matter. However, there is some concern that, as cytoarchitectonic criteria of great apes were not used, that the volume of the PFC as defined in relation to the genu of the corpus callosum is underestimated in these species (Sherwood et al., 2005). Additionally, when the human data from this study are regressed to a best-fit line that is based only on the great ape data, rather than all of the nonhuman primate data, white matter in the PFC is at best, only slightly larger in humans than expected. Also of note is a recent study examining nonhuman anthropoids, including great apes, which reported that the hyperscaling of both the frontal lobe and the whole neocortex to the rest of the brain is primarily due to frontal white matter volumes (Smaers et al., 2010, 2011).

Comparative work on PFC subdivisions: Volumetric, DTI, and minicolumn studies

Although the frontal lobe as a whole does not seem to have been differentially enlarged throughout human evolution, there is evidence for its reorganization, as some regions with known functional correlates are either bigger or smaller in the human brain than expected when compared with the same region in great apes. Several of these functional subdivisions of the PFC that are homologous across humans and great apes have been examined histologically in a comparative context (Fig. 3).

Limbic frontal cortex, BA 13, occupies a portion of the orbitofrontal cortex and is part of the neural substrate underlying emotional reactions to social stimuli (Damasio and Van Hoesen, 1983). It is found in the posterior orbitofrontal cortex and shares strong reciprocal connections with the insular, temporal polar, and parahippocampal cortices, as well as with basal forebrain structures like the ventral striatum, nucleus basalis of Meynert, and amygdala (Nauta, 1962; Van Hoesen, 1981). BA 13 has been identified across humans, chimpanzees, bonobos, gorillas, orangutans, and gibbons, and its volume was estimated in all species using stereological analyses (Semendeferi et al., 1998). BA 13 is conserved in its structure, and features such as size

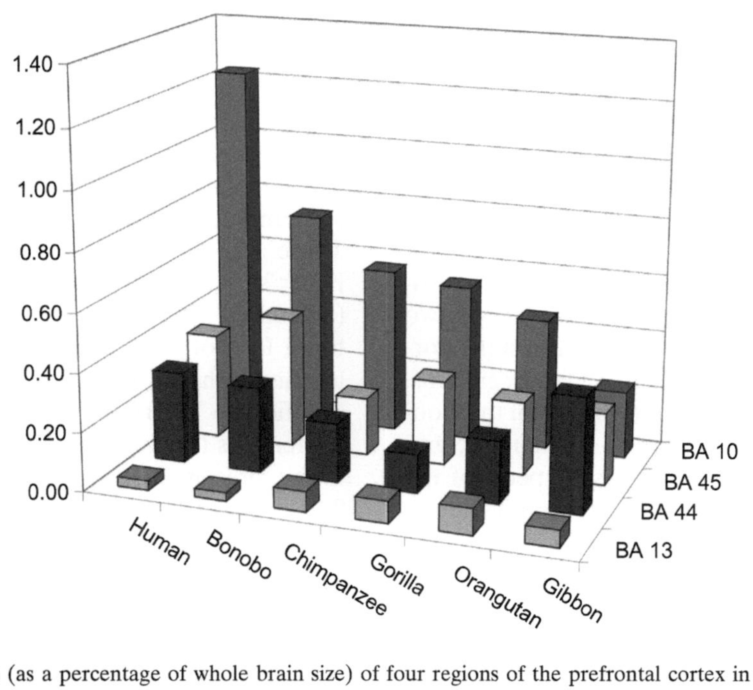

Fig. 3. Relative volumes (as a percentage of whole brain size) of four regions of the prefrontal cortex in humans and great apes. Data from Schenker (2007) and Semendeferi et al. (1998, 2001).

of cortical layers, density of neurons, and space available for connections are similar across hominoids, with only subtle differences present. However, in contrast to the homogeneity found in its organization, variation is present in the relative size of this cortical area, when it is considered as a percentage of total brain volume. Area 13 occupies a smaller percentage of the human and bonobo brains than it does in other hominoid species; the absolute size of BA 13 is quite similar among humans and all of the great apes, save the bonobos (Semendeferi et al., 1998). Its volume is not correlated with the volume of the brain as a whole. It is parsimonious to suggest that BA 13 was part of the Plio-Pleistocene hominoid and hominid brain; BA 13 may have hypothetically occupied a restricted area in the most posterior parts of the medial orbital gyrus and the posterior orbital gyrus, with structural features similar to those present in extant species.

BA 10, which lies at the most anterior aspect of the PFC, is a region of association cortex known to be involved in higher cognitive functions, such as planning future actions and decision-making (Fuster, 2008). BA 10, also called the frontal pole in hominoids, has been identified across humans and most of the apes (Semendeferi et al., 2001). Area 10 has similar cytoarchitectonic features among hominoids, and it forms the entirety of the frontal pole in humans, bonobos, chimpanzees, orangutans, and gibbons, but its presence is not yet established in gorillas. It has two components in the macaque brain: one on the dorsolateral aspect and one on the orbital. In gibbons, area 10 occupies only the orbital sector of the frontal pole, while in chimpanzees, orangutans, and humans, it occupies both sectors. Aspects of the frontal pole's organization vary slightly across hominoid species, including the relative width of its cortical layers and the space available for connections. BA 10 is larger in humans than in apes both in absolute terms and relative to the rest of the brain. Based on least squares regression, the expected volume for BA 10 in an ape brain of human size is little more than half of the observed volume. Supragranular layers II and III also have more space between neurons in the human brain, possibly for connections with other higher-order association areas, a hypothesis lent further support by recent findings regarding minicolumn size in the human PFC, which are discussed in more detail later in this chapter (Semendeferi et al., 2011). This suggests that the neural substrates supporting cognitive functions associated with this part of the cortex enlarged and became specialized during hominid evolution. In the great apes, BA 10 expanded from its restricted orbital location and came to occupy the entire frontal pole in hominoids.

Broca's area, or BAs 44 and 45, comprises part of the inferior frontal gyrus in the human brain. These regions are involved in language production, particularly linguistic motor control, sequencing, planning, syntax, and phonological processing (Broca, 1861; Damasio et al., 2004; Price, 2000). Given their association with language production, the question of their presence and homology in nonhuman primates is of obvious interest. Based on cytoarchitectonic criteria, both regions have been identified in the inferior frontal gyrus of chimpanzees, bonobos, gorillas, and orangutans (Schenker et al., 2008) and display similar cytoarchitectonic characteristics in all hominoid species examined, including humans. There are no relative volumetric differences in Broca's area between humans and the apes (Schenker, 2007). In humans, there is a distinct trend for both BA 44 and 45 to be larger in the left hemisphere than the right; this trend reaches significance in BA 44 for males and in BA 45 for females (Uylings et al., 2006). This asymmetry is not present in chimpanzees (Schenker et al., 2010), suggesting that Broca's area in the left hemisphere expanded in relative size during human evolution, possibly as an adaptation for our species' language abilities. The arcuate fasciculus, a white matter fiber tract that connects regions in the dorsolateral frontal cortex to language regions in the temporal cortex, is more complex in humans than in chimpanzees (Rilling et al., 2008a) and also exhibits a population-level leftward asymmetry in humans, though not in chimpanzees (Glasser and Rilling, 2008).

Thus, although the PFC as a whole is not differentially enlarged over the course of human brain evolution (Smaers et al., 2011), it seems that there is strong evidence for reorganization within the PFC. In addition to volumetric and asymmetry differences among subdivisions, spacing distance between neurons in layer III also sets the human PFC apart. The human frontal pole and Broca's area, BA 44/45, both exhibit differences in columnar organization when compared with the same regions in great apes (Fig. 4). Spacing distance between neurons (HSD, or horizontal spacing distance) and gray level ratio (GLR that measures the area fraction occupied by cell bodies) have been measured in both of these regions, across a histological sample of humans and apes. The combination of increased HSD and decreased GLR values can be used to identify the presence of wider minicolumns in the cortex, signifying enlarged intracolumnar and intercolumnar neuropil space in layer III (Buxhoeveden et al., 1996, 2001; see Semendeferi et al., 2011 for an in-depth discussion of the methods).

HSD is significantly larger in humans than in the great apes in both regions of Broca's area, while GLR is lower in humans than in all of the great apes (Schenker et al., 2008), indicating wider minicolumns. However, relative to brain size, humans have narrower minicolumns than great apes in both regions. Wider minicolumns have likewise been found in the human frontal pole, or BA 10, when it is compared with apes (Semendeferi et al., 2011), based on HSD and GLR measurements. Spacing distance, as measured by HSD, in the human frontal pole, in particular, stood out by being 30% larger than in the frontal pole of the other species. However, as in BA 44/45, when these measurements are placed within the context of overall brain size, humans have relatively narrower minicolumns in BA 10 than do apes.

Within the human brain, the frontal pole also has wider minicolumns than BA 3 (primary somatosensory cortex) or BA 17 (primary visual

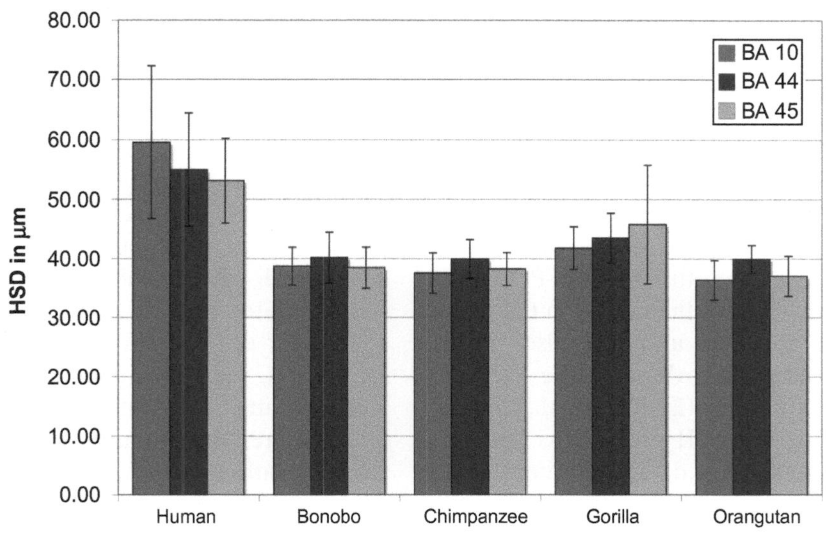

Fig. 4. Horizontal spacing distance (µm) in three regions of the prefrontal cortex (BA 10, 44, and 45) in all species examined. Data from Schenker et al. (2008) and Semendeferi et al. (2011).

cortex), although this pattern does not appear in any apes (Semendeferi et al., 2011). In apes, BA 4 (primary motor cortex) possesses wider minicolumns than BA 10, although HSD was not significantly different between BA 10 and BA 4 in humans. Minicolumns were found to be wider in dorsolateral, medial, and orbital regions of the PFC than in BA 17 in normal human adult brains (Buxhoeveden et al., 2006), which supports the hypothesis that minicolumns are wider throughout the PFC in humans. Another comparison (Casanova et al., 2006) also reports a very similar pattern to those previously described; in that study's control humans, interneuronal distance is reported to be highest in PFC region BA 9, followed closely by BA 4, then BA 3 and BA 17.

This recent research suggests that neurons in layer III are significantly more widely spaced throughout the PFC in humans than they are in great apes, while spacing in sensorimotor and visual cortex is similar in humans and apes. One interpretation of the functional significance of absolutely wider minicolumns, such as those noted in this and the previously mentioned studies, is that they are associated with being more generalized processors (Gustafsson, 1997, 2004). Minicolumns largely comprise pyramidal neurons in layer III, along with their myelinated axons and apical dendrites (DeFelipe, 2005; Peters and Sethares, 1996). The human PFC is also known to exhibit more complex dendritic branching than visual cortex (Elston et al., 2006). Interspecific differences in dendritic structure have also been noted; pyramidal cells in the human PFC are more branched and spinous than those in the temporal and occipital lobes and are also more branched and spinous than those in the PFC of macaques and marmosets (Elston et al., 2001). In the human PFC, layer III projections possess more branched and spinous dendritic arbors than in temporal, occipital, or parietal cortex (Elston et al., 2001; Jacobs et al., 2001; Petanjek et al., 2008). The long-range cortico-cortical projections of layer IIIc neurons (Lewis et al., 2002), in particular, are thought to be critical to working memory and other higher-order cognitive processes in primates (Elston et al., 2006; Fuster, 2000b), suggesting that the reported differences in dendritic tree structure are related to cognitive differences (Zeba et al., 2008). These findings may be indicative of some degree of reorganization characteristic of the human PFC in general, and possibly the frontal lobe as a whole. Human minicolumns are reported to be wider in the lateral superior temporal cortex (BA 22) than in chimpanzees (Buxhoeveden et al., 2001), which suggests that wider minicolumns may be a human trait throughout association cortex beyond the PFC.

A novel class of neurons, Von Economo neurons (VENs), has been identified in the anterior cingulate cortex and frontal insula in humans (Allman et al., 2002; Fajardo et al., 2008) and great apes (Allman et al., 2010; Nimchinsky et al., 1999), though not in other nonhuman primates (Nimchinsky et al., 1999). Frontoinsular cortex and the anterior cingulate project to the frontal pole, other parts of frontal and insular cortex, the septum, and the amygdala. These specialized large projection neurons are also present in several other species of large-brained social mammals, including elephants (Hakeem et al., 2009) and cetaceans (Butti et al., 2009), leading to the proposal that they are the result of convergent evolution in large-brained mammals that require rapid computation of social information (Allman et al., 2010). Also of note is the fact that VENs appear most numerous on the crowns of gyri, which when combined with the finding that gyral white matter is expanded in humans (Schenker et al., 2005) suggests that gyral areas of the PFC may have undergone specific changes during human evolution.

In the realm of neurotransmitters, differences in innervation have been found in the PFC of humans and chimpanzees, both when they are compared to the PFC of macaques, and when compared to primary motor cortex in all three species. In humans and chimpanzees, both BA 9 and 32, which are involved in working memory and theory of mind, respectively, possessed a higher number of dopaminergic afferents in layers III, V, and VI (Raghanti et al., 2008a) and greater density of serotonin transporter-immunoreactive axons in layers V and VI (Raghanti et al., 2008b); for a lengthier discussion of this line of work, see Chapter 11.

Thus, the past 15 years of research into comparative neuroanatomy support the idea (Allen, 2009; Semendeferi et al., 2002) that human brain evolution is characterization by distinct changes in the local circuitry and interconnectivity of the PFC. In particular, modifications throughout the human PFC include increased gyral white matter, a relatively smaller BA 13, a relatively larger BA 10, and greater spacing between layer III neurons in BA 10, 44, and 45. Microstructural changes shared among humans and our closest relatives, the African apes, include VENs.

Pathology

The PFC is affected in a number of conditions and disorders. Here we discuss the impact of autism and schizophrenia on the PFC and frontal lobe in terms of histological and microscopic studies. The late maturation of the PFC, as detailed in the section "Development," makes it particularly vulnerable to developmental disorders (Bradshaw and Sheppard, 2000; Dumontheil et al., 2008; Ghika, 2008). The dorsolateral PFC and anterior cingulate cortex are two regions of the PFC that are affected in both autism and schizophrenia. Collectively, along with lateral orbital PFC, interconnected regions of the basal ganglia, and the supplementary motor area, these regions are called the frontostriatal system, and they work together to subserve many of the cognitive capacities that characterize the human species (Goldman-Rakic, 1988).

Autism spectrum disorder (ASD) is a neurodevelopmental disorder characterized by deficits in social cognition, theory of mind, language, communication, and by stereotypical patterns of behavior (Geschwind, 2009). ASD has a complex and often inconsistent neuropathological profile (Salmond et al., 2007; Schmitz and Rezaie, 2008), likely due to the "spectrum" aspect of the disorder; the phenotype of ASD is heterogeneous, and it is probable that there are significant interindividual differences in the samples examined.

Children with ASD typically display larger brains than average throughout their infancy and toddler years, but by school age these global differences disappear, and brain volumes are normal or even slightly smaller than normal (Courchesne et al., 2011; Redcay and Courchesne, 2005). This trajectory is strongly suggestive of prenatal or quite early postnatal factors playing the determining role in the development of this neuropathology. Head circumference measurements, which are highly correlated with whole brain volumes early in development, are notably larger in infants and very young children who are diagnosed (sometimes later, due to their age) with ASD (Courchesne et al., 2004; Dementieva et al., 2005). This is suspected to be the result of abnormal white and gray matter developmental processes; while head circumference is normal or even below normal in early infancy, it then reliably increases to the 84th percentile between 6 and 14 months. Between the second and third years of life, 90% of autistic children examined in one study had head circumferences larger than average (Courchesne et al., 2001).

These global differences are so robust that young autistic and nonautistic brains can be distinguished solely on cerebral and cerebellar volumes, with 95% accuracy (Akshoomoff et al., 2004). Overgrowth is most marked in frontal and temporal lobes, and far less so in occipital (Carper et al., 2002). However, this differential enlargement in the ASD brain stops after early childhood (Dawson et al., 2007), and autopsies of adult individuals with ASD report average brain weight in the vast majority of cases

(Courchesne et al., 1999), while one even reported a reduced frontal lobe volume in adults (Schmitz et al., 2007). A recent analysis of over 500 MRIs of autistic individuals from age 1 to 50 years confirms the pattern seen in autopsies (Courchesne et al., 2011).

Several structural MRI studies from the past decade corroborate the existence of abnormalities in the frontal regions. Carper and Courchesne (2005) divided the frontal lobe into dorsolateral, medial frontal, precentral gyrus, and orbitofrontal regions in children with ASD and measured their respective volumes. The dorsolateral and medial frontal cortices were significantly enlarged in individuals aged 2–5 years when compared with controls. Frontal sulci are shifted anteriorly in older autistic children, a finding consistent with early overgrowth in this region (Levitt et al., 2003). Development of the dorsolateral PFC seems particularly disrupted in autism; there is an increase in size of 48% from ages 2 to 9 in normal controls, but only a 10% increase in age-matched autistic children. Thus, overgrowth occurs quite early in autistic children, and there is subsequently a striking lack of the age-related increase in volume seen in normal developmental processes, as detailed in the section "Development." Accordingly, the frontal cortex of autistic adults contains abnormally high levels of proapoptotic molecules and decreased amounts of antiapoptotic ones (Araghi-Niknam and Fatemi, 2003).

When gray and white matter volumes within the frontal lobe are considered separately, gray matter is significantly larger in individuals with ASD regardless of the age of the sample; autistic individuals in these studies ranged from 7–15 years of age in Palmen et al. (2005) to 13–29 years of age in Hazlett et al. (2006). Structural MRI studies have also noted decreases of gray matter in the left inferior frontal gyrus in young adults (Abell et al., 1999). However, some argue that it is white matter volume, not gray matter volume, that is unique in autistic children; white matter tracts in the PFC of autistic children from 5 to 11 are reported to be 36% larger than those in normal controls (Herbert et al., 2004) and myelination occurs prematurely throughout the frontal cortex (Ben Bashat et al., 2007). DTI has noted white matter abnormalities in both dorsolateral and medial PFC (Barnea-Goraly et al., 2004).

Regardless of whether gray matter, white matter, or both experience abnormal development in autistic children, the initial overgrowth has been presumed to lead to abnormal connectivity patterns forming early in development. These abnormalities in neural networks in turn result in the behaviors that characterize ASD. This supposition has been borne out by more recent MRI studies of autistic individuals, which note that executive function deficits observed in autistic children are not correlated to gross measures such as dorsolateral PFC volume (Griebling et al., 2010). Further evidence for abnormalities in neurogenesis and neuronal migration comes from the recent discovery that there are poorly defined boundaries between gray and white matter in the frontal lobe (Avino and Hutsler, 2010) and that the dorsolateral PFC, in particular, has less clear lamination (Mukaetova-Ladinska et al., 2004).

Although histological studies of young children with autism are rare (Bauman and Kemper, 2005), those that exist confirm abnormalities in both gray and white matter volumes during development. While these volumetric abnormalities occur throughout much of the brain, the frontal lobes, in particular, exhibit a noteworthy enlargement of both gray and white matter in toddlers (Carper et al., 2002). There are more spindle cells, specialized pyramidal cells involved in social information processing, in the frontal lobes of autistic children (Santos et al., 2010), although this difference may not exist in adult brains (Kennedy et al., 2007). Postmortem histological studies have likewise not produced a consensus on what the neuropathology of autism looks like in adults. An early histological study found few abnormalities in the neocortex; there were smaller and more closely packed neurons and less distinct laminar architecture in anterior cingulate in eight of the nine brains (Kemper and Bauman, 1998).

By far, the most common type of histological study on autistic brains is minicolumnar analysis, which was described in more detail earlier in this article. An examination of columnar organization has revealed intriguing regional differences within the PFC in both autistic children and adults. The columnar organization of dorsolateral PFC (BA 9) is disrupted in adults with ASD; minicolumns are unusually narrow compared to controls (Buxhoeveden et al., 2006; Casanova et al., 2002, 2006). There is also increased neuronal density in BA 9, suggesting that there are more neurons than expected in that region. Although neuropil space was reduced in dorsolateral PFC (BA 9) in autistic brains, this difference was not significant. There is significantly greater neuropil space in autistic brains in the frontal pole (BA 10) and anterior cingulate (BA 24) than in normal, age-matched controls (Casanova et al., 2006). There were no differences seen in orbitofrontal cortex (BA 11) or the part of Broca's area examined in the same study (BA 44). However, when the frontal cortex is parceled into dorsolateral, mesial, and orbital regions, narrower minicolumns can be seen throughout the entire frontal cortex, especially the dorsolateral and orbitofrontal sectors (Buxhoeveden et al., 2006). Neither of the aforementioned studies noticed any differences between autistic and normal brains in visual or sensorimotor cortex. Minicolumnar pathology has been suggested as an important characteristic of a number of developmental and psychological disorders, including ASD and schizophrenia (Casanova and Tillquist, 2008).

There have been many attempts over the years to discern reliable biomarkers for early detection of autism (Pierce et al., 2009), but most candidates are not present in all or even most autistic individuals examined. However, VENs, specialized projection neurons discussed above in this article, appear to be involved in autism; in one study, autistic brains fell into two groups, where VENs were either present in significantly higher numbers or significantly lower numbers than in controls (Simms et al., 2009). Additionally, there are hints that immune dysfunction is common in autism. Microglia, glial cells crucial to immune response, have been found to exhibit abnormalities in autistic individuals; microglia were active in the dorsolateral PFC in 70% of 13 cases, and also displayed increased density and somal enlargement (Morgan et al., 2010).

There is not a strong consensus among the data regarding what abnormalities are present in the brains of schizophrenics, nor where these abnormalities are located. Autopsies of schizophrenics have reported decreases in total brain weight (Brown et al., 1986; Bruton et al., 1990; Pakkenberg, 1987) and a number have reported reduced head circumference in infants, indicative of diminished total brain volume (McNeil et al., 1993). However, only 22% of 50 structural MRI studies found differences between the whole brain, primarily with brain size decreased (Shenton et al., 2001). One possible explanation for this degree of incongruity is that any volumetric variation in the whole cortex may simply be too small to be detected via MRI; one meta-analysis of total brain volumes in 58 studies found that mean cerebral volume is 2% smaller in schizophrenics (Wright et al., 2000). The majority of studies report ventricular enlargement as well as decreased hippocampal volume (Arnold, 1999; Harrison, 1999; Honea et al., 2005).

This divergence of opinion extends to the frontal lobe and the PFC. In a meta-analysis of structural MRI literature, 60% of 50 studies found some difference between the frontal lobes of schizophrenic and normal control brains (Shenton et al., 2001). Once again, volumetric differences in a region as large as the frontal lobe may be undetectable by MRI studies (Shenton et al., 2001); one histological study reported reduction in PFC cortical thickness of 8%, which is noteworthy but not statistically significant (Selemon et al., 1998). There are few structural MRI studies of schizophrenic individuals that address frontal subdivisions, but those that do describe volumetric abnormalities in a host of brain regions, including the inferior frontal gyrus and the medial frontal gyrus in the PFC; over half of the studies examined in a meta-analysis report volume deficits in the left hemisphere for both of these regions (Honea et al., 2005). Hundred percent of 12 studies that looked specifically at superior temporal gyrus gray matter report dissimilarities (Shenton et al., 2001), while the majority of studies (10 out of 15) that looked at both gray matter and white matter in the same region also reported differences between schizophrenic brains and controls.

Cytoarchitectonic findings are similarly divided. In a meta-analysis (Selemon, 2001), four studies report increased neuronal density in BA 9 in schizophrenic brains (Rajkowska et al., 1998; Rajkowska-Markow et al., 1999; Selemon et al., 1995, 2003), while one reported increased neuronal density in BA 46 (Selemon et al., 1998). Three studies reported no differences in neuronal density in BA 24 (Benes, 1991; Cotter et al., 2000; Kalus et al., 1997), while later studies have found a reduction of neuronal density specifically in calbindin-binding neurons in BA 24 (Cotter et al., 2002). Another study that examined BA 10 found decreased small neuron density but increased pyramid density (Benes et al., 1991). The anterior cingulate cortex in schizophrenics is characterized by some as having smaller and more widely spaced neurons in layer II (Benes and Bird, 1987; Benes et al., 1987). Decreased neuronal density has also been reported in the anterior cingulate (Benes et al., 1986; Benes et al., 1991) and dorsolateral PFCs (Benes et al., 1986). However, other studies report that there is in fact an increase in the density of smaller neurons in dorsolateral PFC, along with a decrease in the neuropil space (Rajkowska et al., 1998; Selemon et al., 1998). Decreased neuronal size has also been observed in both the anterior cingulate (Benes et al., 1986) and in the dorsolateral PFC (Rajkowska et al., 1998).

There is decreased neuronal density throughout several regions of the PFC, while some regions exhibit increased cellular density. The regions with decreased density include primary motor cortex (BA 4), the frontal pole (BA 10), and the anterior cingulate, BA 24 (Benes et al., 1986), although another study reported no differences in neuronal density in motor cortex (Arnold et al., 1995). Smaller and more dispersed neurons have also been reported for BA 24 (Benes and Bird, 1987). Throughout dorsolateral PFC (BA 9 and 46), neuronal density is increased (Goldman-Rakic and Selemon, 1997) by 17% in BA 9 (Selemon et al., 1995) and 21% in 46 (Selemon et al., 1998). Minicolumnar analysis of dorsolateral PFC also reports increased cell density in schizophrenic BA 9 (Casanova et al., 2008). Other studies examining the density of neurons in dorsolateral PFC (BA 9) found no differences in the brains of schizophrenic individuals (Akbarian et al., 1995), while others report decreased neuronal density in the same region (Selemon et al., 1995). Pyramidal neurons in layer III of frontal cortex display significantly decreased density of dendritic spines (Garey et al., 1998), which may explain the loss of cortical volume reported in some regions of the frontal lobe in schizophrenics without a concomitant loss in neuron number. This hypothesis is further supported by reduced synaptophysin protein in dorsolateral PFC, a finding which implies reduced presence of synapses (Glantz and Lewis, 1997; Perrone-Bizzozero et al., 1996).

The microstructural abnormalities that characterize the frontal pole, dorsolateral PFC, and anterior cingulate cortex in schizophrenia do not seem to extend to all regions of the PFC; an examination of BA 9 and 44, while confirming a 12% increase in neuronal density in BA 9, did not find any differences regarding neuronal density, glial density, cortical thickness, or somal size in BA 44 (Selemon et al., 2003).

Conclusion

The human brain is the largest primate brain, but accumulating evidence suggests that absolute size may not be the only variable that sets humans apart from other primates. Specific reorganizational events in neural circuitry took place, either as a result of adjusting to increases in total brain size or as adaptive responses to specific selection pressures (Krubitzer and Kaas, 2005; Krubitzer and Kahn, 2003). While the human frontal lobe is not enlarged in humans relative to apes, there are indeed significant volumetric and microstructural differences within the PFC. Minicolumns, the vertical arrangement of neurons that is a vestige of the radial migration of neurons during development, are wider in humans than in great apes in all PFC regions where they have been examined (BA 10, 44, and 45), but not in sensorimotor cortex or visual cortex. Additionally, minicolumns in these late-developing PFC regions are the most widely spaced in the human cortex, while motor cortex possesses the most widely spaced minicolumns in all of the apes. It is only after the split from the last common ancestor with the chimpanzees that PFC neuronal spacing became the largest, compared to sensorimotor and visual cortex in the human brain and compared to PFC in the other apes.

There are other important volumetric differences in the human PFC that suggest reorganization during human evolution. Gyral white matter is enlarged in the human frontal lobe compared to the great apes, although human frontal lobe gray matter volume is not. Limbic frontal area 13 is relatively smaller in humans than it is in great apes, while frontal pole (BA 10) is relatively larger. There are no relative volumetric differences in Broca's area, BA 44 and 45, but chimpanzees do not exhibit the leftward asymmetry that characterizes human Broca's area. Thus, it appears that frontal pole and left Broca's area expanded in humans after the split with the African apes, while area 13 diminished.

Living in complex environments has been recognized as a considerable factor in the evolution of primate cognition (Byrne, 2007; Rilling et al., 2008b; Whiten, 2010). As discussed in the Introduction, the PFC is crucial for normal executive and social-emotional functioning, a suite of cognitive abilities that humans needed for navigating both complex social groups and changeable, hazardous environments throughout their evolution. Normal frontal lobe development and function are also compromised in several neurological and psychiatric disorders, including autism. We believe that a phylogenetically recent reorganization of frontal cortical circuitry took place (involving an increase in size of some regions, the decrease of others, and increased neuronal spacing distance) that may be critical to the emergence of human-specific executive and social-emotional functions. Relatedly, a developmental pathology in these same systems underlies many neurological disorders, including autism, which involves disturbances in both executive and socio-emotional functioning. Anatomically, autism is characterized by early overgrowth, and then diminution, throughout the PFC.

The PFC exhibits a notably lengthened development, and is one of the last regions of the brain to complete maturation, as based on anatomical indices including cortical thickness, gray matter volume, white matter volume, neurogenesis, synaptic density, and degree of dendritic development. The majority of these developmental processes follow a U-shaped trajectory, with an initial peak and then decline. In general, development proceeds in a caudal to rostral fashion, with the most anterior regions maturing the latest. However, a few exceptions exist; the most rostral area of the PFC, the frontal pole, matures earlier than lateral PFC, as do the ventromedial regions. Just as the unusually prolonged development of the human PFC translates into increased vulnerability to disorders such as autism, it may also enable some of the microstructural species differences described in this chapter.

In conclusion, the field of comparative neuroanatomy holds great promise in its potential to aid in the elucidation of the cognitive difference between humans and other primates, as well as when these differences may have arisen on an evolutionary timescale. Conventional wisdom holds that there is a grade shift in cognitive abilities between humans and extant great apes (Tomasello, 1998; Tomasello and Rakoczy, 2003), suggesting that many traits are the sole province of human intelligence, and evolved sometime after chimpanzee and hominin lineages diverged from a common ancestor. At this time, it seems that evolution in human ancestors was accompanied by discrete modifications in local circuitry and interconnectivity of selected parts of the brain. These modifications may have also predisposed humans to a number of neurological and psychological disorders, including autism and schizophrenia.

Acknowledgments

Funding: The Chancellor's Interdisciplinary Collaboratory Scholarship and the Kavli Institute for Brain and Mind at UCSD, and NIH grant PPG HD33113-14A1.

References

Abell, F., Krams, M., Ashburner, J., Passingham, R. E., Friston, K., Frackowiak, R., et al. (1999). The neuroanatomy of autism: A voxel-based whole brain analysis of structural scans. *NeuroReport*, *10*, 1647–1651.

Akbarian, S., Kim, J. J., Potkin, S. G., Hagman, J. O., Tafazzoli, A., Bunney, W. E., et al. (1995). Gene expression for glutamic acid decarboxylase is reduced without loss of neurons in prefrontal cortex of schizophrenics. *Archives of General Psychiatry*, *52*, 258–266.

Akshoomoff, N., Lord, C., Lincoln, A. J., Courchesne, R. Y., Carper, R. A., Townsend, J., et al. (2004). Outcome classification of preschool children with autism spectrum disorders using mri brain measures. *Journal of the American Academy of Child and Adolescent Psychiatry*, *43*, 349–357.

Allen, J. S. (2009). *The lives of the brain: Human evolution and the organ of mind.* Cambridge, MA: The Belknap Press of Harvard University Press.

Allman, J. M., Hakeem, A., & Watson, K. (2002). Two phylogenetic specializations in the human brain. *The Neuroscientist*, *8*, 335–346.

Allman, J. M., Tetreault, N. A., Hakeem, A. Y., Manaye, K. F., Semendeferi, K., Erwin, J. M., et al. (2010). The von Economo neurons in frontoinsular and anterior cingulate cortex in great apes and humans. *Brain Structure and Function*, *214*, 495–517.

Amunts, K., Schleicher, A., Ditterich, A., & Zilles, K. (2003). Broca's region: Cytoarchitectonic asymmetry and developmental changes. *The Journal of Comparative Neurology*, *465*, 72–89.

Araghi-Niknam, M., & Fatemi, S. H. (2003). Levels of bcl-2 and p53 are altered in superior frontal and cerebellar cortices of autistic subjects. *Cellular and Molecular Neurobiology*, *23*, 945–952.

Arnold, S. E. (1999). Neurodevelopmental abnormalities in schizophrenia: Insights from neuropathology. *Development and Psychopathology*, *11*, 439–456.

Arnold, S. E., Franz, B. R., Gur, R. C., Gur, R. E., Shapiro, R. M., Moberg, P. J., et al. (1995). Smaller neuron size in schizophrenia in hippocampal subfields that mediate cortical-hippocampal interactions. *The American Journal of Psychiatry*, *152*, 738–748.

Avino, T. A., & Hutsler, J. J. (2010). Abnormal cell patterning at the cortical gray-white matter boundary in autism spectrum disorders. *Brain Research*, *1360*, 138–146.

Baddeley, A. (1992). Working memory. *Science*, *255*, 556–559.

Barbas, H., & Pandya, D. N. (1989). Architecture and intrinsic connections of the prefrontal cortex in the rhesus-monkey. *The Journal of Comparative Neurology*, *286*, 353–375.

Barkovich, A. J. (2005). Magnetic resonance techniques in the assessment of myelin and myelination. *Journal of Inherited Metabolic Disease*, *28*, 311–343.

Barnea-Goraly, N., Lotspeich, L. J., & Reiss, A. L. (2004). Similar white matter aberrations in children with autism and their unaffected siblings a diffusion tensor imaging study using tract-based spatial statistics. *Archives of General Psychiatry*, *67*, 1052–1060.

Barton, R. A. (2001). The coordinated structure of mosaic brain evolution. *The Behavioral and Brain Sciences*, *24*, 281–282.

Barton, R. A., & Harvey, P. H. (2000). Mosaic evolution of brain structure in mammals. *Nature*, *405*, 1055–1058.

Bauman, M. L., & Kemper, T. L. (2005). Neuroanatomic observations of the brain in autism: A review and future directions. *International Journal of Developmental Neuroscience*, *23*, 183–187.

Bechara, A., Damasio, H., & Damasio, A. R. (2000). Emotion, decision making and the orbitofrontal cortex. *Cerebral Cortex*, *10*, 295–307.

Becker, L. E., Armstrong, D. L., Chan, F., & Wood, M. M. (1984). Dendritic development in human occipital cortical neurons. *Developmental Brain Research*, *13*, 117–124.

Beer, J. S., Heerey, E. A., Keltner, D., Scabini, D., & Knight, R. T. (2003). The regulatory function of self-conscious emotion: Insights from patients with orbitofrontal damage. *Journal of Personality and Social Psychology*, *85*, 594–604.

Ben Bashat, D., Kronfeld-Duenias, V., Zachor, D. A., Ekstein, P. M., Hendler, T., Tarrasch, R., et al. (2007). Accelerated maturation of white matter in young children with autism: A high b value DWI study. *NeuroImage*, *37*, 40–47.

Benes, F. M. (1991). Evidence for neurodevelopment disturbances in anterior cingulate cortex of postmorterm schizophrenic brains. *Schizophrenia Research*, *5*, 187–188.

Benes, F. M., & Bird, E. D. (1987). An analysis of the arrangement of neurons in the cingulate cortex of schizophrenic patients. *Archives of General Psychiatry*, *44*, 608–616.

Benes, F. M., Davidson, J., & Bird, E. D. (1986). Quantitative cytoarchitectural studies of the cerebral cortex of schizophrenics. *Archives of General Psychiatry*, *43*, 31–35.

Benes, F. M., Majocha, R., Bird, E. D., & Marotta, C. A. (1987). Increased vertical axon numbers in cingulate cortex of schizophrenics. *Archives of General Psychiatry*, *44*, 1017–1021.

Benes, F. M., McSparren, J., Bird, E. D., Sangiovanni, J. P., & Vincent, S. L. (1991). Deficits in small interneurons in prefrontal and cingulate cortices of schizophrenic and schizoaffective patients. *Archives of General Psychiatry*, *48*, 996–1001.

Blinkov, S., & Glezer, I. (1968). *Das zentralnervensystem in zahlen und tabellen*. Jena: Fischer Trans.

Bradshaw, J. L., & Sheppard, D. M. (2000). The neurodevelopmental frontostriatal disorders: Evolutionary adaptiveness and anomalous lateralization. *Brain and Language*, *73*, 297–320.

Broca, P. (1861). Remarques sur le siege de la faculte du langage articule suivies d'une observation d'aphemie. *Bulletin de la Societe Anatomique de Paris*, *36*, 330–357.

Brodmann, K. (1909). *Vergleichende localisationslehre der grosshirnrinde in ihren prinzipien dargestellt auf grund des zellenbaues*. Leipzig: Barth.

Brodmann, K. (1912). Neue ergebnisse uber die vergleichende histologische lokalisation der grosshirnrinde mit besonderer berucksichtigung des stirn-hirns. *Anatomischer Anzeiger (Jena)*, *41*, 157–216.

Brown, R., Colter, N., Corsellis, J. A. N., Crow, T. J., Frith, C. D., Jagoe, R., et al. (1986). Postmortem evidence of structural brain changes in schizophrenia—Differences in brain weight, temporal horn area, and parahippocampal gyrus compared with affective disorder. *Archives of General Psychiatry*, *43*, 36–42.

Bruton, C. J., Crow, T. J., Frith, C. D., Johnstone, E. C., Owens, D. G. C., & Roberts, G. W. (1990). Schizophrenia and the brain—A prospective cliniconeuropathological study. *Psychological Medicine*, *20*, 285–304.

Bunge, S. A., Dudukovic, N. M., Thomason, M. E., Vaidya, C. J., & Gabrieli, J. D. E. (2002). Immature frontal lobe contributions to cognitive control in children: Evidence from fmri. *Neuron*, *33*, 301–311.

Bush, E. C., & Allman, J. (2004). The scaling of frontal cortex in primates and carnivores. *Proceedings of the National Academy of Sciences of the United States of America*, *101*, 3962–3966.

Butti, C., Sherwood, C. C., Hakeem, A. V., Allman, J. M., & Hof, P. R. (2009). Total number and volume of von Economo neurons in the cerebral cortex of cetaceans. *The Journal of Comparative Neurology*, *515*, 243–259.

Buxhoeveden, D. P., Lefkowitz, W., Loats, P., & Armstrong, E. (1996). The linear organization of cell columns in human and nonhuman anthropoid tpt cortex. *Anatomy and Embryology*, *194*, 23–36.

Buxhoeveden, D. P., Semendeferi, K., Buckwalter, J., Schenker, N., Switzer, R., & Courchesne, E. (2006). Reduced minicolumns in the frontal cortex of patients with autism. *Neuropathology and Applied Neurobiology*, *32*, 483–491.

Buxhoeveden, D. P., Switala, A. E., Roy, E., Litaker, M., & Casanova, M. F. (2001). Morphological differences between minicolumns in human and nonhuman primate cortex. *American Journal of Physical Anthropology*, *115*, 361–371.

Byrne, R. W. (2007). Clues to the origin of the human mind from primate observational field data. *Japanese Journal of Animal Psychology*, *57*, 1–14.

Cabeza, R., Anderson, N. D., Locantore, J. K., & McIntosh, A. R. (2002). Aging gracefully: Compensatory brain activity in high-performing older adults. *NeuroImage*, *17*, 1394–1402.

Carper, R. A., & Courchesne, E. (2005). Localized enlargement of the frontal cortex in early autism. *Biological Psychiatry*, *57*, 126–133.

Carper, R. A., Moses, P., Tigue, Z. D., & Courchesne, E. (2002). Cerebral lobes in autism: Early hyperplasia and abnormal age effects. *NeuroImage*, *16*, 1038–1051.

Casanova, M. F., Buxhoeveden, D. P., Switala, A. E., & Roy, E. (2002). Minicolumnar pathology in autism. *Neurology*, *58*, 428–432.

Casanova, M. F., Kreczmanski, P., Trippe, J., Switala, A., Heinsen, H., Steinbusch, H. W. M., et al. (2008). Neuronal distribution in the neocortex of schizophrenic patients. *Psychiatry Research*, *158*, 267–277.

Casanova, M. F., & Tillquist, C. R. (2008). Encephalization, emergent properties, and psychiatry: A minicolumnar perspective. *The Neuroscientist*, *14*, 101–118.

Casanova, M. F., van Kooten, I., Switala, A. E., van Engeland, H., Heinsen, H., Steinbusch, H. W. M., et al. (2006). Abnormalities of cortical minicolumnar organization in the prefrontal lobes of autistic patients. *Clinical Neuroscience Research*, *6*, 127–133.

Casey, B. J., Galvan, A., & Hare, T. A. (2005). Changes in cerebral functional organization during cognitive development. *Current Opinion in Neurobiology*, *15*, 239–244.

Caviness, V. S., Kennedy, D. N., Richelme, C., Rademacher, J., & Filipek, P. A. (1996). The human brain age 7–11 years: A volumetric analysis based on magnetic resonance images. *Cerebral Cortex*, *6*, 726–736.

Chugani, H. T., & Phelps, M. E. (1986). Maturational changes in cerebral function in infants determined by f-18 dg positron emission tomography. *Science*, *231*, 840–843.

Chugani, H. T., & Phelps, M. E. (1991). Imaging human brain development with positron emission tomography. *Journal of Nuclear Medicine*, *32*, 23–26.

Coffey, C. E. (1994). Anatomic imaging of the aging human brain: Computer tomography and magnetic resonance imaging. In *The American Psychiatric Press textbook of geriatric neuropsychiatry* (pp. 159–194). Washington DC: American Psychiatric Press.

Conel, J. L. (1939). *Postnatal development of the human cerebral cortex: The cortex of the newborn.* Cambridge, MA: Harvard University Press.

Conel, J. L. (1941). *Postnatal development of the human cerebral cortex: The cortex of the one-month infant.* Cambridge, MA: Harvard University Press.

Conel, J. L. (1947). *Postnatal development of the human cerebral cortex: The cortex of the three-month old infant.* Cambridge, MA: Harvard University Press.

Conel, J. L. (1951). *Postnatal development of the human cerebral cortex: The cortex of the six-month old infant.* Cambridge, MA: Harvard University Press.

Conel, J. L. (1955). *Postnatal development of the human cerebral cortex: The cortex of the fifteen-month old infant.* Cambridge, MA: Harvard University Press.

Conel, J. L. (1959). *Postnatal development of the human cerebral cortex: The cortex of the twenty-four-month infant.* Cambridge, MA: Harvard University Press.

Conel, J. L. (1963). *Postnatal development of the human cerebral cortex: The cortex of forty-eight-month infant.* Cambridge, MA: Harvard University Press.

Conel, J. L. (1967). *Postnatal development of the human cerebral cortex: The cortex of the seventy-two-month infant.* Cambridge, MA: Harvard University Press.

Cotter, D., Landau, S., Beasley, C., Stevenson, R., Chana, G., MacMillan, L., et al. (2002). The density and spatial distribution of GABAergic neurons, labelled using calcium binding proteins, in the anterior cingulate cortex in major depressive disorder, bipolar disorder, and schizophrenia. *Biological Psychiatry*, *51*, 377–386.

Cotter, D., Stevenson, R. L., Kerwin, R., Beasley, C. L., & Everall, I. P. (2000). No alterations in the density of parvalbumin or calretinin immunoreactive neurons in the anterior cingulate cortex (ba 24b/c) in schizophrenia. *Schizophrenia Research*, *41*, A175.

Courchesne, E., Campbell, K., & Solso, S. (2011). Brain growth across the life span in autism: Age-specific changes in anatomical pathology. *Brain Research*, *1380*, 138–145.

Courchesne, E., Chisum, H. J., Townsend, J., Cowles, A., Covington, J., Egaas, B., et al. (2000). Normal brain development and aging: Quantitative analysis at in vivo MR imaging in healthy volunteers. *Radiology*, *216*, 672–682.

Courchesne, E., Karns, C. M., Davis, H. R., Ziccardi, R., Carper, R. A., Tigue, Z. D., et al. (2001). Unusual brain growth patterns in early life in patients with autistic disorder—An MRI study. *Neurology*, *57*, 245–254.

Courchesne, E., Muller, R. A., & Saitoh, O. (1999). Brain weight in autism: Normal in the majority of cases, megalencephalic in rare cases. *Neurology*, *52*, 1057–1059.

Courchesne, E., Redcay, E., & Kennedy, D. (2004). The autistic brain: Birth through adulthood. *Current Opinion in Neurobiology*, *17*, 489–496.

Cowell, P. E., Sluming, V. A., Wilkinson, I. D., Cezayirli, E., Romanowski, C. A. J., Webb, J. A., et al. (2007). Effects of sex and age on regional prefrontal brain volume in two human cohorts. *European Journal of Neuroscience*, *25*, 307–318.

Damasio, H., Tranel, D., Grabowski, T., Adolphs, R., & Damasio, A. (2004). Neural systems behind word and concept retrieval. *Cognition*, *92*, 179–229.

Damasio, A. R., & Van Hoesen, G. (1983). Emotional disturbances associated with focal lesions of the frontal lesions of the frontal lobe. In K. Heilman & P. Satz (Eds.), *Neuropsychology of human emotion: Recent advances* (pp. 85–110). New York: Guilford.

Dawson, G., Estes, A., Munson, J., Schellenberg, G., Bernier, R., & Abbott, R. (2007). Quantitative assessment of autism symptom-related traits in probands and parents: Broader phenotype autism symptom scale. *Journal of Autism and Developmental Disorders*, *37*, 523–536.

De Carli, C., Murphy, D. G. M., Gillette, J. A., Haxby, J. V., Teichberg, D., Schapiro, M. B., et al. (1994). Lack of age-related differences in temporal lobe volume of very healthy adults. *American Journal of Neuroradiology*, *15*, 689–696.

DeFelipe, J. (2005). Reflections of the structure of the cortical minicolumn. In M. Casanova (Ed.), *Neocortical modularity and the cell minicolumn* (pp. 57–92). New York: Nova Biomedical.

Dementieva, Y. A., Vance, D. D., Donnelly, S. L., Elston, L. A., Wolpert, C. M., Ravan, S. A., et al. (2005). Accelerated head growth in early development of individuals with autism. *Pediatric Neurology*, *32*, 102–108.

Diamond, A. D. (2001). Frontal lobe involvement in cognitive changes during early development. *Brain and Cognition*, *47*, 11.

Dobbing, J., & Sands, J. (1973). Quantitative growth and development of human brain. *Archives of Disease in Childhood*, *48*, 757–767.

Dumontheil, I., Burgess, P. W., & Blakemore, S. J. (2008). Development of rostral prefrontal cortex and cognitive and behavioural disorders. *Developmental Medicine and Child Neurology*, *50*, 168–181.

Durston, S., Thomas, K. M., Yang, Y. H., Ulug, A. M., Zimmerman, R. D., & Casey, B. J. (2002). A neural basis for the development of inhibitory control. *Developmental Science, 5*, F9–F16.

Elston, G. N., Benavi-Despiccione, R., & DeFelipe, J. (2001). The pyramidal cell in cognition: A comparative study in human and monkey. *Journal of Neuroscience, 21* RC163, 1–5.

Elston, G. N., Benavides-Piccione, R., Elston, A., Zietsch, B., Defelipe, J., Manger, P., et al. (2006). Specializations of the granular prefrontal cortex of primates: Implications for cognitive processing. *Anatomical Record Part A: Discoveries in Molecular Cellular and Evolutionary Biology, 288A*, 26–35.

Fajardo, C., Escobar, M. I., Buritica, E., Arteaga, G., Umbarila, J., Casanova, M. F., et al. (2008). Von Economo neurons are present in the dorsolateral (dysgranular) prefrontal cortex of humans. *Neuroscience Letters, 435*, 215–218.

Fellows, L. K. (2007a). Advances in understanding ventromedial prefrontal function—The accountant joins the executive. *Neurology, 68*, 991–995.

Fellows, L. K. (2007b). The role of orbitofrontal cortex in decision making—A component process account. In: G. Schoenbaum, J. A. Gottfried, E. A. Murray & S. J. Ramus (Eds.), *Linking affect to action: Critical contributions of the orbitofrontal cortex*, (Vol. 1121), (pp. 421–430). Oxford: Blackwell Publishing.

Fields, R. D., & Stevens-Graham, B. (2002). Neuroscience—New insights into neuron-glia communication. *Science, 298*, 556–562.

Finlay, B. L., Darlington, R. B., & Nicastro, N. (2001). Developmental structure in brain evolution. *The Behavioral and Brain Sciences, 24*, 263–308.

Flechsig, P. (1901). Developmental myelogenetic localisation of the cerebral cortex in the human subject. *Lancet, 2*, 1027–1029.

Flechsig, P. (1920). *Anatomie des menschlichen gehirns und rückenmarks auf myelogenetischer grundlage.* Leipzig: Thieme.

Fuster, J. M. (2000a). Executive frontal functions. *Experimental Brain Research, 133*, 66–70.

Fuster, J. M. (2000b). Memory networks in the prefrontal cortex. *Progress in Brain Research, 122*, 309–316.

Fuster, J. M. (2001). The prefrontal cortex—An update: Time is of the essence. *Neuron, 30*, 319–333.

Fuster, J. M. (2002). Frontal lobe and cognitive development. *Journal of Neurocytology, 31*, 373–385.

Fuster, J. M. (2008). *The prefrontal cortex* (4th ed.). London: Academic Press.

Garey, L. J., Ong, W. Y., Patel, T. S., Kanani, M., Davis, A., Mortimer, A. M., et al. (1998). Reduced dendritic spine density on cerebral cortical pyramidal neurons in schizophrenia. *Journal of Neurology, Neurosurgery, and Psychiatry, 65*, 446–453.

Geschwind, D. H. (2009). Advances in autism. *Annual Review of Medicine, 60*, 367–380.

Ghashghaei, H. T., & Barbas, H. (2002). Pathways for emotion: Interactions of prefrontal and anterior temporal pathways in the amygdala of the rhesus monkey. *Neuroscience, 115*, 1261–1279.

Ghika, J. (2008). Paleoneurology: Neurodegenerative diseases are age-related diseases of specific brain regions recently developed by Homo sapiens. *Medical Hypotheses, 71*, 788–801.

Giedd, J. N., Blumenthal, J., Jeffries, N. O., Castellanos, F. X., Liu, H., Zijdenbos, A., et al. (1999). Brain development during childhood and adolescence: A longitudinal MRI study. *Nature Neuroscience, 2*, 861–863.

Giedd, J. N., Castellanos, F. X., Blumenthal, J., & Rapoport, J. L. (1999). Sexual dimorphism of human brain development: Ages 3 to 18. *Biological Psychiatry, 45*, 370.

Giedd, J. N., Snell, J. W., Lange, N., Rajapakse, J. C., Casey, B. J., Kozuch, P. L., et al. (1996). Quantitative magnetic resonance imaging of human brain development: Ages 4–18. *Cerebral Cortex, 6*, 551–560.

Glantz, L. A., & Lewis, D. A. (1997). Reduction of synaptophysin immunoreactivity in the prefrontal cortex of subjects with schizophrenia—Regional and diagnostic specificity (vol 54, pg 660, 1997). *Archives of General Psychiatry, 54*, 943–952.

Glasser, M. F., & Rilling, J. K. (2008). DTI tractography of the human brain's language pathways. *Cerebral Cortex, 18*, 2471–2482.

Gogtay, N., Giedd, J. N., Lusk, L., Hayashi, K. M., Greenstein, D., Vaituzis, A. C., et al. (2004). Dynamic mapping of human cortical development during childhood through early adulthood. *Proceedings of the National Academy of Sciences of the United States of America, 101*, 8174–8179.

Goldman-Rakic, P. S. (1988). Topography of cognition—Parallel distributed networks in primate association cortex. *Annual Review of Neuroscience, 11*, 137–156.

Goldman-Rakic, P. S., & Selemon, L. D. (1997). Functional and anatomical aspects of prefrontal pathology in schizophrenia. *Schizophrenia Bulletin, 23*, 437–458.

Grady, C. L. (2000). Functional brain imaging and age-related changes in cognition. *Biological Psychology, 54*, 259–281.

Griebling, J., Minshew, N. J., Bodner, K., Libove, R., Bansal, R., Konasale, P., et al. (2010). Dorsolateral prefrontal cortex magnetic resonance imaging measurements and cognitive performance in autism. *Journal of Child Neurology, 25*, 856–863.

Gustafsson, L. (1997). Inadequate cortical feature maps: A neural circuit theory of autism. *Biological Psychiatry, 42*, 1138–1147.

Gustafsson, L. (2004). Comment on "Disruption in the inhibitory architecture of the cell minicolumn: Implications for autism". *The Neuroscientist*, *10*, 189–191.

Habib, M., Demonet, J. F., & Frackowiak, R. (1996). Cognitive neuroanatomy of language: Contribution of functional cerebral imaging. *Revue Neurologique*, *152*, 249–260.

Hakeem, A. Y., Sherwood, C. C., Bonar, C. J., Butti, C., Hof, P. R., & Allman, J. M. (2009). Von Economo neurons in the elephant brain. *Anatomical Record: Advances in Integrative Anatomy and Evolutionary Biology*, *292*, 242–248.

Harrison, P. J. (1999). The neuropathology of schizophrenia—A critical review of the data and their interpretation. *Brain*, *122*, 593–624.

Hazlett, H. C., Poe, M. D., Gerig, G., Smith, R. G., & Piven, J. (2006). Cortical gray and white brain tissue volume in adolescents and adults with autism. *Biological Psychiatry*, *59*, 1–6.

Herbert, M. R., Ziegler, D. A., Makris, N., Filipek, P. A., Kemper, T. L., Normandin, J. J., et al. (2004). Localization of white matter volume increase in autism and developmental language disorder. *Annals of Neurology*, *55*, 530–540.

Hill, J., Inder, T., Neil, J., Dierker, D., Harwell, J., & Van Essen, D. (2010). Similar patterns of cortical expansion during human development and evolution. *Proceedings of the National Academy of Sciences of the United States of America*, *107*, 13135–13140.

Holloway, R. L. (1968). The evolution of the primate brain: Some aspects of quantitative relations. *Brain Research*, *7*, 121–172.

Holloway, R. L. (2001). Does allometry mask important brain structure residuals relevant to species-specific behavioral evolution? *The Behavioral and Brain Sciences*, *24*, 286.

Honea, R., Crow, T. J., Passingham, D., & Mackay, C. E. (2005). Regional deficits in brain volume in schizophrenia: A meta-analysis of voxel-based morphometry studies. *The American Journal of Psychiatry*, *162*, 2233–2245.

Huttenlocher, P. R. (1990). Morphometric study of human cerebral cortex development. *Neuropsychologia*, *26*, 517–527.

Huttenlocher, P. R., & Dabholkar, A. S. (1997). Regional differences in synaptogenesis in human cerebral cortex. *The Journal of Comparative Neurology*, *387*, 167–178.

Huttenlocher, P. R., & De Courten, C. (1987). The development of synapses in striate cortex of man. *Human Neurobiology*, *6*, 1–9.

Jacobs, B., Driscoll, L., & Schall, M. (1997). Life-span dendritic and spine changes in areas 10 and 18 of human cortex: A quantitative Golgi study. *The Journal of Comparative Neurology*, *386*, 661–680.

Jacobs, B., Schall, M., Prather, M., Kapler, E., Driscoll, L., Baca, S., et al. (2001). Regional dendritic and spine variation in human cerebral cortex: A quantitative golgi study. *Cerebral Cortex*, *11*, 558–571.

Jernigan, T. L., Archibald, S. L., Fennema-Notestine, C., Gamst, A. C., Stout, J. C., Bonner, J., et al. (2001). Effects of age on tissues and regions of the cerebrum and cerebellum. *Neurobiology of Aging*, *22*, 581–594.

Jurado, M. B., & Rosselli, M. (2007). The elusive nature of executive functions: A review of our current understanding. *Neuropsychology Review*, *17*, 213–233.

Kalus, P., Senitz, D., & Beckmann, H. (1997). Cortical layer i changes in schizophrenia: A marker for impaired brain development? *Journal of Neural Transmission*, *104*, 549–559.

Kemper, T. L. (Ed.), (1994). *Neuroanatomical and neuropathological changes during aging and in dementia*. New York: Oxford University.

Kemper, T. L., & Bauman, M. (1998). Neuropathology of infantile autism. *Journal of Neuropathology and Experimental Neurology*, *57*, 645–652.

Kennedy, D. P., Semendeferi, K., & Courchesne, E. (2007). No reduction of spindle neuron number in frontoinsular cortex in autism. *Brain and Cognition*, *64*, 124–129.

Klingberg, T., Vaidya, C. J., Gabrieli, J. D. E., Moseley, M. E., & Hedehus, M. (1999). Myelination and organization of the frontal white matter in children: A diffusion tensor MRI study. *NeuroReport*, *10*, 2817–2821.

Knickmeyer, R. C., Gouttard, S., Kang, C. Y., Evans, D., Wilber, K., Smith, J. K., et al. (2008). A structural MRI study of human brain development from birth to 2 years. *Journal of Neuroscience*, *28*, 12176–12182.

Krubitzer, L., & Kaas, J. (2005). The evolution of the neocortex in mammals: How is phenotypic diversity generated? *Current Opinion in Neurobiology*, *15*, 444–453.

Krubitzer, L., & Kahn, D. M. (2003). Nature versus nurture revisited: An old idea with a new twist. *Progress in Neurobiology*, *70*, 33–52.

Lebel, C., Caverhill-Godkewitsch, S., & Beaulieu, C. (2010). Age-related regional variations of the corpus callosum identified by diffusion tensor tractography. *NeuroImage*, *52*, 20–31.

Lenroot, R. K., & Giedd, J. N. (2006). Brain development in children and adolescents: Insights from anatomical magnetic resonance imaging. *Neuroscience and Biobehavioral Reviews*, *30*, 718–729.

Levitt, P. (2003). Structural and functional maturation of the developing primate brain. *Journal of Pediatrics*, *143*, S35–S45.

Levitt, J. G., Blanton, R. E., Smalley, S., Thompson, P. M., Guthrie, D., McCracken, J. T., et al. (2003). Cortical sulcal maps in autism. *Cerebral Cortex*, *13*, 728–735.

Lewis, D. A., Melchitzky, D. S., & Burgos, G. G. (2002). Specificity in the functional architecture of primate prefrontal cortex. *Journal of Neurocytology*, *31*, 265–276.

MacBride, T., Arnold, S. E., & Gur, R. C. (1999). A comparative volumetric analysis of the prefrontal cortex in human and baboon MRI. *Brain, Behavior and Evolution, 54,* 159–166.

Masliah, E., Mallory, M., Hansen, L., Deteresa, R., & Terry, R. D. (1993). Quantitative synaptic alterations in the human neocortex during normal aging. *Neurology, 43,* 192–197.

McNeil, T. F., Cantorgraae, E., & Cardenal, S. (1993). Prenatal cerebral development in individuals at genetic risk for psychosis—Head size at birth in offspring of women with schizophrenia. *Schizophrenia Research, 10,* 1–5.

Milham, M. P., Erickson, K. I., Banich, M. T., Kramer, A. F., Webb, A., Wszalek, T., et al. (2002). Attentional control in the aging brain: Insights from an fMRI study of the stroop task. *Brain and Cognition, 49,* 277–296.

Miller, E. K. (2001). Prefrontal cortex and the neural basis of executive functions. *Brain and Cognition, 47,* 22.

Miller, E. K., & Cohen, J. D. (2001). An integrative theory of prefrontal cortex function. *Annual Review of Neuroscience, 24,* 167–202.

Morgan, J. T., Chana, G., Pardo, C. A., Achim, C., Semendeferi, K., Buckwalter, J., et al. (2010). Microglial activation and increased microglial density observed in the dorsolateral prefrontal cortex in autism. *Biological Psychiatry, 68,* 368–376.

Mrzljak, L., Uylings, H. B. M., Kostovic, I., & Vaneden, C. G. (1988). Prenatal development of neurons in the human prefrontal cortex 1A qualitative Golgi study. *The Journal of Comparative Neurology, 271,* 355–386.

Mrzljak, L., Uylings, H. B. M., Kostovic, I., & Vaneden, C. G. (1992). Prenatal development of neurons in the human prefrontal cortex. 2. A quantitative Golgi study. *The Journal of Comparative Neurology, 316,* 485–496.

Mukaetova-Ladinska, E. B., Arnold, H., Jaros, E., Perry, R., & Perry, E. (2004). Depletion of map2 expression and laminar cytoarchitectonic changes in dorsolateral prefrontal cortex in adult autistic individuals. *Neuropathology and Applied Neurobiology, 30,* 615–623.

Muller, N. G., Machado, L., & Knight, R. T. (2002). Contributions of subregions of the prefrontal cortex to working memory: Evidence from brain lesions in humans. *Journal of Cognitive Neuroscience, 14,* 673–686.

Nagy, Z., Westerberg, H., & Klingberg, T. (2004). Maturation of white matter is associated with the development of cognitive functions during childhood. *Journal of Cognitive Neuroscience, 16,* 1227–1233.

Nauta, W. J. (1962). Some efferent connections of the prefrontal cortex in the monkey. In J. M. Warren & K. Akert (Eds.), *The frontal granular cortex and behavior* (pp. 397–409). New York: McGraw-Hill Book Company.

Nimchinsky, E. A., Gilissen, E., Allman, J. M., Perl, D. P., Erwin, J. M., & Hof, P. R. (1999). A neuronal morphologic type unique to humans and great apes. *Proceedings of the National Academy of Sciences of the United States of America, 96,* 5268–5273.

Olesen, P. J., Nagy, Z., Westerberg, H., & Klingberg, T. (2003). Combined analysis of DTI and fMRI data reveals a joint maturation of white and grey matter in a fronto-parietal network. *Cognitive Brain Research, 18,* 48–57.

Pakkenberg, B. (1987). Postmortem study of chronic schizophrenic brains. *The British Journal of Psychiatry, 151,* 744–752.

Pakkenberg, B., & Gundersen, H. J. G. (1997). Neocortical neuron number in humans: Effect of sex and age. *The Journal of Comparative Neurology, 384,* 312–320.

Palmen, S., Pol, H. E. H., Kemner, C., Schnack, H. G., Durston, S., Lahuis, B. E., et al. (2005). Increased gray-matter volume in medication-naive high-functioning children with autism spectrum disorder. *Psychological Medicine, 35,* 561–570.

Passingham, R. E. (1973). Anatomical differences between the neocortex of man and other primates. *Brain, Behavior and Evolution, 7,* 337–359.

Passingham, R. E. (2002). The frontal cortex: Does size matter? *Nature Neuroscience, 5,* 190–192.

Paus, T., Zijdenbos, A., Worsley, K., Collins, D. L., Blumenthal, J., Giedd, J. N., et al. (1999). Structural maturation of neural pathways in children and adolescents: In vivo study. *Science, 283,* 1908–1911.

Perrone-Bizzozero, N. I., Sower, A. C., Bird, E. D., Benowitz, L. I., Ivins, K. J., & Neve, R. L. (1996). Levels of the growth-associated protein gap-43 are selectively increased in association cortices in schizophrenia. *Proceedings of the National Academy of Sciences of the United States of America, 93,* 14182–14187.

Petanjek, Z., Judas, M., Kostovic, I., & Uylings, H. B. M. (2008). Lifespan alterations of basal dendritic trees of pyramidal neurons in the human prefrontal cortex: A layer-specific pattern. *Cerebral Cortex, 18,* 915–929.

Peters, A., & Sethares, C. (1996). Myelinated axons and the pyramidal cell modules in monkey primary visual cortex. *The Journal of Comparative Neurology, 365,* 232–255.

Pfefferbaum, A., Mathalon, D. H., Sullivan, E. V., Rawles, J. M., Zipursky, R. B., & Lim, K. O. (1994). A quantitative magnetic resonance imaging study of changes in brain morphology from infancy to late adulthood. *Archives of Neurology, 51,* 874–887.

Pierce, K., Glatt, S. J., Liptak, G. S., & McIntyre, L. L. (2009). The power and promise of identifying autism early: Insights from the search for clinical and biological markers. *Annals of Clinical Psychiatry, 21,* 132–147.

Price, C. J. (2000). The anatomy of language: Contributions from functional neuroimaging. *Journal of Anatomy, 197,* 335–359.

Rabinowicz, T., de Courten-Myers, G. M., Petetot, J. M. C., Xi, G. H., & de los Reyes, E. (1996). Human cortex development: Estimates of neuronal numbers indicate major loss late during gestation. *Journal of Neuropathology and Experimental Neurology, 55,* 320–328.

Raghanti, M. A., Stimpson, C. D., Marcinkiewicz, J. L., Erwin, J. M., Hof, P. R., & Sherwood, C. C. (2008a). Cortical dopaminergic innervation among humans, chimpanzees, and macaque monkeys: A comparative study. *Neuroscience*, *155*, 203–220.

Raghanti, M. A., Stimpson, C. D., Marcinkiewicz, J. L., Erwin, J. M., Hof, P. R., & Sherwood, C. C. (2008b). Differences in cortical serotonergic innervation among humans, chimpanzees, and macaque monkeys: A comparative study. *Cerebral Cortex*, *18*, 584–597.

Rajkowska, G., Selemon, L. D., & Goldman-Rakic, P. S. (1998). Neuronal and glial somal size in the prefrontal cortex—A postmortem morphometric study of schizophrenia and Huntington disease. *Archives of General Psychiatry*, *55*, 215–224.

Rajkowska-Markow, G., Miguel-Hidalgo, J. J., Wei, J., & Stockmeier, C. A. (1999). Reductions in glia distinguish orbitofrontal region from dorsolateral prefrontal cortex in schizophrenia. *Society for Neuroscience Abstracts*, *25*, 818.

Rakic, P. (1988). Defects of neuronal migration and the pathogenesis of cortical malformations. *Progress in Brain Research*, *73*, 15–37.

Rakic, P., Bourgeois, J. P., & Goldmanrakic, P. S. (1994). Synaptic development of the cerebral cortex—Implications for learning, memory, and mental illness. *Progress in Brain Research*, *102*, 227–243.

Raz, N. (Ed.), (1996). In: *Neuroanatomy of aging brain: Evidence from structural MRI*New York: Plenum Press.

Raz, N., Ghisletta, P., Rodrigue, K. M., Kennedy, K. M., & Lindenberger, U. (1997). Trajectories of brain aging in middle-aged and older adults: Regional and individual differences. *NeuroImage*, *51*, 501–511.

Raz, N., Ghisletta, P., Rodrigue, K. M., Kennedy, K. M., & Lindenberger, U. (2010). Trajectories of brain aging in middle-aged and older adults: Regional and individual differences. *NeuroImage*, *51*, 501–511.

Redcay, E., & Courchesne, E. (2005). When is the brain enlarged in autism? A meta-analysis of all brain size reports. *Biological Psychiatry*, *58*, 1–9.

Reuter-Lorenz, P. A. (2002). New visions of the aging mind and brain. *Trends in Cognitive Sciences*, *6*, 394–400.

Rilling, J. K. (2001). Allometric departures for the human brain provide insights into hominid brain evolution. *The Behavioral and Brain Sciences*, *24*, 292–293.

Rilling, J. K., Glasser, M. F., Preuss, T. M., Ma, X. Y., Zhao, T. J., Hu, X. P., et al. (2008). The evolution of the arcuate fasciculus revealed with comparative DTI. *Nature Neuroscience*, *11*, 426–428.

Rilling, J. K., King-Casas, B., & Sanfey, A. G. (2008). The neurobiology of social decision-making. *Current Opinion in Neurobiology*, *18*, 159–165.

Rosenberg, K., & Trevathan, W. (1995). Bipedalism and human birth: The obstetrical dilemma revisited. *Evolutionary Anthropology*, *4*, 161–168.

Salat, D. H., Kaye, J. A., & Janowsky, J. S. (1999). Prefrontal gray and white matter volumes in healthy aging and Alzheimer disease. *Archives of Neurology*, *56*, 338–344.

Salat, D. H., Stangl, P. A., Kaye, J. A., & Janowsky, J. S. (1999). Sex differences in prefrontal volume with aging and Alzheimer's disease. *Neurobiology of Aging*, *20*, 591–596.

Salmond, C. H., Vargha-Khademl, F., Gadian, D. G., de Haan, M., & Baldeweg, T. (2007). Heterogeneity in the patterns of neural abnormality in autistic spectrum disorders: Evidence from ERP and MRI. *Cortex*, *43*, 686–699.

Samuelsen, G. B., Larsen, K. B., Bogdanovic, N., Laursen, H., Graem, N., Larsen, J. F., et al. (2003). The changing number of cells in the human fetal forebrain and its subdivisions: A stereological analysis. *Cerebral Cortex*, *13*, 115–122.

Santos, M., Uppal, N., Butti, C., Wicinski, B., Schmeidler, J., Giannakopoulos, P., et al. (2010). Von economo neurons in autism: A stereologic study of the frontoinsular cortex in children. *Brain Research*, *1380*, 206–217.

Schade, J. P., & Van Groenou, W. B. (1961). Structural organization of human cerebral cortex. *Acta Anatomica*, *47*, 74–111.

Scheff, S. W., Price, D. A., & Sparks, D. L. (2001). Quantitative assessment of possible age-related change in synaptic numbers in the human frontal cortex. *Neurobiology of Aging*, *22*, 355–365.

Schenker, N. M. (2007). *Comparative analysis of Broca's area in hominoids.* Ph.D. Dissertation, University of California, San Diego, La Jolla.

Schenker, N. M., Buxhoeveden, D. P., Blackmon, W. L., Amunts, K., Zilles, K., & Semendeferi, K. (2008). A comparative quantitative analysis of cytoarchitecture and minicolumnar organization in Broca's area in humans and great apes. *The Journal of Comparative Neurology*, *510*, 117–128.

Schenker, N. M., Desgouttes, A. M., & Semendeferi, K. (2005). Neural connectivity and cortical substrates of cognition in hominoids. *Journal of Human Evolution*, *49*, 547–569.

Schenker, N. M., Hopkins, W. D., Spocter, M. A., Garrison, A. R., Stimpson, C. D., Erwin, J. M., et al. (2010). Broca's area homologue in chimpanzees (Pan troglodytes): Probabilistic mapping, asymmetry, and comparison to humans. *Cerebral Cortex*, *20*, 730–742.

Schmitz, N., Daly, E., & Murphy, D. (2007). Frontal anatomy and reaction time in autism. *Neuroscience Letters*, *412*, 12–17.

Schmitz, C., & Rezaie, P. (2008). The neuropathology of autism: Where do we stand? *Neuropathology and Applied Neurobiology*, *34*, 4–11.

Schoenemann, P. T., Sheehan, M. J., & Glotzer, L. D. (2005). Prefrontal white matter volume is disproportionately larger in humans than in other primates. *Nature Neuroscience*, *8*, 242–252.

Selemon, L. D. (2001). Regionally diverse cortical pathology in schizophrenia: Clues to the etiology of the disease. *Schizophrenia Bulletin*, *27*, 349–377.

Selemon, L. D., Mrzljak, J., Kleinman, J. E., Herman, M. M., & Goldman-Rakic, P. S. (2003). Regional specificity in the neuropathologic substrates of schizophrenia—A morphometric analysis of Broca's area 44 and area 9. *Archives of General Psychiatry*, *60*, 69–77.

Selemon, L. D., Rajkowska, G., & Goldmanrakic, P. S. (1995). Abnormally high neuronal density in the schizophrenic cortex—A morphometric analysis of prefrontal areas 9 and occipital area 17. *Archives of General Psychiatry*, *52*, 805–818.

Selemon, L. D., Rajkowska, G., & Goldman-Rakic, P. S. (1998). Elevated neuronal density in prefrontal area 46 in brains from schizophrenic patients: Application of a three-dimensional, stereologic counting method. *The Journal of Comparative Neurology*, *392*, 402–412.

Semendeferi, K., Armstrong, E., Schleicher, A., Zilles, K., & Van Hoesen, G. W. (1998). Limbic frontal cortex in hominoids: A comparative study of area 13. *American Journal of Physical Anthropology*, *106*, 129–155.

Semendeferi, K., Armstrong, E., Schleicher, A., Zilles, K., & Van Hoesen, G. W. (2001). Prefrontal cortex in humans and apes: A comparative study of area 10. *American Journal of Physical Anthropology*, *114*, 224–241.

Semendeferi, K., & Damasio, H. (2000). The brain and its main anatomical subdivisions in living hominoids using magnetic resonance imaging. *Journal of Human Evolution*, *38*, 317–332.

Semendeferi, K., Damasio, H., & Frank, R. (1997). The evolution of the frontal lobes: A volumetric analysis based on three-dimensional reconstructions of magnetic resonance scans of human and ape brains. *Journal of Human Evolution*, *32*, 375–388.

Semendeferi, K., Lu, A., Schenker, N., & Damasio, H. (2002). Humans and great apes share a large frontal cortex. *Nature Neuroscience*, *5*, 272–276.

Semendeferi, K., Teffer, K., Buxhoeveden, D., Park, M. S., Bludau, S., Amunts, K., et al. (2011). Spatial organization of neurons in the frontal pole sets humans apart from great apes. *Cerebral Cortex*, *21*, 1485–1497.

Shankle, W. R., Rafii, M. S., Landing, B. H., & Fallon, J. H. (1999). Approximate doubling of numbers of neurons in postnatal human cerebral cortex and in 35 specific cytoarchitectural areas from birth to 72 months. *Pediatric and Developmental Pathology*, *2*, 244–259.

Shaw, P., Kabani, N. J., Lerch, J. P., Eckstrand, K., Lenroot, R., Gogtay, N., et al. (2008). Neurodevelopmental trajectories of the human cerebral cortex. *Journal of Neuroscience*, *28*, 3586–3594.

Shenton, M. E., Dickey, C. C., Frumin, M., & McCarley, R. W. (2001). A review of MRI findings in schizophrenia. *Schizophrenia Research*, *49*, 1–52.

Sherwood, C. C., Holloway, R. L., Semendeferi, K., & Hof, P. R. (2005). Is prefrontal white matter enlargement a human evolutionary specialization? *Nature Neuroscience*, *8*, 537–538.

Sidman, R. L., & Rakic, P. (1973). Neuronal migration, with special reference to developing human brain—Review. *Brain Research*, *62*, 1–35.

Simms, M. L., Kemper, T. L., Timbie, C. M., Bauman, M. L., & Blatt, G. J. (2009). The anterior cingulate cortex in autism: Heterogeneity of qualitative and quantitative cytoarchitectonic features suggests possible subgroups. *Acta Neuropathologica*, *118*, 673–684.

Smaers, J. B., Schleicher, A., Zilles, K., & Vinicius, L. (2010). Frontal white matter volume is associated with brain enlargement and higher structural connectivity in anthropoid primates. *PLoS One*, *5*, 6.

Smaers, J. B., Steele, J., Case, C. R., Cowper, A., Amunts, K., & Zilles, K. (2011). Primate prefrontal cortex evolution: Human brains are the extreme of a lateralized ape trend. *Brain, Behavior, and Evolution*, *77*, 67–78.

Solbakk, A. K., Alpert, G. F., Furst, A. J., Hale, L. A., Oga, T., Chetty, S., et al. (2008). Altered prefrontal function with aging: Insights into age-associated performance decline. *Brain Research*, *1232*, 30–47.

Sowell, E. R., Delis, D., Stiles, J., & Jernigan, T. L. (2001). Improved memory functioning and frontal lobe maturation between childhood and adolescence: A structural MRI study. *Journal of International Neuropsychological Society*, *7*, 312–322.

Sowell, E. R., Thompson, P. M., Holmes, C. J., Batth, R., Jernigan, T. L., & Toga, A. W. (1999). Localizing age-related changes in brain structure between childhood and adolescence using statistical parametric mapping. *NeuroImage*, *9*, 587–597.

Sowell, E. R., Thompson, P. M., Holmes, C. J., Jernigan, T. L., & Toga, A. W. (1999). In vivo evidence for post-adolescent brain maturation in frontal and striatal regions. *Nature Neuroscience*, *2*, 859–861.

Sowell, E. R., Thompson, P. M., Leonard, C. M., Welcome, S. E., Kan, E., & Toga, A. W. (2004). Longitudinal mapping of cortical thickness and brain growth in normal children. *Journal of Neuroscience*, *24*, 8223–8231.

Sowell, E. R., Thompson, P. M., & Toga, A. W. (2004). Mapping changes in the human cortex throughout the span of life. *The Neuroscientist*, *10*, 372–392.

Sowell, E. R., Trauner, D. A., Gamst, A., & Jernigan, T. L. (2002). Development of cortical and subcortical brain structures in childhood and adolescence: A structural MRI study. *Developmental Medicine and Child Neurology*, *44*, 4–16.

Stiles, J., & Jernigan, T. L. (2010). The basics of brain development. *Neuropsychology Review*, *20*, 327–348.

Stone, V. E., Baron-Cohen, S., & Knight, R. T. (1998). Frontal lobe contributions to theory of mind. *Journal of Cognitive Neuroscience*, *10*, 640–656.

Striedter, G. F. (2005). *Principles of brain evolution.* Sunderland, MA: Sinauer Associates, Inc.

Thompson, R. A., & Nelson, C. A. (2001). Developmental science and the media: early brain development. *American Psychologist, 56*, 5–15.

Thorpe, S. J., & Fabre-Thorpe, M. (2001). Seeking categories in the brain. *Science, 291*, 260–263.

Tomasello, M. (1998). Uniquely primate, uniquely human. *Developmental Science, 1*, 1–16.

Tomasello, M., & Rakoczy, H. (2003). What makes human cognition unique? From individual to shared to collective intentionality. *Mind and Language, 18*, 121–147.

Travis, K., Ford, K., & Jacobs, B. (2005). Regional dendritic variation in neonatal human cortex: A quantitative Golgi study. *Developmental Neuroscience, 27*, 277–287.

Uylings, H. B. M., Groenewegen, H. J., & Kolb, B. (2003). Do rats have a prefrontal cortex? *Behavioural Brain Research, 146*, 3–17.

Uylings, H. B. M., Jacobsen, A. M., Zilles, K., & Amunts, K. (2006). Left-right asymmetry in volume and number of neurons in adult Broca's area. *Cortex, 42*, 652–658.

Uylings, H. B. M., & Van Eden, C. (Eds.), (1990). In: *Qualitative and quantitative comparison of the prefrontal cortex in rat and in primates, including humans*, (Vol. 85). New York: Elsevier.

Van Essen, D. (1997). A tension-based theory of morphogenesis and compact wiring in the central nervous system. *Nature, 385*, 313–318.

Van Hoesen, G. (1981). The differential distribution, diversity, and sprouting of cortical projections to the amygdala in the rhesus monkey. In Y. Ben-Ari (Ed.), *The amygdaloid complex: INSERM symposium no. 20* (pp. 77–90). New York: Elsevier/North Holland Biomedical Press.

Wagner, A. D., Maril, A., Bjork, R. A., & Schacter, D. L. (2001). Prefrontal contributions to executive control: fMRI evidence for functional distinctions within lateral prefrontal cortex. *NeuroImage, 14*, 1337–1347.

Waldemar, G. (1995). Functional brain imaging with spect in normal aging and dementia—Methodological, pathophysiological, and diagnostic aspects. *Cerebrovascular and Brain Metabolism Reviews, 7*, 89–130.

Watson, R. E., DeSesso, J. M., Hurtt, M. E., & Cappon, G. D. (2006). Postnatal growth and morphological development of the brain: A species comparison. *Birth Defects Research, 77*, 471–484.

Whiten, A. (2010). The scope of culture in chimpanzees, humans and ancestral apes. *Philosophical transactions of the Royal Society of London. Series B, Biological sciences, 366*, 997–1007.

Wright, I. C., Rabe-Hesketh, S., Woodruff, P. W. R., David, A. S., Murray, R. M., & Bullmore, E. T. (2000). Meta-analysis of regional brain volumes in schizophrenia. *The American Journal of Psychiatry, 157*, 16–25.

Yakovlev, P., & LeCours, A. (1967). The myelogenetic cycles of regional maturation of the brain. In A. Minkowski (Ed.), *Regional development of the brain in early life* (pp. 3–65). Oxford: Blackwell.

Zeba, M., Jovanov-Milosevic, N., & Petanjek, Z. (2008). Quantitative analysis of basal dendritic tree of layer IIIc pyramidal neurons in different areas of adult human frontal cortex. *Collegium Antropologicum, 32*, 161–169.

Zysset, S., Huber, O., Samson, A., Ferstl, E. C., & von Cramon, D. Y. (2003). Functional specialization within the anterior medial prefrontal cortex: A functional magnetic resonance imaging study with human subjects. *Neuroscience Letters, 335*, 183–186.

M. A. Hofman and D. Falk (Eds.)
Progress in Brain Research, Vol. 195
ISSN: 0079-6123

CHAPTER 10

Minicolumn size and human cortex

Daniel P. Buxhoeveden*

College of Social Work and Department of Anthropology, University of South Carolina, Columbia, SC, USA

Abstract: Minicolumns in primates are small when compared with those of other mammals, both in absolute and relative terms. The data suggest that minicolumns in the earliest primates were especially narrow and increased in accordance with encephalization so that the largest minicolumns in this mammalian order are found in apes and humans. Among the evolutionary strategies that led to the successful human brain was a combination of enhanced cortical volume based on increases in the number of ontogenetic units, along with enlargement of the individual minicolumns. However, continued encephalization of the large human brain presents serious problems that may limit future growth. When further increases in brain size can no longer be sustained, the alternative for further adaptations will have to be done at the level of brain organization. A downsizing of minicolumns may be among those responses. This has the advantage of permitting increases in the number of processing units without adding surface area. However, it is argued that narrow minicolumns process information differently, which raises questions about the relation between minicolumn size and behavior. There is evidence that minicolumns may be smaller in extant humans within selected populations, and the implications of this are briefly considered.

Keywords: minicolumns; radial unit hypothesis; hominids; encephalization.

Introduction

For the purposes of this chapter, the minicolumn is defined as the vertical orientation of cortical organization, both anatomically and physiologically, at a spatially small scale ranging between 20 and 80 μm in width. Physiological, metabolic, anatomical, and theoretical modeling work have provided evidence for their functional capabilities as well as how they might perform their tasks (Amirikian and Georgopoulos, 2003; Favorov and Kelly, 1996; Hasselmo, 2005; Johansson and Lansner, 2007; Kohn et al., 2002; Lucke, 2004; Lucke and von der Malsburg, 2004; Mountcastle, 1997, 2003; Rao et al., 1999; Sugimoto et al., 1997). Nonetheless, there remains much to learn about them (Rakic, 2008) and debates about their definition, ubiquity, and functionality remain

*Corresponding author.
Tel.: +1-803-777-4460; Fax: 803 777-0259
E-mail: buxhoeve@gwm.sc.edu

DOI: 10.1016/B978-0-444-53860-4.00010-6

(Catania, 2002; Jones, 2000; Krieger et al., 2007; Rockland and Ichinohe, 2004; Swindale, 1990). It is difficult to provide a rigid definition of the minicolumn that will hold up to all the findings in the literature. One way around this is to define the minicolumn as a common template rather than a stereotypical component found in all brains and regions (Buxhoeveden and Casanova, 2002; Mountcastle, 2003; Silberberg et al., 2002).

Vernon Mountcastle (2003) noted:

> "The important point is that columnar organization depends upon a certain set of properties common to all neurons in the elementary unit, but that other properties may vary between different neurons in the same minicolumn."

Mountcastle provides a conceptual basis to variability upon the basic template by stating that "differences in afferent input are convolved with different intrinsic operations in different cortical areas to produce what we call different functions."

Silberberg et al. (2002) also conclude that despite the great range in microcircuitry, stereotypical features exist at multiple levels indicating a deterministic basis for them and suggest that all neocortical microcircuits may be subtle variations of a common template (see also Jin et al., 2001; Kisvarday et al., 2002; Kozloski et al., 2001). Thus, a broader conception of the minicolumn is to see it as a "template" for a shared set of properties of a given set of neurons across several or more lamina. It seems to be a general principle that cortical neurons with similar stimulus selection properties are found in close proximity to each other (Reich et al., 2001) and the minicolumn is the vertical component in that association. The minicolumn is a subset of organizational units nested within the cortex such as the cortical column and larger hypercolumn. It is possible that the minicolumn is the more fundamental one upon which others are built around but that is one of the questions still debated.

Encephalization and organization

In the field of paleoneurology, there has been a long debate about encephalization and reorganization of the brain and how these relate to computational capacity. The questions centered around the relative importance of enlarging the brain versus its internal organization and especially how these relate to the emergence of human-specific behavior. The discovery of *Homo floresiensis*, with its combination of a chimpanzee-sized brain and human-like behavior, reinforces the crucial role of brain organization when contrasted to size alone. An analogy that may help depict the relationship of brain size, organization, and processing capacity is the use of the internal combustion engine. The engine of a current Formula One racing car has a displacement of 2.4l, the same size as that found in many passenger car engines. However, in the former, the engines produces over 800 hp, while a street engine of that size typically produces somewhere around 160–180 hp. The differences, which extend to other things like torque, the power band, responsiveness, engine speed, and so on, are the result of design changes based on the basic template of the internal combustion piston engine. And as in evolution, both types of engines are responses to specific needs and can only be understood accordingly. The Formula engine would be a terrible and unreliable choice for daily driving, and the passenger engine would be a complete failure in Formula One.

If the design principles remain similar and both engines are enlarged, then each will generate more power, and the amount of the increase will depend on the internal configuration already in place. Enlargement and internal design are highly interrelated (Preuss, 2001). A few years ago when Formula One engines had a 3-l displacement, they were approaching 1000 hp. A 3-l passenger car does not come close. So what a brain is capable of must be judged on the combination of its organization in the context of its volume.

The ontogenetic column

The mechanisms responsible for how the cortex enlarged in evolution were made known by the seminal work of Rakic (1972, 1978, 1988). In this model, the necessary fundamentals are found to explain how the cortex enlarges and how that same process creates an environment conducive to reorganization, whether it be the subtle forms responsible for biological variability or something more significant.

"Thus, mutation of a regulatory gene(s) that controls the timing and ratio of symmetric and asymmetric modes of cell divisions in the proliferative zone, coupled with radial constraints in the distribution of migrating neurons, could create an expanded cortical plate with enhanced capacity for establishing new patterns of connectivity that are validated through natural selection" (Rakic, 1995).

The role of the ontogenetic column in brain evolution is vital for any modern study of paleoneurology and provides a model in which encephalization and reorganization are closely linked. The genesis of the cortex occurs in the ventricles by a series of symmetrical and asymmetrical divisions (Rakic and Kornack, 2001). In the first phase, cells located in the ventricular zone produce two additional progenitor cells with each mitotic cell division (Rakic, 1988). This symmetrical division is responsible for the number of founder cells which controls the total number of ontogenetic columns that will be produced in the cortex. According to the radial unit hypothesis, it is the number of these ontogenetic columns that determines the cortical surface area (Rakic and Kornack, 2001). At some point, progenitor cells begin to divide asymmetrically, producing one daughter cell that becomes a neuron and will move out into the cortical plate, and which will not undergo further division. The second phase is responsible for the number of cells within a column and the thickness of the cortex. Several clones of neurons that share a common site of origin in the ventricular zone use a common migratory pathway along the fascicles of the radial glial cells to settle within the same column in the cortical plate (Rakic, 2003). Radial glial cells create long fascicles that extend from the ventricular zone to the top of the cortical plate so that they span the entire width of the cerebral wall during corticoneurogenesis. Newborn nerve cells use these to traverse the cortical plate. Though there are small differences between radial glial cells among mammals, overall they are very similar in morphology and chemistry.

However, some cortical interneurons do not originate from the ventricular zone and migrate in a radial fashion. In rodents, this is most notable as the majority of cortical interneurons originate from the ganglionic eminence of the ventral telecephalon and migrate tangentially to the cortical plate (Marín and Rubenstein, 2001). In mice, up to 25% of all cortical neurons migrate non-radially, whereas in human, this percentage is less than 10% of the total (Letinic et al., 2002). Thus, there are taxonomic specializations associated with this process.

The total amount of radial units that will be present in the cortex is controlled during embryogenesis by a few regulatory genes, while the final pattern and size of cytoarchitectonic regions are thought to be the work of a different set of genes (Rakic and Kornack, 2001). The final configuration of columns within a cytoarchitectonic area is therefore the result of the genetic influences described above and epigenetic factors, such as interactions of cells, inhibitory neurons, and afferent systems. It is clear to see that alterations in these genes or their influences can have profound effects on the cortex. The increase in founder cell number is exponential and not linear, so that a small prolongation of cell division or changes in length of the cell cycle would result in significant increases in the number of ontogenetic units produced. This is a key to encephalization as well as reorganization of the cortex.

In summary, the process of encephalization that occurred in mammalian evolution is thought to have arisen from the addition of more ontogenetic units which is the basis for increased cortical surface area (Rakic and Kornack, 2001). Ontogenetic column number determines cortical surface area, whereas cell number influences cortical depth. Since surface area has increased a 1000-fold (comparing mouse to human), while cortical depth has only increased around three to four times, the major impetus for cortical enlargement has been the addition of new ontogenetic units. Therefore, it can be expected that the addition of more ontogenetic cell columns should normally result in an increase in cortical surface area and white matter.

The minicolumn in postnatal cortex is arguably derived from the cell columns of the fetal cortex (Fig. 1; Buxhoeveden et al., 1996; Curtetti et al., 2002; Krmpotic-Nemanic et al., 1984; Lohmann and Koppen, 1995; LoTurco et al., 1991; Ong and Carey, 1990; Peinado et al., 1993; Vercelli et al., 2004). Further evidence of this is found in the development of the early cortex, where prospective pyramidal neurons are clustered into vertical columns and coupled by gap junctions (LoTurco et al., 1991; Peinado et al., 1993) and the specificity of excitatory synapses within ontogenetic columns (Yu et al., 2009). Apical dendrites and myelinated axons also bundle their fibers in the earliest stages of development (Fig. 2).

Minicolumns in primate evolution

The "size" of a minicolumn is usually measured according to the horizontal spacing that separates vertical arrays of pyramidal cells, bundled axons, or apical dendrites from pyramidal cells of layers III and V. It has been shown that the layer V apical dendrite bundles, the myelinated axons of layers III and V pyramidal cells, and the neurons themselves, as well as the bundled axons of double-bouquet cells (in primates), display a one-to-one correspondence. In the cortex of the fetal brain, neurons are packed tightly together, and during development, the "noncell" space between them increases both in the vertical and horizontal axes. The neocortex in its earliest stages presents an almost perfect linearity of neurons packed closely together (Fig. 1), but dispersion of these highly linear groups is brought about by factors such as afferent and efferent fibers, neuropil expansion, cell growth, and laminar specializations. A study of cell column development in humans showed that the neuropil space increases disproportionately to the column size during development (Buxhoeveden et al., 1996; Buxhoeveden, unpublished data) so that the increase in neuropil space is what accounts for the majority of the enlargement of minicolumns in development (Fig. 3).

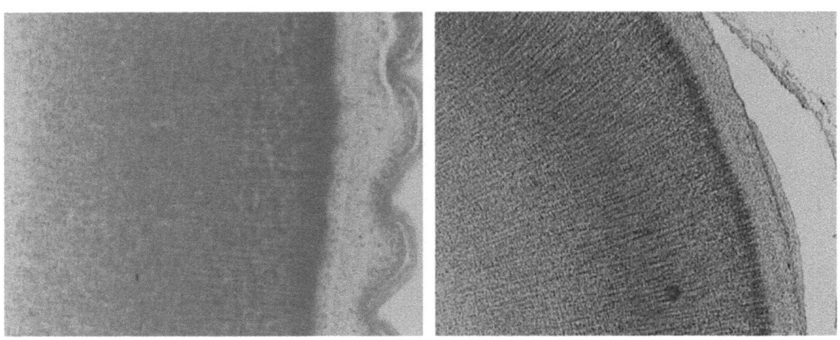

Fig. 1. Left: Cell columns in human fetal brain before the onset of lamination. These are highly linear arrays of neurons with very little neuropil space between them. Right: Minicolumns in human fetal brain at gestational age of approximately 28 weeks. The highly linear configuration is obvious despite the onset of early lamination. Neuropil space is becoming more notable in the pyramidal layers.

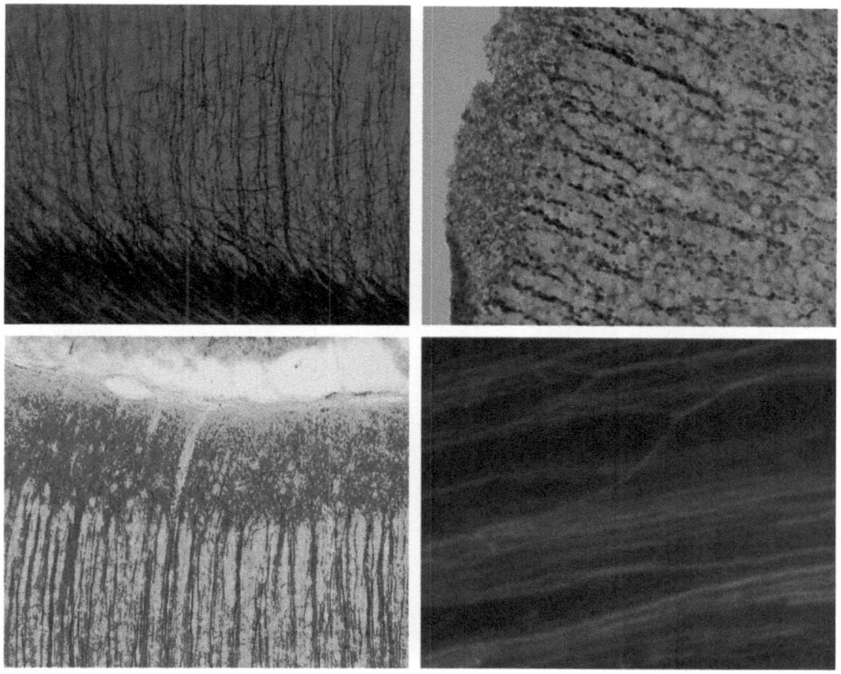

Fig. 2. Top left: Immature myelinated axons are clearly bundled in developing rat cortex. Top right: MAP 2 (microtube-associated protein 2) stain showing apical dendrites are already bundled in day 1 rat brain. Bottom left: MAP 2 stain showing thick apical dendrite bundles in rat brain at 21days. Note the massive dendritic branching in layer I. Image was color enhanced to highlight stain for apical dendrites which are presented as green in the micrograph. Bottom right: Immunoflourescent image of apical dendrite bundles in day 21 rat. Near top of the picture can be seen two apical dendrites that fork and appear to send a branch to neighboring bundles. These suggest that minicolumns are highly interconnected with each other with multiple levels of functionality based on composites of the individual minicolumn. (For color version of this figure, the reader is referred to the Web version of this chapter.)

Certain general facts have emerged concerning the size of minicolumns that appear fairly secure. The first is that neither they are the same size nor do they contain identical components between species, within species, or within regions of the same brain (Buxhoeveden and Casanova, 2002, 2005; Herculano-Houzel et al., 2008). The second is that there is no particular correlation between column size and brain size for animals with a *diverse* evolutionary history (Buxhoeveden and Casanova, 2002). This means there is no way to predict minicolumns size based on cortical surface area across the higher levels of taxonomic divisions. However, within the primate order, there is a notable correlation between the size of the cortex and column size which suggests that, among a related taxonomic group, a loose correspondence between cortical volume and column size may be expected. Presumably, this is because animals with a shared evolutionary history share a basic minicolumn configuration, and all primates share the same scaling for neuronal density (Sarko et al., 2009). If this is the case, it also supports the argument that minicolumns are not generic units that are interchangeable across different animals.

As noted above, the primary basis for encephalization has been the addition of more ontogenetic units resulting from mutational events during symmetrical cell division. However,

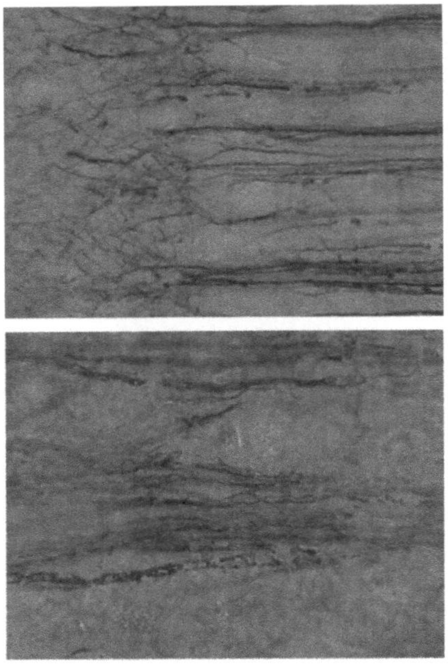

Fig. 3. Top: Horizontal view of apical dendrite bundles in 16-day rat cortex . The left section of the image shows layer I, and the terminal branches of the afferents can be seen. The area immediately below is layer II which in a cell soma stain appears noncolumnar. However, as can be seen here, the minicolumn is not defined primarily by the appearance of neurons but by their vertical interconnections and functional organization (Buxhoeveden and Casanova, 2005). The space between the bundles is the neuropil space as defined by apical dendrite bundles. Bottom: High magnification look at apical dendrite bundles in 21-day-old rat. The image provides a look at the individual fibers that constitute the bundle. Taken at 1000× total magnification with oil immersion.

a secondary and more subtle determinant of surface area has to do with the size of the minicolumns. If minicolumns average 40 μm in cortex "A" and 50 μm in cortex "B," given that each has the same number of minicolumns, the latter cortex will have more surface area. It must be emphasized that changes in surface area attributed to minicolumns size will necessarily be restricted. The size range is limited and cannot account for the manifold differences witnessed in primate evolution. The distinction between human and anthropoid apes shows that cortex is 3–4× greater in humans, and 14× larger than that of the macaque monkey. Therefore, changes in surface area resulting from changes in the size of minicolumns are a factor for within-species variation only.

The relative stability in minicolumn size (20–80 μm across all species tested) compared to the 1000-fold increase in surface area may be because the basic elements that constitute a minicolumn place constraints on how small or large they can be. These include neurons, apical dendrites, myelinated efferents, double-bouquet bundles; thalamic, callosal, and cortico-cortical input; dendritic arborizations and spines, synaptic contacts, as well as glia. These require a minimum size to function and, at the same time, can only get so large. As a result, their dimensions stay more similar while cortical surface area expands.

Minicolumn size, like brain volume, can be described in absolute or relative terms and is usually more meaningful when understood within the context of cortical volume. A small minicolumn in the human brain is still comparatively large and within the range of what would be found in the ape or monkey cortex. What defines a minicolumn as small or large must be considered in context of the available cortical space in which they are found.

Are minicolumns in the primate order smaller than expected for their size?

The data accumulated so far support a hypothesis that minicolumns are smaller in primates when compared to other land mammals. They are small in absolute measurements (in the case of the smaller brained species) and in relative terms for the larger species. Though it may be counterintuitive, the very large cortex of apes and humans contain minicolumns that approximate the size of those found in very small-brained mammals. This means that apes with brain weights of 200–500 g and humans with brain weights of 1300+ share minicolumn sizes with small-brained mammals with brain weights as small as 4 g.

Small-brained primates, like *Saimiri* with a brain weight around 25g, have minicolumns that are smaller than those of mice, even though the latter may have brain weights around 0.4g (Table 1; Buxhoeveden and Casanova, 2002, 2005).

These results are significant because they demonstrate that neither do minicolumns continue to increase in size merely because the cortex does nor do they have to be small because they are situated inside a small brain, all of which implies a functional role to minicolumn size. Unfortunately, lacking a more exhaustive and systematic body of data, we cannot go too much further with these findings. While there is a good sample size for the primate order, it is still incomplete and there is a paucity or total absence of data for minicolumn size in the larger brain mammals, something needed to clarify the place of human and primate minicolumn size within the larger mammalian class. The only other large brain mammal for which there is data are the cetaceans (Table 1), but since they are not land mammals and the configuration of their cortex is dramatically different from a land mammal, comparisons are difficult to make.

Table 1. Typical minicolumn sizes reported in the literature or measured by the author

	Brain weight (g)	Minicolumn (μm)	Minicolumns in V1
Primates			
Saimiri	25	20	N/A
OWM[a]	70–100+	30–40	23–30
Apes	250–500	33–40+	30
Human	1350	40–60+	30+
Land mammals			
Mouse	0.4	22–30	26
Rat	2	40–60	30–60
Hedgehog	4	32	N/A
Rabbit	10	40–50	40–50
Cat	30	50–60	56
Cetaceans	350–3000	20–30	20

(1) There is no correlation between minicolumn size and brain volume among highly diverse taxonomic categories. (2) Within the primate order, a correlation is present. (3) Minicolumns in primates are either absolutely small or small for their brain size compared to the other land mammals. (4) Cetaceans display a strategy of employing very small minicolumns in a cortex that is very large but also very thin. For more specific details regarding minicolumn size across species, see Buxhoeveden and Casanova (2002, 2005).

[a]Old-world monkeys. Generally includes different species of Macaques as well as baboons. The great apes include the Orangutan, Chimpanzee, and Gorilla.

Mechanisms have been proposed to explain how minicolumns change their size. Some of the models derive from neural network theory where self-organizing networks like minicolumns are formed when lateral feedback synaptic strength is a function of lateral distance as shaped by the Mexican hat model. If the inhibitory synaptic strengths increase, the columns become narrower while the reverse is also true (Favorov and Kelly, 1994a,b; Gustafsson, 1997). This can be expected to occur during development. It is also possible for columnar organization to emerge without the usual lateral excitatory–inhibitory feedback mechanism. A basic organization can be laid down before the lateral feedback connections are developed so that, when they do arise, they fine-tune or maintain the columnar organization. Others have reported that neural columns would be narrower if levels of nitric oxide (NO) were reduced so that, given the same stimulus drive, the column size varied according to the level of NO (Gally et al., 1990; Krekelberg and Taylor, 1996). It is also found to be involved in the metasynaptic organization of the frontal cortex in primate but had no effect in visual cortex.

There may be another mechanism for changing the size of minicolumns that is directly related to the production of additional ontogenetic units. Since each minicolumn contains approximately 80–100 neurons (except in V1), the addition of a few hundred more minicolumns implies thousands of more neurons that must find connections. If there is no increase in the amount of afferent input for them to connect with, then the distribution to each column will be altered. This is of course a highly simplified model that does not take account of many other factors. The result would be a decrease in the neuropil space which translates into a narrowing of their

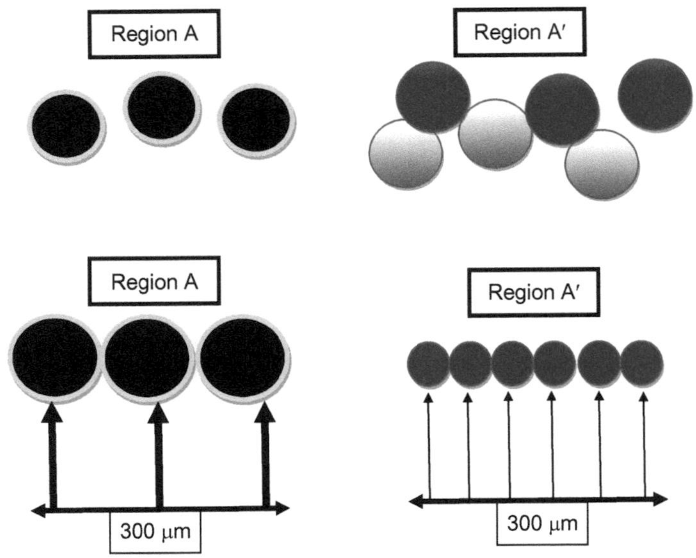

Fig. 4. Top left: Region A represents the normal configuration of ontogenetic units for a given region. Top right: Region A′ represents the same region undergoing a significant increase in the number of ontogenetic units produced. Bottom left: Region A after maturation of the cortex. Bottom right: Region A′ after maturation. Since the available afferent input is the same for region A and A′, competition between units results in sharing of the afferent input to more minicolumns than in region A. The decreased neuropil space results in smaller minicolumns in A′. However, there is no change in the size of the cortical column and therefore in surface area, but information processing has been altered.

width (Fig. 4). Because of the decrease in the overall spacing of minicolumns, several possibilities emerge in regard with surface area. One is that any increase that might have resulted from the production of more ontogenetic units is offset by the smaller size of each. Depending on how small they become, surface area might diminish, remain the same, or increase slightly. In order for the surface area to increase noticeably, what is required is a continuation of this process so that any diminution of column size is eventually offset by the sheer number of additional units.

If adding many more ontogenetic units without a matched increase in input decreases the size of the adult units, then it can be seen that increasing the afferent input relative to each ontogenetic unit would result in more neuropil space and therefore wider columns. Stasis in the size of the minicolumn would presumably occur when there are no mutational events creating additional ontogenetic units, no enhancement of afferent input, or when the addition of more ontogenetic units is matched by increases from incoming afferents. This picture is most likely the case when changes occur slowly over many generations and does not involve other factors that might contribute to minicolumn size (above). Finally, the rate at which additional ontogenetic units are produced is pivotal to the dynamics of column size, since a few additional column units could more easily be absorbed into the system with less impact on size and wiring.

Minicolumn size and cortical organization

Aside from surface area, differences in the size of minicolumns would have no functional significance unless it can be shown that these differences reflect alterations in the wiring of the cortex. Minicolumns in the primary visual cortex of primates are especially small when compared to those of other mammals or other regions of primate cortex.

Minicolumns in the visual cortex of small-brained rodents and cats are estimated to be three times larger (when extrapolated to three dimensions) than those of primates (Peters and Sethares, 1996, 1997; Peters and Yilmaz, 1993), so the difference is dramatic and not subtle. In a series of studies comparing primary visual cortex of primates to cats and rodents, numerous differences were found in their basic anatomy. In the primate, each small minicolumn was considered more specialized and was more heavily interconnected. The conclusion of the authors was that the primate model represents a very different approach to the processing of visual information, one that reflects the greater complexity of primate vision. The argument is that the very small columns are an adaptation for highly complex processing.

It has been proposed that smaller minicolumns enhance resolution because the input signals are distributed to more units, each of which processes a smaller part of the input (Gustafsson, 1997, 2004). This enables the brain to concentrate on more specific aspects of the signal and is thought to provide better discrimination and more focused attention. For example, a given afferent terminal of 300 μm that contains minicolumns with an average size of 50 μm will have the information distributed to six minicolumns (in a two-dimensional model used for simplicity), whereas if the minicolumns in the same area were 30 μm in size, the same input would be distributed to 10 minicolumns (Fig. 5). The disadvantage of small minicolumns is that each can process less information than a larger one because they contain less neuropil space within them. This is offset by having more interconnections between minicolumns.

Studies conducted decades ago link differences in minicolumn size directly to changes in the wiring of the cortex (Seldon, 1981a,b, 1982). In a study of the lateralization of minicolumns in human auditory cortex, it was shown that minicolumns in the right hemisphere were smaller than those on the left in homologous areas. The changes in size between the two hemispheres resulted in a different configuration of the afferent, efferent, and intrinsic connections. Narrower minicolumns in the right hemisphere received the same number of incoming afferent inputs so that the signals were distributed among more minicolumns on the right than in the left

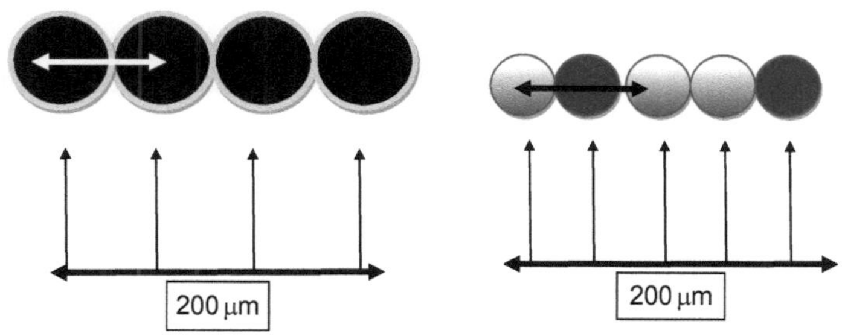

Fig. 5. Diagram depicting two cortical columns of the same size. Left: Cortical column contains four minicolumns with a mean width of 50 μm each. Right: This cortical column contains five minicolumns with a mean width of 40 μm. The width of the afferent terminals that dictate the size of the cortical column is the same for both. Small vertical arrows indicate individual thalamic afferents to each minicolumn. Small horizontal arrows within the columns depict the interconnections between minicolumns. The minicolumns on the left interconnect with two minicolumns, while the ones on the right connect with three. The combination of potential changes in the ratio between long distance and local connections, subcortical and cortical, the altered afferent and efferent connectivity between minicolumns, the reduction in the amount of input signals arriving in each minicolumn results in significant changes in the way signals are processed. The afferent terminals are distributed to more minicolumns in the column on the right, thus reducing the amount of information to each minicolumn but this is offset by greater sharing of information among the units. Changes like this would affect the internal wiring but not result in encephalization. This is the kind of event that may have resulted in reorganization of the neocortex prior to the onset of noticeable cortical enlargement.

hemisphere, suggesting increased resolution or specificity. Within a cortical column, minicolumns in the right hemisphere received input from more minicolumns because their horizontal dendrites (though the same width as those on the left) extended to more minicolumns. In regard with efferent outputs, the same circumstances were in evidence. The number of columns contacted by the dendrite of a single column within a radius of 150μm was about 450 in the right hemisphere compared to 375 in the left hemisphere. These changes occurred without any alteration in the size of the afferent terminals or the horizontal dendritic branches and were solely the result of the size of the columns. It is a reflection of what can transpire simply by changing one part of the anatomical template without changing others.

Further, the addition of many more ontogenetic units can change the ratio of connection types based on whether they are local, subcortical, or long distance. If the highest density of synaptic connections for a given neuron is found within a relatively short distance of a parent neuron (Budd and Kisvarday, 2001; Elston, 2000; Elston and Rosa, 2000), this means that additional ontogenetic columns reciprocally connect to each other and become a major source for synaptogenesis (and in that way also contribute to their survival). It also implies that these connections would remain somewhat constant in number by comparison to other inputs. If there is no increase in subcortical input, then the ratio of these inputs in relation to the local ones would change. Similar kinds of redistributions are envisioned for long distance and contralateral connections. Modifications like these may contribute to normal biological variability as well as open up opportunities for more significant changes.

Increased minicolumn size in hominids

It appears to be the case that primate evolution began with small brain animals containing small minicolumns. Over time, the body size increased, cortical volume and the number of regions increased, and so did the size of minicolumns (Semendeferi et al., 2011). This is supported by the data which show that the smallest-brained primates display the smallest minicolumns, the intermediate ones are intermediate in size, and the apes and humans have the largest minicolumns. This relationship has some overlap due to biological variability, regional specificity, and lateralization. There is also overlap between minicolumn size among some of the largest apes and humans, depending on the region, though the largest primate minicolumns are always found in human cortex (Semendeferi et al., 2011).

While the enlargement of minicolumns along with encephalization may seem intuitive or expected, there is no evidence that this necessarily follows. Though the comparative data are limited, we do know as fact that brains many times smaller than even the smallest anthropoid primates can have very large minicolumns. It cannot be assumed that minicolumns always enlarge with cortical volume. We do not know enough about other taxonomic categories to make that claim.

The pattern of encephalization witnessed during hominid evolution shows that the hominid brain remained basically stable in size over the first 4 million years, with significant increases occurring in the past 2 million years. The selection pressures behind the more rapid increases are evident in the archeological record. This can be described as resulting from a positive feedback loop between behavioral and neural complexity. It may also be easier for encephalization to occur at a more rapid rate in a brain that has already attained a sufficiently large cortex. What that size is would be difficult to judge, but clearly the ape size brain of the australopithecines should have been a good starting point. By virtue of their size, large brains contain more ontogenetic units and more cortical regions which result in a target-rich environment where additional columns added to a new brain can establish reciprocal connections. The implication is that the process of encephalization would proceed more slowly in a small brain and become facilitated as the brain enlarged. While there was a rapid increase in brain size for hominids beginning with the first genus *Homo*, and accelerating rapidly with *Homo erectus*, the past few hundred thousand years represent a slow down if not a stasis regarding encephalization. Why this occurred is discussed below.

A time for downsizing?

The picture presented above is that the strategy in evolution has been to enlarge the surface area and the size of the minicolumns in response to challenges for greater neural complexity. Selection pressures have been favoring a human brain that is better at processing more complex forms of intellectual and abstract thinking. The past few thousand years have witnessed the emergence of new forms of mental skills, such as complex auditory and written languages, higher mathematics, philosophy and analytical thinking, engineering and science, music and the arts, and so on, most of which either represent new mental capacities or are so far improved from earlier forms that the similarity is minimal. All of this occurred rather quickly in a biological system that was shaped for millions of years in an environment radically different from modern society. Given this history, the question is, will the strategies used in evolution that got us to where we are today continue to work in the future?

The answer in regard with encephalization is probably no. Brain size in hominids appears to have reached a point of stasis, with cranial capacity remaining essentially unchanged over the past few hundred thousand years, and if anything, it is reduced from that achieved by the Neanderthal. There are very practical reasons why encephalization in human brains is no longer the best option. To begin with, the brain is one of many organs integrated into a whole body. It does not exist separately and as such it is an expensive organ, comprising about 1–2% of total body weight while consuming in the neighborhood of 20% of the bodies energy. Large brains require high metabolic energy and alterations had to be made in hominid evolution to cool them so as to prevent heat stroke, something that necessitated diverting more blood flow to the brain from other organs of the body (Falk, 1986, 2009a). The size of the human pelvis presents a very serious challenge to further cortical enlargement, and the large head size of human infants is a cause of many of the difficulties associated with child birth. Hominids (probably beginning with *H. erectus*) already made an adaptation for this by changing the primate development pattern of doubling the brain size from birth to adult, to a two-step process in which there is a doubling of size from birth to the first year, and another from that period to adulthood. Whether yet another accommodation can be made is difficult to imagine.

Another problem is that as the cortex enlarges, it results in a disproportionate increase in white matter (Hofman, 2001; see Chapter 18) thereby increasing the total volume of the cerebrum with a potential to encroach on subcortical space and decreasing efficiency. Areas that need to communicate get moved further apart which increases the distance between neurons. The effect is a slowing of communication, a requirement for larger neurons (which are more vulnerable to pathology), and thick myelinated long distance axons, all of which further taxes the energy demands of the body. It should be noted that organizing neurons with similar functional properties into modules such as minicolumns is a way to improve efficiency and overcome some of the disadvantages of brain enlargement.

Given these constraints on significant brain growth in humans, the options remaining are to either radically reconfigure the basic template of the mammalian brain or optimize the internal organization based on what is currently available. The process of evolution is such that the first option is highly doubtful, and if it were to occur, it would not be in anyone's foreseeable lifetime.

The second, however, is what evolution excels at, which is making changes in response to environmental influences and demands, based on what it has to work with. This could take place at all levels, including molecular organization, anatomy, physiology, or chemistry.

What the future holds for minicolumn size depends on a balance between the need to restrain cortical volume while favoring enhancements in computational capacity. The undeniable advantage provided by small minicolumns is the ability to pack more units into less cortical space, and if that is a high priority, then this may be the trend and there is some evidence that this may be occurring (below).

What is less clear about downsizing minicolumns from the current norm concerns the functional and behavioral outcomes. If making minicolumns smaller also means they process information differently, then downsizing may not be a particularly successful strategy even if it limits cortical expansion, or it may be lead to increasing variability in human behavior. It must be asked if small minicolumns are more advantageous for humans then why were minicolumns enlarged in the association cortex during evolution, and it is worth noting that the largest minicolumns in human brain tend to be in areas that are important to human-specific behavior such as frontal cortex and left hemisphere auditory-association cortex.

The answer to this question may be that small minicolumns are advantageous for certain forms of human behavior, behavior that is being selected for due to the radical changes wrought by modern human society, science, and technology over the centuries. The small minicolumns in primate primary visual cortex cannot be used as a model for association cortex. Visual cortex must be seen in the context of what it represents; a highly specialized region that displays many anatomical differences not found in other regions of the neocortex. Their configuration cannot easily be transposed to association neocortex. This does not mean that all small minicolumns have to be designed in the manner of those in visual cortex, but it is a warning not to extrapolate from visual cortex to associational cortex.

A very large brain is ideal in regard with computational complexity and flexibility because it can have many minicolumn units and the size of each can be larger than in a small brain. The advantage of a larger minicolumn is that it can do more processing than a smaller one. The argument that larger minicolumns are more generalized processors, while smaller ones are more specialized because incoming signals can be broken down into more segments, must be understood within the context of the cortical volume. Certainly, if two regions of the cortex are identical in surface area, and one has minicolumns that are significantly smaller, then the smaller ones may be more specialized. This appears to be the strategy behind the use of lateralization in human auditory cortex. However, the larger minicolumns in humans are capable of more specificity and more complex processing because each region of human cortex contains many more units and each one can process more information. In addition, the subtle and complex fine tuning available within the circuits of each minicolumn gives them the capacity to alter incoming signals as needed. Smaller minicolumns are arguably more necessary in a smaller brain than a larger one. The example from lateralization is that both large and small minicolumns in combination serve a function. It seems nonsensical to suggest that the larger minicolumns in the left hemisphere do not process information as well as those on the right, or vice versa. They do so differently.

Small minicolumns in modern humans

There is an emerging body of research that documents the presence of smaller than normal minicolumns in certain extant humans. The bulk of these studies derive from research in the autistic spectrum disorder (Buxhoeveden et al., 2006;

Casanova et al., 2002a, 2003). This spectrum encompasses severely impaired individuals at one end, highly intelligent individuals at the other known as high-functioning autistics, and those with Asperger's syndrome. All of them display variances from "normal" human behavior at some very fundamental levels. From a physical anthropology viewpoint, some of these are striking as they challenge the most fundamental definitions of the human, language, socialization, and theory of mind, all of which is either lacking or is deficient. In Asperger's syndrome, there are serious problems with socialization (but not language). However, there is an ability to think visually rather than with words and many of the high-functioning autistics and those with Asperger's syndrome tend to display above average intelligence.

Research to date has found that in individuals with autism and Asperger's syndrome, minicolumns are smaller in regions associated with higher cognitive functions, such as the frontal lobes, while the primary sensory and motor areas are normal (Buxhoeveden et al., 2006; Casanova et al., 2003). Because these are generally considered to be pathologies, it is difficult what to make of the small minicolumns except to link them as part of the disorder (Tommerdahl, et al., 2008). One option is that there may be several variants of small minicolumns in humans, where some are linked to pathology and others are not. Another is that when only some of the genes associated with these disorders are involved, it may result in either no pathology or a reduction in the severity. Various traits associated with these disorders are said to be present in people who do not otherwise manifest symptoms of autism or Asperger's. These are individuals that can be very successful but demonstrate certain traits that are typical of the spectrum disorder. Many people that excelled in intellectual achievements are claimed to have displayed some of these traits (Boso et al., 2010; Fitzgerald, 2002; Lyons and Fitzgerald, 2005; Treffert, 2009; Treffert and Wallace, 2002). For example, Einstein acquired language very late (about 4–5years old) and had other behaviors associated with these traits, including a preference for thinking in visual images even though his cortex was otherwise generally unexceptional in size or other gross anatomical features (Falk, 2009b). A recent paper even argues that there may be evidence of gliogenesis in Einstein's brain that is the kind found in autism (Yuan, 2009). If the focus is on these individuals, and perhaps Asperger's and some high-functioning autistics, a picture emerges that suggests a variation in the "normal" template of the human brain, perhaps associated with genes that are otherwise associated with dysfunctional anomalies.

Taking this, a step further are tantalizing data from a very small group which displays smaller than normal minicolumns without pathology. This study found that minicolumns were smaller than controls in three supernormals (people who display exceptional intelligence and are otherwise normal) and were in fact in the same size range as Asperger's and autistic individuals (Casanova et al., 2007). The very small sample size makes it difficult to extrapolate further; however, it is well-known that many geniuses did not have larger than normal brains, and some had even smaller than normal ones, something expected if the minicolumns were smaller. In contrast, studies conducted in the brains of Down syndrome individuals found the minicolumns to be normal to large in size, so that when coupled with a significantly smaller cortical volume compared to controls presents a picture of fewer processing units (Buxhoeveden and Casanova, 2004). A study of minicolumns in a Dysletic subject reported they were normal in size (Casanova et al., 2002b). What these studies do is rule out small minicolumns as somehow associated with all pathologies, and more significantly, it associates small minicolumns with a certain form of behavior. Since the definition of minicolumn size is a relative one, the nature of the control group is critical in future studies and should include greater information and detail regarding behavioral traits.

Concluding remarks

It is possible that there is no link between the small minicolumns in the supernormals and the spectrum disorders that the cause of each is separate. Were we able to know more about the details of the internal circuitry, they would be found to be different. What we are more certain of is that encephalization in human brains appears to be in stasis. We also know that the size of minicolumns can say something about how information is processed and that there is a subgroup of extant humans that display minicolumns that are smaller than the average. Some of them are associated with pathology, while some are not. Something missing from our database, and which would be vital in putting the pieces together, is whether the average minicolumn size in extant humans is the same as it was in fossil humans such as *H. erectus*, the archaic *Homo sapiens*, or the anatomically modern *H. sapiens*. Were they larger, smaller, or similar in size compared to the present? If we knew that, then we would have a much better handle on the questions posed in this chapter. It is possible that minicolumns were larger in the past and have been getting progressively smaller in response to the intellectual challenges posed by societies in the past few thousand years.

It is also fairly certain that if no further increases in cortical volume can be expected, it becomes imperative to make the existing brain tissue more efficient. The demand for intellectual complexity is increasing as never before. From the perspective of the minicolumn, it would seem that downsizing is one viable option but can it be done without changing human behavior? There is no reason to suspect that the mutational events that lead to the production of more ontogenetic units will cease, which means that additional units will either continue to add cortical volume, a trend that has a limited future, they will conform to the existing cortical space by downsizing, or simply fail to sustain themselves due to excessive competition and cell death. Another option is that the current size minicolumn will undergo further subtle reorganization or efficiency will be increased at other levels of brain biology.

The range for normal brain size in extant humans is somewhere between 1000 and over 2000cc which is more than brain volume has been for most of hominid evolution. This huge variability in brain size allows for many forms of internal variances in the number and size of minicolumns. The situation is similar with minicolumn size where the range for the size of minicolumns found in extant humans is greater than the absolute size of minicolumns in the smallest monkey brain. Large brains, while having their limitations and taxing bodily resources, nonetheless, afford possibilities closed to small brains (Herculano-Houzel, 2009).

It is reasonable to suspect that variable size minicolumns work well in the human condition since adaptation to modern society calls for many different kinds of skills. The societal options open to a monkey or ape are highly restricted compared to a human, and because of this there is less variance in selection pressures regarding neurological design in these nonhuman primates. The challenges and niches available for a monkey brain are narrow compared to the almost limitless diverse forms of intelligence open to human populations in the modern world. This means that to the extent minicolumn size has any relation to how minds think, the way information is processed, it can be expected that they will always be highly variable in humans.

What appears certain is that continued infinite growth (in respect to spatial expansion) of the cortex is not viable. We have discovered this in regard with human exploitation of limited resources on our planet, and it is true of the brain as well. The trick will be to meet the needs for kinds of human intelligence demanded of the extant and future human population, without detrimentally changing our behavior so as to make it maladaptive, all the while keeping the size of the brain in check.

References

Amirikian, B., & Georgopoulos, A. P. (2003). Modular organization of directionally tuned cells in the motor cortex: Is there a short-range order? *Proceedings of the National Academy of Sciences of the United States of America*, *14*, 12474–12479.

Boso, M., Emanuel, E., Prestori, F., Politi, P., Barale, F., & D'Angelo, E. (2010). Autism and genius: is there a link? The involvement of central brain loops and hypotheses for functional testing. *Functional Neurology*, *25*, 15–20.

Budd, L. M. L., & Kisvarday, Z. F. (2001). Local lateral connectivity of inhibitory clutch cells in layer 4 of cat visual cortex (area 17). *Experimental Brain Research*, *140*, 245–250.

Buxhoeveden, D., & Casanova, M. F. (2002). The minicolumn and evolution of the brain. *Brain, Behavior and Evolution*, *60*, 25–51.

Buxhoeveden, D., & Casanova, M. F. (2004). Accelerated maturation in brains of patients with Down's syndrome. *Journal of Intellectual Disability Research*, *48*, 705–706.

Buxhoeveden, D., & Casanova, M. F. (2005). The cell column in comparative anatomy, Chap 5. Encephalization, minicolumns, and hominid evolution, Chap 6. In M. F. Casanova (Ed.), *Neocortical modularity and the cell minicolumn.* Hauppauge, New York: Nova Science Publishers.

Buxhoeveden, D. P., Lefkowitz, W., Loats, P., & Armstrong, E. (1996). The linear organization of cell columns in human and nonhuman anthropoid Tpt cortex. *Anatomy and Embryology*, *194*, 23–36.

Buxhoeveden, D. P., Semendeferi, K., Schenker, N., Switzer, R., & Courchesne, E. (2006). Reduced minicolumns in the frontal cortex of patients with autism. *Neuropathology and Applied Neurobiology*, *32*, 483–491.

Casanova, M. F., Buxhoeveden, D. P., Switala, A. E., & Roy, E. (2002a). Minicolumnar pathology in autism. *Neurology*, *58*, 428–432.

Casanova, M. F., Buxhoeveden, D. P., Cohen, M., Switala, A. E., & Roy, E. L. (2002b). Minicolumnar pathology in dyslexia. *Annals of Neurology*, *52*, 108–110.

Casanova, M., Buxhoeveden, D., & Gomez, E. (2003). Disruption in the inhibitory architecture of the cell minicolumn: Implications for autism. *The Neuroscientist*, *9*, 496–507.

Casanova, M. F., Switala, A. E., Trippe, J., & Fitzgerald, M. (2007). Comparative minicolumnar morphometry of three distinguished scientists. *Autism*, *11*, 557–569.

Catania, K. C. (2002). Barrels, stripes and fingerprints in the brain-implications for theories of cortical organization. *Journal of Neurocytology*, *31*, 347–358.

Curtetti, R., Garbossa, D., & Vercelli, A. (2002). Development of dendritic bundles of pyramidal neurons in the rat visual cortex. *Mechanism of Ageing and Development*, *123*, 473–479.

Elston, G. N. (2000). Pyramidal cells of the frontal lobe: All the more spinous to think with. *The Journal of Neuroscience*, *20*, 1–4.

Elston, G. N., & Rosa, M. G. P. (2000). Pyramidal cells, patches, and cortical columns: A comparative study of infragranular neurons in TEO, TE, and the superior temporal polysensory areas of the macaque monkey. *The Journal of Neuroscience*, *20*, 1–5.

Falk, D. (1986). Evolution of cranial blood drainage in hominids: Enlarged occipital/marginal sinuses and emissary foramina. *American Journal of Physical Anthropology*, *70*, 311–324.

Falk, D. (2009a). Brain evolution: The radiator theory. In L. R. Squire (Ed.), *Encylopedia of brain science*, (Vol. 2), (pp. 347–352). Oxford: Academic Press.

Falk, D. (2009b). New information about Albert Einstein's brain. *Frontiers in Evolutionary Neuroscience*, *1*, 3.

Favorov, O. V., & Kelly, G. (1994a). Minicolumnar organization within somatosensory cortical segregates I: Development of afferent connections. *Cerebral Cortex*, *4*, 408–427.

Favorov, O. V., & Kelly, G. (1994b). Minicolumnar organization within somatosensory cortical segregates II: Emergent functional properties. *Cerebral Cortex*, *4*, 428–442.

Favorov, O. V., & Kelly, D. G. (1996). Stimulus–response diversity in local neuronal populations of the cerebral cortex. *NeuroReport*, *7*, 2293–2301.

Fitzgerald, M. (2002). Misdiagnosis of Asperger syndrome as anankastic personality disorder. *Autism*, *6*, 435.

Gally, J. A., Montague, P. R., Reeke, G. N., Jr., & Edelman, G. M. (1990). The NO hypothesis: Possible effects of a short-lived, rapidly diffusible signal in the development and function of the nervous system. *Proceedings of the National Academy of Sciences of the United States of America*, *87*, 3547–3551.

Gustafsson, L. (1997). Inadequate cortical feature maps: A neural circuit theory of autism. *Biological Psychiatry*, *42*, 1138–1147.

Gustafsson, L. (2004). Comment on "Disruption in the inhibitory architecture of the cell minicolumn: Implications for autism". *The Neuroscientist*, *10*, 189–191.

Hasselmo, M. E. (2005). A model of prefrontal cortical mechanisms for goal-directed behavior. *Journal of Cognitive Neuroscience*, *7*, 1115–1129.

Herculano-Houzel, S. (2009). The human brain in numbers: A linearly scaled-up primate brain. *Frontiers in Human Neuroscience*, *3*, 31.

Herculano-Houzel, S., Collins, C. E., Wong, P., Kaas, J. H., & Lent, R. (2008). The basic nonuniformity of the cerebral cortex. *Proceedings of the National Academy of Sciences of the United States of America*, *26*, 1.

Hofman, M. A. (2001). Brain evolution in hominids: Are we at the end of the road? In D. Falk & K. R. Gibson (Eds.), *Evolutionary anatomy of the primate cerebral cortex* (pp. 113–127). Cambridge: Cambridge University Press.

Jin, X., Mathers, P. H., Szabo, G., Katarova, Z., & Agmon, A. (2001). Vertical bias in dendritic trees of non-pyramidal neocortical neurons expressing GAD67-GFP in vitro. *Cerebral Cortex*, *11*, 666–678.

Johansson, C., & Lansner, A. (2007). Towards cortex sized artificial neural systems. *Neural Networks*, *1*, 48–61.

Jones, E. G. (2000). Microcolumns in the cerebral cortex. *Proceedings of the National Academy of Sciences of the United States of America*, *97*, 5019–5021.

Kisvarday, Z. E., Ferecsko, A. S., Kovacs, K., Buzas, P., Budd, J. M., & Eysel, U. T. (2002). One axon-multiple functions: Specificity of lateral inhibitory connections between large basket cells. *Journal of Neurocytology*, *31*, 255–264.

Kohn, A., Metz, C., Tommerdahl, M. A., & Whitsel, B. L. (2002). Stimulus-evoked modulation of sensorimotor pyramidal neuron EPSPs. *Journal of Neurophysiology*, *88*, 3331–3347.

Kozloski, J., Hamzei-Sichani, F., & Yuste, R. (2001). Stereotyped position of local synaptic targets in neocortex. *Science*, *293*, 868–972.

Krekelberg, B., & Taylor, J. G. (1996). Nitric oxide in cortical map formation. *Journal of Chemical Neuroanatomy*, *10*, 191–196.

Krieger, P., Funer, T., & Sakmann, B. (2007). Synaptic connections between Layer 5B pyramidal neurons in mouse somatosensory cortex are independent of apical dendrite bundling. *The Journal of Neuroscience*, *27*, 11473–11482.

Krmpotic-Nemanic, J., Kostovic, I., & Nemanic, D. (1984). Prenatal and perinatal development of radial cell columns in the human auditory cortex. *Acta Oto-Laryngologica*, *97*, 489–495.

Letinic, K., Zoncu, N., & Rakic, P. (2002). Origin of GABAergic neurons in the human neocortex. *Nature*, *417*, 645–649.

Lohmann, H., & Koppen, H. J. (1995). Postnatal development of pyramidal dendritic and axonal bundles in the visual cortex of the rat. *Journal für Hirnforschung*, *36*, 101–111.

LoTurco, J. J., Blanton, M. G., & Kriegstein, A. R. (1991). Initial expression and endogenous activation of NMDA channels in early neocortical development. *The Journal of Neuroscience*, *3*, 792–799.

Lucke, J. (2004). Hierarchical self-organization of minicolumnar receptive fields. *Neural Networks*, *17*, 1377–1389.

Lucke, J., & von der Malsburg, C. (2004). Rapid processing and unsupervised learning in a model of the cortical macrocolumn. *Neural Computation*, *16*, 501–533.

Lyons, V., & Fitzgerald, M. (2005). Early memory and autism. *Journal of Autism and Developmental Disorders*, *35*, 683.

Marín, O., & Rubenstein, J. L. (2001). A long, remarkable journey: Tangential migration in the telencephalon. *National Review of Neuroscience*, *2*, 780–790.

Mountcastle, V. B. (1997). The columnar organization of the neocortex. *Brain*, *120*, 701–722.

Mountcastle, V. B. (2003). Introduction. Computation in cortical columns. *Cerebral Cortex*, *13*, 2–4.

Ong, W. Y., & Carey, L. J. (1990). Neuronal architecture of the human temporal cortex. *Anatomy and Embryology*, *181*, 351–369.

Peinado, A., Yuste, R., & Katz, L. C. (1993). Gap junctional communication and then development of local circuits in neocortex. *Cerebral Cortex*, *3*, 488–498 (review).

Peters, A., & Sethares, C. (1996). Myelinated axons and the pyramidal cell modules in monkey primary visual cortex. *The Journal of Comparative Neurology*, *365*, 232–255.

Peters, A., & Sethares, C. (1997). The organization of double bouquet cells in monkey striate cortex. *Journal of Neurocytology*, *26*, 779–797.

Peters, A., & Yilmaz, E. (1993). Neuronal organization in area 17 of cat visual cortex. *Cerebral Cortex*, *3*, 49–68.

Preuss, T. (2001). The discovery of cerebral diversity: An unwelcome scientific revolution. In D. Falk & K. R. Gibson (Eds.), *Evolutionary anatomy of the primate cerebral cortex* (pp. 138–164). Cambridge: Cambridge University Press.

Rakic, P. (1972). Mode of cell migration to the superficial layers of fetal monkey cortex. *The Journal of Comparative Neurology*, *145*, 61–84.

Rakic, P. (1978). Neuronal migration and conduct guidance in primate telencephalon. *Postgraduate Medical Journal*, *54*, 25–40.

Rakic, P. (1988). The specification of cerebral cortical areas: The radial unit hypothesis. *Science*, *241*, 928–931.

Rakic, P. (1995). A small step for the cell, a giant leap for mankind: A hypothesis of neocortical expansion during evolution. *Trends in Neuroscience*, *18*, 383–388.

Rakic, P. (2003). Developmental and evolutionary adaptations of cortical radial glia. *Cerebral Cortex*, *13*, 541–549.

Rakic, P. (2008). Confusing cortical columns. *Proceedings of the National Academy of Sciences of the United States of America*, *105*, 12099–12100.

Rakic, P., & Kornack, D. R. (2001). Neocortical expansion and elaboration during primate evolution: A view from neuroembryology. In D. Falk & K. R. Gibson (Eds.), *Evolutionary anatomy of the primate cerebral cortex* (pp. 30–56). Cambridge: Cambridge University Press.

Rao, S. G., Williams, G. V., & Goldman-Rakic, P. S. (1999). Isodirectional tuning of adjacent interneurons and pyramidal cells during working memory: Evidence for microcolumnar organization in PFC. *Journal of Neurophysiology*, *81*, 1903–1916.

Reich, D. S., Mechler, F., & Victor, J. D. (2001). Independent and redundant information in nearby cortical neurons. *Science*, *294*, 2566–2568.

Rockland, K. S., & Ichinohe, N. (2004). Some thoughts on cortical minicolumns. *Experimental Brain Research*, *158*, 265–277.

Sarko, D. K., Catania, K. C., Leitch, D. B., Kaas, J. H., & Herculano-Houzel, S. (2009). Cellular scaling rules of insectivore brains. *Frontiers in Neuroanatomy*, *3*, 8.

Seldon, H. L. (1981a). Structure of human auditory cortex. I. Cytoarchitectonics and dendritic distributions. *Brain Research*, *229*, 277–294.

Seldon, H. L. (1981b). Structure of human auditory cortex. II. Axon distributions and morphological correlates of speech perception. *Brain Research*, *229*, 295–310.

Seldon, H. L. (1982). Structure of human auditory cortex. III. Statistical analysis of dendritic trees. *Brain Research*, *249*, 211–221.

Semendeferi, K., Teffer, K., Buxhoeveden, D. P., Park, M. S., Bludau, S., Amunts, K., et al. (2011). Spatial organization of neurons in the frontal pole sets humans apart from great apes. *Cerebral Cortex*, *21*, 1485–1497.

Silberberg, G., Gupta, A., & Markram, H. (2002). Stereotypy in neocortical microcircuits. *Trends in Neuroscience*, *25*, 227–230 (review).

Sugimoto, S., Sakurada, M., Horikawa, J., & Taniguchi, I. (1997). The columnar and layer-specific response properties of neurons in the primary auditory cortex of Mongolian gerbils. *Hearing Research*, *112*, 175–185.

Swindale, N. V. (1990). Is the cerebral cortex modular? *Trends in Neuroscience*, *13*, 487–492.

Tommerdahl, M., Tannan, V., Holden, J. K., & Baranek, G. T. (2008). Absence of stimulus-driven synchronization effects on sensory perception in autism: Evidence for local underconnectivity? *Behavior and Brain Function*, *24*, 4–19.

Treffert, D. A. (2009). The savant syndrome: an extraordinary condition. A synopsis: past, present, future. *Philosophical Transactions of the Royal Society of London B: Biological Sciences*, *364*, 1351–1357.

Treffert, D. A., & Wallace, G. L. (2002). Islands of genius. Artistic brilliance and a dazzling memory can sometimes accompany autism and other developmental disorders. *Scientific American*, *286*, 76–85.

Vercelli, A. E., Garbossa, D., Curtetti, R., & Innocenti, G. M. (2004). Somatodendritic minicolumns of output neurons in the rat visual cortex. *The European Journal of Neuroscience*, *20*, 495–502.

Yu, Y. C., Bultje, R. S., Wang, X., & Shi, S. H. (2009). Specific synapses develop preferentially among sister excitatory neurons in the neocortex. *Nature*, *458*, 501–504.

Yuan, T. F. (2009). Einstein's brain: gliogenesis in autism? *Medical Hypotheses*, *72*, 753.

M. A. Hofman and D. Falk (Eds.)
Progress in Brain Research, Vol. 195
ISSN: 0079-6123

CHAPTER 11

Human brain evolution writ large and small

Chet C. Sherwood*[,†], Amy L. Bauernfeind[†], Serena Bianchi[†], Mary Ann Raghanti[‡] and Patrick R. Hof[§,¶]

[†] *Department of Anthropology, The George Washington University, Washington, DC, USA*
[‡] *Department of Anthropology and School of Biomedical Sciences, Kent State University, Kent, OH, USA*
[§] *Fishberg Department of Neuroscience and Friedman Brain Institute, Mount Sinai School of Medicine, New York, NY, USA*
[¶] *New York Consortium in Evolutionary Primatology, New York, NY, USA*

Abstract: Human evolution was marked by an extraordinary increase in total brain size relative to body size. While it is certain that increased encephalization is an important factor contributing to the origin of our species-specific cognitive abilities, it is difficult to disentangle which aspects of human neural structure and function are correlated by-products of brain size expansion from those that are specifically related to particular psychological specializations, such as language and enhanced "mentalizing" abilities. In this chapter, we review evidence from allometric scaling studies demonstrating that much of human neocortical organization can be understood as a product of brain enlargement. Defining extra-allometric specializations in humans is often hampered by a severe lack of comparative data from the same neuroanatomical variables across a broad range of primates. When possible, we highlight evidence for features of human neocortical architecture and function that cannot be easily explained as correlates of brain size and, hence, might be more directly associated with the evolution of uniquely human cognitive capacities.

Keywords: pyramidal neuron; cortical area; chimpanzee; great ape.

Human brain evolution writ large

The most obvious and distinctive evolutionary specialization of the modern human brain is its exceptionally large size. Although elephants and whales have larger brains than humans, allometric scaling analyses have demonstrated that humans are the most encephalized of all mammals (Jerison, 1973; Martin and Harvey, 1985), with a brain that is more than three times larger than would be expected for a primate at the same body mass (Holloway, 1979). This disproportionate

*Corresponding author.
Tel.: +1-202-994-6346; Fax: +1-202-994-6097
E-mail: sherwood@gwu.edu

DOI: 10.1016/B978-0-444-53860-4.00011-8

growth of the brain in the human lineage is a relatively recent phenomenon, having increased dramatically in the past 2.5 million years (Falk et al., 2000; Holloway et al., 2004). Although it is apparent that large brain size is a hallmark of human cognitive and cultural evolution, consensus is lacking on the selection pressures driving encephalization. Among the hypotheses put forward, it has been proposed that the complexity of social interaction, with a greater focus on cooperation and learning from others (Boyd et al., 2011), as well as deception (Byrne and Whiten, 1988), might have played a role.

While difficult to test, these hypothesized causes of encephalization are appealing because they attempt to explain the evolutionary benefits associated with such a prominent feature of human neuroanatomy. Given the high energetic costs of growing and maintaining neural tissue (Aiello and Wheeler, 1995; Chugani and Phelps, 1986), it is reasonable to conclude that there must be decisive fitness benefits associated with increased investment in brain mass beyond what is minimally necessary for a given body size. Indeed, it has been demonstrated across taxa that there is a trade-off between relative brain mass and other metabolically expensive tissues, as well as the extent and timing of life history stages (Barrickman et al., 2008; Barton and Capellini, 2011; Deaner et al., 2002). Some have proposed that encephalization (Jerison, 1973) or greater total numbers of neurons (Herculano-Houzel, 2011) can be taken as a satisfactory, or singular, explanation for our cognitive capacities. Among primates, correlations have been found between relative brain size and an enormously diverse range of variables, including exercise capacity (Raichlen and Gordon, 2011), the total amount of visual input as measured by the size of the optic canal (Kirk, 2006), the extent of stereoscopy as indicated by the degree of orbital convergence (Barton, 2004), behavioral innovation (Reader and Laland, 2002), sociality (Shultz and Dunbar, 2010a), and executive function (Shultz and Dunbar, 2010b). The adage, "a theory that explains everything really explains nothing" seems apt.

To some extent, explaining human cognitive uniqueness as merely a by-product of encephalization, absolute brain size, or total numbers of neurons reflects the infancy of studies in evolutionary neuroscience. If the goal is to understand the distinctive neural bases of the specific behavioral abilities that are unique to humans, then how can any unitary variable explain such a multifaceted suite of characteristics? Although still a source of debate, there appears to be a growing consensus that human cognition is most unique in (1) the representational understanding of one's own and other's mental states, such as beliefs, desires, and goals—that is, "theory of mind" or "mentalizing"—and (2) syntactically ordered symbolic communication in the form of language (Hauser et al., 2002; Herrmann et al., 2007; Suddendorf et al., 2009). Even if brain size can be understood as a major contributor to human cognitive uniqueness in these respects, it would still be necessary to learn more about how this single large variable translates to smaller-scale differences that can be interpreted in terms of the development of connectivity, the integration and signaling of neurons, and the flow of information within the central nervous system. Further, advances from modern behavioral neuroscience and clinical neuropsychology show that dramatic differences in behavior, including social cognition and language, can be mediated by subtle microstructural and molecular changes in brain organization (Arnold and Breedlove, 1985; Craig and Halton, 2009; Donaldson and Young, 2008; Robinson and Becker, 1986; Rosenzweig and Bennett, 1996), most often in the absence of any major difference in brain size.

At present, however, we have only a rudimentary understanding of the anatomical, functional, and energetic consequences of increased brain size or encephalization in human evolution. In large part, this is because only a few studies have yet probed the differences between human brains and those of our close relatives, the great apes, in any detail. Consequently, many questions remain unanswered. For example, the extent to which particular modifications of neuronal morphology and cell type distributions regularly scale up with increases in brain size is poorly known; it is unclear whether all regions of the cerebral cortex tend to increase in size at the same rate, or if there are particular areas that grow disproportionately in correlation with brain size evolution; and it has not been determined how the connectivity of the cerebral cortex varies with brain size. While theoretical models exist to predict some of these scaling regularities across mammals (Kaas, 2000; Striedter, 2005), the absence of sufficient comparative data from hominoid primates precludes clear assessment of whether human neural organization is largely determined by increased brain size, or alternatively, whether certain features have emerged as departures from allometric expectations.

Within this framework, we will discuss changes in neocortical architecture in the evolution of the human brain. Specifically, we will review evidence for microstructural modifications of neocortical structure as reflected in cellular distributions and neuronal morphology, and we will consider their implications for energetics. Other excellent recent reviews discuss the allometric scaling of larger anatomical components of the brains among humans and other primates (Rilling, 2006; Schoenemann, 2006).

Scaling regularities and the human brain writ small: Cellular distributions and morphology

One way of exploring how the human brain differs from other primates is to determine whether features of neocortical architecture deviate from predictions derived from other smaller-brained primates, with the aim of identifying scaling regularities and evolutionary specializations (de Sousa et al., 2010a; Rilling, 2006; Rilling and Insel, 1999b; Rilling and Seligman, 2002; Semendeferi and Damasio, 2000; Semendeferi et al., 1998, 2001, 2002; Sharma et al., 2010; Sherwood et al., 2004, 2005a,b, 2006, 2007, 2010). In many ways, after taking overall brain size into account, comparative evidence indicates that human neuroanatomy is not unexpected. For example, it has been shown that the total number of neurons in the neocortex of humans closely matches expectations for a primate of the same brain size (Azevedo et al., 2009), that total neocortical white matter and corpus callosum size are predicted by scaling (Bush and Allman, 2003; Rilling and Insel, 1999a), and that the frontal cortex of humans is not any larger than expected for brain size (Bush and Allman, 2004; Semendeferi et al., 2002). Human neocortical architecture at the histological level may also be examined from the perspective of allometric scaling. For example, the ratio of glial cells to neurons and the proportion of different subtypes of inhibitory GABAergic interneurons in the dorsolateral prefrontal cortex (Brodmann's area 9) of humans have been shown to be explained by scaling predictions (Sherwood et al., 2006, 2010). Thus, at present, scaling exponents have been calculated for a number of macro- and microscopic neocortical variables, including neuron number and density, cortical thickness and surface area, white matter volume, number of brain areas, number of synapses per neurons, synaptic density, cell soma size, and axon diameter (for review see Changizi, 2001). Accordingly, many anatomical characteristics of the human brain are likely to be closely intertwined with an increase in overall size. Table 1 summarizes results from several studies that have examined whether human neuroanatomical structure is predicted by allometric scaling equations derived from smaller-brained primates.

While scaling regularities may reflect underlying biophysical, computational, or biochemical laws that operate to maintain functional equivalence across variation in brain size (as is probably the case for increases in white matter and

Table 1. Percent difference from allometric predictions for various neuroanatomical variables in humans

Variable	Percent difference from allometric prediction	Independent variable used in the prediction	Nonhuman species used in the prediction equation	Human sample size	Data source
Volumes of structures					
Neocortex (gray and white matter)	+9%	Brain	Anthropoids	n=6	Rilling (2006), data from Rilling and Insel (1999b)
Frontopolar cortex (area 10)	+6%	Brain	Hominoids	n=1	Holloway (2002), data from Semendeferi et al. (2001)
Primary visual cortex (area 17)	−121%	Brain	Primates	n=1	Holloway (2002), data from Stephan et al. (1981)
Hippocampus	−6%	Brain	Primates	n=1	Holloway (2002), data from Stephan et al. (1981)
Striatum	−69%	Brain	Primates	n=1	Holloway (2002), data from Stephan et al. (1981)
Thalamus	−39%	Brain	Primates	n=1	Holloway (2002), data from Stephan et al. (1981)
Lateral geniculate nucleus	−144%	Brain	Primates	n=1	Holloway (2002), data from Stephan et al. (1981)
Cerebellum	−20%	Brain	Anthropoids	n=6	Rilling (2006), data from Rilling and Insel (1999b)
Midbrain	−14%	Brain	Primates	n=1	Holloway (2002), data from Stephan et al. (1981)
Medulla	−74%	Brain	Primates	n=1	Holloway (2002), data from Stephan et al. (1981)
Trigeminal motor nucleus	−17%	Medulla	Primates	n=5	Sherwood et al. (2005a)
Facial motor nucleus	+3%	Medulla	Primates	n=5	Sherwood et al. (2005a)
Hypoglossal nucleus	+24%	Medulla	Haplorhines	n=4	Sherwood et al. (2005a)
Cellular organization					
Glia-neuron ratio in DLPFC (area 9)	−5%	Brain mass	Anthropoids	n=6	Sherwood et al. (2006)
CB-ir interneuron density in DLPFC (area 9)	−40%	Total neuron density	Anthropoids	n=6	Sherwood et al. (2010)
CR-ir interneuron density in DLPFC (area 9)	−35%	Total neuron density	Anthropoids	n=6	Sherwood et al. (2010)
PV-ir neuron density in DLPFC (area 9)	−43%	Total neuron density	Anthropoids	n=6	Sherwood et al. (2010)
Number of TH-ir neurons in locus coeruleus	+3% −126%	Medulla vol Neocortex vol	Catarrhines	n=4	Sharma et al. (2010)
Number of ChAT-ir neurons in nbM	−110% −104%	Brain mass Neocortex vol	Anthropoids	n=2	Raghanti et al. (2011b)

DLPFC, dorsolateral prefrontal cortex; CB, calbindin; CR, calretinin; PV, parvalbumin; ChAT, choline acetyltransferase; ir, immunoreactive; nbM, nucleus basalis of Meynert.
All variables represent volumes unless otherwise indicated. All percent differences between observed and predicted values are based on contemporary species data, not independent contrasts. We include percent differences only from studies where this value was calculated and reported. Other scaling analyses have also examined whether humans are within the 95% prediction intervals generated from nonhuman data, but do not calculate a percent difference between the observed and predicted values. We excluded those studies from this table.

glia–neuron ratios), deviations from scaling predictions provide strong evidence for particular evolutionary specializations, as has been shown for the reduction of primary visual cortex (de Sousa et al., 2010a) and the expansion of the temporal lobe (Rilling and Seligman, 2002).

With regard to scaling regularities and evolutionary specializations, pyramidal cell structure deserves particular attention. The excitatory pyramidal cells constitute approximately 70% of the neurons of the neocortex, with variation among regions and species. Morphologically, they are characterized by a triangular-shaped soma, an axon for long-distance projection, one apical dendrite extending toward the pial surface, and a skirt of dendritic arbors that sample inputs from adjacent cells. Since the early observations by Ramon y Cajal (1909), it had become apparent that the morphology of pyramidal cells varies across species, with the greatest degree of variation being observed in the structure of the dendritic arbors. To a certain extent, this variation may be explained by scaling with body and brain size (Purves, 1988; Wittenberg and Wang, 2007). As brains become larger, neuron density is reduced and greater dendritic arbors may be needed to maintain connections with neighboring cells (Purves, 1988). While scaling exponents have been proposed for neuronal features such as cell body size and axon diameter (Changizi, 2001), less is known about the scaling properties of dendrites.

In fact, recent studies examining dendritic morphology across different cortical areas indicate that pyramidal cell size may depend more on regional specializations than solely on brain size (Elston et al., 2001). In particular, it has been shown that some features of dendrites (i.e., dendritic spine density), but not others (i.e., branching patterns, cell size, and total spine number), correlate with overall brain size (Elston et al., 2001). Moreover, examination of regional differences in the pyramidal cell phenotype of humans and nonhuman primates has revealed substantial variation across cortical areas, with peaks of branching complexity and spine density typically observed in the prefrontal region (Fig. 1; Elston, 2000; Elston et al., 2001; Jacobs et al., 2001). As greater dendritic branching and higher spine densities allow sampling and integration of a larger number of inputs from neighboring cells, these regional differences may reflect functional

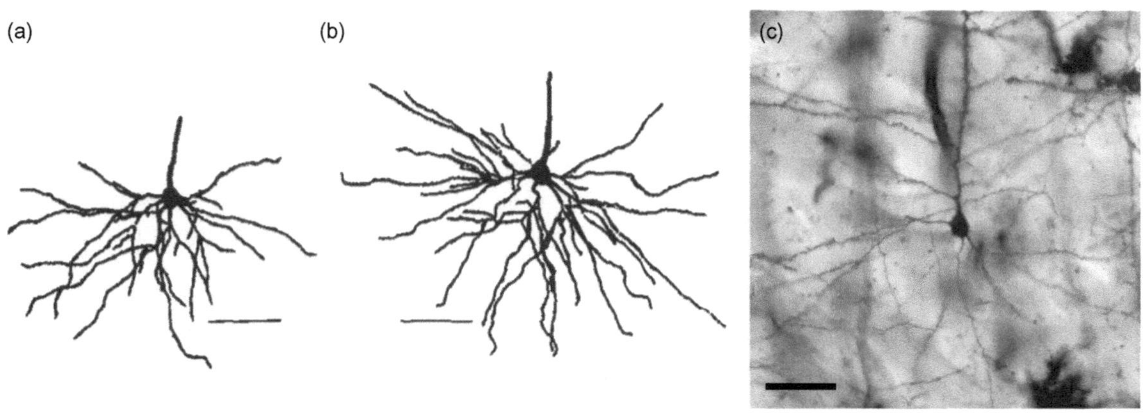

Fig. 1. Morphology of layer III pyramidal neurons in a human and a chimpanzee as revealed by Golgi impregnation. (a) Tracing of a neuron from primary motor cortex (area 4) in a human. (b) Tracing of a neuron from frontopolar cortex (area 10) in the same human individual as in (a). (c) Photomicrograph of a neuron in frontopolar cortex (area 10) of a chimpanzee. In the human, note the greater extent of dendritic branching in the region of the prefrontal cortex as compared to the primary motor cortex. Tracings of human neurons modified from Jacobs et al. (2001). Scale bar=100 μm.

differences in neuronal computational power (Elston, 2000; Elston et al., 2001; Jacobs et al., 2001). Interestingly, when compared to macaque monkeys, the closest human relative yet examined for comparative studies of pyramidal morphology, the prefrontal cortex of humans exhibits a greater degree of branching complexity as well as spine density and total spine number (Elston et al., 2001, 2006). By contrast, pyramidal neurons of the primary visual cortex do not demonstrate as much species-specific variation among primates (Elston et al., 2006). Comparative studies of neuropil in the neocortical gray matter (which is primarily occupied by dendrites, axon, and synapses) across hominoid primates also show increased neuropil selectively in the frontopolar cortex of humans (Semendeferi et al., 2011). Taken together, these data support the suggestion that increased connectivity within the prefrontal cortex is an evolutionary specialization of the human neocortex (Elston, 2003). Because there is not a significant correlation between the fraction of neuropil in the neocortex (as measured in areas 4, 10, 13, 17, and 18) and brain size in hominoids (de Sousa et al., 2010b; Sherwood and Hof, 2007), the evidence for increased connectivity in the human prefrontal cortex may be an evolutionary change that cannot be accounted for by models of scaling.

One possibility that needs further testing, however, is whether regional differences in dendritic morphology straightforwardly reflect variation in neuronal density and neuropil fraction across cortical areas. As shown by a recent study of galagos, macaques, and baboons, neuronal density varies up to five times within species, with the highest density being observed in the primary visual cortex (Collins et al., 2010). Moreover, interpreting the functional implications of structural differences may be further complicated by the fact that, although cell morphology does not correlate with cortical area size, increasing or decreasing the dimension of neurons may have different impacts on small or large areas (Kaas, 2000).

The neuroanatomy of cognitive specializations: Comparing cortical area size and neurotransmission between humans and apes

In addition to investigations of scaling regularities that might account for the evolution of human brain organization, other analyses that make the narrow comparison of neocortical areas between humans and our close relatives, the great apes, also provide important insight. Although such comparison would ideally incorporate data from a greater number of primate species to examine brain size-related correlations, this is not always practical. Nonetheless, these studies provide an essential step to locate changes in the brain which might relate to the recently evolved cognitive specializations of humans by focusing on particular cortical areas that are known to be involved with language and cognition. Broca's area and Wernicke's area in the left hemisphere, for example, are principal nodes in a network that is strongly activated during language processing (Friederici et al., 2006). Regions of the medial prefrontal cortex and along the posterior superior temporal sulcus have been consistently found to be involved in "theory of mind" tasks (Saxe, 2006). Identifiable homologues exist in other primates for all the basic circuitry that is employed for language function and "theory of mind" in humans (Petrides and Pandya, 1994, 2002; Petrides et al., 2005; Preuss and Goldman-Rakic, 1991a,b). Therefore, understanding the neural basis for the origin of these functions requires examining how evolution has modified these regions or their connections in humans.

One important characteristic of the functional control of language in humans is that it is strongly lateralized to the left hemisphere in most people (Toga and Thompson, 2003). This functional dominance is associated with certain anatomical asymmetries, such as an increased size of the planum temporale on the left (Geschwind and Levitsky, 1968; Toga and Thompson, 2003). To examine the coevolution of the brain and language, Broca's area (area 44 and area 45) and

Wernicke's area (the posterior part of area 22; also called area Tpt) in chimpanzees have been studied using magnetic resonance imaging, histological data, and associated behavioral information (Hopkins et al., 2008, 2010; Schenker et al., 2010; Sherwood et al., 2003a, 2010; Spocter et al., 2010). Using cytoarchitecture to define cortical areas, it has been shown that humans and chimpanzees are similar in showing a population-level bias toward leftward asymmetry of Wernicke's area (area Tpt) volume (Spocter et al., 2010). By contrast, Broca's area (area 44 and area 45) of chimpanzees does not display human-like lateralization (Schenker et al., 2010). In addition, when the size of these cortical areas is compared between humans and chimpanzees, it is evident that Wernicke's area has increased in size only proportionately with overall neocortex in human evolution, whereas Broca's area has increased in size to a substantially greater extent (Table 2; Fig. 2). In fact, of the cortical areas measured for their volume in both humans and chimpanzees so far, area 44 in the left hemisphere displays the greatest enlargement in humans. These results allow us to speculate that the uniqueness of human language function may depend more critically on modifications to the processing capacity of the inferior frontal cortex for syntax, lexical retrieval, and other hierarchically ordered information but is built upon more ancestral functions of the superior temporal cortex (area Tpt) in phonological processing.

At present, only a small number of regions in both humans and chimpanzees have been measured for their size based on cytoarchitecture (Table 2). Interestingly, from the cortical areas that have been compared, a similar pattern of disproportionate regional expansion seems to also characterize differences in the size of cortical areas between human and macaques, as well as the heterogeneous growth of the cerebral cortex that occurs over the course of postnatal development in humans (Hill et al., 2010; Fig. 2). This suggests that the difference in cortical area sizes that evolved in humans after branching from the rest of the ape lineage may be an extension of common scaling trends across primate phylogeny that are mediated by conserved developmental mechanisms (see Chapter 4). More detailed parcellations and volumetric comparisons of the temporal cortex and inferior parietal cortex in apes are warranted and may provide insight into additional cortical modifications that have contributed to the evolution of human language and other cognitive functions.

Table 2. Comparison of fold-differences in the sizes of neocortical areas between humans and chimpanzees

Structure	Human versus chimpanzee fold-difference	Data source
Brain	3.6	Holloway (1996)
Neocortical gray	4.0	Rilling and Insel (1999b)
Area 44 (left)	6.6	Amunts et al. (1999), Schenker et al. (2010)
Area 10 (right)	6.3	Semendeferi et al. (2001)
Area 45 (left)	6.0	Amunts et al. (1999), Schenker et al. (2010)
Area 45 (right)	5.0	Amunts et al. (1999), Schenker et al. 2010
Area Tpt (left)	4.2	Galaburda and Sanides (1980), Spocter et al. (2010)
Area 44 (right)	4.1	Amunts et al. (1999), Schenker et al. (2010)
Area 6 (hemisphere unknown)	2.5	Glezer 1958[a]
Area Tpt (right)	2.0	Galaburda and Sanides (1980), Spocter et al. (2010)
Area 17 (left)	1.8	de Sousa et al. (2010a)
Area 13 (right)	1.4	Semendeferi et al. (1998)
Area 4 (hemisphere unknown)	0.8	Glezer (1958)[a]

[a]All comparisons are based on volumes, except for the data from Glezer (1958) which provide surface area measurements.

In addition to clarifying which regional size differences appear to be most significant in human brain evolution, recent research has also revealed human-specific changes to the microstructure of connections and distributions of

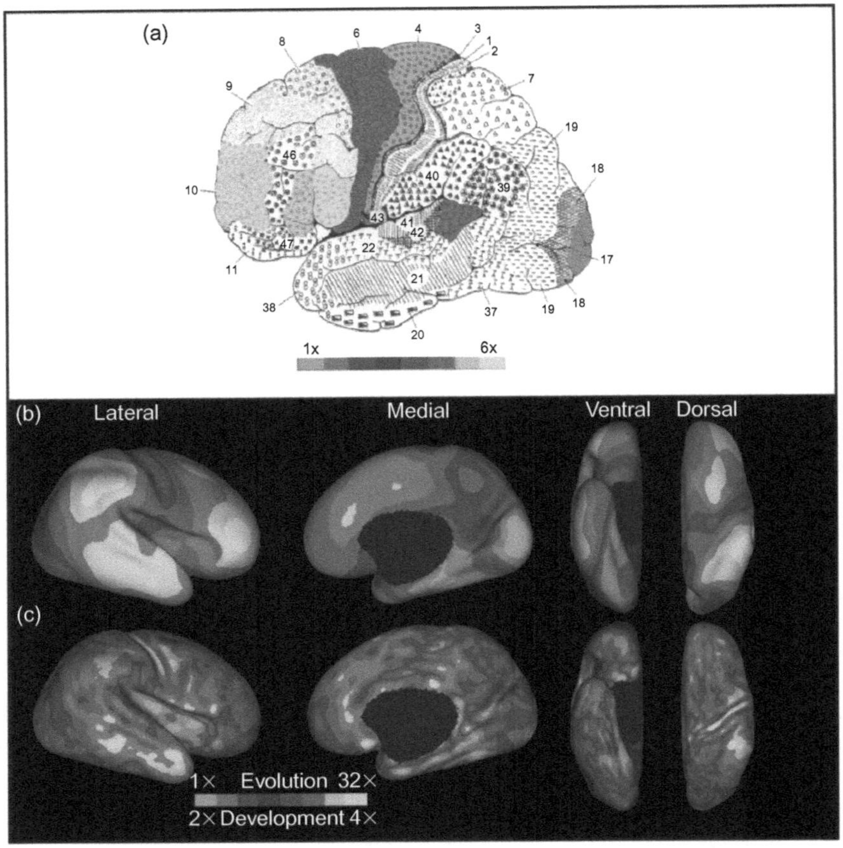

Fig. 2. Maps illustrating the disproportionate expansion of the neocortex in human evolution and development. (a) Fold-difference in size of cortical areas between humans and chimpanzees. Values were taken from Table 1 and overlaid on Brodmann's map of the cerebral cortex. All data are from the left hemisphere except where only right hemisphere or unknown were available. (b) Fold-difference in surface area of cerebral cortex between humans and macaque monkeys expressed relative to the total neocortical size difference between species, modified from Hill et al. (2010). Right hemisphere is shown. (c) Fold-difference in surface area of cerebral cortex between neonatal and adult humans, modified from Hill et al. (2010). Right hemisphere is shown.

neuron types. The supply of neurotransmitter afferents to the cerebral cortex, originating from subcortical neuron populations located in the basal forebrain and brainstem, plays a key role in the balance of excitation and inhibition underlying information processing in the neocortex. These neurotransmitters (e.g., dopamine, norepinephrine, acetylcholine, and serotonin) have diffuse termination zones and have long-term effects on the processing characteristics of postsynaptic cells by interacting with multiple G-protein-linked receptor subtypes. Severe psychiatric disturbances in human patients have been shown to correlate with abnormalities in the synthesis, transport, and metabolism of these neuromodulatory systems (for review see Briand et al., 2007). By using immunohistochemical staining to identify individual axon fibers in combination with stereologic quantification, it has been shown that the extrinsic supply of serotonergic, dopaminergic, and cholinergic axons to the prefrontal cortex has been selectively altered in humans and chimpanzees as compared to macaque monkeys (Raghanti et al., 2008a–c). In both humans and chimpanzees, a greater axonal length density of fibers that are immunoreactive for these neurotransmitters innervates prefrontal cortex (areas 9 and 32) in a layer- and species-specific manner (Fig. 3). Because these neurotransmitter systems are involved in behavioral flexibility, attention, and learning, it is tempting to speculate that this evolutionary shift might contribute to the some of the cognitive abilities that are exclusively shared between ourselves and the great apes, such as increased behavioral inhibition, enhanced attention to the gaze of others, greater social tolerance, diffusion of social learning through regional traditions, and a capacity for self-awareness (Barth et al., 2005; Beran and Evans, 2006; Boesch, 1993; Evans and Beran, 2007a,b; Suddendorf and Whiten, 2001).

Notably, comparative allometric scaling analyses have revealed that the subcortical neuron populations that provide neurotransmitter innervation to the prefrontal cortex do not show a corresponding relative increase in humans, despite the enlargement of human cerebral cortex. Surprisingly, for both the locus coeruleus (supplying norepinephrine) and the nucleus basalis of Meynert (supplying acetylcholine), the human subcortical neuron populations actually are smaller than expected relative to brain size and neocortical volume in comparisons involving large samples of nonhuman primates (Table 1; Raghanti et al., 2011b; Sharma et al., 2010). Taken together, these results indicate that modifications in the anatomy of these neurotransmitter systems in human evolution involved alterations in terminal axon patterns, independent of correlated changes in numbers of neurons in the basal forebrain nuclei themselves.

Additional studies focusing on intrinsic sources of neurotransmitters within the cortex have also revealed that there is a significant degree of variability in both the density and distribution of cortical neurons immunoreactive for various

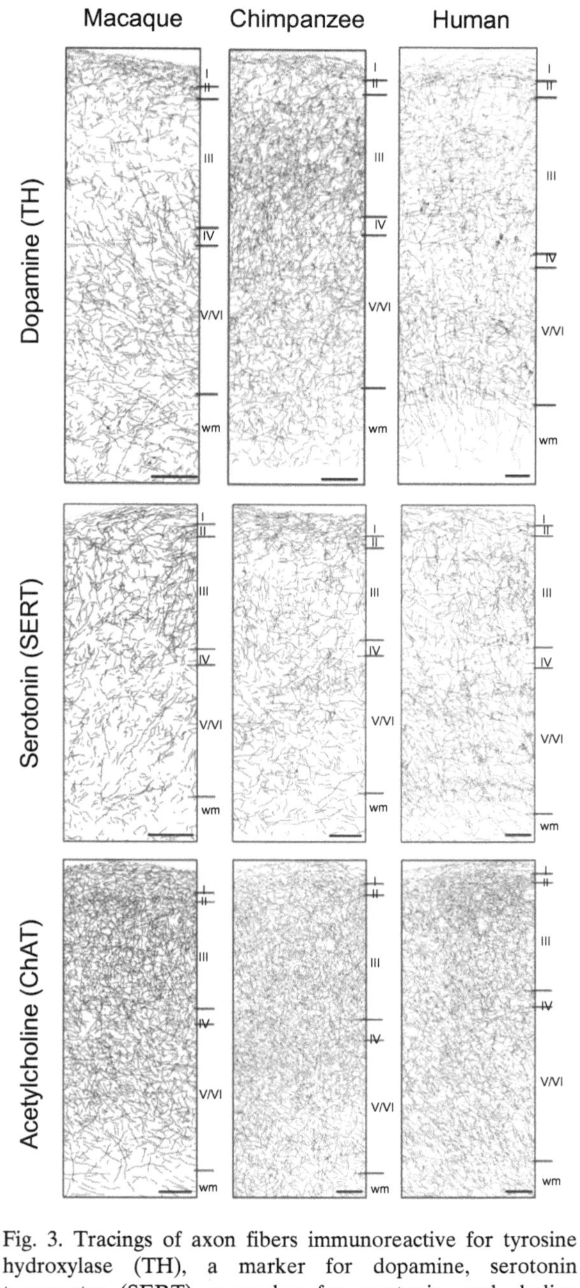

Fig. 3. Tracings of axon fibers immunoreactive for tyrosine hydroxylase (TH), a marker for dopamine, serotonin transporter (SERT), a marker for serotonin, and choline acetyltransferase (ChAT), a marker for acetylcholine, in dorsolateral prefrontal cortex (area 9) of macaque monkeys, chimpanzees, and humans. Images modified from Raghanti et al. (2008a–c). Scale bar=250µm.

neurotransmitters among primates. As with neurotransmitter-immunoreactive axons, these populations of cortical neurons are also selectively targeted in human neuropathologies (Fukuda et al., 1999; Marui et al., 2003; Nihei and Kowall, 1993). Tyrosine hydroxylase-immunoreactive neurons are found sparsely in the cerebral cortex of several mammals; however, humans are the only species that possess these neurons distributed throughout the entire cerebral cortex, with the highest densities occurring in the dorsolateral prefrontal cortex and anterior cingulate cortex (Benavides-Piccione and DeFelipe, 2007; Kohler et al., 1983). However, while present in the siamang, these cells are conspicuously and consistently absent among the great apes (chimpanzee, bonobo, gorilla, and orangutan), indicating that this neurochemical phenotype has independently evolved in humans (Raghanti et al., 2009). It is unclear why this biochemical expression pattern has reemerged in humans, after having been previously lost from the great ape lineage.

Neuropeptide Y-immunoreactive cortical cells are another class of neurons that are present in human and nonhuman primates, including the great apes. As with the neuron populations expressing tyrosine hydroxylase, there are species-specific patterns of variation, but quantitative analyses do not reveal anything unique about their densities or distributions in humans (Raghanti et al., 2011a).

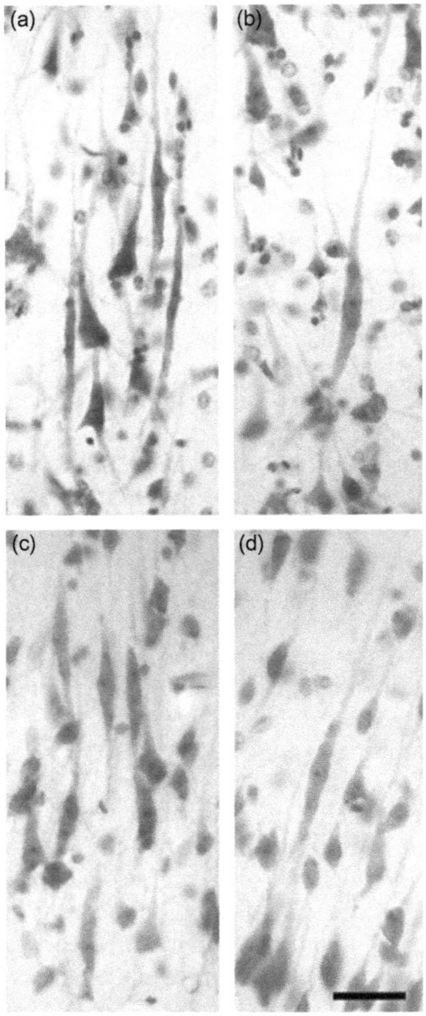

Fig. 4. von Economo neurons (VENs) as revealed by Nissl staining from layer V of anterior cingulate cortex in great apes and humans. (a) human, (b) chimpanzee, (c) bonobo, and (d) gorilla. Scale bar=100 μm.

The emergence of neuronal specializations for social cognition: VENs

Another interesting feature of the hominoid cerebral cortex is the presence of VENs (Fig. 4; Allman et al., 2010; Nimchinsky et al., 1995, 1999; Seeley et al., 2012; von Economo, 1926). VENs are projection neurons located principally in layer V of the anterior cingulate and frontoinsular cortices and, in more limited numbers, in the superior frontal cortex (area 9; Fajardo et al., 2008). Current data suggest that VENs represent a specialized neuronal type with a characteristic morphology that evolved only in a restricted number of species, most likely from a population of pyramidal neurons present in ancestral mammals (Butti and Hof, 2010; Butti et al., 2011). VENs, which are especially numerous in the hominoid lineage, are particularly vulnerable in neuropsychiatric conditions in which social and emotional skills are characteristically affected. Moreover, recent evidence on the neurochemical profile, morphologic features, and laminar and regional distribution of VENs suggests that the functional specificity of this neuronal population could be critically involved in autonomic regulation.

VENs are generally larger than layer V pyramidal neurons and their somatic volume is strongly correlated with the encephalization quotient, unlike that of pyramidal cells (Nimchinsky et al., 1999). In adults, VENs are more abundant in the right hemisphere (Allman et al., 2010), possibly reflecting asymmetries in the organization of afferents from the autonomic nervous system (Craig, 2005). Their densities, however, are low in all species in which they occur, representing only a few percent of the total number of pyramidal neurons (Allman et al., 2010). VENs have been shown to be enriched in nonphosphorylated epitopes of neurofilament proteins, similar to large pyramidal neurons (Nimchinsky et al., 1995), and to express several markers such as dopamine D3 receptor, vasopressin 1a receptor, activating transcription factor 3, interleukin-4 receptor α chain, neuromedin B, gastrin-releasing peptide, and disrupted on schizophrenia-1, in higher levels than neighboring pyramidal cells (Allman et al., 2010, 2011; Stimpson et al., 2011).

The function of VENs remains poorly understood. Nonetheless, it is interesting that VENs are affected in a number of neuropsychiatric illnesses that present impairments of social and communication skills, emotionality, morality, and self-awareness. They are severely lost in the behavioral variant of frontotemporal dementia and in agenesis of the corpus callosum (Kaufman et al., 2008; Kim et al., 2012; Seeley et al., 2006), exhibit decreased densities in schizophrenia (Brüne et al., 2010), and show abnormal cortical distribution and increased number in young children with autism (Santos et al., 2011), as well as increased densities in suicide victims with psychosis (Brüne et al., 2011). The specific localization of VENs in cortical areas in which information on the physiological state of the body is used to guide behavioral choices (Craig, 2009), their involvement in diseases in which social conduct is dramatically affected, their richness in markers such as bombesin-related peptides, and their position in a layer characteristically sending subcortical projections (Brodal, 1978; Glickstein et al., 1985) suggest a role for VENs in cortico-autonomic pathways, supporting the original intuition of von Economo (1926) on the involvement of VENs in autonomic function. The fact that they have also independently emerged in other large-brained social mammals, such as elephants and whales (Butti et al., 2009; Hakeem et al., 2009; Hof and Van der Gucht, 2007), strongly suggests that their development is mediated by mechanisms related to brain size scaling.

Energetics and microstructural changes in human neocortical evolution

Kleiber's law states that the mass-specific energetic cost of an organ declines with increases in total mass (Kleiber, 1961). Accordingly, because the energy required for the maintenance and development of neural tissue is correlated with the overall mass of the brain, greater metabolic efficiency of the human brain would be predicted because of its large size. However, contrary to this expectation, recent comparative studies of gene expression indicate that the human cerebral cortex is characterized by an upregulation of RNA transcripts involved in energy metabolism compared to the cortex of great apes (Caceres et al., 2003; Fu et al., 2011; Khaitovich et al., 2008; Preuss et al., 2004; Uddin et al., 2004). Moreover, the observation that genes coding for synaptic function are also highly expressed in human samples suggests that the human neocortex may be more metabolically expensive as a way to support its underlying cytoarchitecture and greater demand for synaptic signaling (Uddin et al., 2004). However, because these studies were performed with homogenates of tissue, it is still unknown which cellular and biochemical modifications are especially responsible for the apparent increased mass-specific metabolic demand of human neocortical tissue.

Studies have found that the metabolic expense of the rodent brain is tightly correlated with neural activity associated with the most prevalent excitatory neurotransmitter in the brain, glutamate (Attwell and Laughlin, 2001; Sokoloff, 1977). Among the most energetically expensive aspects of neural activity are the depolarization of a neuron's membrane potential following chemical stimulation at a synapse and presynaptic neurotransmitter release and recycling (Lennie, 2003). Together, these processes account for over 60% of the cost of neural activity and localize the majority of metabolic expense to the excitatory glutamatergic synapse (Raichle and Mintun, 2006). If it is assumed that the primate brain allocates energy in the same proportions to neural functions as the rodent brain, it would be expected that cortical regions with a higher density of excitatory synapses, or those regions more consistently stimulated by excitatory inputs, would be more energetically expensive to maintain throughout adulthood. Thus, greater metabolic requirement may be predicted in regions such as in the human prefrontal cortex and other association cortical areas, which have been shown to exhibit greater connectivity (Elston et al., 2006; Jacobs et al., 2001; Semendeferi et al., 2011).

Additionally, studies using radiolabeled glucose to track energy uptake in the brain have found evidence for a default mode network (DMN), a group of regions including the medial prefrontal and medial parietal cortices that are active even during times of cognitive "rest" (Raichle and Snyder, 2007; Raichle et al., 2001). Although similar patterns and high levels of activity have been found during rest in chimpanzees (Rilling et al., 2007) and macaques (Vincent et al., 2007), recent studies indicate that regions activated as part of the DMN in humans are the same regions that first exhibit the pathophysiology unique to Alzheimer's disease, a uniquely human illness (Vaishnavi et al., 2010). If the consistently high levels of activation in the DMN predispose humans to neurological diseases not seen in other primates, a case for the uniqueness of neural activity and energy expenditure in the human lineage is stronger than ever. Thus, human neural connectivity and the biochemical processes involved in energy metabolism may set humans apart from other primates in a manner not readily predicted by brain size.

Conclusions

Only by examining the detailed architecture and function of the human neocortex in a comparative perspective will it be possible ultimately to uncover the neurobiological basis of human behavioral distinctiveness. It is to be expected that some modifications to neocortical organization will be consequences of overall brain-size expansion. Others features, however, may prove to have emerged independent of encephalization. The complexity of the human mind, of course, is not likely to be explained solely by either brain size or by statistical residuals from allometric scaling equations. Rather, the human brain phenotype is constructed dynamically in ontogeny through the interaction of uniquely modified genes that regulate neuronal proliferation (i.e., the size of parts of the brain), cell migration (i.e., the histological organization of the brain), cell adhesion, and axon guidance (i.e., connectivity), as well as the synthesis and turnover of chemicals involved in signaling and energy utilization (Gilbert et al., 2005; Konopka and Geschwind, 2010). Many exciting discoveries have already been made identifying genes that control these processes and which have undergone adaptive evolution in human descent (Dorus et al., 2004; Enard et al., 2002; Evans et al., 2004; Konopka et al., 2009; McLean et al., 2011; Popesco et al., 2006; Uddin et al., 2008a,b). The complexity with which these gene networks are regulated through development and in response to environmental factors provides further opportunities for biocultural interactions to affect the phenotype and function of modern human brains. We are only now scratching the surface and beginning to reveal what makes the human brain phenotype truly special. This effort necessitates interdisciplinary comparative analyses involving a wide variety of molecular, histological, and imaging techniques that include data from our closest kin, the great apes.

Acknowledgments

This work was supported by the National Science Foundation (BCS-0515484, BCS-0549117, BCS-0827531, BCS-0921079, DGE-0801634), the National Institutes of Health (NS42867), and the James S. McDonnell Foundation (22002078 to P. R. H. and C. C. S., and a Scholar Award in Understanding Human Cognition to C. C. S.).

References

Aiello, L. C., & Wheeler, P. (1995). The expensive-tissue hypothesis: The brain and the digestive system in human and primate evolution. *Current Anthropology*, *36*, 199–211.

Allman, J. M., Tetreault, N. A., Hakeem, A. Y., Manaye, K. F., Semendeferi, K., Erwin, J. M., et al. (2010). The von Economo neurons in frontoinsular and anterior cingulate cortex in great apes and humans. *Brain Structure & Function*, *214*, 495–517.

Allman, J. M., Tetreault, N. A., Hakeem, A. Y., Manaye, K. F., Semendeferi, K., Erwin, J. M., et al. (2011). The von Economo neurons in the frontoinsular and anterior cingulate cortex. *Annals of the New York Academy of Sciences*, *1225*, 59–71.

Amunts, K., Schleicher, A., Burgel, U., Mohlberg, H., Uylings, H. B., & Zilles, K. (1999). Broca's region revisited: Cytoarchitecture and intersubject variability. *The Journal of Comparative Neurolology*, *412*, 319–341.

Arnold, A. P., & Breedlove, S. M. (1985). Organizational and activational effects of sex steroids on brain and behavior: A reanalysis. *Hormones and Behavior*, *19*, 469–498.

Attwell, D., & Laughlin, S. B. (2001). An energy budget for signaling in the grey matter of the brain. *Journal of Cerebral Blood Flow and Metabolism*, *21*, 1133–1145.

Azevedo, F. A., Carvalho, L. R., Grinberg, L. T., Farfel, J. M., Ferretti, R. E., Leite, R. E., et al. (2009). Equal numbers of neuronal and nonneuronal cells make the human brain an isometrically scaled-up primate brain. *The Journal of Comparative Neurology*, *513*, 532–541.

Barrickman, N. L., Bastian, M. L., Isler, K., & van Schaik, C. P. (2008). Life history costs and benefits of encephalization: A comparative test using data from long-term studies of primates in the wild. *Journal of Human Evolution*, *54*, 568–590.

Barth, J., Reaux, J. E., & Povinelli, D. J. (2005). Chimpanzees' (*Pan troglodytes*) use of gaze cues in object-choice tasks: Different methods yield different results. *Animal Cognition*, *8*, 84–92.

Barton, R. A. (2004). Binocularity and brain evolution in primates. *Proceedings of the National Academy of Sciences of the United States of America*, *101*, 10113–10115.

Barton, R. A., & Capellini, I. (2011). Maternal investment, life histories, and the costs of brain growth in mammals. *Proceedings of the National Academy of Sciences of the United States of America*, *108*, 6169–6174.

Benavides-Piccione, R., & DeFelipe, J. (2007). Distribution of neurons expressing tyrosine hydroxylase in the human cerebral cortex. *Journal of Anatomy*, *211*, 212–222.

Beran, M. J., & Evans, T. A. (2006). Maintenance of delay of gratification by four chimpanzees (*Pan troglodytes*): The effects of delayed reward visibility, experimenter presence, and extended delay intervals. *Behavioural Processes*, *73*, 315–324.

Boesch, C. (1993). Aspects of transmission of tool-use in wild chimpanzees. In K. R. Gibson & T. Ingold (Eds.), *Tools, language and cognition in human evolution* (pp. 171–183). Cambridge: Cambridge University Press.

Boyd, R., Richerson, P. J., & Henrich, J. (2011). Colloquium Paper: The cultural niche: Why social learning is essential for human adaptation. *Proceedings of the National Academy of Sciences of the United States of America*, *108*(Suppl. 2), 10918–10925.

Briand, L. A., Gritton, H., Howe, W. M., Young, D. A., & Sarter, M. (2007). Modulators in concert for cognition: Modulator interactions in the prefrontal cortex. *Progress in Neurobiology*, *83*, 69–91.

Brodal, P. (1978). The corticopontine projection in the rhesus monkey. Origin and principles of organization. *Brain*, *101*, 251–283.

Brüne, M., Schobel, A., Karau, R., Benali, A., Faustmann, P. M., Juckel, G., et al. (2010). Von Economo neuron density in the anterior cingulate cortex is reduced in early onset schizophrenia. *Acta Neuropathologica*, *119*, 771–778.

Brüne, M., Schobel, A., Karau, R., Faustmann, P. M., Dermietzel, R., Juckel, G., et al. (2011). Neuroanatomical correlates of suicide in psychosis: The possible role of von economo neurons. *PloS One*, *6*, e20936.

Bush, E. C., & Allman, J. M. (2003). The scaling of white matter to gray matter in cerebellum and neocortex. *Brain, Behavior and Evolution, 61*, 1–5.

Bush, E. C., & Allman, J. M. (2004). The scaling of frontal cortex in primates and carnivores. *Proceedings of the National Academy of Sciences of the United States of America, 101*, 3962–3966.

Butti, C., & Hof, P. R. (2010). The insular cortex: A comparative perspective. *Brain Structure & Function, 214*, 477–493.

Butti, C., Raghanti, M. A., Sherwood, C. C., & Hof, P. R. (2011). The neocortex of cetaceans: Cytoarchitecture and comparison with other aquatic and terrestrial species. *Annals of the New York Academy of Sciences, 1225*, 47–58.

Butti, C., Sherwood, C. C., Hakeem, A. Y., Allman, J. M., & Hof, P. R. (2009). Total number and volume of Von Economo neurons in the cerebral cortex of cetaceans. *The Journal of Comparative Neurology, 515*, 243–259.

Byrne, R. W., & Whiten, A. (1988). *Machiavellian intelligence: Social expertise and the evolution of intellect in monkeys, apes, and humans.* Oxford: Clarendon Press.

Caceres, M., Lachuer, J., Zapala, M. A., Redmond, J. C., Kudo, L., Geschwind, D. H., et al. (2003). Elevated gene expression levels distinguish human from non-human primate brains. *Proceedings of the National Academy of Sciences of the United States of America, 100*, 13030–13035.

Changizi, M. A. (2001). Principles underlying mammalian neocortical scaling. *Biological Cybernetics, 84*, 207–215.

Chugani, H. T., & Phelps, M. E. (1986). Maturational changes in cerebral function in infants determined by 18FDG positron emission tomography. *Science, 231*, 840–843.

Collins, C. E., Airey, D. C., Young, N. A., Leitch, D. B., & Kaas, J. H. (2010). Neuron densities vary across and within cortical areas in primates. *Proceedings of the National Academy of Sciences of the United States of America, 107*, 15927–15932.

Craig, A. D. (2005). Forebrain emotional asymmetry: A neuroanatomical basis? *Trends in Cognitive Sciences, 9*, 566–571.

Craig, A. D. (2009). How do you feel—Now? The anterior insula and human awareness. *Nature Reviews Neuroscience, 10*, 59–70.

Craig, I. W., & Halton, K. E. (2009). Genetics of human aggressive behaviour. *Human Genetics, 126*, 101–113.

de Sousa, A. A., Sherwood, C. C., Mohlberg, H., Amunts, K., Schleicher, A., MacLeod, C. E., et al. (2010a). Hominoid visual brain structure volumes and the position of the lunate sulcus. *Journal of Human Evolution, 58*, 281–292.

de Sousa, A. A., Sherwood, C. C., Schleicher, A., Amunts, K., MacLeod, C. E., Hof, P. R., et al. (2010b). Comparative cytoarchitectural analyses of striate and extrastriate areas in hominoids. *Cerebral Cortex, 20*, 966–981.

Deaner, R. O., Barton, R. A., & van Schaik, C. P. (2002). Primate brains and life histories: Renewing the connection. In P. M. Kappeler & M. E. Pereira (Eds.), *Primate life histories and socioecology* (pp. 233–265). Chicago: Univeristy of Chicago Press.

Donaldson, Z. R., & Young, L. J. (2008). Oxytocin, vasopressin, and the neurogenetics of sociality. *Science, 322*, 900–904.

Dorus, S., Vallender, E. J., Evans, P. D., Anderson, J. R., Gilbert, S. L., Mahowald, M., et al. (2004). Accelerated evolution of nervous system genes in the origin of Homo sapiens. *Cell, 119*, 1027–1040.

Elston, G. N. (2000). Pyramidal cells of the frontal lobe: All the more spinous to think with. *Journal of Neuroscience, 20*, RC95.

Elston, G. N. (2003). Cortex, cognition and the cell: New insights into the pyramidal neuron and prefrontal function. *Cerebral Cortex, 13*, 1124–1138.

Elston, G. N., Benavides-Piccione, R., & DeFelipe, J. (2001). The pyramidal cell in cognition: A comparative study in human and monkey. *Journal of Neuroscience, 21*, RC163.

Elston, G. N., Benavides-Piccione, R., Elston, A., Zietsch, B., DeFelipe, J., Manger, P., et al. (2006). Specializations of the granular prefrontal cortex of primates: Implications for cognitive processing. *The Anatomical Record, 288*, 26–35.

Enard, W., Przeworski, M., Fisher, S. E., Lai, C. S., Wiebe, V., Kitano, T., et al. (2002). Molecular evolution of FOXP2, a gene involved in speech and language. *Nature, 418*, 869–872.

Evans, P. D., Anderson, J. R., Vallender, E. J., Gilbert, S. L., Malcom, C. M., Dorus, S., et al. (2004). Adaptive evolution of ASPM, a major determinant of cerebral cortical size in humans. *Human Molecular Genetics, 13*, 489–494.

Evans, T. A., & Beran, M. J. (2007a). Chimpanzees use self-distraction to cope with impulsivity. *Biology Letters, 3*, 599–602.

Evans, T. A., & Beran, M. J. (2007b). Delay of gratification and delay maintenance by rhesus macaques (*Macaca mulatta*). *The Journal of General Psychology, 134*, 199–216.

Fajardo, C., Escobar, M. I., Buritica, E., Arteaga, G., Umbarila, J., Casanova, M. F., et al. (2008). Von Economo neurons are present in the dorsolateral (dysgranular) prefrontal cortex of humans. *Neuroscience Letters, 435*, 215–218.

Falk, D., Redmond, J. C., Jr., Guyer, J., Conroy, C., Recheis, W., Weber, G. W., et al. (2000). Early hominid brain evolution: A new look at old endocasts. *Journal of Human Evolution, 38*, 695–717.

Friederici, A. D., Bahlmann, J., Heim, S., Schubotz, R. I., & Anwander, A. (2006). The brain differentiates human and non-human grammars: Functional localization and structural connectivity. *Proceedings of the National Academy of Sciences of the United States of America, 103*, 2458–2463.

Fu, X., Giavalisco, P., Liu, X., Catchpole, G., Fu, N., Ning, Z. B., et al. (2011). Rapid metabolic evolution in human prefrontal cortex. *Proceedings of the National Academy of Sciences of the United States of America, 108*, 6181–6186.

Fukuda, T., Takahashi, J., & Tanaka, J. (1999). Tyrosine hydroxylase-immunoreactive neurons are decreased in number in the cerebral cortex of Parkinson's disease. *Neuropathology*, *19*, 10–13.

Galaburda, A., & Sanides, F. (1980). Cytoarchitectonic organization of the human auditory cortex. *The Journal of Comparative Neurology*, *190*, 597–610.

Geschwind, N., & Levitsky, W. (1968). Human brain: Left-right asymmetries in temporal speech region. *Science*, *161*, 186–187.

Gilbert, S. L., Dobyns, W. B., & Lahn, B. T. (2005). Genetic links between brain development and brain evolution. *Nature Reviews Genetics*, *6*, 581–590.

Glezer, I. I. (1958). Area relationships in the precentral region in a comparative-anatomical series of primates. *Arkhiv anatomii, gistologii i émbriologii*, *2*, 26.

Glickstein, M., May, J. G., 3rd, & Mercier, B. E. (1985). Corticopontine projection in the macaque: The distribution of labelled cortical cells after large injections of horseradish peroxidase in the pontine nuclei. *The Journal of Comparative Neurology*, *235*, 343–359.

Hakeem, A. Y., Sherwood, C. C., Bonar, C. J., Butti, C., Hof, P. R., & Allman, J. M. (2009). Von Economo neurons in the elephant brain. *The Anatomical Record*, *292*, 242–248.

Hauser, M. D., Chomsky, N., & Fitch, W. T. (2002). The faculty of language: What is it, who has it, and how did it evolve? *Science*, *298*, 1569–1579.

Herculano-Houzel, S. (2011). Brains matter, bodies maybe not: The case for examining neuron numbers irrespective of body size. *Annals of the New York Academy of Sciences*, *1225*, 191–199.

Herrmann, E., Call, J., Hernandez-Lloreda, M. V., Hare, B., & Tomasello, M. (2007). Humans have evolved specialized skills of social cognition: The cultural intelligence hypothesis. *Science*, *317*, 1360–1366.

Hill, J., Inder, T., Neil, J., Dierker, D., Harwell, J., & Van Essen, D. (2010). Similar patterns of cortical expansion during human development and evolution. *Proceedings of the National Academy of Sciences of the United States of America*, *107*, 13135–13140.

Hof, P. R., & Van der Gucht, E. (2007). Structure of the cerebral cortex of the humpback whale, *Megaptera novaeangliae* (Cetacea, Mysticeti, Balaenopteridae). *The Anatomical Record*, *290*, 1–31.

Holloway, R. L. (1979). Brain size, allometry, and reorganization: Toward a synthesis. In M. E. Hahn, C. Jensen & B. C. Dudek (Eds.), *Development and evolution of brain size: Behavioral implications* (pp. 59–88). New York: Academic Press.

Holloway, R. L. (1996). Toward a synthetic theory of human brain evolution. In J. P. Changeux & J. Chavaillon (Eds.), *Origins of the human brain* (pp. 42–54). Oxford: Clarendon Press.

Holloway, R. L. (2002). Brief communication: How much larger is the relative volume of area 10 of the prefrontal cortex in humans? *American Journal of Physical Anthropology*, *118*, 399–401.

Holloway, R. L., Broadfield, D. C., & Yuan, M. S. (2004). *The human fossil record. Brain endocasts—The paleoneurological evidence* (Vol. 3). New York: Wiley.

Hopkins, W. D., Taglialatela, J. P., Meguerditchian, A., Nir, T., Schenker, N. M., & Sherwood, C. C. (2008). Gray matter asymmetries in chimpanzees as revealed by voxel-based morphometry. *NeuroImage*, *42*, 491–497.

Hopkins, W. D., Taglialatela, J. P., Nir, T., Schenker, N. M., & Sherwood, C. C. (2010). A voxel-based morphometry analysis of white matter asymmetries in chimpanzees (*Pan troglodytes*). *Brain, Behavior and Evolution*, *76*, 93–100.

Jacobs, B., Schall, M., Prather, M., Kapler, E., Driscoll, L., Baca, S., et al. (2001). Regional dendritic and spine variation in human cerebral cortex: A quantitative Golgi study. *Cerebral Cortex*, *11*, 558–571.

Jerison, H. J. (1973). *Evolution of the brain and intelligence.* New York: Academic Press.

Kaas, J. H. (2000). Why is brain size so important: Design problems and solutions as neocortex gets bigger or smaller. *Brain and Mind*, *1*, 7–23.

Kaufman, J. A., Paul, L. K., Manaye, K. F., Granstedt, A. E., Hof, P. R., Hakeem, A. Y., et al. (2008). Selective reduction of Von Economo neuron number in agenesis of the corpus callosum. *Acta Neuropathologica*, *116*, 479–489.

Khaitovich, P., Lockstone, H. E., Wayland, M. T., Tsang, T. M., Jayatilaka, S. D., Guo, A. J., et al. (2008). Metabolic changes in schizophrenia and human brain evolution. *Genome Biology*, *9*, R124.

Kim, E. J., Sidhu, M., Gaus, S. E., Huang, E. J., Hof, P. R., Miller, B. L., et al. (2012). Selective frontoinsular von Economo neuron and fork cell loss in early behavioral variant frontotemporal dementia. *Cerebral Cortex*, doi:10.1093/cercor/bhr1004.

Kirk, E. C. (2006). Visual influences on primate encephalization. *Journal of Human Evolution*, *51*, 76–90.

Kleiber, M. (1961). *The fire of life: An introduction to animal energetics.* New York: Wiley.

Kohler, C., Everitt, B. J., Pearson, J., & Goldstein, M. (1983). Immunohistochemical evidence for a new group of catecholamine-containing neurons in the basal forebrain of the monkey. *Neuroscience Letters*, *37*, 161–166.

Konopka, G., Bomar, J. M., Winden, K., Coppola, G., Jonsson, Z. O., Gao, F., et al. (2009). Human-specific transcriptional regulation of CNS development genes by FOXP2. *Nature*, *462*, 213–217.

Konopka, G., & Geschwind, D. H. (2010). Human brain evolution: Harnessing the genomics (r)evolution to link genes, cognition, and behavior. *Neuron*, *68*, 231–244.

Lennie, P. (2003). The cost of cortical computation. *Current Biology*, *13*, 493–497.

Martin, R. D., & Harvey, P. H. (1985). Brain size allometry: Ontogeny and phylogeny. In W. L. Jungers (Ed.), *Size and scaling in primate biology* (pp. 147–173). New York: Plenum.

Marui, W., Iseki, E., Kato, M., & Kosaka, K. (2003). Degeneration of tyrosine hydroxylase-immunoreactive neurons in the cerebral cortex and hippocampus of patients with dementia with Lewy bodies. *Neuroscience Letters, 340*, 185–188.

McLean, C. Y., Reno, P. L., Pollen, A. A., Bassan, A. I., Capellini, T. D., Guenther, C., et al. (2011). Human-specific loss of regulatory DNA and the evolution of human-specific traits. *Nature, 471*, 216–219.

Nihei, K., & Kowall, N. W. (1993). Involvement of NPY-immunoreactive neurons in the cerebral cortex of amyotrophic lateral sclerosis patients. *Neuroscience Letters, 159*, 67–70.

Nimchinsky, E. A., Gilissen, E., Allman, J. M., Perl, D. P., Erwin, J. M., & Hof, P. R. (1999). A neuronal morphologic type unique to humans and great apes. *Proceedings of the National Academy of Sciences of the United States of America, 96*, 5268–5273.

Nimchinsky, E. A., Vogt, B. A., Morrison, J. H., & Hof, P. R. (1995). Spindle neurons of the human anterior cingulate cortex. *The Journal of Comparative Neurology, 355*, 27–37.

Petrides, M., Cadoret, G., & Mackey, S. (2005). Orofacial somatomotor responses in the macaque monkey homologue of Broca's area. *Nature, 435*, 1235–1238.

Petrides, M., & Pandya, D. N. (1994). Comparative architectonic analysis of the human and macaque frontal cortex. In F. Boller & J. Grafman (Eds.), *Handbook of neuropsychology* (pp. 17–58). Amsterdam: Elsevier.

Petrides, M., & Pandya, D. N. (2002). Comparative cytoarchitectonic analysis of the human and the macaque ventrolateral prefrontal cortex and corticocortical connection patterns in the monkey. *European Journal of Neuroscience, 16*, 291–310.

Popesco, M. C., Maclaren, E. J., Hopkins, J., Dumas, L., Cox, M., Meltesen, L., et al. (2006). Human lineage-specific amplification, selection, and neuronal expression of DUF1220 domains. *Science, 313*, 1304–1307.

Preuss, T. M., Caceres, M., Oldham, M. C., & Geschwind, D. H. (2004). Human brain evolution: Insights from microarrays. *Nature Reviews Genetics, 5*, 850–860.

Preuss, T. M., & Goldman-Rakic, P. S. (1991a). Architectonics of the parietal and temporal association cortex in the strepsirhine primate *Galago* compared to the anthropoid primate *Macaca*. *The Journal of Comparative Neurology, 310*, 475–506.

Preuss, T. M., & Goldman-Rakic, P. S. (1991b). Myelo- and cytoarchitecture of the granular frontal cortex and surrounding regions in the strepsirhine primate *Galago* and the anthropoid primate *Macaca*. *The Journal of Comparative Neurology, 310*, 429–474.

Purves, D. (1988). *Body and brain. A trophic theory of neural connections*. Cambridge, MA: Harvard University Press.

Raghanti, M. A., Conley, T., Erwin, J. M., Hof, P. R., & Sherwood, C. C. (2011a). Neuropeptide Y-immunoreactive neurons in the cerebral cortex of humans and other anthropoid primates. In: *Society for neuroscience meeting abstract planner 2011*.

Raghanti, M. A., Simic, G., Watson, S., Stimpson, C. D., Hof, P. R., & Sherwood, C. C. (2011b). Comparative analysis of the nucleus basalis of Meynert among primates. *Neuroscience, 184*, 1–15.

Raghanti, M. A., Spocter, M. A., Stimpson, C. D., Erwin, J. M., Bonar, C. J., Allman, J. M., et al. (2009). Species-specific distributions of tyrosine hydroxylase-immunoreactive neurons in the prefrontal cortex of anthropoid primates. *Neuroscience, 158*, 1551–1559.

Raghanti, M. A., Stimpson, C. D., Marcinkiewicz, J. L., Erwin, J. M., Hof, P. R., & Sherwood, C. C. (2008a). Cholinergic innervation of the frontal cortex: Differences among humans, chimpanzees, and macaque monkeys. *The Journal of Comparative Neurology, 506*, 409–424.

Raghanti, M. A., Stimpson, C. D., Marcinkiewicz, J. L., Erwin, J. M., Hof, P. R., & Sherwood, C. C. (2008b). Cortical dopaminergic innervation among humans, chimpanzees, and macaque monkeys: A comparative study. *Neuroscience, 155*, 203–220.

Raghanti, M. A., Stimpson, C. D., Marcinkiewicz, J. L., Erwin, J. M., Hof, P. R., & Sherwood, C. C. (2008c). Differences in cortical serotonergic innervation among humans, chimpanzees, and macaque monkeys: A comparative study. *Cerebral Cortex, 18*, 584–597.

Raichle, M. E., MacLeod, A. M., Snyder, A. Z., Powers, W. J., Gusnard, D. A., & Shulman, G. L. (2001). A default mode of brain function. *Proceedings of the National Academy of Sciences of the United States of America, 98*, 676–682.

Raichle, M. E., & Mintun, M. A. (2006). Brain work and brain imaging. *Annual Review of Neuroscience, 29*, 449–476.

Raichle, M. E., & Snyder, A. Z. (2007). A default mode of brain function: A brief history of an evolving idea. *NeuroImage, 37*, 1083–1090.

Raichlen, D. A., & Gordon, A. D. (2011). Relationship between exercise capacity and brain size in mammals. *PloS One, 6*, e20601.

Ramon y Cajal, S. (1909). *Histologie du système nerveux de l'homme et des vertébrés*. Paris: Maloine (pp. 859–888).

Reader, S. M., & Laland, K. N. (2002). Social intelligence, innovation, and enhanced brain size in primates. *Proceedings of the National Academy of Sciences of the United States of America, 99*, 4436–4441.

Rilling, J. K. (2006). Human and nonhuman primate brains: Are they allometrically scaled versions of the same design? *Evolutionary Anthropology, 15*, 65–77.

Rilling, J. K., Barks, S. K., Parr, L. A., Preuss, T. M., Faber, T. L., Pagnoni, G., et al. (2007). A comparison of resting-state brain activity in humans and chimpanzees. *Proceedings of the National Academy of Sciences of the United States of America*, *104*, 17146–17151.

Rilling, J. K., & Insel, T. R. (1999a). Differential expansion of neural projection systems in primate brain evolution. *Neuroreport*, *10*, 1453–1459.

Rilling, J. K., & Insel, T. R. (1999b). The primate neocortex in comparative perspective using magnetic resonance imaging. *Journal of Human Evolution*, *37*, 191–223.

Rilling, J. K., & Seligman, R. A. (2002). A quantitative morphometric comparative analysis of the primate temporal lobe. *Journal of Human Evolution*, *42*, 505–533.

Robinson, T. E., & Becker, J. B. (1986). Enduring changes in brain and behavior produced by chronic amphetamine administration: A review and evaluation of animal models of amphetamine psychosis. *Brain Research*, *396*, 157–198.

Rosenzweig, M. R., & Bennett, E. L. (1996). Psychobiology of plasticity: Effects of training and experience on brain and behavior. *Behavioural Brain Research*, *78*, 57–65.

Santos, M., Uppal, N., Butti, C., Wicinski, B., Schmeidler, J., Giannakopoulos, P., et al. (2011). von Economo neurons in autism: A stereologic study of the frontoinsular cortex in children. *Brain Research*, *1380*, 206–217.

Saxe, R. (2006). Uniquely human social cognition. *Current Opinion in Neurobiology*, *16*, 235–239.

Schenker, N. M., Hopkins, W. D., Spocter, M. A., Garrison, A. R., Stimpson, C. D., Erwin, J. M., et al. (2010). Broca's area homologue in chimpanzees (*Pan troglodytes*): Probabilistic mapping, asymmetry, and comparison to humans. *Cerebral Cortex*, *20*, 730–742.

Schoenemann, P. T. (2006). Evolution of the size and functional areas of the human brain. *Annual Review of Antropology*, *35*, 379–406.

Seeley, W. W., Carlin, D. A., Allman, J. M., Macedo, M. N., Bush, C., Miller, B. L., et al. (2006). Early frontotemporal dementia targets neurons unique to apes and humans. *Annals of Neurology*, *60*, 660–667.

Seeley, W. W., Merkle, F. T., Gaus, S. E., Craig, A. D., Allman, J. M., & Hof, P. R. (2012). Distinctive neurons of the anterior cingulate and frontoinsular cortex: A historical perspective. *Cerebral Cortex*, doi:10.1093/cercor/bhr1005.

Semendeferi, K., Armstrong, E., Schleicher, A., Zilles, K., & Van Hoesen, G. W. (1998). Limbic frontal cortex in hominoids: A comparative study of area 13. *American Journal of Physical Anthropology*, *106*, 129–155.

Semendeferi, K., Armstrong, E., Schleicher, A., Zilles, K., & Van Hoesen, G. W. (2001). Prefrontal cortex in humans and apes: A comparative study of area 10. *American Journal of Physical Anthropology*, *114*, 224–241.

Semendeferi, K., & Damasio, H. (2000). The brain and its main anatomical subdivisions in living hominoids using magnetic resonance imaging. *Journal of Human Evolution*, *38*, 317–332.

Semendeferi, K., Lu, A., Schenker, N., & Damasio, H. (2002). Humans and great apes share a large frontal cortex. *Nature Neuroscience*, *5*, 272–276.

Semendeferi, K., Teffer, K., Buxhoeveden, D. P., Park, M. S., Bludau, S., Amunts, K., et al. (2011). Spatial organization of neurons in the frontal pole sets humans apart from great apes. *Cerebral Cortex*, *21*, 1485–1497.

Sharma, Y., Xu, T., Graf, W. M., Fobbs, A., Sherwood, C. C., Hof, P. R., et al. (2010). Comparative anatomy of the locus coeruleus in humans and nonhuman primates. *The Journal of Comparative Neurology*, *518*, 963–971.

Sherwood, C. C., Broadfield, D. C., Holloway, R. L., Gannon, P. J., & Hof, P. R. (2003a). Variability of Broca's area homologue in African great apes: Implications for language evolution. *Anatomical Record*, *271A*, 276–285.

Sherwood, C. C., & Hof, P. R. (2007). The evolution of neuron types and cortical histology in apes and humans. In *The evolution of primate nervous systems*. In: T. M. Preuss & J. H. Kaas (Eds.), *Evolution of nervous systems*, (Vol. 4), (pp. 355–378). Oxford: Academic Press.

Sherwood, C. C., Hof, P. R., Holloway, R. L., Semendeferi, K., Gannon, P. J., Frahm, H. D., et al. (2005a). Evolution of the brainstem orofacial motor system in primates: A comparative study of trigeminal, facial, and hypoglossal nuclei. *Journal of Human Evolution*, *48*, 45–84.

Sherwood, C. C., Holloway, R. L., Erwin, J. M., & Hof, P. R. (2004). Cortical orofacial motor representation in Old World monkeys, great apes, and humans. II. Stereologic analysis of chemoarchitecture. *Brain, Behavior and Evolution*, *63*, 82–106.

Sherwood, C. C., Holloway, R. L., Semendeferi, K., & Hof, P. R. (2005b). Is prefrontal white matter enlargement a human evolutionary specialization? *Nature Neuroscience*, *8*, 537–538.

Sherwood, C. C., Raghanti, M. A., Stimpson, C. D., Bonar, C. J., de Sousa, A. A., Preuss, T. M., et al. (2007). Scaling of inhibitory interneurons in areas V1 and V2 of anthropoid primates as revealed by calcium-binding protein immunohistochemistry. *Brain, Behavior and Evolution*, *69*, 176–195.

Sherwood, C. C., Raghanti, M. A., Stimpson, C. D., Spocter, M. A., Uddin, M., Boddy, A. M., et al. (2010). Inhibitory interneurons of the human prefrontal cortex display conserved evolution of the phenotype and related genes. *Proceedings of the Royal Society. B, Biological Sciences*, *277*, 1011–1020.

Sherwood, C. C., Stimpson, C. D., Raghanti, M. A., Wildman, D. E., Uddin, M., Grossman, L. I., et al. (2006). Evolution of increased glia-neuron ratios in the human frontal cortex. *Proceedings of the National Academy of Sciences of the United States of America*, *103*, 13606–13611.

Shultz, S., & Dunbar, R. (2010a). Encephalization is not a universal macroevolutionary phenomenon in mammals but is associated with sociality. *Proceedings of the National Academy of Sciences of the United States of America, 107*, 21582–21586.

Shultz, S., & Dunbar, R. I. (2010b). Species differences in executive function correlate with hippocampus volume and neocortex ratio across nonhuman primates. *Journal of Comparative Psychology, 124*, 252–260.

Sokoloff, L. (1977). Relation between physiological function and energy metabolism in the central nervous system. *Journal of Neurochemistry, 29*, 13–26.

Spocter, M. A., Hopkins, W. D., Garrison, A. R., Bauernfeind, A. L., Stimpson, C. D., Hof, P. R., et al. (2010). Wernicke's area homologue in chimpanzees (*Pan troglodytes*) and its relation to the appearance of modern human language. *Proceedings of the Royal Society. B, Biological Sciences, 277*, 2165–2174.

Stephan, H., Frahm, H. D., & Baron, G. (1981). New and revised data on volumes of brain structures in insectivores and primates. *Folia Primatologica, 35*, 1–29.

Stimpson, C. D., Tetreault, N. A., Allman, J. M., Jacobs, B., Butti, C., Hof, P. R., et al. (2011). Biochemical specificity of von economo neurons in hominoids. *American Journal of Human Biology, 23*, 22–28.

Striedter, G. F. (2005). *Principles of brain evolution.* Sunderland, MA: Sinauer Associates, Inc. Publishers.

Suddendorf, T., Addis, D. R., & Corballis, M. C. (2009). Mental time travel and the shaping of the human mind. *Philosophical Transactions of the Royal Society. B, Biological Sciences, 364*, 1317–1324.

Suddendorf, T., & Whiten, A. (2001). Mental evolution and development: Evidence for secondary representation in children, great ages, and other animals. *Psychological Bulletin, 127*, 629–650.

Toga, A. W., & Thompson, P. M. (2003). Mapping brain asymmetry. *Nature Reviews Neuroscience, 4*, 37–48.

Uddin, M., Goodman, M., Erez, O., Romero, R., Liu, G., Islam, M., et al. (2008a). Distinct genomic signatures of adaptation in pre- and postnatal environments during human evolution. *Proceedings of the National Academy of Sciences of the United States of America, 105*, 3215–3220.

Uddin, M., Opazo, J. C., Wildman, D. E., Sherwood, C. C., Hof, P. R., Goodman, M., et al. (2008b). Molecular evolution of the cytochrome c oxidase subunit 5A gene in primates. *BMC Evolutionary Biology, 8*, 8.

Uddin, M., Wildman, D. E., Liu, G., Xu, W., Johnson, R. M., Hof, P. R., et al. (2004). Sister grouping of chimpanzees and humans as revealed by genome-wide phylogenetic analysis of brain gene expression profiles. *Proceedings of the National Academy of Sciences of the United States of America, 101*, 2957–2962.

Vaishnavi, S. N., Vlassenko, A. G., Rundle, M. M., Snyder, A. Z., Mintun, M. A., & Raichle, M. E. (2010). Regional aerobic glycolysis in the human brain. *Proceedings of the National Academy of Sciences of the United States of America, 107*, 17757–17762.

Vincent, J. L., Patel, G. H., Fox, M. D., Snyder, A. Z., Baker, J. T., Van Essen, D. C., et al. (2007). Intrinsic functional architecture in the anaesthetized monkey brain. *Nature, 447*, 83–86.

von Economo, C. (1926). Eine neue Art Spezialzellen des Lobus cinguli and Lobus insulae. *Zschr Ges Neurol Psychiat, 100*, 706–712.

Wittenberg, G. M., & Wang, S. S.-H. (2007). Evolution and scaling of dendrites. In G. Stuart, N. Spruston & M. Häusser (Eds.), *Dendrites* (pp. 43–67). New York: Oxford University Press.

M. A. Hofman and D. Falk (Eds.)
Progress in Brain Research, Vol. 195
ISSN: 0079-6123

CHAPTER 12

Hominin paleoneurology: Where are we now?

Dean Falk*

School for Advanced Research, Santa Fe, NM, USA
Department of Anthropology, Florida State University, Tallahassee, FL, USA

Abstract: Hominin paleoneurology is the subfield of paleoanthropology that investigates brain evolution in human ancestors. For over a century, paleoneurologists have focused on analyses of cranial capacities (as surrogates for brain size) and endocranial casts (endocasts), which are prepared from the interiors of fossilized braincases and reproduce details of external brain morphology. This review discusses recent improvements in our understanding of hominin brain evolution in terms of brain size, sulcal patterns, and cortical shape features. To the extent possible, the evolution of neurological reorganization is assessed in light of findings from paleoneurology. In order to make inferences about cognitive evolution, paleoneurologists interpret their data within a framework that incorporates behavioral information from comparative primatological studies and findings from comparative neuroanatomical and medical imaging investigations. Advances in our knowledge about the evolution of the prefrontal cortex (Brodmann's area 10) provide an example of a productive synthesis of comparative neuroanatomical and behavioral research with investigations of the fossil record of hominin endocasts.

Keywords: brain shape; brain size; endocast; lunate sulcus; neurological reorganization; paleoneurology; sulcal patterns.

Introduction

Hominin paleoneurologists study fossilized skulls and casts of their braincases (endocasts) to investigate the evolution of the brain and cognition in our ancestors. Although endocasts sometimes occur naturally, they are traditionally prepared by casting the insides of braincases with latex, or some other molding material. In recent years, however, it has become common to acquire "virtual endocasts" electronically by using 3D imaging techniques, such as computed tomography (Falk, 2004). Virtual endocasts are easier to reconstruct, manipulate, and measure than traditionally prepared ones.

*Corresponding author.
Tel.: +1-850-644-7016; Fax: +1-850-645-3200
E-mail: dfalk@fsu.edu

DOI: 10.1016/B978-0-444-53860-4.00012-X

Endocasts reproduce a good deal of information about the brain, including its general shape, and details of some of its associated blood vessels, cranial nerves, and cranial sutures. With luck, endocasts also reproduce information about the convolutions of the brain that were imprinted on the inner walls of the braincase during life. The convolutions, or folds of gray matter on the brain's surface, consist of bulges (gyri) and the grooves (sulci) that separate them. Sulcal patterns have been a focus of hominin paleoneurology for over a century, although the amount of information about them that is reproduced on hominin endocasts is usually quite limited, as described below. Partly because of this, and partly for historical reasons (Falk, 2009b), hominin paleoneurology is a highly contentious field (Falk, 2011).

Brain size, which is a less controversial topic than hominin sulcal patterns, is also an important parameter for assessing hominin brain evolution. A longstanding debate continues about the relative importance of the evolution of brain size versus that of the internal reorganization of the brain's connections, components, and neurochemistry (neurological reorganization). Below, I describe current findings about hominin brain evolution that paleoneurologists have gleaned from comparisons of the skulls and endocasts of apes and hominins. The evolution of brain size is discussed first, followed by speculation about the mode and tempo of neurological reorganization as indicated by sulcal patterns in two parts of the brain and certain details of brain shape.

Brain size

Brain size is estimated by measuring the cranial capacities of fossil skulls or, alternatively, the volumes of their endocasts (in cubic centimeters, cm^3). Ideally, cranial capacities should be decreased by a corrective factor to compensate for the volume of fluids and meninges that occupy the braincase along with the brain. It is quite common, however, for cranial capacities to be used without correction as proxies for brain size. By analyzing cranial capacities and estimates of body size (often based on postcranial remains), earlier researchers hypothesized that both the absolute mass of the brain and its size relative to body mass (relative brain size, RBS) increased independently during the evolution of the major clades of primates, as well as during the evolution of other mammals (Jerison, 1973; Radinsky, 1979).

More recent quantitative analyses have verified that selective pressures for enlarged brains began early in primate evolution but have also revealed that brain size decreased independently in some branches of old world monkeys, new world monkeys, and strepsirhines (Montgomery et al., 2010). (As an aside, Montgomery et al. (2010) analyzed brain and body size in the tiny type specimen for *Homo floresiensis* (LB1) and concluded that the data fit within the broader context of primate phylogeny (Falk et al., 2009).) Larger-bodied primate species tend to have smaller measures of RBS than smaller-bodied ones, although there are exceptions such as extremely large-brained *Homo*. Partly for this reason, numerous analytical techniques and indices that "subtract" allometric scaling associated with body size from brain size (e.g., encephalization quotient, EQ; index of progression, IP) have been developed to quantify the extent of encephalization in mammals, including nonhuman and human primates (see Falk, 2007a for review).

Recent studies suggest that brain mass is more indicative of advanced cognitive abilities in primates than measures that control for body size, such as IP and EQ. Thus, "the functional integration of different brain regions is so strong that the brain as a whole is a relevant unit for cognitive performance" (Deaner et al., 2007:121; Herculano-Houzel, 2009, Chapter 15). The emerging preference for data based on brain mass is not surprising in light of problems inherent in constructing and using indices that control for body size. These problems include difficulties in identifying appropriate reference groups for baseline data, challenges in selecting exponents for regression equations, and results that (because of artifacts that are inherent in the methodology) tend to overestimate indices for smaller-bodied species and underestimate them for larger-bodied ones (see Falk, 2007a for details).

Another emerging trend is a preference for absolute over RBS as the best indicator of advanced cognition. For example, "Only in terms of absolute mass and the rate of change in absolute mass has the increase in brain size been exceptional along the terminal branch leading to humans. Once scaling effects with body mass have been accounted for the rate of increase in relative brain mass remains high but is not exceptional" (Montgomery et al., 2010:11). Researchers studying neurogenesis have reached a similar conclusion: "The most likely brain alteration resulting from selection for any behavioral ability may be a coordinated enlargement of the entire nonolfactory brain" (Finlay and Darlington, 1995:1578).

Human brains are large

Human cranial capacities (and brains) are, by far, the largest of all the primates. As shown in Fig. 1, when extreme outliers are excluded, human cranial capacities vary from around 1100 to 1700 cm^3, and they are completely separated from those of the great apes. (One researcher who included outliers reported cranial capacities for normal humans that ranged from 790 to 2350 cm^3 (Dart, 1956)!). A figure of 1350–1400 cm^3 commonly appears in the literature as an estimate for the mean cranial capacity in living people. This is about three times the size of the mean estimate of 450 cm^3 for australopithecines (Falk et al., 2000). Various workers have also shown that the volumes of the brains (and, separately, neocortices) of living people are, on average, three times the size predicted for nonhuman primates that are scaled to the same body size as humans (Passingham, 1973, 1975; Stephan et al., 1970). This frequently cited observation is consistent with the following conclusion based on comparative behavioral and neuroanatomical data: "The most practical measure for distinguishing intelligence and predicting the presence of humanlike mental skills in hominid fossils is absolute brain size" (Gibson, 2001:92). So whether or not one examines absolute brain size or RBS,

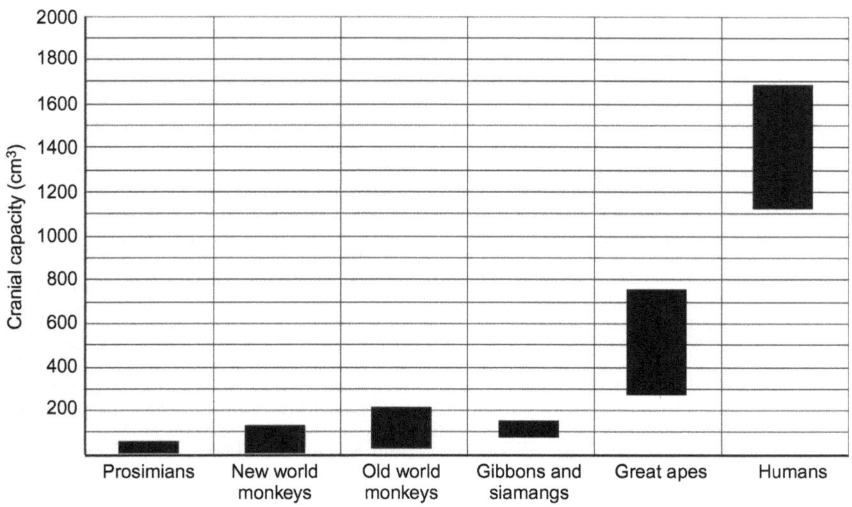

Fig. 1. Cranial capacities as approximations of the ranges of brain sizes in extant primates. Reproduced from Falk (2007a) with permission.

it appears that the mass of the brain increased approximately threefold in the lineage leading from *Australopithecus* to extant *Homo*. Further, the old adage that absolute brain size suddenly "took off" in *Homo* around 2 million years ago needs revision in light of relatively new discoveries of fossil hominins, which suggest that brain size began to increase considerably earlier in the *Australopithecus* ancestors of *Homo* (Falk, 2004, 2007a).

Conclusion regarding brain size

From the above brief discussion, it is understandable that numerous researchers advocate brain size as the most important parameter that changed during hominin brain evolution. However, brains evolved not only in size but also in their neurological organization. Other researchers, therefore, favor neurological reorganization as the most important aspect of hominin brain evolution. As Stephen Jay Gould observed a decade ago, the debate about the relative importance of brain size versus neurological reorganization is based on a false dichotomy (Gould, 2001). Both are important. Studying neurological reorganization is trickier than assessing brain size evolution because most of the changes related to internal wiring, relative sizes of different parts of the brain, and neurochemistry are not revealed in braincases or on endocasts. Nonetheless, investigators are able to obtain hints about neurological reorganization from the sulcal patterns and shape features that, with luck, are reproduced on hominin endocasts.

Neocortical reorganization of sulcal patterns

Human cerebral cortices have a greater number of sulci than those of apes, which is associated with several factors including allometric scaling of the cortical surface relative to brain volume (Jerison, 1973, 1975), an increase in the number of cortical areas that developed during primate brain evolution (Kaas, 2000; Kaas and Preuss, 2008) and constraints related to the evolution of cortical wiring (Hofman, 2001, Chapter 18). The additional sulci in humans are mostly unnamed (Connolly, 1950). Although ape and human brains share most of their named sulci and fissures, the configuration of sulci that appear on the external cortical surface in two regions of the brain is derived in humans compared to apes (and monkeys): (1) the caudal lateral border of the orbitofrontal cortex (Fig. 2) and (2) the rostral border of primary visual cortex (V1 or BA 17). Because cortical reorganization in these parts of the human brain is associated with changed sulcal patterns, the relevant sulci have been investigated on ape and hominin endocasts with an eye toward gaining insight into the pattern and timing of cortical reorganization during hominin evolution, as well as its relationship to brain enlargement. However, the discussion of sulcal patterns on hominin endocasts has been, and continues to be, highly controversial (Holloway, 2008), partly for historical reasons (Falk, 2011).

Sulcal pattern difference 1

As detailed by Connolly (1950), the lateral border of the caudal part of the frontal lobe of all genera of great apes is consistently incised by a fronto-orbital sulcus (*fo*) that courses caudally on the orbital surface toward the temporal pole (see diagram for *Pan* in Fig. 2). This is never the case for human brains. Instead, human brains typically manifest a sulcal pattern in which two rami of the Sylvian fissure (*R'*, *R*) delimit the rostral and caudal boundaries of the *pars triangularis* (BA 45) (which, in the left hemisphere, is part of Broca's speech area consisting of BA 45 and BA 44, see diagram for *Homo* in Fig. 2). Although sulci often fail to delimit cytoarchitectonic regions reliably (Amunts et al., 2007; but see Fischl et al., 2008), these two branches of the Sylvian fissure

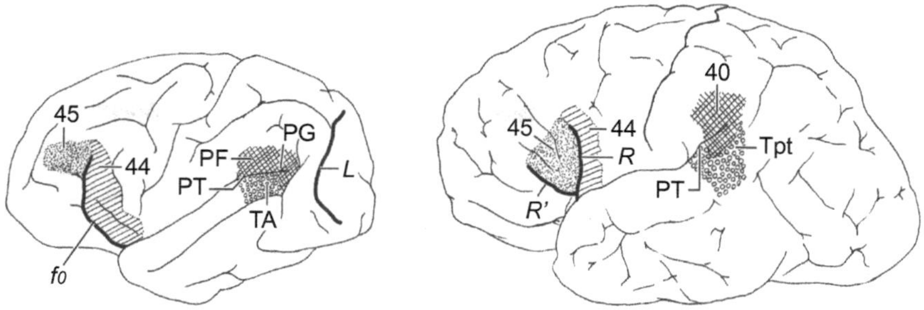

Fig. 2. The classic language areas in left hemispheres of humans and their proposed homologs in chimpanzees. In humans, Brodmann's area (BA) 44 and BA 45 constitute Broca's speech area, while PT (planum temporale), Tpt (temporoparietal), and BA 40 are part of Wernicke's receptive area for language. Proposed homologs of human BA 40 and Tpt with chimp areas PF/PG (inferior parietal lobule) and TA (part of temporal lobe), respectively, are based on cytoarchitectonic and functional similarities and are tentative. The sulci associated with Broca's speech area in the left hemisphere of humans form a distinctive pattern (as do the sulci in the same position on the right) that differs from the sulcal pattern in the frontal lobes of monkeys and apes. The fronto-orbital (*fo*) and lunate (*L*) sulci of the chimpanzee brain and the two sulci delimiting the *pars triangularis* (*R'*, *R*, anterior horizontal and anterior ascending rami, respectively, of the Sylvian fissure) in the human frontal lobe are thickened for illustrative purposes. See Falk (2007b) for details. Figure modified after Falk (2007a) and Schenker et al. (2008); © Dean Falk, reproduced with permission.

bear a predictable relationship to the free surfaces of BA 45 and BA 44 in human brains (Amunts et al., 1999). This indicates that they are potentially good landmarks when studying hominin endocasts (which reproduce only the surface of the cortex): Thus, "there are regions, i.e., the free surfaces of the triangular and opercular parts, in which the probability is very high of localizing areas 45 and 44, respectively" (Amunts et al., 1999:339).

Connolly hypothesized that *fo* does not appear on the lateral surface of human brains because it was displaced caudally by the opercularization of the frontal lobe as brain size increased during hominin brain evolution (Connolly, 1950:330). According to Connolly, *fo* became buried within the brains of humans, where it became part of the anterior limiting sulcus of the insula. The distinction between the sulcal patterns in this part of the frontal lobes of apes and humans is consistent and has paleoneurological significance because of the association of the human pattern with neurological reorganization related to language (Falk, 1983; Tobias 1987). Significantly, an apelike *fo* is present on the ape-sized natural endocast of Taung (the type specimen for *Australopithecus africanus*) (Dart, 1929; Falk, 1980, 2009b).

Figure 2 has an important implication for paleoneurology. In human brains, the rostral part of Broca's speech area (area 45) and the area that borders it ventrally (area 47, not labeled in Fig. 2) together form a slight bulge, which has been called "Broca's cap." Some workers equate this with a bulge that appears in the same general region on ape brains. Cytoarchitectonic evidence, however, reveals that these bulges are not equivalent. In chimpanzees, the bulge is formed by area 44 and sometimes part of area 45 (Sherwood et al., 2003) instead of areas 45 and 47, and of course, apes do not have speech (Falk, 2007b). One should therefore be cautious about inferring that a bulge in this general location on an ape brain or on a small early hominin endocast is equivalent to Broca's cap of humans. What is needed to interpret an early hominin endocast in this region is information about the precise sulcal pattern. Does the endocast have an apelike *fo*? If not, does it reproduce two sulci that suggest the presence of a *pars triangularis*? One cannot always determine the answer to these questions from hominin endocasts, but they should be asked.

Sulcal pattern difference 2

The second sulcal pattern that is relevant for hominin paleoneurology concerns the lateral representation of the primary visual cortex (V1), which is relatively smaller and located noticeably more caudally in humans than the homologous area in monkeys and apes (Connolly, 1950). In nonhuman anthropoids, the rostral border of V1 is approximated by a large crescent-shaped sulcus (*L* in *Pan*, Fig. 2) formerly called the *Affenspalte* (ape sulcus). At the beginning of the twentieth century, Grafton Elliot Smith hypothesized that V1 of humans was bordered by a homologous sulcus, the name of which he changed to "lunate sulcus" (*L*) in keeping with its recognition in humans (Smith, 1903, 1904a,b). Long ago, Smith hypothesized that, as hominin brains enlarged and evolved, the lateral representation of V1 and its bordering *L* were displaced caudally by expansion of the adjacent parieto–occipito-temporal association cortices.

Smith's protégé, Raymond Dart, picked up on his hypothesis in 1925 when he identified and illustrated what he thought was a caudally displaced *L* on the ape-sized endocast of Taung (Fig. 3). Based solely on this identification, Dart concluded that Taung's brain was neurologically advanced toward a human condition because it had relatively expanded nearby association cortices that displace *L* caudally (Dart, 1925). Unfortunately, Dart had incorrectly identified the lambdoid suture of the skull (which had been reproduced on the endocast) as *L*—an identification that Dart's colleagues, including Smith, were skeptical about (Falk, 2009b). Unpublished materials in the Raymond Dart collection of the archives of the University of Witwatersrand reveal that Dart identified 14 additional sulci on the Taung endocast in addition to the two that he identified in his 1925 publication, and that he

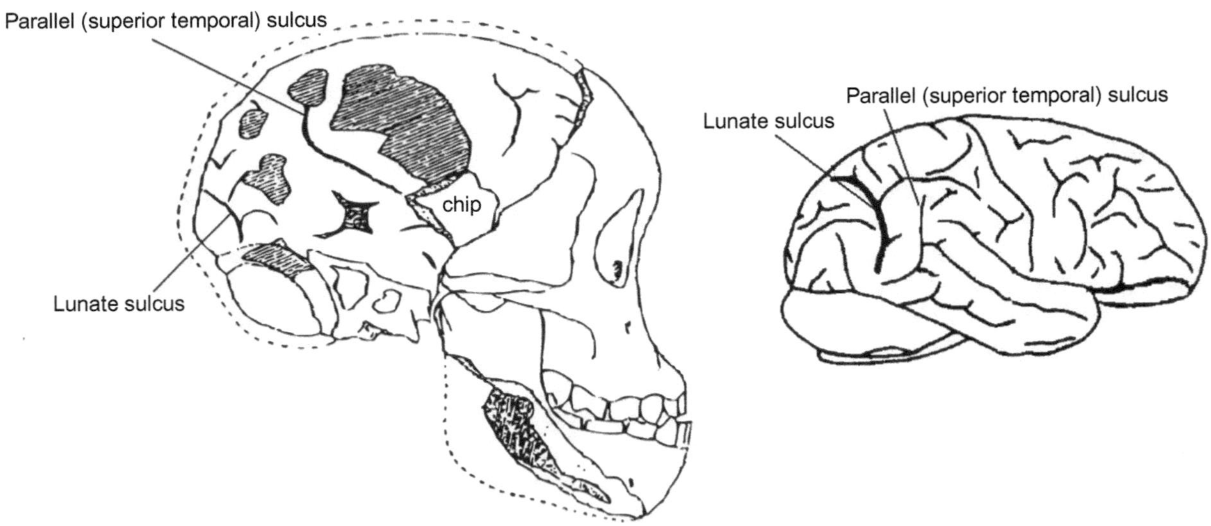

Fig. 3. Dart's (1925) illustration of the right side of the Taung natural endocast, facial fragment, and jaw compared to the right side of a chimpanzee brain. The feature Dart identified as the lunate sulcus is actually the lambdoid suture. Notice that the lunate sulcus on the chimpanzee brain (thickened for illustrative purposes) is more rostally located than the suture that Dart misidentified as the lunate sulcus on Taung. Reproduced from Falk (2011) with permission.

knew he had a lunate sulcus problem (Falk, 2009b). In a previously unpublished illustration (reproduced in Falk, 2009b), Dart illustrated the lambdoid suture on the Taung endocast for the first (and perhaps only) time and hypothesized that a portion of Taung's *L* coursed directly underneath it (Falk, 2009b). Although most of his contemporaries did not accept Dart's identification of *L*, his incorrect interpretation of that feature on the Taung endocast has been used in recent years to bolster the hypothesis that caudal regions of the hominin brain became reorganized prior to reorganization of other areas (mosaic brain evolution) and before evolutionary brain expansion in hominins (Barton and Harvey, 2000; Holloway, 2001; Holloway and Kimbel, 1986). We now know from his unpublished manuscript that Dart, on the other hand, favored a global rather than mosaic view of neurological reorganization (Falk, 2009b).

Current findings regarding the lunate sulcus

Through the years, lunate sulci of humans have been described as shorter, more variable in their configurations, and appearing less frequently than in the other anthropoids (Connolly, 1950; Ono et al., 1990). A recent study provides welcome, if unsurprising, quantitative support for the observation that humans have relatively reduced primary visual cortex compared to other anthropoids and that the volume of V1 in apes is predictable from the position of *L* (de Sousa et al., 2010; see also Fischl et al., 2008). The authors concluded that "the position of the lunate sulcus on fossil endocasts may provide information about brain organization" (de Sousa et al., 2010). However, another study that used high resolution MRI to assess the presence/absence of *L* in 110 adult humans revealed that the rare occurrences of sulci in, or near, the occipital lobes that superficially resemble those of ape lunate sulci were discontinuous beneath the surface and did not approximate the rostral border of V1 (Allen et al., 2006). In other words, there is little, if any, evidence in support of the view that contemporary humans have lunate sulci. It, thus, appears that *L* was lost at some undetermined time during hominin brain evolution. If so, a *lack* of lunate sulci on the brains of hominins, although difficult to verify because this sulcus does not reproduce well on hominoid, including human, endocasts (Le Gros Clark et al., 1936; Connolly, 1950), is the derived condition associated with cortical reorganization.

The only australopithecine endocast that is currently hypothesized to reproduce an "unmistakenly posterior placement" of *L* is that from Stw 505 (*A. africanus*) from Sterkfontein (Holloway et al., 2004). Because of this one endocast, the authors conclude that neurological reorganization occurred in caudal parts of early hominin brains prior to reorganization in other parts of the brain, and prior to brain enlargement. For various reasons, I am unconvinced that the feature identified as *L* on the Stw 505 endocast is that sulcus. If *L* was lost during human brain evolution, as seems likely from Allen et al. (2006), the hypothesis of a derived caudally located *L* in ape-sized australopithecine brains requires that this sulcus was an ancestral retention that was displaced caudally from an apelike location in conjunction with a (derived) differential expansion of association cortices just rostral to it, but with no overall increase in brain size. Another requirement is that, after being displaced caudally in small-brained hominins, *L* was subsequently lost in conjunction with the increase in overall brain size in *Homo*. To me, the hypothesis that *L* was located relatively caudally in early hominins is not parsimonious and, so far, it lacks convincing paleoneurological support from endocasts.

The evolution of cortical sulci

What alternative hypothesis might explain the evolution (or devolution) of the lunate sulcus?

A broad approach to this conundrum is to ask why the cerebral cortex remains smooth in smaller-brained species, yet becomes highly convoluted in larger-brained ones (Van Essen, 1997). The evolution of mammalian, including primate, cortical folding patterns probably entailed many factors, including alterations in the durations of neurogenesis (Finlay and Darlington, 1995). As noted, it has also been associated with optimization of neurological wiring patterns (Hofman, 2001, Chapter 18; Kaas, 2000; Kaas and Preuss, 2008) and an increase in the number of cortical areas with increasing brain size (Kaas and Preuss, 2008; Preuss, 2007a,b). At an allometric level, "convolutions increase with brain size primarily because the expansion of the cortical sheet outpaces the minimal area needed to envelop the underlying cerebral volume" (Van Essen, 1997:314; see also Jerison, 1973). Van Essen's tension-based theory of the formation of convolutions and sulci during brain development takes these various factors into account (Van Essen, 1997, 2007; Van Essen and Dierker, 2007) and is helpful for elucidating how L might have been lost during hominin evolution.

Van Essen hypothesizes that the development of gyral and sulcal patterns during prenatal and perinatal development is mediated by mechanical tensions along the axons as cortical–cortical connections are formed (Fig. 4). Thus, as neurons

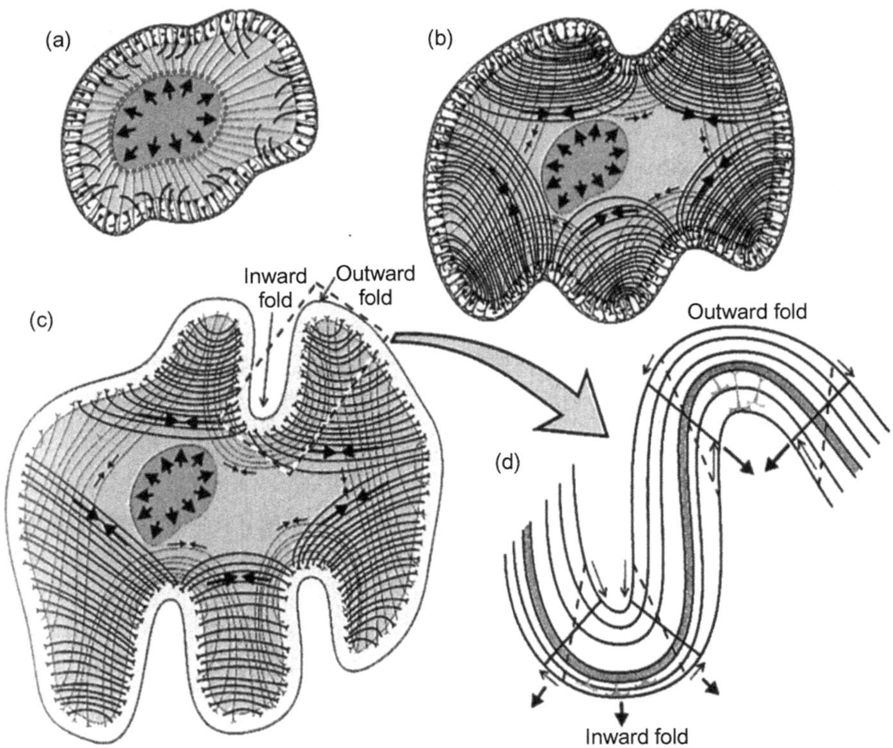

Fig. 4. In Van Essen's illustration of his tension-based theory of how convolutions develop, tightly interconnected regions of the cortical surface begin to swell and change the external shape of the cortex before the sulci that separate them are completely formed. Reproduced from Van Essen (1997) with permission. (For color version of this figure, the reader is referred to the Web version of this chapter.)

migrate to the cortical plate and make connections there, tensions between strongly interconnected areas pull them together creating an externally protruding gyrus. Sulci, on the other hand, are inward folds that separate regions that have weak interconnections. Accordingly, "consistency in folding may reflect consistent patterns of connectivity among nearby areas" (Van Essen and Dierker, 2007: 212). Van Essen's hypothesis may also shed light on sulcal patterns that are relatively variable. "If cortical sulcal patterns are reflective of the tension of subcortical and corticocortical axonal projections (Van Essen, 1997), then it may be that the variability in the location of a cortical area relates to the degree of heterogeneity in its pattern of connectivity" (Fischl et al., 2008: 1978). An important implication of Van Essen's model is that the shape of the brain's surface changes as gyri and sulci develop and that these changes may begin to take place due to tensions from areas that are expanding (becoming strongly interconnected) (Fig. 4b) *before* sulci are fully formed (Fig. 4c). As discussed below, this possibility may have important implications for hominin paleoneurology.

Van Essen's hypothesis suggests that *L* may have been lost during hominin evolution because of changing patterns of cortical interconnections associated with the posterior and medial displacement of visual cortex. The lunate sulcus in monkeys and apes separates strongly interconnected visual areas from bordering association cortices, with which the former are relatively weakly connected (Van Essen, 1997). It seems likely that, as hominin brains increased in size and became neurologically reorganized, the strength of the interconnections between visual areas and the bordering association cortices increased in conjunction with the increase in the absolute and relative size of the latter (Falk and Gibson, 2001; de Sousa et al., 2010). The lunate sulcus may, thus, have disappeared in the ancestors of humans because the regions it formerly separated became more strongly interconnected with each other as the cortex reorganized.

Summary and conclusion regarding sulcal patterns

The options are very limited for gleaning information about the evolution of cortical folding patterns from fossil hominin endocasts. Ape and human brains consistently differ in their named sulci in only two parts of the cerebral cortex. In both cases, expanded association cortices appear to have displaced adjacent regions caudally as the cerebral cortex enlarged and reorganized during hominin brain evolution: In the frontal lobes, two new sulci (*R'*, *R*) appeared in humans that approximate the borders of the *pars triangularis* of Broca's speech area in the left hemisphere (and its homologous area in the right hemisphere) as the apelike *fo* was displaced caudally beneath the exterior surface of the brain. The second area entailed enlargement of the parieto–occipito-temporal association cortices, which displaced the primary visual cortex caudally. Unlike the first region, however, the evolution of this part of the hominin cerebral cortex entailed the loss of a major sulcus, *L*, as the primary visual cortex became more strongly interconnected with bordering association cortices (Allen et al., 2006; de Sousa et al., 2010).

Unfortunately, *L* does not reproduce well on endocasts from either apes or humans (Connolly, 1950). The fact that sulcal patterns of humans are derived both rostrally and caudally suggests that hominin brain evolution entailed global reorganization of the cerebral cortex (Dart, 1929; Falk, 2009b), contrary to the assertion of "mosaic brain evolution" in which the caudal portion of the brain is asserted to have evolved before other regions (Barton, 2001; Holloway, 2001; de Sousa et al., 2010). Ever since Dart misidentified the lambdoid suture for *L* on the Taung endocast (Dart, 1925, 1929; Falk, 2009b), assessment of the presence and location of *L* on early hominin endocasts has been muddied by paleopolitics (Falk, 2011). Although *fo* reproduces better on ape endocasts than *L*, it has received considerably less attention in the paleoneurological literature (Falk, 2009b). It would be wonderful if hominin endocasts reproduced crystal clear sulcal patterns, but they do not. They do, however, reveal a good deal of information about shape features of the cerebral cortex that appears to be associated with cortical reorganization.

Neocortical reorganization of endocast (brain) shape

Although a good deal of attention has been given to the evolution of brain size, and some attention has been focused on the evolution of sulcal patterns, researchers are just beginning to apply imaging and geometric morphometric techniques to the study of brain shape changes during hominoid ontogeny and phylogeny (Chapter 13). Results show that brain shapes of humans and chimpanzees (as reproduced on virtual endocasts) are distinctive for each species at birth, and for each, they continue to change dynamically during infancy and childhood (Dosenbach et al., 2010; Neubauer et al., 2010; Ventrice, 2011). Human infants experience an early shape globularization of their brains that does not occur in chimpanzees before or after birth, which has been interpreted as a uniquely human trait that may be related to the evolved cortical reorganization that underpins derived human behaviors and cognitive abilities (Neubauer et al., 2010). Asymmetries in the gross brain shape of humans have also been associated with such traits, including language and handedness (see below).

Brains of anthropoid primates are functionally lateralized, which is superficially manifested in gross difference in the shapes of the two cerebral hemispheres. As is well known, cerebral lateralization is especially marked in humans, in whom the neurological substrates for language and right-handedness usually depend largely on the left hemisphere, whereas processing of more holistic endeavors such as musical activities is largely the domain of the right hemisphere (Falk, 2010; Chapter 6). In keeping with this, shape asymmetries of the whole brain, known as petalias, are more dramatic in humans than in the other higher primates (LeMay et al., 1982). Additionally, human brains are derived compared to apes and early hominins in the gross shapes of certain parts of their brains (Falk et al., 2000).

Petalias

Asymmetrical brain shape is the norm for adult humans, in whom the most frequent petalia pattern, known as the Yakovlevian torque, combines a more protuberant and wider right frontal lobe with a more protuberant and wider left occipital lobe (Fig. 5) (Galaburda et al., 1978; Chiu and Damasio, 1980; LeMay, 1984; Toga and

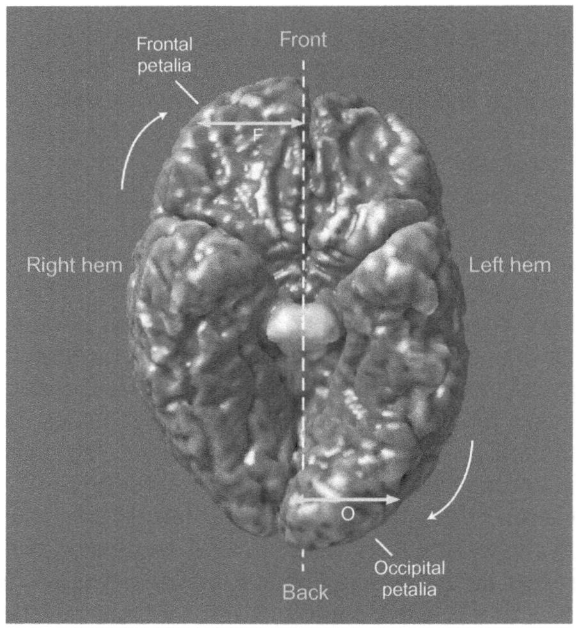

Fig. 5. The most common petalia pattern in the human brain, consisting of a right frontal and left occipital petalia. This rendering of the inferior surface of a human brain is from a magnetic resonance imaging (MRI) scan that has been exaggerated to illustrate the typical human petalia pattern and Yakovlevian torque. Reproduced from Toga and Thompson (2003), courtesy of Dr. Arthur W. Toga and Dr. Paul M. Thompson, Laboratory of Neuro Imaging at UCLA. (For color version of this figure, the reader is referred to the Web version of this chapter.)

Thompson, 2003; Kivilevitch et al., 2010). This right frontal, left occipital petalia pattern is statistically correlated with right-handedness, whereas the reversed left frontal, right occipital petalia pattern is associated to some degree with left-handedness (LeMay, 1977; LeMay et al., 1982). The latter is especially true for left-handed women, particularly if the degree of the petalias is extreme (Bear et al., 1986). As noted, these petalia patterns exist to a lesser degree in nonhuman primates and early hominins (LeMay et al., 1982). Accordingly, the relatively extreme torques in human brains is viewed as the result of a prolonged evolutionary trend for brain lateralization (Falk, 2009a, 2010).

Ventrice (2011) has recently observed that the ontogenetic development of human petalia patterns is a dynamic process during which shape torques change directions. According to Ventrice, brain shape of infants and juveniles are typically characterized by left frontal and right occipital petalias, which is the reverse of the most common adult pattern. This surprising new finding needs confirmation, but should be kept in mind when interpreting petalia patterns from australopithecine infants or juveniles such as the Dikika infant (*Australopithecus afarensis*) and Taung juvenile (*A. africanus*). On a technical note, because of shape torques, the midline of brains meanders a bit and the common practice of mirror-imaging missing parts of hominin endocasts around an estimated midsagittal plane is bound to introduce some error in both shape and brain size estimates. One way to minimize reconstruction error is to use automated computer programs for establishing the most optimal midsagittal plane (Falk and Clarke, 2007).

Shape of the lobes

As noted, Van Essen's tension-based theory of the formation of convolutions suggests that evolutionary changes in the patterns of neurological connections influenced local shapes of the cerebral cortex, which paved the way for the formation of sulci separating less interconnected regions (as well as the reverse process in which sulci may have disappeared as previously separated areas became increasingly interconnected) (Van Essen, 1997). Findings regarding sulcal patterns and endocast shapes of two different genera of fossil hominins (*Paranthropus* and *Australopithecus*) that lived contemporaneously in Africa between approximately 2.6 and 1.9 million years ago are consistent with this hypothesis (Berger et al., 2010; Falk et al., 2000). As far as I have been able to determine from their endocasts, the brain size and sulcal patterns of both groups were similar and apelike (Falk, 2009b; Falk et al., 2000). Brain shape, however, differed markedly between the two genera. The robust australopithecines (*Paranthropus*) are thought not to have been directly ancestral to *Homo*, which is consistent with certain apelike features of their endocasts compared to those of *Australopithecus*—the genus that is believed to have given rise to *Homo* (Berger et al., 2010).

Endocasts of *Paranthropus* were primitive in their relatively pointed frontal lobes (when seen in dorsal view) compared to *Australopithecus*, which had frontal lobes that were more squared off at the rostral lateral borders (Falk et al., 2000) (Fig. 6). Consequently, the overall perimeter of *Paranthropus* endocasts has a teardrop shape compared to *Australopithecus* endocasts. The orbital surfaces of the frontal lobes of *Australopithecus* are also expanded ventrally compared to the flatter orbital surfaces of *Paranthropus*. It is noteworthy that the frontal lobes of *Australopithecus* are elongated rostrally in a region that corresponds to Brodmann's area 10 (BA 10) in both apes and humans. When viewed basally, *Australopithecus* endocasts have temporal poles that are expanded and pointed rostrolaterally compared to the relatively stubby temporal poles of *Paranthropus* and African apes (for illustrations and further details, see Falk et al., 2000).

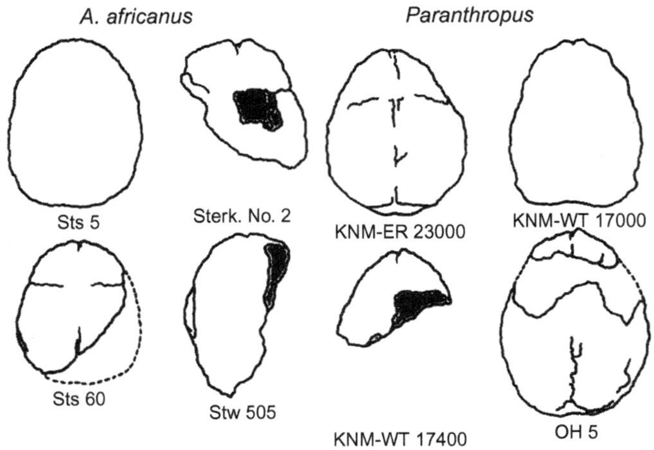

Fig. 6. Shape differences between similarly sized endocasts of *Paranthropus* and *Australopithecus africanus*, seen from dorsal view (with the frontal lobes located superiorly). Specimen numbers are next to endocasts, some of which are fragmentary. Compared to *Australopithecus*, endocasts of *Paranthropus* have more pointed frontal lobes, which give the overall perimeter of their endocasts a teardrop shape. Rather than being pointed, the frontal lobes of *Australopithecus* are broader, with sides that are more squared off laterally. Reproduced with permission from Falk et al. (2000).

Raymond Dart's observations of the Taung endocast that recently came to light also suggest that the shape of the prefrontal cortex was derived toward a human condition in *A. africanus* compared to apes (Dart, 1929; Falk, 2009b). Additionally, Dart observed that the occipital pole of *Australopithecus* projected caudally relative to the cerebellar pole, which is another feature that may be derived in *Australopithecus* compared to *Paranthropus* and apes (Falk et al., 2009). (For the sake of completion, Dart also believed that the caudal lateral portion of Taung's temporal lobe was expanded and somewhat derived toward a human condition, although I have not compared this feature in different australopithecines (Falk, 2009b).

Interestingly, *Paranthropus* endocasts reproduce an enlarged occipital/marginal venous sinus caudally, as do most, if not all, of the scorable specimens belonging to *A. afarensis* (Falk et al., 1995). This feature has been observed in Taung (Tobias and Falk, 1988) and possibly in a fragmentary occipital fragment (Stw 187a) (Lockwood and Tobias, 2002) among the available *A. africanus* specimens. Since brain sizes were very similar in the two genera of australopithecines (Falk et al., 2000), their different blood drainage patterns, as well as the derived cortical shape features of *Australopithecus*, were not the result of allometric scaling. With respect to the latter, and consistent with Van Essen's hypothesis, it is reasonable to speculate that certain neurological regions may have become more interconnected and derived toward the human condition in *Australopithecus*, thus causing the noted shape changes, although these had not become pronounced enough to cause changes in their sulcal patterns—at least to an extent that can presently be inferred from endocasts.

Conclusion regarding endocast (brain) shape

Most of the information about hominin brain evolution that paleoneurologists can reliably glean from endocasts is limited to details about brain size and the gross shape of the cerebral hemispheres (including asymmetries) and lobes of the brain. Unfortunately, although sulcal patterns in the caudal lateral part of the frontal lobes and near the rostral lateral borders of primary visual cortex are potentially informative, they do not reproduce well on hominin endocasts and have, therefore, been subject to intense controversy (Falk, 2011). In other words, apart from gross size and shape of the brain, endocasts offer few hints about the trajectory of brain evolution in hominins.

Comparative neuroanatomical studies: Implications for hominin paleoneurology

In order to assess more fully the nature of the evolved neurological substrates that underpin human cognitive abilities, paleoneurologists must turn to findings from comparative neuroanatomy, neurochemistry, genetics, and functional imaging studies such as those discussed in this volume (Falk, 2010). Although a thorough review of the relevant literature is beyond the scope of this chapter, some studies that have particularly important implications regarding neurological features that seem to be evolutionarily advanced in humans are briefly described here. To begin, and as noted above, the most obvious derived characteristic of human brains is that they are absolutely and relatively large, averaging about three times the size for australopithecines as well as three times the size expected for nonhuman primates of similar body size (Passingham, 1973, 1975; Stephan et al., 1970). Despite their large mass, however, the "quest for uniqueness" in human brains has been frustrated because

> "the human brain has the number of neurons that is expected of a primate brain of its size; a cerebral cortex that is exactly as large as expected for a primate brain of [its size]; just as many neurons as expected in the cerebral cortex for the size of this structure; and, despite having a relatively large cerebral cortex ..., this enlarged cortex holds just the same proportion of brain neurons in humans as do other primate cortices.... This final observation calls for a reappraisal of the view of brain evolution that concentrates on the expansion of cerebral cortex and its replacement with a more integrated view of coordinate evolution of cellular composition, neuroanatomical structure, and function of cerebral cortex and cerebellum"
> (Herculano-Houzel, 2009:10)

According to Herculano-Houzel (2009; Chapter 15), what is unique about the human brain is that humans have the largest absolute number of neurons among primates and probably other animals. This fits nicely with the fact that people also have the largest brains, by far, of any primate.

As the research of Herculano-Houzel (2009) and Herculano-Houzel et al. (2010) illustrates, the search for advanced brain features in humans has become less focused on gross anatomy and more concerned with cytoarchitecture, neuronal connections, and functions at the cellular level. At one point, for example, the relative size of the human frontal lobe was believed to be differentially large. However, Semendeferi and her colleagues have demonstrated that the overall size of human frontal lobes is not greater than expected for brains of their size. Instead, it now appears that alterations in internal wiring and differential enlargement occurred during hominin evolution in certain subareas of the prefrontal cortex including BA 10 (Semendeferi and Damasio, 2000; Semendeferi et al., 2001, 2002), while other areas such as Brodmann's area 13 (BA 13, part of the limbic system) decreased in relative size (Semendeferi et al., 1998). Human prefrontal cortex is especially important for higher cognitive processing in humans, in keeping with the finding that differential expansion of white matter (Schoenemann et al., 2005) and pronounced gyrification (Armstrong et al., 1993; Rilling, 2006) have also been described for this part of the brain.

Because the relative size of human BA 10 is twice that of both bonobos and chimpanzees,

Semendeferi (1994) suggested that this area of the cerebral cortex increased in relative size at some point along the line from the first hominins to the early representatives of the genus *Homo*. Recently, she and her colleagues compared the spacing organization of neurons in layer III in frontopolar (BA 10), primary motor (BA 4), primary somatosensory (BA 3), and primary visual cortex (BA 17) in ape and human brains (Semendeferi et al., 2011) (Fig. 7). Their results strongly suggest that the horizontal spacing distance (HSD) between neurons increased in BA 10 (but not the three other areas) in hominins after they split from the ancestors of chimpanzees in a manner that facilitated complex interconnectivity and information processing (Fig. 7). Interestingly, similar histological findings have also been reported for human BA 44/45 (Broca's area) (Schenker et al., 2008), which raises the fascinating possibility that the human prefrontal cortex was widely reorganized during hominin cognitive evolution (Semendeferi et al., 2011).

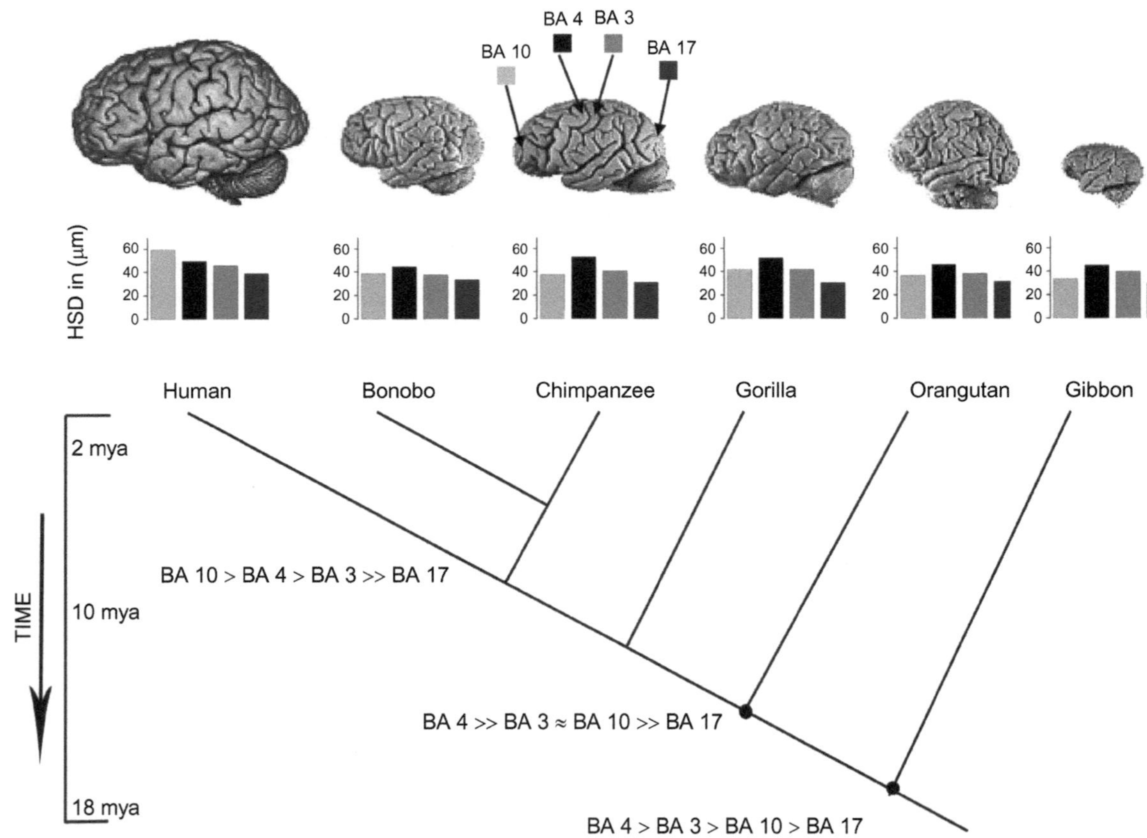

Fig. 7. Cladogram showing left lateral views of the human and ape brains and the relative degree of horizontal spacing distance (HSD) of neurons in four areas of their cerebral cortices (BA 10, BA 4, BA 3, BA 17). Symbols: >, greater HSD between neurons; >>, statistically significant greater HSD between neurons; ≈, HSD about the same. After human and chimpanzee lineages split, the HSD of BA 10 in humans became the largest (indicating more complex connectivity) compared with the three other cortical areas in the human brain and compared with BA 10 in the apes. Reproduced with permission from Semendeferi et al. (2011).

The size and organization of the human frontal pole clearly stands out from that of apes and its functions constitute one of the most fascinating puzzles in cognitive neuroscience (Burgess et al., 2005; Gilbert et al., 2006). This area has been implicated in a range of activities including "watchfulness," remembering to carry out intended activities, aspects of recollection, anticipating the future, multitasking, switching between externally versus internally oriented thoughts, and integrating limbic input to arousal, motivation, and intentions (Burgess et al., 2005; Koechlin and Hyafil, 2007; Tucker and Holmes, 2011). Thus, a key adaptive advantage of an evolved frontopolar cortex may have been "an ability to pursue long-term behavioral plans and at the same time respond to demands of the physical or social environments... the frontopolar cortex may have played an even more critical role in the gradual formation of complex behavioral and cognitive routines such as tool use in individuals and societies, that is, in human creativity rather than complex decision-making and reasoning" (Koechlin and Hyafil, 2007:598).

Concluding remarks

I have spent some time reviewing the literature on BA 10 because it is an excellent example of research that is beginning to shed light on the evolution of advanced cognitive abilities in hominins based on a synthesis of findings from paleoneurology and comparative neuroanatomy. As we have seen, the shapes of the frontal lobes that are reproduced in the frontopolar region on endocasts of *Australopithecus* and *Paranthropus* appear expanded toward a human condition in the former but not the latter. This observation is consistent with Semendeferi et al.'s (2011) hypothesis that an increase in the horizontal spacing between neurons and an associated increased complexity in connectivity occurred in BA 10 of our ancestors' brains at some point after our lineage split from that of chimpanzees. When, exactly, this change began during the approximately 7 million years of hominin evolution is unknown. However, the comparative paleoneurological evidence regarding brain size, frontal lobe shape, and sulcal patterns (including an apelike fronto-orbital sulcus in *Australopithecus*) suggests that the early stages of prefrontal cortical evolution may have been underway in the *Australopithecus* population(s) that gave rise directly to *Homo*. Indeed, Raymond Dart would have embraced this hypothesis, as shown by his 1929 unpublished monograph, which languishes in the archives of the University of Witwatersrand.

References

Allen, J. S., Bruss, J., & Damasio, H. (2006). Looking for the lunate sulcus: A magnetic resonance imaging study in modern humans. *The Anatomical Record. Part A, Discoveries in Molecular, Cellular, and Evolutionary Biology*, *288*, 867–876.

Amunts, K., Schleicher, A., Burgel, U., Mohlberg, H., Uylings, H. B., & Zilles, K. (1999). Broca's region revisited: Cytoarchitecture and intersubject variability. *The Journal of Comparative Neurology*, *412*, 319–341.

Amunts, K., Schleicher, A., & Zilles, K. (2007). Cytoarchitecture of the cerebral cortex—More than localization. *NeuroImage*, *37*, 1061–1065 discussion 1066–1068.

Armstrong, E., Zilles, K., & Schleicher, A. (1993). Cortical folding and the evolution of the human brain. *Journal of Human Evolution*, *25*, 387–392.

Barton, R. A. (2001). The coordinated structure of mosaic brain evolution (commentary). *The Behavioral and Brain Sciences*, *24*, 281–282.

Barton, R. A., & Harvey, P. H. (2000). Mosaic evolution of brain structure in mammals. *Nature*, *405*, 1055–1058.

Bear, D., Schiff, D., Saver, J., Greenberg, M., & Freeman, R. (1986). Quantitative analysis of cerebral asymmetries. Fronto-occipital correlation, sexual dimorphism and association with handedness. *Archives of Neurology*, *43*, 598–603.

Berger, L. R., De Ruiter, D. J., Churchill, S. E., Schmid, P., Carlson, K. J., Dirks, P. H., et al. (2010). *Australopithecus sediba*: A new species of *Homo*-like Australopith from South Africa. *Science*, *328*, 195–204.

Burgess, P. W., Simons, J. S., Dumontheil, I., & Gilbert, S. J. (2005). The gateway hypothesis of rostral prefrontal cortex (area 10) function. In J. Duncan, P. Mcleod & L. Phillips (Eds.), *Measuring the mind: Speed, control, and age* (pp. 215–246). Oxford: Oxford University Press.

Chiu, H. C., & Damasio, A. R. (1980). Human cerebral asymmetries evaluated by computed tomography. *Journal of Neurology, Neurosurgery, and Psychiatry*, *43*, 873–878.

Connolly, J. C. (1950). *External morphology of the primate Brain.* Springfield, IL: C. C. Thomas.

Dart, R. A. (1925). *Australopithecus africansus*: The man-ape of South Africa. *Nature, 115*, 195–199.

Dart, R. A. (1929). *Australopithecus africanus: And His Place in Human Origins.* Johannesburg: University of Witwatersrand Archives unpublished manuscript.

Dart, R. A. (1956). The relationships of brain size and brain pattern to human status. *South African Journal of Medical Sciences, 21*, 23–45.

de Sousa, A. A., Sherwood, C. C., Mohlberg, H., Amunts, K., Schleicher, A., Macleod, C. E., et al. (2010). Hominoid visual brain structure volumes and the position of the lunate sulcus. *Journal of Human Evolution, 58*, 281–292.

Deaner, R. O., Isler, K., Burkart, J., & Van Schaik, C. (2007). Overall brain size, and not encephalization quotient, best predicts cognitive ability across non-human primates. *Brain, Behavior and Evolution, 70*, 115–124.

Dosenbach, N. U., Nardos, B., Cohen, A. L., Fair, D. A., Power, J. D., Church, J. A., et al. (2010). Prediction of individual brain maturity using fMRI. *Science, 329*, 1358–1361.

Falk, D. (1980). A reanalysis of the South African australopithecine natural endocasts. *American Journal of Physical Anthropology, 53*, 525–539.

Falk, D. (1983). The Taung endocast: A reply to Holloway. *American Journal of Physical Anthropology, 60*, 479–489.

Falk, D. (2004). Hominin brain evolution—New century, new directions. *Collegium Antropologicum, 28*(Suppl. 2), 59–64.

Falk, D. (2007a). Evolution of the primate brain. In W. Henke & I. Tattersall (Eds.), *Handbook of palaeoanthropology Vol. 2: Primate evolution and human origins* (pp. 1133–1162). Heidelberg: Springer-Verlag.

Falk, D. (2007b). *The evolution of Broca's area. IBRO History of Neuroscience.* http://www.ibro.info/Pub_Main_Display.asp?LC_Docs_ID=3145.

Falk, D. (2009a). *Finding Our Tongues: Mothers, Infants & The Origins of Language.* New York: Pertheus (Basic Books).

Falk, D. (2009b). The natural endocast of Taung (Australopithecus africanus): Insights from the unpublished papers of Raymond Arthur Dart. *Yearbook of Physical Anthropology, 52*, 49–65.

Falk, D. (2010). Evolution of the brain, cognition, and speech. In C. Larson (Ed.), *A companion to biological anthropology* (pp. 258–271). Oxford: Blackwell Publishing.

Falk, D. (2011). *The fossil chronicles: How two controversial discoveries changed our view of human evolution.* Berkeley: University of California Press.

Falk, D., & Clarke, R. (2007). Brief communication: New reconstruction of the Taung endocast. *American Journal of Physical Anthropology, 134*, 529–534.

Falk, D., Gage, T., Dudek, B., & Olson, T. R. (1995). Did more than one species of hominid coexist before 3.0 ma?: Evidence from blood and teeth. *Journal of Human Evolution, 29*, 591–600.

Falk, D., & Gibson, K. R. (Eds.), (2001). *Evolutionary anatomy of the primate cerebral cortex.* Cambridge, UK: Cambridge University Press.

Falk, D., Redmond, J. C., Jr., Guyer, J., Conroy, G. C., Recheis, W., Weber, G. W., et al. (2000). Early hominid brain evolution: A new look at old endocasts. *Journal of Human Evolution, 38*, 695–717.

Falk, D., Hildebolt, C., Smith, K., Morwood, M. J., Sutikna, T., Jatmiko, et al. (2009). LB1's virtual endocast, microcephaly, and hominin brain evolution. *Journal of Human Evolution, 57*, 597–607.

Finlay, B. L., & Darlington, R. B. (1995). Linked regularities in the development and evolution of mammalian brains. *Science, 268*, 1578–1584.

Fischl, B., Rajendran, N., Busa, E., Augustinack, J., Hinds, O., Yeo, B. T., et al. (2008). Cortical folding patterns and predicting cytoarchitecture. *Cerebral Cortex, 18*, 1973–1980.

Galaburda, A. M., Lemay, M., Kemper, T. L., & Geschwind, N. (1978). Right-left asymmetries in the brain. *Science, 199*, 852–856.

Gibson, K. R. (2001). Bigger is better: Primate brain size in relationship to cognition. In D. Falk & K. R. Gibson (Eds.), *Evolutionary anatomy of the primate cerebral cortex* (pp. 79–97). Cambridge, UK: Cambridge University Press.

Gilbert, S. J., Spengler, S., Simons, J. S., Steele, J. D., Lawrie, S. M., Frith, C. D., et al. (2006). Functional specialization within rostral prefrontal cortex (area 10): A meta-analysis. *Journal of Cognitive Neuroscience, 18*, 932–948.

Gould, S. J. (2001). Size matters and function counts. In D. Falk & K. R. Gibson (Eds.), *Evolutionary anatomy of the primate cerebral cortex* (pp. xiii–xvii). Cambridge, UK: Cambridge University Press.

Herculano-Houzel, S. (2009). The human brain in numbers: A linearly scaled-up primate brain. *Frontiers in Human Neuroscience, 3*, 31.

Herculano-Houzel, S., Mota, B., Wong, P., & Kaas, J. H. (2010). Connectivity-driven white matter scaling and folding in primate cerebral cortex. *Proceedings of the National Academy of Sciences of the United States of America, 107*, 19008–19013.

Hofman, M. A. (2001). Brain evolution in hominids: Are we at the end of the road? In D. Falk & K. R. Gibson (Eds.), *Evolutionary anatomy of the primate cerebral cortex* (pp. 113–127). Cambridge, UK: Cambridge University Press.

Holloway, R. L. (2001). Does allometry mask important brain structure residuals relevant to species-specific behavioral evolution? *The Behavioral and Brain Sciences, 24*, 286–287.

Holloway, R. L. (2008). The human brain evolving: A personal retrospective. *Annual Review of Anthropology, 37*, 1–37.

Holloway, R. L., Clarke, R. J., & Tobias, P. V. (2004). Posterior lunate sulcus in *Australopithecus africanus*: Was Dart right? *Comptes Rendus, Palevol, 3*, 287–293.

Holloway, R. L., & Kimbel, W. H. (1986). Endocast morphology of Hadar hominid AL 162-28. *Nature, 321*, 536–537.

Jerison, H. J. (1973). *Evolution of the brain and intelligence.* New York: Academic Press.

Jerison, H. J. (1975). Fossil evidence of evolution of the human brain. *Annual Review of Anthropology*, *4*, 27–58.

Kaas, J. H. (2000). Why is brain size so important: Design problems and solutions as neocortex gets bigger or smaller. *Brain and Mind*, *1*, 7–23.

Kaas, J. H., & Preuss, T. M. (2008). Human brain evolution. In L. R. Squire (Ed.), *Fundamental neuroscience* (pp. 1017–1038). (3rd ed.). San Diego: Academic Press.

Kivilevitch, Z., Achiron, R., & Zalel, Y. (2010). Fetal brain asymmetry: In utero sonographic study of normal fetuses. *American Journal of Obstetrics and Gynecology*, *202*, 359 e351–359.e358.

Koechlin, E., & Hyafil, A. (2007). Anterior prefrontal function and the limits of human decision-making. *Science*, *318*, 594–598.

Le Gros Clark, W. E., Cooper, D. M., & Zuckerman, S. (1936). The endocranial cast of the chimpanzee. *The Journal of the Royal Anthropological Institute*, *66*, 249–268.

Lemay, M. (1977). Asymmetries of the skull and handedness. Phrenology revisited. *Journal of the Neurological Sciences*, *32*, 243–253.

LeMay, M. (1984). Radiological, developmental and fossil asymmetries. In N. Geschwind & A. M. Galaburda (Eds.), *Cerebral dominance: The biological foundations* (pp. 26–42). Cambridge, MA: Harvard University Press.

Lemay, M., Billig, M. S., & Geschwind, N. (1982). Asymmetries of the brains and skulls of nonhuman primates. In E. Armstrong & D. Falk (Eds.), *Primate brain evolution: Methods and concepts* (pp. 263–277). New York: Plenum.

Lockwood, C. A., & Tobias, P. V. (2002). Morphology and affinities of new hominin cranial remains from Member 4 of the Sterkfontein Formation, Gauteng Province, South Africa. *Journal of Human Evolution*, *42*, 389–450.

Montgomery, S. H., Capellini, I., Barton, R. A., & Mundy, N. I. (2010). Reconstructing the ups and downs of primate brain evolution: Implications for adaptive hypotheses and *Homo floresiensis*. *BioMed Central Biology*, *8*, 9. doi:10.1186/1741-7007-8-9.

Neubauer, S., Gunz, P., & Hublin, J. J. (2010). Endocranial shape changes during growth in chimpanzees and humans: A morphometric analysis of unique and shared aspects. *Journal of Human Evolution*, *59*, 555–566.

Ono, M., Kubik, S., & Abernathy, C. D. (1990). *Atlas of the cerebral sulci.* New York: Hafner Publishing Company.

Passingham, R. E. (1973). Anatomical differences between the neocortex of man and other primates. *Brain, Behavior and Evolution*, *7*, 337–359.

Passingham, R. E. (1975). Changes in the size and organisation of the brain in man and his ancestors. *Brain, Behavior and Evolution*, *11*, 73–90.

Preuss, T. (2007a). Evolutionary specializations of primate brain systems. In M. Ravosa & M. Dagasto (Eds.), *Primate origins: Adaptations and evolution* (pp. 625–675). New York: Springer.

Preuss, T. M. (2007b). Primate brain evolution in phylogenetic context. In J. H. Kaas & T. M. Preuss (Eds.), *Evolution of nervous systems Vol. 4: The evolution of primate nervous systems* (pp. 1–34). Oxford: Elsevier.

Radinsky, L. B. (1979). *The fossil record of primate brain evolution (49th James Arthur Lecture).* New York: American Museum of Natural History.

Rilling, J. K. (2006). Human and nonhuman primate brains: Are they allometrically scaled versions of the same design? *Evolutionary Anthropology*, *15*, 65–77.

Schenker, N. M., Buxhoeveden, D. P., Blackmon, W. L., Amunts, K., Zilles, K., & Semendeferi, K. (2008). A comparative quantitative analysis of cytoarchitecture and minicolumnar organization in Broca's area in humans and great apes. *The Journal of Comparative Neurology*, *510*, 117–128.

Schoenemann, P. T., Sheehan, M. J., & Glotzer, L. D. (2005). Prefrontal white matter volume is disproportionately larger in humans than in other primates. *Nature Neuroscience*, *8*, 242–252.

Semendeferi, K. (1994). *Evolution of the hominoid prefrontal cortex: A quantitative and image analysis of area 13 and 10. Anthropology.* Iowa City: University of Iowa Ph.D. dissertation.

Semendeferi, K., Armstrong, E., Schleicher, A., Zilles, K., & Van Hoesen, G. W. (1998). Limbic frontal cortex in hominoids: A comparative study of area 13. *American Journal of Physical Anthropology*, *106*, 129–155.

Semendeferi, K., Armstrong, E., Schleicher, A., Zilles, K., & Van Hoesen, G. W. (2001). Prefrontal cortex in humans and apes: A comparative study of area 10. *American Journal of Physical Anthropology*, *114*, 224–241.

Semendeferi, K., & Damasio, H. (2000). The brain and its main anatomical subdivisions in living hominoids using magnetic resonance imaging. *Journal of Human Evolution*, *38*, 317–332.

Semendeferi, K., Lu, A., Schenker, N., & Damasio, H. (2002). Humans and great apes share a large frontal cortex. *Nature Neuroscience*, *5*, 272–276.

Semendeferi, K., Teffer, K., Buxhoeveden, D. P., Park, M. S., Bludau, S., Amunts, K., et al. (2011). Spatial 0rganization of neurons in the frontal pole sets humans apart from great apes. *Cerebral Cortex*, *5*, 1485–1497.

Sherwood, C. C., Broadfield, D. C., Holloway, R. L., Ganon, P. J., & Hof, P. R. (2003). Variability of Broca's area homologue in African great apes: Implications for language evolution. *Anatomical Record*, *271A*, 276–285.

Smith, G. E. (1903). The so-called Affenspalte in the Human (Egyptian) brain. *Anatomischer Anzeiger*, *24*, 74–83.

Smith, G. E. (1904a). The morphology of the occipital region of the cerebral hemisphere in man and the apes. *Anatomischer Anzeiger*, *24*, 436–447.

Smith, G. E. (1904b). The morphology of the retrocalcarine region of the cerebral cortex. *Proceedings of the Royal Society of London*, *73*, 59–65.

Stephan, H., Bauchot, R., & Andy, O. J. (1970). Data on size of the brain and of various brain parts in insectivores and primates. In C. R. Noback & W. Montagna (Eds.), *Advances in primatology, Vol 1: The primate brain.*New York: Appleton-Century-Crofts.

Tobias, P. V. (1987). The brain of *Homo habilis*: A new level of organization in cerebral evolution. *Journal of Human Evolution, 16*, 741–761.

Tobias, P. V., & Falk, D. (1988). Evidence for a dual pattern of cranial venous sinuses on the endocranial cast of Taung (*Australopithecus africanus*). *American Journal of Physical Anthropology, 76*, 309–312.

Toga, A. W., & Thompson, P. M. (2003). Mapping brain asymmetry. *Nature Reviews Neuroscience, 4*, 37–38.

Tucker, D. M., & Holmes, M. D. (2011). Fractures and bindings of consciousness. *American Scientist, 99*, 32–39.

Van Essen, D. C. (1997). A tension-based theory of morphogenesis and compact wiring in the central nervous system. *Nature, 385*, 313–318.

Van Essen, D. C. (2007). Cerebral cortical folding patterns in primates: Why they vary and what they signify. In J. H. Kaas & T. M. Preuss (Eds.), *Evolution of nervous systems*, (Vol. 4), (pp. 267–289). London: Elsevier.

Van Essen, D. C., & Dierker, D. L. (2007). Surface-based and probabilistic atlases of primate cerebral cortex. *Neuron, 56*, 209–225.

Ventrice, F. (2011). *Modern human brain growth and development. Contribution to brain evolution in hominids.* Zurich, Paris: University of Zurich, Museum of Natural History Ph.D. dissertation.

M. A. Hofman and D. Falk (Eds.)
Progress in Brain Research, Vol. 195
ISSN: 0079-6123

CHAPTER 13

Evolution of hominin cranial ontogeny

Christoph P. E. Zollikofer*

Anthropological Institute and Museum, University of Zurich, Zurich, Switzerland

Abstract: Hominin evolution is characterized by two main trends, transition to bipedality and increase in brain size. Fossil evidence shows that both trends had a major impact on the structure and function of the hominin skull. This chapter asks how evolutionary modification of the cranial ontogenetic program led to morphological reorganization of the hominin skull and ultimately to hominin cranial diversity. Three major mechanisms of evolutionary developmental reorganization are proposed: modified prenatal development of the cranial base and face reflects adaptation to bipedality; high rates of neurocranial growth during early postnatal ontogeny are essential to attain large brain sizes; taxon-specific modification of facial development reflects dietary adaptation and—in the genus *Homo*—a general trend toward neoteny.

Keywords: evolutionary developmental paleoanthropology; growth; development; heterochrony; heterotopy; geometric morphometrics.

Introduction

Immature fossil hominin individuals have played a significant role in shaping the ways in which we think about human evolution. The fossilized cranial remains of a child found in 1830 in the cave of Engis near Liège, Belgium, mark the beginning of fossil-based paleoanthropology (Schmerling, 1832–1833). And the Taung child skull (the type specimen of *Australopithecus africanus*) was the first direct fossil evidence for early hominin origins in Africa (Dart, 1925). Inferences drawn from juvenile fossil hominins were received with reservation, however, because immatures do not display the full set of features typically thought to be relevant for the proper definition of a paleospecies. As an effect, Dart's ideas on Taung were only fully accepted in the 1950s (Tobias, 1985), and the Engis child was identified as a Neanderthal more than a century after its discovery (Fraipont, 1936).

Today, fossil hominin cranial ontogeny has become a research field in its own, yielding new and often surprising insights into the evolution of growth and development in the hominin lineage. Evolutionary developmental biology, or

*Corresponding author
Tel.: +41 44 635 5427; Fax +41 44 635 6886
E-mail: zolli@aim.uzh.ch

DOI: 10.1016/B978-0-444-53860-4.00013-1

"evo-devo," investigates how ontogenetic programs evolve. A central question of evo-devo is how ontogenetic programs undergo evolutionary modification, and how these modifications cause the emergence of new phenotypes. Another important question is how the process of ontogeny—which represents a spatiotemporal feature of the phenotype—is shaped through adaptation and selection.

The aim of this chapter is to provide an overview of current trends in evolutionary developmental biology of the skull of fossil hominins. Since direct observation of the development of paleospecies is not possible, fossil hominin evo-devo heavily relies on comparative data from extant taxa. Also, it has to use a variety of methods to compare patterns of ontogeny and to infer processes of evolutionary developmental modification. Following this logic, the chapter introduces concepts and methods of evo-devo, presents evidence from extant species, and then considers what can be inferred from combined extant and fossil data about the evolution of hominin cranial ontogeny.

Fossil hominin evo-devo: Concepts and terminology

The term *hominins* is used here to denote the tribe Hominini (modern humans and their extinct relatives); *panins* is used to denote the tribe Panini (extant taxa of the genus *Pan*—*Pan troglodytes* [chimpanzee] and *Pan paniscus* [bonobo]—and their extinct relatives). The term *skull* denotes the bony elements of the head, comprising the cranium and mandible. In many studies, the cranium and mandible are treated separately, not least because only few fossil hominins are represented by associated cranial and mandibular remains, and because skeletal collections often preserve either cranial or mandibular specimens. However, the skull represents a developmental and functional unit and will be considered here as a whole as far as fossil data are available.

The phyletically oldest parts of the skull encapsulate structures of the nervous system, while the younger parts constitute the scaffold for the dentognathic system. In higher vertebrates, the brain and some of the primary sense organs, such as the vestibulocochlear organs of hearing and equilibrium, the eyes, and the nose, are at least partly enclosed in bony structures. The fossilized remains of these encasements thus convey information about brain and sensory structure and function, and ultimately about the behavior of extinct taxa.

To study the ontogeny of the skull at the phenotypic level, several basic concepts of morphology are used. The term *form* is used to denote the spatial properties of an organismic structure. Form is decomposed into *size* and *shape* (Mosimann, 1988), where size denotes the extent of the structure (e.g., mass, volume, stature, length), while shape denotes spatial (geometric) relationships between subunits (e.g., length/width). Size and shape have both morphometric and biological significance. Body size, for example, is among the most important parameters constraining an individual's morphological, ecophysiological, behavioral, and life history properties. Maxillofacial shape, as another example, characterizes relevant biomechanical (structural–functional) properties of the masticatory system. During ontogeny, most structures increase in size and concomitantly change their shape. The term *growth* is used here to denote increase in size, while *development* is used to denote change in shape. Furthermore, size-dependent change in shape is termed *ontogenetic allometry*.

Another important conceptual framework of evolutionary developmental biology is that of heterochrony and heterotopy (Gould, 1977). *Heterochrony* denotes evolutionary modification of *temporal* properties of an ancestor's ontogenetic program, for example, changes in rates and timing of growth and development, while *heterotopy* denotes modification of *spatial* properties of ontogeny, for example, directions of growth or spatial distribution of bone growth fields.

Heterochrony and heterotopy are descriptive rather than explanatory terms; they facilitate classification and comparative analysis of ontogenetic trajectories, but do not automatically lead to insights into how actual developmental mechanisms were modified during evolution (McKinney, 1999). Nevertheless, comparing ancestral and descendant modes of ontogeny in terms of heterochrony/heterotopy permits establishment of explicit hypotheses about how ontogenetic programs have been modified during evolution (Zollikofer and Ponce de León, 2004).

Heterochronic hypotheses have played a key role in studying the evolution of hominin cranial ontogeny. As proposed by Bolk (1926), and later elaborated by Gould (1977), one specific form of heterochrony, namely, pedomorphy through neoteny, is a key mechanism of the evolution of hominin ontogeny. Pedomorphy means that adults of a descendant species are similar to juveniles of the ancestral species, while neoteny implies low rates of somatic ontogeny compared to sexual maturation (developmental retardation).

The third conceptual framework that is used to analyze cranial evo-devo is that of cranial *modularity* and *integration* (Enlow, 1990; Lieberman, 2011; Mitteroecker and Bookstein, 2008). The skull is a composite structure: components (modules) with specific functions have been added at various points along the evolutionary timeline. As expressed in the functional matrix hypothesis of Moss (1986), cranial modules have their functional, developmental, and genetic identity. At the same time, however, modules are networked through genes, developmental mechanics, and spatial constraints. As can be shown by theoretical network models of cranial ontogeny (Zollikofer and Ponce de León, 2004), and by experimental data (Szabo-Rogers et al., 2010), even slight modification in a single system parameter can have substantial effects on how cranial modules interact with each other. For example, genes primarily affecting the development of the derivatives of the first branchial arch (mandible, maxilla, and malar/middle ear region) have pervasive effects on cranial morphology as a whole (Gitton et al., 2010). Local ontogenetic and functional constraints thus may lead to global structural conflicts, which need to be solved during evolution and development through cranial integration.

As becomes evident from the morphological diversity of hominin skulls, even the highly constrained bauplan of the skull apparently bears enormous potential for evolutionary developmental modification. The processes by which diversity is generated have aptly been characterized as *evolutionary tinkering* (Lieberman et al., 2004). Modification of modular developmental mechanisms occurs at random, but ontogenetic homeostasis, plus the action of natural selection, leads to formation of highly integrated cranial phenotypes.

Figure 1 illustrates cranial modularity and integration with three examples. Major contrasts between the skulls of hominoids and bovines are due to differences in the size and orientation of the face relative to the brain. In the bovine skull, the masticatory system is placed in front of the brain, and it grows to considerably larger sizes than the brain, resulting in decoupling between endocranial and exocranial modules. The evolutionary solution to this developmental problem consists in filling the space between these modules with pneumatized bone ("spandrels" sensu Gould and Lewontin, 1979). The bony crests and lamellae developing between internal and external tables of the cranial vault bones represent a lightweight but mechanically stable structure (which, for example, withstands the high forces acting on the skull during head-to-head confrontation). Also, the large external surface area provides attachment areas for masticatory and nuchal muscles.

Compared to bovines, hominoids have large brains and short (orthognathic) but wide faces with fully convergent orbits. In the gorilla skull, the space between the wide cranial base and the endocranium is filled with pneumatized bone, while attachment areas for the masticatory and nuchal muscles come in the form of external cranial

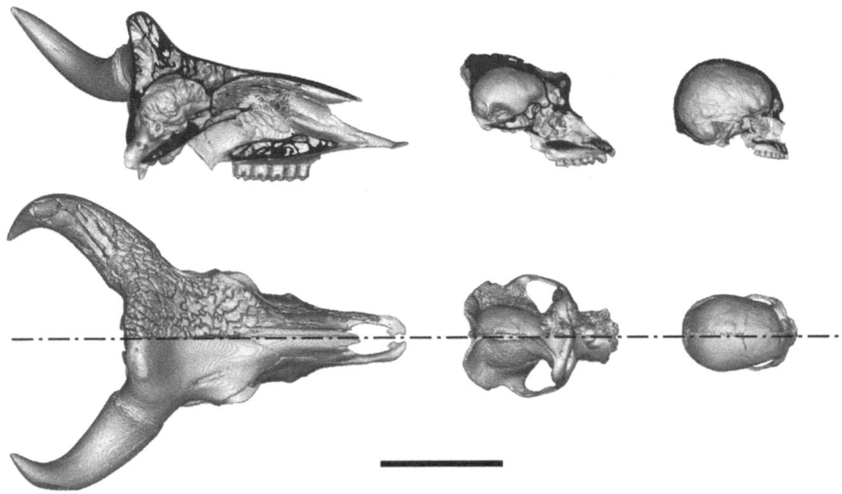

Fig. 1. Crania of a gaur (*Bos gaurus*, Indian bison), a gorilla (*Gorilla gorilla beringei*), and a modern human (*Homo sapiens*); visualization of CT data. Superior views: right and left sides represent external and internal surfaces, respectively. Midsagittal views: black areas indicate cancellous bone. Scale bar is 20cm.

superstructures, which also provide overall mechanical stability. In humans compared to gorillas, the face is reduced, the brain is enlarged, and the skull is balanced on top of the vertebral column. Forces acting on the head are thus comparatively small, and there is ample space for the attachment of the relatively slender masticatory and nuchal muscles, such that cranial superstructures are not required. Overall, Fig. 1 shows that while the standard bauplan of the skull is functionally and developmentally constrained, evolutionary tinkering provides a wide range of solutions to the problem of integrating cranial modules.

Measuring cranial morphology in time and space

To quantify and compare cranial ontogenetic trajectories of different species, measures of cranial geometry and developmental time are needed. Technically, patterns of growth, development, and ontogenetic allometry are best analyzed with methods of geometric morphometrics (GM). GM provides a well-established statistical framework to quantify the form of complex three-dimensional structures with sets of anatomical points of reference, so-called landmarks (Bookstein, 1991; Mitteroecker and Gunz, 2009). Size is measured as centroid size (CS), which is proportional to the standard deviation of a specimen's landmark positions around their center of mass. Specimens are then superimposed with generalized Procrustes analysis to evaluate a mean (consensus) landmark configuration, and each specimen's shape is measured as its deviation from the consensus. These data can be further analyzed with classical multivariate statistics, such as principal components analysis (Jolliffe, 1986), and the results can be visualized as a data scatter in morphospace or as actual shape transformations in physical space (Zollikofer and Ponce de León, 2002).

Time represents the fourth dimension of ontogeny. In primates in general, as well as within the hominoids, dental development is closely correlated with the general pace of ontogeny and life history (Smith and Tompkins, 1995). Also, key events of dental development can be associated with key events of cranial, somatic, and gonadal ontogeny. For example, cessation of brain growth [but not of endocranial volume (ECV) growth; see Fig. 2] approximately coincides with M1 eruption, the adolescent growth spurt with M2 eruption, and onset of reproduction with M3 eruption (Kelley and Schwartz, 2010; Smith and Tompkins, 1995). In accordance with their slow facial ontogeny, the eruption of the permanent molars in humans is delayed compared to chimpanzees, while the eruption of the anterior dentition follows largely similar schedules in both taxa (Zollikofer and Ponce de León, 2010). Overall, the eruption sequence has shifted from M1-I1-I2-M2-(P3, P4)-C-M3 (great apes) to M1-I1-I2-(P3, C, P4)-M2-M3 (genus *Homo*). Calibrating cranial ontogeny with dental eruption schedules thus uses taxon-specific physiological time (life history-related time) as a frame of reference.

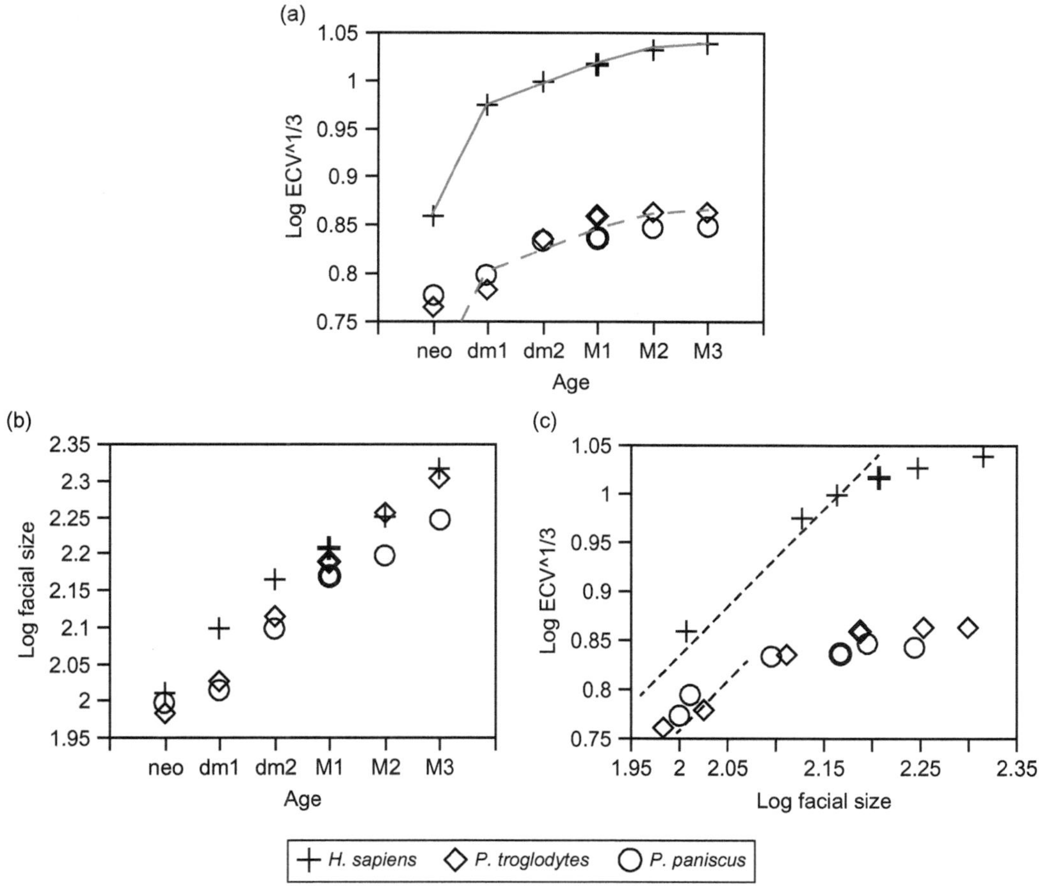

Fig. 2. Brain and facial growth in humans, chimpanzees, and bonobos. Ontogenetic time is calibrated to six stages: neonates, eruption of dm1, dm2, M1 (in bold), M2, and M3 (adult). (a) Endocranial growth trajectories. To facilitate comparisons, the human trajectory (gray line) is superimposed (dashed line) onto the *Pan* trajectories (transposition of log $ECV^{1/3}$ corresponds to scaling of $ECV^{1/3}$). Note close similarity of trajectories during late but not during early postnatal ontogeny. (b) Facial growth trajectories largely coincide. (c) Ontogenetic allometry of endocranial-to-facial growth. Dashed lines indicate isometric growth; note extended isometric trajectory in humans.

The enamel of the tooth crown and the dentin of the root are formed through incremental growth, such that teeth keep a record of circadian enamel and dentine deposition patterns. The resulting incremental growth structures permit detailed comparative analyses of tissue-specific growth trajectories, such as rates and duration of root and crown extension (Smith, 2008). Moreover, in analogy to dendrochronology, a total count of the growth structures in teeth of juvenile specimens permits evaluation of their age at death (Dean, 2006). Overall, relative (dental eruption) and absolute (age-at-death) time scales represent alternative frames of reference to compare rates of growth and development.

Human and chimpanzee cranial ontogeny

Comparative studies of human and panin cranial ontogeny have yielded various insights into how ontogenetic programs have been modified during the evolution of these sister clades (Bastir and Rosas, 2004; Lieberman et al., 2002, 2007; Mitteroecker et al., 2004; Penin et al., 2002; Ponce de León and Zollikofer, 2006). Here, I review the current body of evidence with a focus on basic ontogenetic mechanisms and constraints, which constitute the "ontogenetic raw material" upon which natural selection acts during hominin and panin evolution.

In absence of direct fossil evidence of the last common ancestor (LCA) of hominins and panins (often termed the "human-chimp LCA," HC-LCA), panins are often used as models for the ancestor. This actualistic comparative approach has several limitations. Since the split from the HC-LCA, hominins as well as panins experienced several million years of separate evolution, such that chimpanzees are as far away from the HC-LCA as we are. Nevertheless, for the majority of cranial morphological and ontogenetic features, chimpanzees tend to exhibit less derived character states than modern humans (Lieberman et al., 2002).

Compared to chimpanzees, modern human crania can be characterized by three major derived traits:

1. basal (rather than occipital) position of the foramen magnum and of the nuchal plane,
2. a large, spherically shaped ("globular") braincase overarching the face,
3. a small, anteroposteriorly short, vertically oriented, but relatively wide face.

Thinking in terms of an evolutionary developmental engineer, the following major changes have to be effected in the ontogenetic program of the HC-LCA in order to evolve a modern human skull:

1. cranial base: growth rates and directions have to be modified in order to rotate the foramen magnum basally and to accommodate a large brain on top of a small, retracted face,
2. brain: growth rates have to be upregulated,
3. face: growth rates have to be downregulated, and growth directions modified.

Skull growth

Various approaches have been proposed to compare human and chimpanzee endocranial (and brain) growth trajectories. The major difference between these approaches lies in the method of calibrating the *x* and *y* axes representing developmental time and ECV (used as a proxy of brain size), respectively (Zollikofer and Ponce de León, 2010). When absolute coordinates are used, contrasts between human and chimpanzee endocranial trajectories are considerable. From birth to adulthood, the ECV of chimpanzees expands from an average of 155ccm to 380ccm (2.5×), while that of modern humans expands from 400ccm to 1350ccm (3.3×) (DeSilva and Lesnik, 2006). One striking aspect of these figures is that the ECV of adult chimpanzees is smaller than that of modern human neonates.

When relative scales are used on both axes (e.g., adult ECV=100%; time at reaching adult ECV=100%), human and chimp trajectories largely coincide, and this may be taken as evidence that spatiotemporal up/down-scaling of existing ontogenetic programs is a major evolutionary mechanism to generate novel ontogenetic trajectories (Rice, 1997) (Fig. 2a). If one is interested in the metabolic costs of the growing brain, it is sensible to evaluate specific (instantaneous) growth rates (Leigh, 2004). And finally, ECV may be calibrated to adult average values (100%) and absolute time (years), such that differences during early postnatal brain growth appear more clearly (Hublin and Coqueugniot, 2006).

Depending on which comparative approach is used, human brain growth appears as "fast" or "slow" compared to chimpanzees: human neonates sustain high "fetal" brain growth rates well beyond birth, while in chimpanzees, these rates drop to moderate levels relatively rapidly after birth (Leigh, 2004, 2006). On the other hand, human brain growth is "slow" in the sense that neonate relative to adult brain size at birth is smaller and that adult brain size is reached slightly later in life than in chimpanzees.

Facial growth is more difficult to quantify than endocranial growth, especially because human and chimpanzee faces have clearly different shapes. Compared to chimpanzees, modern human faces are less prognathic (because the front dentition is greatly reduced) and less projecting (because the cranial base is short) such that the face is tucked under the neurocranium (Lieberman et al., 2002). However, human faces are comparatively wide and vertically extended. As an effect, absolute facial size (as measured by CS) is similar in modern humans and chimpanzees, and facial growth trajectories largely overlap (Fig. 2b).

Figure 2 indicates that facial growth characteristics are clearly distinct from neurocranial growth characteristics. While the neurocranium follows a saturation curve, which rapidly levels off around the eruption of M1, the face follows an almost linear growth trajectory from birth to adulthood. Integration of these two distinct modes of growth results in a two-phase trajectory of cranial ontogeny (Fig. 2c). Early postnatal ontogeny is characterized by an almost isometric increase of neurocranial versus facial size, while late ontogeny is characterized by negative ontogenetic allometry of the neurocranium relative to the face (or conversely, by positive facial relative to neurocranial allometry). The cessation of brain growth marks the transition between the two phases; transition occurs earlier in chimpanzees than in humans (Fig. 2c).

Skull development

Chimpanzee and human developmental trajectories have been studied extensively with GM methods over the past years. Research has focused on shape changes in cranial modules such as the face, the mandible, the cranial base, the endocranium, as well as on the cranium as a whole (Bastir and Rosas, 2004; Lieberman et al., 2007; Mitteroecker et al., 2004; Neubauer et al., 2010; Penin et al., 2002; Ponce de León and Zollikofer, 2006). The results of these studies are summarized and complemented here with a comparative GM study of craniomandibular ontogeny [neonate to adult samples of *Homo sapiens* (*N*=53), *P. troglodytes* (*N*=69), *P. paniscus* (*N*=39); landmark definitions and GM methods as in Ponce de León and Zollikofer (2001, 2006)].

Figure 3 represents species-specific craniomandibular ontogenetic trajectories through shape space. Humans and chimpanzees are clearly different already at the time of birth, indicating species-specific prenatal developmental processes (Mitteroecker et al., 2004; Ponce de León and Zollikofer, 2006). After birth, both species follow curved (biphasic) trajectories, where the early postnatal phase corresponds to isometric neurocranial versus facial growth and the late postnatal phase to negative allometry of endocranial versus facial growth (see Fig. 2).

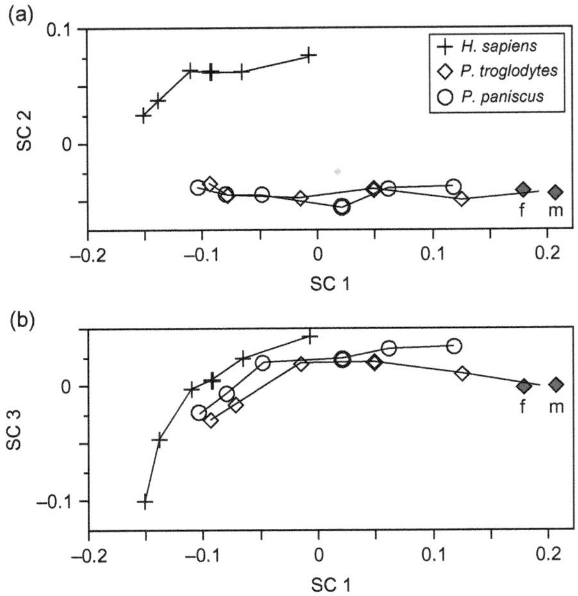

Fig. 3. Skull development in humans, chimpanzees, and bonobos. Ontogenetic trajectories through shape space are represented by mean shapes at birth, at eruption of dm1, dm2, M1 (in bold), M2, and M3 (adult). (a) Bivariate plot of shape components SC1 and SC2. (b) Bivariate plot of shape components SC1 and SC3. SC2 captures major prenatal differences between humans and *Pan* species; SC3 captures shape change during early postnatal development; SC1 captures shape change during late postnatal development.

Postnatal differences between species arise from different lengths of early and late postnatal phases and from slight divergence of trajectories.

Figure 4 visualizes craniomandibular shape differences corresponding to the data of Fig. 3. Already at birth, the human cranium is characterized by a basal position of the foramen magnum, a short anterior but wide posterior dental arcade, and a spherical braincase (Fig. 4a). During early postnatal ontogeny, the face and the neurocranium rotate in opposite directions relative to each other, resulting in vertical orientation of the face, basal shift of the foramen magnum, and flexion of the cranial base (Fig. 4b). As shown earlier (Zollikofer and Ponce de León, 2002), substantial intraspecific variation exists in early postnatal ontogeny, resulting in brachycephalic versus dolichocephalic skull morphologies, and varying degrees of neurocranial globularity (Enlow, 1990; Neubauer et al., 2010). Late postnatal ontogeny is dominated by increasing facial relative to neurocranial size (Fig. 4c). The face becomes more projecting but less vertically inclined, the cranial base is extended such that the foramen magnum is relocated to a more posterior position. Late postnatal ontogeny thus partially reverses the effects of early neurocranial-to-facial reorientation.

The characteristic morphology of the human skull appears to be brought about by specific modifications in three relatively independent ontogenetic mechanisms: (1) modification of prenatal basicranial development leads to basal shift of the foramen magnum and a short but wide dentognathic apparatus; (2) high brain growth rates and late cessation of brain growth lead to flexion of cranial base, basal rotation of foramen magnum, and facial reorientation; (3) reduction of facial growth rates and early cessation of facial growth prevent formation of a projecting face during adolescence.

In comparative analyses of cranial ontogeny, the role of the cranial base has received special interest because it constitutes an interface between the neurocranium and the viscerocranium (Lieberman et al., 2000). Since the cranial base is the first region to ossify in the fetal skull, it is supposed to act as a scaffold that constrains early neurocranial versus facial development (Jeffery and Spoor, 2004). Later on, the cranial base mediates between the growing brain on its superior side and the orbits and nasomaxillary complex on its inferior side. One basicranial variable that has received considerable attention is cranial base flexion (Lieberman, 2011; Lieberman et al., 2000). Compared to chimpanzees, modern human cranial base flexion increases (rather than decreases) during postnatal ontogeny. Cranial base flexion is often seen as exerting major constraints on neurocranial versus

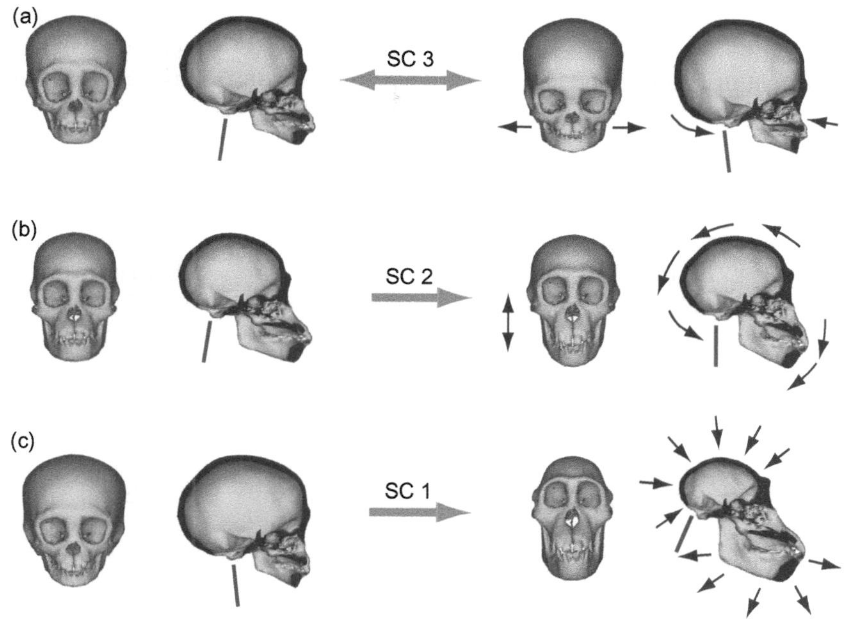

Fig. 4. Visualization of major modes of skull shape variation in humans, chimpanzees, and bonobos (corresponding to shape transformation along shape components SC1–3 in Fig. 3). (a) Prenatal differences between humans and *Pan* species (note differences in foramen magnum position/orientation and in facial morphology). (b) Neurocranial-to-facial reorientation during early postnatal ontogeny. (c) Changes in neurocranial-to-facial size and orientation during late postnatal ontogeny.

facial development, but relationships between cause and effect are complex (Lieberman et al., 2000, 2008). Overall, cranial base flexion reflects changes in the relative size, position, and orientation of the face relative to the braincase (Bastir et al., 2010).

While GM-based comparative analyses of cranial ontogeny reveal taxon-specific differences in relative rates and directions of shape change, the actual developmental mechanisms underlying these patterns can be investigated by direct observation of local growth fields. Ontogeny of the bony elements of the head is mediated by a complex interplay of bone deposition and resorption. Skeletal ontogeny thus acts like a sculptor using a combination of additive and subtractive technologies: while bone is added to the developing skull in some regions, it is removed from other regions. The spatial distribution and temporal activation of depository and resorptive growth fields in the human skull are documented in detail in Enlow's classic textbook on facial growth (Enlow, 1990). Compared to chimpanzees and macaques, human skulls exhibit a derived spatial distribution of growth fields, especially in the face (Enlow, 1990). While the internal (basicranial) surface of the maxilla grows via interosseous bone deposition (sutural bone growth), the external (facial) surface of the maxilla as well as the external symphyseal area of the mandible exhibit resorptive fields. As an effect, the human face grows inferiorly and anteriorly through maxillo-basicranial bone apposition, but forward growth is counteracted by maxillofacial resorptive growth fields. The combination of these processes results in a retracted, vertically oriented face in which the chin represents the most prominent (i.e., least resorbed) part.

Reconstructing fossil hominin cranial ontogenies

Comparison of modern human and chimpanzee cranial ontogenies leads to three evolutionary questions: When did the proposed ontogenetic modifications occur during hominin evolution? How was craniomandibular ontogeny modified? Which selective pressures favored these modifications? Studying craniomandibular ontogeny in fossil hominin species faces several challenges. Fossilized hominin remains are deteriorated, fragmented, and distorted by processes of taphonomy and diagenesis. Moreover, immature skeletal elements are often unfused, adding further difficulties to the reconstruction of a specimen's morphology at the time of death. Computer-assisted paleoanthropology provides a set of methods to efficiently tackle these problems (Zollikofer and Ponce de León, 2005). Following non-invasive 3D data acquisition with computed tomography, computer graphics tools are used to prepare specimens non-invasively, correct the effects of taphonomic deformation, recompose their three-dimensional morphology on the computer screen, and complement missing regions via mirror-imaging and/or interpolation/extrapolation. These virtual skulls can then be submitted to morphometric analysis.

A further challenge is to evaluate the individual age of immature fossil specimens. As mentioned, teeth (if present) are routinely used as absolute and relative age indicators. Synchrotron-based dental microstructural analysis now permits age-at-death estimates of fossil hominins with an error range of a few weeks (Tafforeau and Smith, 2008; Tafforeau et al., 2006).

Evolutionary modifications of cranial ontogeny in the earliest hominins

The earliest phase of hominin evolution is documented by three fossil taxa: *Sahelanthropus*, *Orrorin*, and *Ardipithecus* (Brunet et al., 2002; Senut et al., 2001; White et al., 2009). Following a proposition of White et al. (2009), these paleospecies may be attributed to "stage I" of hominin evolution, which is characterized by facultative bipedalism and reduction of dental sexual dimorphism. The crania of *Sahelanthropus* and *Ardipithecus* indicate that the foramen magnum was at a basal position, and the face somewhat shorter and more vertically oriented than in chimpanzees (Suwa et al., 2009; Zollikofer et al., 2005). The nuchal plane (which is preserved in *Sahelanthropus*) has a more basal orientation than in chimpanzees. Together, these features are indicative of early cranial adaptations to bipedal locomotion. On the other hand, both *Sahelanthropus* and *Ardipithecus* had ECVs in the range of similarly sized great apes. Overall, one may conclude that bipedality is the prime feature of hominin evolution, while the increase in brain size is the secondary feature.

What is the developmental basis of stage I hominin cranial morphologies? GM analyses show that the *Sahelanthropus* cranium is more similar to that of later hominins than to any of the great ape species (Zollikofer et al., 2005). A comparative ontogenetic analysis, however, suggests that postnatal cranial ontogeny was chimpanzee-like (Guy et al., 2005) (Fig. 5). The following hypothesis may thus be proposed: After early cessation of brain growth, *Sahelanthropus* grew a large face, but its distinctive "bipedal" cranial morphology with a low position of the foramen magnum and reduced midfacial projection was already present at birth.

Sahelanthropus has a unique morphology of the supraorbital region. The supraorbital tori are exceptionally massive (similar in size to those of adult male gorillas), but there is no postorbital sulcus, as one would expect in a great ape of similar cranial size (Brunet et al., 2002). One possible explanation of this feature is that, in contrast to extant great apes and later fossil hominins, the "gap" between the endocranium and the supraorbital region was filled with pneumatized bone. This hypothesis receives support from synchrotron tomography-based analysis of the endocranial structures of *Sahelanthropus* (Bienvenu, 2011).

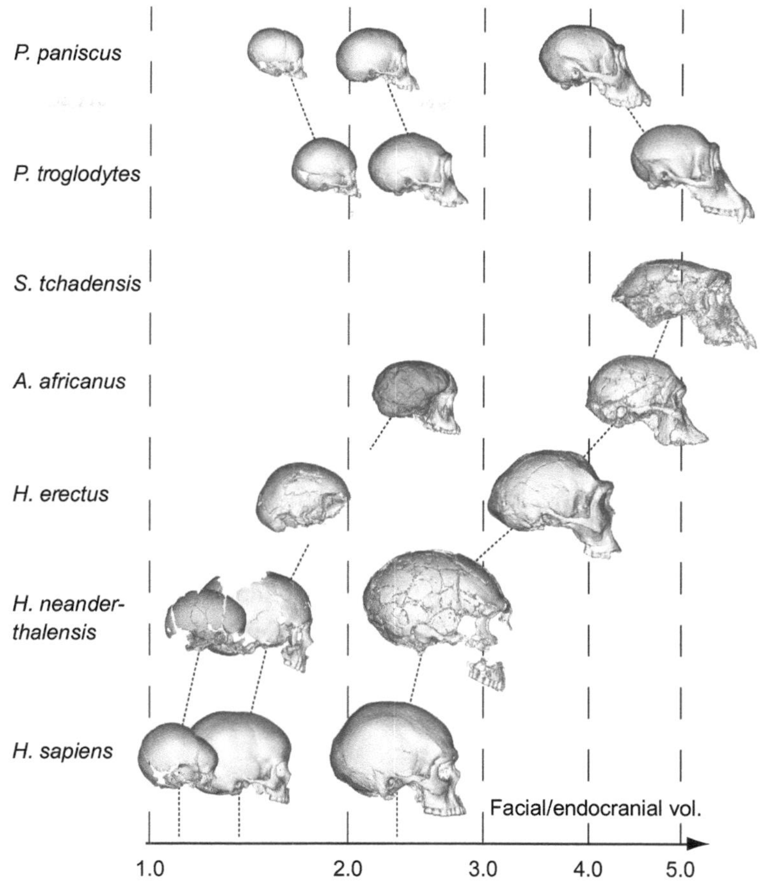

Fig. 5. Cranial ontogeny in hominins and panins. Position of specimens along the horizontal axis indicates cranial shape measured by facial/endocranial volume. Dashed lines connect neonates, infants (before M1 eruption), and adults, respectively. Fossil specimens are *H. neanderthalensis*: Mezmaiskaya 1, Roc de Marsal 1, Amud 1; *H. erectus*: Mojokerto, KNM-ER3733; *A. africanus*: Taung, Sts5; *S. tchadensis*: TM266. Note evolutionary trend toward short ontogenetic trajectories, that is, reduced facial relative to neurocranial growth (modified after Zollikofer and Ponce de León, 2010).

Cranial ontogeny in the australopiths: A theme with ontogenetic variations

The australopiths—a term typically used to subsume Pliocene hominins ascribed to the genera *Australopithecus* and *Paranthropus*—exhibited an enormous evolutionary diversity of forms, both in time and space (Wood and Lonergan, 2008). Australopiths are best characterized as stage II hominins, for example, "Pan-African striding bipeds with a wide ecological niche" (White et al., 2009). The evolutionary developmental foundations of australopithecine radiation are best studied in the framework of functional and ontogenetic constraints that had been established during stage-I hominin evolution. Hominin bipedality led to reorientation of the posterior cranial base toward the horizontal plane, and of the face toward the vertical plane. As outlined above, it is likely that ontogenetic facial-to-basicranial reorientation occurs before birth, such that, during postnatal ontogeny, the face tends to grow downward rather than forward relative to the neurocranium. Accordingly, developmental variation in facial-to-neurocranial growth in the australopiths has effects on craniofacial morphology that are different from patterns of variation between *Pan* species.

A functional constraint arising from striding bipedality is that a large head has to be balanced on top of the vertebral column (encephalization of australopiths was slightly larger than in stage-I hominins). To optimize head stabilization, any projection and/or prognathism of the face must be biomechanically compensated for by the nuchal musculature. To distribute head mass equally around the vertebral axis, the cranial base grows in anteroposterior and mediolateral directions. A wide cranial base is also significant in the context of dentognathic evolution. Australopiths are characterized by a reduced anterior dentition as opposed to increased molar size, resulting in comparatively short but wide dental arcades and jaws.

Early australopiths, which are best documented by *Australopithecus afarensis*, had a *Sahelanthropus*-like neurocranial-to-facial size and orientation. Only an in-depth analysis of the Dikika infant (Alemseged et al., 2006) will provide insights into how brain growth trajectories were modified to reach increased encephalization. During the Pliocene, australopiths underwent substantial radiation, and it is not yet clear how the various taxa are phyletically related to each other. Evo-devo trends seen in these fossils thus might represent parallel evolution. The gracile australopiths—best represented by *A. africanus* and the recently discovered paleopopulation of *Australopithecus sediba* (Berger et al., 2010)—exhibited less facial relative to neurocranial growth (Fig. 5). As a consequence, gracile australopithecine crania are more pedomorphic than those of *A. afarensis*, similar to the contrast between chimpanzees and bonobos. The Taung specimen, with an estimated individual age of 3.8 years, is informative about how the small face of gracile australopiths developed: a comparison of inferred facial-to-neurocranial growth in *A. africanus* and chimpanzees shows that facial expansion from Taung to adulthood is comparable to facial expansion from Taung-aged chimps to adult chimps (McNulty et al., 2006). This indicates that reduction in facial growth must have occurred earlier during life, be it prenatally or early during postnatal ontogeny.

The robust australopithecines, which represent the most specialized group of hominins (Ungar et al., 2008), exhibited a derived pattern of facial development: modification of bone growth fields in the nasal floor resulted in a vertically oriented and superoinferiorly extended face (McCollum, 1999). As mentioned, developmental constraints inherent to the hominin bauplan favor facial enlargement in vertical and transverse directions, but less in anterior direction. In the robust australopiths, this process gave rise to large orthognathic jaws and associated neurocranial superstructures. The pronounced sexual dimorphism in the robust australopithecines may thus be due to differences in rate and/or duration of facial growth during late postnatal ontogeny.

Cranial ontogeny in *Homo erectus*

Various australopith taxa have been proposed as candidate ancestors of *Homo*, and since the discovery of *A. sediba* at the site of Malapa, South Africa, it has become even more difficult to draw a line between australopiths and *Homo* (Berger et al., 2010), and between stage II and stage III hominins (White et al., 2009). The *A. sediba* sample is especially interesting because it most likely represents a paleopopulation comprising immature and adult individuals. Although *A. sediba* comes closest in cranial morphology to early *Homo*, it is not necessarily its direct ancestor, as it might represent one of the many evolutionary developmental variants that led to *Homo*-like morphologies.

In *A. sediba*, a *Homo*-like vertically oriented face is attached to a small australopith-like braincase. The premaxillary suture in the juvenile individual (MH1) is still patent, indicating australopith-like sustained growth of the anterior dental arcade. The combination in *A. sediba* of features previously thought to be characteristic of *Australopithecus* versus *Homo* corroborates the notion that neurocranial and facial modules evolved relatively independently, resulting in a pattern known as mosaic evolution.

Evolutionary tinkering with two fairly independent ontogenetic modules of the skull might have laid the foundations for the evolution of *Homo* and might explain why it is difficult to draw borders between *Homo habilis* and later forms attributed to *Homo rudolfensis*, *Homo ergaster*, and *H. erectus* (Wood and Collard, 1999). Overall, however, the genus *Homo* is best characterized by a shift toward increased encephalization (Holloway, 2008) and facial reorientation (Fig. 5). In early Pleistocene *H. erectus*, ECV varies between 600 and 900ccm. Toward the middle Pleistocene *H. erectus*, ECV approaches the lower range of modern human variation (1000–1250ccm) (Rightmire, 2004; Ruff et al., 1997; Spoor et al., 2007; Vekua et al., 2002). Modes of growth and development of the neurocranium thus underwent significant evolutionary modifications during the Pleistocene.

Only few *H. erectus* specimens are sufficiently well preserved to permit direct assessment of facial size and shape, and facial-to-neurocranial orientation. Frontozygomatic and basicranial dimensions of *H. erectus* calvariae suggest that facial size (and accordingly, facial growth) was as variable as brain size/growth. Overall, it appears that the face of *H. erectus* was more vertically oriented than in earlier hominins, and that this condition was probably brought about by less midfacial growth (reduced prognathism) and more upper facial growth (increased facial projection). The spatial separation of the upper face from the neurocranium results in postorbital constriction and a supraorbital sulcus. The conspicuous occipital and angular tori found in many large *H. erectus* specimens also indicate the presence of large faces, as these superstructures reflect a well-developed nuchal musculature to counterbalance the face. The allometrically expanded supraorbital tori of the large specimens most probably represent pronounced sexual dimorphism (Spoor et al., 2007) rather than biomechanical adaptation (Hylander et al., 1991).

The basal position of the foramen magnum, together with vertical reorientation of the face, leads to anteroposterior compression and flexion of the cranial base. Potential spatial packing problems arising from the enlarged brain are circumvented by widening of the cranial base in transverse direction. This becomes evident by the more transverse orientation of the temporal pyramids in *H. erectus* compared to earlier hominins, and characteristically inclined parietal walls.

Direct fossil evidence for cranial ontogeny in *H. erectus* comes from three well-preserved specimens, the immature Mojokerto neurocranium with an estimated ECV of 663ccm (Coqueugniot et al., 2004) and the complete skulls from Nariokotome (880ccm, ca. 8 years) (Walker and Leakey, 1993) and from Dmanisi (600ccm, 11–13 years) (Vekua et al., 2002).

Since the face of the Mojokerto individual is missing, age at death cannot be estimated from dental data, but a well-founded study using neuro-basicranial developmental markers yielded an estimate of 1.5 years (Coqueugniot et al., 2004). The high ECV of Mojokerto at this early age (230% neonate ECV versus 80% adult ECV) reflects a unique combination of primitive and derived features of the brain growth trajectory: the former figure indicates high postnatal brain growth rates—like in modern humans, while the latter figure indicates early cessation of brain growth—like in australopiths and chimpanzees (Zollikofer and Ponce de León, 2010).

The paleopopulation from Dmanisi, Georgia, is of special relevance for ontogenetic studies because it comprises adolescent to adult and senile individuals (Gabunia et al., 2000; Lordkipanidze et al., 2006; Rightmire et al., 2006; Vekua et al., 2002). An edentulous specimen (D3444) and a mandible with heavily worn teeth (D2600) provide first insights into patterns of dentognathic aging until complete tooth loss and massive alveolar bone resorption. Craniodental aging forms an important part of an individual's ontogeny, and further studies are required to evaluate whether these late ontogenetic processes differed between hominins and great apes.

The Neanderthals and us

The LCA of Neanderthals and us is still unknown, but fossils subsumed under the hypodigm of *Homo heidelbergensis* (such as the Heidelberg mandible, and the Steinheim, Broken Hill, and Petralona crania, as well as the Atapuerca paleopopulation) represent likely candidates (Rightmire, 2008). Transition from *H. erectus* to the LCA of humans and Neanderthals is characterized by a marked increase in brain size (Rightmire, 2004), and this trend is continued in the descendant species. Both Neanderthals and modern humans grow very large brains, but while Neanderthals also grew exceptionally large faces, modern human facial growth rates are clearly reduced (Fig. 5).

Which mechanisms govern these species-specific modes of cranial ontogeny? Comparisons of Neanderthal and human cranial ontogenetic trajectories revealed that Neanderthals were clearly different from modern humans already at birth (Gunz et al., 2010; Ponce de León and Zollikofer, 2001; Ponce de León et al., 2008). Many of the "classical" Neanderthal features are already manifest in the neonates, most notably the large projecting face with an anteriorly positioned (prognathic) dentition, and a low but anteroposteriorly and transversely wide neurocranium (Ponce de León et al., 2008).

Neanderthal–human differences in cranial morphology can tentatively be explained by evolutionary modification of growth fields. For example, in modern humans, bone resorption in the external symphyseal region of the mandible leads to a retracted dental arcade and a prominent chin. In the Neanderthals, the alveolar process overgrows the symphyseal region, resulting in a receding chin (Ponce de León and Zollikofer, 2001). A recent study provided the first direct evidence for different locations (heterotopy) and differential activation (heterochrony) of bone growth fields in Neanderthal versus modern human mandibles (Martinez-Maza et al., 2011), demonstrating that the characteristic retromolar space and square-shaped anterior dental arcade are excavated by the activity of resorptive bone growth fields.

Similar arguments can be applied to explain neurocranial shape differences. The wide but low braincase of Neanderthals probably resulted from hyperactivation of lateral endocranial resorptive growth fields and of corresponding exocranial depository fields (Ponce de León and Zollikofer, 2001). These fields are also present in human crania, but less active, thus giving rise to the narrow but high braincase characteristic for our species.

Postnatal ontogeny of Neanderthals and modern humans deployed along essentially parallel trajectories, indicating largely similar modes of cranial development. However, the Neanderthal trajectory was slightly more extended than that of modern humans, indicating more intense facial relative to neurocranial growth (Ponce de León and Zollikofer, 2001). In heterochronic terms, modern humans are thus more pedomorphic than Neanderthals, and vice versa, Neanderthals are more peramorphic than modern humans. Over the past years, the picture of a general heterochronic shift has received several modifications (Bastir et al., 2007). Comparison of neuroendocranial development in modern humans, Neanderthals, and chimpanzees revealed intriguing differences: modern human neuroendocranial ontogeny is characterized by an early phase of "neurocranial globularization," which is less expressed in the Neanderthals and absent in chimpanzees (Bruner et al., 2003; Gunz et al.,

2010; Neubauer et al., 2010). On the other hand, early brain growth occurred at significantly higher rates in Neanderthals than in modern humans, such that Neanderthals reached larger brain sizes at adulthood than extant modern humans (Ponce de León et al., 2008).

These studies raise questions regarding the causal relationships between module-specific ontogenetic trajectories of the brain, the neurocranium, the dentognathic complex, and the cranium as a whole. One fundamental but still unresolved question is to which extent the growing brain shapes the skull and to which extent the growing skull shapes the brain. Based on the hypothesis that neurocranial and facial ontogenies are integrated, endocranial globularity appears as a consequence of the peculiar spatial relationships between the neurocranium and the face in modern humans. However, if one assumes that the brain and the face are largely independent cranial ontogenetic modules and that endocranial shape reflects brain shape, differences in endocranial development might indicate differences in early brain *development.*

Currently, we can only speculate about the adaptive significance of differences in endocranial ontogenies. Differences in brain growth rates imply different metabolic requirements along the ontogenetic trajectory and might have resulted in different life history profiles (Ponce de León et al., 2008; van Schaik et al., 2006). Differences in endocranial development ultimately might reflect differences in cognitive development.

Synchrotron-based histological analysis of dental ontogeny in Neanderthals and in fossil and extant *H. sapiens* yields additional insights into mechanisms of evolutionary developmental modification in these two hominin sister species (Smith, 2008; Smith et al., 2010). The pace of Neanderthal dental ontogeny was at the fast end of the range of variation seen among modern human populations. Whether this implies a faster pace of Neanderthal life history remains to be clarified by further comparative studies. Still little is known about the relationship between dental eruption schedules and mandibular/maxillary development (Boughner and Hallgrimsson, 2008), and it remains to be investigated to which extent differences in dental developmental schedules among modern human populations (Liversidge, 2003) reflect differences in maxillofacial growth and development and/or differences in life history.

Evolved or diseased, or both?—*Homo floresiensis*

The short-statured, small-brained *H. floresiensis* (Brown et al., 2004) has alternatively been interpreted as an Asian branch of australopiths, an endemic island dwarf form of East Asian *H. erectus*, and a microcephalic form of *H. sapiens* (Aiello, 2010). Each of these hypotheses raises interesting ontogenetic questions. The australopith ancestry hypothesis currently lacks support from fossil evidence in SE Asia, but it appears plausible on evolutionary developmental grounds: comparisons of *H. floresiensis* with the similarly sized *A. sediba* will help elucidate whether the former taxon followed a normal or pathological developmental trajectory.

The island dwarfism hypothesis raises questions as to how an organism is "scaled down" through evolutionary modification of developmental rates. Although miniaturization is widespread among primates, still relatively little is known about its developmental basis and on how it affects different ontogenetic modules (Marroig and Cheverud, 2009). Currently, the best comparative evidence for brain–body size miniaturization comes from dwarf hippos, which indicate that dwarfing does not follow the standard allometric brain–body relationships established for mammals (Weston and Lister, 2009).

The microcephaly hypothesis refers to an extreme case of gene-based developmental modification (Ponting and Jackson, 2005), in which rates and/or duration of brain growth are reduced, but facial growth follows a normal trajectory. As shown above, differential modification of brain relative to facial ontogenetic trajectories has substantial effects on overall cranial shape. Without further data on *H. erectus* early cranial ontogeny and on the cranial ontogeny in modern human microcephalic individuals, it will thus be difficult to discriminate between "normal" and pathological patterns of brain-to-face ontogeny.

Hominin cranial ontogeny and birth

During hominin evolution, the trend toward increased encephalization led to large-brained neonates. On the other hand, the trend toward obligate bipedalism led to narrow hips. Modern humans have reached an evolutionary developmental limit, where neonate heads just fit through the birth canal and where birthing bears considerable risks of the mother and the baby (Rosenberg and Trevathan, 1995). In stark contrast, birth in chimpanzees is unproblematic because neonate heads are small and maternal pelvises wide.

When did the "birth dilemma" arise during hominin evolution, and how were obstetric constraints optimized? Modern human optimization consists in giving birth to relatively immature (secondary altricial) neonates, which maintain high fetal-like growth rates for several months after birth (Martin, 1986). In fossil hominins, direct neonate–maternal evidence only exists for the Neanderthals. Neonate brain size was similar as in modern humans, and Neanderthals also reached obstetric limits, although in a different way than modern humans (Ponce de León et al., 2008; Weaver and Hublin, 2009). Computerized simulation of Neanderthal birth indicates that the wide maternal pelves were adapted to give birth to large-faced neonates (Ponce de León et al., 2008).

When exactly fetal cranial ontogeny came into evolutionary conflict with maternal pelvic morphology is still a matter of debate. There is evidence that australopiths gave birth to relatively large neonates, but it is not yet clear how this is related to perinatal cranial growth and birth mechanisms (reviewed in Berge and Goularas, 2010). A female pelvis from Gona, Ethiopia (ascribed to *H. erectus*), has been taken as evidence for incipient obstetric adaptation to large-headed neonates (Ruff, 2010).

Outlook: Evolutionary developmental paleogenomics

Several genes have been advocated as potential hot spots to explain how differences in cranial developmental programs lead to phenetic differences between modern human, fossil hominin, and chimpanzee crania (Carroll, 2003; Fisher and Scharff, 2009; Gilbert et al., 2005; Ponting and Jackson, 2005; Stedman et al., 2004). The Neanderthal draft genome (Green et al., 2010) now offers new ways to investigate the developmental foundations of the last evolutionary split of the hominins. For example, differences in the *RUNX2* gene (whose mutant alleles cause cleidocraniosynostosis) have been proposed to be involved in the differential development of Neanderthal and modern human facial morphologies.

Paleoanthropology has a long tradition of interpreting new hominin fossil finds as modern human pathologies. While Neanderthals are long established as a proper paleospecies, the case for *H. floresiensis* is still open. Traditionally, evolutionary and pathological explanations of fossil morphologies are seen as opposing domains. Evolutionary developmental paleogenomics can be expected to establish explicit links between these domains, as it brings together insights from ontogenetic studies of fossil hominins with insights from clinical studies of congenital cranial malformations (Richtsmeier et al., 2006; Wilkie and Morriss-Kay, 2001). Congenital malformations represent extreme modifications of the cranial ontogenetic program, which decrease the fitness of the affected individual. Evolutionary modifications of the cranial developmental program, on the other hand, have become genetically fixated because they improved the fitness of the affected individuals. In both cases, however, the same basic set of cranial ontogenetic mechanisms is modified.

References

Aiello, L. C. (2010). Five years of *Homo floresiensis*. *American Journal of Physical Anthropology*, *142*(2), 167–179.

Alemseged, Z., Spoor, F., Kimbel, W. H., et al. (2006). A juvenile early hominin skeleton from Dikika, Ethiopia. *Nature*, *443*, 296–301.

Bastir, M., & Rosas, A. (2004). Facial heights: Evolutionary relevance of postnatal ontogeny for facial orientation and skull morphology in humans and chimpanzees. *Journal of Human Evolution*, *47*(5), 359–381.

Bastir, M., O'Higgins, P., & Rosas, A. (2007). Facial ontogeny in Neanderthals and modern humans. *Proceedings of the Royal Society. Series B, Biological Sciences*, *274*(1614), 1125–1132.

Bastir, M., Rosas, A., Stringer, C., et al. (2010). Effects of brain and facial size on basicranial form in human and primate evolution. *Journal of Human Evolution*, *58*(5), 424–431.

Berge, C., & Goularas, D. (2010). A new reconstruction of Sts 14 pelvis (*Australopithecus africanus*) from computed tomography and three-dimensional modeling techniques. *Journal of Human Evolution*, *58*, 262–272.

Berger, L. R., de Ruiter, D. J., Churchill, S. E., et al. (2010). *Australopithecus sediba*: A new species of *Homo*-Like australopith from South Africa. *Science*, *328*(5975), 195–204.

Bienvenu, T. (2011). *The endocranium of* Sahelanthropus tchadensis *(Hominidae, Upper Miocene of Chad): 3D reconstruction and morphology; comparison with extant and fossil hominoids.* Poitiers: University of Poitiers pp. 237.

Bolk, L. (1926). On the problem of anthropogenesis. *Proceedings of the Section of Sciences, Koninklijke Akademie van Wetenschappen te Amsterdam*, *29*, 465–475.

Bookstein, F. L. (1991). *Morphometric tools for landmark data.* Cambridge: Cambridge University Press.

Boughner, J. C., & Hallgrimsson, B. (2008). Biological spacetime and the temporal integration of functional modules: A case study of dento-gnathic developmental timing. *Developmental Dynamics*, *237*(1), 1–17.

Brown, P., Sutikna, T., Morwood, M. J., et al. (2004). A new small-bodied hominin from the Late Pleistocene of Flores, Indonesia. *Nature*, *431*(7012), 1055–1061.

Bruner, E., Manzi, G., & Arsuaga, J. L. (2003). Encephalization and allometric trajectories in the genus *Homo*: Evidence from the Neandertal and modern lineages. *Proceedings of the National Academy of Sciences of the United States of America*, *100*(26), 15335–15340.

Brunet, M., Guy, F., Pilbeam, D., et al. (2002). A new hominid from the Upper Miocene of Chad, Central Africa. *Nature*, *418*(6894), 145–151.

Carroll, S. B. (2003). Genetics and the making of *Homo sapiens*. *Nature*, *422*(6934), 849–857.

Coqueugniot, H., Hublin, J. J., Veillon, F., Houet, F., & Jacob, T. (2004). Early brain growth in *Homo erectus* and implications for cognitive ability. *Nature*, *431*(7006), 299–302.

Dart, R. A. (1925). *Australopithecus africanus*: The man-ape of South Africa. *Nature*, *115*, 195–199.

Dean, M. C. (2006). Tooth microstructure tracks the pace of human life-history evolution. *Proceedings of the Royal Society. Series B, Biological Sciences*, *273*(1603), 2799–2808.

DeSilva, J., & Lesnik, J. (2006). Chimpanzee neonatal brain size: Implications for brain growth in *Homo erectus*. *Journal of Human Evolution*, *51*(2), 207–212.

Enlow, D. H. (1990). *Facial growth.* Philadelphia: Saunders.

Fisher, S. E., & Scharff, C. (2009). FOXP2 as a molecular window into speech and language. *Trends in Genetics*, *25*(4), 166–177.

Fraipont, C. (1936). Les hommes fossiles d'Engis. *Archives de l'Institut de Paleontologie Humaine*, *16*, 1–51.

Gabunia, L., Vekua, A., Lordkipanidze, D., et al. (2000). Earliest Pleistocene hominid cranial remains from Dmanisi, Republic of Georgia: Taxonomy, geological setting, and age. *Science*, *288*(5468), 1019–1025.

Gilbert, S. L., Dobyns, W. B., & Lahn, B. T. (2005). Genetic links between brain development and brain evolution. *Nature Reviews Genetics*, *6*(7), 581–590.

Gitton, Y., Heude, G., Vieux-Rochas, M., et al. (2010). Evolving maps in craniofacial development. *Seminars in Cell & Developmental Biology*, *21*(3), 301–308.

Gould, S. J. (1977). *Ontogeny and phylogeny.* Cambridge, MA and London: The Belknap Press of Harvard University Press.

Gould, S. J., & Lewontin, R. C. (1979). The spandrels of San Marco and the Panglossian paradigm: A critique of the adaptationist programme. *Proceedings of the Royal Society of London. Series B*, *205*, 581–598.

Green, R. E., Krause, J., Briggs, A. W., et al. (2010). A draft sequence of the Neandertal genome. *Science*, *328*(5979), 710–722.

Gunz, P., Neubauer, S., Maureille, B., & Hublin, J.-J. (2010). Brain development after birth differs between Neanderthals and modern humans. *Current Biology*, *20*(21), R921–R922.

Guy, F., Lieberman, D. E., Pilbeam, D., et al. (2005). Morphological affinities of the *Sahelanthropus tchadensis* (Late Miocene hominid from Chad) cranium. *Proceedings of the National Academy of Sciences of the United States of America*, *102*(52), 18836–18841.

Holloway, R. L. (2008). The human brain evolving: A personal retrospective. *Annual Review of Anthropology*, *37*, 1–19.

Hublin, J. J., & Coqueugniot, H. (2006). Absolute or proportional brain size: That is the question. A reply to Leigh's (2006) comments. *Journal of Human Evolution*, *50*, No. 1, 109–113.

Hylander, W. L., Picq, P. G., & Johnson, K. R. (1991). Masticatory-stress hypotheses and the supraorbital region of primates. *American Journal of Physical Anthropology*, *86* (1), 1–36.

Jeffery, N., & Spoor, F. (2004). Ossification and midline shape changes of the human fetal cranial base. *American Journal of Physical Anthropology*, *123*(1), 78–90.

Jolliffe, I. T. (1986). *Principal component analysis.* Berlin, Heidelberg, New York: Springer.

Kelley, J., & Schwartz, G. T. (2010). Dental development and life history in living African and Asian apes. *Proceedings of the National Academy of Sciences of the United States of America*, *107*(3), 1035–1040.

Leigh, S. R. (2004). Brain growth, life history, and cognition in primate and human evolution. *American Journal of Primatology*, *62*(3), 139–164.

Leigh, S. R. (2006). Brain ontogeny and life history in *Homo erectus*. *Journal of Human Evolution*, *50*(1), 104–108.

Lieberman, D. (2011). *The evolution of the human head.* Cambridge, MA: The Belknap Press of Harvard University Press.

Lieberman, D. E., Ross, C. F., & Ravosa, M. J. (2000). The primate cranial base: Ontogeny, function, and integration. *Yearbook of Physical Anthropology*, *43*, 117–169.

Lieberman, D. E., McBratney, B. M., & Krovitz, G. (2002). The evolution and development of cranial form in *Homo sapiens*. *Proceedings of the National Academy of Sciences of the United States of America*, *99*(3), 1134–1139.

Lieberman, D. E., Krovitz, G. E., & McBratney-Owen, B. (2004). Testing hypotheses about tinkering in the fossil record: The case of the human skull. *Journal of Experimental Zoology. Part B, Molecular and Developmental Evolution*, *302B*(3), 284–301.

Lieberman, D. E., Carlo, J., Ponce de León, M. S., & Zollikofer, C. P. E. (2007). A geometric morphometric analysis of heterochrony in the cranium of chimpanzees and bonobos. *Journal of Human Evolution*, *52*(6), 647–662.

Lieberman, D. E., Hallgrimsson, B., Liu, W., Parsons, T. E., & Jamniczky, H. A. (2008). Spatial packing, cranial base angulation, and craniofacial shape variation in the mammalian skull: Testing a new model using mice. *Journal of Anatomy*, *212*(6), 720–735.

Liversidge, H. (2003). Variation in modern human dental development. In J. L. Thompson, G. E. Krovitz & A. J. Nelson (Eds.), *Patterns of growth and development in the genus Homo*. Cambridge: Cambridge University Press.

Lordkipanidze, D., Vekua, A., Ferring, R., et al. (2006). A fourth hominin skull from Dmanisi, Georgia. *Anatomical Record*, *288A*(11), 1146–1157.

Marroig, G., & Cheverud, J. M. (2009). Size and shape in Callimico and marmoset skulls: Allometry and heterochrony in the morphological evolution of small anthropoids. In S. M. Ford, L. M. Porter & L. C. Davis (Eds.), *The smallest anthropoids* (pp. 331–353). New York: Springer.

Martin, R. D. (1986). Ontogenetic and phylogenetic aspects of human brain size. In M. Sakka (Ed.), *Définition et origines de l'homme*. Paris: CNRS.

Martinez-Maza, C., Rosas, A., García-Vargas, S., Estalrrich, A., & de la Rasilla, M. (2011). *Bone remodelling in Neanderthal mandibles from the El Sidrón site (Asturias, Spain).* Biology Letters, *7*(4), 593–596.

McCollum, M. A. (1999). The robust australopithecine face: A morphogenetic perspective. *Science*, *284*, 301–305.

McKinney, M. L. (1999). Heterochrony: Beyond words. *Paleobiology*, *23*, 149–153.

McNulty, K. P., Frost, S. R., & Strait, D. S. (2006). Examining affinities of the Taung child by developmental simulation. *Journal of Human Evolution*, *51*(3), 274–296.

Mitteroecker, P., Gunz, P., Bernhard, M., Schaefer, K., & Bookstein, F. L. (2004). Comparison of cranial ontogenetic trajectories among great apes and humans. *Journal of Human Evolution*, *46*(6), 679–697.

Mitteroecker, P., & Bookstein, F. (2008). The evolutionary role of modularity and integration in the hominoid cranium. *Evolution*, *62*(4), 943–958.

Mitteroecker, P., & Gunz, P. (2009). Advances in geometric morphometrics. *Evolutionary Biology*, *36*(2), 235–247.

Mosimann, J. E. (1988). Size and shape analysis. In L. Kotz & N. L. Johnson (Eds.), *Encyclopedia of statistical sciences* (pp. 219–239). Dordrecht: Holland.

Moss, M. L. (1986). Newer analytical models of craniofacial growth. *Nova Acta Leopoldina NF*, *58*(262), 17–25.

Neubauer, S., Gunz, P., & Hublin, J. J. (2010). Comparing endocranial ontogenetic trajectories between humans and chimpanzees. *American Journal of Physical Anthropology*, 199.

Penin, X., Berge, C., & Baylac, M. (2002). Ontogenetic study of the skull in modern humans and the common chimpanzees: Neotenic hypothesis reconsidered with a tridimensional Procrustes analysis. *American Journal of Physical Anthropology*, *118*, 50–62.

Ponce de León, M. S., & Zollikofer, C. P. E. (2001). Neanderthal cranial ontogeny and its implications for late hominid diversity. *Nature*, *412*, 534–538.

Ponce de León, M. S., & Zollikofer, C. P. E. (2006). Neanderthals and modern humans—Chimps and bonobos: Similarities and differences in development and evolution. In K. Harvati & T. Harrison (Eds.), *Neanderthals revisited: New approaches and perspectives* (pp. 71–90). New York: Springer.

Ponce de León, M. S., Golovanova, L., Doronichev, V., et al. (2008). Neanderthal brain size at birth provides insights into the evolution of human life history. *Proceedings of the National Academy of Sciences of the United States of America*, *105*, 13764–13768.

Ponting, C., & Jackson, A. P. (2005). Evolution of primary microcephaly genes and the enlargement of primate brains. *Current Opinion in Genetics and Development*, *15*(3), 241–248.

Rice, S. H. (1997). The analysis of ontogenetic trajectories: When a change in size or shape is not heterochrony. *Proceedings of the National Academy of Sciences of the United States of America*, *94*(3), 907–912.

Richtsmeier, J. T., Aldridge, K., DeLeon, V. B., et al. (2006). Phenotypic integration of neurocranium and brain. *Journal of Experimental Zoology. Part B, Molecular and Developmental Evolution*, *306*(4), 360–378.

Rightmire, G. P. (2004). Brain size and encephalization in early to mid-Pleistocene Homo. *American Journal of Physical Anthropology*, *124*(2), 109–123.

Rightmire, G. P., Lordkipanidze, D., & Vekua, A. (2006). Anatomical descriptions, comparative studies and evolutionary significance of the hominin skulls from Dmanisi, Republic of Georgia. *Journal of Human Evolution*, *50*(2), 115–141.

Rightmire, G. P. (2008). Homo in the Middle Pleistocene: Hypodigms, variation, and species recognition. *Evolutionary Anthropology*, *17*, 8–21.

Rosenberg, K., & Trevathan, W. (1995). Bipedalism and human birth: The obstetrical dilemma revisited. *Evolutionary Anthropology*, *4*(5), 161–168.

Ruff, C. (2010). Body size and body shape in early hominins—Implications of the Gona pelvis. *Journal of Human Evolution*, *58*(2), 166–178.

Ruff, C. B., Trinkaus, E., & Holliday, T. W. (1997). Body mass and encephalization in Pleistocene *Homo*. *Nature*, *387* (6629), 173–176.

Schmerling, P.-C. (1832–1833). Sur des cavernes à ossements de la province de Liège. *Bulletin de la Societe Geologique de France*, *3*, 217–222.

Senut, B., Pickford, M., Gommery, D., Mein, P., Cheboi, K., & Coppens, Y. (2001). First hominid from the Miocene (Lukeino Formation, Kenya)Premier hominidÈ du MiocËne (formation de Lukeino, Kenya). *Comptes rendus de l'Académie des Sciences*, *332*(2), 137–144.

Smith, B. H., & Tompkins, R. L. (1995). Toward a life history of the *Hominidae*. *Annual Review of Anthropology*, *24*, 257–279.

Smith, T. M. (2008). Incremental dental development: Methods and applications in hominoid evolutionary studies. *Journal of Human Evolution*, *54*(2), 205–224.

Smith, T. M., Tafforeau, P., Reid, D. J., et al. (2010). Dental evidence for ontogenetic differences between modern humans and Neanderthals. *Proceedings of the National Academy of Sciences of the United States of America*, *107* (49), 20923–20928.

Spoor, F., Leakey, M. G., Gathogo, P. N., et al. (2007). Implications of new early *Homo* fossils from Ileret, east of Lake Turkana, Kenya. *Nature*, *448*, 688–691.

Stedman, H. H., Kozyak, B. W., Nelson, A., et al. (2004). Myosin gene mutation correlates with anatomical changes in the human lineage. *Nature*, *428*(6981), 415–418.

Suwa, G., Asfaw, B., Kono, R. T., Kubo, D., Lovejoy, C. O., & White, T. D. (2009). The *Ardipithecus ramidus* skull and its implications for hominid origins. *Science*, *326*, 68e1–68e7 No. 5949.

Szabo-Rogers, H. L., Smithers, L. E., Yakob, W., & Liu, K. J. (2010). New directions in craniofacial morphogenesis. *Developmental Biology*, *341*(1), 84–94.

Tafforeau, P., Boistel, R., Boller, E., et al. (2006). Applications of X-ray synchrotron microtomography for non-destructive 3D studies of paleontological specimens. *Applied Physics. A, Materials Science & Processing*, *83*(2), 195–202.

Tafforeau, P., & Smith, T. M. (2008). Nondestructive imaging of hominoid dental microstructure using phase contrast X-ray synchrotron microtomography. *Journal of Human Evolution*, *54*(2), 272–278.

Tobias, P. V. (1985). *Dart, Taung and the "missing link"*. Johannesburg: Witwatersrand Univ Press.

Ungar, P. S., Grine, F. E., & Teaford, M. F. (2008). Dental microwear and diet of the Plio-Pleistocene hominin *Paranthropus boisei*. *PLoS One*, *3*(4), e2044.

van Schaik, C., Barrickman, N., Bastian, M., Krakauer, E., & van Noordwijk, M. (2006). Primate life histories and the role of brains. In K. Hawkes & R. R. Pain (Eds.), *The evolution of human life history*. Santa Fe: School of American Reserach Press.

Vekua, A., Lordkipanidze, D., Rightmire, G. P., et al. (2002). A new skull of early *Homo* from Dmanisi, Georgia. *Science*, *297*(5578), 85–89.

Walker, A., & Leakey, R. (1993). *The Nariokotome* Homo erectus *skeleton*. Berlin: Springer.

Weaver, T. D., & Hublin, J. J. (2009). Neandertal birth canal shape and the evolution of human childbirth. *Proceedings of the National Academy of Sciences of the United States of America*, *106*(20), 8151–8156.

Weston, E. M., & Lister, A. M. (2009). Insular dwarfism in hippos and a model for brain size reduction in *Homo floresiensis*. *Nature*, *459*(7243), 85–88.

White, T. D., Asfaw, B., Beyene, Y., et al. (2009). *Ardipithecus ramidus* and the paleobiology of early hominids. *Science*, *326*(5949), 75–86.

Wilkie, A. O. M., & Morriss-Kay, G. M. (2001). Genetics of craniofacial development and malformation. *Nature Reviews Genetics*, *2*, 458–468.

Wood, B., & Collard, M. (1999). The human genus. *Science*, *284*, 65–71.

Wood, B., & Lonergan, N. (2008). The hominin fossil record: Taxa, grades and clades. *Journal of Anatomy*, *212*(4), 354–376.

Zollikofer, C. P. E., & Ponce de León, M. S. (2002). Visualizing patterns of craniofacial shape variation in *Homo sapiens*. *Proceedings of the Royal Society. Series B, Biological Sciences*, *269*, 801–807.

Zollikofer, C. P. E., & Ponce de León, M. S. (2004). Kinematics of cranial ontogeny: Heterotopy, heterochrony, and geometric morphometric analysis of growth models. *The Journal of Experimental Zoology*, *302B*, 322–340.

Zollikofer, C. P. E., & Ponce de León, M. S. (2005). *Virtual reconstruction: A primer in computer-assisted paleontology and biomedicine.* New York: Wiley.

Zollikofer, C. P. E., Ponce de León, M. S., Lieberman, D. E., et al. (2005). Virtual cranial reconstruction of *Sahelanthropus tchadensis*. *Nature*, *434*(7034), 755–759.

Zollikofer, C. P. E., & Ponce de León, M. S. (2010). The evolution of hominin ontogenies. *Seminars in Cell & Developmental Biology*, *21*(4), 441–452.

M. A. Hofman and D. Falk (Eds.)
Progress in Brain Research, Vol. 195
ISSN: 0079-6123

CHAPTER 14

Hominins and the emergence of the modern human brain

Alexandra de Sousa* and Eugénia Cunha

Department of Life Sciences, Forensic Sciences Center,
University of Coimbra, Coimbra, Portugal

Abstract: Evidence used to reconstruct the morphology and function of the brain (and the rest of the central nervous system) in fossil hominin species comes from the fossil and archeological records. Although the details provided about human brain evolution are scarce, they benefit from interpretations informed by interspecific comparative studies and, in particular, human pathology studies. In recent years, new information has come to light about fossil DNA and ontogenetic trajectories, for which pathology research has significant implications. We briefly describe and summarize data from the paleoarcheological and paleoneurological records about the evolution of fossil hominin brains, including behavioral data most relevant to brain research. These findings are brought together to characterize fossil hominin taxa in terms of brain structure and function and to summarize brain evolution in the human lineage.

Keywords: evolution; endocast; fossil hominin; paleoneurology; cognition; paleoarcheology.

Introduction

A major obstacle to understanding the evolution of the human brain is that it is a soft tissue not directly preserved in the fossil record, although information about its size and shape is preserved in natural endocasts and fossilized neurocrania. Recent insights have come from progress in archeology, ancient DNA, and life history. The combination of paleoneurological, paleoarcheological, paleogenetic, and ontogenetic evidence, informed by comparative studies and pathology research, is reviewed here to reconstruct the evolution of the human brain.

"Hominins" include extant humans (herewith referred to as "humans" or "recent anatomically modern *Homo sapiens*—rAMHS") and all extinct species that are more closely related to humans than to any other living taxon; "Panins" include chimpanzees, bonobos, and all fossil species more closely related to them than to humans. The

*Corresponding author.
E-mail: alexandra.allison.sousa@gmail.com

DOI: 10.1016/B978-0-444-53860-4.00014-3

hominin clade dates back to the most recent common ancestor (MRCA) of humans, chimpanzees, and bonobos, via the appearance of the first hominins ca. 8–4 Ma. Anatomically modern *Homo sapiens* (AMHS), encompassing both fossil (fAMHS) and recent (rAMHS) members, emerged from this clade a little over 190 ka (McDougall et al., 2005). The emergence of the human brain is reviewed here with respect to different hominin taxa, based on a speciose taxonomy (Table 1), as described previously (de Sousa and Wood, 2007; see also Wood and Lonergan, 2008).

Interpretations of fossil hominin brain structure and function benefit from ongoing comparative and pathology studies. Comparative samples are used to determine interspecific trends, for example, to estimate the degree of encephalization. Comparisons between living species improve our understanding of how the brain works and indicate any unique attributes of the human brain. Although humans are more closely related to panins than to any other living animals, it is usual to compare them to the "apes," a group defined on the basis of behavioral and anatomical similarities. Apes include the great apes (panins and gorillas in Africa, orangutans in Asia) and the lesser apes (gibbons). Apes and humans together comprise the "hominoids."

Panins (especially chimpanzees) are used as a proxy for the ancestral hominin state, unless indicated otherwise. In order to determine whether a feature is "primitive" or "derived" within the hominin clade, it is necessary to consider the brain morphology of the hominin–panin MRCA, which would have possessed all shared derived features of extant humans, chimpanzees, and bonobos but would lack those features acquired solely along either the panin or the hominin lineages. It is particularly difficult to reconstruct the brain of the hominin–panin MRCA because a panin fossil

Table 1. Approximate date ranges for fossil hominin species, which are defined according to both speciose and short taxonomies

	Speciose taxonomy	Short taxonomy	Approximate date range
Possible hominins	*Sahelanthropus tchadensis*	*Ardipithecus ramidus s.l.*	7 Ma
	Orrorin tugenensis	*Ardipithecus ramidus s.l.*	6 Ma
	Ardipithecus kadabba	*Ardipithecus ramidus s.l.*	5.8–5.2 Ma
	Ardipithecus ramidus	*Ardipithecus ramidus s.l.*	4.5–4.3 Ma
Archaic hominins	*Australopithecus anamensis*	*Australopithecus afarensis s.l.*	4.2–3.9 Ma
	Australopithecus afarensis	*Australopithecus afarensis s.l.*	3.7–3 Ma
	Australopithecus bahrelghazali	*Australopithecus afarensis s.l.*	3.5–3.0 Ma
	Kenyanthropus platyops	*Kenyanthropus platyops*	3.5–3.3 Ma
	Australopithecus africanus	*Australopithecus africanus*	3–2.4 Ma
	Australopithecus sediba	*Australopithecus sediba*	1.95 Ma
Megadont archaic hominins	*Australopithecus garhi*	*Australopithecus garhi*	2.5 Ma
	Paranthropus aethiopicus	*Paranthropus boisei s.l.*	2.5–2.3 Ma
	Paranthropus boisei s.s.	*Paranthropus boisei s.l.*	2.3–1.4 Ma
	Paranthropus robustus	*Paranthropus robustus*	2.0–1.5 Ma
Transitional hominins	*Homo habilis s.s.*	*Homo habilis s.l.*	2.4–1.4 Ma
	Homo rudolfensis	*Homo habilis s.l.*	2.4–1.6 Ma
Premodern *Homo*	*Homo erectus s.s.*	*Homo erectus s.l.*	1.9–1.5 Ma
	Homo ergaster	*Homo erectus s.l.*	1.8 Ma–30 ka
	Homo antecessor	*Homo antecessor*	780–500 ka
	Homo heidelbergensis	*Homo sapiens s.l.*	600–100 ka
	Homo neanderthalensis	*Homo sapiens s.l.*	200–28 ka
	Homo floresiensis	*Homo floresiensis*	74–17 ka
Anatomically modern *Homo*	*Homo sapiens s.s.*	*Homo sapiens s.l.*	195 ka–present

Note: *s.s.*, *sensu stricto*; *s.l.*, *sensu lato*.

record was completely unknown until recently (McBrearty and Jablonski, 2005). So, for practical purposes, the brains and behaviors of panins are assumed to be equivalent to those primitive for hominins.

Traveling from past to present in the hominin fossil record, species appear which are increasingly related to humans (although there exists evidence for at least one other major hominin lineage, the megadont archaic hominins). Hominins which are phylogenetically closest to humans are presumably less panin-like and more humanlike. For this reason, intraspecific comparisons, usually between clinical cases and neurotypical humans, are used to assess variation at low taxonomic levels.

Pathology's contributions to brain evolution research

Pathology (here defined broadly to include the study of all clinically abnormal conditions including injuries, disabilities, disorders, syndromes, and infections) has enlightened our understanding of hominin brain evolution in two major ways. First, pathologies highlight human biological mechanisms. Second, an understanding of the anatomical (and behavioral) manifestations of pathologies is necessary for correctly interpreting the fossil (and archeological) record. These major categories of contributions are described here, and further examples of how pathologies play into interpretations of fossil hominin brains can be found elsewhere in this chapter.

Pathologies mark neural and genetic mechanisms

Pathological studies have made significant contributions to neuroscience. Early brain mapping was conducted by associating behavioral deficits to regions of physical brain damage. For example, Paul Broca found that two stroke patients who had lost speech ability both had damage to the posterior inferior frontal gyrus at autopsy, a region which now bears his name (Dronkers et al., 2007). Similarly, human pathologies are increasingly attributed to genetic causes, such as mutations in gene sequences, or changes in gene copy number. Pathologies which interrupt the function of human-specific behaviors have drawn special attention in brain evolution research. Microcephaly and the severe speech and language disorder of the KE family highlight mechanisms which may be responsible for human-specific brain structure and function, and the possibility that these pathologies are atavisms has been discussed (Fisher and Marcus, 2006; Jackson et al., 2002).

Autism, a neurodevelopmental disorder characterized by impaired social interaction for which a clinical definition is ongoing (Charman et al., 2010; Happe et al., 2006), is referenced in studies of fossil hominin brain structure and function, either as an analogy for developmental differences between closely related species or as a potentially atavistic indication of actual primitive phenotypes. For example, an autistic child lacking language created naturalistic artwork much like that from the Upper Paleolithic, on the basis of which it was suggested that fAMHS could have also lacked fully modern cognition (Humphrey, 1998).

Recognizing pathology in fossils

Pathology has always had a role in paleoanthropology. At the time of the first acknowledgment of a fossil hominin species, *H. neanderthalensis*, the prominent pathologist Rudolf Virchow wrote off the distinct anatomy as features of aging, arthritis, fracture, and rickets (Cartmill and Smith, 2009). Virchow's interpretation was quashed after additional specimens were found to show the same distinct morphology. Pathological explanations continue to be put forward for *H. neanderthalensis*, including vitamin D deficiency (Ivanhoe, 1970) and cretinism (Dobson, 1998), although these hypotheses are not generally accepted.

Most recently, a scatter of pathologies have been proposed to describe the morphology of *H.*

floresiensis, including Laron syndrome (Hershkovitz et al., 2007), cretinism (Obendorf et al., 2008; Oxnard et al., 2010), and microcephaly and dwarfism (Martin et al., 2006), or microcephaly in a pygmoid population (Jacob et al., 2006). All these have subsequently been addressed (Falk et al., 2007, 2009a, b; Jungers et al., 2009). There is much yet to glean from the discovery of *H. floresiensis*, a species which has quickly achieved a special position in neuroscience and paleoanthropology through its challenge to prevailing notions about brain size, organization, and function (Cunha and Silva, 2005). The interpretation of *H. floresiensis* faces similar challenges as with *H. neanderthalensis*, although it is further complicated by the poor preservation of skeletal material and DNA on a tropical island. Identifying pathology also has direct implications for assessing behavior. Specimens of *H. neanderthalensis* and *Homo heidelbergensis* showing advanced age, trauma, and pathology (Bonmati et al., 2010; Gracia et al., 2011) have been interpreted as evidence of humanlike social and cognitive capacities (Spikins et al., 2010).

Fossil hominin brain size

Human mean adult's (21- to 39-year-old) brain weight is 1450g for males and 1290g for females (Fig. 1; Table 2) (Dekaban and Sadowsky, 1978). Chimpanzee mean adolescent and young adult (7–30 years) brain weight is 406g for males and 368 g for females (Herndon et al., 1999). It is expected that the averages used here are higher than in other autopsy samples, which include a majority of individuals with advanced age. In both species, brain weight decreases in older adults (Resnick et al., 2003), for example, Dekaban and Sadowsky (1978)

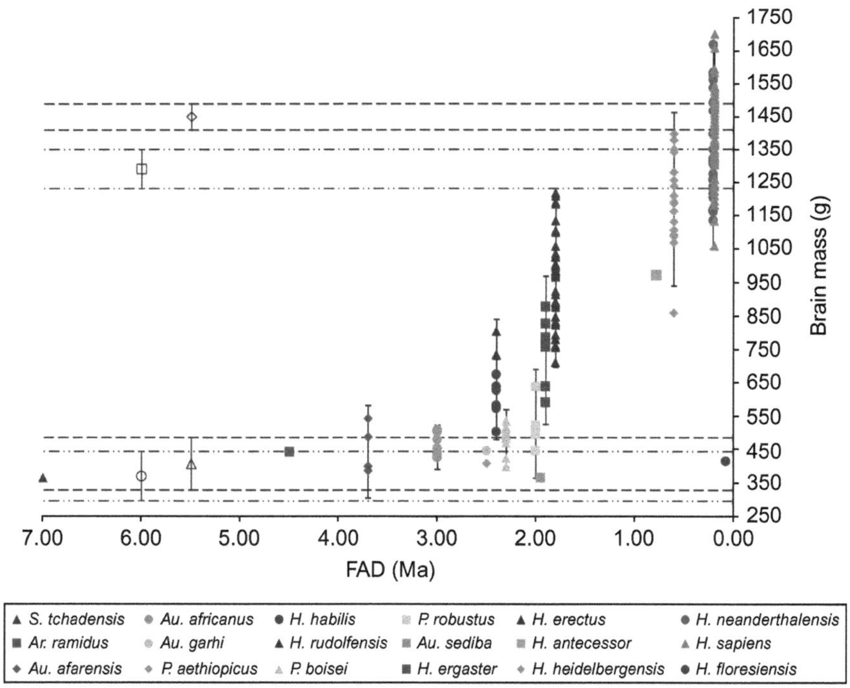

Fig. 1. Chimpanzee male (open triangle) and female (open circle) and rAMHS male (open diamond) and female (open square) brain weight means are plotted, with *Y*-axis bars and dashed lines showing ranges within two standard deviations. Fossil hominin brain weight individual specimen values are plotted with *Y*-axis bars showing range within two standard deviations from the mean. For more information, see Table 2.

Table 2. Absolute and relative brain size values for fossil and extant panin and hominin taxa

Taxon[a]	FAD (mya)	No. endocranial vols. or brain wts.	Mean endocranial vol. (cm^3)	Minimum endocranial vol. (cm^3)	Maximum endocranial vol. (cm^3)	Mean brain wt. (g)[b]	Minimum brain wt. (g)	Maximum brain wt. (g)	Brain wt. SD	Mean body wt. (kg)[c]	EQ[d]
Pan troglodytes (M)		17				406	347	530	39	58	1.65
Pan troglodytes (F)		17				368	308	458	37	43	1.88
rAMHS (M)		351				1450	1343	1526	20	70	5.10
rAMHS (F)		201				1290	1239	1366	30	57	5.35
S. tchadensis	7.0	1	365			363					
Au. afarensis	3.7	5	446	387	550	442	385	542	69	38	2.50
Au. afarensis (M?)	3.7	2	521	492	550	514	486	542	40	45	2.56
Au. afarensis (F?)	3.7	3	396	387	400	393	385	397	7	29	2.69
Au. africanus	3.0	9	460	428	515	455	424	508	33	34	2.78
Au. garhi	2.5	1	450			446					
P. aethiopicus	2.5	1	410			407					
H. habilis	2.4	6	609	509	687	599	503	674	60	33	3.72
H. rudolfensis	2.4	3	776	750	825	758	734	805	41	55	3.21
P. boisei	2.3	10	488	400	545	483	397	537	43	41	2.54
P. robustus	2.0	4	533	450	650	525	446	638	82	36	3.07
Au. sediba	2.0	1	420			363					
H. erectus	1.9	36	991	727	1260	963	712	1218	134	58	3.94
H. ergaster	1.8	6	763	600	900	746	590	877	111	64	2.81
H. ergaster (Africa)	1.8	3	851	804	900	830	785	877	46	64	3.12
H. ergaster (Dmanisi)	1.8	3	675	600	775	662	590	758	86		
H. antecessor	0.78	1	1000			972					
H. heidelbergensis	0.60	21	1242	880	1450	1200	858	1397	131	71	4.21
H. neanderthalensis	0.20	27	1404	1172	1740	1353	1135	1669	153	72	4.67
H. sapiens	0.20	79	1463	1090	1880	1408	1057	1799	124	64	5.30
H. floresiensis	0.07	1	417			414				26	3.10

[a]Sources as follows: Chimpanzee brain and body weights from individuals 7–30 years, from Herndon et al. (1999). rAMHS brain and body weights from adults 21–39 years (except min. and max. brain weights, which are for 20–30 years), from Dekaban and Sadowsky (1978). In both datasets, "brain weight" is taken from fresh autopsy specimens and includes brain tissue as well as leptomeninges and CSF. See Appendix for references of the volumes on which fossil hominin species data are based.
[b]Fossil endocranial volumes were converted into brain weights after Ruff et al. (1997).
[c]Mean body weights are from Skinner and Wood (2006).
[d]After Martin (1981) and Ruff et al. (1997). Extant taxon EQs are means of individual EQs. Fossil taxon sample mean EQs are obtained from each taxon's mean brain weight and mean body weight estimates. EQs obtained by either method are very similar and have been used interchangeably (e.g., Ruff et al., 1997).

reported a 7.4% (~100g) decrease in human brain weight between 20–30 years and 70–80 years. Although in living species males and females often have significantly different brain sizes, it is not possible to know the sex of fossils with certainty, and statistical methods of sexing are not possible for the small samples of early hominin crania. Here, fossil specimens are not sexed but are compared to extant species data in which sex is known (Fig. 1). All statistical comparisons from this dataset (given in text) are derived from a Kruskal–Wallis test of significance.

Previously, absolute brain size has been used to determine a cerebral rubicon criterion for inclusion in the genus *Homo*, variably set between 600 and 800 cm^3 (Leakey et al., 1964). Until recently, absolute brain size was thought to lack biological significance, as it does not give an indication of degree of encephalization, or the number of "extra neurons" (Jerison, 1973; Martin, 1990). However, a recent study has found that absolute brain size predicts cognitive ability in primates, whereas encephalization quotient (EQ, a measure of brain weight relative to body size) does not (Deaner et al., 2007). Many aspects of brain morphology such as brain component volumes and degree of gyrification scale to absolute brain size (Semendeferi and Damasio, 2000; Semendeferi et al., 2002; Weaver, 2005; Zilles et al., 1988, 1989), so it is an important consideration when making comparisons between the morphology of fossil endocasts.

Fossil hominin species mean brain weights and EQs were estimated using allometric equations (Table 1; for a discussion of methods, see de Sousa, 2008; de Sousa and Wood, 2007). We used Ruff et al.'s (1997) formula, based on Martin (1981), and used Ruff et al.'s (1997) calculation for estimating brain weight from endocranial volume, based on Martin (1990).

Fossil hominin brain morphology

Inferences can be made about the size and shape of the brain from natural endocasts (casts of the neurocranial cavity) and fossilized cranial skeletal remains. Convolutional details are notoriously difficult to interpret (Holloway, 1966; Symington, 1916). A feature may be the impression of a sulcus, a blood vessel, or a skeletal suture, or it may be an artifact, and observers may offer genuinely different interpretations of what the same feature represents (Connolly, 1950; Falk, 1980b). Aspects of brain morphology of hominin species inferred from fossil endocranial data are summarized here and in Table 3 (see also Chapter 12).

Not all researchers are convinced that the detailed morphology of endocasts has functional relevance. At one time, paleoneurology took for granted that sulci delimit functional or somatotopic cortical areas (see Radinsky, 1972, and references therein). It is not understood, however, that primate brains exhibit a substantial amount of intraspecific variability in sulcal anatomy and cytoarchitectural boundaries (Geyer et al., 2001, Rademacher et al., 2001, Amunts et al., 2007). There are some cases in which the relationship between a sulcal landmark and functional area border is maintained, at least within a species (Holloway et al., 2003), but in other cases, it varies within species (Sherwood et al., 2003). Extant primates' brains and endocrania are used to make inferences about fossil hominins' brains, but data from apes are rare, so most inferences should be treated as preliminary.

Left-occipital right-frontal petalia

A petalia is a protrusion of one cerebral hemisphere relative to the other. The left-occipital right-frontal (LORF) petalia is an asymmetrical pattern in which there is a wider and more posteriorly protruding left-occipital pole, and a wider and more rostrally protruding right-frontal lobe. The LORF petalia is typical of humans and is statistically related to right-handedness —that is, left-handed and ambidextrous people are more likely to be symmetrical or have the opposite pattern (Le May, 1976). It is not clear whether apes exhibit humanlike petalias. Le May (1976) and

Table 3. Brain morphology of fossil hominin species inferred from fossil endocranial data

Taxon	FAD (Ma)	LORF petalia[a]	Fronto-orbital sulcus[b]	Orbital surface of the frontal lobe[c]	Broca's cap region[d]	Neurocranial globularity[e]	Temporal pole morphology[f]	Lunate sulcus position[g]	Relative size of cerebellum (CQ)[h]
P. troglodytes		P	P	P	P	P	P	P	1.2
rAMHS		H	H	H	H	H	H	H	1
Au. afarensis	3.7	I	–	–	I	–	–	P/H	–
Au. africanus	3	h	P	h	h	–	h	H	0.8
P. aethiopicus	2.5	h	–	P	–	–	P	–	–
H. habilis s.s.	2.4	P	P	–	I	–	–	–	1
H. rudolfensis	2.4	H	H	–	H	–	–	–	0.9
P. boisei s.s.	2.3	h	–	P	–	–	P	H	1
P. robustus	2	I	–	P	–	–	P	H	–
H. erectus s.s.	1.9	H	–	–	H	P	–	H	0.9
H. ergaster	1.8	H	–	–	I	–	–	–	0.9
H. heidelbergensis	0.60	H	–	–	H	P	–	H	0.8
H. neanderthalensis	0.20	H	–	–	H	P	–	H	0.7
H. sapiens s.s.	0.20	H	–	–	H	H	–	H	0.7
H. floresiensis	0.07	H	H	H	I	–		H	–

Notes: –, no relevant evidence; I, insufficient evidence; H, humanlike morphology either described or inferred; h, incipient humanlike morphology either described or inferred; P, panin-like morphology either described or inferred.

Panin-like (P) and humanlike (H) morphology as follows. Refer to text for a more details.

[a]Left-occipital right-frontal petalial pattern (P) infrequent, rarely involves both frontal and occipital lobes; (H) usual.

[b]Fronto-orbital sulcus (P) present; (H) absent.

[c]Orbitofrontal region (P) beak shaped; (H) blunt and expanded.

[d]Asymmetrical Broca's area (P) not asymmetrically enlarged; (H) L>R asymmetry.

[e]Endocast shape (P) "archaic"; (H) globular, suggests expanded parietal.

[f]Temporal pole morphology (P) rounded; (H) expanded in anterior and lateral directions.

[g]Lunate sulcus position (P) anterior (some variability); (H) more posterior.

[h]Mean CQ values, calculated from specimen CQ values (LSR-05 in Weaver, 2001).

Le may et al. (1982) found that petalias are also common in great apes. However, Holloway and de Lacoste-Lareymondie (1982) found them to be less frequent than in humans and rarely involving both the frontal and the occipital lobes but noted a high incidence of left-occipital petalias in gorillas. In a more recent MR study, however, Hopkins and Marino (2000) found that great apes display humanlike right-frontal and left-occipital petalias.

Humanlike LORF petalias are also usual in *H. heidelbergensis*, *H. neanderthalensis*, *H. sapiens* (Holloway et al, 2004a). LORF petalias are also attributed to specimens of *H. rudolfensis* (Holloway, 1983), *H. ergaster* (Begun and Walker, 1993; Holloway et al., 2004a), *H. erectus* (Broadfield et al., 2001; Holloway, 1980; Holloway et al., 2004a). A pronounced but reversed (ROLF) petalia pattern has been described for *H. floresiensis* and may be related to left-handedness (Falk et al., 2005, 2009b). Small LORF petalias have been described for *Australopithecus africanus*, *Paranthropus boisei*, and *Paranthropus aethiopicus* (Holloway et al., 2004a). *H. habilis* specimens all lack humanlike LORF petalias. (Holloway et al., 2004a; Tobias, 1987). There is insufficient evidence for describing petalias in *Australopithecus afarensis*, although there may be evidence of a slight left-occipital petalia in one specimen, AL 333–45 (Holloway et al., 2004a).

Orbital frontal lobe shape

The orbital surface of the frontal lobe is blunt and expanded in humans. In contrast, it is described as beaked and pointed in the African apes. This region corresponds to part of BA10, which is involved in planning future actions, abstract thinking, and undertaking initiatives (Semendeferi et al., 2001). The human BA10 is characterized by increased horizontal spacing between cell bodies (Semendeferi et al., 2011), and a larger volume than expected for an ape of human brain size (but the residual [6%] is less striking than for other regions; Holloway, 2002). The prefrontal cortex in the region of BA10 is expanded in two large convolutions that straddle the rostrodorsal midline between the frontal lobes in *H. floresiensis*, unlike neurotypical (non-microcephalic) *H. sapiens*, in whom BA10 is enlarged but not manifested in such convolutions (Falk, 2009; Falk et al., 2005). This region is also somewhat expanded in *Au. africanus* (Smith, 1927), but not in *P. aethiopicus*, *P. boisei*, and *Paranthropus robustus* (Falk et al., 2000).

Fronto-orbital sulcus

The fronto-orbital (orbitofrontal) sulcus typically incises the orbitolateral border of the frontal lobe of African apes but is not present on human brains (Connolly, 1950; Falk, 1980a; Chapter 12). According to Connolly (1950), opercular expansion of the frontal lobe during hominin evolution covered this sulcus and caused it to shift caudally, where it became part of the anterior limiting sulcus of the insula. The human frontal lobe (Semendeferi and Damasio, 2000; Semendeferi et al., 1997) and its cortex (Semendeferi et al., 2002) have volumes expected for an ape of similar brain size. It has been suggested that the human prefrontal cortex is larger than expected for a primate with a similar sized brain (Deacon, 1997), supported by the finding of increased gyrification in this region (Rilling and Insel, 1999). It has further been suggested that the human prefrontal cortex has a higher than expected white-to-gray matter ratio (Schoenemann et al., 2005). Inferences about the prefrontal cortex are disputed because this region is difficult to delimit (Semendeferi et al., 2002; Sherwood et al., 2005). It is possible that the human frontal lobe scales to brain size, although the prefrontal cortex is enlarged proportionally within it.

Humanlike morphology, in which the fronto-orbital sulcus is absent, has been reported for *H. rudolfensis* (Falk, 1983) and *H. floresiensis* (Falk et al., 2005). An apelike fronto-orbital sulcus has been reported for *Au. africanus* and *H. habilis* (Falk, 1980a, 1983).

Broca's cap

Broca's cap, as seen on endocasts, represents portions of Brodmann's areas (BAs) 47 and 45 as identified on human brains (Broadfield et al., 2001; Chapter 12). Broca's cap overlaps (but does not exactly correspond to) Broca's language area. Broca's area corresponds to BA45 and BA44, (respectively, pars triangularis and pars opercularis of the inferior frontal gyrus; Aboitiz and Garcia, 1997). In the majority of humans, the left hemisphere is dominant for language, and BA44 (but not BA45) on the left hemisphere is asymmetrically enlarged in comparison to the contralateral BA44 (Amunts et al., 1999). Although an enlarged Broca's cap is a characteristic of humans, it might also occur, albeit more rarely, in apes (Holloway, 1996).

Questions persist about whether the homologue of Broca's area in apes exhibits humanlike asymmetry (Cantalupo and Hopkins, 2001; Holloway, 1996; Sherwood et al., 2003). A recent study of minicolumn size in BA44 and BA45 has indicated that apes lack a species-level pattern of asymmetries like that seen in humans (Schenker et al., 2008). Investigators draw attention to humanlike Broca's cap asymmetry in fossil hominins, in particular, in specimens in which the left side is larger than its homologue on the right

(L>R). They also describe overall size and convolutional detail—particularly in fossils where only one hemisphere is present. Because a Broca's area L>R asymmetry is a characteristic of most humans, and most humans are right-handed, it is related to right-handedness and attention is drawn to fossil hominins with both Broca's area L>R and LORF L>R asymmetries.

Attention is drawn to fossil hominins in which there is a pattern of both Broca's area and LORF L>R asymmetries, presumably because these L>R asymmetries are characteristic of humans, and most humans are both right-handed and left-lateralized for language. Broca's area's involvement in hand movement (Fadiga and Craighero, 2006) as well as language is the basis of the mirror system hypothesis which sees laterality for handedness as a precursor to language (Arbib, 2005; Rizzolatti and Arbib, 1998). However, the anatomical pattern of asymmetry in Broca's area is variable and poorly understood in relation to the functional asymmetries of language and handedness (Keller et al., 2009). Further, brain surface morphology does not provide consistent landmarks for identifying the cytoarchitectonic borders of Broca's region, particularly when making comparisons between species (Falk, 2007; Sherwood et al., 2003).

A clearly delimited, humanlike Broca's cap exhibiting L>R asymmetry is usual in *H. heidelbergensis*, *H. neanderthalensis*, and *H. sapiens* (Holloway et al., 2004a). It has also has been reported in specimens of *H. rudolfensis* (Begun and Walker, 1993; Holloway, 1983; Holloway et al., 2004a; Tobias, 1975) and *H. erectus* (Broadfield et al., 2001; Holloway et al., 2004a). The Broca's cap region demonstrates a trend toward a humanlike pattern in *Au. africanus* (Holloway et al., 2004a). Descriptions of the Broca's cap region have been insufficient or inconclusive in *Au. afarensis* (Holloway et al., 2004a), *H. habilis* (Holloway et al., 2004a; Tobias, 1987), *H. ergaster* (Begun and Walker, 1993; Holloway et al., 2004a), and *H. floresiensis* (Falk et al., 2005).

Temporal poles

Falk et al. (2005, 2000) described human endocasts as having temporal poles which are extended in the anterior and lateral directions, whereas African apes have rounded temporal poles. More generally, human temporal lobes are larger in total volume, white matter volume, and surface area than predicted for an ape of similar brain size (Rilling and Seligman, 2002). In humans, the anterior lateral temporal pole, particularly in the left hemisphere, is involved in face recognition and naming (Damasio et al., 1996; Grabowski et al., 2001). The corresponding monkey area, TG, also functions in visual learning and recognition (Horel et al., 1984; Nakamura and Kubota, 1995).

Enlarged temporal lobes in fossil hominins are considered to be humanlike, in contrast to the smaller temporal lobes of apes (Dart, 1940; Falk et al., 2000; Smith, 1927). Falk and others have made two different categories of observations about temporal lobe size, where the information is available. First, they have described the temporal poles of rAMHS and *Au. africanus* as pointed due to the forward projections of the poles beyond the anterior borders of sella turcica, and the distances between the poles. In contrast, in *P. boisei*, *P. robustus*, and *P. aethiopicus*, the temporal lobes are rounded and apelike (Falk et al., 2000).

Second, the unusually wide temporal lobes of *H. floresiensis* are indicated by the overall endocast breadth/width, which exceeds that of any human, ape, or fossil hominin endocast measured, due to the lateral expansion of the caudal part of the temporal lobes (Falk et al., 2005, 2009b). However, consistent with its small cranial capacity, the distance between temporal poles in *H. floresiensis* is only slightly higher than that of *Au. afarensis* and less than that of rAMHS (Falk et al., 2005).

Lunate sulcus position

The lunate sulcus (LS) is within the secondary visual area of apes, close to the anterior border of the primary visual cortex, BA17. The position of the LS has long been used to estimate BA17 size, and LS position correlates with cytoarchitecturally defined total BA17 volume in apes (de Sousa et al., 2010a). In chimpanzees, the LS marks the anterior extent of BA17, even in cases where the LS is in an unusual position for the species (Holloway et al., 2003).

BA17 is the histologically defined cortical area which is most reduced in humans, based on predictions for an ape of similar brain size. Humans have a substantially smaller (−121%) BA17 than expected for a nonhuman primate of similar brain size (Holloway, 1992), a finding which is supported even when phylogeny is controlled for de Sousa et al. (2010a). Although chimpanzees typically have a relatively larger BA17 than do humans, a minority of chimpanzees show repositioning of the LS to a more humanlike posterior position and therefore might also have reduced BA17 volumes (Holloway et al., 2003). Holloway et al. (2003) use this point to argue that the hypothetical panin–hominin MRCA must also have had within its population individuals with reduced primary visual cortices, so one would expect this condition in early hominins such as *Au. afarensis*. The LS may be unique among the cortical sulci visible on endocasts in that it may provide information about the proportion of cortex allocated to distinct functional categories, and provide an estimate of the aforementioned ratio of association to sensory cortex (Holloway, 1966, 1968).

Inferences about LS position are based on observations of the LS itself, and also (or alternatively) on the position of the interparietal sulcus, which abuts the LS. Dart (1925) described a humanlike posteriorly positioned LS on a small, early hominin endocast (the Taung child, *Au. africanus*), and in doing so began a long debate over the identification and location of the LS (Falk, 1980a, 1985; Holloway, 1975; Keith, 1931; Le Gros Clark et al., 1936; Schepers, 1946). In fact, many authors note that it is impossible to know the LS position in Taung with certainty (Falk, 2009; Holloway, 1985; Le Gros Clark, 1947; Tobias, 1991). In contrast, it has been asserted that Stw 505 provides better evidence of a posteriorly positioned LS in *Au. afarensis* (Holloway et al., 2004b). In addition, posteriorly positioned lunate sulci have been inferred from the fossil endocasts of *P. boisei*, *P. robustus*, *H. erectus*, *H. heidelbergensis*, *H. neanderthalensis*, and *H. sapiens* (Holloway et al., 2004a). It is suggested that in *Au. afarensis*, as in chimpanzees, the LS may be variably in a posterior or anterior position (Holloway et al., 2004a, 2003).

Parietal lobe expansion

The relative reduction of BA17 is associated with the relative expansion of the posterior parietal association cortex. The posterior parietal lobe is concerned with several aspects of sensory processing and sensorimotor integration (Hyvarinen, 1981; Lynch, 1980). The superior parietal lobule subcomponent is involved in visuomotor tasks, including finger movements (Shibata and Ioannides, 2001) and visual attention (Yantis et al., 2002). The superior parietal lobule (BA7) functions in spatial cognition and demonstrates differential activation during an Oldowan toolmaking task (Stout et al., 2000). The inferior parietal lobule is involved in language and calculation abilities, and it is greatly expanded in humans compared to monkeys (Simon et al., 2002). Derived human behaviors involving the posterior parietal lobe include enhanced social behavior including communication, toolmaking, and tool-use (Holloway et al., 2004a). It is suggested that the unique globular shape of the neurocranium of AMHS is related to an additional expansion of the parietal lobe and may be associated with the manufacture of more sophisticated tools and refined language ability (Bruner, 2004; Bruner et al., 2003).

Cerebellum size

The cerebellum is well known for its functions in motor control such as coordination, precision, and accurate timing; however, recent attention has been brought to the evolution of the cerebellum because the neocerebellum has been implicated in cognitive tasks, and there is a differential expansion of the lateral cerebellum (largely neocerebellum) in hominoids compared to other anthropoids (MacLeod et al., 2003; Chapter 8). The human cerebellum is smaller than would be expected for an ape of similar brain size (Rilling and Insel, 1998; Semendeferi and Damasio, 2000). A cerebellar quotient (CQ=actual/predicted value) was obtained when rAMHS cerebellar volume (determined from posterior cranial fossa volume) was regressed against brain volume (determined from endocranial capacity) minus cerebellar volume (Weaver, 2005). The difference between human and great ape relative cerebellar volumes is statistically significant, although less dramatic when considered among the range of inferred relative cerebellar volumes of fossil hominins (Weaver, 2005). The relative cerebellum size of rAMHS is estimated to be similar to or larger than that of earlier hominins. However, fAMHS and *H. neanderthalensis* have the smallest mean CQs, indicating a recent increase in relative cerebellum size to that of rAMHS (Weaver, 2005).

Archeological implications for fossil brain function

The fossil hominin archeological record contains a wealth of information about the emergence of human-specific behaviors. Given their special relevance to understanding hominin brain evolution, the earliest evidence for the following human-specific behaviors are discussed here and summarized in Table 4: (1) tool-use, (2) intentionally manufactured stone tools, (3) symmetrical tools, (4) handedness among toolmakers, and (5) symbolic activity (burials, abstract representations, personal ornamentation, and figurative representations).

Tool-use

Comparative studies of material culture have specific implications for understanding fossil hominin taxa because tool-use and production are the earliest (and also the most extensive) category of hominin behavioral evidence. Chimpanzee archeology informs our interpretation of behavior as represented in the archeological record. In particular, information about tool-use in chimpanzees and humans is used to reconstruct the behaviors of the hominin–panin MRCA (Haslam et al., 2009).

Chimpanzees are the apes that use tools most extensively in the wild. Chimpanzees use stone "hammers" to crack open nuts and modify twigs for termite hunting (Goodall, 1986) and have recently been observed to hunt small primates with spears (Pruetz and Bertolani, 2007). As is the case with hominins, tool-use in chimps is socially learned (Whiten and van Schaik, 2007). Chimpanzee nut-cracking leaves a recoverable material record of stone and plant remains (Mercader et al., 2002). The material culture of chimpanzees can be used to reconstruct operational sequences showing their utilization, and these are distinguishable between different chimpanzee cultural groups (Carvalho et al., 2008, 2009). It has been hypothesized that even the earliest hominins used tools, made of stone or other materials, prior to the earliest evidence for stone tool making (Panger et al., 2002). The earliest known evidence of stone tool-use, inferred indirectly on the basis of cuts and percussion related to the consumption of animal tissues, is from before 3.39Ma at Dikika, Ethiopia and is attributed to *Au. afarensis*, which was the only hominin present in the area at that time (McPherron et al., 2010). Although the full extent of tool-use in *Au. afarensis* is uncertain, at present, the evidence supports the conclusion that they used tools. However, whether or not they intentionally made tools is an open question. Unlike humans, chimpanzees are not known to intentionally manufacture stone tools in the wild.

Table 4. Behaviors of fossil hominin species inferred from the archeological record

Taxon	FAD (mya)	Tool-use	Intentional tool manufacture	Handedness (across populations)	Symmetry in tool morphology	Symbolism	Burial	Ornamentation	Abstract representation	Sound-modifying instruments	Figurative representation
P. troglodytes		A									
rAMHS		A									
Au. afarensis	3.7	U									
Au. africanus	3	U	U								
Au. garhi	2.5	U	U								
P. aethiopicus	2.5	U	U								
H. habilis s.s.	2.4	U	U	U							
H. rudolfensis	2.4	U	U								
P. boisei s.s.	2.3	U	U	U							
P. robustus	2	U	U								
H. erectus s.s.	1.9	A	A	A	A						
H. ergaster	1.8	A	A	U	A						
H. antecessor	0.70	A	A		A						
H. heidelbergensis	0.60	A	A	A	A						
H. neanderthalensis	0.20	A	A	A	A	A	A	A	U	U	U
H. sapiens s.s.	0.20	A	A	A	A	A	A	A	A	A	A
H. floresiensis	0.07	A	A								

Notes: A, associated with behavior in archeological record; U, uncertain association, for which it is a candidate species.

In captivity, it is possible to train a bonobo to produce by hard-hammer percussion fragments for food consumption use, using a stone-flaking method less advanced than the Oldowan (Schick et al., 1999; Toth et al., 1993).

Intentionality

Oldowan tools from Gona, Ethiopia, from 2.6 to 2.5 Ma are the oldest direct evidence of stone tool making (Semaw et al., 1997). These tools are not in direct association with any hominin species, although *Australopithecus garhi* occurred contemporaneously a short distance away (Asfaw et al., 1999; de Heinzelin et al., 1999). However, any of several species which occurred at this time might have made them. Early *Homo* is often credited for the earliest stone tool making, and thus starting a tradition that continued in the undisputed toolmaking members of its genus including rAMHS (Plummer, 2004; Toth and Schick, 2009). The earliest hominin specimen associated with tools is AL 666-1, which may be an early member of the genus *Homo* (Kimbel et al., 1996). However, non-*Homo* hominids have also been put forward as potential toolmakers because they are synchronic with lithic industries (de Heinzelin et al., 1999; Kuman and Clarke, 2000), share a trend toward brain size increase (Elton et al., 2001), and may have had hand morphology compatible with tool-use (Ricklan, 1987; Susman, 1994, 1998; but see Tocheri et al., 2008). *Australopithecus garhi*, *Au. africanus*, *K. platyops*, *P. aethiopicus*, *P. boisei*, *P. robustus*, *H. habilis*, and *H. rudolfensis* have all been implicated as potential creators of the earliest Oldowan. The earliest definite toolmaker is *H. ergaster*, a species which is associated with both late Oldowan and early Acheulean (Kuman and Clarke, 2000; Plummer, 2004). In addition, stone tools have been found in association with the skeletal remains of *H. ergaster*, *H. erectus*, *H. antecessor*, *H. heidelbergensis*, *H. neanderthalensis*, *H. floresiensis*, and *H. sapiens*.

Handedness

All human populations are characterized by having a right-handed majority. Handedness in apes is less regular, and although it may occur at the level of populations (Hopkins et al., 2003), the pattern is not consistent across an entire species (Uomini, 2009) and continues to be debated (Hopkins et al., 2011).

Evidence of population-level right-handedness was proposed for an Oldowan assemblage, dated to 1.9–1.4 Ma, from Koobi Fora, Kenya (Toth, 1985). This is not in direct association with any hominin species, although the candidate species are *H. habilis*, *H. ergaster*, and *P. boisei*. On the basis of lithic morphology, population-level right-handedness has been proposed for *H. erectus*, *H. neanderthalensis*, and *H. sapiens* (Uomini, 2009). It has also been suggested for *H. neanderthalensis* and *H. heidelbergensis* on the basis of dental microwear patterns (Frayer et al., 2011a,b; Lozano et al., 2009).

Symmetry

The production of lithics with symmetrical (or otherwise standardized) shape has not been observed in extant species other than *H. sapiens*. However, the cognitive implications of this human-specific behavior are strongly disputed. A technological industry spanning one million years and several continents, the Acheulean, is generally typified according to the existence of bifacially and bilaterally symmetrical teardrop-shaped lithics and is the focus of dispute. It has been suggested that symmetrical standardization of lithics signifies the ability to impose a predetermined form on a piece of stone, and by extrapolation cognitive capacities for planning and perhaps language ability (Gowlett, 2006). However, it has been suggested that Acheulean tools are, for the most part, not particularly symmetrical (Clark and Riel-Salvatore, 2005), vary in form according to raw material and production intensity (McPherron, 2000), and may have been cores rather than end-products (Davidson and Noble, 1993). Further, the manufacture of symmetrical objects is not distinct to humans, as apes may produce radially symmetrical sleeping nests (Wynn and McGrew, 1989). The earliest handaxes are from 1.6 to 1.7 Ma, from eastern and southern Africa, and include 1.7 Ma tools at Konso-Gardula, Ethiopia (Asfaw et al., 1992) and 1.6 Ma tools at Sterkfontein, South Africa (Gibbon et al., 2009). The Acheulean is normally attributed to *H. ergaster* and African *H. erectus*. *H. antecessor*, *H. heidelbergensis*, *H. neanderthalensis*, and early AMHS are also associated with symmetrical handaxes (Clark et al., 2003; McPherron, 2000). In addition, rAMHS and *H. floresiensis* are also associated with symmetrical tool morphologies.

Symbolism

The fabrication of material symbols is unique to humans among extant species. The quintessential symbol system, language, is human specific, but symbolism is defined more broadly to include any arbitrary assignment of meaning to a thing that bears no necessary resemblance to its referent (Pierce, 1932). Aspects of symbolic capacity exist in apes, including language in a captive bonobo (Benson et al., 2002), auditory communication in wild chimpanzees (Boesch, 1991), and juvenile female chimpanzees carrying sticks as if they were infants (Kahlenberg and Wrangham, 2010). This has prompted examination of any particular aspects of symbolic behavior which may in fact be human specific (Deacon, 1997; Mignault, 1985). Because wild chimpanzees are not known to manufacture symbolic artifacts, at least this aspect of symbolic behavior seems to be exclusively hominin.

There is no definitive way in which symbolism is physically represented, making it difficult to identify archeologically. Language is the likely predecessor to all forms of symbolic behavior, but it is not identifiable archeologically until the advent of writing. Burial, ornamentation, and abstract or figurative auditory and visual representations (art and music) are often cited as evidence of early symbolic capacity. However, none of these is necessarily symbolic (Duff et al., 1992); therefore, it is common to look more broadly for evidence for symbolic behavior. It has been argued that symbolic behaviors appeared as a synchronic "package," coincident with a cognitive "revolution" (Klein, 1999, 2003), although most archeologists prefer a model in which attributes of symbolic behavior appear gradually. The timeframe may have encompassed speciation events, and sequential species may have differed in the biological basis of behavior (McBrearty and Brooks, 2000). However, demographic and ecological variables could have contributed to the appearance and patterning of new behaviors even in the absence of biological changes (Powell et al., 2009).

The earliest evidence of probable symbolic behavior of any kind is the use of pigments, presumably in personal ornamentation or art. The earliest known occurrence of pigment use is at Pinnacle Point, South Africa, 164 ka, attributed to *H. sapiens* (Marean et al., 2007).

Burial: The oldest purposeful burials are *H. sapiens* from Qafzeh and Skhul, Israel, 115 ka (Grun et al., 2005). Two species, *H. sapiens* and *H. neanderthalensis*, are associated with purposeful burials (Riel-Salvatore and Clark, 2001).

Ornamentation: More robust evidence of personal ornamentation is the use of shell beads, which occurred in similar patterns at several middle stone age sites in North Africa (Bouzouggar et al., 2007; Vanhaereny et al., 2006) and Sub-Saharan Africa (d'Errico et al., 2005; Henshilwood et al., 2009), and Western Asia (Bar-Yosef Mayer et al., 2009; Vanhaereny et al., 2006). Of these, the earliest are the perforated shells from Skhul, Israel, from 100 to 135 ka, associated with *H. sapiens* (Vanhaereny et al., 2006). Two species, *H. sapiens* and *H. neanderthalensis*, are associated with ornamentation. The latter is associated with beads made of pierced shells and teeth (Zilhão, 2007; Zilhao et al., 2010) and may have used feathers as ornaments (Peresani et al., 2011).

Abstract representation: The earliest evidence of abstract representation are deliberately engraved ochre pieces from Blombos Cave, South Africa, dated to 75ka (Henshilwood et al., 2002, 2009). Two species, *H. sapiens* and *H. neanderthalensis*, are associated with abstract visual and auditory representations (Soressi and d'Errico, 2007), although in the latter species the evidence is far less robust.

Sound-modifying instruments: The oldest undisputed intentionally crafted sound-modifying instruments are bone flutes, thought to be musical instruments, attributed to over 35ka from Hohle Fels, Germany (Conard et al., 2009). Undisputed examples of intentionally crafted musical instruments have been found in association with *H. sapiens* only (d'Errico and Lawson, 2006).

Figurative representation: There are several candidates for the oldest figurative representations (i.e., for which the intentionality is undisputed and which are thus interpreted as "art"), all dating to 60–30ka. There is a great deal of dispute about precise dates, however, and the time frame is relatively narrow, so the earliest examples are provided here. The oldest possible date for figurative representations come from painted slabs depicting animals, in the Apollo 11 Cave, which is dated to 18–34ka but argued to have a date older than 59ka based on their Middle Stone Age context in Namibia (McBrearty and Brooks, 2000, Wendt, 1974, 1975). Also during this time frame, rock paintings depicting animals from Grotte Chauvet, France are dated up to 32 ka, and figurines from Hohle Fels Cave date to older than 30ka (Conard, 2003, 2010). Undisputed examples of figurative art have been found in association with *H. sapiens* only. However, no specific taxon is associated with the earliest figurative representations. It has been suggested that archaic humans such as *H. neanderthalensis* had the potential to create figurative representations and/or that *H. sapiens*' intergroup contact may have inspired their creation (Conard, 2010; d'Errico and Stringer, 2011).

Neuroimaging fossil hominin archeology

An approach which links hominin brains to behaviors entails the neuroimaging of humans engaged in tool-use and toolmaking. Such studies identify regions involved in fossil hominin-like toolmaking. Toolmaking by experienced toolmakers differs from that of inexperienced toolmakers in that there is more activation in regions of language and manual praxis circuits, including parietofrontal regions in both the hemispheres and the right hemisphere homologue of Broca's area (Stout et al., 2008). Also of interest are differences in brain activation according to toolmaking method, demonstrating the differential cognitive demands of different technological industries. For example, the observation of Acheulean toolmaking, compared with Oldowan toolmaking, corresponds to increased activation of left anterior intraparietal and inferior frontal sulci, which are regions of the brain involved in "action understanding" (Stout et al., 2011).

Unfortunately, thus far, it has proved difficult to do functional neuroimaging of tool-use and manufacture in apes. Comparisons between humans and macaques during tool-use (and related activities) may highlight regions of interest for future studies. For example, a rostral sector of the left inferior parietal lobule active during the observation of tool-use by humans, but not macaques, has been identified through functional neuroimaging (Peeters et al., 2009).

The archeological record is not limited to toolmaking, and there exists additional potential for identifying the neural correlates of other human-specific activities known to exist in fossil hominins. Notably, artifacts as early as the Acheulean may provide information about esthetics in early hominins. Further research may reveal how transitions in the fossil hominin archeological record in both the technology and the production of art and ornamentation are related to the neural correlates of esthetic behavior in humans as well as nonhumans.

Fossil brain genetics

Ancient DNA sequences provide evidence which can be used in conjunction with archeological and paleontological evidence to reconstruct the structure and function of fossil hominin brains. Interestingly, human-specific cognition and brain morphology may be the product of contributions from multiple hominin lineages, either directly through genomic contributions or through social and cultural interactions. Recently, it has been suggested that *H. neanderthalensis* may have contributed 1–4% of the genomes of living Eurasian *H. sapiens* (Green et al., 2010) and a separate, genetically defined species, from 50 to 30ka in Denisova cave, Russia, may have contributed 4% of the genomes of living Melanesian *H. sapiens* (Reich et al., 2010). As mentioned in the section on archeology, intergroup contact may have driven behavioral changes at the cultural and/or genetic levels.

Genes associated with neurological conditions give insight into the mechanisms influencing the evolution of brain structure and function. Differences between the genomes of *H. neanderthalensis* and *H. sapiens* have been identified and could potentially be linked to species-specific cognition. Of these, several genes associated with human neurological disorders were inferred to show evidence of positive selection in *H. sapiens* since the time of divergence with *H. neanderthalensis.* These are DYRK1A, a gene associated with Down syndrome; NRG, a gene associated with schizophrenia; and CADPS2 and AUTS2, both associated with autism (Green et al., 2010). Further analyses are needed to indicate the specific functional, and potentially cognitive, significance of any of these mutations.

Two genes related to cognitive function have been specifically investigated in *H. neanderthalensis* because of their probable roles in language and brain size, respectively: FOXP2 and microcephalin (MCPH1). FOXP2 is implicated in a severe form of speech and language disorder which is associated with a heterozygous missense mutation at the locus SPCH1 (Fisher et al., 1998). FOXP2 may be essential for normal language and speech function (Lai et al., 2001; MacDermot et al., 2005). Two amino acid substitutions appeared in FOXP2 in hominins after the divergence from the hominin–panin MRCA and are found in *H. sapiens* (Enard et al., 2002), *H. neanderthalensis* (Krause et al., 2007), and the Denisovans (Hawks, 2011a).

Homozygosis of loss-of-function mutations in MCPH1 causes a condition known as primary microcephaly, associated with severe (three- to fourfold) reduction in brain volume (Jackson et al., 2002). A derived group of haplotypes at the MCPH1 locus (haplogroup D) occurs in 70% of *H. sapiens* but is estimated to have appeared relatively recently (14–62ka) (Evans et al., 2005). It was proposed that the rapid increase of this genetic variant indicates that it was positively selected for in *H. sapiens* and appeared due to admixture with other hominin species—originally suggested to be *H. neanderthalensis*, because of the lower frequency of haplotype D in Sub-Saharan Africa (Evans et al., 2005, 2006). However, in spite of its recent predominance, no particular function is attributed to the novel variant; in fact, a recent study found no evidence linking MCPH1 to brain size evolution (Montgomery et al., 2011). Further, all specimens with known sequences for the gene in *H. neanderthalensis* show the ancestral variant (Green et al., 2010; Lari et al., 2010). The Denisovans also show the ancestral variant (Hawks, 2011b).

Fossil brain ontogeny

Humans and chimpanzees differ in their patterns of brain growth (increase in size) and development (change in shape) (see Chapter 13). Most obviously, there is a difference between the total growth of the brain, indicated by the three- to fourfold larger adult brain volume of humans than chimpanzees. However, this may seem surprising given that the duration of postnatal brain growth is similar in humans and chimpanzees. Several factors explain this. First, humans have larger brains at birth (DeSilva and Lesnik, 2006). Second, postnatally, the human brain grows proportionally more, expanding by a factor of 3.3, compared with 2.5 in chimpanzees (DeSilva and Lesnik, 2006). Third, postnatally, a greater absolute volume of additional brain tissue is produced in humans than in chimpanzees (DeSilva and Lesnik, 2006). Fourth, in the early postnatal period, humans have a higher growth rate (Leigh, 2004). Few juvenile cranial capacities are known, making it very difficult to ascertain anything about brain development in early hominins. The earliest species for which the brain growth trajectory is studied in detail is *H. erectus*. *H. erectus* has been suggested to have a chimpanzee-like brain growth rate based on the juvenile Mojokerto specimen (Coqueugniot et al., 2004), but this has been met with skepticism due to difficulty in accurately aging the specimen, and because the Nariokotome boy suggests a more humanlike growth pattern (DeSilva and Lesnik, 2006; Leigh, 2006). There is better evidence for determining the growth trajectory of *H. neanderthalensis*, which seems to have had growth rates during early infancy which were even higher than those of humans, and resulted in larger adult brain sizes but not in earlier completion of brain growth (Ponce de Leon et al., 2008). This pattern is suggested to also apply to fAMHS because they have large brains, although growth rates have not been demonstrated for them using fossil samples (Ponce de Leon et al., 2008).

Human brains also differ from those of chimpanzees in overall shape (Aldridge, 2011). Developmental patterns differ between humans and chimpanzees during the stage directly after birth in that only humans undergo a "globularization" phase; although, later in development, the patterns of development are similar (Neubauer et al., 2010). *H. neanderthalensis* resembles chimpanzees in that it lacks the AMHS globularization phase (Gunz et al., 2010). In spite of the primitive brain shape development of *H. neanderthalensis*, its growth appears further "derived" along the hominin trend than rAMHS due to a larger adult brain size and a more rapid postnatal brain growth period.

It has been suggested that differences between *H. neanderthalensis* and rAMHS in brain morphology and cognition have parallels in the differences between autistic versus neurotypical individuals (Neubauer et al., 2010). That is, *H. neanderthalensis*, like autistic individuals, has been suggested to have undergone an early postnatal spurt in brain development. The irregular developmental trajectory of autistics results in more short-distance connections and larger absolute brain size, whereas the developmental trajectory of neurotypicals results in longer-distance connections and smaller brain size (Courchesne et al., 2010; Lewis and Elman, 2008).

Bringing together evidence for fossil hominin brain structure and function

Possible hominins

We know very little about the brains and behaviors of possible hominin species, and there is no basis on which they seem to be derived in comparison to a panin ancestral condition. *Sahelanthropus tchadensis* is the earliest possible hominin and it is also the hominin species with the smallest mean brain volume, which falls just below the female chimpanzee mean (but note that specimens of *P. aethiopicus*, *Au. garhi*, and *H. floresiensis* plot around the male chimpanzee mean). The lack of archeological data is consistent with what would be expected for a very early hominin, or, for that matter, a member of any great ape lineage.

Archaic hominins: Reintroducing "Man the Toolmaker"

The notion that the human lineage could be defined according to toolmaking was quashed with the discovery of *Au. afarensis*, which shares humanlike postcranial anatomy and bipedal locomotion but lacks the big brains and behaviors of humans. However, more recent findings indicate that *Au. afarensis* was likely a tool-user and *Au. garhi* was likely a toolmaker. As noted, chimpanzees have their own material culture. These recent findings suggest that tool associations may be expected even for probable hominins, regardless of brain size. This, in turn, suggests that it would be worthwhile to investigate the relationship between early, subtle changes in brain size and organization and material culture. For example, possible differences in brain organization within panins (cf. de Sousa et al., 2010a,b) could be related to tool behaviors.

The earliest detectable aspect of humanlike brain morphology is the reduction of BA17. A posterior LS has been reported for some *Au. afarensis* specimens, although it is variable within the taxon. Given the small sample, it is difficult to tell whether the *Au. afarensis* brain really is derived in the direction of the human brain, or whether it expresses variability similar to that seen in chimpanzees. The *Au. afarensis* mean brain mass is not significantly different from the combined sex sample of chimpanzees ($p=0.093$) nor from the male chimpanzee sample ($p=0.456$), although it is significantly larger than the female chimpanzee sample ($p=0.011$), although the EQs of *Au. afarensis* (2.5) and *Au. africanus* (2.8) are well above those for chimpanzees (male EQ=1.7; female EQ=1.9). Given small samples, one cannot be certain whether this variation is in fact a change from the variation seen in chimpanzees. Further, humanlike endocranial anatomy in *Au. afarensis* might be a preadaptation which only acquires its modern functions in *Au. africanus*, *H. rudolfensis*, or in even later hominins. Behaviorally, *Au. afarensis* could be derived compared to chimpanzees: Like chimpanzees, they may have used tools; however, unlike chimpanzees, it is likely they used tools to butcher animals.

Many aspects of humanlike endocast morphology make an appearance in *Au. africanus*, including (1) evidence for a reduced BA17, (2) frontal lobes that are expanded orbitally and a prefrontal cortex that appears squared off rostrolaterally when viewed dorsally, (3) anteriorly expanded, laterally pointed temporal poles, (4) an incipient LORF petalial pattern, and (5) a humanlike Broca's cap region. Although these features are not as pronounced as in humans, they can be interpreted as being derived in the direction of humans. The reason for their occurrence in this taxon is uncertain but may be influenced by brain size increase, and it is quite possibly related to exceptional preservation of brain morphology in *Au. africanus*. The appearance of several aspects of modern brain morphology in *Au. africanus* complement the fact that this taxon is the first to have a brain size significantly different from chimpanzees. The *Au. africanus* sample is significantly different from the combined sex sample ($p<0.001$), and the male ($p=0.001$) and the female ($p<0.001$) subsamples of chimpanzees. This finding is further evidenced by the *Au. afarensis* EQ (2.5) which is well above that for chimpanzees and equals that of *P. boisei*.

However, the *Au. africanus* brain differs considerably from the human brain, and any similarities are not considered sufficient to suggest humanlike brain structure and function in *Au. africanus*. *Au. africanus* appears later in the fossil record than *Au. afarensis* and there exists some possibility that it could have manufactured Oldowan tools. *Australopithecus* brain size and structure are rarely related to derived cognitive capacities.

Megadont archaic hominins: Potential parallels

The *P. boisei* mean estimated brain weight (483 g) is larger than that of *P. aethiopicus* (407 g) and somewhat larger than those of *Au. africanus* (455 g) and *Au. afarensis* (442 g). Further, the majority of the *P. boisei* specimens fall outside two standard deviations of the male chimpanzee mean. Therefore, it is inferred that *P. boisei* has increased its absolute brain size relative to the primitive condition. The *P. boisei* mean is not significantly different ($p=0.357$) from that of the later occurring *P. robustus* sample, even though the latter attains a much higher maximum value (638 g) and has a much higher mean (525 g). *P. boisei* and *P. robustus* have EQs that are higher than those for male and female chimpanzees. However, the *P. boisei* EQ is smaller than the *Au. africanus* and it is similar to the *Au. afarensis* value. Given the lack of postcranial evidence, one cannot be certain that EQ has increased from *P. aethiopicus* to later *Paranthropus* taxa. These data are, however, consistent with the suggestion of a temporal trend for brain size increase within the *Paranthropus* lineage (Elton et al., 2001).

There is little evidence suggesting humanlike reorganization of the *Paranthropus* brain. In particular, slight LORF petalial patterns are found in *P. aethiopicus* and *P. boisei*, and a posteriorly positioned LS has been identified in *P. boisei*. The evidence does not suggest that the *Paranthropus* brain becomes increasingly humanlike over time, as is the case for *Homo*. Further, *Paranthropus* retains an apelike beak-shaped orbital surface of frontal lobe and rounded temporal poles, differentiating it from *Au. africanus* and AMHS. The humanlike endocranial features seen in *Paranthropus* most likely reflect a shared ancestry with the human lineage. Similarly, brain size increase in *Paranthropus* is probably the continuation of a trend beginning in the AMHS-*Paranthropus* MRCA.

Transitional hominins: Intelligent, assuming we are related...

The more humanlike brain morphology of *H. rudolfensis* is generally taken as evidence of more humanlike cognitive capacities. Most notably, these features are suggestive of language ability and right-handedness—coincident with the first stone tools which apparently were made by right-handed hominins. The LORF petalial pattern and Broca's cap region have become increasingly humanlike in *H. rudolfensis*, the earliest taxon for which there is strong evidence for humanlike brain organization. In addition, *H. rudolfensis* is the earliest taxon not to have a fronto-orbital sulcus (but the evidence is based on very little endocranial morphology). Interestingly, there is no good evidence for a humanlike LORF petalial pattern and a Broca's cap region in *H. habilis*. Instead, there is evidence of an African apelike fronto-orbital sulcus. This is associated with the earliest brain weights that exceed expectations for chimpanzees, and an EQ higher than that of earlier taxa.

Relative brain weight in both *H. habilis* and *H. rudolfensis* is greater than that in *Australopithecus* and *Paranthropus*, and they approach the values for *H. erectus* (EQ=3.9). By the time of the appearance of *H. rudolfensis* and *H. habilis*, both absolute and relative brain size have clearly departed from the *Pan*-like condition. *H. habilis* and *H. rudolfensis* are significantly different in brain weight ($p=0.02$), and the entire range of *H. rudolfensis* values plot above the range of *H. habilis* values. However, *H. habilis* has a higher EQ than *H. rudolfensis* and also has brain weight values outside of those expected for chimpanzees. As yet, it is not possible to tell whether the more humanlike brain morphology of *H. rudolfensis*, compared to *H. habilis* is, or is not, size related.

Early *Homo* has been described as linking several aspects of the brain (absolute size, encephalization trend, humanlike morphology—in particular, related to lateralization) to behaviors (intentionality, handedness). However, the archeological associations are uncertain, and the endocranial evidence is scarce. Doubt has been cast on the *Homo*-status of early *Homo* on the basis of its australopith-like postcrania and dentition (Wood and Collard, 1999), and it has even been suggested that their brains require downsizing (Bromage et al., 2008). More reliable data about the brains and behaviors of early *Homo* would shed much light on definitive aspects of brain evolution in the human lineage.

Premodern Homo*: Making space for* H. floresiensis

H. erectus and *H. ergaster* tend to share the humanlike endocranial features found in *H. rudolfensis*. Absolute and relative brain sizes in *H. floresiensis* are thought to have decreased from the ancestral condition (Brown et al., 2004). *H. floresiensis* had a very small brain (414g), with an EQ (3.0) much lower than that of one of its presumed close relatives, *H. erectus*. Interestingly, its EQ is higher than the one listed here for *H. ergaster* (2.8—includes Dmanisi) and only slightly lower than the EQ for African *H. ergaster* (3.1). Body weight estimates obtained from Dmanisi postcranial remains will refine the *H. ergaster* EQ. Given that *H. ergaster* is thought to have expanded its range outside of Africa, evidence from the relative brain size alone suggest that it rather than *H. erectus* may be the sister-taxon of *H. floresiensis*. If so, this would indicate that EQ did not actually decrease in *H. floresiensis*—solving one of the major puzzles of this taxon (Brown et al., 2004). *H. floresiensis* possesses many features which are derived relative to an apelike condition, although several of these already existed in the *H. erectus*.

Anatomically modern Homo*: When "modern" is not "recent"*

Although fAMHS is included in our species on the basis of anatomy (and possibly DNA; Caramelli et al., 2008) that is indistinguishable from rAMHS, the evidence reviewed here suggest that the brains and behaviors of fAMHS are not identical to those of rAMHS. Parietal lobe expansion that is related to brain globularization (Bruner et al., 2003) and occurs during a novel postnatal developmental stage (Gunz et al., 2010) distinguishes both fossil and recent AMHS from *H. neanderthalensis* and other hominins. However, fAMHS more closely resembles *H. neanderthalensis* than rAMHS in having larger brains (and by extrapolation, a similarly brain growth trajectory) and smaller relative cerebellum sizes. Further, behavioral distinctions between fAMHS and *H. neanderthalensis* are becoming blurred as new studies reveal increasingly early symbolic behaviors, including some associated with *H. neanderthalensis* (d'Errico and Stringer, 2011). In fact, the earliest art is not directly associated with AMHS, or any other hominin species. Finally, the most recent genetic analyses (Green et al., 2010; Reich et al., 2010), as well as some archeological and morphological studies (Duarte et al., 1999, Zilhão, 2006) have suggested admixture between fAMHS and other contemporary species, including *H. neanderthalensis*. The biological and cultural factors which define brain structure and function in rAMHS might not best be revealed by comparing *H. neanderthalensis* to fAMHS. Rather, more comparisons are needed between fAMHS and rAMHS.

Hominin behavior may have changed more dramatically from the early to late Upper Paleolithic than during the Middle to Upper Paleolithic "transition" (Riel-Salvatore and Clark, 2001). Studies making such a distinction might explain why the makers of late Upper Paleolithic cave art resemble rAMHS in drawing animal limb joints (Biederman and Kim, 2008), while early Upper Paleolithic art may more closely approximate autistic rather than neurotypical productions (Humphrey, 1998). Because these comparisons involve members of a single species, they will rely even more heavily on pathology and other interindividual comparisons in neural and psychological sciences.

Conclusions

The data suggest that while fully humanlike brain morphology only occurs in rAMHS, many aspects of human brain morphology are present in earlier forms. Further, encephalization in the human lineage may have begun as early as *Au. afarensis*, and it was more evident in *Au. africanus* (in parallel to the encephalization of *Paranthropus*) and had definitely occurred by the time of the appearance of *H. habilis* and *H. rudolfensis*. Interestingly, brain size increase and the appearance of some aspects of humanlike brain morphology occur in at least two hominin lineages. Both *Paranthropus* and *Homo* have absolutely and relatively significantly larger brains than *Australopithecus*. However, only in *Homo* does brain size increase occur in parallel with the acquisition of humanlike brain morphology. *H. floresiensis* provides striking evidence that within *Homo* brain size and morphology may have become disassociated. New evidence of possible continuity between the brains and behaviors of fAMHS and contemporaneous hominins also indicate that further changes may have taken place more recently in our species. New lines of research are shedding light on the biological and cultural factors resulting in recent human brain structure and function.

Acknowledgment

Fundação para a Ciência e a Tecnologia (SFRH/BPD/43518/2008) supported this research. The chapter was substantially improved by editorial suggestions from Dean Falk and Michel Hofman.

Appendix

Notes for brain size data as used in Table 2.

Fossil hominin endocranial volume measurements were taken from reviews as well other papers. In order to maximize sample size, the approach taken was to all specimens for which an estimate could be found, noting the reliability is variable.

S. tchadensis: Based on TM 266-01-060-1 (Zollikofer et al., 2005).

Ar. ramidus: Based on ARA-VP-6/500 (Suwa et al., 2009).

Au. afarensis s.s.: Based on AL 162-28, AL 288-1, AL 333-105, AL 333-45, AL 444-2 (Holloway et al., 2004a).

Au. africanus: Based on STW 505 (Conroy et al., 1998), MLD 1, STS 19, STS 5, STS 60, STS 71, Taung, Type 2 (Holloway et al., 2004a), MLD 37/38 (Neubauer et al., 2004).

Au. garhi: Based on Bou-VP-12/130 (Holloway et al., 2004a).

P. aethiopicus: Based on KNM-WT 17000 (Holloway et al., 2004a).

P. boisei s.s.: Based on Omo-323-1976-896 (Falk et al., 2000), KNM-ER 406 (Holloway, 1988 [in Falk et al., 2000]), KGA-10-525, KNM-ER 23000, KNM-ER 407, KNM-ER 732, KNM-WT 13750, KNM-WT 17400, OH 5, Omo L-338Y-6 (Holloway et al., 2004a).

P. robustus: Based on TM 1517 (Broom and Robinson, 1948), SK 1585, SK 54, SK 859 (Holloway et al., 2004a).

Au. sediba: Based on MH1 (Berger et al., 2010).

H. rudolfensis: Based on KNM-ER 1470, KNM-ER 1590, KNM-ER 3732 (Holloway et al., 2004a).

H. habilis s.s.: Based on KNM-ER 1813, KNM-ER 1805, OH 7, OH 13, OH 16, OH 24 (Holloway et al., 2004a).

H. erectus s.s.: Based on Modjokerto 1 (Anton, 1997), Sangiran IX (Anton, 2003), Zhoukoudian V (Chiu et al., 1973), Gongwangling 1, Hexian, Narmada 1, Ngandong 1, Ngandong 6 (5), Ngandong 7, Ngandong (10), Ngandong 13 (11), Ngandong 14 (12), Ngawi, OH 9, OH 12, Sambungmacan 1, Sambungmacan 3, Sambungmacan 4, Sangiran 2, Sangiran 3, Sangiran 4, Sangiran 10, Sangiran 12, Sangiran 17, Trinil 2 (1891), Zhoukoudian, Zhoukoudian

I, Zhoukoudian III (E), Zhoukoudian III (L), Zhoukoudian VI (Holloway et al., 2004a), Bou-VP-2/66, Buia (UA 31), Ceprano (Lee and Wolpoff, 2003), Nanjing 1 (Liu et al., 2005), Poloyo PL-1 (Mowbray et al., 2000), Zhoukoudian II (Weidenreich, 1943).

H. ergaster: Based on D2280, D2282, KNM-ER 3733, KNM-ER 3883, KNM-WT 15000 (Holloway et al., 2004a), D2700 (Lee and Wolpoff, 2003).

H. antecessor: Based on ATD-15 (Bermudez de Castro et al., 1997).

H. heidelbergensis: Based on Steinheim (Rightmire, 2004), Florisbad (Aiello and Dean, 1990), Vertesszollos II (Delson et al., 2000), Arago (composite), Atapuerca 4, Atapuerca 5, Atapuerca 6, Bodo, Broken Hill 1, Dali 1, Ehringsdorf, Jinniushan 1, Guomde, Lazaret, Petralona 1, Reilingen, Sale, SAM-PQ-EH1, Swanscombe, Yunxian (1 and 2) (Holloway et al., 2004a), Ndutu 1 (Rightmire, 1983 [in Rightmire, 2004]).

H. neanderthalensis: Based on Krapina 4 (Delson et al., 2000), Ehringsdorf 9, Gibraltar 1, Le Moustier 1 (Grimaud-Hervé, 1997), Amud 1, Biache-Saint Vaast, Engis 2, Ganovce 1, Gibraltar 2, Krapina (4) 2, Krapina (4) 3, Krapina 6, La Chapelle-aux-Saints 1, La Ferrassie 1, La Quina 18, La Quina 5, Monte Circeo I, Neanderthal, Saccopastore I, Saccopastore II, Shanidar 5, Spy I, Spy II, Tabun I, Teshik-Tash 1 (Holloway et al., 2004a), Fontéchevade, Shanidar 1 (Lee and Wolpoff, 2003).

H. sapiens s.s.: Based on Eyasi 1 (Easi) (Conroy, 1997), Kanjera 1 (Brauer, 1984), Border Cave 1, Brno I, Brno II, Brno III, Bruniquel 2, Cap Blanc 1, Chancelade, Combe Capelle, Cro-Magnon I, Cro-Magnon III, Dolni Vestonice 3, Dolni Vestonice 14, Dolni Vestonice 18, Dolni Vestonice 20, Dolni Vestonice 21, Grotte des Enfants 4, Grotte des Enfants 5, Grotte des Enfants 6, Herto 1/16, Jebel Irhoud 1, Jebel Irhoud 2, Kostenki 14, Kostenki 2, LH 18, Liujiang, Minatogawa 1, Minatogawa 2, Minatogawa 4, Mladec 1, Mladec 2, Mladec 5, Nazlet Khater 2, Oberkassel 1, Oberkassel 2, Omo-Kibbish 2, Pataud 1, Pavlov 1, Predmosti 3, Predmosti 4, Predmosti 9, Predmosti 10, Qafzeh 11, Qafzeh 6, Qafzeh 9, San Teodoro 1, San Teodoro 2, San Teodoro 3, San Teodoro 5, Singa 1, Skhul 1, Skhul 4, Skhul 5, Skhul 9, St. Germain-la-Riviere 1, Sungir 1, Sungir 2, Sungir 3, Sungir 5, Veyrier 1, Yinkou, Zhoukoudian (Upper Cave) 1, Zhoukoudian (Upper Cave) 2, Zhoukoudian (Upper Cave) 3 (Holloway et al., 2004a), Eliye Springs 11693, Omo-Kibbish 1, Paderbourne (Lee and Wolpoff, 2003), Arene Candide 1, Arene Candide 1-IP, Arene Candide 2, Arene Candide 4, Arene Candide 5, Asselar, Barma Grande 2 (Ruff et al., 1997), Dolni Vestonice 8, Dolni Vestonice 15, Dolni Vestonice 16 (Schwartz and Tattersall, 2002), BOU-VP-16/1 (White et al., 2003).

H. floresiensis: Based on LB1 (Falk et al., 2005).

Abbreviations

AMHS	anatomically modern *Homo sapiens*
fAMHS	fossil anatomically modern *Homo sapiens*
rAMHS	recent anatomically modern *Homo sapiens*
CQ	cerebellar quotient
EQ	encephalization quotient
FAD	first appearance datum
L>R	left larger than right
LORF	left-occipital right-frontal
LS	lunate sulcus
MRCA	most recent common ancestor

References

Aboitiz, F., & Garcia, R. (1997). The anatomy of language revisited. *Brain Research Reviews*, *30*, 171–183.

Aiello, L., & Dean, C. (1990). *An introduction to human evolutionary anatomy.* London; San Diego: Academic Press.

Aldridge, K. (2011). Patterns of differences in brain morphology in humans as compared to extant apes. *Journal of Human Evolution*, *60*, 94–105.

Amunts, K., Schleicher, A., Burgel, U., Mohlberg, H., Uylings, H. B., & Zilles, K. (1999). Broca's region revisited: Cytoarchitecture and intersubject variability. *The Journal of Comparative Neurology*, *412*, 319–341.

Amunts, K., Schleicher, A., & Zilles, K. (2007). Cytoarchitecture of the cerebral cortex - More than localization. *Neuroimage*, *37*, 1061.

Anton, S. C. (1997). Developmental age and taxonomic affinity of the Mojokerto child, Java, Indonesia. *American Journal of Physical Anthropology*, *102*, 497–514.

Anton, S. C. (2003). Natural history of *Homo erectus*. *Yearbook of Physical Anthropology*, *46*, 126–170.

Arbib, M. A. (2005). From monkey-like action recognition to human language: An evolutionary framework for neurolinguistics. *The Behavioral and Brain Sciences*, *28*, 105–124 (discussion 125–67).

Asfaw, B., Beyene, Y., Suwa, G., Walter, R. C., White, T. D., Woldegabriel, G., et al. (1992). The earliest Acheulean from Konso-Gardula. *Nature*, *360*, 732.

Asfaw, B., White, T., Lovejoy, O., Latimer, B., Simpson, S., & Suwa, G. (1999). *Australopithecus garhi*: A new species of early hominid from Ethiopia. *Science*, *284*, 629–635.

Bar-Yosef Mayer, D. E., Vandermeersch, B., & Bar-Yosef, O. (2009). Shells and ochre in Middle Paleolithic Qafzeh Cave, Israel: Indications for modern behavior. *Journal of Human Evolution*, *56*, 307–314.

Begun, D. R., & Walker, A. (1993). The endocast. In A. Walker & R. Leakey (Eds.), *The Nariokotome Homo erectus skeleton* (pp. 326–358). Cambridge, MA: Harvard University Press.

Benson, J., Greaves, W., O'Donnell, M., & Taglialatela, J. (2002). Evidence for symbolic language processing in a bonobo (*Pan paniscus*). *Journal of Consciousness Studies*, *9*, 33–56.

Berger, L. R., de Ruiter, D. J., Churchill, S. E., Schmid, P., Carlson, K. J., Dirks, P. H., et al. (2010). Australopithecus sediba: A new species of Homo-like australopith from South Africa. *Science*, *328*, 195–204.

Bermudez de Castro, J. M., Arsuaga, J. L., Carbonell, E., Rosas, A., Martinez, I., & Mosquera, M. (1997). A hominid from the lower Pleistocene of Atapuerca, Spain: Possible ancestor to Neandertals and modern humans. *Science*, *276*, 1392–1395.

Biederman, I., & Kim, J. G. (2008). 17,000 years of depicting the junction of two smooth shapes. *Perception*, *37*, 161–164.

Boesch, C. (1991). Symbolic communication in wild chimpanzees? *Human Evolution*, *6*, 81–89.

Bonmati, A., Gomez-Olivencia, A., Arsuaga, J. L., Carretero, J. M., Gracia, A., Martinez, I., et al. (2010). Middle Pleistocene lower back and pelvis from an aged human individual from the Sima de los Huesos site, Spain. *Proceedings of the National Academy of Sciences of the United States of America*, *107*, 18386–18391.

Bouzouggar, A., Barton, N., Vanhaeren, M., D'Errico, F., Collcutt, S., Higham, T., et al. (2007). 82,000-year-old shell beads from North Africa and implications for the origins of modern human behavior. *Proceedings of the National Academy of Sciences of the United States of America*, *104*, 9964–9969.

Brauer, G. (1984). A craniological approach to the origin of anatomically modern *Homo sapiens* in Africa and implications for the appearance of modern Europeans. In F. Spencer (Ed.), *The origin of modern humans: A world survey of the fossil evidence* (pp. 327–410). New York: Alan R. Liss.

Broadfield, D. C., Holloway, R. L., Mowbray, K., Silvers, A., Yuan, M. S., & Marquez, S. (2001). Endocast of Sambungmacan 3 (Sm 3): A new *Homo erectus* from Indonesia. *Anatomical Record*, *262*, 369–379.

Bromage, T. G., McMahon, J. M., Thackeray, J. F., Kullmer, O., Hogg, R., Rosenberger, A. L., et al. (2008). Craniofacial architectural constraints and their importance for reconstructing the early *Homo* skull KNM-ER 1470. *The Journal of Clinical Pediatric Dentistry*, *33*, 43–54.

Broom, R., & Robinson, J. T. (1948). Size of the brain in the ape-man, *Plesianthropus*. *Nature*, *161*, 438.

Brown, P., Sutikna, T., Morwood, M. J., Soejono, R. P., Jatmiko, , Saptomo, E. W., et al. (2004). A new small-bodied hominin from the Late Pleistocene of Flores, Indonesia. *Nature*, *431*, 1055–1061.

Bruner, E. (2004). Geometric morphometrics and paleoneurology: Brain shape evolution in the genus *Homo*. *Journal of Human Evolution*, *47*, 279–303.

Bruner, E., Manzi, G., & Arsuaga, J. L. (2003). Encephalization and allometric trajectories in the genus *Homo*: Evidence from the Neandertal and modern lineages. *Proceedings of the National Academy of Sciences of the United States of America*, *100*, 15335–15340.

Cantalupo, C., & Hopkins, W. D. (2001). Asymmetric Broca's area in great apes. *Nature*, *414*, 505.

Caramelli, D., Milani, L., Vai, S., Modi, A., Pecchioli, E., Girardi, M., et al. (2008). A 28,000 years old Cro-Magnon mtDNA sequence differs from all potentially contaminating modern sequences. *PloS One*, *3*, e2700.

Cartmill, M., & Smith, F. H. (2009). *The human lineage*. Hoboken, NJ: Wiley-Blackwell.

Carvalho, S., Cunha, E., Sousa, C., & Matsuzawa, T. (2008). Chaînes opératoires and resource-exploitation strategies in chimpanzee (*Pan troglodytes*) nut cracking. *Journal of Human Evolution*, *55*, 148–163.

Carvalho, S., Biro, D., McGrew, W. C., & Matsuzawa, T. (2009). Tool-composite reuse in wild chimpanzees (Pan troglodytes): archaeologically invisible steps in the technological evolution of early hominins?. *Animal Cognition*, *12* (Suppl 1), S103–S114.

Charman, T., Jones, C. R., Pickles, A., Simonoff, E., Baird, G., & Happe, F. (2010). Defining the cognitive phenotype of autism. *Brain Research*, *1380*, 10–21.

Chiu, C., Ku, Y., Chang, Y., & Chang, S. (1973). Peking man fossils and cultural remains newly discovered at Choukoutien. *Vertebrata Palasiatica*, *11*, 109–131.

Clark, J. D., Beyene, Y., Woldegabriel, G., Hart, W. K., Renne, P. R., Gilbert, H., et al. (2003). Stratigraphic, chronological and behavioural contexts of Pleistocene Homo sapiens from Middle Awash, Ethiopia. *Nature*, *423*, 747–752.

Clark, G. A., & Riel-Salvatore, J. (2005). Observations on systematics in Paleolithic archaeology. In E. Hovers & S. L. Kuhn (Eds.), *Transitions before the transition* (pp. 29–56). New York: Springer.

Conard, N. J. (2003). Palaeolithic ivory sculptures from southwestern Germany and the origins of figurative art. *Nature*, *426*, 830–832.

Conard, N. J. (2010). Cultural modernity: Consensus or conundrum? *Proceedings of the National Academy of Sciences of the United States of America*, *107*, 7621–7622.

Conard, N. J., Malina, M., & Munzel, S. C. (2009). New flutes document the earliest musical tradition in southwestern Germany. *Nature*, *460*, 737–740.

Connolly, C. J. (1950). *External morphology of the primate brain.* Springfield, IL: C.C. Thomas.

Conroy, G. C. (1997). *Reconstructing human origins: A modern synthesis.* New York: W.W. Norton & Company.

Conroy, G. C., Weber, G. W., Seidler, H., Tobias, P. V., Kane, A., & Brunsden, B. (1998). Endocranial capacity in an early hominid cranium from Sterkfontein, South Africa. *Science*, *280*, 1730–1731.

Coqueugniot, H., Hublin, J. J., Veillon, F., Houet, F., & Jacob, T. (2004). Early brain growth in *Homo erectus* and implications for cognitive ability. *Nature*, *431*, 299–302.

Courchesne, E., Campbell, K., & Solso, S. (2010). Brain growth across the life span in autism: Age-specific changes in anatomical pathology. *Brain Research*, *1380*, 138–145.

Cunha, E., & Silva, H. P. (2005). A recente descoberta paleoantropológica da Ilha das Flores: alguns comentários. *Antropo*, [Online], 10. Available from: http://www.didac.ehu.es/antropo/10/10-3/Cunha.pdf.

D'Errico, F., Henshilwood, C., Vanhaeren, M., & van Niekerk, K. (2005). *Nassarius kraussianus* shell beads from Blombos Cave: Evidence for symbolic behaviour in the Middle Stone Age. *Journal of Human Evolution*, *48*, 3–24.

D'Errico, F., & Lawson, G. (2006). The Sound Paradox. How to assess the acoustic significance of archaeological evidence? In C. S. G. Lawson (Ed.), *Archaeoacustics* (pp. 41–57). Cambridge, UK: McDonald Institute.

D'Errico, F., & Stringer, C. B. (2011). Evolution, revolution or saltation scenario for the emergence of modern cultures? *Philosophical Transactions of the Royal Society B: Biological Sciences*, *366*, 1060–1069.

Damasio, H., Grabowski, T. J., Tranel, D., Hichwa, R. D., & Damasio, A. R. (1996). A neural basis for lexical retrieval. *Nature*, *380*, 499–505.

Dart, R. A. (1925). *Australopithecus africanus*: The man-ape of South Africa. *Nature*, *115*, 195–199.

Dart, R. A. (1940). The status of *Australopithecus*. *American Journal of Physical Anthropology*, *26*, 167–186.

Davidson, I., & Noble, W. (1993). Tools and language in human evolution. In K. Gibson & T. Ingold (Eds.), *Tools, language and cognition in human evolution* (pp. 363–388). Cambridge: Cambridge University Press.

de Heinzelin, J., Clark, J. D., White, T., Hart, W., Renne, P., Woldegabriel, G., et al. (1999). Environment and behavior of 2.5-million-year-old Bouri hominids. *Science*, *284*, 625–629.

de Sousa, A. A. (2008). *Hominoid brain organization: Histometric and morphometric comparisons of visual brain structures.* Ph.D., The George Washington University.

de Sousa, A. A., Sherwood, C. C., Mohlberg, H., Amunts, K., Schleicher, A., Macleod, C. E., et al. (2010a). Hominoid visual brain structure volumes and the position of the lunate sulcus. *Journal of Human Evolution*, *58*, 281–292.

de Sousa, A. A., Sherwood, C. C., Schleicher, A., Amunts, K., Macleod, C. E., Hof, P. R., et al. (2010b). Comparative cytoarchitectural analyses of striate and extrastriate areas in hominoids. *Cerebral Cortex*, *20*, 966–981.

de Sousa, A., & Wood, B. A. (2007). The hominin fossil record and the emergence of the modern human central nervous system. In J. H. Kaas & T. M. Preuss (Eds.), *Evolution of nervous systems, Vol. 4: The evolution of primate nervous systems* (pp. 291–336). Oxford: Academic Press.

Deacon, T. W. (1997). *The symbolic species: The co-evolution of language and the brain.* New York: W.W. Norton & Company.

Deaner, R. O., Isler, K., Burkart, J., & van Schaik, C. (2007). Overall brain size, and not encephalization quotient, best predicts cognitive ability across non-human primates. *Brain, Behavior and Evolution*, *70*, 115–124.

Dekaban, A. S., & Sadowsky, D. (1978). Changes in brain weights during the span of human life: Relation of brain weights to body heights and body weights. *Annals of Neurology*, *4*, 345–356.

Delson, E., Tattersall, I., Couvering, J. A. V., & Brooks, A. S. (2000). *Encyclopedia of human evolution and prehistory.* New York: Garland/Taylor & Francis.

Desilva, J., & Lesnik, J. (2006). Chimpanzee neonatal brain size: Implications for brain growth in *Homo erectus*. *Journal of Human Evolution*, *51*, 207–212.

Dobson, J. E. (1998). The iodine factor in health and evolution. *Geographical Review*, *88*, 3–28.

Dronkers, N. F., Plaisant, O., Iba-Zizen, M. T., & Cabanis, E. A. (2007). Paul Broca's historic cases: High resolution MR imaging of the brains of Leborgne and Lelong. *Brain*, *130*, 1432–1441.

Duarte, C., Mauricio, J., Pettitt, P. B., Souto, P., Trinkaus, E., van der Plicht, H., et al. (1999). The early Upper Paleolithic human skeleton from the Abrigo do Lagar Velho (Portugal) and modern human emergence in Iberia. *Proceedings of the National Academy of Sciences, 96*, 7604–7609.

Duff, A. I., Clark, G. A., & Chadderdon, T. J. O. (1992). Symbolism in the early Paleolithic: A conceptual odyssey. *Cambridge Archaeological Journal, 2*, 211–229.

Elton, S., Bishop, L. C., & Wood, B. (2001). Comparative context of Plio-Pleistocene hominin brain evolution. *Journal of Human Evolution, 41*, 1–27.

Enard, W., Przeworski, M., Fisher, S. E., Lai, C. S., Wiebe, V., Kitano, T., et al. (2002). Molecular evolution of FOXP2, a gene involved in speech and language. *Nature, 418*, 869–872.

Evans, P. D., Gilbert, S. L., Mekel-Bobrov, N., Vallender, E. J., Anderson, J. R., Vaez-Azizi, L. M., et al. (2005). Microcephalin, a gene regulating brain size, continues to evolve adaptively in humans. *Science, 309*, 1717–1720.

Evans, P. D., Mekel-Bobrov, N., Vallender, E. J., Hudson, R. R., & Lahn, B. T. (2006). Evidence that the adaptive allele of the brain size gene microcephalin introgressed into Homo sapiens from an archaic Homo lineage. *Proceedings of the National Academy of Sciences of the United States of America, 103*, 18178–18183.

Fadiga, L., & Craighero, L. (2006). Hand actions and speech representation in Broca's area. *Cortex, 42*, 486–490.

Falk, D. (1980a). A reanalysis of the South African australopithecine natural endocasts. *American Journal of Physical Anthropology, 53*, 525–539.

Falk, D. (1980b). Hominid brain evolution: The approach from paleoneurology. *Yearbook of Physical Anthropology, 23*, 93–107.

Falk, D. (1983). Cerebral cortices of East-African early hominids. *Science, 221*, 1072–1074.

Falk, D. (1985). Apples, oranges, and the lunate sulcus. *American Journal of Physical Anthropology, 67*, 313–315.

Falk, D. (2007). The evolution of Broca's area. *IBRO History of Neuroscience*, [Online]. Available from: http://www.ibro.org/Pub/Pub_Main_Display.asp?LC_Docs_ID=3145.

Falk, D. (2009). The natural endocast of Taung *Australopithecus africanus*: Insights from the unpublished papers of Raymond Arthur Dart. *American Journal of Physical Anthropology, 140*, 49–65.

Falk, D., Hildebolt, C., Smith, K., Jungers, W., Larson, S., Morwood, M., et al. (2009). The type specimen (LB1) of *Homo floresiensis* did not have Laron syndrome. *American Journal of Physical Anthropology, 140*, 52–63.

Falk, D., Hildebolt, C., Smith, K., Morwood, M. J., Sutikna, T., Brown, P., et al. (2005). The brain of LB1, *Homo floresiensis*. *Science, 308*, 242–245.

Falk, D., Hildebolt, C., Smith, K., Morwood, M. J., Sutikna, T., Jatmiko, et al. (2007). Brain shape in human microcephalics and *Homo floresiensis*. *Proceedings of the National Academy of Sciences of the United States of America, 104*, 2513–2518.

Falk, D., Hildebolt, C., Smith, K., Morwood, M. J., Sutikna, T., Jatmiko, et al. (2009). LB1's virtual endocast, microcephaly, and hominin brain evolution. *Journal of Human Evolution, 57*, 597–607.

Falk, D., Redmond, J. C., Jr., Guyer, J., Conroy, C., Recheis, W., Weber, G. W., et al. (2000). Early hominid brain evolution: A new look at old endocasts. *Journal of Human Evolution, 38*, 695–717.

Fisher, S. E., & Marcus, G. F. (2006). The eloquent ape: Genes, brains and the evolution of language. *Nature Reviews. Genetics, 7*, 9–20.

Fisher, S. E., Vargha-Khadem, F., Watkins, K. E., Monaco, A. P., & Pembrey, M. E. (1998). Localisation of a gene implicated in a severe speech and language disorder. *Nature Genetics, 18*, 168–170.

Frayer, D. W., Fiore, I., Lalueza-Fox, C., Radovcic, J., & Bondioli, L. (2011a). Right handed Neandertals: Vindija and beyond. *Journal of Anthropological Sciences, 88*, 113–127.

Frayer, D. W., Lozano, M., Bermudez DE Castro, J. M., Carbonell, E., Arsuaga, J. L., Radovcic, J., et al. (2011b). More than 500,000 years of right-handedness in Europe. Laterality, 1–19.

Geyer, S., Schleicher, A., Schormann, T., Mohlberg, H., Bodegard, A., Roland, P. E., & Zilles, K. (2001). Integration of microstructural and functional aspects of human somatosensory areas 3a, 3b, and 1 on the basis of a computerized brain atlas. *Anatomy and Embryology (Berl), 204*, 351–366.

Gibbon, R. J., Granger, D. E., Kuman, K., & Partridge, T. C. (2009). Early Acheulean technology in the Rietputs Formation, South Africa, dated with cosmogenic nuclides. *Journal of Human Evolution, 56*, 152–160.

Goodall, J. (1986). *The chimpanzees of Gombe: Patterns of behavior.* Boston: Bellknap Press of the Harvard University Press.

Gowlett, J. A. J. (2006). The elements of design form in Acheulian bifaces: Modes, modalities, rules and languag. In N. Goren-Inbar & G. Sharon (Eds.), *Axe age: Acheulian tool-making from quarry to discard* (pp. 203–221). London: Equinox.

Grabowski, T. J., Damasio, H., Tranel, D., Ponto, L. L., Hichwa, R. D., & Damasio, A. R. (2001). A role for left temporal pole in the retrieval of words for unique entities. *Human Brain Mapping, 13*, 199–212.

Gracia, A., Martinez-Lage, J. F., Arsuaga, J. L., Martinez, I., Lorenzo, C., & Perez-Espejo, M. A. (2011). The earliest evidence of true lambdoid craniosynostosis: the case of "Benjamina", a Homo heidelbergensis child. *Child's Nervous System, 26*, 723–727.

Green, R. E., Krause, J., Briggs, A. W., Maricic, T., Stenzel, U., Kircher, M., et al. (2010). A draft sequence of the Neandertal genome. *Science, 328*, 710–722.

Grimaud-Hervé, D. (1997). *L'évolution de l'encéphale chez Homo erectus et Homo sapiens.* Paris: CNRS éditions.

Grun, R., Stringer, C., McDermott, F., Nathan, R., Porat, N., Robertson, S., et al. (2005). U-series and ESR analyses of bones and teeth relating to the human burials from Skhul. *Journal of Human Evolution, 49*, 316–334.

Gunz, P., Neubauer, S., Maureille, B., & Hublin, J. J. (2010). Brain development after birth differs between Neanderthals and modern humans. *Current Biology, 20*, R921–R922.

Happe, F., Ronald, A., & Plomin, R. (2006). Time to give up on a single explanation for autism. *Nature Neuroscience, 9*, 1218–1220.

Haslam, M., Hernandez-Aguilar, A., Ling, V., Carvalho, S., de la Torre, I., Destefano, A., et al. (2009). Primate archaeology. *Nature, 460*, 339–344.

Hawks, J. (2011a). Denisova FOXP2 status. *John Hawks Weblog*, [Online]. Available from: http://johnhawks.net/weblog/reviews/denisova/foxp2-denisova-humanlike-2011.html.

Hawks, J. (2011b). Denisova microcephalin status. *John Hawks Weblog*, [Online]. Available from: http://johnhawks.net/weblog/reviews/denisova/denisova-mcph1-2011.html.

Henshilwood, C. S., D'Errico, F., & Watts, I. (2009). Engraved ochres from the Middle Stone Age levels at Blombos Cave, South Africa. *Journal of Human Evolution, 57*, 27–47.

Henshilwood, C. S., D'Errico, F., Yates, R., Jacobs, Z., Tribolo, C., Duller, G. A., et al. (2002). Emergence of modern human behavior: Middle Stone Age engravings from South Africa. *Science, 295*, 1278–1280.

Herndon, J. G., Tigges, J., Anderson, D. C., Klumpp, S. A., & McClure, H. M. (1999). Brain weight throughout the life span of the chimpanzee. *The Journal of Comparative Neurology, 409*, 567–572.

Hershkovitz, I., Kornreich, L., & Laron, Z. (2007). Comparative skeletal features between Homo floresiensis and patients with primary growth hormone insensitivity (Laron syndrome). *American Journal of Physical Anthropology, 134*, 198–208.

Holloway, R. L. (1966). Cranial capacity, neural reorganization, and hominid evolution—Search for more suitable parameters. *American Anthropologist, 68*, 103–121.

Holloway, R. L. (1968). The evolution of the primate brain: Some aspects of quantitative relations. *Brain Research, 7*, 121–172.

Holloway, R. L. (1975). Early hominid endocasts: Volumes, morphology and significance. In R. Tuttle (Ed.), *Primate functional morphology and evolution*. The Hague: Mouton.

Holloway, R. L. (1980). Indonesian "Solo" (Ngandong) endocranial reconstructions: Some preliminary observations and comparisons with Neandertal and *Homo erectus* groups. *American Journal of Physical Anthropology, 53*, 285–295.

Holloway, R. L. (1983). Human paleontological evidence relevant to language behavior. *Human Neurobiology, 2*, 105–114.

Holloway, R. L. (1985). The past, present, and future significance of the lunate sulcus. In P. V. Tobias, V. Strong & H. White (Eds.), *Hominid evolution: Past, present and future.* New York: A.R. Liss.

Holloway, R. L. (1992). The failure of the gyrification index (GI) to account for volumetric reorganization in the evolution of the human brain. *Journal of Human Evolution, 22*, 163.

Holloway, R. L. (1996). Evolution of the human brain. In A. Lock & C. R. Peters (Eds.), *Handbook of human symbolic evolution*. New York: Clarendon Press.

Holloway, R. L. (2002). Brief communication: How much larger is the relative volume of area 10 of the prefrontal cortex in humans? *American Journal of Physical Anthropology, 118*, 399–401.

Holloway, R. L., Broadfield, D. C., & Yuan, M. S. (2003). Morphology and histology of chimpanzee primary visual striate cortex indicate that brain reorganization predated brain expansion in early hominid evolution. *Anatomical Record, 273A*, 594–602.

Holloway, R. L., Broadfied, D. C., & Yuan, M. (2004a). *The human fossil record, volume three: Brain endocasts—The paleoneurological evidence.* Hoboken, NJ: John Wiley & Sons Inc.

Holloway, R. L., Clarke, R. J., & Tobias, P. V. (2004b). Posterior lunate sulcus in *Australopithecus africanus*: Was Dart right? *Comptes Rendus Palevol, 3*, 287–293.

Holloway, R. L., & De Lacoste-Lareymondie, M. C. (1982). Brain endocast asymmetry in pongids and hominids: Some preliminary findings on the paleontology of cerebral dominance. *American Journal of Physical Anthropology, 58*, 101–110.

Hopkins, W. D., Hook, M., Braccini, S., & Schapiro, S. J. (2003). Population-level right handedness for a coordinated bimanual task in chimpanzees: Replication and extension in a second colony of apes. *International Journal of Primatology, 24*, 677–689.

Hopkins, W. D., & Marino, L. (2000). Asymmetries in cerebral width in nonhuman primate brains as revealed by magnetic resonance imaging (MRI). *Neuropsychologia, 38*, 493–499.

Hopkins, W. D., Phillips, K. A., Bania, A., Calcutt, S. E., Gardner, M., Russell, J., et al. (2011). Hand preferences for coordinated bimanual actions in 777 great apes: Implications for the evolution of handedness in hominins. *Journal of Human Evolution, 60*, 605–611.

Horel, J. A., Voytko, M. L., & Salsbury, K. G. (1984). Visual learning suppressed by cooling the temporal pole. *Behavioral Neuroscience, 98*, 310–324.

Humphrey, N. (1998). Cave art, autism, and the evolution of the human mind. *Cambridge Archaeological Journal, 8*, 165–191.

Hyvarinen, J. (1981). Regional distribution of functions in parietal association area 7 of the monkey. *Brain Research, 206*, 287-303.

Ivanhoe, F. (1970). Was Virchow right about Neandertal? *Nature, 227*, 577–579.

Jackson, A. P., Eastwood, H., Bell, S. M., Adu, J., Toomes, C., Carr, I. M., et al. (2002). Identification of microcephalin, a protein implicated in determining the size of the human brain. *American Journal of Human Genetics, 71*, 136–142.

Jacob, T., Indriati, E., Soejono, R. P., Hsu, K., Frayer, D. W., Eckhardt, R. B., et al. (2006). Pygmoid Australomelanesian *Homo sapiens* skeletal remains from Liang Bua, Flores: Population affinities and pathological abnormalities. *Proceedings of the National Academy of Sciences of the United States of America, 103*, 13421–13426.

Jerison, H. J. (1973). *Evolution of the brain and intelligence.* New York: Academic Press.

Jungers, W. L., Falk, D., Hildebolt, C., Smith, K., Prior, F., Tocheri, M. W., et al. (2009). The hobbits (*Homo floresiensis*) were not cretins. *American Association of Physical Anthropologists, 138*(S48), 244.

Kahlenberg, S. M., & Wrangham, R. W. (2010). Sex differences in chimpanzees' use of sticks as play objects resemble those of children. *Current Biology, 20*, R1067–R1068.

Keith, A. (1931). *New discoveries relating to the antiquity of man.* New York: Williams and Norgate.

Keller, S. S., Crow, T., Foundas, A., Amunts, K., & Roberts, N. (2009). Broca's area: Nomenclature, anatomy, typology and asymmetry. *Brain and Language, 109*, 29–48.

Kimbel, W. H., Walter, R. C., Johanson, D. C., Reed, K. E., Aronson, J. L., Assefa, Z., et al. (1996). Late Pliocene *Homo* and Oldowan tools from the Hadar Formation (Kada Hadar member), Ethiopia. *Journal of Human Evolution, 31*, 549–561.

Klein, R. G. (1999). *The human career: Human biological and cultural origins.* Chicago: University of Chicago Press.

Klein, R. G. (2003). Whither the Neanderthals? *Science, 299*, 1525–1527.

Krause, J., Lalueza-Fox, C., Orlando, L., Enard, W., Green, R. E., Burbano, H. A., et al. (2007). The derived FOXP2 variant of modern humans was shared with Neandertals. *Current Biology, 17*, 1908–1912.

Kuman, K., & Clarke, R. J. (2000). Stratigraphy, artefact industries and hominid associations for Sterkfontein, member 5. *Journal of Human Evolution, 38*, 827–847.

Lai, C. S., Fisher, S. E., Hurst, J. A., Vargha-Khadem, F., & Monaco, A. P. (2001). A forkhead-domain gene is mutated in a severe speech and language disorder. *Nature, 413*, 519–523.

Lari, M., Rizzi, E., Milani, L., Corti, G., Balsamo, C., Vai, S., et al. (2010). The microcephalin ancestral allele in a Neanderthal individual. *PloS One, 5*, e10648.

Le Gros Clark, W. E. (1947). Observations on the anatomy of the fossil Australopithecinae. *Journal of Anatomy, 81*, 300–333.

Le Gros Clark, W. E., Cooper, D. M., & Zuckerman, S. (1936). The endocranial cast of the chimpanzee. *Journal of the Royal Anthropological Institute of Great Britain and Ireland, 66*, 249–268.

Le May, M. (1976). Morphological cerebral asymmetries of modern man, fossil man and nonhuman primates. *Annals of the New York Academy of Sciences, 280*, 349–366.

Le May, M., Billig, M. S., & Geschwind, N. (1982). Asymmetries in the brains and skulls of nonhuman primates. In E. Armstrong & D. Falk (Eds.), *Primate brain evolution: Methods and concepts* (pp. 263–278). New York: Plenum Press.

Leakey, L. S. B., Tobias, P. V., & Napier, J. R. (1964). A new species of genus *Homo* from Olduvai Gorge. *Nature, 202*, 7–9.

Lee, S. H., & Wolpoff, M. H. (2003). The pattern of evolution in Pleistocene human brain size. *Paleobiology, 29*, 186–196.

Leigh, S. R. (2004). Brain growth, life history, and cognition in primate and human evolution. *American Journal of Primatology, 62*, 139–164.

Leigh, S. R. (2006). Brain ontogeny and life history in Homo erectus. *Journal of Human Evolution, 50*, 104–108 (author reply 109–13).

Lewis, J. D., & Elman, J. L. (2008). Growth-related neural reorganization and the autism phenotype: A test of the hypothesis that altered brain growth leads to altered connectivity. *Developmental Science, 11*, 135–155.

Liu, W., Zhang, Y., & Wu, X. (2005). Middle Pleistocene human cranium from Tangshan (Nanjing), Southeast China: A new reconstruction and comparisons with Homo erectus from Eurasia and Africa. *American Journal of Physical Anthropology, 127*, 253–262.

Lozano, M., Mosquera, M., De Castro, J. M. B., Arsuaga, J. L. & Carbonell, E. 2009. Right handedness of *Homo heidelbergensis* from Sima de los Huesos (Atapuerca, Spain) 500,000 years ago. *Evolution and Human Behavior, 30*, 369–376.

Lynch, J. C. (1980). The functional organization of posterior parietal association cortex. *The Behavioral and Brain Sciences, 3*, 485–534.

Macdermot, K. D., Bonora, E., Sykes, N., Coupe, A. M., Lai, C. S., Vernes, S. C., et al. (2005). Identification of FOXP2 truncation as a novel cause of developmental speech and language deficits. *American Journal of Human Genetics, 76*, 1074–1080.

Macleod, C. E., Zilles, K., Schleicher, A., Rilling, J. K., & Gibson, K. R. (2003). Expansion of the neocerebellum in Hominoidea. *Journal of Human Evolution, 44*, 401–429.

Marean, C. W., Bar-Matthews, M., Bernatchez, J., Fisher, E., Goldberg, P., Herries, A. I., et al. (2007). Early human use of marine resources and pigment in South Africa during the Middle Pleistocene. *Nature, 449*, 905–908.

Martin, R. D. (1981). Relative brain size and basal metabolic rate in terrestrial vertebrates. *Nature, 293*, 57–60.

Martin, R. D. (1990). *Primate origins and evolution: A phylogenetic reconstruction.* London/New Jersey: Chapman Hall/ Princeton University Press.

Martin, R. D., Maclarnon, A. M., Phillips, J. L., & Dobyns, W. B. (2006). Flores hominid: New species or microcephalic dwarf? *Anatomical Record, 288*, 1123–1145.

McBrearty, S., & Brooks, A. (2000). The revolution that wasn't: A new interpretation of the origin of modern human behavior. *Journal of Human Evolution, 39*, 453–563.

McBrearty, S., & Jablonski, N. G. (2005). First fossil chimpanzee. *Nature, 437*, 105–108.

McDougall, I., Brown, F. H., & Fleagle, J. G. (2005). Stratigraphic placement and age of modern humans from Kibish, Ethiopia. *Nature, 433*, 733–736.

McPherron, S. P. (2000). Handaxes as a measure of the mental capabilities of early hominids. *Journal of Archaeological Science, 27*, 655–663.

McPherron, S. P., Alemseged, Z., Marean, C. W., Wynn, J. G., Reed, D., Geraads, D., et al. (2010). Evidence for stone-tool-assisted consumption of animal tissues before 3.39 million years ago at Dikika, Ethiopia. *Nature, 466*, 857–860.

Mercader, J., Panger, M., & Boesch, C. (2002). Excavation of a chimpanzee stone tool site in the African rainforest. *Science, 296*, 1452–1455.

Mignault, C. (1985). Transition between sensorimotor and symbolic activities in nursery-reared chimpanzees (*Pan troglodytes*). *Journal of Human Evolution, 14*, 747–758.

Montgomery, S. H., Capellini, I., Venditti, C., Barton, R. A., & Mundy, N. I. (2011). Adaptive evolution of four microcephaly genes and the evolution of brain size in anthropoid primates. *Molecular Biology and Evolution, 28*, 625–638.

Mowbray, K., Marquez, S., & Anton, S. (2000). The newly recovered Poloyo hominin (PL-1) from Java. *American Journal of Physical Anthropology Supplement, 30*, 233.

Nakamura, K., & Kubota, K. (1995). Mnemonic firing of neurons in the monkey temporal pole during a visual recognition memory task. *Journal of Neurophysiology, 74*, 162–178.

Neubauer, S., Gunz, P., & Hublin, J. J. (2010). Endocranial shape changes during growth in chimpanzees and humans: A morphometric analysis of unique and shared aspects. *Journal of Human Evolution, 59*, 555–566.

Neubauer, S., Gunz, P., Mitteroecker, P., & Weber, G. W. (2004). Three-dimensional digital imaging of the partial *Australopithecus africanus* endocranium MLD 37/38. *Canadian Association of Radiologists Journal, 55*, 271–278.

Obendorf, P. J., Oxnard, C. E., & Kefford, B. J. (2008). Are the small human-like fossils found on Flores human endemic cretins? *Proceedings of the Royal Society of London Series B-Biological Sciences, 275*, 1287–1296.

Oxnard, C., Obendorf, P. J., & Kefford, B. J. (2010). Post-cranial skeletons of hypothyroid cretins show a similar anatomical mosaic as *Homo floresiensis. PloS One, 5*, e13018.

Panger, M. A., Brooks, A. S., Richmond, B. G., & Wood, B. (2002). Older than the Oldowan? Rethinking the emergence of hominin tool use. *Evolutionary Anthropology, 11*, 235–245.

Peeters, R., Simone, L., Nelissen, K., Fabbri-Destro, M., Vanduffel, W., Rizzolatti, G., et al. (2009). The representation of tool use in humans and monkeys: Common and uniquely human features. *The Journal of Neuroscience, 29*, 11523–11539.

Peirce, C. S. (1932). The icon, index, and symbol. In C. Hartshorne & P. Weis (Eds.), *The collected papers of Charles Sanders Peirce, Vol. II* (pp. 156–173). Cambridge, MA: Harvard.

Peresani, M., Fiore, I., Gala, M., Romandini, M., & Tagliacozzo, A. (2011). Late Neandertals and the intentional removal of feathers as evidenced from bird bone taphonomy at Fumane Cave 44 ky B.P., Italy. *Proceedings of the National Academy of Sciences of the United States of America, 108*, 3888–3893.

Plummer, T. (2004). Flaked stones and old bones: Biological and cultural evolution at the dawn of technology. *American Journal of Physical Anthropology, 125*, 118–164.

Ponce de Leon, M. S., Golovanova, L., Doronichev, V., Romanova, G., Akazawa, T., Kondo, O., et al. (2008). Neanderthal brain size at birth provides insights into the evolution of human life history. *Proceedings of the National Academy of Sciences of the United States of America, 105*, 13764–13768.

Powell, A., Shennan, S., & Thomas, M. G. (2009). Late Pleistocene demography and the appearance of modern human behavior. *Science, 324*, 1298–1301.

Pruetz, J. D., & Bertolani, P. (2007). Savanna chimpanzees, Pan troglodytes verus, hunt with tools. *Current Biology, 17*, 412–417.

Radinsky, L. (1972). Endocasts and studies of primate brain evolution. In R. Tuttle (Ed.), *The functional and evolutionary biology of primates*. Chicago: Aldine-Atherton Inc.

Rademacher, J., Burgel, U., Geyer, S., Schormann, T., Schleicher, A., Freund, H. J., & Zilles, K. (2001). Variability and asymmetry in the human precentral motor system. A cytoarchitectonic and myeloarchitectonic brain mapping study. *Brain, 124*, 2232–2258.

Reich, D., Green, R. E., Kircher, M., Krause, J., Patterson, N., Durand, E. Y., et al. (2010). Genetic history of an archaic hominin group from Denisova Cave in Siberia. *Nature, 468*, 1053–1060.

Resnick, S. M., Pham, D. L., Kraut, M. A., Zonderman, A. B., & Davatzikos, C. (2003). Longitudinal magnetic resonance imaging studies of older adults: A shrinking brain. *The Journal of Neuroscience, 23*, 3295–3301.

Ricklan, D. E. (1987). Functional anatomy of the hand of *Australopithecus africanus*. *Journal of Human Evolution*, *16*, 643–664.

Riel-Salvatore, J., & Clark, G. A. (2001). Grave markers: Middle and Early Upper Paleolithic burials and the use of chronotypology in contemporary Paleolithic research. *Current Anthropology*, *42*, 449–479.

Rightmire, G. P. (2004). Brain size and encephalization in early to mid-Pleistocene *Homo*. *American Journal of Physical Anthropology*, *124*, 109–123.

Rilling, J. K., & Insel, T. R. (1998). Evolution of the cerebellum in primates: Differences in relative volume among monkeys, apes and humans. *Brain, Behavior and Evolution*, *52*, 308–314.

Rilling, J. K., & Insel, T. R. (1999). The primate neocortex in comparative perspective using magnetic resonance imaging. *Journal of Human Evolution*, *37*, 191–223.

Rilling, J. K., & Seligman, R. A. (2002). A quantitative morphometric comparative analysis of the primate temporal lobe. *Journal of Human Evolution*, *42*, 505–533.

Rizzolatti, G., & Arbib, M. A. (1998). Language within our grasp. *Trends in Neurosciences*, *21*, 188–194.

Ruff, C. B., Trinkaus, E., & Holliday, T. W. (1997). Body mass and encephalization in Pleistocene *Homo*. *Nature*, *387*, 173–176.

Schenker, N. M., Buxhoeveden, D. P., Blackmon, W. L., Amunts, K., Zilles, K., & Semendeferi, K. (2008). A comparative quantitative analysis of cytoarchitecture and minicolumnar organization in Broca's area in humans and great apes. *The Journal of Comparative Neurology*, *510*, 117–128.

Schepers, G. W. H. (1946). The endocranial casts of the South African apemen. In R. Broom & G. W. H. Schepers (Eds.), *The South African fossil apemen: The Australopithecinae*. Pretoria: Transvaal Museum.

Schick, K. D., Toth, N., Garufi, G., Savage-Rumbaugh, E., Rumbaugh, D., & Sevcik, R. (1999). Continuing investigations into the stone tool-making and tool-using capabilities of a Bonobo (*Pan paniscus*). *Journal of Archaeological Science*, *26*, 821–832.

Schoenemann, P. T., Sheehan, M. J., & Glotzer, L. D. (2005). Prefrontal white matter volume is disproportionately larger in humans than in other primates. *Nature Neuroscience*, *8*, 242–252.

Schwartz, J. H., & Tattersall, I. (2002). *The human fossil record, volume 1. Terminology and craniodental morphology of genus Homo (Europe)*. New York: Wiley-Liss.

Semaw, S., Renne, P., Harris, J. W., Feibel, C. S., Bernor, R. L., Fesseha, N., et al. (1997). 2.5-million-year-old stone tools from Gona, Ethiopia. *Nature*, *385*, 333–336.

Semendeferi, K., Armstrong, E., Schleicher, A., Zilles, K., & van Hoesen, G. W. (2001). Prefrontal cortex in humans and apes: A comparative study of area 10. *American Journal of Physical Anthropology*, *114*, 224–241.

Semendeferi, K., & Damasio, H. (2000). The brain and its main anatomical subdivisions in living hominoids using magnetic resonance imaging. *Journal of Human Evolution*, *38*, 317–332.

Semendeferi, K., Damasio, H., Frank, R., & van Hoesen, G. W. (1997). The evolution of the frontal lobes: A volumetric analysis based on three-dimensional reconstructions of magnetic resonance scans of human and ape brains. *Journal of Human Evolution*, *32*, 375–388.

Semendeferi, K., Lu, A., Schenker, N., & Damasio, H. (2002). Humans and great apes share a large frontal cortex. *Nature Neuroscience*, *5*, 272–276.

Semendeferi, K., Teffer, K., Buxhoeveden, D. P., Park, M. S., Bludau, S., Amunts, K., et al. (2011). Spatial organization of neurons in the frontal pole sets humans apart from great apes. *Cerebral Cortex*, *21*(7), 1485–1497.

Sherwood, C. C., Broadfield, D. C., Holloway, R. L., Gannon, P. J., & Hof, P. R. (2003). Variability of Broca's area homologue in African great apes: Implications for language evolution. *Anatomical Record*, *271A*, 276–285.

Sherwood, C. C., Holloway, R. L., Semendeferi, K., & Hof, P. R. (2005). Is prefrontal white matter enlargement a human evolutionary specialization? *Nature Neuroscience*, *8*, 537–538.

Shibata, T., & Ioannides, A. A. (2001). Contribution of the human superior parietal lobule to spatial selection process: An MEG study. *Brain Research*, *897*, 164–168.

Simon, O., Mangin, J. F., Cohen, L., Le Bihan, D., & Dehaene, S. (2002). Topographical layout of hand, eye, calculation, and language-related areas in the human parietal lobe. *Neuron*, *33*, 475–487.

Skinner, M., & Wood, B. (2006). The evolution of modern human life history—A paleontological perspective. In K. Hawkes & R. Paine (Eds.), *The evolution of modern human life history*. Santa Fe: School of American research.

Smith, G. E. (1927). *Evolution of man: Essays*. London: Oxford University Press.

Soressi, M., & D'errico, F. (2007). Pigments, gravures, parures: Les comportements symboliques controversés des Néandertaliens. In B. Vandermeersch & B. Maureille (Eds.), *Les Néandertaliens. Biologie et cultures*, (Vol. 23), (pp. 297–309). Paris: Éditions du CTHS.

Spikins, P. A., Rutherford, H. E., & Needham, A. P. (2010). From hominінity to humanity: Compassion from the earliest archaics to modern humans. *Time and Mind*, *3*, 303–325.

Stout, D., Passingham, R., Frith, C., Apel, J., & Chaminade, T. (2011). Technology, expertise and social cognition in human evolution. *The European Journal of Neuroscience*, *33*, 1328–1338.

Stout, D., Toth, N., Schick, K., & Chaminade, T. (2008). Neural correlates of Early Stone Age toolmaking: Technology, language and cognition in human evolution. *Philosophical Transactions of the Royal Society of London B*, *363*, 1939–1949.

Stout, D., Toth, N., Schick, K., Stout, J., & Hutchins, G. (2000). Stone tool-making and brain activation: Position emission tomography (PET) studies. *Journal of Archaeological Science*, *27*, 1215–1223.

Susman, R. L. (1994). Fossil evidence for early hominid tool use. *Science*, *265*, 1570–1573.

Susman, R. L. (1998). Hand function and tool behavior in early hominids. *Journal of Human Evolution*, *35*, 23–46.

Suwa, G., Asfaw, B., Kono, R. T., Kubo, D., Lovejoy, C. O., & White, T. D. (2009). The *Ardipithecus ramidus* skull and its implications for hominid origins. *Science*, *326*, 68e1–68e7.

Symington, J. (1916). Endocranial casts and brain form: A criticism of some recent speculations. *Journal of Anatomy and Physiology*, *11*, 111–130.

Tobias, P. V. (1975). Brain evolution in the Hominoidea. In R. H. Tuttle (Ed.), *Primate functional morphology and evolution*. The Hague: Mouton.

Tobias, P. V. (1987). The brain of *Homo habilis*: A new level of organization in cerebral evolution. *Journal of Human Evolution*, *16*, 741.

Tobias, P. V. (1991). *Olduvai Gorge, Volume 4. The skulls, endocasts and teeth of Homo habilis*. Cambridge: Cambridge University Press.

Tocheri, M. W., Orr, C. M., Jacofsky, M. C., & Marzke, M. W. (2008). The evolutionary history of the hominin hand since the last common ancestor of *Pan* and *Homo*. *Journal of Anatomy*, *212*, 544–562.

Toth, N. (1985). Archaeological evidence for preferential right-handedness in the Lower and Middle Pleistocene, and its possible implications. *Journal of Human Evolution*, *14*, 607–614.

Toth, N., & Schick, K. (2009). The Oldowan: The tool making of early hominins and chimpanzees compared. *Annual Review of Anthropology*, *38*, 289–305.

Toth, N., Schick, K. D., Savage-Rumbaugh, E. S., Sevcik, R. A., & Rumbaugh, D. M. (1993). *Pan* the tool-maker: Investigations into the stone tool-making and tool-using capabilities of a bonobo (*Pan paniscus*). *Journal of Archaeological Science*, *20*, 81–91.

Uomini, N. T. (2009). The prehistory of handedness: Archaeological data and comparative ethology. *Journal of Human Evolution*, *57*, 411–419.

Vanhaereny, M., D'Errico, F., Stringer, C., James, S. L., Todd, J. A., & Mienis, H. K. (2006). Middle Paleolithic shell beads in Israel and Algeria. *Science*, *312*, 1785–1788.

Weaver, A. G. H. (2001). *The cerebellum and cognitive evolution in Pliocene and Pleistocene hominids*. The University of New Mexico: Unpublished doctoral dissertation.

Weaver, A. H. (2005). Reciprocal evolution of the cerebellum and neocortex in fossil humans. *Proceedings of the National Academy of Sciences of the United States of America*, *102*, 3576–3580.

Weidenreich, F. (1943). The skulls of Sinanthropus pekinensis: A comparative study on a primitive hominid skull. *Palaeontologia Sinica, Series D*, *10*, 1–485.

Wendt, W. E. (1974). "Art mobilier" aus der Apollo 11-Grotte in Südwest-Afrika. *Acta praehistorica et archaeologica*, *5*, 1–42.

Wendt, W. E. (1975). "Art Mobilier" from the Apollo 11 Cave, South West Africa: Africa's oldest dated works of art. *South African Archaeological Bulletin*, *31*, 5–11.

White, T., Asfaw, B., Degusta, D., Gilbert, H., Richards, G. D., Suwa, G., et al. (2003). Pleistocene *Homo sapiens* from Middle Awash, Ethiopia. *Nature*, *423*, 742–747.

Whiten, A., & van Schaik, C. P. (2007). The evolution of animal "cultures" and social intelligence. *Philosophical Transactions of the Royal Society B: Biological Sciences*, *362*, 603.

Wood, B., & Collard, M. (1999). The human genus. *Science*, *284*, 65–71.

Wood, B., & Lonergan, N. (2008). The hominin fossil record: Taxa, grades and clades. *Journal of Anatomy*, *212*, 354–376.

Wynn, T., & McGrew, W. C. (1989). An ape's view of the Oldowan. *Man*, *24*, 383–398.

Yantis, S., Schwarzbach, J., Serences, J. T., Carlson, R. L., Steinmetz, M. A., Pekar, J. J., et al. (2002). Transient neural activity in human parietal cortex during spatial attention shifts. *Nature Neuroscience*, *5*, 995–1002.

Zilhão, J. (2006). Neandertals and moderns mixed, and it matters. *Evolutionary Anthropology*, *15*, 183–195.

Zilhão, J. (2007). The emergence of ornaments and art: An archaeological perspective on the origins of "behavioral modernity" *Journal of Archaeological Research*, *15*, 1–54.

Zilhão, J., Angelucci, D. E., Badal-Garcia, E., D'Errico, F., Daniel, F., Dayet, L., et al. (2010). Symbolic use of marine shells and mineral pigments by Iberian Neandertals. *Proceedings of the National Academy of Sciences of the United States of America*, *107*(3), 1023–1028.

Zilles, K., Armstrong, E., Moser, K. H., Schleicher, A., & Stephan, H. (1989). Gyrification in the cerebral cortex of primates. *Brain, Behavior and Evolution*, *34*, 143–150.

Zilles, K., Armstrong, E., Schleicher, A., & Kretschmann, H. J. (1988). The human pattern of gyrification in the cerebral cortex. *Anatomy and Embryology*, *179*, 173–179.

Zollikofer, C. P. E., Ponce de Leon, M. S., Lieberman, D. E., Guy, F., Pilbeam, D., Likius, A., et al. (2005). Virtual cranial reconstruction of *Sahelanthropus tchadensis*. *Nature*, *434*, 755–759.

SECTION V

Theories of neural organization

M. A. Hofman and D. Falk (Eds.)
Progress in Brain Research, Vol. 195
ISSN: 0079-6123

CHAPTER 15

Neuronal scaling rules for primate brains: The primate advantage

Suzana Herculano-Houzel*

Instituto de Ciências Biomédicas, Universidade Federal do Rio de Janeiro, Brasil and Instituto Nacional de Neurociência Translacional, Rio de Janeiro, Brazil

Abstract: In what concerns cognitive abilities, primates usually outrank other mammals of similar, or even larger, brain size, as illustrated by comparisons between a macaque monkey and a capybara; a chimpanzee and a cow; or a human and a dolphin, whale, or elephant. Such a cognitive advantage is inconsistent with the traditional view of brain scaling in mammalian evolution as a homogeneous phenomenon regarding numbers of neurons and neuronal density, with brains of different sizes viewed as similarly scaled-up or scaled-down versions of a shared basic plan. Here, I will argue, instead, that different neuronal scaling rules apply to different mammalian orders and that the particular rules that apply to primates are such that endow us with an advantage over other mammals that is likely to have important cognitive consequences: a larger number of neurons concentrated per volume in the brain.

Keywords: brain size; allometry; numbers of neurons; numbers of glia; scaling; metabolism; cognition; encephalization.

Introduction

Because brains are made of neurons, it makes intuitive sense to expect larger brains to be made of larger numbers of neurons. If neurons are the computational units of the brain, then larger brains, made of larger numbers of neurons, should have larger computational abilities. By the same token, two brains of comparable size should be made of comparable numbers of neurons and hence have similar cognitive abilities.

A simple observation of mammalian behavior indicates, however, that this logic is flawed. While larger brain size is indeed generally correlated with better cognitive capacities across species of a same order (Deaner et al., 2007; Roth and Dicke, 2005), the correlation breaks down across orders. Compare, for instance, the range and complexity of the cognitive abilities of a macaque

*Corresponding author.
Tel.: +55-21-2562-6390; Fax +55-21-2562-6674
E-mail: suzanahh@gmail.com

DOI: 10.1016/B978-0-444-53860-4.00015-5

monkey and a capybara (~70 g brain mass); a chimpanzee and a cow (~400 g brain mass); or a human, at about 1500 g brain mass, with dolphins, whales, and elephants, endowed with even larger brains of 2–9 kg (Haug, 1987; Herculano-Houzel et al., 2006; Marino, 1998). In such comparisons, primates seem to have a consistent cognitive advantage over other mammals of a similar brain size. Thus, either the logic that larger brains always have more neurons in a homogeneous manner across mammals is flawed, or number of neurons (and brain size, if a valid proxy for number of neurons) is not the most important determinant of cognitive abilities.

The previous view: "All brains scale the same"

For decades, studies on comparative neuroanatomy were largely comparisons of absolute size (volume or mass), relative structure size, or cell densities across species of the different mammalian orders indiscriminately: rats, monkeys, cows, cats, humans, elephants, and whales (Finlay and Darlington, 1995; Haug, 1987; Rockel et al., 1980; Tower and Elliott, 1952; Zhang and Sejnowski, 2000), with the tacit assumption that larger brains are simply homogeneously scaled-up versions of smaller brains. Such studies established that brain size is related to body size by a power law of exponent inferior to 1.0 (Fox and Wilczynski, 1986; Martin, 1981), such that brain size increases at a slower pace than body size, and at different rates across mammalian orders (Marino, 1998). Although primates were shown to have larger brains than rodents and insectivores for a given body mass, they seemed to be reasonably aligned with cetaceans in their body–brain relationship (Fig. 1).

Within the brain, those comparative studies found that the cerebral cortex increases faster in volume than the remaining brain structures, gaining in relative size such that larger brains are more and more dominated by cortex (Frahm et al., 1982). Although the human cerebral cortex is the largest among mammals in its relative size, at 75.5% (Rilling and Insel, 1999), 75.7% (Frahm et al., 1982), or even 84.0% (Hofman, 1988) of the entire brain mass or volume, other animals, primate and nonprimate, are not far behind: the

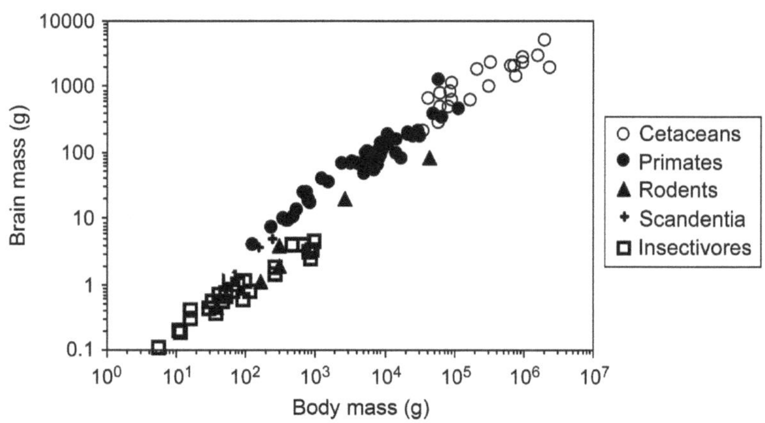

Fig. 1. Brain–body mass relationships among primates, cetaceans, rodents, and insectivores. Each datapoint represents an individual species belonging to one of the groups indicated. Brain mass scales with body mass in ways that can be described as power laws of different exponents for each group: cetaceans, 0.529; insectivores, 0.643; primates, 0.701; rodents, 0.758. Data from Frahm et al. (1982) (insectivores and scandentia), Marino (1998) (primates and cetaceans), and Herculano-Houzel et al. (2006) (rodents).

cerebral cortex represents 73.0% of the entire brain mass in the chimpanzee (Stephan et al., 1981), 74.5% in the horse, and 73.4% in the short-finned whale (Hofman, 1985). Despite the only small advantage, the relative increase in the size of the cerebral cortex has come to be equated with brain evolution and is often offered as an explanation for our cognitive superiority compared to other species (Rakic, 2009).

In comparison, larger brains have isometrically larger cerebella, which accompany almost linearly the size of the cerebral cortex, and retain a stable relative size with increasing brain size, such that larger brains have cerebella of the same relative volume. Depending on whether emphasis was placed on the absolute size or on the relative size of these structures, as proxies for their absolute or relative numbers of neurons, these studies apparently supported either a tendency toward relative expansion of the role of the cerebral cortex in evolution (Clark et al., 2001) or the coordinated evolution of the roles of cerebral cortex and cerebellum (Sultan, 2002)—both of which cannot be simultaneously true.

Comparisons of cell densities were also often made irrespectively of mammalian order, and with a heavy bias on analyses of the cerebral cortex, ever since Franz Nissl, based on visual inspection of the brains of various unrelated species, observed that neurons are distributed more sparsely in larger brains (Nissl, 1898). Further studies soon supported his observation, showing that neuronal density declines in the cerebral cortex as a power function of increasing brain volume across unrelated species with a negative exponent of −0.32 (Tower and Elliott, 1952), and that the decrease in neuronal density with increasing brain size applies to a large group of species comprising from the smallest insectivores (Stolzenburg et al., 1989) to primates, dolphins, elephants, and whales (Garey and Leuba, 1986; Haug, 1987; Tower, 1954). Those studies led to the widespread notions that larger brains as a rule have smaller neuronal densities (and thus presumably larger neurons), accompanied by larger glia/neuron ratios, which express the relative number of glial cells distributed among the neurons (Friede, 1954; Garey and Leuba, 1986; Haug, 1987; Hawkins and Olszewski, 1957; Stolzenburg et al., 1989; Tower, 1954; Tower and Elliott, 1952).

A decreased neuronal density in larger cerebral cortices is attributable to two factors: increased neuronal size (including the neuropil) and increased relative number of the interspersed glial cells. Interestingly, the increased glia/neuron ratio is not accompanied by any major variation in glial density, which has been reported either to vary widely but independently of brain size (Haug, 1987) or to remain stable (Stolzenburg et al., 1989; Tower and Young, 1973) across mammalian species of increasing brain size. Therefore, decreases in neuronal density most likely reflect increases in average neuronal cell size. Because of these supposedly universal neuronal and glial scaling rules, glia are widely said to be the most numerous cell type in the brain (Doetsch, 2003; Nishiyama et al., 2005) and to be 10–50 times more numerous than neurons in humans (Kandel et al., 2000). Evidence for this particular assertion, however, is scant.

Different neuronal scaling rules for the brains of different mammalian orders

One possible explanation for the cognitive advantage of primates over other mammals of similar brain size is their higher degree of encephalization (Jerison, 1977, 1985), that is, the excess of brain mass over what would be expected for a mammal of a given body mass from relationships such as those depicted in Fig. 1. However, the notion that higher encephalization correlates with improved cognitive abilities has been disputed, in favor of absolute numbers of cortical neurons and connections (Roth and Dicke, 2005) or simply absolute brain size (Deaner et al., 2007). One difficulty is that the encephalization quotient, EQ, does not take into account sheer brain size: For example, if EQ were the main determinant of cognitive abilities, small-brained animals with very large encephalization quotients, such as capuchin monkeys, should be expected to be more cognitively able than large-brained but less encephalized animals, such as the gorilla (Marino, 1998). However, the former, smaller, are outranked by the latter in cognitive performance (Deaner et al., 2007).

An alternative explanation for the cognitive superiority of primate species over nonprimate mammals of similar brain size is that all mammalian brains are not made the same, as larger or smaller versions of a same basic plan with proportionately larger or smaller numbers of neurons. In that scenario, similarly sized brains, such as cow and chimpanzee, might contain very different numbers of neurons, just as a very large cetacean brain might contain fewer neurons than a gorilla or even human brain. Testing this hypothesis required using a volume-independent method to obtain data on the numbers of neurons that compose the brains of different mammalian species so that these numbers could be examined against the respective brain volumes.

Such a method was developed in our lab a few years ago: the isotropic fractionator, which allows the nonstereological determination of the absolute number of neuronal and nonneuronal cells in different brain regions (Herculano-Houzel and Lent, 2005). It consists in transforming highly anisotropic brain structures into homogeneous, isotropic suspensions of fixed, free cell nuclei which can then be counted and identified immunocytochemically as neuronal or nonneuronal. The method can be applied either to the brain as a whole or to its dissected parts, such as cerebral cortex or cerebellum, whose respective numbers of cells can next be added up in order to obtain a whole-brain estimate. Estimates of total neuronal and nonneuronal numbers in any brain structure can be obtained in 24h, with a coefficient of variation below 10%. As the estimates obtained are independent of brain mass or volume, they can be used in comparative studies of variation in brain size among species and in studies of phylogenesis, development, adult neurogenesis, and pathology.

We have so far used the isotropic fractionator to compare the numbers of cells that compose the entire adult brain (divided into cerebral cortex [grey and white matter combined], cerebellum [grey and white matter and deep nuclei combined], and rest of brain or RoB, excluding the olfactory bulb) of 28 mammalian species that can be grouped into three large clades: 10 Glires (nine rodents and one lagomorph, which will heretofore referred to as "rodents"; Herculano-Houzel et al., 2006, 2011), 12 primates (including humans; Azevedo et al., 2009; Gabi et al., 2010; Herculano-Houzel et al., 2007) plus the closely related tree shrew (order Scandentia; the tree shrew is, however, not included in the primate dataset for quantification in this review); and five Eulipotyphla (insectivores; Sarko et al., 2009).

Our comparative studies of the neuronal composition of mammalian species showed that, indeed, different neuronal scaling rules apply to the various mammalian orders examined so far. Remarkably, the differences are such that similarly sized cerebral cortices always contain more neurons in primates than in rodents. For instance, while the agouti cortex, at 17.7g, holds 795 million neurons, the slightly smaller owl monkey cortex, at 15.7g and 1.5 billion neurons, contains almost twice as many (Herculano-Houzel et al., 2006, 2007). The larger the mass of the cerebral cortex, the larger the discrepancy in numbers of neurons between rodents and primates (Fig. 2, top). This is because the structure is found to scale in mass as different functions of its number of neurons across the two clades: as a power function of exponent 1.7 in rodents and as a power function of exponent 1.1 in primates that is equally well fitted as a linear function (Gabi et al., 2010). The rodent cerebral cortex, therefore, gains neurons in a very volume-costly manner, while the primate cortex gains neurons more economically in terms of resulting structure volume. The insectivore

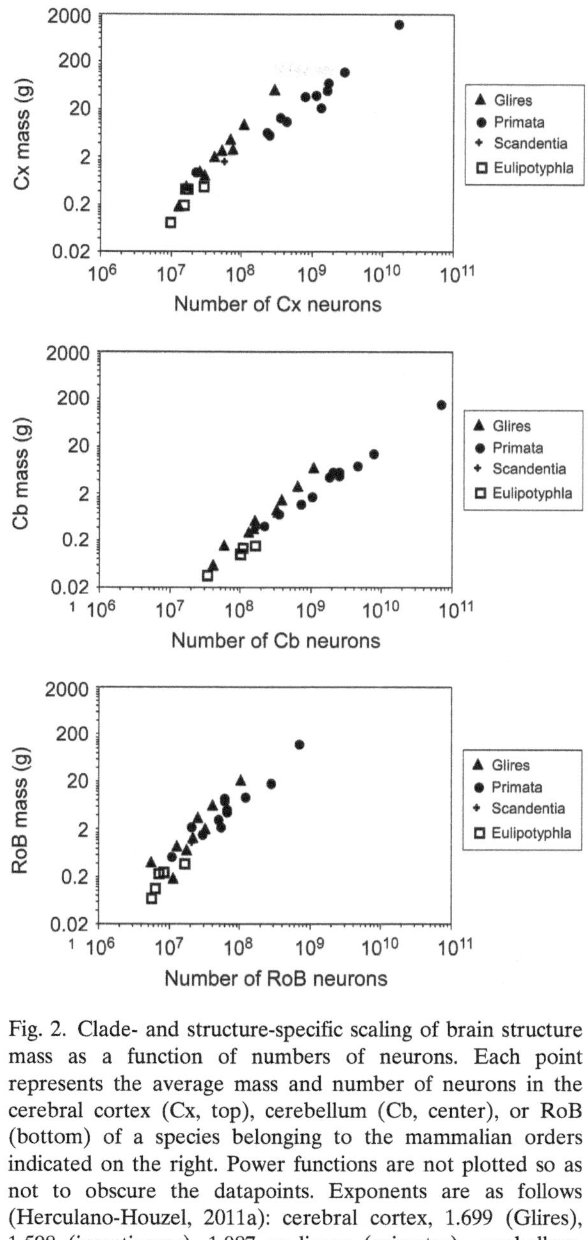

Fig. 2. Clade- and structure-specific scaling of brain structure mass as a function of numbers of neurons. Each point represents the average mass and number of neurons in the cerebral cortex (Cx, top), cerebellum (Cb, center), or RoB (bottom) of a species belonging to the mammalian orders indicated on the right. Power functions are not plotted so as not to obscure the datapoints. Exponents are as follows (Herculano-Houzel, 2011a): cerebral cortex, 1.699 (Glires), 1.598 (insectivores), 1.087 or linear (primates); cerebellum, 1.305 (Glires), 1.028 or linear (insectivores), 0.976 or linear (primates); RoB, 1.568 (Glires), 1.297 (insectivores), 1.198 (or 1.4 when corrected for phylogenetic relatedness in the dataset, primates; Herculano-Houzel, 2011a). Data from Herculano-Houzel et al. (2006, 2007), Sarko et al. (2009), Azevedo et al. (2009), and Gabi et al. (2010).

cerebral cortex, in turn, overlaps with rodents in its neuronal scaling and shares a similarly large allometric exponent of 1.6 (Fig. 2, top), increasing in size very rapidly as it gains neurons (Sarko et al., 2009).

The cerebellum, like the cerebral cortex, gains mass faster than it gains neurons in rodents, with an allometric exponent of 1.3 (Fig. 2, center). In primates and insectivores, in contrast, cerebellar mass scales linearly with the number of cerebellar neurons (Fig. 2, center). As a result, primate and insectivore cerebella are found to contain many more neurons than rodent cerebella of similar mass. For instance, the bonnet monkey cerebellum, at 5.7g, contains 2 billion neurons, almost twice as many as the capybara cerebellum at 6.6 g and 1.2 billion neurons (Herculano-Houzel et al., 2006, 2007); the eastern mole cerebellum, at 0.15g, contains 158 million neurons, over twice as many neurons as the hamster at 0.14g and 61 million neurons (Herculano-Houzel et al., 2006; Sarko et al., 2009).

Interestingly, the relationship between the mass of the remaining brain structures that compose the RoB and its number of neurons does not appear as clearly separated across the three clades, with a much larger overlap among the data points (Fig. 2, bottom) and exponents of 1.568 (rodents), 1.297 (insectivores), and 1.198 (primates, an exponent that increases to 1.4 after correction for phylogenetic relatedness in the dataset; Gabi et al., 2010; Herculano-Houzel, 2011a). This raises the interesting possibility that the scaling rules for the RoB, in contrast to the cerebral cortex and cerebellum, are shared across rodents, primates, and insectivores.

The different neuronal scaling rules reflect the highly diverse scaling of neuronal cell densities across species and structures, contrary to the homogeneous scaling rules assumed in the literature from cross-order comparisons (e.g., Haug, 1987). In rodents and insectivores, the cerebral cortex increases in size with an accompanying steep decrease in neuronal density (Fig. 3, top), which varies with cortical mass raised to the

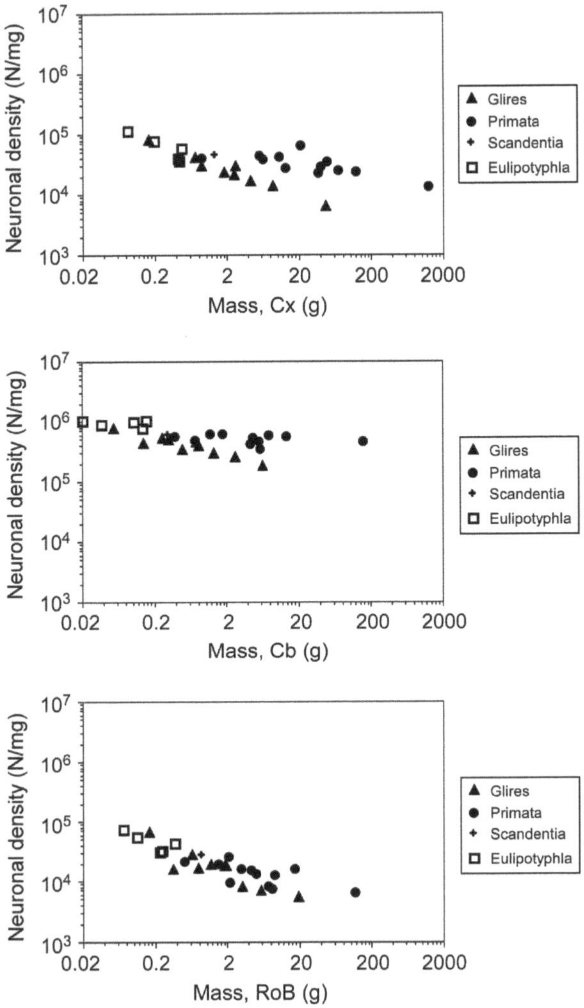

Fig. 3. Neuronal cell densities scale differently across structures and orders but are always larger in primates than in Glires. Each point represents the average structure mass and neuronal cell density (number of neurons/mg) in the cerebral cortex (top), cerebellum (center), or RoB (bottom) of each species. Power functions are not plotted so as not to obscure the datapoints. Exponents are as follows (Herculano-Houzel, 2011a): cerebral cortex, −0.424 (Glires), −0.569 (insectivores), −0.168 (primates); cerebellum, −0.271 (Glires), not significant (insectivores and primates); RoB, −0.467 (Glires), not significant (insectivores), −0.220 (primates). Data from Herculano-Houzel et al. (2006, 2007), Sarko et al. (2009), Azevedo et al. (2009), and Gabi et al. (2010).

power of −0.424 or −0.569, respectively. In primates, there is only a slight decrease in neuronal density in larger cortices, with an exponent of −0.168. As a result, neuronal densities in the cerebral cortex are always larger in primates than in rodents of a similar cortical mass, and the discrepancy becomes larger with increasing cortical mass. On the other hand, the cerebellum scales in size with no significant change in neuronal density in insectivores and primates, and with a more modest decrease in neuronal density in rodents, such that again neuronal densities are always larger in primate than in rodent cerebella of a similar mass (Fig. 3, center). Likewise, only in rodents does an increase in RoB mass correlate significantly with a decrease in neuronal density, which varies with RoB mass raised to the power of −0.467 (Fig. 3, bottom). Consistently with the possibility that the RoB neuronal scaling rules are shared across the three clades, however, the distributions of neuronal densities in this structure are fairly overlapping across the three clades—although, once again, neuronal densities appear larger in the primate RoB compared to most rodents of a similar RoB mass.

An interesting point regards the position of the tree shrew along the distribution of brain size and number of neurons. The tree shrew is currently classified in the order Scandentia, which, together with Rodentia, Lagomorpha, and Primata (as well as the Dermoptera, not analyzed here), composes the superorder Euarchontoglires, whose members shared a common ancestor over 90million years ago (Murphy et al., 2004). The cellular composition of the tree shrew brain can be predicted very well by the primate neuronal scaling rules, deviating on average by only 12.5% of the predicted numbers of neurons in the different brain structures, by 28.3% from the values predicted by the rodent neuronal scaling rules, and by a larger 42.0% from the predictions for insectivores. The good alignment with primates suggests that orders Scandentia and Primata, although considered sister clades, nevertheless share the same neuronal scaling rules—just like the sister orders Rodentia and Lagomorpha (Herculano-Houzel, 2011a; Herculano-Houzel et al., 2011).

Intriguingly, in the distributions of brain structure mass against numbers of neurons, the tree shrew is positioned approximately at the intersection between rodents and primates (see Figs. 2 and 3). Although this placing might be meaningless, as the two distributions are bound to intersect at some point, it raises the interesting possibility that the tree shrew brain is similar to the brain that once belonged to the original ancestral species that gave rise to the extant Euarchontoglires, branching in the two directions that evolved into Glires and Primata, with their clade-specific neuronal scaling rules.

Shared scaling rules for the brain across mammalian orders: Coordinate scaling of numbers of cortical and cerebellar neurons

While larger brains possess relatively larger cerebral cortices, the relative size of the cerebellum fails to increase with brain size (Clark et al., 2001; Fig. 4, top), as if these two structures did not evolve in concert. Assuming that relatively larger structures hold increasingly larger percentages of brain neurons across species, this discrepancy has been used to favor the traditional view that emphasizes the importance of relative neocortex expansion in brain function and evolution (Clark et al., 2001; Hofman, 1985; Jerison, 2007).

Once the numbers of neurons that compose these structures are available for scrutiny, however, we find that the distribution of mass in the brain does not reflect the distribution of neurons across the cerebral cortex and the cerebellum. The cerebral cortex typically contains around 20% of all brain neurons, while the cerebellum holds 70–80% of all brain neurons (Herculano-Houzel, 2009), regardless of brain size and inclusive of the human brain (Azevedo et al., 2009; Fig. 4, center). Neither in primates nor in rodents or insectivores is the relative number of neurons in either the

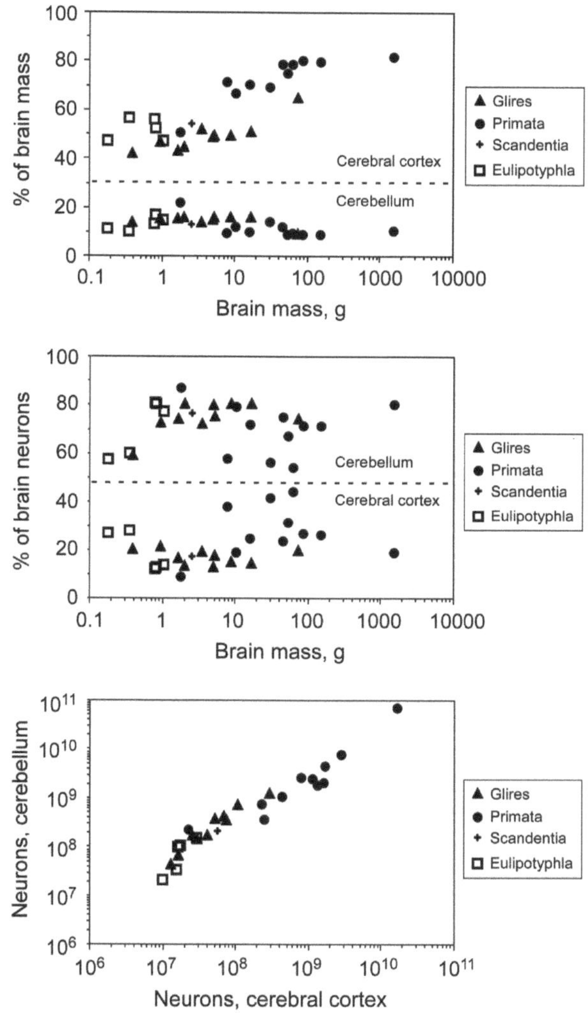

Fig. 4. Numbers of neurons scale coordinately across the cerebral cortex and cerebellum, despite the increase in relative cortical but not cerebellar mass, with increasing brain mass. Top, percentage of brain mass contained in the cerebral cortex or cerebellum of each species, plotted against total brain mass. Center, percentage of all brain neurons contained in the cerebral cortex or cerebellum of each species, plotted against total brain mass. Bottom, total number of cerebellar neurons plotted against total number of neurons in the cerebral cortex for each species. Each point represents the average values per species. The slope of the linear function that relates the number of cerebellar neurons to the number of cerebral cortical neurons is 4.2. Data from Herculano-Houzel et al. (2006, 2007), Sarko et al. (2009), Azevedo et al. (2009), and Gabi et al. (2010).

cerebral cortex or the cerebellum (calculated as a percentage of all brain neurons) correlated with the respective relative mass of the structure (Spearman correlation, all p-values>0.05). This indicates that a relatively larger cerebral cortex does not hold relatively more neurons in larger brains (Herculano-Houzel, 2010).

Strikingly, the cerebral cortex and the cerebellum gain neurons coordinately across the 28 species (Fig. 4, bottom), as originally reported for the initial dataset of 19 species (Herculano-Houzel, 2010), at an average rate of 4.2 neurons in the cerebellum to every neuron in the cerebral cortex, despite the different scaling rules that apply to cerebral cortex and cerebellum across the three clades (Fig. 2). Such a coordinated addition of neurons to the two structures is compatible with the modern view of the integrated function of cerebral cortex and cerebellum, and supports the notion that the two structures are subject to similar selective pressures and evolve concertedly (Balsters et al., 2010; Ramnani et al., 2006; Whiting and Barton, 2003). The coordinated scaling of their numbers of neurons is masked by their different mass scaling relationships, given that the cerebral cortex increases in mass as it gains neurons with a higher exponent than the cerebellum, as described above, such that the relative mass of the cerebral cortex increases in larger primate and Glire brains, while the relative mass of the cerebellum does not vary systematically. Primates, therefore, are similar to at least insectivores and rodents, among mammals, in the coordinate scaling of numbers of neurons in the cerebral cortex and cerebellum in evolution, despite any changes in the relative size of these structures.

Shared scaling rules for the brain across mammalian orders: Nonneuronal cells

In contrast to the clade-specific rules that apply to how the cerebral cortex and cerebellum scale in size as they gain neurons, the rules that govern the addition of nonneuronal (other) cells to the brain appear to be shared not only across clades but also across brain structures. As shown in Fig. 5 (top), the cerebral cortex, cerebellum, and RoB scale in size as similar, overlapping power functions of their respective numbers of other cells raised to exponents of 0.9–1.1 (or as linear functions of their numbers of other cells). As a result of the approximately linear relationship

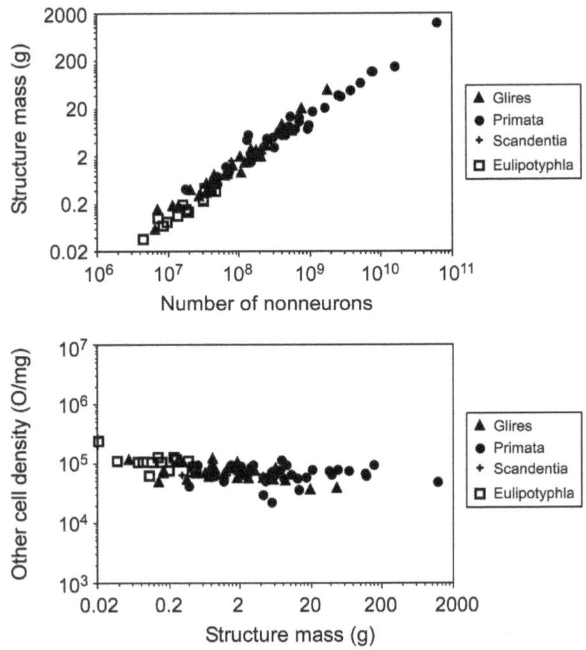

Fig. 5. Other cell scaling rules are shared across both clades and structures. Top, each point represents the average structure mass and number of other (nonneuronal) cells in the cerebral cortex, cerebellum, or RoB of a species belonging to the mammalian orders indicated on the right. Bottom, each point represents the average other cell density (in cells/mg of tissue) and the average mass of the same structures and species as above. All structures are plotted together so as to illustrate their overlapping of other cell scaling rules and other cell density. Exponents are as follows (Herculano-Houzel, 2011a): cerebral cortex, 1.132 (Glires), 1.143 (insectivores), 1.036 (primates); cerebellum, 1.002 (Glires), 1.094 (insectivores), 0.873 (primates); RoB, 1.073 (Glires), 0.926 (insectivores), 1.065 (primates). Data from Herculano-Houzel et al. (2006, 2007), Sarko et al. (2009), Azevedo et al. (2009), and Gabi et al. (2010).

between brain structure mass and number of other cells, we find that the three structures share a similar range of densities of other cells, which do not correlate significantly with variations in structure mass (Fig. 5, bottom).

Nonneuronal (other) cells include all glial cell types, ependymal cells, and endothelial cells. Because the latter are estimated to amount to at most 5% of all brain cells, given the small relative volume of brain vasculature (Lawers et al., 2008), and because the relative number of ependymal cells is most likely very small, the other/neuronal cell ratio serves as an upper limit of the glia/neuron ratio and provides a reasonable approximation of its actual value. For simplicity, the other/neuronal ratio in our sample will heretofore be referred to as the glia/neuron ratio.

Contrary to commonly assumed in the literature (Marino, 2006), we find no general trend for larger brains (or brain structures) to have larger glia/neuron ratios. Because of the different combinations of shared nonneuronal scaling rules and clade- and structure-specific neuronal scaling rules, the glia/neuron ratio is found to increase together with structure size only in the cerebral cortex, cerebellum, and RoB of rodents, and in the cerebral cortex of insectivores (Fig. 6, top) while varying nonsystematically with structure mass across primate brains. This difference means that while neurons represent decreasing percentages of all brain cells across rodents of increasing brain size, neurons represent about 50% of all brain cells in all primates examined so far, including in the human brain (Azevedo et al., 2009; Gabi et al., 2010; Herculano-Houzel et al., 2007), contrary to the often quoted line according to which there would be "10 times more neuron than glia in the human brain" (for instance, Barres and Allen, 2009; Kandel et al., 2000).

Remarkably, however, we find that the variation in the glia/neuron ratio accompanies decreasing neuronal density in a manner that is overlapping across all structures and species, including primates (Fig. 6, bottom), and therefore appears to obey a shared scaling rule that, like the addition of nonneuronal cells to brain tissue, is clade- and structure-non-specific. Because of the inverse relationship between neuronal density and average neuronal size, this finding suggests that the glia/neuron ratio is directly related to average neuronal size: the larger the average neuronal size in a structure, whatever the species, primate or nonprimate, the larger the glia/neuron ratio in the structure.

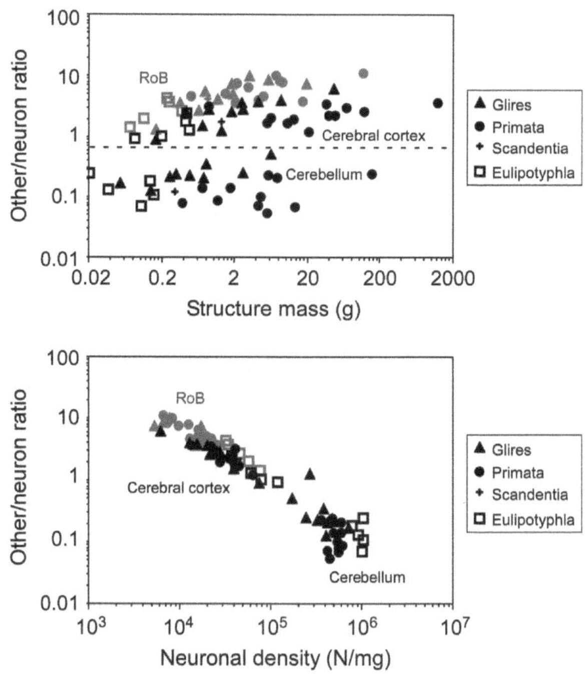

Fig. 6. Glia/neuron ratio scales differently across structures and orders with structure mass but scales homogeneously with neuronal density. Each point represents the average other cell/neuron ratio (which approximates the glia/neuron ratio) and structure mass (top) or neuronal density (bottom) in the cerebral cortex (grey), cerebellum (black), or RoB (black) of a species. Notice that in contrast to the scattered distribution across species and structures in the top graph, datapoints are aligned across species and structures in the bottom plot, suggesting that it is smaller neuronal densities (i.e., larger average neuronal cell size), not larger structure mass, that is accompanied by a larger glia/neuron ratio. Data from Herculano-Houzel et al. (2006, 2007), Sarko et al. (2009), Azevedo et al. (2009), and Gabi et al. (2010).

The primate advantage: More neurons in the same volume

The comparative analysis of the neuronal and nonneuronal scaling rules that apply to the brains of different mammalian orders, although limited so far to rodents, insectivores, and primates, already points to a cellular basis for the cognitive advantage of primates over other mammals of similar brain mass in the total number of brain neurons, regardless of brain size. This putative advantage results from the fact that, among primates, increases in brain mass, or in the mass of individual structures such as the cerebral cortex or cerebellum, are not accompanied by systematic changes in average neuronal cell size (as gauged by decreased neuronal density, given that glial cell density does not scale)—while in rodents, and in the insectivore cerebral cortex, increases in numbers of neurons are accompanied by increased average neuronal size. This results in an inflationary scaling of the brains of rodents, compared to a much more volume-economical scaling of primate brains. For instance, while a 10-fold increase in the number of neurons in the brains of rodents leads to a 32-fold larger brain, in the brains of primates, a similar increase leads only to an equivalent 10-fold larger brain (Herculano-Houzel, 2011a). Because of the higher neuronal densities in primate compared to rodent brains of equivalent size (for instance, about 40,000 neurons/mg in the cortex of *Aotus*, against 12,000 neurons/mg in the agouti) while maintaining similar other cell densities, primate brains hold more neurons than rodent brains of equivalent size. Previous analyses showing even smaller neuronal densities in the cerebral cortex of whales and elephants (Tower, 1954) suggest that this trend will be confirmed in the comparison between great ape and human brains and cetacean and elephant brains, despite the larger size of the latter. Speculatively, the estimate of neuronal density in the gray matter of the cerebral cortex of the whale and the elephant at a low figure of about 7000 neurons/mm^3 (Tower, 1954) suggests that these brains conform to scaling rules that are much closer to those that apply to rodents than to the primate scaling rules. In the latter case, the brains of these animals would be predicted to have 212–241 billion neurons; in the former case, however, they would have only 22–24 billion neurons (Herculano-Houzel, 2009). It may turn out, therefore, these very large brains are composed of far fewer neurons than the human brain, despite their size, thanks to the distinct, economical scaling rules that apply to primates in general (and not to humans in particular). Remarkably, we find that the same scaling rules apply to monkeys, humans, and great apes (Herculano-Houzel and Kaas, 2011), which implies that extinct hominin brains were also built according to the same neuronal scaling rules that are observed today among primate brains.

The larger number of neurons per unit volume presumably endows primates with a larger computational capacity than rodent brains of equivalent size. This type of evolutionary change allowed primate brains to accumulate large numbers of neurons without becoming prohibitively large: A macaque brain of 6.4 billion neurons built with the neuronal scaling rules that apply to rodents would, for example, weigh 575 g, instead of its actual 87 g. These findings suggest that the divergence of primate evolution away from the common ancestor with rodents involved mechanisms that favored either a reduction in average neuronal cell size or the ability to add neurons to the brain without making them larger, for instance, through circuitry changes that favored local connectivity (Herculano-Houzel et al., 2010).

The evolutionary implications of the clade- and structure-specific neuronal scaling rules with putatively universal glial scaling rules are intriguing: in mammalian brain evolution, it appears that neurons (supposedly of each of the various neuronal cell types, although that remains to be examined) have been largely free to vary in size across structures and species, while glial cells, however variable in their morphology (Barres,

2008; Walz, 2000), do not quite vary in size across species or even across structures, maintaining very similar properties among mammals (Mishima and Hirase, 2010; Picker et al., 1981), and even in amphibia (Kuffler et al., 1966). These findings indicate that glial cell evolution is severely constrained, which, in turn, suggests that glial cells as a whole perform such a fundamental job that their structure and function can hardly be messed with. This is in agreement with the intricate functional and metabolic interactions between neurons and glia that have been found to apply to human and rat brains alike (Magistretti et al., 1999; Shen et al., 1999; Sibson et al., 1998). Indeed, the shared scaling of brain size with numbers of glial cells suggests that the glial characteristics that apply today to extant brains were already present in the common ancestor to the current 28 species, over 90million years ago (Murphy et al., 2004).

We find that primates share with all mammals examined so far the coordinate scaling of numbers of neurons in the cerebral cortex and the cerebellum. However, primates can be distinguished from rodents and insectivores by the scaling of the joint number of neurons in these two structures relative to the RoB, which includes all remaining structures from the brainstem to the basal ganglia. If the number of neurons in the RoB can be considered to indicate the scaling of the number of neurons available for body-related functions (Herculano-Houzel, 2011b), then any faster scaling of the number of neurons in the cerebral cortex and the cerebellum with the number of neurons in the RoB might be considered a reasonable approximation of the scaling of numbers of neurons in excess of those necessary to deal simply with the bodily functions, and therefore available to add complexity and flexibility to behavior.

The RoB contains at most 6.5% of all brain neurons in the 12 primate species analyzed so far, including humans (Azevedo et al., 2009; Gabi et al., 2010; Herculano-Houzel et al., 2007). In rodents and insectivores, we find that the number of neurons in the RoB varies within the same range as in nonhuman primates, between 5 and 108 million neurons (Herculano-Houzel et al., 2006, 2011; Sarko et al., 2009). In rodents, the RoB represents from 5% to 22% of all brain neurons; in insectivores, from 6% to 15%. Thus, the vast majority of brain neurons are found in the ensemble of cerebral cortex and cerebellum of these animals.

In rodents, insectivores and primates, the percentage of RoB neurons relative to the total number of brain neurons becomes significantly smaller with increasing brain mass or number of neurons, while the relative number of neurons in the ensemble of cerebral cortex and cerebellum increases in larger brains (regression to power law, all values of $p<0.05$; Fig. 7, top). The increase in the joint relative number of cerebral cortical and cerebellar neurons with increasing brain mass is more readily noticeable in rodents and insectivores than in primates, with exponents of 0.016, 0.059, and 0.006, respectively (95% confidence intervals, 0.004–0.028, 0.019–0.098, and 0.002–0.009; Fig. 7, top). Remarkably, however, primates exhibit a much larger ratio of neurons in the cerebral cortex and cerebellum to the number of neurons in the RoB than rodents and insectivores: as seen in Fig. 7 (bottom), while rodent and insectivore brains have, respectively, 6–18 times and 6–16 times more neurons in the ensemble of cerebral cortex and cerebellum than in the RoB, primate brains have 20–124 times more neurons in the cerebral cortex and cerebellum than in the RoB (Herculano-Houzel, 2011a,b).

Thus, if the scaling of the number of RoB neurons can be considered a proxy for the scaling of the number of neurons directly required for bodily functions, the faster scaling of neurons in the ensemble of cerebral cortex and cerebellum relative to the RoB with increasing brain size suggests that larger brains tend to have increasingly larger numbers of cerebral cortical and cerebellar neurons in excess of those required to deal directly with bodily functions and therefore

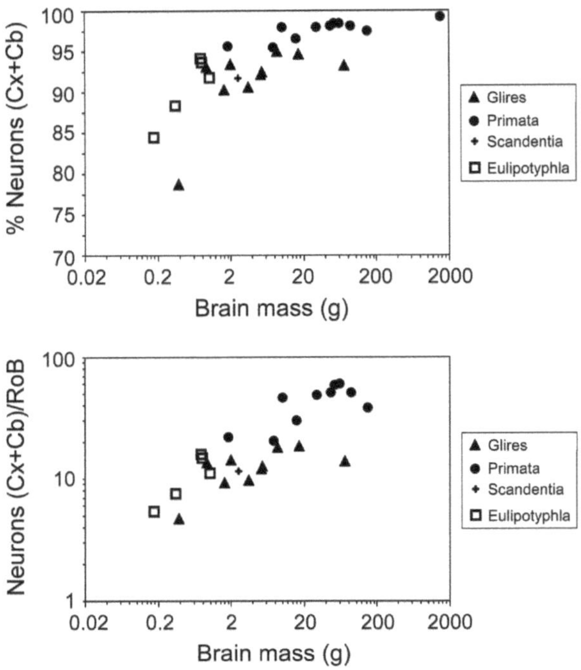

Fig. 7. Relative increase of the joint number of neurons in the cerebral cortex and cerebellum in larger brains, but at different rates across mammalian orders. Average percentage of brain neurons located in the ensemble of cerebral cortex and cerebellum (top) and average ratio between numbers of neurons in the ensemble of cerebral cortex and cerebellum and in the RoB (bottom) are plotted for each species as a function of average brain mass. For a same brain mass, primates have a much larger percentage of brain neurons located in the ensemble of cerebral cortex and cerebellum than rodents and insectivores. Data from Herculano-Houzel et al. (2006, 2007), Sarko et al. (2009), Azevedo et al. (2009), and Gabi et al. (2010).

available to operate the body with increasing complexity and flexibility, which should offer a cognitive advantage. Notice that this increasing ratio of "excess neurons," thus defined, is a different function of brain size across mammalian orders and is, so far, most remarkable in primate brains, which enjoy a much larger ratio of "excess neurons" than rodents of similar brain mass (Fig. 7, bottom). In the present view, the larger the number of neurons in excess of that required to operate the body, the more complex and flexible the behavior of an animal can be expected to be, and thus the larger its cognitive abilities.

Still, the simple total number of brain neurons is likely to be a more practical proxy of cognitive abilities, if not its main limiting factor. First, the number of neurons found outside the ensemble of cerebral cortex and cerebellum, which we suggest that might reflect the number of neurons directly related to bodily functions, is relatively very small in the brains examined so far (see above). Second, given that a same ratio of cortex–cerebellum to RoB neurons can correspond to a much larger absolute number of neurons in the cerebral cortex and cerebellum of one species than in another, and given the relatively small number of body-related neurons in the RoB, the absolute number of neurons in the ensemble of cerebral cortex and cerebellum is likely to be the most limiting determinant of cognitive abilities. Finally, given that the RoB might share neuronal scaling rules across different mammalian brains (Herculano-Houzel, 2011a), varying in size across possibly all mammalian species as a similar function of its number of neurons but not tightly related to body mass (Burish et al., 2010), the cognitive abilities of a species might be simply a function of its total number of brain neurons, an increasing fraction of which is found in the cerebral cortex and cerebellum in larger brains (Herculano-Houzel, 2011b).

Regarding how cognitive abilities compare across animals of different mammalian orders, species, and brain sizes, perhaps the most important realization that arises from studying numbers of neurons in different species is that body size might not be a relevant parameter for defining a species' behavioral performance. Rather, absolute numbers of brain neurons might be the key factor (Fig. 8). This is in line with the recent finding that cognitive abilities in nonhuman primates correlate best with absolute brain size (and hence with absolute number of brain neurons; Deaner et al., 2007; Gabi et al., 2010) and agrees with the general observation that

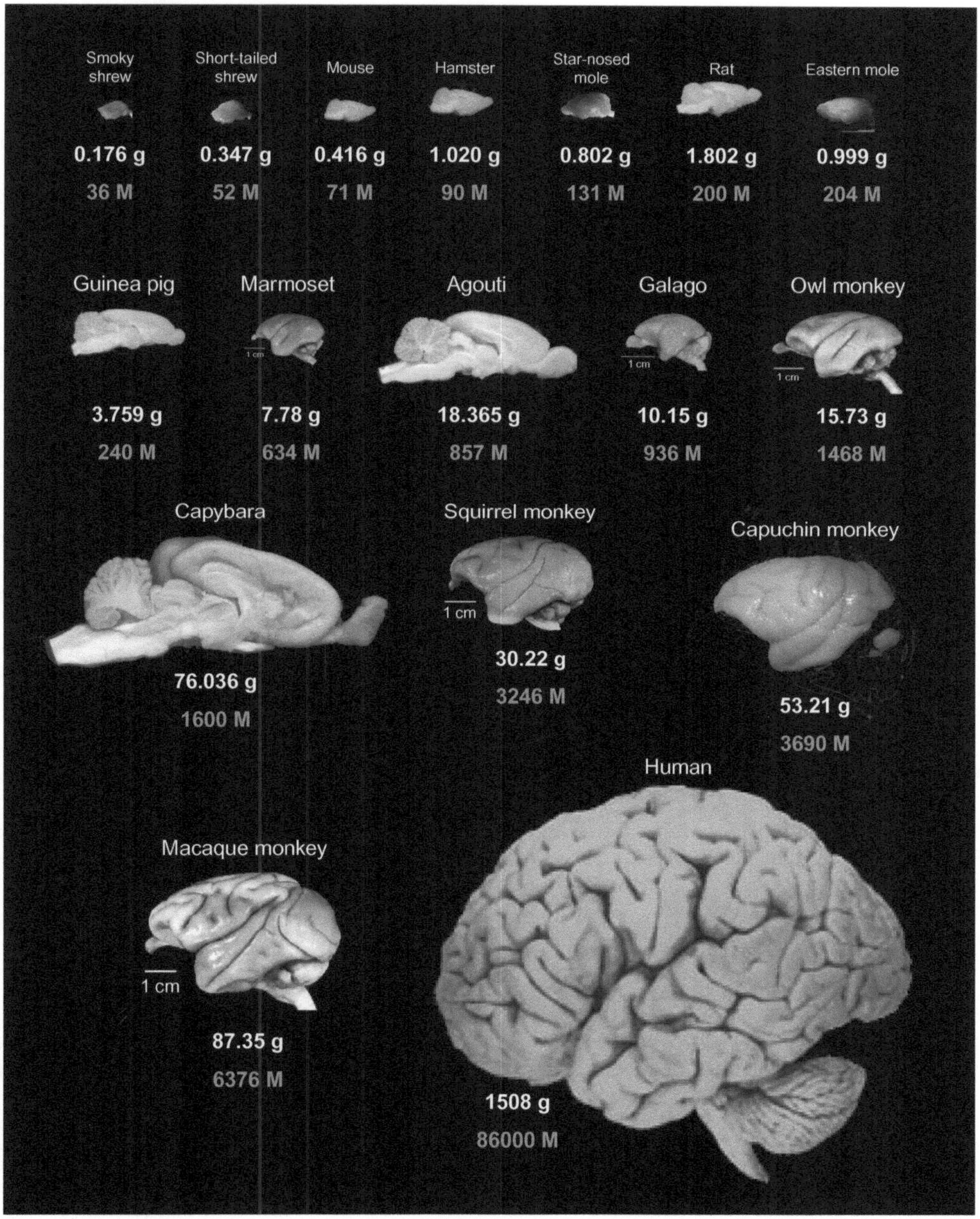

Fig. 8. Primates, rodents, and insectivores ranked by numbers of brain neurons (red, in millions). Average brain mass for each species is also shown (in grams). Notice the difference in size between rodent and primate brains containing comparable numbers of neurons, such as agouti and galago, and capybara and owl monkey. All brains represented to scale. Primate brains reproduced from www.brainmuseum.org. Figure reproduced, with permission, from Herculano-Houzel (2009). (For interpretation of the references to color in this figure legend, the reader is referred to the Web version of this chapter.)

primates tend to be more cognitively able than others with a similar brain size. Most important, considering total numbers of brain neurons as a key determinant of cognitive abilities is a testable working hypothesis. Now that these numbers can be readily examined across species, it will be fascinating to determine whether the primate advantage compared to rodents, described here, also extends to other mammalian orders with even larger brains—but, possibly, still fewer neurons than in primate brains, for a similar brain size.

Acknowledgments

Thanks to Jon Kaas and Roberto Lent for continued support and encouragement, and to our many collaborators for their involvement in acquiring the data on which this review is based. This work is supported by grants from CNPq, FAPERJ, MCT, and the James S. McDonnell Foundation.

References

Azevedo, F. A., Carvalho, L. R., Grinberg, L. T., Farfel, J. M., Ferretti, R. E., Leite, R. E., et al. (2009). Equal numbers of neuronal and nonneuronal cells make the human brain an isometrically scaled-up human brain. *The Journal of Comparative Neurology*, *513*, 532–541.

Balsters, J. H., Cussans, E., Diedrichsen, J., Phillips, K. A., Preuss, T. M., Rilling, J. K., et al. (2010). Evolution of the cerebellar cortex: The selective expansion of prefrontal-projecting cerebellar lobules. *NeuroImage*, *49*, 2045–2052.

Barres, B. A. (2008). The mystery and magic of glia: A perspective on their roles in health and disease. *Neuron*, *60*, 430–440.

Barres, B. A., & Allen, N. (2009). Glia—More than just brain glue. *Nature*, *457*, 675–677.

Burish, M. J., Peebles, J. K., Tavares, L., Baldwin, M., Kaas, J. H., & Herculano-Houzel, S. (2010). Cellular scaling rules for primate spinal cords. *Brain, Behavior and Evolution*, *76*, 45–59.

Clark, D. A., Mitra, P. P., & Wang, S. S. (2001). Scalable architecture in mammalian brains. *Nature*, *411*, 189–193.

Deaner, R. O., Isler, K., Burkart, J., & van Schaik, C. (2007). Overall brain size, and not encephalization quotient, best predicts cognitive ability across non-human primates. *Brain, Behavior and Evolution*, *70*, 115–124.

Doetsch, F. (2003). The glial identity of neural stem cells. *Nature Neuroscience*, *6*, 1127–1134.

Finlay, B. L., & Darlington, R. B. (1995). Linked regularities in the development and evolution of mammalian brains. *Science*, *268*, 1578–1584.

Fox, J. H., & Wilczynski, W. (1986). Allometry of major CNS divisions: Towards a reevaluation of somatic brain-body scaling. *Brain, Behavior and Evolution*, *28*, 157–169.

Frahm, H. D., Stephan, H., & Stephan, M. (1982). Comparison of brain structure volumes in insectivora and primates. I. Neocortex. *Journal für Hirnforschung*, *23*, 375–389.

Friede, R. (1954). Der quantitative Anteil der Glia an der Cortexentwicklung. *Acta Anatomica*, *20*, 290–296.

Gabi, M., Collins, C. E., Wong, P., Torres, L. B., Kaas, J. H., & Herculano-Houzel, S. (2010). Cellular scaling rules for the brains of an extended number of primate species. *Brain, Behavior and Evolution*, *76*, 32–44.

Garey, L. J., & Leuba, G. (1986). A quantitative study of neuronal and glial numerical density in the visual cortex of the bottlenose dolphin: Evidence for a specialized subarea and changes with age. *The Journal of Comparative Neurology*, *247*, 491–496.

Haug, H. (1987). Brain sizes, surfaces, and neuronal sizes of the cortex cerebri: A stereological investigation of man and his variability and a comparison with some mammals (primates, whales, marsupials, insectivores, and one elephant). *The American Journal of Anatomy*, *180*, 126–142.

Hawkins, A., & Olszewski, J. (1957). Glia/nerve cell index for cortex of the whale. *Science*, *126*, 76–77.

Herculano-Houzel, S. (2009). The human brain in numbers: A linearly scaled-up primate brain. *Frontiers in Human Neuroscience*, *3*, 31.

Herculano-Houzel, S. (2010). Coordinated scaling of cortical and cerebellar numbers of neurons. *Frontiers in Neuroanatomy*, *4*, 12.

Herculano-Houzel, S. (2011a). Not all brains are made the same: New views on brain scaling in evolution. *Brain, Behavior and Evolution*, *78*, 22–36.

Herculano-Houzel, S. (2011b). Brains matter, bodies maybe not: The case for examining neuron numbers irrespectively of body size. *Annals of the New York Academy of Sciences*, *1225*, 191–199.

Herculano-Houzel, S., Collins, C., Wong, P., & Kaas, J. H. (2007). Cellular scaling rules for primate brains. *Proceedings of the National Academy of Sciences of the United States of America*, *104*, 3562–3567.

Herculano-Houzel, S., & Kaas, J. H. (2011). Great ape brains conform to the primate scaling rules: Implications for hominin evolution. *Brain, Behavior and Evolution*, *77*, 33–44.

Herculano-Houzel, S., & Lent, R. (2005). Isotropic fractionator: A simple, rapid method for the quantification of total cell and neuron numbers in the brain. *The Journal of Neuroscience, 25*, 2518–2521.

Herculano-Houzel, S., Mota, B., & Lent, R. (2006). Cellular scaling rules for rodent brains. *Proceedings of the National Academy of Sciences of the United States of America, 103*, 12138–12143.

Herculano-Houzel, S., Mota, B., Wong, P., & Kaas, J. H. (2010). Connectivity-driven white matter scaling and folding in primate cerebral cortex. *Proceedings of the National Academy of Sciences of the United States of America, 107*, 19008–19013.

Herculano-Houzel, S., Ribeiro, P. F. M., Campos, L., da Silva, A. V., Torres, L. B., Catania, K. C., & Kaas, J. H. (2011). Updated neuronal scaling rules for the brains of Glires (rodents/lagomorphs). *Brain, Behavior and Evolution, 78*, 302–314.

Hofman, M. A. (1985). Size and shape of the cerebral cortex in mammals. I. The cortical surface. *Brain, Behavior and Evolution, 27*, 28–40.

Hofman, M. A. (1988). Size and shape of the cerebral cortex in mammals. II. The cortical volume. *Brain, Behavior and Evolution, 32*, 17–26.

Jerison, H. J. (1977). The theory of encephalization. *Annals of the New York Academy of Sciences, 299*, 146–160.

Jerison, H. J. (1985). Animal intelligence as encephalization. *Philosophical Transactions of the Royal Society of London. Series B, Biological Sciences, 308*, 21–35.

Jerison, H. J. (2007). How can fossils tell us about the evolution of the neocortex? In J. Kaas (Ed.), *Evolution of nervous systems: A comprehensive reference*, (Vol. 3). Oxford: Elsevier, 1–12.

Kandel, E., Schwartz, J., & Jessel, T. (2000). *Principles of neural science* (4th ed.). New York: McGraw-Hill (p. 20).

Kuffler, S. W., Nicholls, J. G., & Orkand, R. K. (1966). Physiological properties of glial cells in the central nervous system of amphibia. *Journal of Neurophysiology, 29*, 768–787.

Lawers, F., Cassot, F., Lauwers-Cances, V., Puwanarajah, P., & Duvernoy, H. (2008). Morphometry of the human cerebral cortex microcirculation: General characteristics and space-related profiles. *NeuroImage, 39*, 936–948.

Magistretti, P. J., Pellerin, L., Rothman, D. L., & Shulman, R. G. (1999). Energy on demand. *Science, 283*, 496–497.

Marino, L. (1998). A comparison of encephalization between odontocete cetaceans and anthropoid primates. *Brain, Behavior and Evolution, 51*, 230–238.

Marino, L. (2006). Absolute brain size: Did we throw the baby out with the bathwater? *Proceedings of the National Academy of Sciences of the United States of America, 103*, 13563–13564.

Martin, R. D. (1981). Relative brain size and basal metabolic rate in terrestrial vertebrates. *Nature, 293*, 57–60.

Mishima, T., & Hirase, H. (2010). In vivo intracellular recording suggests that gray matter astrocytes in mature cerebral cortex and hippocampus are electrophysiologically homogeneous. *The Journal of Neuroscience, 30*, 3093–3100.

Murphy, W. J., Pevzner, P. A., & O'Brien, S. J. (2004). Mammalian phylogenomics comes of age. *Trends in Genetics, 20*, 631–639.

Nishiyama, A., Yang, Z., & Butt, A. (2005). Astrocytes and NG2-glia: What's in a name? *Journal of Anatomy, 207*, 687–693.

Nissl, F. (1898). Nervenzellen und graue Substanz. *Muenchener Medisches Wochenschrift, 45*, 988–992 (1023–1029, 1060–1062).

Picker, S., Pieper, C. F., & Goldring, S. (1981). Glial membrane potentials and their relationship to [K+]o in man and guinea pig. A comparative study of intracellularly marked normal, reactive, and neoplastic glia. *Journal of Neurosurgery, 55*, 347–363.

Rakic, P. (2009). Evolution of the neocortex: A perspective from developmental biology. *Nature Reviews Neuroscience, 10*, 724–735.

Ramnani, N., Behrens, T. E., Johansen-Berg, H., Richter, M. C., Pinsk, M. A., Andersson, J. L., et al. (2006). The evolution of prefrontal inputs to the cortico-pontine system: Diffusion imaging evidence from macaque monkeys and humans. *Cerebral Cortex, 16*, 811–818.

Rilling, J. K., & Insel, T. R. (1999). The primate neocortex in comparative perspective using magnetic resonance imaging. *Journal of Human Evolution, 37*, 191–223.

Rockel, A. J., Hiorns, R. W., & Powell, T. P. (1980). The basic uniformity in structure of the neocortex. *Brain, 103*, 221–244.

Roth, G., & Dicke, U. (2005). Evolution of the brain and intelligence. *Trends in Cognitive Sciences, 9*, 250–257.

Sarko, D. K., Catania, K. C., Leitch, D. B., Kaas, J. H., & Herculano-Houzel, S. (2009). Cellular scaling rules of insectivore brains. *Frontiers in Neuroanatomy, 3*, 8.

Shen, J., Petersen, K. F., Bekar, K. L., Brown, P., Nixon, T. W., Mason, G. F., et al. (1999). Determination of the rate of the glutamate/glutamine cycle in the human brain by in vivo 13C NRM. *Proceedings of the National Academy of Sciences of the United States of America, 96*, 8235–8240.

Sibson, N. R., Chankar, A., Mason, G. F., Rothman, D. L., Bekar, K. L., & Shulman, R. G. (1998). Stoichiometric coupling of brain glucose metabolism and glutamatergic neuronal activity. *Proceedings of the National Academy of Sciences of the United States of America, 95*, 316–321.

Stephan, H., Frahm, H., & Baron, G. (1981). New and revised data on volumes of brain structures in insectivores and primates. *Folia Primatologica, 35*, 1–29.

Stolzenburg, J. U., Reichenbach, A., & Neumann, M. (1989). Size and density of glial and neuronal cells within the cerebral neocortex of various insectivorian species. *Glia*, *2*, 78–84.

Sultan, F. (2002). Analysis of mammalian brain architecture. *Nature*, *415*, 133–134.

Tower, D. B. (1954). Structural and functional organization of mammalian cerebral cortex; the correlation of neurone density with brain size; cortical neurone density in the fin whale (Baleanoptera physalus L.) with a note on the cortical neurone density in the Indian elephant. *The Journal of Comparative Neurology*, *101*, 19–51.

Tower, D. B., & Elliott, K. A. (1952). Activity of acetylcholine system in cerebral cortex of various unanesthetized mammals. *The American Journal of Physiology*, *168*, 747–759.

Tower, D. B., & Young, D. M. (1973). The activities of butyrylcholinesterase and carbonic anhydrase, the rate of anaerobic glucolysis, and the question of a constant density of glial cells in cerebral cortices of various mammalian species from mouse to whale. *Journal of Neurochemistry*, *20*, 269–278.

Walz, W. (2000). Controversy surrounding the existence of discrete functional classes of astrocytes in the adult grey matter. *Glia*, *31*, 95–103.

Whiting, B. A., & Barton, R. A. (2003). The evolution of the cortico-cerebellar complex in primates: Anatomical connections predict patterns of correlated evolution. *Journal of Human Evolution*, *44*, 3–10.

Zhang, K., & Sejnowski, T. J. (2000). A universal scaling law between gray matter and white matter of cerebral cortex. *Proceedings of the National Academy of Sciences of the United States of America*, *97*, 5621–5626.

M. A. Hofman and D. Falk (Eds.)
Progress in Brain Research, Vol. 195
ISSN: 0079-6123

CHAPTER 16

Self-organization and interareal networks in the primate cortex

Henry Kennedy[†,*] and Colette Dehay[‡]

[†] *Inserm U846, Stem Cell and Brain Research Institute, Bron, France*
[‡] *Université de Lyon, Université Lyon I, Lyon, France*

Abstract: Variability of gene expression of cortical precursors may partially reflect the operation of the gene regulatory network and determines the boundaries of the state space within which self-organization of the cortex can unfold. In primates, including humans, the outer subventricular zone, a primate-specific germinal zone, generates a large contingent of the projection neurons participating in the interareal network. The number of projection neurons in individual pathways largely determines the network properties as well as the hierarchical organization of the cortex. Mathematical modeling of cell-cycle kinetics of cortical precursors in the germinal zones reveals how multiple control loops ensure the generation of precise numbers of different categories of projection neurons and allow partial simulation of cortical self-organization. We show that molecular manipulation of the cell cycle of cortical precursors shifts the trajectory of the cortical precursor within its state space, increases the diversity in the cortical lineage tree, and explores changes in phylogenetic complexity. These results explore how self-organization underlies the complexity of the cortex and suggest evolutionary mechanisms.

Keywords: corticogenesis; development; neocortex; macaque; monkey; connections; pathways; proliferation; cell cycle; outer subventricular zone; evolution.

Introduction

Under nonequilibrium conditions, natural occurring fluctuations amongst the multiple components of an emerging complex system may undergo reorganization toward a dynamic equilibrium. Under dissipative conditions with a maintained energy input, dynamic equilibrium allows complex spatiotemporal patterning, and the system is said to self-organize. Self-organization underlies the complexity found in both the biological and nonbiological world and therefore is expected to play a decisive role in the development of the cerebral cortex (Halley and Winkler, 2008a).

*Corresponding author.
Tel.: +33-472913476; Fax: +33-472913461
E-mail: henry.kennedy@inserm.fr

DOI: 10.1016/B978-0-444-53860-4.00016-7

In biocomplex systems, it is not evident what is meant by the natural occurring fluctuations that will be finely tuned by the self-organization process. Recent work in stem cell biology goes some way to suggesting what could be the initial conditions that support self-organization. A population of stem cells derived from a single cell were shown to exhibit a bell-shaped curve of protein expression (Chang et al., 2008). Flow cytometry analysis shows that the observed wide dispersion of protein levels is due to slow fluctuations over days of gene expression and may contribute to stemness. While phenotypic variability may specifically contribute to stem cell biology, it also may have profound implications for understanding cell fate and diversity. Understanding the origins of this nongenetic phenotypic heterogenity can best be approached in the framework of the epigenetic landscape (Huang, 2009).

Waddington coined the term epigenetic landscape that has been interpreted by Stewart Kaufmann as an *attractor landscape* in the high-dimensional genome-wide gene expression state space (Kauffman, 1993; Fig. 1). The landscape topography reflects the substructure of the state space where each point $S(t)$ represents a network state of the gene regulatory network, reflecting the collective expression of the genome via the activity states of the N genes x_i:

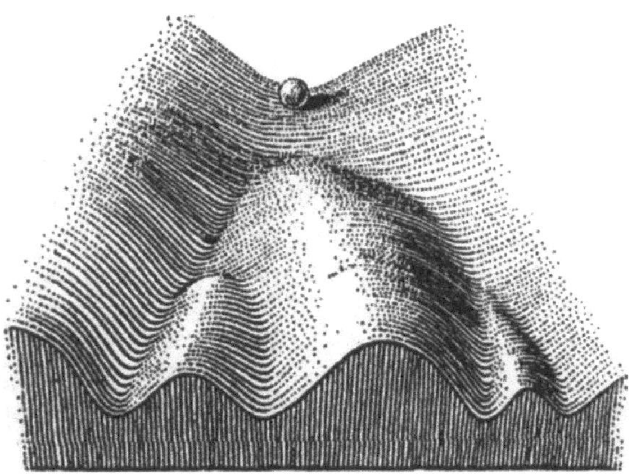

Fig. 1. The epigenetic landscape of Waddington.

$$S(t) = [x_1(t), x_2(t), \ldots, x_N(t)]$$

Conflicting network interactions mean that some configurations of gene activity $S(t)$ are more stable and constitute valleys while others are unstable (hill tops) (Fig. 1). Kauffman proposed that the valleys are attractor states and correspond to a distinct subclass of model gene networks defining cell types. These basins of attraction drain the slopes in their neighborhood, and the collection of valleys and hilltops constitutes an N-dimensional space, which is explored by the dynamics of the gene regulatory network (Brazhnik et al., 2002).

There are a number of different approaches one can adopt to study cortical development. One is to describe the cellular events, usually *in vitro*, of various developmental stages. This approach is often coupled to the study of gene expression and ultimately with the modification of the expression of one or more genes. While this molecular level of investigation is extremely powerful, there are numerous occasions where the removal or overexpression of a gene will have little effect on the developmental process where the gene is expressed, thereby testifying on the one hand to the complexity of the developmental process, and on the other indicating the limitations of the reductionist approach. An alternative but complementary approach is to record events during development in such a way as to be able to develop a quantitative database sufficiently detailed to allow a simulation of the developmental process itself (van Ooyen, 2011; Zubler and Douglas, 2009). This requires adopting an experimental methodology that makes it possible to obtain reliable count data. This provides a valuable means of testing the cellular mechanisms and relevant molecular data but more importantly to link specific aspects of the developmental process to key functional features of the mature cortex. But above all, simulating development can provide insight into the integration of the multiple interactions of diverse components and therefore provide a deeper understanding of the complexity of the cortex (van Ooyen, 2011).

Here, we set out to review processes that may underlie the self-organization of the cortex. This involves briefly describing the complexity of cortical function before proceeding to identify essential aspects of the architecture supporting this function. We describe a number of developmental studies that outline the regulatory control of cortical projection neuron production, and how this regulatory control acts on the phenotypic variability of cortical stem cells.

Self-organization in ontogeny, phylogeny, and the role of postnatal experience

The cortex detects statistical regularities in the environment by sensing local and global correlations of neuronal activity. In this way, invariant characteristics of the world can be inferred as illustrated by color constancy and object segmentation. This notion of cortical function tempers efforts to understand the brain uniquely by studying all of its component independently of the environment and is relevant in a deep sense to understanding both the development and function of the cortex. It suggests that for understanding function, it is necessary to consider the statistical features of the environment that the brain is able to detect and to directly link the structural/functional properties of the brain to its perceptual capacities (Shepard, 2001). For understanding development, detection of the statistical features is expected to largely depend on self-organization (Kennedy and Dehay, 1993). This is upheld by recent evidence that shows that during cortical development there is a progressive optimization of the internal model of area V1 to the statistics of the natural visual environment that is achieved by adjusting the weights of connections (Berkes et al., 2011).

To understand self-organization, we need to take into account the environmental factors that have molded its phylogenetic history. This suggests that self-organization and natural selection are two facets of the single evolutionary process (Halley and Winkler, 2008b; Kauffman, 2000). The self-organization process means that corticogenesis cannot be understood uniquely in terms of molecular prespecification but must also take into account the internal and external environmental factors that modulate organization as cortical development unfolds.

The developing sensory apparatus produces environmental information from which the brain needs to extract behaviorally relevant patterns. By rewiring or reweighting connections, it tunes itself to or learns about coherent (and presumably relevant) patterns in its input. This unsupervised classification procedure is used to generate self-organized maps. There is an evidence that the neuronal mechanisms of ontogenetic self-organization actually persist into adulthood when they mediate adaptive changes in learning and memory.

The species as a whole is subject to environmental patterns that exert pressure through natural selection that could promote the development of suitable circuits and processing modules, that are tuned to the exigencies that led to survival of the current generation (Geisler and Diehl, 2002). The proposed process carries the prediction that corticogenesis even at very early stages of development is influenced by extrinsic factors, echoing earlier stages of phylogeny. During the early 1970s, there were considerable efforts to show that corticogenesis is significantly shaped by extrinsic factors related to the sensory periphery (Van der Loos, 1977). This work was largely supported by the observation that visual experience plays an important role in the elaboration of the functional architecture of the primary visual cortex (LeVay et al., 1980; Thompson et al., 1983). This understanding of corticogenesis was later referred to as protocortex theory but has been largely superseded by protomap theory which postulates that corticogenesis is driven by intrinsic molecular mechanisms. While in recent years there has been overwhelming evidence in favor of a genetic specification of cortical areas, this evidence does not invalidate the numerous instances of so-called afferent specification of the cortex and points to the need for a reappraisal of self-organization (Kennedy and Dehay, 1993; Killackey, 1990; O'Leary, 1989; Sur and Rubenstein, 2005).

Because self-organizing biological systems are initially in a poorly differentiated state and by definition respond over time to changing signals from the environment, one might expect that they would be characterized by prolonged maturational processes in relation to their complexity. The phenomena of self-organization has been traditionally linked to Hebbian plasticity by which competitive modification of synaptic strength underlies experience-dependent self-organization of the functional properties of the visual cortex, so leading to postnatal plasticity in the orientation and ocular dominance domains as well as the analysis of visual movement (Kennedy and Orban, 1983; LeVay et al., 1980; Thompson et al., 1983).

Self-organization and the OSVZ-SGL model

Compared to the other species, the primate cortex is characterized by an overrepresented supragranular layer (SGL) compartment whose neurons are dedicated to forming local connections as well as the transfer of information between cortical areas (Fig. 2). In monkey cortex, the SGL is generated by a primate-specific germinal zone during *in utero* development. Later postnatally, there is a remodeling of the connectivity of the SGL when the animal is visually exploring its environment (Barone et al., 1995, 1996; Kennedy et al., 1989). A further argument in favor of self-organization in the SGL is a dependency on activity for correct development. This has been shown to be the case for corticocortical pathways since immature cortical pathways are highly susceptible to manipulation of the ascending pathways (Dehay et al., 1986, 1989). In this review, we examine the possibility that environmental factors might contribute to shaping the function of these cortical layers and that this might be a characteristic feature of primate cortical development.

In the six-layered cortex, the SGLs are dedicated to forming interareal connections in contrast to the infragranular layers, which contain projection neurons that project to subcortical structures as well as to cortical areas. It is

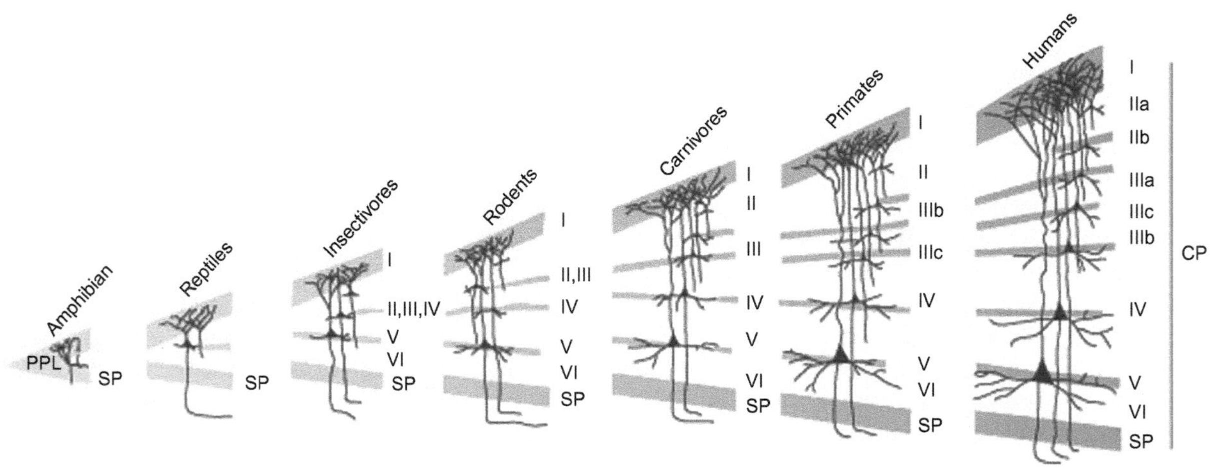

Fig. 2. Evolution of the supragranular layers of the cortex. Supragranular layers (SGL), generated late in neurogenesis, are greatly expanded in the primate cerebral cortex, especially in humans. In primates, SGL neurons form local patchy connections and feedforward long-distance corticocortical connections. (Reprinted with permission from Hill and Walsh, 2005.)

generally accepted that the physiological function of the cortex is shaped by the pattern of interareal connections. Here, we focus on the SGLs of the cortex that are thought to undergo selective expansion in primates (Dehay and Kennedy, 2007). In a first instance, we review how connection weight determines the general network and hierarchical features of the cortical network. We then show how the stringent control of the proliferative program of the cortical progenitors in the germinal zone determines the exact numbers of the different projection neurons that constitute the cortical network. Because the control of cortical precursors comes under multiple environmental influences and determines the structural features of the cortex, it is best understood in the context of self-organization.

The pyramidal projection neurons of the neocortex are generated by at least five different progenitor cell types located in the germinal zones lining the ventricle. The first progenitors in the cortex are the neuroepithelial cells (NECs), which give rise to radial glial progenitors (RGPs), the grandmother cell of cortical development as they give rise directly or indirectly to all cortical projection neurons (Heins et al., 2002; Malatesta et al., 2003). Both NECs and RGPs divide at the apical surface and form the ventricular zone, which is the first formed germinal zone. Iain Smart observed, as far back as 1973, that relatively early on crowding of precursors due to queuing at the apical surface is relieved by some precursors shifting to undergo mitosis at the top of the ventricular zone (Smart, 1973). These basal dividing precursors were referred to as intermediate progenitors (IP), shown to express the transcription factor Tbr2 which distinguished them from the NECs and RGPs, which express prominin 1 and the transcription factor Pax6 (Götz et al., 1998; Hartfuss et al., 2001; Malatesta et al., 2000; Miyata et al., 2004; Noctor et al., 2004). Finally, a short neural precursor (SNP) was identified in the early ventricular zone, which like the RGPs undergoes apical or near apical division (Gal et al., 2006).

A major research breakthrough occurred when it was shown that the RGPs are not only part of a glial scaffolding but also constitute multipotent cortical progenitors (Malatesta et al., 2000; Noctor et al., 2001, 2002). There is further heterogeneity of RGP in the embryonic primate, where a fraction cease dividing and function as migration scaffolding for several months before reinitiating proliferation and generating astrocytes (Rakic, 2003; Schmechel and Rakic, 1979).

There is a major difference in the cellular composition of the primate outer subventricular zone (OSVZ) with respect to the rodent SVZ. Whereas in the rodent all RGP nuclei are restricted to the VZ, RGPs somata are morphologically identified in the OSVZ of the primate (Fietz et al., 2010; Fish et al., 2008; Hansen et al., 2010; Levitt et al., 1981; Lukaszewicz et al., 2005). There is evidence that precursors of the primate OSVZ express Pax6, which characterizes RGP identity in the rodent (Fish et al., 2008).

A number of observations link production of infragranular layers to VZ and SVZ to production of SGL. Although SVZ are derived from VZ precursors, there are clear differences in gene expression between the two precursor pools and these differences correlate with distinct neuronal progeny. For instance, Otx1 and Fez1 are both expressed in VZ precursors and downregulated in SVZ and then subsequently upregulated in subsets of deep-layer neurons (Arlotta et al., 2005; Chen et al., 2005a,b; Frantz et al., 1994; Molyneaux et al., 2005). Further, both Otx1 and Fez1 play a crucial role in specifying the axonal projections of subsets of lower layer neurons. Recent studies in mice show that several transcription factors (Cux2, Tbr2, Satb2, and Nex) (Britanova et al., 2005; Nieto et al., 2004; Pinto et al., 2009; Wu et al., 2005; Zimmer et al., 2004) as well as the noncoding RNA Svet-1 (Tarabykin et al., 2001) are selectively expressed in both the SVZ and upper layer neurons. This congruency of expression of genes first in SVZ progenitors and subsequently in supragranular neurons together with time-lapse microscopy observations suggest that the SVZ gives rise to upper layer neurons (Noctor et al., 2004; Tarabykin et al., 2001; Zimmer et al., 2004).

Consistent with these findings, distinct molecular mechanisms have been identified for the specification of infra- and supragranular neuronal lineages. Studies from mutant mice show that Ngn1 and Ngn2 activity are required for the specification of a subset of infragranular neurons but not for the specification of SGL neurons. In contrast, Pax 6 and Tlx, two genes required for the formation of the SVZ (Nieto et al., 2004; Roy et al., 2004; Zimmer et al., 2004), are synergistically involved in the specification of SGL (Schuurmans et al., 2004). It therefore can be hypothesized that the selective expansion of the SGL compartment in the primate cortex results from modifications of the Pax6/Tlx related specification without modification of the Ngn specification mechanisms (Schuurmans et al., 2004).

In apparent contrast to the above-mentioned links between the SVZ and the SGL is the observation that in rodent the IPs contribute neurons to all cortical layers (Haubensak et al., 2004; Kowalczyk et al., 2009; Miyata et al., 2004; Noctor et al., 2004; Shen et al., 2006). This has led to the suggestion that rodent IPs are universal generators of neurons, and in this sense the separation of the germinal zones into VZ and SVZ might be considered arbitrary, and instead we should distinguish between apical (NECs and RGPs) and basal dividing progenitors (IPs). However, while there is overwhelming evidence that IPs contribute neurons to all cortical layers there is also evidence that IPs in the VZ and SVZ display important differences. Tbr2+ IP-like progenitors undergo apical or subapical mitosis and constitute as much as 5–20% of the total Tbr2+ population (Kowalczyk et al., 2009) and are reported to resemble the short radial SNPs (Gal et al., 2006). The distinction between rodent IPs in the VZ and SVZ is supported by the following. First, whereas the IPs in the VZ have radial morphologies, those in the SVZ and the interface between SVZ and VZ have multipolar morphologies as first described in 1973 by Smart (Kowalczyk et al., 2009; Smart, 1973). Second, while time-lapse observations suggest that IPs undergo only one or two proliferative divisions, there could be a difference in the tendency of IPs in the SVZ to undergo more frequent proliferative divisions, given the decrease of the neurogenic fraction in basal divisions during corticogenesis (Kowalczyk et al., 2009). Further investigations are needed to determine if the different categories of IP cells have distinct cell-cycle kinetics (Arai et al., 2011). If this should be the case, this would further the distinction of IPs in the VZ and SVZ. Under these circumstances, the fact that IPs contribute neurons to all layers fails to invalidate the observed links between SVZ and SGL. In the primate visual cortex, more than 75% of cortical neurons destined for the upper layers originate from SVZ precursors (Lukaszewicz et al., 2005).

In the primate, there is an important elaboration of the SVZ to form an imposing outer component OSVZ which is not observed in the rodent (Smart et al., 2002; Fig. 3). The OSVZ is a distinct histological structure that is bounded by two clearly defined fiber pathways (the outer and inner fiber layers) that have no parallel in the developing rodent or carnivore. The homologous structure of the rodent SVZ in the primate could be the ISVZ, which like the rodent SVZ is in contact with the VZ. The OSVZ exhibits a number of unique features. Contrarily to what is observed in the rodent, where the VZ is the major germinal compartment throughout corticogenesis, the primate VZ declines rapidly during the course of corticogenesis. This decline is associated with an early appearance of the SVZ followed by the OSVZ. This primate-specific organization, first described in the monkey (Smart et al., 2002), has also been subsequently observed in the developing human cortex (Fietz et al., 2010; Hansen et al., 2010; Zecevic et al., 2005).

Enlargement of the SVZ precursor pool in the primate might correspond to an evolutionary adaptive mechanism ensuring the increased

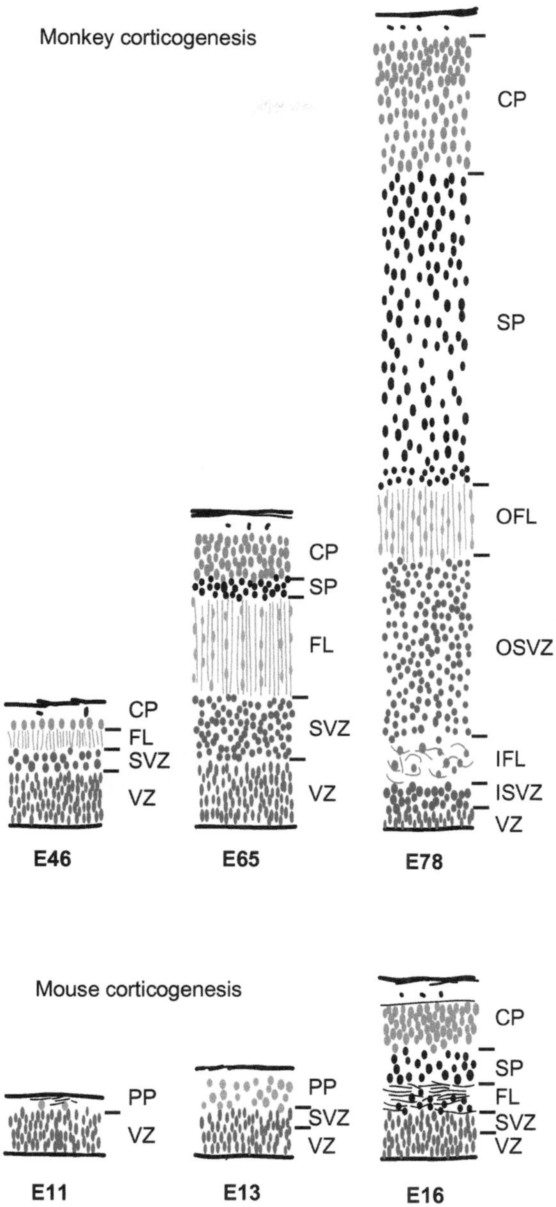

Fig. 3. Comparison of human and mouse germinal zones at equivalent developmental stages. These drawings are transects through presumptive area 17 in monkey and mouse dorsal cortex at comparable developmental stages. The depth of each layer is drawn to a common scale. In the primate, an early appearing outer fiber layer (OFL) forms a major landmark. The ventricular zone (VZ) declines progressively after E65. The subventricular zone (SVZ), by contrast, increases progressively in depth and by E72 is divided into an inner subventricular zone (ISVZ) and outer subventricular zone (OSVZ) by an intruding inner fiber layer (IFL). The increase in the OSVZ is particularly important between E65 and E72 and occurs as the VZ declines. CP, cortical plate; SP, subplate (adapted from Smart et al., 2002). (For color version of this figure, the reader is referred to the Web version of this chapter.)

neuronal output necessary to build a more highly developed neocortex involving a pronounced cytological complexification of the SGL (Dehay et al., 1993; Lukaszewicz et al., 2005; Smart et al., 2002). The rodent SVZ is only partially self-sustaining and instead has to receive a constant supply of precursors from the VZ (Haubensak et al., 2004; Miyata et al., 2004; Noctor et al., 2004; Reznikov et al., 1997; Wu et al., 2005). In primates, self-renewal (i.e., precursor divisions leading to an increase in numbers of precursors) would appear to be much more pronounced in the OSVZ than in rodent SVZ (Lukaszewicz et al., 2005; Smart et al., 2002).

The determination of neuron numbers composing either a cortical area or a cortical layer has been shown by a technique, referred to as the mitotic history of the neuron (Dehay and Kennedy, 2007), to crucially depend on the modulation of the mode of division where high rates of proliferative division lead to increases in the production of neurons (Polleux et al., 1997, 1998). Because cytoarchitecture is characterized by differences in the numbers of neurons in individual layers, arealization and lamination are two sides of the same coin. Hence, the spatial and temporal modulation of the frequency of proliferative divisions of cortical precursors determine the cytoarchitecture of the cortex (Dehay and Kennedy, 2007).

While the improved understanding of the IP precursor does not invalidate the suggested links between SVZ and SGL referred to above, it does give an improved conceptual framework for understanding the organizational principles underlying corticogenesis, as postulated by

Hevner and coworkers (Pontious et al., 2008). These authors have identified three competing but in our view not mutually exclusive hypotheses: the upper layer hypothesis (ULH), the intermediate progenitor hypothesis (IPH), and the radial amplification hypothesis (RAH).

The ULH points to the evidence linking SVZ and SGL (Britanova et al., 2005; Nieto et al., 2004; Tarabykin et al., 2001; Zimmer et al., 2004) and argues that a distinct population of IPs generate the SGL. Further, it is postulated that the population of IPs in question reside in the SVZ. The major objection of the Hevner group against the ULH is that IPs have been shown to contribute to all cortical layers (Pontious et al., 2008); however, as we argue above, there are reasonable grounds to suggest that IPs in the VZ and SVZ may constitute distinct precursor types, and in any case, the ULH remains an accurate description of the correlations between SVZ morphogenesis, gene expression, and the production of the upper layer neurons during middle to late stages of corticogenesis (Pontious et al., 2008). As discussed above, modulation of mode of division of cortical precursors has been implicated in determining cytoarchitecture via specifying the laminar thickness and hence arealization (Dehay and Kennedy, 2007). Because IP is the major precursor involved in amplification of neuron number, these findings implicate a special role of the IP both in determining laminar thickness and areal extent of cortex and this is what Pontious et al. referred to as the IP-hypothesis (Pontious et al., 2008). The increased expansion of the occipital pole of the fetal monkey is thought to be related to the extensive size of the OSVZ in this region and this suggests a link between gyrification and IPs (Lukaszewicz et al., 2006; Smart et al., 2002), which has been further elucidated in the ferret (Kriegstein et al., 2006; Martinez-Cerdeno et al., 2006). Because IPs respond to extrinsic signals released by the thalamic fibers (Dehay et al., 2001) and to genes that control specification of areas and growth (Bedford et al., 2005; Cappello et al., 2006; Chen et al., 2006; Holm et al., 2007; Land and Monaghan, 2003; Quinn et al., 2007; Roy et al., 2004; Schuurmans et al., 2004; Yun et al., 2004; Zhou et al., 2006), IPH postulates that IP amplification determines the arealization during both development and evolution (Cheung et al., 2007; Molnar et al., 2006).

The lack of evidence for IPs amplification in mice influencing expansion of cortical surface area is the major argument put forward by Pontious et al. (2008) in favor of the RAH. In contrast to IPs, factors influencing RGPs have been shown to influence cortical surface area (Chenn and Walsh, 2002; Hevner, 2005; Inglis-Broadgate et al., 2005; Kuida et al., 1998). However, the objection that changes in abundance of IPs cannot impact on surface area ignores two important findings. First, removal of thalamic input to the cortex via bilateral *in utero* enucleation drastically reduces the dimensions of cortical area 17 (Dehay et al., 1989, 1991; Rakic, 1988). Because the removal of inputs to the cortex occurred after the early phase of expansion of the RGPs, it would seem to be a direct consequence of alteration of the proliferation of IPs. This would seem to argue against IPH and in favor of RAH. However, this understanding of the RAH is based on the erroneous notion that RGPs undergo early symmetrical division to establish the pool of RGPs prior to the onset of neuron production. In fact, it is now established that RGPs proliferation occurs continuously throughout corticogenesis (Kowalczyk et al., 2009; Miyata et al., 2004; Noctor et al., 2004). Hence, factors that influence IPs could also be influencing RGPs either directly or indirectly. That such influences might exist is illustrated in recent experiments aimed at accelerating the cell cycle by selective reduction of G1 via transfection of cyclin E in rodents (Pilaz et al., 2009). The cell-cycle acceleration of Pilaz et al. (2009) confirmed the hypothesis of the control of mode of division by cell-cycle duration (Calegari and Huttner, 2003; Dehay et al., 2001; Gotz and Huttner, 2005; Lukaszewicz et al., 2002, 2005).

The cell-cycle acceleration study of Pilaz et al. (2009) was carried out in mouse and showed that reduction of G1 led to an increase in the frequency of proliferative divisions where both daughter cells reenter the cell cycle accompanied by a decrease in differentiative neurogenic divisions where daughter cells quit the cell cycle to become neurons. This cell-cycle reentry led to a transient expansion of the precursor pool followed by a subsequent surge in neuron production. Interestingly, the increase in cell-cycle reentry of the Pax6+ RGPs was very short lived and did not appear to have any consequence on the dimensions of the RGPs pool nor the cortex. This contrasted with the effects of cell-cycle acceleration on the Tbr2+ population, which showed a much bigger amplification leading to a marked and relatively long-lived increase in the dimensions of the SVZ followed by an increase in the production of SGL neurons and the thickness of the SGLs. This "primatization" of the rodent cortex via a selective increase in the SVZ coupled with an expansion of the SGL suggests a conceptual exploration of the IPH and the RUH. Cell-cycle acceleration experiments by Calegari's group revealed a 300% increase in cortical surface area (Lange et al., 2009), which could suggest that the increased IP population has a feedback control on proliferation in the RGPs population, which causes the increase in cortical size. Certainly, such an influence of the IPs on the RGPs is required to explain the claims of the role of IPs in gyrification (Kriegstein et al., 2006; Lukaszewicz et al., 2006; Martinez-Cerdeno et al., 2006; Smart et al., 2002). For instance in the E80 fetal monkey at the onset of SGL production, there is a very prominent OSVZ which decreases in size concomitantly with the increased rates of neuron production and the very rapid growth of the occipital pole and the formation of the lunate sulcus. During this period of ballooning out of the occipital pole, the cortical plate of the incipient area 17 is actually thinner than the presumptive area 18 despite the fact that later in the adult it will house a greater number of neurons in its thickness (Lukaszewicz et al., 2006). During this period of rapid occipital pole growth, the maintenance of the VZ will require an increase in rates of proliferation and implies that there is a concerted mitogenic effect relayed from the OSVZ down to the VZ.

Neuron number specifies cortical networks

The study of complex networks has had an important impact on a wide range of scientific fields including social sciences, physics, and biology. In fact, any situation involving the interaction of numerous components to form complex systems with emergent properties can be investigated usefully with a graph theoretic approach. In this way, understanding connectivity has made important contributions to phenomena as diverse as molecular interactions, metabolic pathways, ecological food webs, and the brain (Sporns, 2011).

A number of studies have applied graph theory to investigate the network of connections linking cortical areas. A number of studies have used the database of Felleman and Van Essen based on the compilation of data across some 350 papers concerning the connectivity of 31 cortical areas (Felleman and Van Essen, 1991). Malcolm Young and his colleagues used multidimensional scaling in order to obtain a topological model of the cortex (Young, 1992). Jouve used a similar approach modified in such a way as to infer connections that had not been tested (Jouve et al., 1998). This showed that the frontal eye field, a prefrontal area involved in directing attention, had a surprisingly central position in the cortical graph (Fig. 4).

Studies using a database which has been compiled from the published literature suffer from the presence of numerous connections that have not been tested. To overcome this, we have carried out an exhaustive study of the inputs to 26 cortical areas (Markov et al., 2011a). This enables us to determine the percentage of connections that exist in comparison to the theoretical maximum connectivity. This is referred to as the density of the graph. Network studies using previous

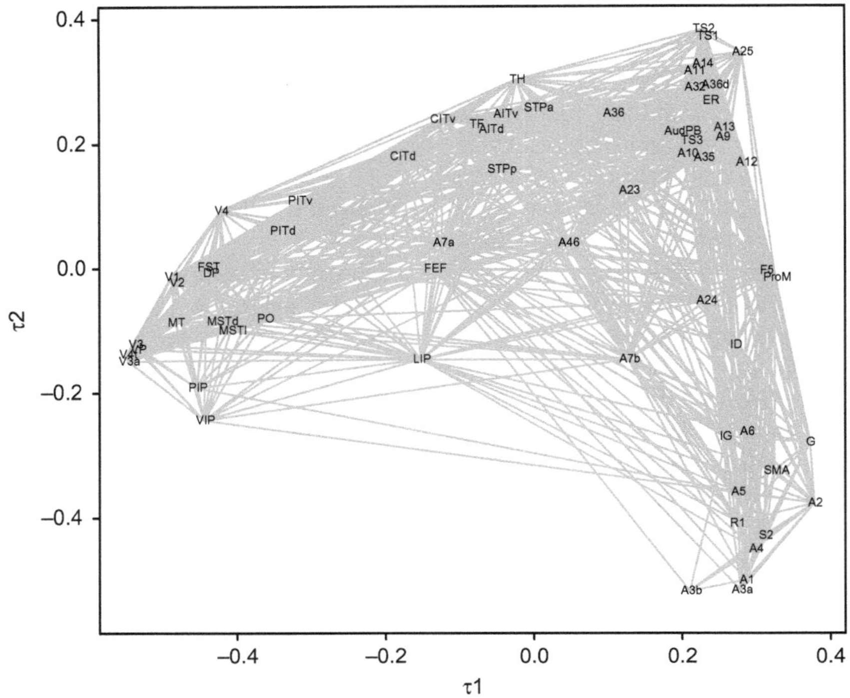

Fig. 4. Topological model of cortex. Using Jouve's algorithm (Jouve et al., 1998) on a database of Barone et al. (2000). The network organization of the cortex can be represented as a graph, where cortical areas are nodes and the pathways between areas are the links between nodes. In this model, areas are closer according to the similarity of their connectivity.

published database have chosen to ignore the untested connections and this has led them to conclude that the density of the graph is between 12% and 30% (Honey et al., 2007).

In our study, we found that the density was considerably higher, in the region of 60–70%. At such high densities, the binary (i.e., connected/not connected) features of the graph reveal little specificity. This makes it necessary to consider the weights of connections between areas.

If the specificity of the cortical graph is derived from the weights of the connections between areas, then one would predict that connection weight between two areas will be consistent across individuals. Few studies have attempted to characterize the interindividual variability. One exception is a study by Scannell et al. (2000) who reported in collated data a 100-fold difference in the strength of a given corticocortical pathway across individuals. Because the differences in experimental procedure in different laboratories could largely contribute to variability observed by Scannel, we set out to make repeat injections in areas V1, V2, and V4 across individuals (Markov et al., 2011b). The results showed that connections formed by cortical neurons (i) are highly local, 80% of the connections are restricted to the cortical area and do not transverse more than 2–3 mm; (ii) of the 20% that project out of the injected area about two-thirds are to the neighboring area. This leaves about 5% of the connections to form the extrinsic associative connectivity to be shared amongst the 25–50 cortical pathways that are formed by each cortical area. It is in the extrinsic pathways that we were able to show that weights spanned a considerable distance, over 5 units of log, and that the weights were highly consistent across individuals (Falchier et al., 2002; Markov et al., 2011b; Vezoli et al., 2004). These results show that the connectivity profile of an individual cortical area is highly specific and that it is the weight of the connections that will largely determine the specificity of the cortical graph. Further, we were able to show that there is a weight–distance relationship so that the nearest connections are the strongest and the weakest connections cover the greatest distance (Markov et al., 2011a,b). While one can choose to ignore these weak long-distance connections, it is important to note that they interconnect very different sorts of cortical areas and that their role in synchronization could mean that they have important roles in long-distance coordination of brain activity. In a more philosophical vein, one needs to remember the strength of weak connections in graph theory as a whole (Csermely, 2006).

Neuron number specifies cortical hierarchy

Much of our understanding of cortical function comes from the work in the visual cortex where the stimulus response function has been most extensively studied. Hubel and Wiesel's pioneering work showed that the receptive field structure of neurons in the visual cortex is progressively elaborated exhibiting simple, then complex, and finally hypercomplex features. This observation leads these authors to postulate that cortex processes afferent information through a feedforward (FF) hierarchy of progressive abstract detectors (Hubel and Wiesel, 1968). Anatomical studies showed that FF pathways originate from SGL and terminate in layer 4, while feedback (FB) pathways originate from infragranular layers and terminate outside of layer 4 (Kennedy and Bullier, 1985). In the early 1990s, David Van Essen's group showed that pairwise comparison of these connections allowed the construction of a model of the cortex that captured many features including the dorsal and ventral streams of the visual cortices as well as a strict hierarchical organization which extended up to the prefrontal cortex (Felleman and Van Essen, 1991; Fig. 2a). Malcolm Young's group performed a statistical analysis of the Van Essen database and confirmed the basic features of the Van Essen hierarchy, including the ventral and dorsal streams. However, while the Young et al. organization appeared to be hierarchical, they found that it was highly indeterminate; in fact they found over 150,000 equally plausible solutions to the hierarchy.

Indeterminacy in the Van Essen model stems from the fact that there is no indication of hierarchical distance between nodes coupled to the fact that there are numerous parallel pathways in addition to the dorsal and ventral streams. An anatomical solution to hierarchical distance is provided by the fact that long distance FF pathways arise uniquely from the SGL, and that as distance diminishes there is an increasingly important contribution to the projection from the infragranular layers (Fig. 5). Likewise, long-distance FB projections originate uniquely from infragranular layers and as distance is reduced there are increasing contributions from SGL (Barone et al., 2000; Dehay et al., 1986; Kennedy and Bullier, 1985). Estimating the contributions of the SGL and infragranular layers to a given pathway involves quantitative estimations of numbers of neurons and defines the SLN% (fraction of supragranular neurons) for a given pathway (SLN%=number of SGL neurons/numbers of SGL + infragranular layer neurons). Accurate SLN values make it possible to construct a determinate model of the cortical hierarchy (Fig. 5).

These findings show that the connectivity signature of a cortical area is defined by the individual strengths of 25–80 cortical areas that project to it, the hierarchical distance of each of these areas as reflected by its SLN, and the numerical strength of the individual pathways. This would suggest that the physiological function of, or the range of information processing performed by the target

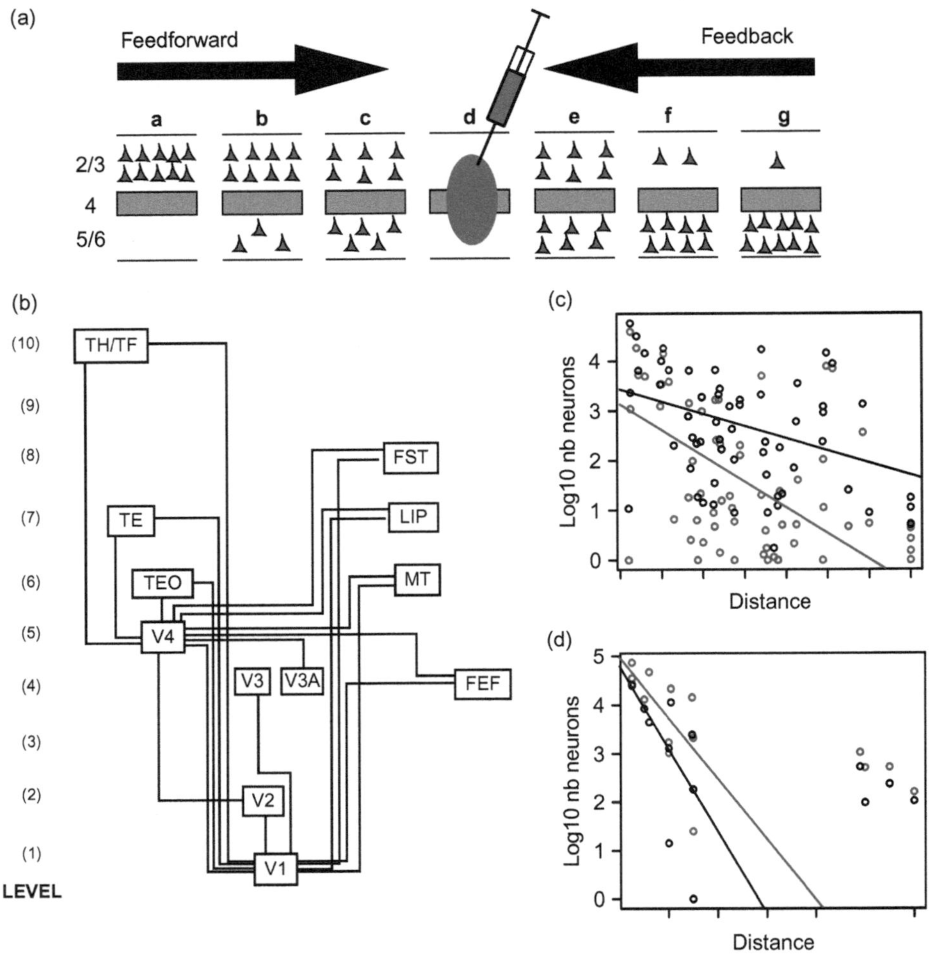

Fig. 5. Laminar distribution and number of projection neurons determine hierarchy according to the operation of a distance rule. (a) Cartoon illustrating the distribution of labeled neurons in feedforward and feedback projections after injection of a retrograde tracer in the target area. Each area exhibits a specific SLN value, which determines its hierarchical distance from the target area. Distant feedforward projections have SLN values of 100 (e.g., area a). More proximal areas have lower SLN values. Distant feedback projections have SLN values of 0 (e.g., area g). More proximal feedback areas have higher SLN values. SLN values of 40–60% correspond to lateral connections (areas c and e). (b) Hierarchical model of cortical areas connected to the target area according to the relationship between the SLN% and the distance rule (Barone et al., 2000) (Modified from Felleman and Van Essen, 1991). (c) Feedback projections in the infragranular layers extend over long distances (black circles) and over short distances in the supragranular layers (red circles). (d) Feedforward projections in the supragranular layers extend over long distances (red circles) while feedforward projections in the infragranular layers extend over short distances (black circles). (For interpretation of the references to color in this figure legend, the reader is referred to the Web version of this chapter.)

area, is constrained by the particular profile of its inputs. Projections originating from SGL terminate in layer 4 and recurrent local circuits amplify the input signals before relaying them to the output neurons in the upper and lower layers (Douglas et al., 1995). The output of the cortex is modulated by the infragranular layer projections to layer 1 (Cauller, 1995) and to a lesser extent to layers 5 and 6. In this way, FF pathways construct the receptive field properties of the target and area, while the FB pathways modulate the features of these receptive fields and the output of the target area.

What are the factors that set up the changes in SLN? One possibility is the way pathways in the cortex are highly dependent on distance. Elsewhere we have shown that there is an important exponential distance rule such that projections have low and neighboring projections high weights (Markov et al., 2011a). We have examined how the weights of labeled neurons in individual layers are influenced by distance. In feedback connections, the weight of layer 6 neurons falls off very much slower than does the weight of SGLs (Fig. 5c), so that long distance feedback connections have low SLN values. Conversely, in FF pathways, the weights of infragranular layers fall off rapidly while those of SGLs fall rapidly so that long-distance FF pathways have high SLN values (Fig. 5d). These results show that the weight relations of a cortical area are globally constrained by an exponential distance rule; at a finer level a multiple distance rule acting independently on different layers establishes the hierarchical interareal relationships.

How does the cortex generate precise numbers of neurons?

The coordinated regulation of two cardinal cell-cycle parameters of cortical precursors determines neuronal production via the regulation of the size of the precursor pool: the duration of the cell cycle and the relative frequency of cell-cycle reentry compared with cell-cycle exit. *In vivo* and *ex vivo* analysis of the cell-cycle regulation of OSVZ precursors of the primate visual cortex has shed light on the molecular correlates of area-specific differences in proliferation that underlie area-specific differences in the thickness of SGLs.

Compared to area 18 OSVZ precursors, area 17 OSVZ precursors are characterized by both a shorter cell-cycle duration, due to a reduction of the G1 phase, and an increased relative frequency of cell-cycle reentry. These areal differences on OSVZ precursor cell-cycle regulation are associated with significant differences in the level of expression of molecular regulators of the G1/S transition p27kip1 and CyclinE. The *ex vivo* up and down modulation of their level of expression significantly affects cell-cycle reentry and the rate of cell-cycle progression and stresses the role of the G1 phase regulation in corticogenesis (Lukaszewicz et al., 2005, 2006). Mathematical modeling of the observed differences in both rates of cell-cycle reentry and in G1 phase duration show that the combined variation of these two parameters are sufficient to generate the enlarged SGL that distinguishes area 17 from area 18. These results show that variations of G1 phase duration and the coordinated variation in mode of division contribute directly to regulate neuron number (Lukaszewicz et al., 2005).

In vitro work on mouse cortical precursors (Dehay et al., 2001) indicates that thalamic afferents control corticogenesis by modulating rates of proliferation. Embryonic thalamic axons release a mitogenic factor that increases the proliferative capacity of mouse cortical precursors during generation of SGL by decreasing the G1 duration and by promoting cell-cycle reentry (Dehay et al., 2001). In the monkey, the LGN axons that selectively project onto the OSVZ of area 17 could be responsible for the temporally and spatially restricted stimulation of proliferation that results in the transient upsurge of the size of SGL precursor pool in area 17 (Dehay et al., 1993; Lukaszewicz et al., 2005; Smart et al., 2002). There are also *in vivo* indications in primates that embryonic thalamic axons could impact on areal size and specification via an early influence on neuron production during cortical neurogenesis (Dehay et al., 1989, 1991). Because thalamic axons are precisely targeted onto distinct cortical areas (reviewed in Lopez-Bendito and Molnar, 2003), they will be able to differentially affect rates of precursor proliferation and of neuron production across the germinal zones and therefore determine local areal cytoarchitectonic features.

As we have shown above, the numerical values of the different projections neuron phenotypes determine the hierarchy and network features of the cortex. There is considerable evidence that the phenotypic fate is sealed during the final round of mitosis in the germinal zones and that the timing of this event is highly significant (Chen et al., 2005a,b; Molyneaux et al., 2005; Polleux et al., 2001; Shen et al., 2006). It seems that fate specification is entirely premigratory as neurons that are destined for a given cortical layer and which end up in the inappropriate layer do not acquire the connectivity of the inappropriate layer (McConnell, 1988; Polleux et al., 2001).

The timing of the differentiative division that generates a given neuron type is directly involved not only in the specification but also in determining the number of neurons generated. Delaying the final differentiative division of a precursor pool leads to increasing the number of proliferative divisions, which in turn leads to its amplification and hence expansion of the numbers of neurons generated (Pilaz et al., 2009; Polleux et al., 1997). Mathematical modeling of these events shows a high predictability of the numbers of neurons generated confirming the relationship between the kinetics of the cycling precursor and the cytoarchitecture of the cortex (Lukaszewicz et al., 2005; Pilaz et al., 2009).

There are numerous feedback loops in the neural epithelium that regulate the rate of proliferation in the germinal zones (Fig. 6). In the reeler mutant, neuroblast migration is perturbed so that the last generated neurons end up deep in the cortex inverting the normal outside first inside last neurogenic gradient. Despite these changes in the laminar location of projection neurons, the laminar and areal time tables are strictly conserved in this mutant. However, this contrasts with the kinetics of the precursors and the determination of cell fate that are both highly perturbed confirming and extending previous findings of regulatory feedback loops from the cortical plate to the germinal zones (for a review,

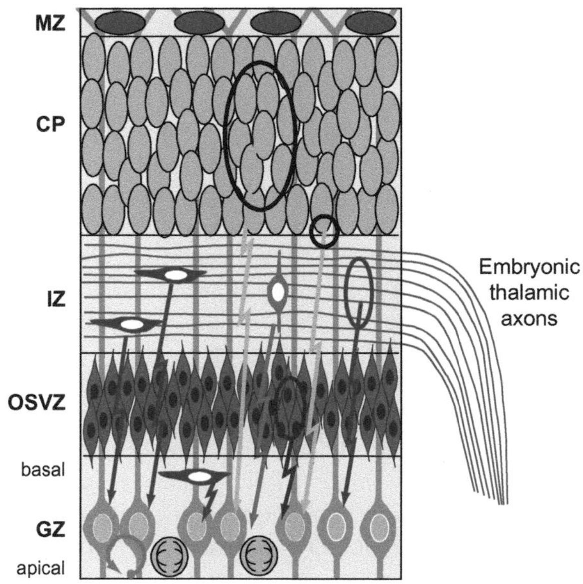

Fig. 6. Schematic diagram of different extrinsic influences affecting precursor proliferation dynamics and cell fate. The different cellular compartments of the embryonic cortex are thought to provide signals that modulate the proliferation and fate of cortical precursors in germinal zones. Regulatory feedback includes influences from the postmitotic pyramidal neurons of the cortical plate (CP) (Polleux et al., 1998, 2001) (blue cells), specifically from the lower layers (Viti et al., 2003) from newborn cortical neurons migrating in the intermediate zone (IZ) compartment (gray cell) and local feedback signaling from the germinal zone (GZ) precursors (circular arrow) (Lien et al., 2006). Extracortical extrinsic signals are provided by tangentially migrating interneurons generated in the ventral telencephalon (green cells) and by ingrowing embryonic thalamic axons (magenta) that have been shown to influence cell-cycle kinetics of precursors (Dehay et al., 2001). We hypothesize in the present report that IP precursors in the OSVZ send mitotic signals to the VZ. MZ, marginal zone. (For interpretation of the references to color in this figure legend, the reader is referred to the Web version of this chapter.)

see, Polleux et al., 1997). The other major regulatory pathways are between the OSVZ and the GZ as hypothesized above.

Gain-of-function experiment exploring self-organization

The attractor or epigenetic landscape considers that the cell phenotype is defined in a state space in which the cell has a specific trajectory. This leads to phenotypic variability which allows complex cellular interactions that underlie the emergence of biocomplexity. The peaks and saddles separating the valleys in the attractor landscape can be transverse via forced gene expression. This we have explored in the above-mentioned experiments where we have caused an acceleration of the cell cycle of cortical precursors during the production of SGL. This gain-of-function experiment led to an expansion of the mouse precursor pool resulting in an increased production of the SGL of the mouse. The increased size of the SVZ and the SGL of the mouse appears as a form of primatization of the rodent. Interestingly, with the expansion of the SVZ precursor pool, there is also an important increase in the fraction of SVZ precursors that coexpress Pax6 and Tbr2. In mouse, very few precursors coexpress the two transcription factors while this is a characteristic feature of the macaque OSVZ.

One can interprete these experimental results in terms of a modification of the normal regulatory control of mouse corticogeneis. One can hypothesize that the gain-of-function experiment described above is essentially revealing what would be the consequence of an increase in gain of thalamic regulation of corticogenesis and that this could be mimicking the expansion of cortex during evolution. Earlier studies have shown that thalamic afferents release a mitogen that shortens the cell cycle via reduction of the G1 phase, leading to an increase in cell-cycle reentry and an increase in the precursor pool (Dehay et al., 2001). This conclusion is in accordance with the demonstration that removal of thalamic fibers has little effect on rodent development and an important effect in prenatal monkey development (Dehay et al., 1989, 1991, 1996; Rakic, 1988). The finding that interests us here is that the experimental gain-of-function experiment resulting from cell-cycle acceleration induces the expression of Pax6 in precursors that would normally be uniquely expressing Tbr2. Understanding this experimental result in the framework of the Kauffman hypothesis suggests what has happened is that there is a modification of the attractor landscape of the mouse cortical precursor via the G1 reduction, that mimics what is happening during normal development under the influence of the thalamic control of corticogenesis, suggesting that the regulatory loops are shaping the attractor landscape of the cortical precursor cells via modification of the gene regulatory network (Kauffman, 1993). These results explore the idea that self-organization of the brain is tightly linked to its evolution involves cell–cell interactions with far reaching consequences and where phenotypic variability plays a crucial role.

Conclusion

A recent study in *Nature* explored experimentally the concept of self-organization. The group of Yoshiki Sasai showed that a homogeneous population of embryonic stem cell-derived neural epithelium spontaneously underwent optic cup morphogenesis in 3D culture (Eiraku et al., 2011). These authors were able to report the spontaneous formation of epithelial vesicles patterned along their proximal–distal axis, exhibiting interkinetic nuclear migration and generating a stratified neural retinal tissue following the steps of normal development *in vivo*. According to our definition of self-organization in the section "Introduction," we would predict that there is an oscillatory state in the homogeneous pool of stem cells, possibly here involving *wnt* signaling, that underwent spatial–temporal modulation to drive the cell population to form optic vesicles *in vitro*. If future regenerative medicines protocols are going to follow this strategy, they will aim to control the identity of the initial oscillatory state so as to obtain the desired organ. While this research heralds enormous potential for regenerative medicine, it also tells us something very deep about normal development. The fact that a simple cell culture system can undergo intrinsic self-organization to generate a complex structure without external cues emphasizes the importance that self-organization has in the normal developmental process.

The brain is a remarkable computational system. Unlike conventional computers, it does not rely on an external agency for its construction and programming. Instead, the entire circuitry is self-constructed by replication and interaction of the germinal cells and their derived neuronal types. Unlike the majority of tissues that emphasize local three-dimensional organization where cells contact their neighbors, the CNS is characterized by complex connectional topologies over very large spatial scales. The underlying need for this organization is due to the fact that information processing is finally about selective communication between particular processors. Such functions can be represented as a graph-like topology composed of processing nodes (single or populations of neurons), and their connecting communication edges (axons).

Acknowledgments

This work was supported by FP6-2005 IST-1583 (HK), FP7-2007 ICT-216593 (HK-CD), ANR-05-NEUR-088 (HK), ANR-06-NEUR-CMMCS (CD). We thank Nikola Markov, who made important contributions to understanding the role of a distance rules in determining cortical hierarchy, and Kenneth Knoblauch, for teaching us the importance of count data for understanding structure.

References

Arai, Y., Pulvers, J. N., Haffner, C., Schilling, B., Nusslein, I., Calegari, F., et al. (2011). Neural stem and progenitor cells shorten S-phase on commitment to neuron production. *Nature Communications*, *2*, 154.

Arlotta, P., Molyneaux, B. J., Chen, J., Inoue, J., Kominami, R., & Macklis, J. D. (2005). Neuronal subtype-specific genes that control corticospinal motor neuron development *in vivo*. *Neuron*, *45*, 207–221.

Barone, P., Batardiere, A., Knoblauch, K., & Kennedy, H. (2000). Laminar distribution of neurons in extrastriate areas projecting to visual areas V1 and V4 correlates with the hierarchical rank and indicates the operation of a distance rule. *The Journal of Neuroscience*, *20*, 3263–3281.

Barone, P., Dehay, C., Berland, M., Bullier, J., & Kennedy, H. (1995). Developmental remodeling of primate visual cortical pathways. *Cerebral Cortex*, *5*, 22–38.

Barone, P., Dehay, C., Berland, M., & Kennedy, H. (1996). Role of directed growth and target selection in the formation of cortical pathways: Prenatal development of the projection of area V2 to area V4 in the monkey. *The Journal of Comparative Neurology*, *374*, 1–20.

Bedford, L., Walker, R., Kondo, T., van Cruchten, I., King, E. R., & Sablitzky, F. (2005). Id4 is required for the correct timing of neural differentiation. *Developmental Biology*, *280*, 386–395.

Berkes, P., Orban, G., Lengyel, M., & Fiser, J. (2011). Spontaneous cortical activity reveals hallmarks of an optimal internal model of the environment. *Science*, *331*, 83–87.

Brazhnik, P., de la Fuente, A., & Mendes, P. (2002). Gene networks: How to put the function in genomics. *Trends in Biotechnology*, *20*, 467–472.

Britanova, O., Akopov, S., Lukyanov, S., Gruss, P., & Tarabykin, V. (2005). Novel transcription factor Satb2 interacts with matrix attachment region DNA elements in a tissue-specific manner and demonstrates cell-type-dependent expression in the developing mouse CNS. *European Journal of Neuroscience*, *21*, 658–668.

Calegari, F., & Huttner, W. B. (2003). An inhibition of cyclin-dependent kinases that lengthens, but does not arrest, neuroepithelial cell cycle induces premature neurogenesis. *Journal of Cell Science*, *116*, 4947–4955.

Cappello, S., Attardo, A., Wu, X., Iwasato, T., Itohara, S., Wilsch-Brauninger, M., et al. (2006). The Rho-GTPase cdc42 regulates neural progenitor fate at the apical surface. *Nature Neuroscience*, *9*, 1099–1107.

Cauller, L. (1995). Layer I of primary sensory neocortex: Where top-down converges upon bottom-up. *Behavioral Brain Research*, *71*, 163–170.

Chang, H. H., Hemberg, M., Barahona, M., Ingber, D. E., & Huang, S. (2008). Transcriptome-wide noise controls lineage choice in mammalian progenitor cells. *Nature*, *453*, 544–547.

Chen, L., Liao, G., Yang, L., Campbell, K., Nakafuku, M., Kuan, C. Y., et al. (2006). Cdc42 deficiency causes Sonic hedgehog-independent holoprosencephaly. *Proceedings of the National Academy of Sciences of the United States of America, 103*, 16520–16525.

Chen, J. G., Rasin, M. R., Kwan, K. Y., & Sestan, N. (2005). Zfp312 is required for subcortical axonal projections and dendritic morphology of deep-layer pyramidal neurons of the cerebral cortex. *Proceedings of the National Academy of Sciences of the United States of America, 102*, 17792–17797.

Chen, B., Schaevitz, L. R., & McConnell, S. K. (2005). Fezl regulates the differentiation and axon targeting of layer 5 subcortical projection neurons in cerebral cortex. *Proceedings of the National Academy of Sciences of the United States of America, 102*, 17184–17189.

Chenn, A., & Walsh, C. A. (2002). Regulation of cerebral cortical size by control of cell cycle exit in neural precursors. *Science, 297*, 365–369.

Cheung, A. F., Pollen, A. A., Tavare, A., DeProto, J., & Molnar, Z. (2007). Comparative aspects of cortical neurogenesis in vertebrates. *Journal of Anatomy, 211*, 164–176.

Csermely, P. (2006). *Weak Links: Stabilizers of complex systems from protein to social networks.* Berlin: Springer.

Dehay, C., Giroud, P., Berland, M., Killackey, H., & Kennedy, H. (1996). Contribution of thalamic input to the specification of cytoarchitectonic cortical fields in the primate: Effects of bilateral enucleation in the fetal monkey on the boundaries, dimensions, and gyrification of striate and extrastriate cortex. *The Journal of Comparative Neurology, 367*, 70–89.

Dehay, C., Giroud, P., Berland, M., Smart, I., & Kennedy, H. (1993). Modulation of the cell cycle contributes to the parcellation of the primate visual cortex. *Nature, 366*, 464–466.

Dehay, C., Horsburgh, G., Berland, M., Killackey, H., & Kennedy, H. (1989). Maturation and connectivity of the visual cortex in monkey is altered by prenatal removal of retinal input. *Nature, 337*, 265–267.

Dehay, C., Horsburgh, G., Berland, M., Killackey, H., & Kennedy, H. (1991). The effects of bilateral enucleation in the primate fetus on the parcellation of visual cortex. *Developmental Brain Research, 62*, 137–141.

Dehay, C., & Kennedy, H. (2007). Cell-cycle control and cortical development. *Nature Reviews Neuroscience, 8*, 438–450.

Dehay, C., Kennedy, H., & Bullier, J. (1986). Callosal connectivity of areas V1 and V2 in the newborn monkey. *The Journal of Comparative Neurology, 254*, 20–33.

Dehay, C., Savatier, P., Cortay, V., & Kennedy, H. (2001). Cell-cycle kinetics of neocortical precursors are influenced by embryonic thalamic axons. *The Journal of Neuroscience, 21*, 201–214.

Douglas, R. J., Koch, C., Mahowald, M., Martin, K. A., & Suarez, H. H. (1995). Recurrent excitation in neocortical circuits. *Science, 269*, 981–985.

Eiraku, M., Takata, N., Ishibashi, H., Kawada, M., Sakakura, E., Okuda, S., et al. (2011). Self-organizing optic-cup morphogenesis in three-dimensional culture. *Nature, 472*, 51–56.

Falchier, A., Clavagnier, S., Barone, P., & Kennedy, H. (2002). Anatomical evidence of multimodal integration in primate striate cortex. *The Journal of Neuroscience, 22*, 5749–5759.

Felleman, D. J., & Van Essen, D. C. (1991). Distributed hierarchical processing in the primate cerebral cortex. *Cerebral Cortex, 1*, 1–47.

Fietz, S. A., Kelava, I., Vogt, J., Wilsch-Brauninger, M., Stenzel, D., Fish, J. L., et al. (2010). OSVZ progenitors of human and ferret neocortex are epithelial-like and expand by integrin signaling. *Nature Neuroscience, 13*, 690–699.

Fish, J. L., Dehay, C., Kennedy, H., & Huttner, W. B. (2008). Making bigger brains-the evolution of neural-progenitor-cell division. *Journal of Cell Science, 121*, 2783–2793.

Frantz, G. D., Weimann, J. M., Levin, M. E., & McConnell, S. K. (1994). Otx1 and Otx2 define layers and regions in developing cerebral cortex and cerebellum. *The Journal of Neuroscience, 14*, 5725–5740.

Gal, J. S., Morozov, Y. M., Ayoub, A. E., Chatterjee, M., Rakic, P., & Haydar, T. F. (2006). Molecular and morphological heterogeneity of neural precursors in the mouse neocortical proliferative zones. *The Journal of Neuroscience, 26*, 1045–1056.

Geisler, W. S., & Diehl, R. L. (2002). Bayesian natural selection and the evolution of perceptual systems. *Philosophical Transactions of the Royal Society of London. Series B, Biological Sciences, 357*, 419–448.

Gotz, M., & Huttner, W. B. (2005). The cell biology of neurogenesis. *Nature Reviews Molecular Cell Biology, 6*, 777–788.

Götz, M., Stoykova, A., & Gruss, P. (1998). Pax6 controls radial glia differentiation in teh cerebral cortex. *Neuron, 21*, 1031–1044.

Halley, J. D., & Winkler, D. A. (2008a). Consistent concepts of self-organization and self-assembly. *Complexity, 14*, 10–17.

Halley, J. D., & Winkler, D. A. (2008b). Critical-like self-organization and natural selection: Two facets of a single evolutionary process? *Biosystems, 92*, 148–158.

Hansen, D. V., Lui, J. H., Parker, P. R., & Kriegstein, A. R. (2010). Neurogenic radial glia in the outer subventricular zone of human neocortex. *Nature, 464*, 554–561.

Hartfuss, E., Galli, R., Heins, N., & Gotz, M. (2001). Characterization of CNS precursor subtypes and radial glia. *Developmental Biology, 229*, 15–30.

Haubensak, W., Attardo, A., Denk, W., & Huttner, W. B. (2004). Neurons arise in the basal neuroepithelium of the early mammalian telencephalon: A major site of neurogenesis. *Proceedings of the National Academy of Sciences of the United States of America, 101*, 3196–3201.

Heins, N., Malatesta, P., Cecconi, F., Nakafuku, M., Tucker, K. L., Hack, M. A., et al. (2002). Glial cells generate neurons: The role of the transcription factor Pax6. *Nature Neuroscience*, *5*, 308–315.

Hevner, R. F. (2005). The cerebral cortex malformation in thanatophoric dysplasia: Neuropathology and pathogenesis. *Acta Neuropathologica*, *110*, 208–221.

Hill, R. S., & Walsh, C. A. (2005). Molecular insights into human brain evolution. *Nature*, *437*, 64–67.

Holm, P. C., Mader, M. T., Haubst, N., Wizenmann, A., Sigvardsson, M., & Gotz, M. (2007). Loss- and gain-of-function analyses reveal targets of Pax6 in the developing mouse telencephalon. *Molecular and Cellular Neuroscience*, *34*, 99–119.

Honey, C. J., Kotter, R., Breakspear, M., & Sporns, O. (2007). Network structure of cerebral cortex shapes functional connectivity on multiple time scales. *Proceedings of the National Academy of Sciences of the United States of America*, *104*, 10240–10245.

Huang, S. (2009). Non-genetic heterogeneity of cells in development: More than just noise. *Development*, *136*, 3853–3862.

Hubel, D. H., & Wiesel, T. N. (1968). Receptive fields and functional architecture of monkey striate cortex. *The Journal of Physiology*, *195*, 215–243.

Inglis-Broadgate, S. L., Thomson, R. E., Pellicano, F., Tartaglia, M. A., Pontikis, C. C., Cooper, J. D., et al. (2005). FGFR3 regulates brain size by controlling progenitor cell proliferation and apoptosis during embryonic development. *Developmental Biology*, *279*, 73–85.

Jouve, B., Rosenstiehl, P., & Imbert, M. (1998). A mathematical approach to the connectivity between the cortical visual areas of the macaque monkey. *Cerebral Cortex*, *8*, 28–39.

Kauffman, S. A. (1993). *The origins of order.* New York/Oxford: Oxford University Press.

Kauffman, S. A. (2000). *Investigations.* New York: Oxford University Press.

Kennedy, H., & Bullier, J. (1985). A double-labeling investigation of the afferent connectivity to cortical areas V1 and V2 of the macaque monkey. *The Journal of Neuroscience*, *5*, 2815–2830.

Kennedy, H., Bullier, J., & Dehay, C. (1989). Transient projections from the superior temporal sulcus to area 17 in the newborn macaque monkey. *Proceedings of the National Academy of Sciences of the United States of America*, *86*, 8093–8097.

Kennedy, H., & Dehay, C. (1993). Cortical specification of mice and men. *Cerebral Cortex*, *3*, 27–35.

Kennedy, H., & Orban, G. A. (1983). Response properties of visual cortical neurons in cats reared in stroboscopic illumination. *Journal of Neurophysiology*, *49*, 686–704.

Killackey, H. P. (1990). Neocortical expansion: An attempt towards relating phylogeny and ontogeny. *Journal of Cognitive Neuroscience*, *2*, 1–17.

Kowalczyk, T., Pontious, A., Englund, C., Daza, R. A., Bedogni, F., Hodge, R., et al. (2009). Intermediate neuronal progenitors (basal progenitors) produce pyramidal-projection neurons for all layers of cerebral cortex. *Cerebral Cortex*, *19*, 2439–2450.

Kriegstein, A., Noctor, S., & Martinez-Cerdeno, V. (2006). Patterns of neural stem and progenitor cell division may underlie evolutionary cortical expansion. *Nature Reviews Neuroscience*, *7*, 883–890.

Kuida, K., Haydar, T. F., Kuan, C. Y., Gu, Y., Taya, C., Karasuyama, H., et al. (1998). Reduced apoptosis and cytochrome c-mediated caspase activation in mice lacking caspase 9. *Cell*, *94*, 325–337.

Land, P. W., & Monaghan, A. P. (2003). Expression of the transcription factor, tailless, is required for formation of superficial cortical layers. *Cerebral Cortex*, *13*, 921–931.

Lange, C., Huttner, W. B., & Calegari, F. (2009). Cdk4/cyclinD1 overexpression in neural stem cells shortens G1, delays neurogenesis, and promotes the generation and expansion of basal progenitors. *Cell Stem Cell*, *5*, 320–331.

LeVay, S., Wiesel, T. N., & Hubel, D. H. (1980). The development of ocular dominance columns in normal and visually deprived monkeys. *The Journal of Comparative Neurology*, *191*, 1–51.

Levitt, P., Cooper, M. L., & Rakic, P. (1981). Coexistence of neuronal and glial precursor cells in the cerebral ventricular zone of the fetal monkey: An ultrastructural immunoperoxidase analysis. *The Journal of Neuroscience*, *1*, 27–39.

Lien, W. H., Klezovitch, O., Fernandez, T. E., Delrow, J., & Vasioukhin, V. (2006). alphaE-catenin controls cerebral cortical size by regulating the hedgehog signaling pathway. *Science*, *311*, 1609–1612.

Lopez-Bendito, G., & Molnar, Z. (2003). Thalamocortical development: How are we going to get there? *Nature Reviews Neuroscience*, *4*, 276–289.

Lukaszewicz, A., Cortay, V., Giroud, P., Berland, M., Smart, I., Kennedy, H., et al. (2006). The concerted modulation of proliferation and migration contributes to the specification of the cytoarchitecture and dimensions of cortical areas. *Cerebral Cortex*, *16*(Suppl. 1), i26–i34.

Lukaszewicz, A., Savatier, P., Cortay, V., Giroud, P., Huissoud, C., Berland, M., et al. (2005). G1 phase regulation, area-specific cell cycle control, and cytoarchitectonics in the primate cortex. *Neuron*, *47*, 353–364.

Lukaszewicz, A., Savatier, P., Cortay, V., Kennedy, H., & Dehay, C. (2002). Contrasting effects of basic fibroblast growth factor and neurotrophin 3 on cell cycle kinetics of mouse cortical stem cells. *The Journal of Neuroscience*, *22*, 6610–6622.

Malatesta, P., Hack, M. A., Hartfuss, E., Kettenmann, H., Klinkert, W., Kirchhoff, F., et al. (2003). Neuronal or glial progeny: Regional differences in radial glia fate. *Neuron*, *37*, 751–764.

Malatesta, P., Hartfuss, E., & Gotz, M. (2000). Isolation of radial glial cells by fluorescent-activated cell sorting reveals a neuronal lineage. *Development*, *127*, 5253–5263.

Markov, N. T., Ercsey-Ravasz, M. M., Gariel, M. A., Dehay, C., Knoblauch, A., Toroczkai, Z., et al. (2011). The tribal networks of the cerebral cortex. In L. M. Chalupa, N. Berardi, M. Caleo, L. Galli-Resta & T. Pizzorusso (Eds.), *Cerebral plasticity* (pp. 275–290). Cambridge, MA: MIT Press.

Markov, N. T., Misery, P., Falchier, A., Lamy, C., Vezoli, J., Quilodran, R., et al. (2011). Weight consistency specifies regularities of macaque cortical networks. *Cerebral Cortex*, *21*, 1254–1272.

Martinez-Cerdeno, V., Noctor, S. C., & Kriegstein, A. R. (2006). The role of intermediate progenitor cells in the evolutionary expansion of the cerebral cortex. *Cerebral Cortex*, *16*(Suppl. 1), i152–i161.

McConnell, S. K. (1988). Fates of visual cortical neurons in the ferret after isochronic and heterochronic transplantation. *The Journal of Neuroscience*, *8*, 945–974.

Miyata, T., Kawaguchi, A., Saito, K., Kawano, M., Muto, T., & Ogawa, M. (2004). Asymmetric production of surface-dividing and non-surface-dividing cortical progenitor cells. *Development*, *131*, 3133–3145.

Molnar, Z., Metin, C., Stoykova, A., Tarabykin, V., Price, D. J., Francis, F., et al. (2006). Comparative aspects of cerebral cortical development. *European Journal of Neuroscience*, *23*, 921–934.

Molyneaux, B. J., Arlotta, P., Hirata, T., Hibi, M., & Macklis, J. D. (2005). Fezl is required for the birth and specification of corticospinal motor neurons. *Neuron*, *47*, 817–831.

Nieto, M., Monuki, E. S., Tang, H., Imitola, J., Haubst, N., Khoury, S. J., et al. (2004). Expression of Cux-1 and Cux-2 in the subventricular zone and upper layers II-IV of the cerebral cortex. *The Journal of Comparative Neurology*, *479*, 168.

Noctor, S. C., Flint, A. C., Weissman, T. A., Dammerman, R. S., & Kriegstein, A. R. (2001). Neurons derived from radial glial cells establish radial units in neocortex. *Nature*, *409*, 714–720.

Noctor, S. C., Flint, A. C., Weissman, T. A., Wong, W. S., Clinton, B. K., & Kriegstein, A. R. (2002). Dividing precursor cells of the embryonic cortical ventricular zone have morphological and molecular characteristics of radial glia. *The Journal of Neuroscience*, *22*, 3161–3173.

Noctor, S. C., Martinez-Cerdeno, V., Ivic, L., & Kriegstein, A. R. (2004). Cortical neurons arise in symmetric and asymmetric division zones and migrate through specific phases. *Nature Neuroscience*, *7*, 136–144.

O'Leary, D. D. M. (1989). Do cortical areas emerge from a protocortex? *Trends in Neurosciences*, *12*, 400–406.

Pilaz, L. J., Patti, D., Marcy, G., Ollier, E., Pfister, S., Douglas, R. J., et al. (2009). Forced G1-phase reduction alters mode of division, neuron number, and laminar phenotype in the cerebral cortex. *Proceedings of the National Academy of Sciences of the United States of America*, *106*, 21924–21929.

Pinto, L., Drechsel, D., Schmid, M. T., Ninkovic, J., Irmler, M., Brill, M. S., et al. (2009). AP2gamma regulates basal progenitor fate in a region- and layer-specific manner in the developing cortex. *Nature Neuroscience*, *12*, 1229–1237.

Polleux, F., Dehay, C., Goffinet, A., & Kennedy, H. (2001). Pre- and post-mitotic events contribute to the progressive acquisition of area-specific connectional fate in the neocortex. *Cerebral Cortex*, *11*, 1027–1039.

Polleux, F., Dehay, C., & Kennedy, H. (1998). Neurogenesis and commitment of corticospinal neurons in reeler. *The Journal of Neuroscience*, *18*, 9910–9923.

Polleux, F., Dehay, C., Moraillon, B., & Kennedy, H. (1997). Regulation of neuroblast cell-cycle kinetics plays a crucial role in the generation of unique features of neocortical areas. *The Journal of Neuroscience*, *17*, 7763–7783.

Pontious, A., Kowalczyk, T., Englund, C., & Hevner, R. F. (2008). Role of intermediate progenitor cells in cerebral cortex development. *Developmental Neuroscience*, *30*, 24–32.

Quinn, J. C., Molinek, M., Martynoga, B. S., Zaki, P. A., Faedo, A., Bulfone, A., et al. (2007). Pax6 controls cerebral cortical cell number by regulating exit from the cell cycle and specifies cortical cell identity by a cell autonomous mechanism. *Developmental Biology*, *302*, 50–65.

Rakic, P. (1988). Defects of neuronal migration and the pathogenesis of cortical malformations. *Progress in Brain Research*, *73*, 15–37.

Rakic, P. (2003). Developmental and evolutionary adaptations of cortical radial glia. *Cerebral Cortex*, *13*, 541–549.

Reznikov, K., Acklin, S. E., & van der Kooy, D. (1997). Clonal heterogeneity in the early embryonic rodent cortical germinal zone and the separation of subventricular from ventricular zone lineages. *Developmental Dynamics*, *210*, 328–343.

Roy, K., Kuznicki, K., Wu, Q., Sun, Z., Bock, D., Schutz, G., et al. (2004). The Tlx gene regulates the timing of neurogenesis in the cortex. *The Journal of Neuroscience*, *24*, 8333–8345.

Scannell, J. W., Grant, S., Payne, B. R., & Baddeley, R. (2000). On variability in the density of corticocortical and thalamocortical connections. *Philosophical Transactions of the Royal Society of London. Series B, Biological Sciences*, *355*, 21–35.

Schmechel, D. E., & Rakic, P. (1979). Arrested proliferation of radial glial cells during midgestation in rhesus monkey. *Nature*, *277*, 303–305.

Schuurmans, C., Armant, O., Nieto, M., Stenman, J. M., Britz, O., Klenin, N., et al. (2004). Sequential phases of

cortical specification involve Neurogenin-dependent and -independent pathways. *EMBO Journal, 23*, 2892–2902.
Shen, Q., Wang, Y., Dimos, J. T., Fasano, C. A., Phoenix, T. N., Lemischka, I. R., et al. (2006). The timing of cortical neurogenesis is encoded within lineages of individual progenitor cells. *Nature Neuroscience, 9*, 743–751.
Shepard, R. N. (2001). Perceptual-cognitive universals as reflections of the world. *The Behavioral and Brain Sciences, 24*, 581–601 discussion 652–571.
Smart, I. H. M. (1973). Proliferative characteristics of the ependymal layer during the early development of the mouse neocortex: A pilot study based on recording the number, location and plane of cleavage of mitotic figures. *Journal of Anatomy, 116*, 67–91.
Smart, I. H., Dehay, C., Giroud, P., Berland, M., & Kennedy, H. (2002). Unique morphological features of the proliferative zones and postmitotic compartments of the neural epithelium giving rise to striate and extrastriate cortex in the monkey. *Cerebral Cortex, 12*, 37–53.
Sporns, O. (2011). *Networks of the brain.* Cambridge, MA: MIT Press.
Sur, M., & Rubenstein, J. L. (2005). Patterning and plasticity of the cerebral cortex. *Science, 310*, 805–810.
Tarabykin, V., Stoykova, A., Usman, N., & Gruss, P. (2001). Cortical upper layer neurons derive from the subventricular zone as indicated by Svet1 gene expression. *Development, 128*, 1983–1993.
Thompson, I. D., Kossut, M., & Blakemore, C. (1983). Development of orientation columns in cat striate cortex revealed by 2-deoxyglucose autoradiography. *Nature, 301*, 712–715.
Van der Loos, H. (1977). Structural changes in the cerebral cortex upon modification of the periphery: Barrels in somatosensory cortex. *Philosophical Transactions of the Royal Society of London. Series B, Biological Sciences, 278*, 373–376.
van Ooyen, A. (2011). Using theoretical models to analyse neural development. *Nature Reviews Neuroscience, 12*, 311–326.
Vezoli, J., Falchier, A., Jouve, B., Knoblauch, K., Young, M., & Kennedy, H. (2004). Quantitative analysis of connectivity in the visual cortex: Extracting function from structure. *The Neuroscientist, 10*, 476–482.
Viti, J., Gulacsi, A., & Lillien, L. (2003). Wnt regulation of progenitor maturation in the cortex depends on Shh or fibroblast growth factor 2. *The Journal of Neuroscience, 23*, 5919–5927.
Wu, S. X., Goebbels, S., Nakamura, K., Nakamura, K., Kometani, K., Minato, N., et al. (2005). Pyramidal neurons of upper cortical layers generated by NEX-positive progenitor cells in the subventricular zone. *Proceedings of the National Academy of Sciences of the United States of America, 102*, 17172–17177.
Young, M. P. (1992). Objective analysis of the topological organization of the primate cortical visual system. *Nature, 358*, 152–155.
Yun, K., Mantani, A., Garel, S., Rubenstein, J., & Israel, M. A. (2004). Id4 regulates neural progenitor proliferation and differentiation *in vivo*. *Development, 131*, 5441–5448.
Zecevic, N., Chen, Y., & Filipovic, R. (2005). Contributions of cortical subventricular zone to the development of the human cerebral cortex. *The Journal of Comparative Neurology, 491*, 109–122.
Zhou, C. J., Borello, U., Rubenstein, J. L., & Pleasure, S. J. (2006). Neuronal production and precursor proliferation defects in the neocortex of mice with loss of function in the canonical Wnt signaling pathway. *Neuroscience, 142*, 1119–1131.
Zimmer, C., Tiveron, M. C., Bodmer, R., & Cremer, H. (2004). Dynamics of Cux2 expression suggests that an early pool of SVZ precursors is fated to become upper cortical layer neurons. *Cerebral Cortex, 14*, 1408–1420.
Zubler, F., & Douglas, R. (2009). A framework for modeling the growth and development of neurons and networks. *Frontiers in Computational Neuroscience, 3*, 25.

M. A. Hofman and D. Falk (Eds.)
Progress in Brain Research, Vol. 195
ISSN: 0079-6123

CHAPTER 17

Neural wiring optimization

Christopher Cherniak*

Committee for Philosophy and the Sciences, Department of Philosophy, University of Maryland, College Park, MD, USA

Abstract: Combinatorial network optimization theory concerns minimization of connection costs among interconnected components in systems such as electronic circuits. As an organization principle, similar wiring minimization can be observed at various levels of nervous systems, invertebrate and vertebrate, including primate, from placement of the entire brain in the body down to the subcellular level of neuron arbor geometry. In some cases, the minimization appears either perfect, or as good as can be detected with current methods. One question such best-of-all-possible-brains results raise is, what is the map of such optimization, does it have a distinct neural domain?

Keywords: Adjacency Rule; *Caenorhabditis elegans*; cerebral cortex; component placement optimization; Size Law; Steiner tree; volume minimization; wirelength minimization.

Introduction

Neuroconnectivity architecture sometimes shows virtually perfect network optimization, rather than just network satisficing. Long-range connections are a critically constrained resource in the brain, hence, there may be great selective pressure to optimize finely their deployment. The formalism of scarcity of interconnections is network optimization theory, which characterizes efficient use of limited connection resources. The field matured decades ago for microcircuit design, typically to minimize the total length of wire needed to make a given set of connections among components. For layout of neural components, such connection minimization has been reported for the nematode nervous system (Cherniak, 1994a), rat amygdala and olfactory cortex (Cherniak and Rodriguez-Esteban, 2010), cat sensory cortex, and macaque visual cortex (Cherniak et al., 2004). Corresponding arbor optimization also applies for some types of dendrites and axons (Cherniak et al., 1999). Results for more primitive nervous systems help fill in some of the evolutionary trajectory of neural optimization phenomena.

Such optimality contrasts with the familiar picture for biological design, of only moderately good engineering: for example, the first chapter

*Corresponding author.
Tel.: +1-301-405-5689; Fax: +301-405-5690
E-mail: cherniak@umd.edu

DOI: 10.1016/B978-0-444-53860-4.00017-9

of *Descent of Man* (Darwin, 1871) enumerated many instances of rudimentary structures in humans that are no longer in use (although the neural examples are in fact functional). Instead, it is almost as if neural connections had an unbounded cost. When this simple "save wire" idea is treated as a generative principle for nervous system organization, it turns out to have some applicability: To an extent, across evolutionary levels, wire-minimization yields brain structure.

A caveat is that, in general, network optimization problems are easy to state, but vastly computationally costly to solve exactly. These connection cost-minimization problems are a major hurdle of microcircuit design and are known to be NP-complete (nondeterministic polynomial time complete), that is, *de facto* intractable (Garey and Johnson, 1979). Computation costs of solving problems of comparatively small size typically grow exponentially, to cosmic scale: exactly solving some could consume more space and/or time than exists in the known Universe. The archetypal example of an NP-complete problem is Traveling Salesman: For a given set of points on a map, simply find the shortest roundtrip tour.

Neuron arbor optimization

The basic concept of an optimal tree is as follows: Given a set of loci in 3D space, find the minimum-cost tree that interconnects them, for instance, the set of interconnections of least total volume. If branches are allowed to join at points other than the given terminal loci (the "leaves" and "root"), the minimum tree is of the most economical type, a Steiner tree. If the synapse sites and origin of a dendrite or axon are treated in this way, optimization of the dendrite or axon can be evaluated. Approximately planar arbors in 2D space are simpler to analyze. The most important feature of naturally occurring arbors—neuronal, vascular, plant, water drainage networks, etc.—is that, unlike much manufactured circuitry, for each internodal junction, trunk cost (e.g., diameter) is higher than branch costs.

Local trees

When such Y-junctions are examined in isolation, positioning of the junction sites shows minimization of total volume cost (vs. surface area or length) to within about 5% of optimal (Cherniak, 1990, 1992; 7 of 25 datasets were from primates). In turn, the relation of branch diameters to trunk diameter fits a simple fluid-dynamical model for minimization of walldrag of internal laminar flow in a tree of tubes: Dendrites and axons act like flowing water.

Global trees

This Y-tree cost-minimization constitutes local optimization. Only one interconnection pattern or topology is involved. Such small-scale optimization does not entail larger-scale optimization, where local trade-offs are often required. When more complex portions of a total arbor are analyzed, optimization becomes a global problem, with an exponentially exploding number of alternative possible interconnection topologies. For example, a nine-terminal tree already has 135,135 alternative topologies, each of which must be generated and costed to verify the best solution (see Fig. 1). Neuron arbor samples, each with three internodal Y-junctions and a distribution of different topologies, minimize their volume to within about 5% of optimal (Cherniak et al., 1996, 1999). This optimality performance is consistent for dendrites (rabbit retina ganglion and amacrine cells, and cat retina ganglion cells) and also for some types of axons (intrinsic and extrinsic mouse thalamus). One of eight datasets was from primates.

Topology

The pattern for natural arbors, living and nonliving, is that more costly topologies are more common, cheapest ones are rarest. However, the

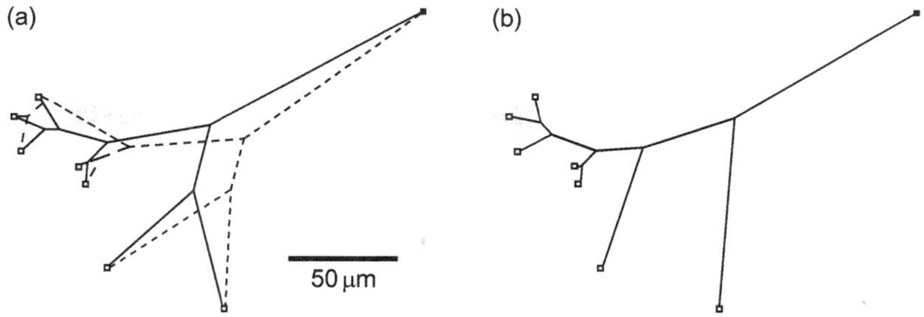

Fig. 1. Actual versus optimal neuron arbors, mouse thalamus extrinsic axon, ascending reticular formation. The arbor best fits a minimized-volume model. (a) Wireframe representation of eight-terminal subtree of observed arbor. Actual tree, with actual topology in its actual embedding, appears in broken lines. Optimal embedding with respect to volume minimization of the actual topology is superimposed in solid lines. The cost in volume of the actual arbor exceeds that of the optimized embedding of its topology by 2.20%. (b) "Best of all possible topologies" connecting the given terminal loci: the optimal topology with respect to volume, optimally embedded. The volume cost of the actual arbor exceeds that of the optimal topology by 2.47%. Only 10 of the 10,395 possible alternative topologies here have lower total volume costs, when optimally embedded, than the actual topology (reprinted with permission from Cherniak et al., 1999).

most costly optimally embedded "pessimal" topologies have relatively little higher cost than the cheapest ones. In this sense, to a first approximation, "topology does not matter". Consequently, neuron arbor anatomy behaves like flowing water, and waterflow in turn acts like a tree composed of weights and pulleys (rather than springs). Fluid dynamics drives fluid statics, that is, vector mechanics. Hence, "instant arbors, just add water," that is, neuroanatomy from physics.

Component placement optimization

Another key problem in microcircuit design is component placement optimization (also characterized as a quadratic assignment problem). Given a set of interconnected components, find the location of the components on a 2D surface that minimizes total cost of connections (e.g., wirelength). A familiar example is siting of computer chips on a motherboard. Again, this concept seems to account for aspects of neuroanatomy at multiple hierarchical levels.

Why the brain is in the head is a one-component placement problem. That is, given the fixed loci of receptors and muscles, positioning the brain as far forward in the body axis as possible minimizes total nerve connection costs to and from the brain, because more sensory and motor connections go to the anterior than to the posterior of the body. This seems to hold for the vertebrate series (e.g., humans) and also for invertebrates with sufficient cephalization to possess a main nervous system concentration (e.g., nematodes).

Caenorhabditis elegans

As for arbors, multiple-component problems again generally require exponentially exploding costs for exact solutions; for an *n*-component system, *n*! alternative layouts must be searched. A typical neural wiring optimization result is for placement of the 11 ganglionic components of the nervous system of the roundworm *C. elegans*, with ~1000 interconnections. This nervous system is the first to be completely mapped (Wood, 1988), which enables fair approximation of connection wirelengths. When all 39,916,800 alternative possible ganglion layouts are generated, the actual layout turns out in fact to be the one with minimum total wirelength (Cherniak, 1991, 1994a, 2003a).

Some optimization mechanisms provide convergent support for this finding: A simple genetic algorithm, with wirecost as fitness measure so that shorter wirelength worm layouts are more likely to survive, will rapidly and robustly converge upon the actual optimal layout (Cherniak et al., 2002; see Fig. 2). Also, a force-directed placement algorithm, with each connection approximated as a weights-and-pulleys mechanism (non-Hooke's law, i.e., not a spring) acting between ganglion components, attains the actual layout as a minimum-energy state, with little local-minima trapping (Cherniak et al., 2002; see Fig. 3). Each of these wire-minimization mechanisms operates top-down: that is, each proceeds from nervous system connections to positioning of neural components; vice versa is not necessary.

There is statistical evidence that this brain-as-microchip framework also extends in the worm down to the level of clustering of individual neurons into ganglionic groups, and to soma positioning within ganglia to reduce connection costs (Cherniak, 1994a).

Cortex

The wiring-minimization approach can be applied to placement of functional areas of the vastly more complex mammalian cerebral cortex. In contrast to the fixed character of neural development in invertebrates such as nematodes, it is striking that optimization also holds for malleable mammal brains. Since wirelengths and branch patterns of corticocortical connections are difficult to estimate, one strategy is instead to explore a simpler measure of connection cost, conformance of a layout to a wire-conserving Adjacency Rule:

> If components *a* & *b* are connected,
> then *a* & *b* are adjacent.

Exhaustive search of all possible layouts is still required to identify the cheapest one(s). A promising calibration is that the actual minimum-wire layouts of the nematode ganglia are among the top layouts with fewest violations of the Adjacency Rule. One primate cortex example is that, for 17 core visual areas of macaque cortex,

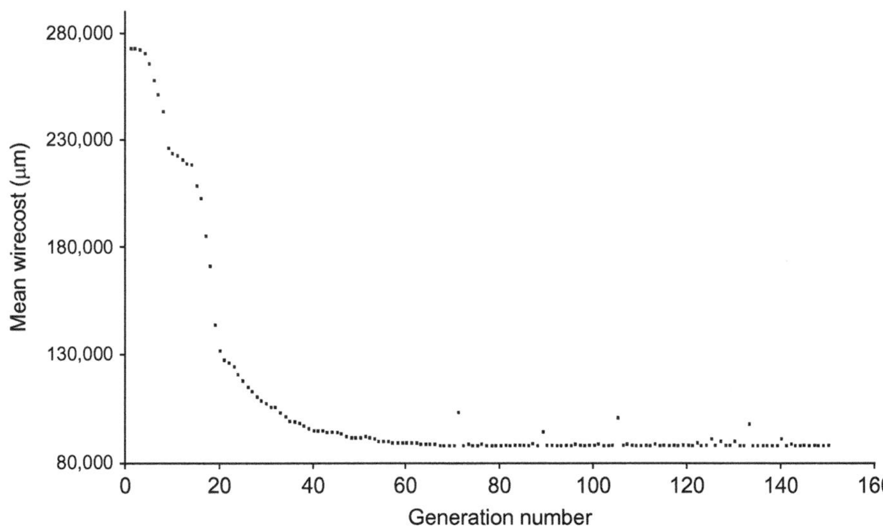

Fig. 2. A simple genetic algorithm, GenAlg, rapidly and reliably finds the optimal layout of *C. elegans* ganglia among 11! alternatives. Fitness measure for survival is total wirelength of individual nervous system. The initial population of this run was only 10 individuals, all with reverse of actual ordering of ganglia (reprinted with permission from Cherniak, 2005).

```
                                                   Input: actual.mtx

                         T E N S A R A M A

Head                                                Tail
0     0     1     1     2     2     3     3     4     4     5
0     5     0     5     0     5     0     5     0     5     0   Tetrons

 PH    (100.000000)
   AN    (300.000000)
    RNG   (440.000000)
     DO    (506.000000)
     LA    (564.000000)
       VN    (744.000000)
         RV    (948.000000)
                  VCa  (1856.000000)
                                      VCp  (3856.000000)
                                               PA    (4726.000000)
                                                DR    (4810.000000)
                                                LU    (4884.000000)

Final layout popped out after:  100,000 iterations
Tension constant = 0.010000
Total wirecost = 87802.750000 µm
```

Fig. 3. Runscreen for a force-directed placement algorithm, Tensarama, for optimizing layout of ganglia of the nematode *C. elegans*, that is, minimizing total length of interconnections. This vector-mechanical simulation represents each of the roundworm's ~1000 interconnections as a weights-and-pulleys mechanism (as opposed to a spring) acting upon the horizontally movable ganglia "PH," "AN," etc. Connections themselves do not appear on runscreen nor fixed components such as sensors and muscles. The screendump here shows the final configuration of the system after 100,000 iterations (reupdate cycles for forces and locations). The system has terminated with the global minimum-cost positioning of the ganglia (using about 8.7cm total of wire), which is also the actual layout. In this way, physics suffices to generate this neuroanatomical structure, out of ~40 million alternative possible configurations (reprinted with permission from Cherniak et al., 2002).

the actual layout of this subsystem ranks in the top 10^{-7} layouts best fitting this adjacency costing. For 15 visual areas of cat cortex, the actual layout ranks in the top 10^{-6} of all layouts (Cherniak, 2003b; Cherniak et al., 2003, 2004; see also Cherniak, 1991, 2003a; Young, 1992). Other examples include rat olfactory cortex and amygdala (Cherniak and Rodriguez-Esteban, 2010; see Fig. 4).

Size Law

In general, a Size Law seems to apply to cases with such local–global trade-offs. If a complete system is in fact perfectly optimized, then the smaller the portion of it considered by itself, the poorer the optimization appears. Or, to reverse the reasoning:

> The larger the proportion of a total optimal system that the evaluated subsystem is,
> the better its optimization.

A Size Law applies to each of the above cortex systems (see Fig. 5). For the largest systems studied (visual, auditory, and somatosensory areas of cat cortex), there is evidence along these lines of optimization approaching limits of current detectability by brute-force sampling techniques. A similar Size Law pattern also appears to hold for Steiner tree optimization of neuron arbor topologies (cf. Fig. 1). The overall picture then is of limited connections deployed very well, a predictive success story. The significance of ultra-fine

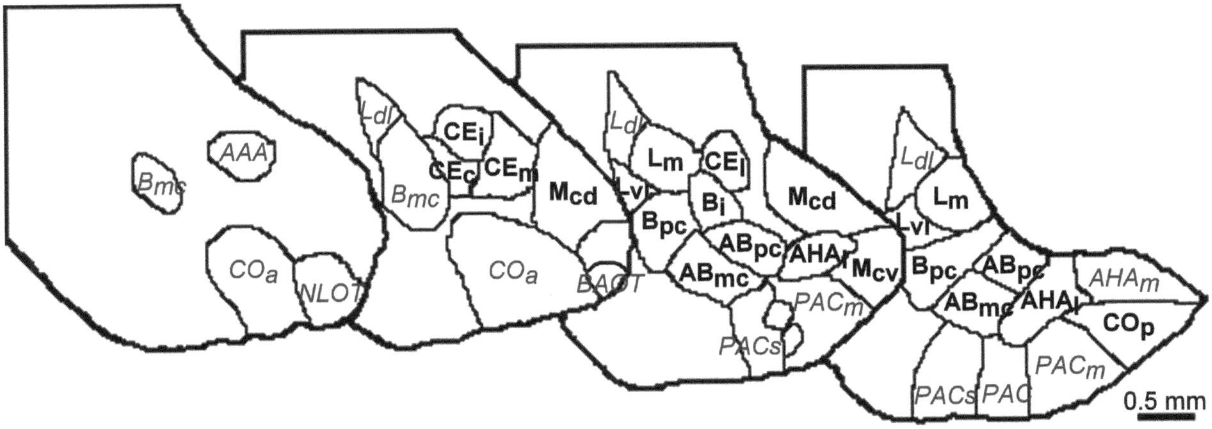

Fig. 4. Rat amygdala represented as a stack of slices: 3D topological interrelations among its nuclei. Fourteen contiguous core components for optimization analysis of their layout are shown with **boldface** labels; immediately surrounding edge areas are in *italic*. For interconnections, and successive subset sizes 1–14, see Table 2, Cherniak and Rodriguez-Esteban (2010).

neural optimization remains an open question. Levels of connection optimization in the nervous system seem unlike levels of optimization elsewhere in organisms.

Related optimization results

Some other recently reported instances of biological network optimization provide perspective on the above neural optimization cases. For example, an amoeboid organism, the plasmodium of the slime mold *Physarum polycephalum*, is capable of solving a maze, that is, not just finding some path across a labyrinth, but a shortest path through it to food sources (Nakagaki et al., 2000). Generating such a minimum-length solution is a network optimization feat for any simple creature. However, it should be noted that this shortest path problem is not computationally intractable; in particular, it is not NP-complete (Garey and Johnson, 1979). "Greedy algorithms" can solve it and also can be implemented as simple vector-mechanical "tug of war" processes. Nonetheless, that a slime mold can optimize a path through a network converges with observations of network optimization in nervous system anatomy. The latter results entail solution of computationally complex (i.e., NP-complete) problems. Such consilience lends support to the neuroanatomical findings.

Path optimization by social insects has also been reported. For example, wood ants (*Formica aquilonia*) form complex tree structures as foraging paths; however, their length by itself is not minimized (Buhl et al., 2009). Bumblebees (*Bombus terrestris*) satisfactorily solve similar Traveling Salesman foraging problems among food sources (Lihoreau et al., 2010), and Argentine ants (*Linepithima humile*) can find efficient tree structures interconnecting their nests (Latty et al., 2011). However, the networks involved have only five or less nodes.

Some critiques have appeared related to the main wiring optimization result reported here for *C. elegans*, that the actual layout of its ganglia has the minimum total wirelength of all 11! alternative possible configurations. Since primate cortex optimization results depend upon the soundness of the nematode results, we review the critiques of the latter. For the exhaustive searches of worm layouts, we employed the

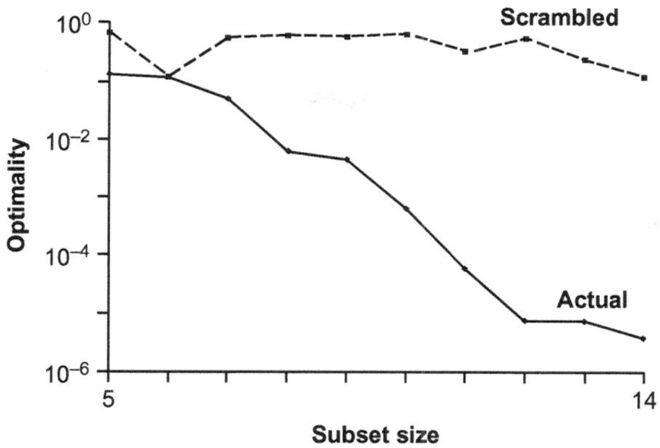

Fig. 5. Rat amygdala layout optimization analysis. Plot of optimality rankings for a series of nested subsets of the 14-area core (see Fig. 4) shown by solid line. For comparison, a randomly generated layout with areas' relative positions scrambled and their interconnections preserved is similarly analyzed for a succession of progressively larger subsets (dashed line). A layout is scored in terms of its violations of the Adjacency Rule. Each nested compact subset is compared with all possible alternative layouts of that subset for Adjacency Rule optimality. A Size Law trend—increasing optimality with increased subsystem size—is apparent for the actual layout, but not for its scrambled control version. For the best-fit line for optimality of the subset series of the actual layout, r^2=0.96, p<0.0001. Layout optimality rank for the complete amygdala system analyzed is in top 3.9×10^{-6} of all possible layouts of the full 14 areas, comparable to cat and macaque visual cortex (Cherniak et al., 2004) (reprinted with permission from Cherniak and Rodriguez-Esteban, 2010).

simplest cost metric, a linear function of total wirelength; this cost measure also performs well for a force-directed placement simulation, and as fitness measure for a genetic algorithm (Cherniak et al., 2002; see also Figs. 2 and 3 above).

Chklovskii (2004) proposes that wiring cost scales instead as sum of squares of wirelengths, and offers a quadratic minimization analysis of ganglion position. However, among perturbation analyses we had also performed to explore the optimization landscape was an exhaustive search of all 11! layouts with connection cost instead as wirelength squared. The worm's actual layout then drops in rank, from optimal # 1 to # 404,959; the actual layout costs 21% more than the cheapest. So, compared to wirecost=wirelength, a wirelength squared model does not do well in terms of goodness of fit. And, in fact, Chklovskii concludes that this model does not predict actual order of all ganglia, in particular, for the dorsorectal (DR) ganglion. (For comparison, see, e.g., DR position in Fig. 3 above.)

Chen et al. (2006) extend the Chklovskii (2004) connection-minimization model from ganglia down to the level of individual neuron positioning in *C. elegans*. Again, a wirelength squared measure for connection cost is employed. At this finer scale of anatomy, for our analyses of neuron arbors (Cherniak et al., 1999), we had found that such a Hooke's Law model, where connections behave like springs, similarly did not perform well compared to a simple linear cost model. In addition, Chen et al. principally employ a "dedicated-wire" model, where a neuron cell body must have a separate connection to each of its synapses, rather than a more realistic "shared-wire" model. One germane calibration is that we had performed another series of exhaustive searches of all 11! layouts without any actual shared connections at the ganglion level. The actual ganglion layout then drops in rank, from # 1 to # 2,948,807; with the redundancy of these dedicated connections, the actual layout now costs 38% more than the least costly one. Further, the more shared connections allowed—the more permitted branchings—the better the actual layout performs.

Chen et al. (2006) calculated neuron positions that minimize their quadratic cost function. A caveat for such analytic solutions (see also Chklovskii, 2004) is that, as mentioned, this optimization problem is in fact NP-complete, that is, generally not exactly solvable without exhaustive search. (For examples of local-minimum trapping for ganglion layouts, see Fig. 6 of Cherniak et al., 2002 and Fig. 8.5 of Cherniak, 2009.). Chen et al. do not address the NP-complete character of the wiring problem. Chen et al. conclude that some neurons show strong deviation from the "optimal" placement model; total wiring cost of the actual configuration is nearly four times greater than that of their optimized layout. Compared to the linear minimization account for the ganglia, performance of this quadratic minimization model for neurons turns attention back to how connections were costed.

Contemporaneous with Chen et al. (2006), Kaiser and Hilgetag (2006) also argued that the actual layouts of macaque cortex areas and of *C. elegans* neurons were not in fact minimum-wirelength configurations. We focus on similar questions that arise for this analysis of the worm nervous system. To reiterate, in combinatorial network optimization theory, the component placement optimization problem is, given a set of components and their interconnections, find a siting of components that yields minimum total cost; length of particular individual connections is not an issue. Kaiser and Hilgetag report that rearranging positions of 277 of the worm's 302 neurons can yield an alternative network with total wiring cost reduced by 48%. However, as for Chen et al. (2006) above, how multiple synapses from a neuron fiber are dealt with in alternate placements again needs examining. As an instance, Kaiser and Hilgetag raise the question of accurate representation of sensory and motor connection costs in their alternative neuron layouts. We had performed another search of all 11! alternative ganglion layouts, with only muscle connections deleted. The actual layout's rank then dropped from # 1 to # 63,161, with 10% greater wirecost than the optimal. Again, interpretation of alternate layout connection-costing would benefit from clarification.

Thus, similar questions seem to remain about meaningfulness of both the Chen et al. and Kaiser–Hilgetag neuron placement optimization results. Also relevant here is the observed neuron wiring-minimization pattern mentioned earlier, that if two *C. elegans* neurons are connected, they tend strongly to be clustered in the same ganglion. Further, within ganglia, antero-posterior siting of somata conforms significantly to a connection-length minimization model (Cherniak, 1994a, 1995). In addition, at a yet finer scale, we reported volume minimization of dendrite and axon arbors (Cherniak et al., 1999).

Finally, Klyachko and Stevens (2003) have reported that layout of functional areas of macaque prefrontal cortex is optimal, in that the actual placement of the 11 areas minimizes total wirelength of their known interconnections. Along lines of Cherniak (2003b) and Cherniak et al. (2003, 2004), we reanalyzed the Carmichael and Price (1996) prefrontal neuroanatomy used here, employing instead simple conformance to the Adjacency Rule as a connection cost measure, as discussed above for cat and macaque visual cortex, etc. An exhaustive search of alternative placements showed that the actual layout of the prefrontal areas then ranked in the best 2×10^{-5} of all possible alternative layouts. In our earlier adjacency-cost analysis of macaque visual cortex, the actual layout of a core subset of 11 areas had ranked in the top 1.07×10^{-5} of all layouts. So, connection optimization of prefrontal cortex areas seems to agree with our results for visual cortex.

Mapping neural optimization

Mechanisms of neural optimization are best understood against the background that the key problems of network optimization theory are NP-complete; hence, exact solutions in general are computationally intractable. For example, blind trial and error exhaustive search for the minimum-wiring layout of a 50-component system (such as all areas of a mammalian cerebral cortex hemisphere), even at a physically unrealistic rate of one layout per picosecond, would still require more than the age of the Universe (Cherniak, 1994b). Instead, even evolution must exploit "quick and dirty" approximation/probabilistic heuristics.

One such possible strategy discernible above is optimization for free, directly from physics. That is, as some structures develop, physical principles cause them automatically to be optimized. Evidence was reviewed above for arbor optimization via fluid dynamics, and for roundworm ganglion layout optimization via force-directed placement. The worm layout is among the most complex biological structures known to be derivable in this way for free, directly from simple physical processes, without intervention by DNA mechanisms.

For processes of component placement optimization, a chicken–egg question arises of whether components begin in particular loci and make connections, or instead start with their interconnections and then adjust their positions, or some mix of both causal directions. However, it is worth noting that both a force-directed placement algorithm for ganglion layout and genetic algorithms for layout of ganglia and of cortex areas suggest that simple "connections→placement" optimization processes can suffice.

Wiring optimization is of course subject to many basic constraints and so cannot be ubiquitous in the nervous system or elsewhere; the question is where it does in fact occur, and how good it is. Trade-offs of local optimality for better global cost minimization of a total system (as Fig. 2, Cherniak et al. (2004) illustrates) are one way in which global optimization can be obscured.

The very fact of neural resource limitations appears to drive "save wire" fine-grained minimization of connections. Another part of the functional role of such optimization may be the picture here of the prebiotic pervading the biotic: "physics→optimization→neuroanatomy." Perhaps, such an economical means of self-organizing complex structure generation eases transmissibility through the "genomic bottleneck," that is, the limited information carrying capacity of DNA. This constitutes a thesis of "Non-Genomic Nativism," that significant complex biological structure is not encoded in DNA, but instead derives from basic physical principles (Cherniak, 2005).

Such an account is an innateness hypothesis: There is inborn structure—not only at the abstract cognitive level (e.g., of linguistic competence) but also at the brain hardware level. The harmony of physics and neuroanatomy yielding optimization is an instance of self-organizing biological structure. For such an account, the blank slate of the nervous system is in fact instead preformatted—however, not via the genome, but by the underlying physical and mathematical order of the Universe (see Chomsky, 2005). A division of labor holds between the genome and this underlying order.

The "connective tissue" minimization findings suggest optimization of neural layouts to a level at least in the best one millionth of all layouts. And this across much of the evolutionary trajectory, from nematode to macaque—another dimension of convergent confirmation of neural optimization.

In addition, the Size Law raises the possibility of extrapolation, that larger neural systems that take into account more connected components may in fact be attaining even better cost minimization. And, in fact, Cherniak (2003b) and Cherniak et al. (2004) include results for the 39-component cat sensory cortex system (visual, auditory, and somatosensory), where optimization falls in the top one billionth of all layout possibilities. This begins to approach some of the most precise confirmed predictions known in science, such as those of quantum electrodynamics (e.g., Peskin and Schroeder, 1995). Such a best-in-a-billion optimization model seems a predictive success story.

Yet, against the familiar background of biological satisficing, this neural minimizing may appear gratuitous. There are many other competing design desiderata besides "save wire." Extreme connection minimization itself in turn stands in need of further explanation. In his discussion of neural wiring economy, Sporns (2010) concludes that brain connectivity optimization to minimize wirecost is unlikely; instead, brain wiring is a compromise of many factors. Such views, of course, are familiar; Gould (1980), along Darwinian lines, is a contemporary *locus classicus* for the idea that evolution yields improvised *ad hoc* tinkering, not ideal design.

However, none of Gould's examples are neural. Optimization to physical limits has long been reported also for stimulus amplitude sensitivity of sensory systems for vision, hearing, olfaction, etc. (e.g., Cherniak et al., 2002). So, an emerging picture might suggest exploring the conjecture of a neural/nonneural divide for the scope of optimization prevalence. Sensors would fall just on the neural side of such a boundary. Next questions include, what are other domains of optimality, and why would neural systems tend more to be organized in this different way from many other biological systems?

References

Buhl, J., Hicks, K., Miller, E. R., Persley, S., Alinvi, O., & Sumpter, D. J. (2009). Shape and efficiency of wood ant foraging networks. *Behavioral Ecology and Sociobiology*, *63*, 451–460.

Carmichael, S., & Price, J. (1996). Connectional networks within the orbital and medial prefrontal cortex of macaque monkeys. *The Journal of Comparative Neurology*, *371*, 179–207.

Chen, B., Hall, D., & Chklovskii, D. (2006). Wiring optimization can relate neuronal structure and function. *Proceedings of the National Academy of Sciences of the United States of America*, *103*, 4723–4728.

Cherniak, C. (1990). *Local network optimization in the brain.* UM Computer Science Technical Report CS-TR-2498, UMIACS-TR-90-90 (July 1990).

Cherniak, C. (1991). *Component placement optimization in the brain.* UM Computer Science Technical Report CS-TR-2711, UMIACS-TR-91-98 (July 1991).

Cherniak, C. (1992). Local optimization of neuron arbors. *Biological Cybernetics*, *66*, 503–510.

Cherniak, C. (1994a). Component placement optimization in the brain. *The Journal of Neuroscience*, *14*, 2418–2427.

Cherniak, C. (1994b). Philosophy and computational neuroanatomy. *Philosophical Studies*, *73*, 89–107.

Cherniak, C. (1995). Neural component placement. *Trends in Neurosciences*, *18*, 522–527.

Cherniak, C. (2003a). *Network optimization in the brain [I, 1991]: From* C. elegans *to cerebral cortex.* UM Computer Science Technical Report CS-TR-4524, UMIACS-TR-2003-92 (September 2003).

Cherniak, C. (2003b). *Network optimization in the brain [II, 2000]: Cerebral cortex layout.* UM Computer Science Technical Report CS-TR-4525, UMIACS-TR-2003-93 (September 2003).

Cherniak, C. (2005). Innateness and brain-wiring optimization: Non-genomic nativism. In A. Zilhao (Ed.), *Cognition, evolution, and rationality*. London: Routledge.

Cherniak, C. (2009). Brain wiring optimization and non-genomic nativism. In M. Piattelli-Palmarini, J. Uriagereka & P. Salaburu (Eds.), *Of minds and language*. New York: Oxford.

Cherniak, C., Changizi, M., & Kang, D. (1996). *Large-scale optimization of neuron arbors.* UM Computer Science Technical Report CS-TR-3708, UMIACS-TR-96-78 (November 1996).

Cherniak, C., Changizi, M., & Kang, D. (1999). Large-scale optimization of neuron arbors. *Physical Review E*, *59*, 6001–6009.

Cherniak, C., Mokhtarzada, Z., & Nodelman, U. (2002). Optimal-wiring models of neuroanatomy. In G. Ascoli (Ed.), *Computational neuroanatomy: Principles and methods* (pp. 71–82). Totowa, NJ: Humana.

Cherniak, C., Mokhtarzada, Z., Rodriguez-Esteban, R., & Changizi, B. (2003). *Global optimization of cerebral cortex layout (I).* UM Computer Science Technical Report CS-TR-4534, UMIACS-TR-2003-102.

Cherniak, C., Moktarzada, Z., Rodriguez-Esteban, R., & Changizi, B. (2004). Global optimization of cerebral cortex layout. *Proceedings of the National Academy of Sciences of the United States of America*, *101*, 1081–1086.

Cherniak, C., & Rodriguez-Esteban, R. (2010). Information processing limits on generating neuroanatomy: Global optimization of rat olfactory cortex and amygdala. *Journal of Biological Physics*, *36*, 45–52.

Chklovskii, D. (2004). Exact solution for the optimal neuronal layout problem. *Neural Computation*, *16*, 2067–2078.

Chomsky, N. (2005). Three factors in language design. *Linguistic Inquiry*, *36*, 1–22.

Darwin, C. (1871). *The descent of man.* London: John Murray.

Garey, M., & Johnson, D. (1979). *Computers and intractability: A guide to NP-completeness.* San Francisco: W. H. Freeman.

Gould, S. (1980). *The panda's thumb.* New York: W. W. Norton.

Kaiser, M., & Hilgetag, C. (2006). Nonoptimal component placement, but short processing paths, due to long-distance projections in neural systems. *PLoS Computational Biology*, *2*, e95.

Klyachko, V., & Stevens, C. (2003). Connectivity optimization and the positioning of cortical areas. *Proceedings of the National Academy of Sciences of the United States of America*, *100*, 7937–7941.

Latty, T., Ramsch, K., Ito, K., Nakagaki, T., Sumpter, D., Middendorf, M., et al. (2011). Structure and formation of ant transportation networks. *Journal of the Royal Society Interface*, *8*(62), 1298–1306. doi:10.1098/rsif.2010.0612.

Lihoreau, M., Chittka, L., & Raine, N. (2010). Travel optimization by foraging bumblebees through readjustments of traplines after discovery of new feeding locations. *The American Naturalist*, *176*, 744–757.

Nakagaki, T., Yamada, H., & Toth, A. (2000). Maze-solving by an amoeboid organism. *Nature*, *407*, 470.

Peskin, M., & Schroeder, D. (1995). *Introduction to quantum field theory*. Boulder: Westview.

Sporns, O. (2010). *Networks of the brain*. Cambridge: MIT.

Wood, W. (Ed.), (1988). In: *The nematode* Caenorhabditis elegans. Cold Spring Harbor, New York: Cold Spring Harbor Laboratory.

Young, M. (1992). Objective analysis of the topological organization of the primate cortical visual system. *Nature*, *358*, 152–155.

M. A. Hofman and D. Falk (Eds.)
Progress in Brain Research, Vol. 195
ISSN: 0079-6123

CHAPTER 18

Design principles of the human brain: An evolutionary perspective

Michel A. Hofman*

Netherlands Institute for Neuroscience, Royal Netherlands Academy of Arts and Sciences, Amsterdam, The Netherlands

Abstract: The evolution of the brain in mammals has been accompanied by a reorganization of the brain as a result of differential growth of certain brain regions. Consequently, the geometry of the brain, and especially the size and shape of the cerebral cortex, has changed notably during evolution. Comparative studies of the cerebral cortex suggest that there are general architectural principles governing its growth and evolutionary development and that the primate neocortex is uniformly organized and composed of neural processing units. We are beginning to understand the geometric, biophysical, and energy constraints that have governed the evolution of these neuronal networks. In this review, some of the design principles and operational modes will be explored that underlie the information processing capacity of the cerebral cortex in primates, and it will be argued that with the evolution of the human brain we have nearly reached the limits of biological intelligence.

Keywords: brain evolution; cortical columns; neural networks; fractals; cortical design; human neocortex.

Introduction

During the past decades, considerable progress has been made in explaining the evolution of brain size in mammals in terms of physical and adaptive principles (see, e.g., Hofman, 2003; Lefebvre et al., 2004; Macphail and Bolhuis, 2001; Roth and Dicke, 2005). In addition, a quantitative approach to the comparative morphology of the brain has made it possible to identify and formalize empirical regularities in the diversity of brain design, especially in the geometry of the cerebral cortex (e.g., Changizi, 2001, 2007; Clark et al., 2001; Hofman, 1989). Though many aspects of brain evolution still remain unexplained, these comparative investigations, using scaling methods and mathematical models, have provided new insights into the evolutionary dynamics of the brain and its morphological constraints. The object of this chapter is to present

*Corresponding author.
Tel.: +31-20-566-5500; Fax: +31-20-696-6121
E-mail: m.hofman@nin.knaw.nl

DOI: 10.1016/B978-0-444-53860-4.00018-0

current perspectives on primate brain evolution, especially in humans, and to examine some hypothetical organizing principles that underlie the brain's complex organization. Some of the design principles and operational modes that underlie the information processing capacity of the cerebral cortex in primates will be explored, and it will be argued that with the evolution of the human brain we have nearly reached the limits of biological intelligence.

Evolution and geometry of the cerebral cortex

The evolution of the brain in mammals has been accompanied by a reorganization of the brain as a result of differential growth of certain brain regions. Consequently, the geometry of the brain, and especially the size and shape of the cerebral cortex, has changed notably since the late Cretaceous (Jerison, 1973). The evolutionary expansion of the cerebral cortex, indeed, is among the most distinctive morphological features of mammalian brains. Particularly in species with large brains, and most notably in great apes and marine mammals, the brain becomes disproportionately composed of this cortical structure (Nieuwenhuys, 1994a,b; Northcutt and Kaas, 1995; Striedter, 2004; Welker, 1990; Fig. 1).

Comparisons among mammals show that the surface area of the cortical sheet varies by more than 5 orders of magnitude, while the thickness of the cortex varies by less than one order of magnitude (Allman, 1990; Hofman, 1989; Welker, 1990). Therefore, evolutionary changes in the cerebral cortex have occurred mainly parallel to the cortical surface (tangentially) and have been sharply constrained in the vertical (radial) dimension, which makes it especially well suited for the elaboration of multiple projections and mapping systems. A mosaic of functionally specialized areas have indeed been found in the mammalian cortex, some of the functions being remarkably diverse (Kaas, 1993, 2008; Krubitzer, 1995, 2007). At the lower processing levels of the cortex, these maps bear a fairly simple topographical relationship to the world, but in higher areas precise topography is sacrificed for the mapping of more abstract functions. Here, selected aspects of the sensory input are combined in ways that are likely to be relevant to the animal.

Using modern anatomical tracing methods, physiological recordings, and mapping studies, it has been established that each sensory modality is mapped several times in different areas, with about a dozen representations of the visual world and a half a dozen each of auditory inputs and somatosensory sensations. In fact, the maps differ in the attributes of the stimulus represented, in how the field is emphasized, and in the types of computations performed. Clearly, the specifications of all these representations mean that functional maps can no longer be considered simply as hard-wired neural networks. They are much more flexible than previously thought and are continually modified by feedback and lateral interactions. These dynamic changes in maps, which seem likely to result from local interactions and modulations in the cortical circuits, provide the plasticity necessary for adaptive behavior and learning. Although species vary in the number of cortical areas they possess, and in the patterns of connections within and between areas, the structural organization of the primate neocortex is remarkably similar. It has been estimated, for example, that within 1 mm^3 of human cortex there are about 50,000 neurons that contain 150 m of dendrites and 100 m of axons and that these neurites have about 50×10^6 synapses (Cherniak, 1990).

Neural mechanisms of cortical folding

It is now well established that the cerebral cortex forms as a smooth sheet populated by neurons that proliferate at the ventricular surface and migrate outward along radial glial fibers (for reviews, see Cheung et al., 2007; Rakic, 2009). Why then does the cortex remain smooth

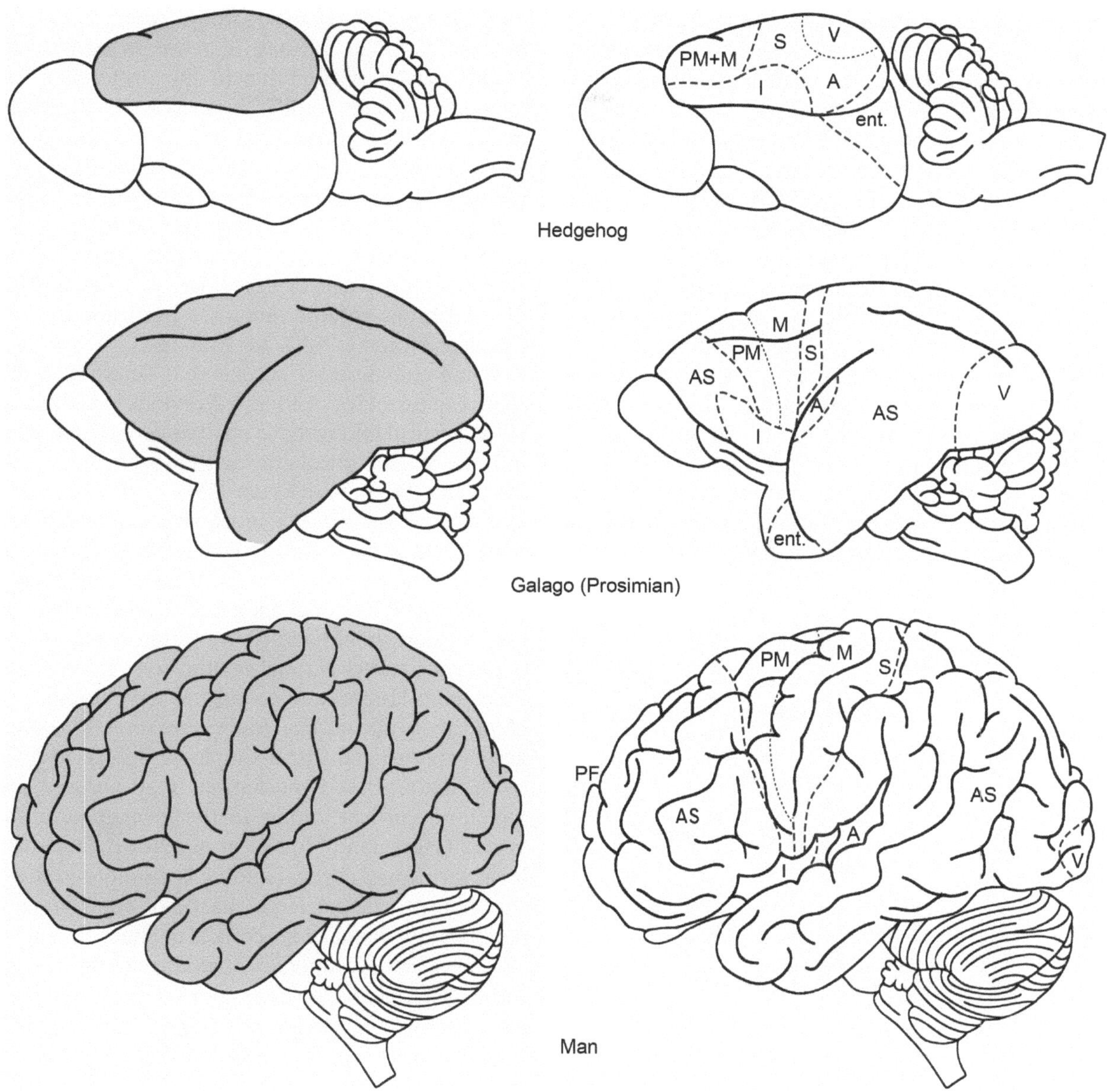

Fig. 1. Lateral views of the brains of some mammals to show the evolutionary development of the neocortex (gray). In the hedgehog almost the entire neocortex is occupied by sensory and motor areas. In the prosimian *Galago* the sensory cortical areas are separated by an area occupied by association cortex (AS). A second area of association cortex is found in front of the motor cortex. In man these anterior and posterior association areas are strongly developed. A, primary auditory cortex; AS, association cortex; Ent, entorhinal cortex; I, insula; M, primary motor cortex; PF, prefrontal cortex; PM, premotor cortex; S, primary somatosensory cortex; V, primary visual cortex. Modified with permission from Nieuwenhuys (1994b). (For color version of this figure, the reader is referred to the Web version of this chapter.)

(lissencephalic) in some species, particularly those with small brains, yet become highly convoluted (gyrencephalic) in others, particularly those with large brains? The primary reason is that, from a brain volume of about 3–4 cm^3 onward, the surface area of the cortex increases disproportionally with brain size (Hofman, 1989; Fig. 2).

In mammals with convoluted brains, among which are almost all primates, the cortical surface area, rather than being proportional to the two-thirds power of geometric similarity, is nearly a linear function of brain volume (Hofman, 1985a, 1989). It means that if a mouse brain (volume = 0.5 cm^3) were scaled up as the two-thirds power to the size of the human brain (volume = 1400 cm^3), it would have a cortical surface of only about 480 cm^2. The actual surface area of the human cortex, however, is about 2000 cm^2, which is more than four times larger than would be predicted assuming geometric similarity, indicating

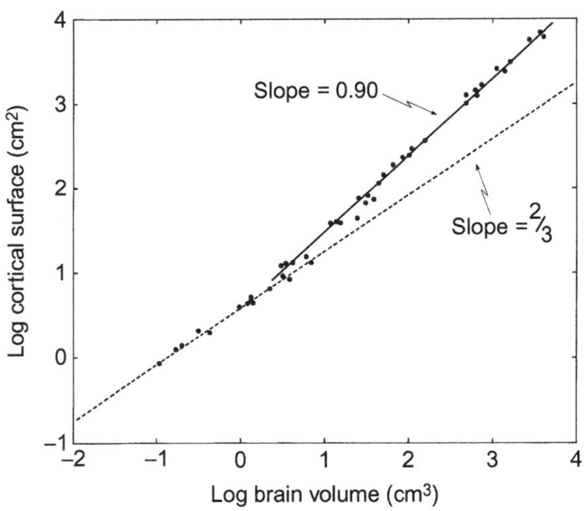

Fig. 2. Total cortical surface area as a function of brain volume in terrestrial mammals. Logarithmic scale. The slope of the regression line is 0.90 ± 0.012, representing the surface–volume relationship for convoluted brains. Note that the cortical surface area of species with convoluted brains (area > 10 cm^2), as in most primates, is nearly a linear function of brain volume, rather than being proportional to the 2/3 power of geometric similarity. Modified with permission from Hofman (1991).

that mammalian brains change their shape by becoming folded as they increase in size. Most of this bias is attributable to the preference for tangential versus radial expansion. Differences in the duration of neurogenesis, which increases more rapidly with brain size for the cerebral cortex than for subcortical areas (Finlay et al., 2001; Rakic, 1995), lead to a systematic increase in the ratio of the cortical to subcortical regions. The volume of gray matter expressed as a percentage of total brain volume increases from about 25% for insectivores to 50% for humans.

When convolutions occur, what determines the spatial pattern of folding? Previous hypotheses about cortical folding have emphasized mechanisms intrinsic to the cortical gray matter (for a review, see Hofman, 1989). Van Essen (1997) suggested that extrinsic factors are the more important and that tension along axons in the white matter is the primary driving force for cortical folding. By keeping the aggregate length of axonal and dendritic wiring low, tension should contribute to the compactness of neural circuitry throughout the cortex. Recently, Herculano-Houzel and colleagues have found that connectivity and cortical folding are directly related across species and that a simple model based on a white matter-based mechanism may account for increased cortical folding in the primate cerebral cortex (Herculano-Houzel et al., 2010). Thus, the local wiring and cortical folding is a simple strategy that helps to fit the large sheet-like cortex into a compact space and keeps cortical connections short. An important evolutionary advantage of this design principle is that it enables brains to be more compact and faster with increasing size (Harrison et al., 2002; Karbowski, 2003).

Scaling of primate neocortex

Analysis of the cerebral cortex in anthropoid primates revealed that the volume of the neocortex is highly predictable from absolute brain size (Finlay and Darlington, 1995; Hofman, 1989, 2007; Zhang and Sejnowski, 2000). The volume of the neocortical gray matter, containing local networks of neurons that are wired by dendrites and mostly nonmyelinated axons, is basically a linear function of brain volume, whereas the mass of long-range axons, forming the underlying white matter volume, increases disproportionately with brain size (Fig. 3). As a result, the volume of gray matter expressed as a percentage of total brain volume is about the same for all anthropoid primates.

The relative white matter volume, on the other hand, increases with brain size, from 9% in pygmy marmosets (*Cebuella pygmaea*) to about 35% in humans, the highest value in primates (Hofman, 1989). The nonlinear nature of this process is further emphasized by plotting the relative volume of white matter as a function of brain size (Fig. 4). The high correlation between both variables ensures that the curve, and its confidence limits, can be used for predictive purposes to estimate the volume of white matter relative to brain volume for a hypothetical primate. The model, for example, predicts a white matter volume of about 1470 cm^3 for an anthropoid primate with a brain volume of 3000 cm^3. In other words, in such a large brained primate, white matter would comprise about half of the entire brain volume, compared to one-third in modern man.

Volumetric measurements of gray and white matter in the neocortex of anthropoid primates have shown that the "universal scaling law" of neocortical gray to white matter applies separately for frontal and nonfrontal lobes and that changes in the frontal (but not nonfrontal) white matter volume are associated with changes in other parts of the brain, including the basal ganglia, a group of subcortical nuclei functionally linked to executive control (Smaers et al., 2010). These comparative analyses indicate that the evolutionary process of neocorticalization in primates is mainly due to the progressive expansion of the axonal mass that implement global communication, rather than to the increase in the number of cortical neurons and the importance of high

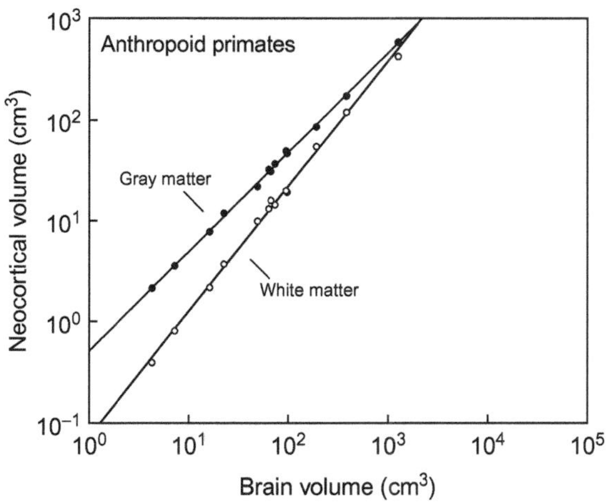

Fig. 3. Volumes of cerebral gray and white matter as a function of brain volume in anthropoid primates, including humans. Logarithmic scale. The slopes of the regression lines are 0.985 ± 0.009 (gray matter) and 1.241 ± 0.020 (white matter). Note the difference in the rate of change between gray matter (neural elements) and white matter (neural connections) as brain size increases. Reproduced with permission from Hofman (2001b).

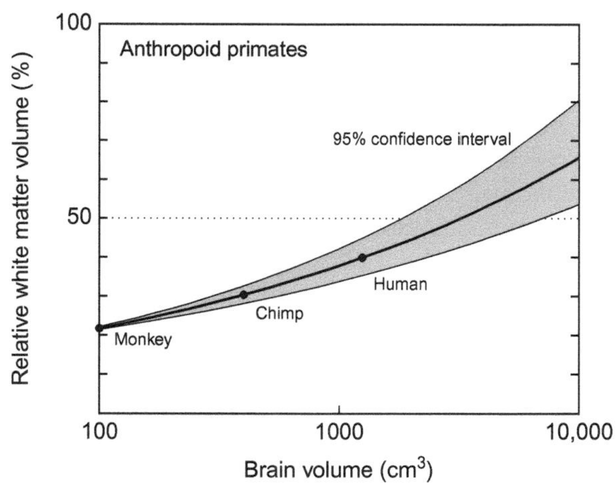

Fig. 4. Relative white matter volume as a function of brain volume in anthropoid primates. Semilogarithmic scale. The proportion of white matter increases with brain size, from 22% in a monkey brain of 100 cm^3 to about 65% in a hypothetical primate with a brain size of 10,000 cm^3. Modified with permission from Hofman (2001b).

neural connectivity in the evolution of brain size in anthropoid primates.

Wen and Chklovskii (2005) have shown that the competing requirements for high connectivity and short conduction delay may lead naturally to the observed architecture of the mammalian neocortex. Obviously, the brain functionally benefits from high synaptic connectivity and short conduction delays. A magnetic resonance imaging study, further, focusing specifically on the prefrontal cortex, has shown that the volume of the white matter underlying prefrontal areas is disproportionately larger in humans than in other primates (Schoenemann et al., 2005). It suggests that the connectional elaboration of the prefrontal cortex, which mediates such important behavioral domains as planning, aspects of language, attention, and social and temporal information processing, has played a key role in human brain evolution.

Fractal geometry of the primate brain

Many objects in nature, including snowflakes, coastlines, clouds, bronchial trees, and neural networks, are neither completely random nor completely ordered (Mandelbrot, 1982; West and Goldberger, 1987). Among these disorderly objects the mammalian brain has long been a paradigm of biological complexity, challenging comparative morphologists to capture the richness of its geometry in mathematical models (e.g., Changizi, 2001, 2007; Cherniak, 1995; Cherniak et al., 2004; Chklovskii et al., 2002; Hofman, 1989; Jerison, 1979). During the past decades, considerable progress has been made in identifying and formalizing empirical regularities in the diversity of brain design. One of the major outcomes of these investigations is that brains are geometrically constrained and that the central nervous system in mammals could evolve within the context of a limited number of potential forms.

From these studies it also became evident that, whereas in small brained species the cortical volume expands by virtue of a combined increase in surface area and cortical thickness, the increase of the cortical volume in larger species is almost entirely due to a disproportionate expansion of the cortical surface area. It is the increase of the cortical surface area beyond that expected for geometrically similar objects of different volumes, as we have seen, which creates the need to cortical folding (Hofman, 1989; Jerison, 1982; Todd, 1986). Consequently, the brains of larger species, like primates, are convoluted and irregular and are not well described by the ideal constructs of Euclidean geometry. Mandelbrot (1982) coined the word "fractal" to identify this group of complex geometric forms and developed the concept of fractal scaling to describe their organized variability. An important feature of fractal objects is that they are invariant, in a statistical sense, over a wide range of scales, a property that is known as scaling.

Fractal scaling and convoluted brains

The principal idea underlying scaling is that, although biological systems may evolve by rules distinct from these governing the development of a physical system, they cannot violate basic physical principles. One of the standard problems of classical scaling is that for any series of similar objects the surface area is proportional to the square of a length dimension, whereas the volume is proportional to the cube. According to this geometric principle, also known as Galileo's principle of "similitude," surface area is proportional to the two-thirds power of volume, or in its generalized form

$$\text{area} = k(\text{volume})^{D/3} \tag{1}$$

where k is a scaling constant and $D=2$, the topological or Euclidean dimension of geometric similarity.

To determine the surface dimension D for mammalian brains, the empirical exponents were related to the generalized surface–volume relation

Table 1. Allometric scaling of the cerebral cortex against brain volume in mammals with convoluted brains

Covariate	(Sub)Orders	Species	Intercept, b	Standard major axis, α ($\pm$SD)	r
Outer cortical surface area (cm^2)	7	18	4.47	0.726$\pm$0.021	0.994
Total cortical surface area (cm^2)	9	23	3.77	0.901$\pm$0.022	0.994
Mean cortical thickness (cm)	7	16	0.106	0.129$\pm$0.019	0.832
Cortical volume (gray matter, cm^3)	7	16	0.520	0.982$\pm$0.017	0.998
Cortical volume (white matter, cm^3)	7	16	0.046	1.280$\pm$0.045	0.991
Cortical volume (total, cm^3)	9	21	0.396	1.099$\pm$0.028	0.994

The model is $\log y = \log b + \alpha \log x$, where x is the variate (i.e., brain volume in cm^3), y is the covariate in centimeter units, $\log b$ is the log-y-intercept at $\log x = 0$, and α is the slope of the standard major axis. The strength of the relationship is reflected by Pearson's correlation coefficient, r. Reproduced with permission from Hofman (1991).

(Eq. (1)). The allometric equations are given in Table 1. The surface dimension of nonconvoluted brains coincides with the Euclidean dimension (Hofman, 1985a), indicating that mammals with a smooth cerebral cortex satisfy the geometric scaling model, very similar to the spherical surface area of an Euclidean hemisphere, where area= 3.84 $(\text{volume})^{2/3}$ (see Fig. 2). Convoluted brains, on the other hand, with their surface dimension of $D=2.70$, have a fractal dimension far above the expected value of standard dimensional analysis (Table 2; Hofman, 1991).

By Mandelbrot's definition, a fractal is any object or process of which the dimension, given by the equation

$$D = \log N / \log(1/r) \qquad (2)$$

strictly exceeds its topological or Euclidean dimension. It means that every set with a non-integer D, as in convoluted brains, is a fractal. To derive this formula for dimension, consider a straight line of unit length divided into N segments of length r (r=ratio vector). By definition, $N \times r = 1$ and so $r = 1/N$. Analogously, a unit square contains N pieces of edge length r in an area of 1, so $r = 1/(N^{1/2})$ and a unit cube N pieces of edge length r in a volume of 1, so $r = 1/(N^{1/3})$. Denoting dimension by D, the relationship can be generalized to $r = 1/(N^{1/D})$, and the equation can be solved for the fractal dimension D.

Table 2. The fractal geometry of convoluted brains

Cortical parameter	Topological dimension	Empirical dimension	Difference (p-value)
Cortical thickness	1	0.39$\pm$0.057	$<$0.001
Outer cortical surface	2	2.18$\pm$0.062	$<$0.02
Total cortical surface	2	2.70$\pm$0.065	$<$0.001
Cortical gray matter	3	2.95$\pm$0.050	NS
Cortical white matter	3	3.84$\pm$0.134	$<$0.001
Total cortical volume	3	3.30$\pm$0.084	$<$0.01

The model is based on the equation: $y = bx^{D/3}$, where y is the cortical structure, x is brain volume, b is an allometric constant, and D is the topological (Euclidean), respectively, empirical (fractal) dimension of the structure. The fractal dimensions are given as mean$\pm$SD. NS, not significant. Reproduced with permission from Hofman (1991).

Dimension D is called the fractal dimension because it is not necessarily integer.

In general, D is the number that tells us something about the overall structure and complexity of an object. The empirical area–volume relation, for example, found for convoluted brains indicates that the cortical surface is partly space-filling and that its surface area fractally evolves into a volume, or that its volume, by fractal folding, attains the properties of an area (Hofman, 1991). We will return to the neural principles that may account for the fractal geometry for convoluted brains in the next section.

Fractal scaling of neocortex

From anatomical studies, we know that the primate neocortex is made up of distinct neural networks, which are organized in columnar arrays stretched out through the depth of the cortex (Buxhoeveden and Casanova, 2002a; Goldman-Rakic, 1984; Jones, 1981; Mountcastle, 1978; Powell, 1981; see Chapter 10). These modules, in turn, are linked together into larger neural networks by coalescing adjacent sets of columnar units. These networks, which also have a columnar arrangement, are found through the cerebral cortex (Buxhoeveden and Casanova, 2002a,b; Rakic, 2002, 2007), so that we may consider columnar arrangements of neural elements as a general organizational framework of the primate cortex.

The arrangements of these clusters of neurons in vertical columns perpendicular to the pial surface, however, leads to a geometric dilemma with the evolutionary expansion of the brain. It is this dilemma—the requirement of the cortical surface area to keep abreast with the volumetric growth of the brain itself—which creates the need for cortical folding. Studies in mammals have shown that in species with convoluted brains the mass of interconnective nerve fibers, forming the underlying white matter, is proportional to the 1.28 power of brain volume (Hofman, 1988, 1991; see also Table 1), meaning that the cortical white matter is a fractal system. As a result, the total cortical surface area (S_{cortex}), including all gyri and sulci, scales approximately as the two-thirds power of the white matter volume (V_{white}) or

$$S_{\mathrm{cortex}} = 36.4(V_{\mathrm{white}})^{2/3} \qquad (3)$$

In other words, the surface area of the cerebral cortex, and with that the total number of neuronal columns, is geometrically similar with the amount of white matter, that is, with the number and length of the interconnective nerve fibers. In small species with nonconvoluted brains, a similar relationship was found between the cortical surface area and the mass of myelinated nerve fibers (Hofman, 1991).

Apparently, the fractal geometry of the mammalian brain is a consequence of the design of the cerebral cortex, in which each cortical module, containing a large number of neurons, is connected to its environment by a specific number of axons. Here, we have an analogy to Rent's rule for computer geometry (Landman and Russo, 1971), which says that the number of components (C) in each module of complex computer circuits is related to the number of terminals (T) according to $T \sim C^{D/D_{\mathrm{T}}}$ where D is the fractal dimension and D_{T} is the Euclidean dimension. In a spatial circuit, where the components are organized in columnar units, while they are in contact with the outside by their surfaces, as in the cortex, $D_{\mathrm{T}}=3$ and D is somewhere between 2 and 3. The ratio D/D_{T} increases with the degree of parallelism that is present in the design (Landman and Russo, 1971; Mandelbrot, 1982).

Therefore, a fractal dimension of $D=2.70$, as found for convoluted brains, suggests a high degree of parallel processing to take place in the cerebral cortex (see, e.g., Ballard, 1986) and emphasizes the processing and/or transfer of information across cortical regions in highly corticalized mammals, such as monkeys and apes, rather than within regions. To reach the state of integral parallelism in which each neural component has its own terminal, the length and number of the interconnective axons must be reduced in order to set limits to the axonal mass.

Design principles of neural organization

Explaining how the brain works requires information about how it is organized structurally. We have seen that the evolutionary changes in the cerebral cortex have occurred mainly parallel to the cortical surface (tangentially) and have been sharply constrained in the vertical (radial) dimension. This tremendous increase in the cortical surface without a comparable increase in its thickness during mammalian evolution has been explained in the context of the radial-unit hypothesis of cortical development (for reviews, see Rakic, 2007, 2009). According to this model, neocortical expansion is the result of changes in proliferation kinetics that increase the number of radial columnar units without changing the number of neurons within each unit significantly. Therefore, the evolutionary expansion of the neocortex in primates is mainly the result of an increase in the number of radial columns.

If the neuron can be regarded as the "atomic" unit of function in the nervous system, then the "molecular" unit of information processing is in a way akin to the neuronal network. In particular, the primate neocortex has been found to be uniformly organized and to be composed of such neural processing units interacting over fairly short distances (Buxhoeveden and Casanova, 2002a; Cherniak, 1995; Douglas and Martin, 2004; Hofman, 1985b, 2007; Mountcastle, 1978, 1997). It appears that the module for information processing in the neocortex consists of a functional neuronal micronetwork with a columnar structure that has the capability of quite sophisticated spatial-temporal firing patterns. These processing units operate as prewired neural assemblies where individual neurons are configured to execute complex transactions.

Their widespread occurrence, further, qualifies them to be considered as fundamental building blocks in neural evolution (for reviews, see Buxhoeveden and Casanova, 2002b; Mountcastle, 1997; Rockland, 2010). Neocortical columns are functional and morphological units. It has become evident that these cortical circuits integrate at higher levels of information processing, as a result of the hierarchical organization of the brain, thus enabling the system to combine dissimilar views of the world. It implies that if we seek the neural basis of biological intelligence, including mind-like properties and consciousness, we can hardly localize it in a specific region of the brain, but must suppose it to involve all those regions through whose activity an organism is able to construct an adequate model of its external world, perhaps it may even encompass the entire neo- and subcortical network.

Cortical circuits: Architecture and wiring

It is evident that these neocortical columns are functional and morphological units whose architecture may have been under selective evolutionary pressure in different mammalian lineages in response to encephalization and specializations of cognitive abilities. We are beginning to understand some of the geometric, biophysical, and energy constraints that have governed the evolution of these neural networks (e.g., Chklovskii et al., 2002; Felleman and Van Essen, 1991; Klyachko and Stevens, 2003; Laughlin and Sejnowski, 2003; Rockland, 2010). To operate efficiently within these constraints, nature has optimized the structure and function of these processing units with design principles similar to those used in electronic devices and communication networks. In fact, the basic structural uniformity of the cerebral cortex suggests that there are general architectural principles governing its growth and evolutionary development (Cherniak, 1995; Hofman, 2001a, 2007; Rakic, 1995).

It has been postulated that these neural processing units have spatial dimensions depending on the number of local circuit neurons and that both the number and size of cortical modules increase with increasing brain size (Hofman, 1985b, 1991; Prothero, 1997). In the 1980s, stereological investigations in mammals already indicated that the number of cortical neurons beneath a given surface area of cortex is not the same in all species, especially in those with large brains (see, e.g., Haug, 1987). Recent studies in primates have shown that the number of neurons underneath a unit area of cortical surface is not constant and varies linearly with neuronal density, a parameter that is neither related to cortical size nor to the total number of neurons (Herculano-Houzel, 2009; Herculano-Houzel et al., 2008; Wang et al., 2008). These studies indicate that the cortical column varies both in size and number of neurons, which is in accordance with predictions based on computational models (Hofman, 1985b).

Comparative studies further indicate that variability in subtle subcomponents of the columnar organization in human and nonhuman primates, such as the composition of the interneuron subtypes, are a primary source of interspecific differences in minicolumn morphology among species (Raghanti et al., 2010). Humans deviate from other primates in having a greater width of minicolumns in specific cortical areas, especially in the prefrontal cortex, owing to constituents of the peripheral neuropil space (Buxhoeveden and Casanova, 2002a; Semendeferi et al., 2011). These findings support the idea (Allen, 2009; Semendeferi et al., 2002; Chapter 9) that human evolution, after the split from the common ancestor with chimpanzees, was accompanied by discrete modifications in local circuitry and interconnectivity of selected parts of the brain. The differences in columnar diameter among primates, however, are only minor compared to the dramatic variation in overall cortex size. Thus, it seems that the main cortical change during evolution has presumably been an increase in the number, rather than the size, of these neural circuits.

Although the details of the interpretation of the columnar organization of the neocortex are still controversial (for recent reviews, see Da Costa and Martin, 2010; Rockland, 2010; Chapter 10), it is evident that the primate neocortex is characterized by the hierarchical organization of groups of neurons. To group neurons into clusters interacting over relatively short distances allows these groups to inform as many adjacent clusters of neurons about the state of the "emitting" cluster with as little as possible redundancy of information. Figure 5 shows a simple schematic diagram illustrating the effect of increasing the number of functional cortical units on the number of interconnections. When the units are connected to all others by separate fibers and when each additional unit becomes connected with each of the already existing ones, then the number of connections (C) is related to the number of units (U) according to the equation: $C=U(U-1)$, which is nearly equivalent to $C=U^2$. In such a system, the number of connections increases much faster than the number of units. Generally, the growth of connections to units is a factorial function of the number of units in a fully connected network and a linear function of the number of units in a minimally connected network.

Neural networks and cognition

Recently, we have shown that in species with convoluted brains the fraction of mass devoted to

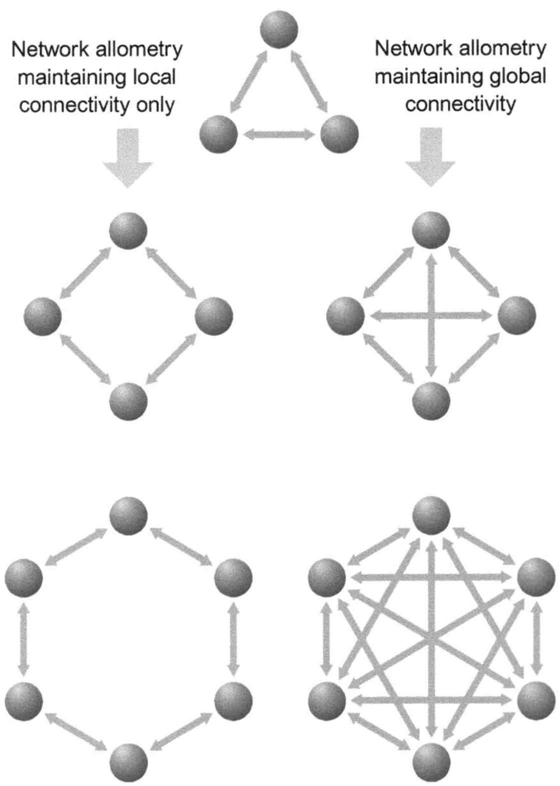

Fig. 5. The problem of network allometry is represented by two neural circuits that exhibit local and global connectivity, respectively. These diagrams depict that the number of connections (C) grows much faster than the number of units (U) in a fully connected network: $C=U(U-1)$ than in a binary system, where the growth of connections is a linear function of the number of units. Reproduced with permission from Hofman (2008).

wiring seems to increase much slower than that needed to maintain a high degree of connectivity between the modular units (Hofman, 2003, 2007). These findings are in line with a model of neuronal connectivity (Deacon, 1990; Ringo, 1991) which says that as brain size increases there must be a corresponding fall in the fraction of neurons with which any neuron communicates directly. The reason for this is that if a fixed percentage of interconnections is to be maintained in the face of increased neuron number, then a large fraction of any brain size increase would be spent maintaining such degree of wiring, while the increasing axon length would reduce neural computational speed (Ringo et al., 1994). The human brain, for example, has an estimated interconnectivity of the order of 10^3, based on data about the number of modular units and myelinated nerve fibers (Fig. 6). This implies that each cortical module is connected to a thousand other modules and that the mean number of processing steps, or synapses, in the path interconnecting these modules, is about two.

Recently, Herculano-Houzel et al. (2010) have shown that in primates the mass of the white matter scales linearly across species with its number of nonneuronal cells, which is expected to be proportional to the total length of myelinated axons in the white matter. They found that the surface area of the white matter increases with the number of neurons in the gray matter, N, according to $N^{0.87}$, not $N^{1.0}$, indicating that connectivity decreases in larger cerebral cortices as a slowly diminishing fraction of neurons. Decreased connectivity in the brain is compatible with previous suggestions that neurons in the cerebral cortex are connected as a small-world network and should slow down the increase in global conduction delay in cortices with larger numbers of neurons (Sporns et al., 2004, 2007; Wang et al., 2008; Fig. 7).

Once the brain has grown to a point where the bulk of its mass is in the form of connections, then further increases (as long as the same ratio in interconnectivity is maintained) will be unproductive. Increases in number of units will be balanced by decreased performance of those units due to the increased conduction time. This implies that large brains may tend to show more specialization in order to maintain processing capacity. Indeed, an increase in the number of distinct cortical areas with increasing brain size has been reported (Kaas, 2000; Striedter, 2004; Welker, 1990). It may even explain why large-brained species may develop some degree of brain lateralization as a direct consequence of size. If there is evolutionary pressure on certain functions that require a high degree of local processing and sequential control, such as linguistic communication in human brains, these will have a strong tendency to develop in one hemisphere (Aboitiz et al., 2003; Ringo et al., 1994).

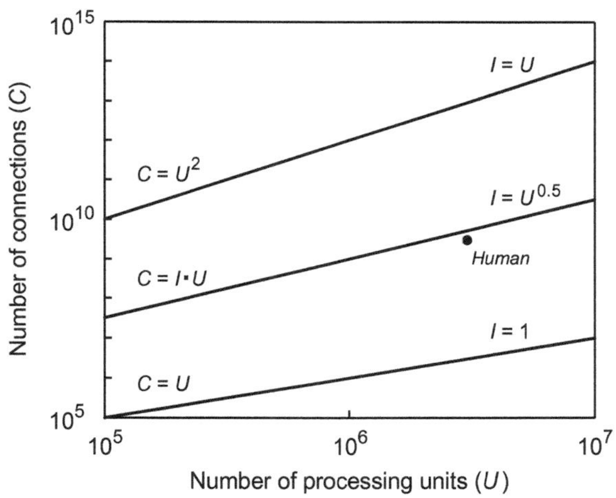

Fig. 6. Number of connections (C) as a function of the number of processing units (U) in a neural network. Logarithmic scale. In a binary system, with an interconnectivity of $I=1$ the growth of connections is a linear function of the number of units, and in a fully connected network ($I=U$) both variables are related according to a power function (cf. Fig. 5). The human cerebral cortex, with an estimated interconnectivity of about 10^3, lies somewhere between these extremes, and close to the line for $I=U^{0.5}$. It implies that in humans the number of myelinated axons scales to the 1.5 power of the number of modular processing units. Reproduced with permission from Hofman (2001b).

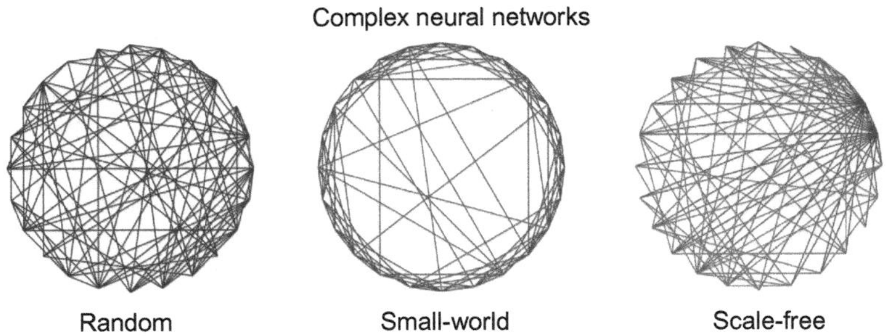

Fig. 7. Organizational principles of random, small-world, and scale-free networks. *Structural* cortical networks are neither completely connected with each other nor randomly linked, instead their connections have small-world attributes with path lengths that are close to those of equivalent random networks but with significantly higher degrees of local clustering. *Functional* cortical networks, on the other hand, exhibit both scale-free attributes with power law degree distributions, as well as small-world attributes. Modified with permission from Sporns et al. (2004). (For color version of this figure, the reader is referred to the Web version of this chapter.)

Biological limits to information processing

The primate brain has evolved from a set of underlying structures that constrain its size, and the amount of information it can store and process. If the ability of an organism to process information about its environment is a driving force behind evolution, then the more information a system, such as the brain, receives, and the faster it can process this information, the more adequately it will be able to respond to environmental challenges and the better will be its chances of survival (Hofman, 2003). The limit to any intelligent system, therefore, lies in its abilities to process and integrate large amounts of sensory information and to compare these signals with as many memory states as possible, and all that in a minimum of time. It implies that the functional capacity of a neuronal structure is inherently limited by its neural architecture and signal processing time (see, e.g., Changizi and Shimojo, 2005; Hofman, 2001a; Laughlin and Sejnowski, 2003).

The processing or transfer of information across cortical regions, rather than within regions, in large-brained primates can only be achieved by reducing the length and number of the interconnective axons in order to set limits to the axonal mass (Fig. 8). The *number* of interconnective fibers can be reduced, as we have seen, by compartmentalization of neurons into modular circuits in which each module, containing a large number of neurons, is connected to its neural environment by a small number of axons. The *length* of the interconnective fibers can be reduced by folding the cortical surface and thus shortening the radial and tangential distances between brain regions. Local wiring—preferential connectivity between nearby areas of the cortex—is a simple strategy that helps keep cortical connections short. In principle, efficient cortical folding could further reduce connection length, in turn reducing white matter volume and conduction times (Chklovskii et al., 2004; Scannell et al., 1995; Young, 1993). Thus, the development of the cortex does seem to coordinate folding with connectivity in a way that could produce smaller and faster brains.

Recently, Wang et al. (2008) have shown that there are functional trade-offs in white matter axonal scaling in mammals. They found that the composition of white matter shifts from compact, slow-conducting, and energetically expensive unmyelinated axons to large, fast-conducting, and energetically inexpensive myelinated axons. The fastest axons have conduction times of 1–5 ms

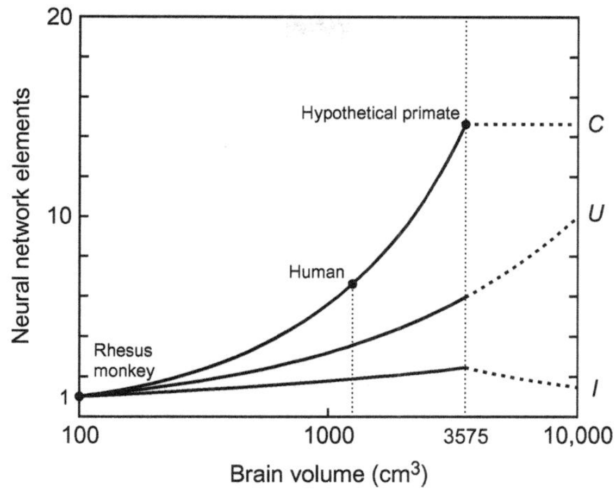

Fig. 8. The number of connections (*C*), cortical processing units (*U*), and level of interconnectivity (*I*) in the primate neocortex as a function of brain size. Semilogarithmic scale. Values are normalized to one at a brain volume of $100\,cm^3$, the size of a monkey brain. Note that the number of myelinated axons increases much faster than the number of cortical processing units (see also Fig. 5). The human cerebrum, for example, contains $6\times$ more myelinated axons than that of a rhesus monkey, whereas the number of cortical processing units is only three times larger. Dashed lines show the potential evolutionary pathway of these neural network elements in primates with very large brains, that is, beyond the hypothetical upper limit of the brain's processing power (see text and Fig. 9). Note that a further exponential growth in the number of cortical processing units, without a corresponding increase in the axonal mass, will lead to a decrease in connectivity between these units and thus to more local wiring.

across the neocortex and <1 ms from the eye to the brain, suggesting that in select sets of communicating fibers, large brains reduce transmission delays and metabolic firing costs at the expense of increased volume. Delays and potential imprecision in cross-brain conduction times are especially great in unmyelinated axons, which may transmit information via firing rate rather than precise spike timing. In the neocortex, axon size distributions can account for the scaling of per-volume metabolic rate and suggest a maximum supportable firing rate, averaged across all axons, of 7 ± 2 Hz. Clearly, the white matter architecture must follow a limited energy budget to optimize both volume and conduction time.

Another way to keep the aggregate length of axonal and dendritic wiring low, and with that the conduction time and metabollic costs, is to increase the degree of cortical folding. A major disadvantage of this evolutionary strategy, however, is that an increase in the relative number of gyri can only be achieved by reducing the gyral width. At the limit, the neurons in the gyri would be isolated from the remainder of the nervous system, since there would no longer be any opening for direct contact with the underlying white matter. Prothero and Sundsten (1984) therefore introduced the concept of the gyral "window," which represents the hypothetical plane between a gyrus and the underlying white matter through which nerve fibers running to and from the gyral folds must pass. According to this hypothesis, there would be a brain size where the gyral "window" area has an absolute maximum. A further increase in the size of the brain beyond that point, that is, at $2800\,cm^3$, would increase the cortical surface area, but the "window" would decrease, leading to a lower degree of neuronal integration and an increase in response time.

The remarkably high correlation between gray matter, white matter, and brain size in anthropoid primates ensures that the proposed model can be used for predictive purposes to estimate the volume of white matter relative to brain volume for a hypothetical primate (Hofman, 2001b). Model studies of the growth of the neocortex at different brain sizes, using a conservative scenario, revealed that with a brain size of about $3500\,cm^3$ the total volume of the subcortical areas (i.e., cerebellum, brain stem, diencephalon, etc.) reaches a maximum value (Fig. 9). Increasing the size of the brain beyond that point, following the same design principle, would lead to a further increase in the size of the neocortex, but to a reduction of the subcortical volume. Consequently, primates with very large brains (e.g., over 5 kg) may have a declining capability for neuronal integration despite their larger number of cortical neurons.

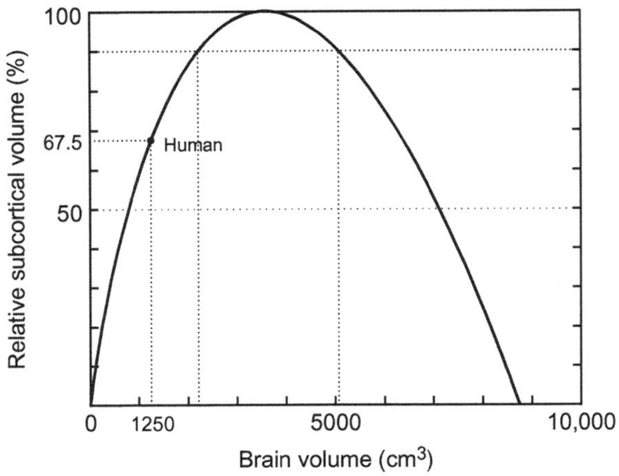

Fig. 9. Relative subcortical volume as a function of brain volume. The predicted subcortical volume (i.e., brain volume—predicted neocortex volume) must be zero at zero brain size. Likewise, the subcortical volume will be zero when the brain is exclusively composed of cortical gray and white matter. At a brain size of 3575cm^3, the subcortical volume has a maximum (see also Fig. 8). The maximum simulated value for the subcortical volume (366cm^3) is then taken as 100%. The larger the brain grows beyond this critical size, the less efficient it will become. Assuming constant design, it follows that this model predicts an upper limit to the brain's processing power. Modern humans are at about two-third of that maximum. Modified with permission from Hofman (2001b).

Limits to human brain evolution

A progressive enlargement of the hominid brain started by about 2–2.5 million years ago, probably from a bipedal, australopithecine form with a brain size comparable to that of a modern chimpanzee (see, e.g., Falk, 2004, 2007; Robson and Wood, 2008). Since then, a threefold increase in endocranial volume has taken place, leading to one of the most complex and efficient structures in the animated universe, the human brain. The human brain contains about 100 billion neurons, more than 100,000km of interconnections, and has an estimated storage capacity of 1.25×10^{12} bytes (Cherniak, 1990; Hofman, 2001a). In view of the central importance placed on brain evolution in explaining the success of our species, one may wonder whether there are physical limits that constrain its processing power and evolutionary potential.

The human brain has evolved from a set of underlying structures that constrain its size, and the amount of information it can store and process. In fact, there are a number of related factors that interact to limit brain size, factors that can be divided into two categories: (1) energetic constraints and (2) neural processing constraints (see, e.g., Herculano-Houzel, 2009; Wang et al., 2008).

Energetic limits

The human brain generates about 15 W in a well-insulated cavity of about 1500cm^3. From an engineering point of view, removal of sufficient heat to prevent thermal overload could be a significant problem. But the brain is actively cooled by blood and not simply by heat conduction from the surface of the head. So the limiting factor is how fast the heat can be removed from the brain by blood flow. It has been suggested by Falk (1990) and others that the evolution of a "cranial radiator" in hominids helped provide additional cooling to delicate and metabolically expensive parts of the brain, such as the cerebral cortex. This vascular cooling mechanism would have served as a "prime releaser" that permitted brain size to increase dramatically during human evolution. So, to increase cooling efficiency in a larger brain, either the blood must be cooler when it first enters the structure or the flow-rate must be increased above current levels.

Another factor related to blood flow has to do with the increasing energy requirements of a larger brain, a problem that is exacerbated by the high metabolic cost of this organ. It is unlikely, however, that the rate of blood flow or the increasing volume used by the blood vessels in the brain—in human about 4%—constrain its potential size. A bigger brain is metabolically possible because our cardiovascular system could evolve to transport more blood at greater pressure to meet the increased demand. This should not be taken to imply that thermal and metabolic mechanisms play no role at all in setting limits to brain size. Ultimately, energetic considerations will dictate and restrict the size of any neuron-based system, but as theoretical analyses indicate, thermal and metabolic factors alone are unlikely to constrain the potential size of our brain until it has increased to at least 10 times its present size (Cochrane et al., 1995).

Neural-processing limits

The limit to any neural system lies in its ability to process and integrate large amounts of information in a minimum of time and therefore its functional capacity is inherently limited by its neural architecture and signal processing time. The scaling model of the geometry of the neocortex, for example, predicts an absolute upper limit to primate brain size (Hofman, 2001b; Fig. 9). Without a radical change in the macroscopic organization of the brain, however, this hypothetical limit will never be approached, since at that point (ca. 8750cm^3) the brain would consist entirely of cortical neurons, and their interconnections, leaving no space for any other brain structure.

Cochrane et al. (1995) looked at the different ways in which the brain could evolve to process more information or work more efficiently. They argue that the human brain has (almost) reached the limits of information processing that a neuron-based system allows and that our evolutionary potential is constrained by the delicate balance maintained between conduction speed, pulse width, synaptic processing time, and neuron density. By modeling the information processing capability per unit time of a human-type brain as a function of interconnectivity and axonal conduction speed, they found that the human brain lies about 20–30% below the optimal, with the optimal processing ability corresponding to a brain about twice the current volume. Any further enhancement of human brain power would require a simultaneous improvement of neural organization, signal processing, and thermodynamics. Such a scenario, however, is an unrealistic biological option and must be discarded because of the trade-off that exists between these factors.

Of course, extrapolations based on brain models, such as the ones used in the present study, implicitly assume a continuation of brain developments that are on a par with growth rates in the past. One cannot exclude the possibility of new structures evolving in the brain, or a higher degree of specialization of existing brain areas, but within the limits of the existing "Bauplan" there does not seem to be an incremental improvement path available to the human brain. At a brain size of about 3500cm^3, corresponding to a brain volume two to three times that of modern man, the brain seems to reach its maximum processing capacity. The larger the brain grows beyond this critical size, the less efficient it will become, thus limiting any improvement in cognitive power.

Concluding remarks

The evolution of the neocortex in primates is mainly characterized by the development and multiplication of clusters of neurons which are strongly interconnected and in physical proximity. Since these clusters of neurons are organized in vertical columns, an increase in the number and complexity of the neuronal networks will be reflected by an expansion of the cortical surface area beyond that expected for geometric similar brains. As a result the cortical surface area fractally evolves into a volume with increasing brain size.

It is evident that the potential for brain evolution results not from the unorganized aggregation of neurons but from cooperative association by the self-similar compartmentalization and hierarchical organization of neural circuits and the invention of fractal folding, which reduces the interconnective axonal distances. The competing requirements for high connectivity and short conduction delay may lead naturally to the observed architecture of the primate neocortex. Obviously, the brain functionally benefits from high synaptic connectivity and short conduction delays. The design of the primate brain is such that it may perform a great number of complex functions with a minimum expenditure of energy and material both in the performance of the functions and in the construction of the system. In general, there will be a number of adequate designs for an object, which, for practical purposes, will all be equivalent.

The brains of birds and mammals, for example, differ quite considerably in their geometry and structural organization but do not differ in the performance of their biological functions. The similarity in brain design among primates, on the other hand, indicates that brain systems among related species are internally constrained and that the primate brain could only evolve within the context of a limited number of potential forms. It means that internal factors of brain design may be the primary determinants constraining the evolution of the brain and that geometric similarity among species in the functional organization of the brain may be derived from a common ancestor rather than being immediately evolved in response to specific environmental conditions.

References

Aboitiz, F., Lopez, J., & Mortiel, J. (2003). Long distance communication in the human brain: Timing constraints for interhemispheric sychrony and the origin of brain lateralization. *Biological Research*, *36*, 89–99.

Allen, J. S. (2009). *The lives of the brain: Human evolution and the organ of mind.* Cambridge, MA: Belknap Press.

Allman, J. M. (1990). Evolution of neocortex. In E. G. Jones & A. Peters (Eds.), *Cerebral cortex*, *8A*, (pp. 269–283). New York: Plenum Press.

Ballard, D. H. (1986). Cortical connections and parallel processing—Structure and function. *The Behavioral and Brain Sciences*, *9*, 67–120.

Buxhoeveden, D. P., & Casanova, M. F. (2002a). The minicolumn hypothesis in neuroscience. *Brain*, *125*, 935–951.

Buxhoeveden, D. P., & Casanova, M. F. (2002b). The minicolumn and the evolution of the brain. *Brain, Behavior and Evolution*, *60*, 125–151.

Changizi, M. A. (2001). Principles underlying mammalian neocortical scaling. *Biological Cybernetics*, *84*, 207–215.

Changizi, M. A. (2007). Scaling the brain and its connections. In J. H. Kaas (Ed.), *Evolution of nervous systems*, *3*, (pp. 167–180). New York: Academic Press.

Changizi, M. A., & Shimojo, S. (2005). Parcellation and area-area connectivity as a function of neocortex size. *Brain, Behavior and Evolution*, *66*, 88–98.

Cherniak, C. (1990). The bounded brain: Toward quantitative neuroanatomy. *Journal of Cognitive Neuroscience*, *2*, 58–66.

Cherniak, C. (1995). Neural component placement. *Trends in Neurosciences*, *18*, 522–527.

Cherniak, C., Mokhtarzada, Z., Rodriguez-Esteban, R., & Changizi, K. (2004). Global optimization of cerebral cortex layout. *Proceedings of the National Academy of Sciences of the United States of America*, *101*, 1081–1086.

Cheung, A. F., Pollen, A. A., Tavare, A., DeProto, J., & Molnár, Z. (2007). Comparative aspects of cortical neurogenesis in vertebrates. *Journal of Anatomy*, *211*, 164–176.

Chklovskii, D. B., Mel, B. W., & Svoboda, K. (2004). Cortical rewiring and information storage. *Nature*, *431*, 782–788.

Chklovskii, D. B., Schikorski, T., & Stevens, C. F. (2002). Wiring optimization in cortical circuits. *Neuron*, *34*, 341–347.

Clark, D. A., Mitra, P. P., Wang, S. S.-H., et al. (2001). Scalable architecture in mammalian brains. *Nature*, *411*, 189–192.

Cochrane, P., Winter, C. S., & Hardwick, A. (1995). *Biological limits to information processing in the human brain. Internet: http://www.cochrane.org.uk/opinion/archive/articles.phd.*

Da Costa, N. M., & Martin, K. A. C. (2010). Whose cortical column would that be? *Frontiers in Neuroanatomy*, *4*, 16 doi:10.3389/fnana.2010.00016.

Deacon, T. W. (1990). Rethinking mammalian brain evolution. *American Zoologist*, *30*, 629–705.

Douglas, R. J., & Martin, K. A. (2004). Neuronal circuits of the neocortex. *Annual Review of Neuroscience*, *27*, 419–451.

Falk, D. (1990). Brain evolution in *Homo*: The "radiator" theory. *The Behavioral and Brain Sciences*, *13*, 333–381.

Falk, D. (2004). *Braindance: New discoveries about human origins and brain evolution* (Revised and expanded ed.). Gainesville, FL: University Press of Florida.

Falk, D. (2007). Evolution of the primate brain. In W. Henke & I. Tattersall (Eds.), *Handbook of palaeoanthropology*, *2*, (pp. 1133–1162). Berlin: Springer Verlag.

Felleman, D. J., & Van Essen, D. C. (1991). Distributed hierarchical processing in the primate cerebral cortex. *Cerebral Cortex*, *1*, 1–47.

Finlay, B. L., & Darlington, D. B. (1995). Linked regularities in the development and evolution of mammalian brains. *Science*, *268*, 1578–1584.

Finlay, B. L., Darlington, D. B., & Nicastro, N. (2001). Developmental structure in brain evolution. *The Behavioral and Brain Sciences*, *24*, 263–278.

Goldman-Rakic, P. S. (1984). Modular organization of prefrontal cortex. *Trends in Neurosciences*, *7*, 419–429.

Harrison, K. H., Hof, P. R., & Wang, S. S.-H. (2002). Scaling laws in the mammalian neocortex: Does form provide clues to function? *Journal of Neurocytology*, *31*, 289–298.

Haug, H. (1987). Brain sizes, surfaces, and neuronal sizes of the cortex cerebri: A stereological investigation of man and his variability and comparison with some mammals (primates, whales, marsupials, insectivores and one elephant. *The American Journal of Anatomy*, *180*, 126–142.

Herculano-Houzel, S. (2009). The human brain in numbers: A linearly scaled-up primate brain. *Frontiers in Human Neuroscience*, *3*, 1–11.

Herculano-Houzel, S., Collins, C. E., Wong, P., Kaas, J. H., & Lent, R. (2008). The basic nonuniformity of the cerebral cortex. *Proceedings of the National Academy of Sciences of the United States of America*, *105*, 12593–12598.

Herculano-Houzel, S., Mota, B., Wong, P., & Kaas, J. H. (2010). Connectivity-driven white matter scaling and folding in primate cerebral cortex. *Proceedings of the National Academy of Sciences of the United States of America*, *107*, 19008–19013.

Hofman, M. A. (1985a). Size and shape of the cerebral cortex in mammals. Part I. The cortical surface. *Brain, Behavior and Evolution*, *27*, 28–40.

Hofman, M. A. (1985b). Neuronal correlates of corticalization in mammals: A theory. *Journal of Theoretical Biology*, *112*, 77–95.

Hofman, M. A. (1988). Size and shape of the cerebral cortex in mammals. Part II. The cortical volume. *Brain, Behavior and Evolution*, *32*, 17–26.

Hofman, M. A. (1989). On the evolution and geometry of the brain in mammals. *Progress in Neurobiology*, *32*, 137–158.

Hofman, M. A. (1991). The fractal geometry of convoluted brains. *Journal für Hirnforschung*, *32*, 103–111.

Hofman, M. A. (2001a). Evolution and complexity of the human brain: Some organizing principles. In G. Roth & M. F. Wullimann (Eds.), *Brain evolution and cognition* (pp. 501–521). New York: Wiley.

Hofman, M. A. (2001b). Brain evolution in hominids: Are we at the end of the road. In D. Falk & K. R. Gibson (Eds.), *Evolutionary anatomy of the primate cerebral cortex* (pp. 113–127). Cambridge: Cambridge University Press.

Hofman, M. A. (2003). Of brains and minds: A neurobiological treatise on the nature of intelligence. *Evolution and Cognition*, *9*, 178–188.

Hofman, M. A. (2007). Brain evolution and intelligence in primates. In S. Watanabe & M. A. Hofman (Eds.), *Integration of comparative neuroanatomy and cognition* (pp. 33–53). Tokyo: Keio University Press.

Hofman, M. A. (2008). Neural networks and cognition: An evolutionary approach. *Japanese Journal of Cognitive Neuroscience*, *10*, 1–4.

Jerison, H. J. (1973). *Evolution of the brain and intelligence.* New York: Academic Press.

Jerison, H. J. (1979). The evolution of diversity in brain size. In M. E. Hahn, C. Jensen & B. C. Dudek (Eds.), *Development and evolution of brain size—Behavioral implications* (pp. 30–60). New York: Academic Press.

Jerison, H. J. (1982). Allometry, brain size, cortical surface and convolutedness. In E. Armstrong & D. Falk (Eds.), *Primate brain evolution* (pp. 77–84). New York: Plenum Press.

Jones, E. G. (1981). Anatomy of cerebral cortex—Columnar input–output organization. In F. O. Schmitt, E. G. Worden, G. Adelman & S. Dennis (Eds.), *The organization of the cerebral cortex* (pp. 1–19). Cambridge, MA: MIT Press.

Kaas, J. H. (1993). Evolution of multiple areas and modules within neocortex. *Perspectives on Developmental Neurobiology*, *1*, 101–107.

Kaas, J. H. (2000). Why is brain size so important: Design problems and solutions as neocortex gets bigger or smaller. *Brain and Mind*, *1*, 7–23.

Kaas, J. H. (2008). The evolution of the complex sensory and motor systems of the human brain. *Brain Research Bulletin*, *75*, 384–390.

Karbowski, J. (2003). How does connectivity between cortical areas depend on brain size? Implications for efficient computation. *Journal of Computational Neuroscience*, *15*, 347–356.

Klyachko, V. A., & Stevens, C. F. (2003). Connectivity optimization and the positioning of cortical areas. *Proceedings of the National Academy of Sciences of the United States of America*, *100*, 7937–7941.

Krubitzer, L. (1995). The organization of neocortex in mammals: Are species differences really so different? *Trends in Neurosciences*, *18*, 408–417.

Krubitzer, L. (2007). The magnificent compromise: Cortical field evolution in mammals. *Neuron*, *56*, 201–208.

Landman, B. S., & Russo, R. L. (1971). On a pin versus block relationship for partitions of logic graphs. *IEEE Trends in Computation*, *20*, 1469–1479.

Laughlin, S. B., & Sejnowski, T. J. (2003). Communication in neural networks. *Science*, *301*, 1870–1874.

Lefebvre, L., Reader, S. M., & Sol, D. (2004). Brains, innovations and evolution in birds and primates. *Brain, Behavior and Evolution*, *63*, 233–246.

Macphail, E. M., & Bolhuis, J. J. (2001). The evolution of intelligence: Adaptive specializations *versus* general process. *Biological Reviews*, *76*, 341–364.

Mandelbrot, B. B. (1982). *The fractal geometry of nature.* San Francisco: Freeman.

Mountcastle, V. B. (1978). An organizing principle for the cerebral function—The unit module and the distributed system. In F. O. Schmitt (Ed.), *The mindful brain* (pp. 7–50). Cambridge, MA: MIT Press.

Mountcastle, V. B. (1997). The columnar organization of the brain. *Brain, 120*, 701–722.

Nieuwenhuys, R. (1994a). The neocortex: An overview of its evolutionary development, structural organization and synaptology. *Anatomy and Embryology, 190*, 307–337.

Nieuwenhuys, R. (1994b). The human brain: An introductory survey. *Medicamundi, 39*, 64–79.

Northcutt, R. G., & Kaas, J. H. (1995). The emergence and evolution of mammalian cortex. *Trends in Neurosciences, 18*, 373–379.

Powell, T. P. S. (1981). Certain aspects of the intrinsic organisation of the cerebral cortex. In O. Pompeiano & C. D. Marsan (Eds.), *Brain mechanisms and perceptual awareness* (pp. 1–19). New York: Raven Press.

Prothero, J. W. (1997). Cortical scaling in mammals: A repeating units model. *Journal of Brain Research, 38*, 195–207.

Prothero, J. W., & Sundsten, J. W. (1984). Folding of the cerebral cortex in mammals: A scaling model. *Brain, Behavior and Evolution, 24*, 152–167.

Raghanti, M. A., Spocter, M. A., Butti, C., Hof, P. R., & Sherwood, C. C. (2010). A comparative perspective on minicolumns and inhibitory GABAergic interneurons in the neocortex. *Frontiers in Neuronanatomy, 4*, 3 doi:10.3389/neuro.05.003.2010.

Rakic, P. (1995). A small step for the cell, a giant leap for mankind: A hypothesis of neocortical expansion during evolution. *Trends in Neurosciences, 18*, 383–388.

Rakic, P. (2002). Evolving concepts of cortical radial and areal specification. *Progress in Brain Research, 136*, 265–280.

Rakic, P. (2007). The radial edifice of cortical architecture: From neuronal silhouettes to genetic engineering. *Brain Research Reviews, 55*, 204–219.

Rakic, P. (2009). Evolution of the neocortex: A perspective from developmental biology. *Nature Reviews Neuroscience, 10*, 724–735.

Ringo, J. L. (1991). Neuronal interconnection as a function of brain size. *Brain, Behavior and Evolution, 38*, 1–6.

Ringo, J. L., Doty, R. W., Demeter, S., & Simard, P. Y. (1994). Time is of the essence: A conjecture that hemispheric specialization arises from interhemispheric conduction delay. *Cerebral Cortex, 4*, 331–343.

Robson, S. L., & Wood, B. (2008). Hominin life history: Reconstruction and evolution. *Journal of Anatomy, 212*, 394–425.

Rockland, K. S. (2010). Five points on columns. *Frontiers in Neuroanatomy, 4*, 22. doi:10.3389/fnana.2010.00022.

Roth, G., & Dicke, U. (2005). Evolution of the brain and intelligence. *Trends in Cognitive Sciences, 9*, 250–257.

Scannell, J. W., Blakemore, C. J., & Young, M. P. (1995). Analysis of connectivity in the cat cerebral cortex. *Journal of Neuroscience, 15*, 1463–1483.

Schoenemann, P. T., Sheehan, M. J., & Glotzer, I. D. (2005). Prefrontal white matter volume is disproportionately larger in humans than in other primates. *Nature Neuroscience, 8*, 242–252.

Semendeferi, K., Lu, A., Schenker, N., & Damasio, H. (2002). Humans and great apes share a large frontal cortex. *Nature Neuroscience, 5*, 272–276.

Semendeferi, K., Teffer, K., Buxhoeveden, D. P., Park, M. S., Bludau, S., Amunts, K., Travis, K., & Buckwalter, J. (2011). Spatial organization of neurons in the frontal pole sets humans apart from great apes. *Cerebral Cortex*, (in press).

Smaers, J. B., Schleicher, A., Zilles, K., & Vinicius, L. (2010). Frontal white matter volume in anthropoid primates. *PloS One, 5*, e9123. doi:10.1371/journal.pone.0009123.

Sporns, O., Chilavo, D. R., Kaiser, M., & Hilgetag, C. C. (2004). Organization, development and function of complex brain networks. *Trends in Cognitive Sciences, 8*, 418–425.

Sporns, O., Honey, C. J., & Kötter, R. (2007). Identification and classification of hubs in brain networks. *PloS One, 2*, e1049. doi:10.1371/journal.pone.0001049.

Striedter, G. F. (2004). *Principles of brain evolution.* Sunderland, MA: Sinauer Associates.

Todd, P. H. (1986). *Intrinsic geometry of biological surface growth. Lecture Notes in Biomathematics 67.* Berlin: Springer Verlag.

Van Essen, D. C. (1997). A tension-based theory of morphogenesis and compact wiring in the central nervous system. *Nature, 385*, 313–318.

Wang, S. S.-H., Shultz, J. R., Burish, M. J., Harrison, K. H., Hof, P. R., Towns, L. C., Wagers, M. W., & Wyatt, K. D. (2008). Functional trade-offs in white matter axonal scaling. *Journal of Neuroscience, 28*, 4047–4056.

Welker, W. (1990). Why does cerebral cortex fissure and fold? A review of determinants of gyri and sulci. In In: E. G. Jones & A. Peters (Eds.), *Cerebral cortex*, (Vol. 8B), (pp. 23–136). New York: Plenum Press.

Wen, Q., & Chklovskii, D. B. (2005). Segregation of the brain into gray and white matter: A design minimizing conduction delays. *PLoS Computational Biology, 1*, 617–630.

West, B. J., & Goldberger, A. L. (1987). Physiology in fractal dimensions. *American Scientist, 75*, 354–365.

Young, M. P. (1993). The organization of neural systems in the primate cerebral cortex. *Proceedings of the Royal Society of London B, 252*, 13–18.

Zhang, K., & Sejnowski, T. J. (2000). A universal scaling law between gray matter and white matter of cerebral cortex. *Proceedings of the National Academy of Sciences of the United States of America, 97*, 5621–5626.

SECTION VI

Cognition: From neuron to behavior

M. A. Hofman and D. Falk (Eds.)
Progress in Brain Research, Vol. 195
ISSN: 0079-6123

CHAPTER 19

Primate encephalization

Louis Lefebvre*

Department of Biology, McGill University, Montréal, QC, Canada

Abstract: Encephalization is a concept that implies an increase in brain or neocortex size relative to body size, size of lower brain areas, and/or evolutionary time. Here, I review 26 large-scale comparative studies that provide robust evidence for five lifestyle correlates of encephalization (group living, a large home range, a high-quality diet, a strong reliance on vision, arboreal and forest dwelling), six cognitive correlates (better performance in captive tests, more tactical deception, innovation, tool use, social learning, all subsumed in part by general intelligence), one life history correlate (a longer lifespan), two evolutionary correlates (a high rate of change in microcephaly genes, an increase in brain size over macroevolutionary time), as well as three trade-offs (a slower juvenile development, a higher metabolic rate, sexually selected dimorphism). Of the 26 different encephalization measures used in these studies, corrected neocortex size, either with a ratio or a residual, is the most popular structural correlate of the functional variables, while residual brain size is the measure associated with the greatest number of them. Controversies remain on corrected or absolute measures of neural structure size, concerted versus mosaic evolution of brain parts and specialized versus domain-general brain structures and cognitive processes.

Keywords: primate; encephalization; cognition; neocortex; brain.

Introduction

Brain size shows a strong positive relationship with body size over a large set of animal species, but some species clearly have brains that are much larger than expected, given their body size. Behaviors that would be considered intelligent in humans have been observed in several of the larger-brained species in different animal classes. Over the past three million years, endocranial volume has increased dramatically in our hominin lineage, as has evidence for key cognitive innovations like biface tool manufacture and fire. Taken together, these three observations have given rise to the idea that something about brain enlargement, once the allometric effects of growth have been removed, has coevolved with cognition. Encephalization is the concept born

*Corresponding author.
Tel.: 1-514-398-6457; Fax: 1-514-398-65069
E-mail: louis.lefebvre@mcgill.ca

DOI: 10.1016/B978-0-444-53860-4.00019-2

from this idea. In comparative biology, the concept describes the difference between animals in the amount of neuronal mass or volume available beyond some value predicted by body size. In paleontology, it refers to increases in brain size observed over evolutionary time in some taxa. In neuroanatomy, it describes the relative increase in size of higher brain structures like the neocortex[1] and pallium in classes like birds and mammals compared to the subpallium in more "primitive" clades like fish or reptiles. Anatomical traits like folding are also sometimes described in similar terms, with highly gyrified brains seen as more encephalized than smoother ones. In this chapter, I first discuss some of the controversies surrounding the concept of encephalization, and then review 26 large-scale comparative studies that identify key correlates of primate encephalization.

Problems with the concept of encephalization

Encephalization is not a simple descriptive concept. It is a relative one that implies a comparison over time, a comparison between taxa, and/or a comparison of some neural structures (so-called higher) with other neural structures (so-called lower) or with the whole body. It is also a concept that mixes two levels of explanation: the structural level of neurons, brains, and bodies and the functional level of information processing. Finally, as evidenced by the terms used in the first paragraph—"something about brain enlargement has coevolved with cognition"—the concept can be fuzzy. What exactly is the "something" in the brain that has coevolved with "cognition"? In fact, what is "cognition" and how can it be operationalized? Is the relative nature of encephalization justified and the removal of the allometric growth component in brain size necessary?

These questions are particularly relevant in primates because recent work in neuroanatomy (Burish et al., 2010; Gabi et al., 2010; Herculano-Houzel et al., 2007) and comparative psychology (Deaner et al., 2007) on this order has raised the possibility that absolute measures of brain and neocortex size may be more relevant than the traditional relative ones. Second, it is in primates that we have the widest array of operational measures of cognition on a wide sample of species, as well as attempts to synthesize this array into a common framework based on general intelligence (Deaner et al., 2006; Reader et al., 2011). The operational measures go from learning and problem-solving tests in captivity (Johnson et al., 2002; Riddell and Corl, 1977) to taxonomic counts of tactical deception (Byrne and Whiten, 1990), innovation, tool use, social learning (Reader and Laland, 2002), and extractive foraging (Reader et al., 2011) taken mostly from the wild. The literature on primates also includes good quantitative estimates of lifestyles where complex cognition might provide a selective advantage, for example, group living (Dunbar, 1998; Dunbar and Shultz, 2007a) or foraging for high-quality dispersed food such as fruit (Barton, 1996; Clutton-Brock and Harvey, 1980; Fish and Lockwood, 2003). Third, primates feature one of the best neuroanatomical databases (Isler et al., 2008; Stephan et al., 1981) to test structural (Barton and Harvey, 2000; Clark et al., 2001; Finlay et al., 2001; Yopak et al., 2010) and functional (Barton, 1998; Dunbar, 1998; Reader et al., 2011) hypotheses.

Based on current knowledge, the structural basis of encephalization in primates can be described as a series of nested scaling relationships that link numbers of neurons and glia, cortical white and gray matter mass, neocortical volume, whole brain, body, and spinal cord mass. The current consensus is that (1) numbers of neurons and glia scale isometrically (i.e., 1 to 1) with neocortex mass as well as brain mass,

[1] I use the term "neocortex" throughout, rather than "isocortex," because it is the term most often used in the literature I cover here.

but cell density does not (Gabi et al., 2010; Herculano-Houzel et al., 2007); (2) neocortical gray matter scales isometrically (and white matter with positive allometry, i.e., slope>1) with mass of the rest of the brain (Barton and Harvey, 2000); (3) frontal white and gray matter have a positive allometric relationship with neocortical mass (Smaers et al., 2010), as does (4) brain mass with spinal cord neuron numbers (Burish et al., 2010); (5) brain mass has a negative allometric relationship (i.e., slope<1) with body mass, with different slopes at different taxonomic levels (Isler et al., 2008; Pagel and Harvey, 1989). In primates (but not in other mammalian orders: rodents: Herculano-Houzel et al., 2006; insectivores: Sarko et al., 2009), brain mass is therefore equivalent to neuron numbers due to the 1 to 1 relationship between them and the lack of relationship of mass and volume with cell density.

Brain size versus control mechanisms

The main functional prediction that corresponds to these structural scaling rules is that cognitive benefits, accruing in certain ecological conditions and traded off against some costs, have coevolved with encephalization at one or more of the neuroanatomical scales mentioned above. One problem with this prediction is that two levels of explanation, as well as two traditions of empirical testing, are used. Knowing which brain area's size is best correlated with a measure of cognition and knowing what brain processes control this cognitive ability are distinct questions. Traditionally, the questions have been addressed with different techniques and, while their answers are mutually relevant, they involve different levels of explanation. To answer the mechanistic control question, one has to identify the key neuronal events that lead to performance differences in cognitive tasks. The events, which could be localized or distributed over many parts of the brain, involve neurotransmitters, receptors, enzymes that affect neurotransmitter metabolization, as well as blood oxygenation changes that follow neuronal activity. The mapping of blood oxygenation changes that is achieved with magnetic resonance imaging goes some way toward linking neuronal events with the identification of brain areas involved in particular cognitive activities, but they do not answer the question of size differences between brain areas across different species. For example, a consensus seems to be developing among neuroscientists (Deary et al., 2010) that tasks with strong loadings on general intelligence involve a distributed network of at least 14 brain areas in humans (review by Jung and Hier, 2007 of 37 imaging studies involving 1557 subjects). Whether the activation of the 14 areas during general intelligence tasks translates into more neurons, and thus, a greater volume in each of the 14 brain parts is another matter. How the 14 brain parts would evolve to different sizes in different species due to the selective advantages of more versus less general intelligence is also unknown. Would it be through concerted evolution (Finlay et al., 2001) of the entire zone encompassing the 14 areas? Would it be via mosaic evolution (Barton and Harvey, 2000) of the 14-area network only? Would it be through 14 separate evolutionary events each affecting a different area? The distinction between size evolution and proximal control, along with the unresolved question of how the two levels are linked, needs to be taken into account in all our thinking about brain-intelligence coevolution.

Theories of encephalization originate from the simple observation that the brain as a whole, as well as areas that take up a large proportion of the brain such as the avian pallium and the mammalian neocortex, is many times larger in some species than in others. These empirical observations warrant scientific study, in the same way that research on body size has long been a legitimate field in ecology and evolution (the search topic "body size evolution" currently yields 7275 articles on the Web of Science). The fact that bodies are made up of many parts that are selected for different purposes has not led critics to doubt the validity of the whole field, but this has sometimes happened for brain size. In particular, recent work on spatial memory and bird song has led to the belief in some circles that the idea of encephalization is meaningless because it refers to too broad an anatomical structure (Healy and Rowe, 2007). However, in the case of both spatial memory and bird song, strong mechanistic research programs had long identified key neural centers involved in the storage of specialized information about space and learned vocalizations. When Sherry and Vaccarino (1989) lesioned the hippocampus of chickadees and found that spatial memory of previously stored food was lost, their focus on that brain structure was based on decades of lesion (Morris et al., 1982) and electrophysiological research (e.g., place cells, O'Keefe and Dostrovsky, 1971). Similar work by Nottebohm et al. (1976) had also identified nuclei like HVC, RA (robust nucleus of the arcopallium), and area X as specialized centers for oscine song, as well as demonstrated neurogenesis coincident with song learning (Paton and Nottebohm, 1984). Sherry's serendipitous discovery (Sherry, 2011) that the chickadee hippocampus was much larger than that of the nonstoring canary whose brain atlas had been used to guide the lesion surgery of Sherry and Vaccarino (1989) was posterior to the strong research tradition identifying this brain structure as a key center for spatial memory. In a similar manner, the discovery by DeVoogd et al. (1993) that the size of nucleus HVC (but not area X) was proportional to the size of the song repertoire of oscine species came several years after Nottebohm's pioneering mechanistic work.

The study of encephalization does not have the same history or luck as that of spatial memory and bird song. The question "why is a corvid brain so large?" is much older than the data showing that crows can manufacture tools (Hunt, 1996) and magpies can recognize themselves in a mirror (Prior et al., 2008). The brain centers and neural events that are necessary for mirror image recognition and tool manufacture in birds are also currently unknown. We are therefore obliged to use a top-down approach rather than the bottom-up program that spatial memory and bird song researchers were fortunate enough to have at their disposal. We start with the old observation of brain and body size covariation by Dubois and Lapicque (Gayon, 2000, for a historical perspective), and then seek cognitive and lifestyle correlates of this variation. If (1) these correlates are independent of each other, (2) each one is more strongly tied to size variation in one brain area than in others, and (3) structural studies of brain organization demonstrate strict mosaic evolution of areas on the basis of functional specialization, then the concept of encephalization loses much of its relevance because it is not specific enough. If instead the cognitive and lifestyle measures covary and size variation in brain areas is concerted with variation in the whole brain through common evo–devo processes, then the idea of encephalization is more useful.

Bearing these caveats in mind, what is the current state of the literature on species differences in structural and functional correlates of encephalization in primates? In this chapter, I deal only with large-scale (either all primates or all haplorrhines) comparative analyses that examine phylogenetically controlled correlates of encephalization. I do not include analyses that focus only on a single clade such as apes or New World or Old World monkeys; I also do not include analyses that add *Homo sapiens* to the primate database. Tables 1–4 summarize the evidence taken from 26 studies that focus on the whole brain or its major "higher" divisions, the telencephalon and the neocortex. Other brain areas that are usually not considered in encephalization research have also been subject to comparative analyses featuring correlations between cognitive functions and structure size (the cerebellum: Lewis and Barton, 2004; Dunbar and Shultz, 2007b; the amygdala: Lewis and Barton, 2006; the hippocampus: Lewis and Barton, 2006; Shultz and Dunbar, 2010a; the main

Table 1. Lifestyle correlates of encephalization

Encephalization measure	Lifestyle correlate				
	DQ	HR	HU	GS	VI
Brain vol or mass				x•	
Res brain vol ag. body	••x	••	•	•xx	•••
EQ					
Neonatal brain vol					
Tel vol				x	
Res tel vol ag. body				x	
Res tel vol ag. brain				x	
Res tel ag. ROB				•	
Neo vol				x	
Neo+striatum vol					
NV neo vol					
Res neo vol ag. body				x	
Res neo vol ag. brain	x	•	x	••x	
Res neo vol ag. ROB	•x	•	•	•••	••
Res neo ag. medulla	•*	•	•	•	
Res NV neo ag. body		•		x	
Res NV neo ag. ROB		x		••	
Res NV neo ag. ROB ag. body	•			•	
Neo/brain				•	
Neo/ROB	x	x		•••	
Neo/ROB, ag. brain					
Neo+stri/bstem	x	x		•	
Neo+stri/bstem, ag. body					
Neo/ROB−cereb	x	x		x	
NV neo/ROB	x	xx		••	
NV neo/ROB−cereb	x	x		•	

Encephalization abbreviations: vol, volume; res, residual or result of partial correlation; EQ, encephalization quotient; neo, neocortex; NV, nonvisual; ag. body, regressed against body mass; tel, telencephalon; ROB, rest of brain; stri, striatum; cereb, cerebellum; bstem, brainstem.
DQ, diet quality: Barton (1996) and Fish and Lockwood (2003).
HR, home range: Deaner et al. (2000) and Walker et al. (2006).
HU, habitat use: Dunbar and Shultz (2007b).
GS, group size: Barton (1996), Deaner et al. (2000), Dunbar (1998), Joffe and Dunbar (1997), Lehmann and Dunbar (2009), Lindenfors (2005), Lindenfors et al. (2007), and Walker et al. (2006).
VI, visual input: Barton (1998, 2004) and Kirk (2006).
•: Result, $p<0.05$; •*: result, $p=0.06$; x: result, ns.

Table 2. Cognitive correlates of encephalization

Encephalization measure	Cognitive correlate					
	TD	IN	TU	SL	G	TR
Brain vol or mass						••
Res brain vol ag. body						x
EQ						x
Neonatal brain vol						
Tel vol						
Res tel vol ag. body						
Res tel vol ag. brain						
Res tel ag. ROB						
Neo vol	•				•	••
Neo+striatum vol		•	•	•		
NV neo vol						•
Res neo vol ag. body						x
Res neo vol ag. brain						
Res neo vol ag. ROB					x	x
Res neo ag. medulla						
Res NV neo ag. body						
Res NV neo ag. ROB						
Res NV neo ag. ROB ag. body						
Neo/brain						
Neo/ROB	•				•	••*
Neo/ROB, ag. brain						
Neo+stri/bstem		•	•	•	•	
Neo+stri/bstem, ag. body		•*	•*			
Neo/ROB−cereb						
NV neo/ROB						
NV neo/ROB−cereb						

Encephalization abbreviations as in Table 1.
TD, tactical deception: Byrne and Corp (2004).
IN, TU, SL, innovation, tool use, social learning: Reader and Laland (2002).
G, general intelligence: Reader et al. (2011).
TR, test rankings: Deaner et al. (2007) and Shultz and Dunbar (2010a).
•: Result, $p<0.05$; •*: result, $p=0.06$; x: result, ns.

and accessory olfactory bulbs: Barton, 2006b; the striatum: Graham, 2011; the primary visual cortex: Barton, 1996), but they are not covered in the tables.

The 26 studies all yield linear associations between some structural measure of encephalization (over two dozen different measures) and a cognitive variable, a trade-off, or a lifestyle. The studies suggest that more encephalized primate species or genera tend to (1) eat a higher quality diet, (2) have larger home ranges, (3) are arboreal and live in closed forests, (4) occur in larger groups, and

Table 3. Trade-offs of encephalization

Encephalization measure	Trade-off		
	JD	MR	SS
Brain vol or mass			
Res brain vol ag. body	••	••	•
EQ			
Neonatal brain vol			
Tel vol			
Res tel vol ag. body			
Res tel vol ag. brain			
Res tel ag. ROB			•
Neo vol			
Neo+striatum vol			
NV neo vol			
Res neo vol ag. body			
Res neo vol ag. brain	x	•	x
Res neo vol ag. ROB	•	•	x
Res neo ag. medulla	•	•*	
Res NV neo ag. body			
Res NV neo ag. ROB			
Res NV neo ag. ROB ag. body			
Neo/brain			
Neo/ROB			•
Neo/ROB, ag. brain			•
Neo+stri/bstem			
Neo+stri/bstem, ag. body			
Neo/ ROB−cereb			
NV neo/ROB	•		
NV neo/ROB−cereb	•		

Encephalization abbreviations as in Table 1.
JD, juvenile development: Barrickman et al. (2008) and Walker et al. (2006).
SD, sexual dimorphism: Lindenfors et al. (2007) and Schillaci (2006, 2008).
MR, metabolic rate: Isler and van Schaik (2006).
•: Result, $p<0.05$; •*: result, $p=0.06$; x: result, ns.

Table 4. Evolutionary correlates of encephalization

Encephalization measure	Evolutionary correlate		
	LS	MG	ET
Brain vol or mass		x	
Res brain vol ag. body	••x	•	•
EQ		x	
Neonatal brain vol		•	
Tel vol			
Res tel vol ag. body			
Res tel vol ag. brain			
Res tel ag. ROB			
Neo vol		x	
Neo+striatum vol			
NV neo vol			
Res neo vol ag. body			
Res neo vol ag. brain		x	
Res neo vol ag. ROB		•	x
Res neo ag. medulla		•	
Res NV neo ag. body			
Res NV neo ag. ROB			
Res NV neo ag. ROB ag. body			
Neo/brain			
Neo/ROB		•	
Neo/ROB, ag. brain			
Neo+stri/bstem	•		
Neo+stri/bstem, ag. body			
Neo/ROB−cereb	x		
NV neo/ROB	x		
NV neo/ROB−cereb	x		

Encephalization abbreviations as in Table 1.
LS, life span: Barrickman et al. (2008); Walker et al. (2006).
MG, microcephaly genes: Ali and Meier (2008); Montgomery et al. (2011).
ET, evolutionary time: Shultz and Dunbar (2010b).
•: Result, $p<0.05$; •*: result, $p=0.06$; x: result, ns.

(5) use more binocular visual input than less encephalized primates (Table 1). They also show (6) more tactical deception, (7) innovation, (8) tool use, (9) social learning, (10) general intelligence, and (11) better performance in captive tests, than do less encephalized primates (Table 2). Encephalization is traded off against (12) a slower juvenile development, (13) a higher metabolic rate, and (14) a greater degree of sexually selected dimorphism (Table 3). More encephalized primates have (15) a longer lifespan. Over evolutionary time, there has been (16) an increase in primate encephalization, along with (17) a high rate of change in some of the genes associated with whole brain and neocortex size (Table 4). In the following sections, I summarize the evidence linking one or more encephalization measure with the 17 functional variables. Many of the comparative studies test several neural measures. The tables include all tests of all neural measures, whether they lead to significant (indicated by a "•") or non-significant (indicated by an "x") results. The text in

the sections leaves out many of these details, so as not to overburden the reading.

Lifestyles associated with encephalization (Table 1)

Historically, studies on lifestyles favoring encephalization preceded those in which direct tests of cognitive measures were used. The comparative psychology work of Riddell and Corl (1977) was contemporaneous with the earlier ecological studies of Clutton-Brock and Harvey (1977) but was not as influential (currently 22 citations in the Web of Science, compared to 570). It is the discovery by Clutton-Brock and Harvey (1980) that frugivorous primates have a larger brain than do folivorous ones that gave an impetus to ecological studies of encephalization. Similar work on other orders suggested that dietary reliance on multiple foods (omnivory) that are hard to find (fruit, vertebrate prey) might select for enlarged brains in many mammals.

Many early studies of encephalization did not remove the potential pseudoreplication effects of common ancestry on their comparative data. The advent of phylogenetic corrections improved comparative work and provided a means to distinguish between phylogenetically confounded versus repeated independent coevolution of brain enlargement and cognition. Barton (1996) and Fish and Lockwood (2003) confirmed with phylogenetic controls the dietary trends reported by Clutton-Brock and Harvey (1980). They also used quantitative indices instead of categorical measures of diet; in Barton's case, the measure was percent fruit in the diet, while in Fish and Lockwood's, the diet quality index included fruit, meat, and leaves.

The abundance and the spatial and temporal distribution of fruit are, on average, more difficult to track than that of leaves. This is the cognitive challenge that is assumed to be behind the relatively large brain of frugivorous primates. Food that is patchy and whose ripeness has to be tracked in space and time might also have to be searched for over a wide range. It is therefore logical that primates with larger brains should also have larger home ranges. Deaner et al. (2000) have confirmed with phylogenetic corrections and two different measures of relative neocortex size the earlier finding by Clutton-Brock and Harvey (1980) that home range is positively associated with residual brain size. Two other habitat use variables have been examined by Dunbar and Shultz (2007b), occurrence in open/mixed versus closed forest habitats, as well as terrestrial versus arboreal locomotion between feeding and resting sites. In univariate analyses, arboreal and closed forest species showed a large residual brain and neocortex size.

Another key lifestyle variable hypothesized to be associated with encephalization is group living, which was first tested by Sawaguchi and Kudo (1990). The assumption here is that a larger brain or neocortex can process a larger amount of social information resulting from the alliances, networks, and dominance relationships that increase, presumably in a nonlinear manner, with the number of individuals in a primate group. Dunbar and others have confirmed with phylogenetic corrections that neocortex size is associated with several features of sociality: social group size (Barton, 1996; Dunbar, 1992), number of females in the group (Lindenfors, 2005), grooming clique size (Kudo and Dunbar, 2001), frequency of coalitions (Dunbar and Shultz, 2007a), and network connectivity (Lehmann and Dunbar, 2009). Lindenfors et al. (2007) suggest that female primates, but not males, show the relationship between neocortex size and sociality. Instead, sexual selection for large size in males is more strongly associated with the size of limbic structures involved in aggression (see the section "Trade-offs").

Barton's recent work has focused on correlates of specialized brain parts, in accordance with his views on mosaic evolution of functionally linked areas (Barton, 1999, 2006a, 2007; Barton and Harvey, 2000; Whiting and Barton, 2003).

Barton and colleagues (Lewis and Barton, 2004, 2006) have found that the size of the cerebellum, amygdala, and hypothalamus are associated with social play (as is the size of the striatum, Graham, 2011); the size of the main olfactory bulb with diet and diurnal versus nocturnal activity; and the size of the accessory olfactory bulb with social and mating systems (Barton, 2006b). More relevant to the question of encephalization are Barton's (1998) findings that the size of specialized visual areas (primary visual cortex, parvocellular and magnocellular layers of the lateral geniculate nucleus) and the number of neurons in them are positively correlated with allometrically corrected brain size. Barton (2004) also shows that degree of binocular convergence correlates both with the size of specialized visual areas in the brain, as well as relative size of the neocortex and whole brain, adding further support to the idea that vision was a major factor in primate encephalization.

Attempts to test lifestyle variables together, to see if one is a better predictor than the other or if they covary in ways that suggest they are not independent, have produced mixed results. Joffe and Dunbar (1997) have shown that visual areas of the cortex (striate cortex and lateral geniculate nucleus) have a poorer relationship with social group size than do nonvisual areas, suggesting that vision *per se*, contrary to Barton's ideas, did not play a major role in the encephalization of the social brain. Barton (1996) showed that social group size and percent fruit in the diet predict independent portions of the variance in relative neocortex size. Deaner et al. (2000) have found that either social variables such as group size or ecological variables such as home range are the only significant correlate of allometrically corrected brain size depending on the method used for the correction. Walker et al. (2006) report a similar result using stepwise regressions. When residual brain size is the encephalization measure, only home range size is significant, while group size and percent fruit in the diet are not. When the encephalization measures are neocortex ratios (calculated in five different ways), only group size now enters the multiple regressions, with home range and percent fruit dropping out. Reader et al. (2011) report that lifestyle variables, whether social (group size) or dietary (percent fruit in diet, number of food categories in diet), load together on the second component of a PCA in which the main component regroups five cognitive measures. The correlations between the social and the dietary variables are weakly positive, varying between 0.14 and 0.25 (Reader et al., 2011). Finally, Dunbar and Shultz (2007b) conducted multivariate and path analyses on several lifestyle and life history variables that were significant predictors of residual whole brain and neocortex size in univariate analyses. In the final multivariate model, residual brain size was best predicted by diet and lifespan, while the effects of group size, home range size, and habitat use did not contribute significantly to the model. Residual neocortex size was best predicted by group size and lifespan, with the other lifestyle variables dropping out. In the path analysis, the relationship between group size and residual neocortex was direct and bidirectional, as was the relationship between residual brain size and lifespan. In contrast, the relationship between residual brain size and diet was indirect and included several unidirectional intermediates, with metabolic rate driving both diet and brain size. Intriguingly, body size was only indirectly driving brain size in the analysis, via its unidirectional effects on metabolic rate and lifespan.

Cognitive correlates of encephalization (Table 2)

Frugivory and sociality seem to have a robust relationship with primate encephalization. However, these are lifestyles in which enhanced cognition might be an advantage, but they are not cognitive variables *per se*. If, for instance, one measured species differences in the number of conspecific faces primates can memorize, this would provide a direct test of the cognitive differences that are constrained by a smaller social brain. In the absence of such experiments, only direct quantitative estimates of cognitive abilities provide the missing link between lifestyles and encephalization.

Riddell and Corl (1977) were the first to show in a large data set that species differences in "cerebral development indices" correlated with performance on a variety of tasks. Deaner and colleagues (Deaner et al., 2006, 2007; Johnson et al., 2002), using phylogenetic corrections, generalized this finding by ranking with a Bayesian procedure the different primate genera on all available comparative tasks studied in captivity. Deaner et al. (2007) then compared these ranks to eight different measures of encephalization. Absolute (log-transformed) size of the whole brain and the neocortex were the only significant correlates of cognitive performance on independent contrasts between primate genera; neocortex ratio showed borderline significance, but other relative measures, whether corrected by body size or size of the rest of the brain, did not. Reader et al. (2011) assessed the relationship between the general intelligence factor they extracted from eight cognitive and lifestyle measures and the experimental data from captive studies described by Riddell and Corl and Deaner and colleagues. They found significant correlations in both cases. This supports the idea that cognitive tasks given in captivity are ecologically valid measures of cognitive differences found in the field, as well as the idea that measures taken both in the field and in captivity are to some extent controlled by general intelligence.

Historically, Byrne and Whiten (1990) were the first to examine taxonomic differences in the frequency of a cognitive ability, tactical deception, in the wild and in captivity. The idea here is that the number of times human observers witness the use of a particular type of cognition in particular species can serve as a quantitative measure of taxonomic variation in that ability. Byrne and Whiten were careful to correct their data for potential biases that might inflate observations in well-studied or more visible species. The anecdotal nature of their data, a method that had been more of less banned from comparative psychology since the days of E.L. Thorndike, was also extensively discussed (see peer commentaries included with Whiten and Byrne, 1988). Byrne and Whiten (1990) focused on cases of "Machiavellian" intelligence involving social manipulation and tactical deception. Byrne and Corp (2004) then showed that deception frequency per species, corrected for research effort, was positively correlated with both absolute and relative size of the neocortex.

The taxonomic count technique was then generalized by Reader and Laland (2002) to cases of innovation, tool use, and social learning. All three of these measures were shown to correlate with neocortex ratio. Recently, Reader et al. (2011) have reexamined the cognitive measures they used in their earlier article. They added a new measure, extractive foraging, as well as Byrne and Whiten's data on tactical deception frequency. These five cognitive measures were then submitted to PCA to see if they all loaded on one general factor or if the social (social learning, tactical deception) measures loaded on a separate factor from the nonsocial ones (tool use, extractive foraging). All five measures loaded strongly on a first component that explained 65% of the variance; this result is compatible with the idea that there is a general intelligence factor (*g*) behind the measures. More interestingly, the five cognitive measures all loaded together on the same factor even when three lifestyles variables (group size, percent fruit in diet, and number of food categories in diet) were added to the factor analysis. The lifestyle variables all loaded on a second, independent factor. What these data suggest is not only that general intelligence might be an important part of primate cognition but also that the distinction mentioned above between lifestyle correlates of brain size and cognitive measures *per se* might be more important than the oft-cited difference between social and nonsocial intelligence (see also Overington et al., 2008). Finally, the positive correlations between Reader et al.'s *g* and the results of the captive tests analyzed by Riddell and Corl (1977) and Deaner et al. (2006) argue against a common bias to taxonomic counts based on quantified anecdotes. Such a bias might lead to positive intercorrelations between taxonomic counts, but should not also produce correlations with the results of laboratory tests. These might instead be biased by species' responses to captivity and testing by humans, but the positive relationship with field counts argues against this possibility.

Trade-offs (Table 3)

A solitary folivore is predicted to profit less from an enlarged neocortex than an omnivore living in a large group. Lifestyles, however, are not the only contexts that affect the evolution of complex cognition and large brains. Life history is also thought to be important, in particular the slow–fast continuum of developmental traits. Based on this continuum, more encephalized primates should have small litters, as well as long periods of gestation, lactation, and juvenile growth. These are costs, as they imply increased parental investment and a delay in reproductive maturity. However, the lengthened lifespan that also goes with the slow–fast continuum can compensate for the sexual maturation delay by increasing the duration of the reproductive period. Several researchers (e.g., Allman et al., 1993; Deaner et al., 2003; Hofman, 1983, 1993; Sacher and Staffeldt, 1974) have examined one or more of these life history traits, often with varying results (see Table 1 in Barrickman et al., 2008 for a review). Based on a large data set taken mostly from field studies, the analysis of Barrickman et al. (2008) supports most of the predictions from the slow–fast view of encephalization. Taking into account body size and phylogeny, primate brain size is positively associated with length of the juvenile period and age at first reproduction. Gestation length is also associated with brain size, but time to weaning is not. The advantage that counterbalances these developmental costs is a lengthened life span (Table 4; Dunbar and Shultz, 2007b; Barrickman et al., 2008).

Many papers on primate (especially human) encephalization mention the metabolic costs that a large brain represents. These costs can be met in large-brained species either by increasing metabolic rate or by reducing the energetic costs associated with other organs, for instance digestive ones (Aiello and Wheeler, 1995). Both can be achieved via an improvement in diet quality, which increases caloric intake as well as digestibility. Isler and collaborators (Isler and van Schaik, 2006; Isler et al., 2008), using two different samples of primate data, have confirmed the predicted relationship between basal metabolic rate and brain size, with phylogeny and body size controlled for, as have Dunbar and Shultz (2007b).

As humans, we might think that an increase in cognitive efficiency is always a good thing, but in some primate species, variance in reproductive success might be more strongly affected by noncognitive factors, to the point of actually selecting against encephalization. The finding of Lindenfors et al. (2007) that limbic areas involved in aggression were associated with group size and dimorphism in male primates, but not in females, hints at such an effect. If a male, emigrating from its natal troop, competes with other males via intense individual aggression, traits such as body size, canine length, and fighting ability might be more important than cognitive abilities that would allow alliance management, tactical deception, large grooming networks, and kin recognition in circumstances where individual aggression is less important. From an "expensive tissue" perspective, this might also create trade-offs between structural investment in brain versus canine and muscle tissue. One operational measure of the intensity of male competition is sexual dimorphism, which can be quite large in some species, such as mandrills. As predicted, Schillaci (2006) reports a significant negative relationship between brain size and degree of dimorphism (Schillaci, 2006), as well as a relationship between brain size and mating system, with monogamous primates having larger brains than polygynous ones (Schillaci, 2008). Coherent with this view is the fact that the correlation between male rank and mating success in polygamous primates is negatively associated with neocortex size: the stronger the reproductive skew in favor of high ranking males, the smaller the neocortex ratio (Pawlowski et al., 1998).

A second possible trade-off involving sexual selection is suggested by Pitnick et al.'s (2006) finding that in bats, sperm competition has led to positive selection on testis size, with a structural trade-off on other "expensive tissues" negatively affecting brain size (see, however, Dechmann and Safi, 2009). Two studies (Lemaître et al., 2009; Schillaci, 2006) have now shown that, although the data confirm Pitnick et al.'s idea in echolocating bats, the prediction is not supported in primates. This negative result is coherent with the data on mating systems: sperm competition is low when only one male copulates with a female, whether the mating system is strict monogamy or strict polygyny. It is when females copulate with several males that sperm competition is highest, a system that is associated with intermediate-sized brains in Schillaci's (2008) analysis. The prediction on sperm competition thus leads the larger brains of the monogamous primates to cancel out the effect of the smaller brains of the polygynous ones in the statistical comparison with the intermediate-sized brains of the multimale/multifemale species.

Are some encephalization measures better than others?

Authors of individual studies often argue that the encephalization measure they are using is the most appropriate one; they sometimes strengthen their argument by showing that alternative size measures show either a poorer or a nonsignificant correlation with the cognitive variable they are testing. Though this approach is defendable, I will concentrate here on the trends in the entire set of tables rather than on single cases. Comparisons of encephalization indices are tricky, especially if different measures are derived from different techniques. For example, correcting whole brain size by body size and neocortex size by medulla size might make the second index look better simply because its correction factor has less measurement error, individual variability, and noncognitive selection pressures favoring a larger or a smaller body. Figure 1 illustrates this point using data on 43 extant nonhuman primate species taken from Stephan et al. (1981).

Regressing log neocortex size against log size of the medulla yields almost the same trends as regressing log whole brain size (minus the medulla) against log medulla (Fig.1a), leading to very similar residuals (Fig. 1b). However, regressing log size of the brain (minus the medulla) against log body size leads to results that show much more variation (Fig. 1c). Phylogenetic corrections might change these results slightly, but not alter the overall, qualitative conclusions. If a neocortex index based on a brainstem control is a much better predictor of a given cognitive measure than is brain size regressed against body mass, this should thus not automatically be taken to mean that extra-cortical areas are not involved in a particular cognitive process. More work is clearly needed on phylogenetically corrected data to compare the different encephalization measures.

The first obvious trend in Tables 1–4 is the very large number of encephalization measures. For corrected neocortex size only, there are 14. Half of these corrections are done with residuals from regressions, while the other half are done with ratios. Some of the neocortex measures use the entire structure, while others subtract the primary visual areas from the rest of the neocortex, and others still add the striatum. It is the structure used as the independent variable in the regressions and the denominator in the ratios that varies the most: it goes from body mass to volume

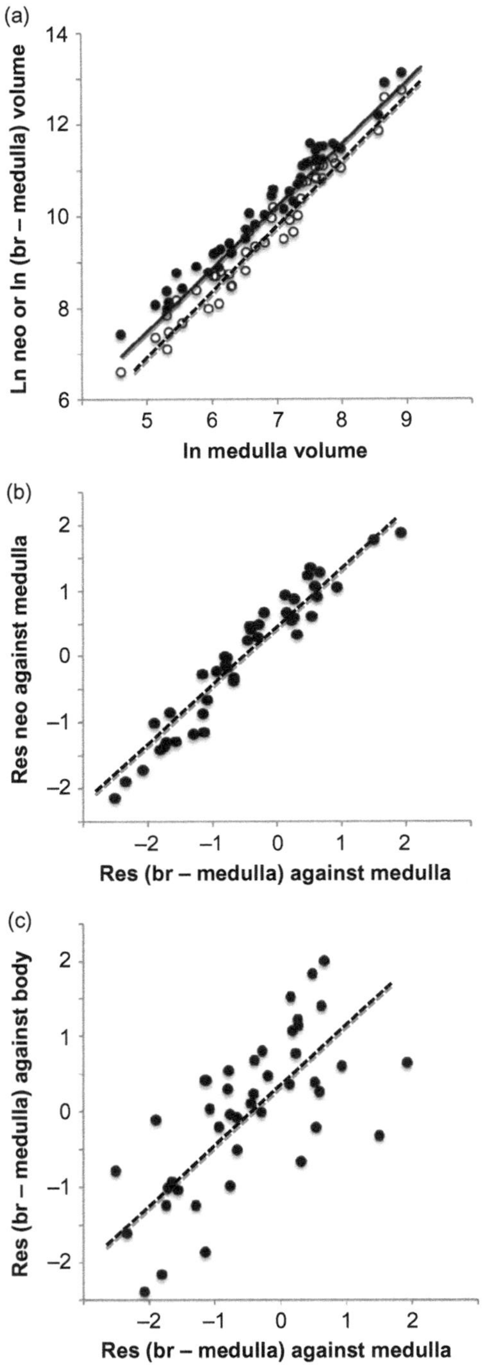

Fig 1. (a) Ln volume of the neocortex (open circles) and ln volume of the brain minus the medulla (filled circles) plotted against ln volume of the medulla. (b) Residual of neocortex volume regressed against medulla volume plotted against residual of brain minus medulla volume regressed against medulla volume. (c) Residual of brain minus medulla volume regressed against medulla volume plotted against residual of brain minus medulla volume regressed against body volume. Abbreviations as in Table 1.

of the whole brain, to that of the brain minus the neocortex or the brain minus the neocortex and the cerebellum, to the brainstem (mesencephalon plus medulla oblongata) to the medulla alone. However, before one concludes that some encephalization measures are better correlates of a particular functional variable than others, we need to know more about the indices themselves. Figure 2 illustrates this point: different ways of calculating relative size of the neocortex yield very different results. It matters little whether size of the neocortex is regressed against that of the whole brain or against that of the rest of the brain (see close relationship in Fig. 2a). However, other indices give discordant results: residual neocortex size regressed against the rest of the brain has a weak positive relationship with neocortex ratio over size of the whole brain (Fig. 2b), but a very poor relationship with neocortex ratio over size of the medulla (Fig. 2c). In turn, the relationship between the two neocortex ratios is strong, but nonlinear (Fig. 2d). What is clearly needed in the future is a comparative study that examines the similarities and differences between the different measures of encephalization, before they are used to test any functional predictions. Failing this, structural differences between encephalization measures may confound any apparent difference in the correlates of different functional variables.

Over all entries in the tables, measures of telencephalon size, whether absolute or corrected, are by far the least popular (six results) and least successful (more nonsignificant results, four, than significant ones, two). Residual brain size is much more successful: of the 24 results that use it, 18 show a significant relationship and only 6 a

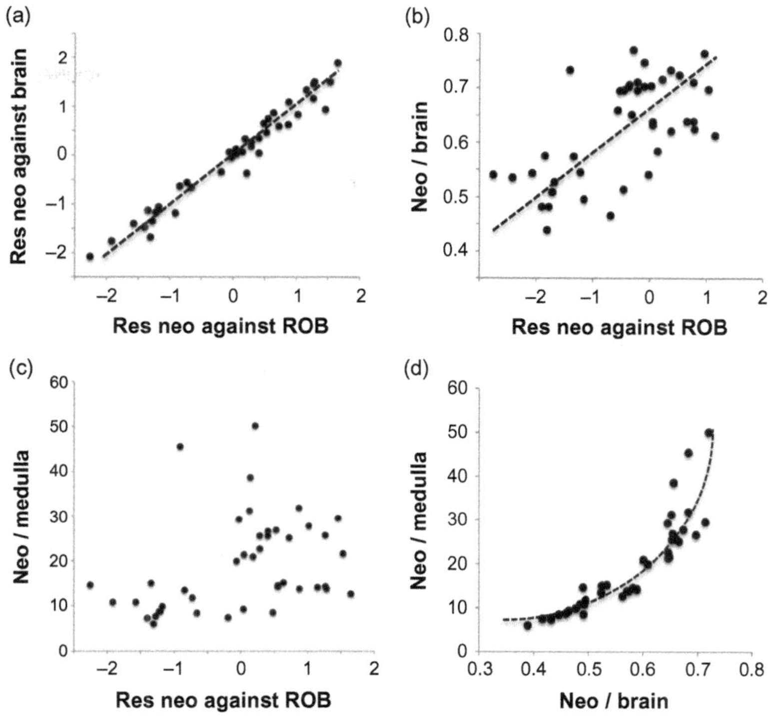

Fig. 2. (a) Residual neocortex volume regressed against volume of the rest of the brain plotted against residual neocortex volume regressed against whole brain volume. (b) Residual neocortex volume regressed against volume of the rest of the brain plotted against neocortex ratio over the whole brain. (c) Residual neocortex volume regressed against volume of the rest of the brain plotted against neocortex ratio over the medulla. (d) Neocortex ratio over the whole brain plotted against neocortex ratio over the medulla. Abbreviations as in Table 1.

nonsignificant one. Neocortex size is the most popular: it appears in 91 results, 59 of which are significant. An obvious caveat on these numbers is that some very productive research groups have, over the years, contributed multiple papers with significant relationships between a particular functional predictor and a particular structure, inflating the trends in the tables.

Over the 17 functional predictors of encephalization in Tables 1–4 (column headings), residual brain size shows at least one significant association with 11 of them. Residual neocortex size regressed against the rest of the brain is significantly associated with eight functional predictors, while neocortex ratio is associated with six. Absolute neocortex volume, in one form or the other (alone or with the volume of the striatum added or the volume of the primary visual cortex removed), is significantly associated with all six cognitive variables in Table 2, but with none of the other predictors in Tables 1, 3, and 4.

The trends in Table 1–4 suggest that both the whole brain and the neocortex, but not the telencephalon, are relevant neuroanatomical levels to test predictors of encephalization. They also suggest that both corrected and absolute neocortex volumes are of interest. Are these results coherent with genetic and evo–devo approaches to encephalization? Several genes that, in their abnormal form cause human microcephaly, have recently been studied in hominid lineages (Evans et al., 2004; Kouprina et al., 2004) and in comparative analyses on wider primate data sets (Ali and Meier, 2008; Montgomery et al., 2011; see Chapter 2 of this volume). Alternative evo–devo processes have also been invoked to account for either concerted evolution of all brain parts via differences in embryological neurogenesis ("late is large," Finlay et al., 2001) or mosaic coevolution of adaptively specialized sets of brain areas (Barton and Harvey, 2000; Whiting and Barton, 2003). Does the current consensus on microcephaly genes and concerted versus mosaic evolution allow us to decide between the whole brain and the neocortex as the most appropriate neuroanatomical level? Does it tell us whether absolute size of the neocortex is more relevant than corrected size?

The problem is that there is no consensus on these points, at least for the moment. Proponents of concerted and mosaic evolution focus on different parts of the variance in brain component size and use different methods to test their predictions. Of the four main papers on brain part evolution published in *Science* or *Nature*, Finlay and Darlington (1995) compare log-transformed absolute volumes of brain parts and find concerted evolution. Barton and Harvey (2000) partial out the size of the rest of the brain for each brain part and find mosaic evolution. Clark et al. (2001) transform each brain part into fractions of the total brain and find scalable taxon-specific cerebrotypes. de Winter and Oxnard (2001) use multivariate analysis on ratios of each brain part divided by the volume of the medulla, then again by the volume of the neocortex and find clusters of unrelated taxa that share similar niches. One or more of these conclusions might well be correct, but the differences in data transformations used in the studies might also constrain the realm of possible results that can be obtained.

An example of the different effects of data transformations is given in Figure 3.

The column on the left features the cerebellum, and the column on the right, the hippocampus, all plotted against the Finlay and Darlington (1995) measure of ln absolute brain part volume. The first line (a and b) features the transformation used by Barton and Harvey (2000), the second line (c and d) the transformation used by Clark et al. (2001), and the third (e and f) and fourth (g and h) lines the transformations used by de Winter and Oxnard (2001). What the figure clearly shows is that the transformations treat the brain part data in very different ways. In half of the eight cases (a, b, c, and f), pairs of transformations are uncorrelated over the different primate species. In one case (e), they show a tight nonlinear positive relationship, while in the other three (d, g, and h), they show a loose negative linear one. From the top and bottom lines of the figure, one could conclude that the cerebellum and hippocampus show similar trends, while from the middle two lines, that they show divergent trends.

What is needed in this debate is more comparative work on embryological neurogenesis in different brain areas of different primate species, similar to what is being done in birds by Striedter and Charvet, (2008, 2009; Charvet and Striedter, 2009). If developmental schedules are concerted and conserved, this should be detectable in the growth trends of different brain areas in different species. If instead brain parts develop as taxon-specific, functionally related mosaic pieces, this should also be obvious in embryonic growth. If strict mosaic evolution prevails, then there is no reason to expect that whole brain size should be relevant to functional predictions about encephalization. In contrast, concerted evolution would imply that both whole brain and brain part size would correlate with cognitive measures.

As far as microcephaly genes are concerned, the three most relevant studies also show contradictory results (Table 4). Evans et al. (2004, 2006) had suggested that the microcephaly genes ASPM and MCPH1 have evolved at a faster rate in lineages leading from the last common ancestor of apes to modern humans than in other lineages, but the data from Montgomery et al. (2011) do not support this idea. Instead, they find adaptive variation across all primate lineages. Ali and Meier (2008) had also linked adaptive evolution

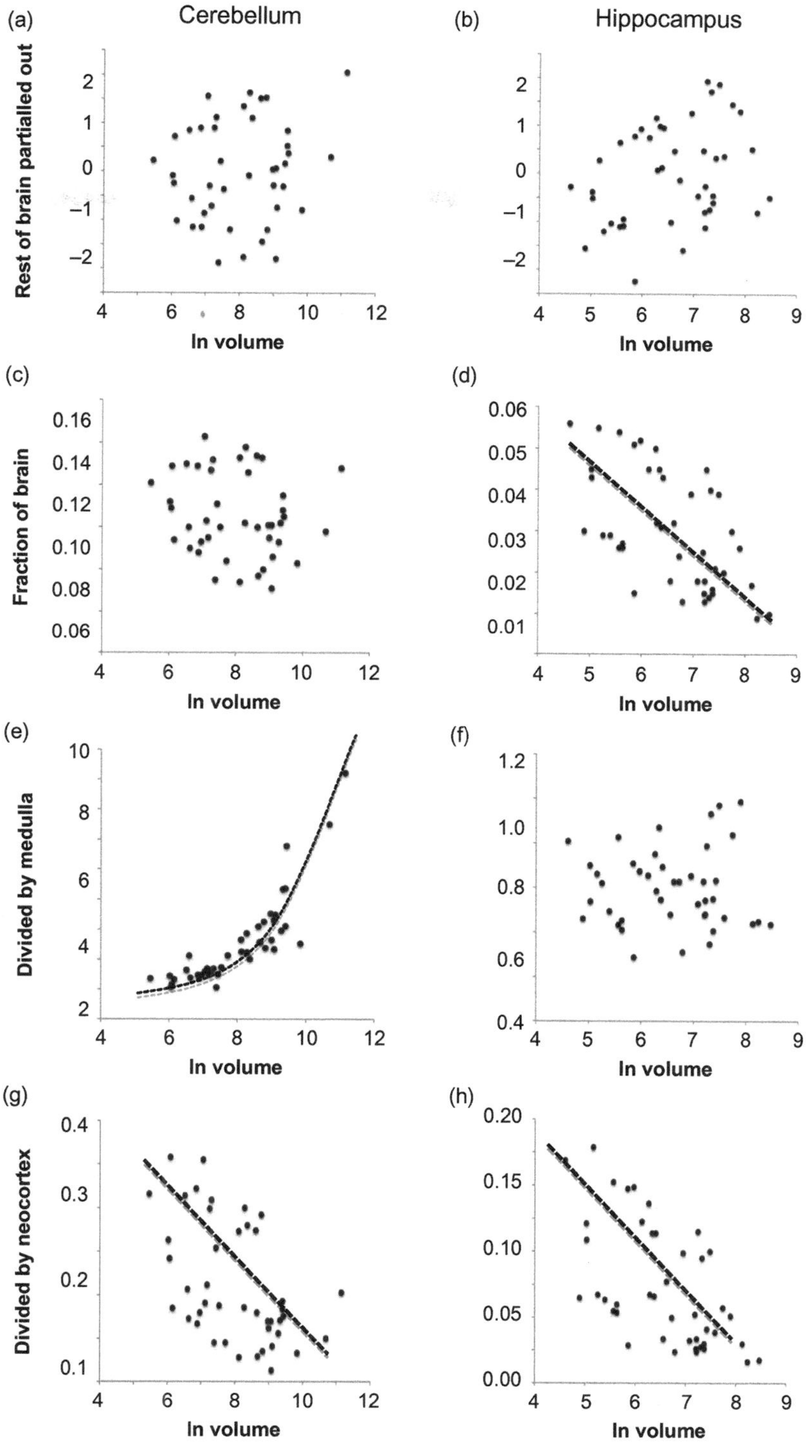

Fig. 3. Left column: ln volume of the cerebellum plotted (a) against ln volume of the cerebellum with volume of the rest of the brain partialled out; (c) against volume of the cerebellum divided by volume of the whole brain; (e) against volume of the cerebellum divided by volume of the medulla; (g) against volume of the cerebellum divided by volume of the neocortex. Right column: ln volume of the hippocampus plotted (b) against ln volume of the hippocampus with volume of the rest of the brain partialled out; (d) against volume of the hippocampus divided by volume of the whole brain; (f) against volume of the hippocampus divided by volume of the medulla; (h) against volume of the hippocampus divided by volume of the neocortex. Abbreviations as in Table 1.

of ASPM to changes in relative neocortex size, but Montgomery et al. (2011) again find little support for this in a wider sample of primates. The strongest relationship in Montgomery et al. (2011) is that between the rate of coding change in the microcephaly genes ASPM and CDK5RAP2 and the absolute size of neonatal primate brains. As Montgomery et al. conclude, this is the relationship that one would predict on causal bases, as microcephaly genes are involved in embryological neurogenesis.

If genes associated with encephalization seem to be selected in primates and if rates of genetic change correlate with brain size differences, is there evidence for brain size changes over macroevolutionary time? Shultz and Dunbar (2010b) have looked at this question in several mammalian lineages. They show that primates have the highest encephalization slope over time of the six mammalian orders they tested (Table 4). At the suborder level, anthropoids have the highest slope, with Strepsirrhines also showing a significant positive slope. Other mammalian orders, for instance, Insectivora, do not show this directional trend toward enlarged brains over evolutionary time.

Conclusion

Primates are by far the best-studied clade in terms of brains and cognition, in part because of the interest in humans shown by neuropsychologists and paleoanthropologists. Robust evidence is now available for several lifestyle, cognitive, and life history correlates of brain and neocortex size. Promising new avenues of research are opening in molecular genetics, with the possibility that some brain regions might be differentially imprinted by the paternal or maternal genome (Keverne et al., 1996; Wilkinson et al., 2007). The main controversies in the field seem to be over the encephalization measures that should be used and whether encephalization results from concerted or mosaic evolution. More embryological work is needed to resolve these issues, beyond the current correlational approaches.

With the current evidence, it is difficult to reject the concept of encephalization as vague and misleading because it deals with the whole brain and neocortex instead of specialized brain areas. First and foremost, the size of the brain and neocortex, whether relative or absolute, varies enormously between clades: a marmoset brain contains 63 million neurons (Herculano-Houzel et al., 2007), while a human brain contains 86 billion (Azevedo et al., 2009). In itself, this variation is worthy of study. Brain evolution is clearly not happening only between specialized areas trading off in volume within unvarying cranial constraints but also in overall brain size and neuron numbers. Second, evidence is mounting in primates (Deaner et al., 2006; Reader and Laland, 2002; Reader et al., 2011), but also in birds (Lefebvre et al., 2004), that the many positive correlations observed between cognitive measures across taxa may in part be subsumed by general intelligence. Third, neuroscientists are increasingly interested in distributed networks of multiple, functionally related brain areas involved in several processes, in contrast to the strict modular view that was dominant a few years ago (Bressler and Menon, 2010). Finally, the trends in Tables 1–4 suggest that both corrected whole brain size and corrected, as well as absolute, neocortex size are robust correlates of several functional variables. In the current state of affairs, the "something about brain enlargement" that was alluded to in the first paragraph of this chapter cannot be pinpointed to only one neuroanatomical level.

The question of absolute versus corrected measures of brain and neocortex size is also difficult to resolve. In Tables 1, 3, and 4, corrected measures seem to be the most successful predictors of functional variables, but good neuroanatomical (Herculano-Houzel et al., 2007) and genetic (Montgomery et al., 2011) arguments have been made in favor of absolute measures, which seem to be better predictors of the cognitive correlates in Table 2. Primates differ from most other clades in that the largest-bodied species, the great apes, are also the largest-brained ones, as well as the ones that show the most complex cognition. In contrast, in groups like birds, cetaceans, and dinosaurs, large brains are found in very large-bodied taxa that show little of the lifestyle or cognitive correlates of encephalization, for example, ostriches, baleen whales, and sauropods. It is only when body-size corrections are made that smaller taxa like corvids, dolphins, and theropods surpass their very large-bodied cousins in relative brain size and reveal the predicted associations with lifestyle and cognition. Whether primates are an exception to a general trend (see Burness et al., 2001; Smith et al., 2010 for the relationship between diet and large size) needs to be determined, taking into account the possibility that the primate equivalent of a baleen whale or a sauropod—a very large-bodied species whose diet would not favor complex cognition—might have recently gone extinct. One candidate here might be *Gigantopithecus*, a 550-kg ape that disappeared 300,000 years ago and whose diet, estimated from dental remains, might have been dominated by bamboo and other high-fiber vegetable foods (Kupczik and Dean, 2008; Wang, 2009).

Another key point for the future is the reconciliation of the macro-anatomical perspective and comparative approach used by researchers interested in encephalization and the much finer techniques used in proximal studies of cognitive processes, which work at the level of single cells, neurotransmitters, receptors, and genes. Bridging this gap in methods and perspectives is crucial. One example of a combined approach is the comparative study of the neuropeptides involved in social and reproductive behavior. The peptides are relatively conserved over several classes, taking slightly different chemical forms in birds, mammals, and fish (Donaldson and Young, 2008; Goodson and Thompson, 2010). Techniques are available to map their receptor distribution in different parts of the brain, as well as identify and manipulate the biochemical (Goodson et al., 2009b) and genomic (Ferguson et al., 2000; Young et al., 1999) differences that cause behavioral differences. Finally, good comparative work taking into account common ancestry and independent evolutionary events has been done on several species that vary in their social behavior (Goodson et al., 2006); similar comparative analyses have also been done on midbrain dopamine neuron numbers (Goodson et al., 2009a). This integration of approaches, which is also used in contemporary studies of bird song, brings together molecular genetics, neuroscience, ecology, behavior, and evolution. With the added insights of embryology, it is an example of possible directions in which research on encephalization might go.

Acknowledgments

I am grateful to Neeltje Boogert, Laura Chouinard-Thuly, Sarah Overington, Daniel Sol, and Romain Willemet for comments on an earlier draft, as well as to NSERC (Canada) and FQRNT (Québec) for funding.

References

Aiello, L. C., & Wheeler, P. (1995). The expensive-tissue hypothesis: The brain and the digestive system in human and primate evolution. *Current Anthropology*, *36*, 199–221.

Ali, F., & Meier, R. (2008). Positive selection in ASPM is correlated with cerebral cortex evolution across primates but not with whole-brain size. *Molecular Biology and Evolution*, *25*, 2247–2250.

Allman, J., McLaughlin, T., & Hakeem, A. (1993). Brain weight and life-span in primate species. *Proceedings of the National Academy of Sciences of the United States of America*, *90*, 118–122.

Azevedo, F. A. C., Carvalho, L. R. B., Grinberg, L. T., Farfel, J. M., Ferretti, R. E. L., Leite, R. E. P., et al. (2009). Equal numbers of neuronal and nonneuronal cells make the human brain an isometrically scaled-up primate brain. *The Journal of Comparative Neurology*, *513*, 532–541.

Barrickman, N. L., Bastian, M. L., Isler, K., & van Schaik, C. P. (2008). Life history costs and benefits of encephalization: A comparative test using data from long-term studies of primates in the wild. *Journal of Human Evolution*, *54*, 568–590.

Barton, R. A. (1996). Neocortex size and behavioural ecology in primates. *Proceedings of the Royal Society. Series B: Biological Sciences*, *263*, 173–177.

Barton, R. A. (1998). Visual specialization and brain evolution in primates. *Proceedings of the Royal Society. Series B: Biological Sciences*, *265*, 1933–1937.

Barton, R. A. (1999). The evolutionary ecology of the primate brain. In P. C. Lee (Ed.), *Comparative primate socioecology* (pp. 167–203). Cambridge: Cambridge University Press.

Barton, R. A. (2004). Binocularity and brain evolution in primates. *Proceedings of the National Academy of Sciences of the United States of America*, *101*, 10113–10115.

Barton, R. A. (2006a). Primate brain evolution: Integrating comparative, neurophysiological, and ethological data. *Evolutionary Anthropology*, *15*, 224–236.

Barton, R. A. (2006b). Olfactory evolution and behavioral ecology in primates. *American Journal of Primatology*, *68*, 545–558.

Barton, R. A. (2007). Evolutionary specialization in mammalian cortical structure. *Journal of Evolutionary Biology*, *20*, 1504–1511.

Barton, R. A., & Harvey, P. H. (2000). Mosaic evolution of brain structure in mammals. *Nature*, *405*, 1055–1058.

Bressler, S. L., & Menon, V. (2010). Large-scale brain networks in cognition: Emerging methods and principles. *Trends in Cognitive Sciences*, *14*, 277–290.

Burish, M. J., Peebles, J. K., Baldwin, M. K., Tavares, L., Kaas, J. H., & Herculano-Houzel, S. (2010). Cellular scaling rules for primate spinal cords. *Brain, Behavior and Evolution*, *76*, 45–59.

Burness, G. P., Diamond, J., & Flannery, T. (2001). Dinosaurs, dragons, and dwarfs: The evolution of maximal body size. *Proceedings of the National Academy of Sciences of the United States of America*, *98*, 14518–14523.

Byrne, R. W., & Corp, N. (2004). Neocortex size predicts deception rate in primates. *Proceedings of the Royal Society. Series B: Biological Sciences*, *271*, 1693–1699.

Byrne, R., & Whiten, A. (1990). Tactical deception in primates: The 1990 database. *Primate Report*, *27*, 1–101.

Charvet, C. J., & Striedter, G. F. (2009). Developmental origins of mosaic brain evolution: Morphometric analysis of the developing zebra finch brain. *The Journal of Comparative Neurology*, *514*, 203–213.

Clark, D. A., Mitra, P. P., & Wang, S. S. H. (2001). Scalable architecture in mammalian brains. *Nature*, *411*, 189–193.

Clutton-Brock, T. H., & Harvey, P. H. (1977). Primate ecology and social-organization. *Journal of Zoology*, *183*, 1–39.

Clutton-Brock, T. H., & Harvey, P. H. (1980). Primates, brain and ecology. *Journal of Zoology*, *190*, 309–323.

Deaner, R., Barton, R., & van Schaik, C. (2003). Primate brains and life histories: Renewing the connection. In P. Kappeler & M. Pereira (Eds.), *Primate life histories and socioecology* (pp. 233–265). Chicago: University of Chicago Press.

Deaner, R. O., Isler, K., Burkart, J., & van Schaik, C. P. (2007). Overall brain size, and not encephalization quotient, best predicts cognitive ability across non-human primates. *Brain, Behavior and Evolution*, *70*, 115–124.

Deaner, R. O., Nunn, C. L., & van Schaik, C. P. (2000). Comparative tests of primate cognition: Different scaling methods produce different results. *Brain, Behavior and Evolution*, *55*, 44–52.

Deaner, R. O., van Schaik, C. P., & Johnson, V. (2006). Do some taxa have better domain-general cognition than others? A meta-analysis of nonhuman primate studies. *Evolutionary Psychology*, *4*, 149–196.

Deary, I. J., Penke, L., & Johnson, W. (2010). The neuroscience of human intelligence differences. *Nature Reviews Neuroscience*, *11*, 201–211.

Dechmann, D. K. N., & Safi, K. (2009). Comparative studies of brain evolution: A critical insight from the Chiroptera. *Biological Reviews*, *84*, 161–172.

DeVoogd, T. J., Krebs, J. R., Healy, S. D., & Purvis, A. (1993). Relations between song repertoire size and the volume of brain nuclei related to song: Comparative evolutionary analyses amongst oscine birds. *Proceedings of the Royal Society. Series B: Biological Sciences*, *254*, 75–82.

de Winter, W., & Oxnard, C. E. (2001). Evolutionary radiations and convergences in the structural organization of mammalian brains. *Nature*, *409*, 710–714.

Donaldson, Z. R., & Young, L. R. (2008). Oxytocin, vasopressin, and the neurogenetics of sociality. *Science*, *322*, 900–904.

Dunbar, R. I. M. (1992). Neocortex size as a constraint on group size in primates. *Journal of Human Evolution*, *20*, 469–493.

Dunbar, R. I. M. (1998). The social brain hypothesis. *Evolutionary Anthropology*, *6*, 178–190.

Dunbar, R. I. M., & Shultz, S. (2007a). Evolution in the social brain. *Science*, *317*, 1344–1347.

Dunbar, R. I. M., & Shultz, S. (2007b). Understanding primate brain evolution. *Philosophical Transactions of the Royal Society of London. Series B, Biological Sciences*, *362*, 649–658.

Evans, P. D., Anderson, J. R., Vallender, E. J., Gilbert, S. L., Malcom, C. M., Dorus, S., et al. (2004). Adaptive evolution of ASPM, a major determinant of cerebral cortical size in humans. *Human Molecular Genetics*, *13*, 489–494.

Evans, P. D., Vallender, E. J., & Lahn, B. T. (2006). Molecular evolution of the brain size regulator genes CDK5RAP2 and CENPJ. *Gene*, *375*, 75–79.

Ferguson, J. N., Young, L. J., Hearn, E. F., Matzuk, M. M., Insel, T. R., & Winslow, J. T. (2000). Social amnesia in mice lacking the oxytocin gene. *Nature Genetics*, *25*, 284–288.

Finlay, B. L., & Darlington, R. B. (1995). Linked regularities in the development and evolution of mammalian brains. *Science*, *268*, 1578–1584.

Finlay, B. L., Darlington, R. B., & Nicastro, N. (2001). Developmental structure in brain evolution. *The Behavioral and Brain Sciences*, *24*, 263–308.

Fish, J. L., & Lockwood, C. A. (2003). Dietary constraints on encephalization in primates. *American Journal of Physical Anthropology*, *120*, 171–181.

Gabi, M., Collins, C. E., Wong, P., Torres, L. B., Kaas, J. H., & Herculano-Houzel, S. (2010). Cellular scaling rules for the brains of an extended number of primate species. *Brain, Behavior and Evolution, 76*, 32–44.

Gayon, J. (2000). History of the concept of allometry. *American Zoologist, 40*, 748–758.

Goodson, J. L., David Kabelik, D., Aubrey, M., Kelly, A. M., & Jacob Rinaldi, J. (2009a). Midbrain dopamine neurons reflect affiliation phenotypes in finches and are tightly coupled to courtship. *Proceedings of the National Academy of Sciences of the United States of America, 106*, 8737–8742.

Goodson, J. L., Evans, A. K., & Wang, Y. (2006). Neuropeptide binding reflects convergent and divergent evolution in species-typical group sizes. *Hormones and Behavior, 50*, 223–236.

Goodson, J. L., Schrock, S. E., Klatt, J. D., Kabelik, D., & Kingsbury, M. A. (2009b). Mesotocin and nonapeptide receptors promote estrildid flocking behavior. *Science, 325*, 862–866.

Goodson, J. L., & Thompson, R. R. (2010). Nonapeptide mechanisms of social cognition, behavior and species-specific social systems. *Current Opinion in Neurobiology, 20*, 784–794.

Graham, K. L. (2011). Coevolutionary relationship between striatum size and social play in nonhuman primates. *American Journal of Primatology, 73*, 314–322.

Healy, S. D., & Rowe, C. (2007). A critique of comparative studies of brain size. *Proceedings of the Royal Society. Series B: Biological Sciences, 274*, 453–464.

Herculano-Houzel, S., Collins, C. E., Wong, P., & Kaas, J. H. (2007). Cellular scaling rules for primate brains. *Proceedings of the National Academy of Sciences of the United States of America, 104*, 3562–3567.

Herculano-Houzel, S., Mota, B., & Lent, R. (2006). Cellular scaling rules for rodent brains. *Proceedings of the National Academy of Sciences of the United States of America, 103*, 12138–12143.

Hofman, M. A. (1983). Energy-metabolism, brain size and longevity in mammals. *The Quarterly Review of Biology, 58*, 495–512.

Hofman, M. A. (1993). Encephalization and the evolution of longevity in mammals. *Journal of Evolutionary Biology, 6*, 209–227.

Hunt, G. R. (1996). Manufacture and use of hook-tools by New Caledonian crows. *Nature, 379*, 249–251.

Isler, K. E., Kirk, C., Miller, J. M. A., Albrecht, G. A., Gelvin, B. R., & Martin, R. D. (2008). Endocranial volumes of primate species: Scaling analyses using a comprehensive and reliable data set. *Journal of Human Evolution, 55*, 967–978.

Isler, K., & van Schaik, C. P. (2006). Metabolic costs of brain size evolution. *Biology Letters, 2*, 557–560.

Joffe, T. H., & Dunbar, R. I. M. (1997). Visual and socio-cognitive information processing in primate brain evolution. *Proceedings of the Royal Society. Series B: Biological Sciences, 264*, 1303–1307.

Johnson, V. E., Deaner, R. O., & van Schaik, C. P. (2002). Bayesian analysis of rank data with application to primate intelligence experiments. *Journal of the American Statistical Association, 97*, 8–17.

Jung, R. E., & Hier, R. J. (2007). The parieto-frontal integration theory (p-fit) of intelligence: Converging neuroimaging evidence. *The Behavioral and Brain Sciences, 30*, 135–187.

Keverne, E. B., Martel, F. L., & Nevison, C. M. (1996). Primate brain evolution: Genetic and functional considerations. *Proceedings of the Royal Society. Series B: Biological Sciences, 262*, 689–696.

Kirk, E. C. (2006). Visual influences on primate encephalization. *Journal of Human Evolution, 51*, 76–90.

Kouprina, N., Pavlicek, A., Mochida, G. H., Solomon, G., Gersch, W., Yoon, Y. H., et al. (2004). Accelerated evolution of the ASPM gene controlling brain size begins prior to human brain expansion. *PLoS Biology, 2*, 653–663.

Kudo, H., & Dunbar, R. I. M. (2001). Neocortex size and social network size in primates. *Animal Behaviour, 62*, 711–722.

Kupczik, K. M., & Dean, C. (2008). Comparative observations on the tooth root morphology of Gigantopithecus blacki. *Journal of Human Evolution, 54*, 196–204.

Lefebvre, L., Reader, S. M., & Sol, D. (2004). Brain, innovation and evolution in birds and primates. *Brain, Behavior and Evolution, 63*, 233–246.

Lehmann, J., & Dunbar, R. I. M. (2009). Network cohesion, group size and neocortex size in female-bonded Old World primates. *Proceedings of the Royal Society. Series B: Biological Sciences, 276*, 4417–4422.

Lemaître, J. F., Ramm, S. A., Barton, R. A., & Stockley, P. (2009). Sperm competition and brain size evolution in mammals. *Journal of Evolutionary Biology, 22*, 2215–2221.

Lewis, K. P., & Barton, R. A. (2004). Playing for keeps—Evolutionary relationships between social play and the cerebellum in nonhuman primates. *Human Nature, 15*, 5–21.

Lewis, K. P., & Barton, R. A. (2006). Amygdala size and hypothalamus size predict social play frequency in nonhuman primates: A comparative analysis using independent contrasts. *Journal of Comparative Psychology, 120*, 31–37.

Lindenfors, P. (2005). Neocortex evolution in primates: The 'social brain' is for females. *Biology Letters, 1*, 407–410.

Lindenfors, P., Nunn, C. L., & Barton, R. A. (2007). Primate brain architecture and selection in relation to sex. *BMC Biology, 5*, 20.

Montgomery, S. H., Capellini, I., Venditti, C., Barton, R. A., & Mundy, N. I. (2011). Adaptive evolution of four microcephaly genes and the evolution of brain size in anthropoid primates. *Molecular Biology and Evolution, 28*, 625–638.

Morris, R. G. M., Garrud, P., Rawlins, J. N. P., & O'Keefe, J. (1982). Place navigation impaired in rats with hippocampal lesions. *Nature, 297*, 681–683.

Nottebohm, F., Stokes, T. M., & Leonard, C. M. (1976). Central control of song in canary, Serinus canarius. *The Journal of Comparative Neurology*, *165*, 457–486.

O'Keefe, J., & Dostrovsky, J. (1971). The hippocampus as a spatial map. Preliminary evidence from unit activity in the freely moving rat. *Brain Research*, *34*, 171–175.

Overington, S. E., Dubois, F., & Lefebvre, L. (2008). Resource unpredictability drives both generalism and social foraging: A game theoretical model. *Behavioral Ecology*, *19*, 836–841.

Pagel, M. D., & Harvey, P. H. (1989). Taxonomic differences in the scaling of brain on body-weight among mammals. *Science*, *244*, 1589–1593.

Paton, J. A., & Nottebohm, F. (1984). Neurons generated in the adult brain are recruited into functional circuits. *Science*, *225*, 1046–1048.

Pawlowski, B., Lowen, C. B., & Dunbar, R. I. M. (1998). Neocortex size, social skills and mating success in primates. *Behaviour*, *135*, 357–368.

Pitnick, S., Jones, K. E., & Wilkinson, G. S. (2006). Mating system and brain size in bats. *Proceedings of the Royal Society. Series B: Biological Sciences*, *273*, 719–724.

Prior, H., Schwarz, A., & Gunturkun, O. (2008). Mirror-induced behavior in the magpie (Pica pica): Evidence of self-recognition. *PLoS Biology*, *6*, 1642–1650 e202.

Reader, S. M., Hager, Y., & Laland, K. N. (2011). The evolution of primate general and cultural intelligence. *Philosophical Transactions of the Royal Society of London. Series B, Biological Sciences*, *366*, 1017–1027.

Reader, S. M., & Laland, K. N. (2002). Social intelligence, innovation and enhanced brain size in primates. *Proceedings of the National Academy of Sciences of the United States of America*, *99*, 4436–4441.

Riddell, W. I., & Corl, K. G. (1977). Comparative investigation of the relationship between cerebral indices and learning abilities. *Brain, Behavior and Evolution*, *14*, 385–398.

Sacher, G., & Staffeldt, E. (1974). Relation of gestation time to brain weight for placental mammals: Implications for the theory of vertebrate growth. *The American Naturalist*, *108*, 593–615.

Sarko, D. K., Catania, K. C., Leitch, D. B., Kaas, J. H., & Herculano-Houzel, S. (2009). Cellular scaling rules of insectivore brains. *Frontiers in Neuroanatomy*, *3*, 1–12.

Sawaguchi, T., & Kudo, H. (1990). Neocortical development and social-structure in primates. *Primates*, *31*, 283–289.

Schillaci, M. A. (2006). Sexual selection and the evolution of brain size in primates. *PLoS One*, *1*, 1–15, e62.

Schillaci, M. A. (2008). Primate mating systems and the evolution of neocortex size. *Journal of Mammalogy*, *89*, 58–63.

Sherry, D. F. (2011). The hippocampus of food-storing birds. *Brain, Behavior and Evolution*, *78*, 133–135.

Sherry, D. F., & Vaccarino, A. L. (1989). Hippocampus and memory for food caches in black-capped chickadees. *Behavioral Neuroscience*, *103*, 308–318.

Shultz, S., & Dunbar, R. I. M. (2010a). Species differences in executive function correlate with hippocampus volume and neocortex ratio across nonhuman primates. *Journal of Comparative Psychology*, *124*, 252–260.

Shultz, S., & Dunbar, R. I. M. (2010b). Encephalization is not a universal macroevolutionary phenomenon in mammals but is associated with sociality. *Proceedings of the National Academy of Sciences of the United States of America*, *107*, 21582–21586.

Smaers, J. B., Schleicher, A., Zilles, K., & Vinicius, L. (2010). Frontal white matter volume is associated with brain enlargement and higher structural connectivity in anthropoid primates. *PLoS One*, *5*, e9123.

Smith, F. A., Boyer, A. G., Brown, J. H., Costa, D. P., Dayan, T., Ernest, S. K. M., et al. (2010). The evolution of maximum body size of terrestrial mammals. *Science*, *330*, 1216–1219.

Stephan, H., Frahm, H., & Baron, G. (1981). New and revised data on volumes of brain structures in insectivores and primates. *Folia Primatologica*, *35*, 1–29.

Striedter, G. F., & Charvet, C. J. (2008). Developmental origins of species differences in telencephalon and tectum size: Morphometric comparisons between a parakeet (Melopsittacus undulatus) and a quail (Colinus virgianus). *The Journal of Comparative Neurology*, *507*, 1663–1675.

Striedter, G. F., & Charvet, C. J. (2009). Telencephalon enlargement by the convergent evolution of expanded subventricular. *Biology Letters*, *5*, 134–137.

Walker, R., Burger, O., Wagner, J., & Von Rueden, C. R. (2006). Evolution of brain size and juvenile periods in primates. *Journal of Human Evolution*, *51*, 480–489.

Wang, W. (2009). New discoveries of *Gigantopithecus blacki* teeth from Chuifeng Cave in the Bubing Basin, Guangxi, south China. *Journal of Human Evolution*, *57*, 229–240.

Whiten, A., & Byrne, R. W. (1988). Tactical deception in primates. *The Behavioral and Brain Sciences*, *11*, 233–244.

Whiting, B. A., & Barton, R. A. (2003). The evolution of the cortico-cerebellar complex in primates: Anatomical connections predict patterns of correlated evolution. *Journal of Human Evolution*, *44*, 3–10.

Wilkinson, L. S., Davies, W., & Isles, A. R. (2007). Genomic imprinting effects on brain development and function. *Nature Reviews Neuroscience*, *8*, 832–843.

Yopak, K. E., Lisney, T. J., Darlington, R. B., Collin, S. P., Montgomery, J. C., & Finlay, B. L. (2010). A conserved pattern of brain scaling from sharks to primates. *Proceedings of the National Academy of Sciences of the United States of America*, *107*, 12946–12951.

Young, L. J., Nilsen, R., Waymire, K. G., MacGregor, G. R., & Insel, T. R. (1999). Affiliative response to vasopressin in mice expressing the v1a receptor from a monogamous vole. *Nature*, *400*, 766–768.

M. A. Hofman and D. Falk (Eds.)
Progress in Brain Research, Vol. 195
ISSN: 0079-6123

CHAPTER 20

Evolution of the brain and intelligence in primates

Gerhard Roth* and Ursula Dicke

Brain Research Institute, University of Bremen, Bremen, Germany

Abstract: Primates are, on average, more intelligent than other mammals, with great apes and finally humans on top. They generally have larger brains and cortices, and because of higher relative cortex volume and neuron packing density (NPD), they have much more cortical neurons than other mammalian taxa with the same brain size. Likewise, information processing capacity is generally higher in primates due to short interneuronal distance and high axonal conduction velocity. Across primate taxa, differences in intelligence correlate best with differences in number of cortical neurons and synapses plus information processing speed. The human brain stands out by having a large cortical volume with relatively high NPD, high conduction velocity, and high cortical parcellation. All aspects of human intelligence are present at least in rudimentary form in nonhuman primates or some mammals or vertebrates except syntactical language. The latter can be regarded as a very potent "intelligence amplifier."

Keywords: brain size; cortex size; cortical neuron number; information processing capacity; brain–body relationship; unique brain properties.

Introduction

Primates are commonly believed to be more intelligent than other animals and humans to be the most intelligent creatures on earth. Further, these alleged differences in intelligence are linked by neuroscientists and psychologists to differences in structure and function of the respective brains. In this chapter, we will critically review these assumptions.

In the first part of this chapter, we will ask how nonhuman intelligence can be defined and compared to human intelligence and discuss the findings on the respective forms and degrees of intelligence found in primate taxa. We will then investigate to what degree possible differences in intelligence can be related to properties of the brains of the taxa compared. In the second,

*Corresponding author.
Tel.: 0049 21862950; Fax: 0049 21862969
E-mail: gerhard.roth@uni-bremen.de

DOI: 10.1016/B978-0-444-53860-4.00020-9

shorter, part we will ask, to what degree correlations between forms of intelligence and brain properties found in primates can be applied to nonprimate mammals and to other groups of vertebrates considered to be highly intelligent, for example, some bird taxa, before we draw a general conclusion. Given the restricted space, it is impossible to cite all relevant literature. Therefore, in the following, we mainly confer to recent literature on intelligent behavior that enlarges the view of comparative facets.

Definition and putative causes of animal intelligence

In humans, intelligence is commonly defined as the sum of mental capacities such as abstract thinking, understanding, communication, reasoning, learning and memory formation, action planning, and problem solving. Usually, human intelligence is measured by intelligence tests and expressed in intelligence quotient (IQ) values expressing different contents (e.g., visual–spatial, verbal, numerical). There is a popular distinction proposed by Cattell (1963) between fluid and crystallized intelligence, where fluid intelligence is considered to be closely related to general intelligence "g" (Spearman, 1904) as a broad ability to reason, form concepts, and solve problems using unfamiliar information or novel procedures, while crystallized intelligence includes the breadth and depth of a person's acquired knowledge, the ability to communicate one's knowledge and to reason using previously learned experiences.

Evidently, such a definition and measurement of intelligence cannot be applied directly to nonhuman animals, because any test depending on verbalization is inapplicable. A number of comparative and evolutionary psychologists and cognitive ecologists converge on the view that mental or behavioral flexibility is a good measure of intelligence culminating in the appearance of novel solutions not part of the animal's normal repertoire (Byrne, 1995; Gibson et al., 2001; Gould, 2003; Roth and Dicke, 2005).

Supposed differences in intelligence among primate taxa

In the following, we concentrate on the most frequently used paradigms for measuring animal intelligence, including tool use and tool fabrication, gaze following, imitation, intentional action and social learning, quantity representation, object permanence, metacognition/metamemory, theory of mind (ToM), and language.

Tool use and tool fabrication

Ring-tailed lemurs have recently been reported to successfully manipulate a puzzle feeder in the wild (Kendal et al., 2010), which is the only known case of lemur tool use in the wild. In captivity, however, lemur manipulatory skills with novel objects are roughly comparable with those of some New and Old World monkeys. The gray mouse lemur mastered to open boxes in different ways including the use of reversed images, and aye-ayes demonstrated basic understanding of features of tools by solving a can-pulling task (cf. Fichtel and Kappeler, 2010). Systematic tool use including limited forms of tool making is found in the capuchin monkey (Ottoni and Izar, 2008; Visalberghi et al., 2009). Iriki and Sakura (2008) argue that in Japanese monkeys, latent cognitive abilities are widened by exposure to a proper environment. Further, the successful training of tool use induced physiological, anatomical, and molecular-genetic changes in the brains of the animals.

Chimpanzees are known to fabricate and use a wide range of complex tools and have been shown to vary in their tool use at many levels. In chimpanzee populations, tool kits consist of about 20 types of tools for various functions. Only chimpanzees appear to be able to use one type of raw material to make different kinds of tools or make one kind of tool from different raw materials. They use tool sets in a sequential order, make use of composite tools and combine tools to a single working unit (McGrew, 2010; Sanz and Morgan, 2009). In this context, chimpanzees and orangutans exhibit insightful problem solving (Mendes et al., 2007; Osvath and Osvath, 2008). They are engaged in action planning, mentally pre-experience an upcoming event, are able to select objects needed for a much-delayed future in tool use.

Gaze following

Lemurs preferentially orient their eyes toward other lemurs and mirror the attentional state of others in their social group (Shepherd and Platt, 2008). A study on gaze orientation and object-choice used a color photo of a conspecific as model, oriented with eyes and head to a right- or left-sided reward (Ruiz et al., 2009). The response of the lemurs to the models' gaze significantly influenced their choice behavior. The authors define this as gaze priming. In monkeys, a mentalistic understanding of the observing animal about the others' visual target remains an open question. In long-tailed macaques, gaze following was accompanied by frequent check-looks and was significantly more frequent in response to a signal of fear and submission than to a neutral facial expression (Goossens et al., 2008). Capuchin and spider monkeys spontaneously followed a human experimenter's gaze, and capuchin monkeys followed the gaze around barriers, but neither capuchin nor spider monkeys displayed any "looking back" behavior (Amici et al., 2009) and, thus, might lack perspective-taking. Marmosets showed high proficiency in extrapolating gaze direction, but failed to show context-independent perspective-taking (Burkart and Heschl, 2007). At present evidence, monkeys appear to deal with a directed gaze without understanding visual perspective. Great apes are able to track gaze to hidden targets and look back to the human experimenter, when they do not find a target (Bräuer et al., 2005; Tomasello et al., 2007). However, great apes use both head and eye direction in gaze following, while human infants are much more attuned to the eyes.

Imitation, intentional action, and social learning

Bates and Byrne (2010) classified several types of imitation. Response facilitation is found in a wide range of animals and means that seeing an action "primes" the individual to do the same. Also, imitation occurs, when a social signal is conveyed; this type of social mimicry may depend on the action copied and the motivation behind the copying. Further, contextual imitation, found in monkeys and apes, includes learning to employ an action already in the repertoire. Production imitation stands for learning a new motor skill by observation and further comprises program-level and rational imitation. In the former, fine detail is unimportant as long as the right result is obtained, while in the latter form an understanding of the logic of how actions achieve their ends is present. Chimpanzees and other great apes show imitative abilities beyond those of other primates. The recent view is that great apes display program-level imitation, explicit recognition of imitation, rational imitation, are capable of mentalizing about others and have some understanding of intentionality and causality.

It is unclear whether copying an expert's use of a rule rather than just copying a certain motor behavior found in macaques (Subiaul et al., 2004) evidences contextual or production imitation at the monkey level. In macaques, posterior parietal and frontal areas including the much discussed "mirror neurons" in frontal area F5 are dedicated to the execution and recognition of meaningful hand reaching and grasping as well as facial movements (Rizzolatti and Craighero, 2004), but their significance for imitation remains unclear.

Chimpanzees are able to distinguish between an experimenter who is either unwilling or unable to give them food. Hence, they do not simply perceive the behavior of others but also interpret it

(Call et al., 2004). Recently, capuchin monkeys were shown to distinguish between intentional agents and unintentional objects (Phillips et al., 2009).

Learning from others' mistakes is as important as copying others' actions. Apes and children differ in the social learning mechanisms they use in problem solving. The tool use of a human demonstrator to retrieve an invisible reward from a puzzle box was reproduced by chimpanzees imitating the overall structure of the task. In the visible condition, chimpanzees ignored the irrelevant actions in favor of a more efficient, emulative technique, while children employed imitation to solve the task in both conditions, at the expense of efficiency (Horner and Whiten, 2005). Capuchin monkeys were unable to spontaneously compensate failures of a human demonstrator, who showed the monkeys an action to open or fail to open a baited box. However, when a conspecific was watched and failed to open the box, the other monkey successfully opened it (Kuroshima et al., 2008). Monkeys were able to refer to the outcome of the others' action as well as to the others' action *per se*, which suggests that monkeys, like humans and great apes, may understand the meaning of others' actions in social learning.

Quantity representation

Lemurs are capable of controlling their impulsive gesture toward a larger option, when selection of a smaller quantity of food is rewarded with a larger one. They also learned to associate a graphic representation of the reward with the corresponding quantity, even though only one subject consistently selected the representation of the smaller quantity to be rewarded with the larger quantity of food (Genty and Roeder, 2011). The fundaments of abstraction appear to be present in prosimians. Nevertheless, numerical discrimination is superior in monkeys and apes. Capuchin monkeys are able to judge larger quantities of two sets contrasting up to five items in food-choice experiments (Evans et al., 2009). Quantity-based judgments for two sets with up to 10 items were tested in rhesus monkeys and great apes. Rhesus monkeys selected the larger of the two sequentially presented sets reliable when one set had fewer or more than four items (Beran, 2007), whereas great apes did so, even when the quantities were large and the numerical distance between them was small (Hanus and Call, 2007).

Object permanence

Object permanence is divided into six major stages according to its gradual development in humans (Piaget, 1954). At stage 4, human infants are able to mentally represent and retrieve an object hidden in a single hiding place. When the object is visibly placed into a new hiding place, infants continue to search the initial location. Stage 5 of object permanence characterizes the ability to find an object that has been hidden successively in multiple locations, whereas at stage 6 direct perception of an object is no longer required to infer an object's location. Lemurs successfully found objects (raisins) being visible displaced and thus fulfill stage 5; they are capable of understanding and mentally representing visible displacements (Deppe et al., 2009). Lemurs and squirrel monkeys did not correctly locate objects during invisible displacements (stage 6). Tamarins, however, accurately selected visibly displaced items and were successful at finding objects in invisible displacement tasks (Neiworth et al., 2003). The ability to locate an invisible moving object has consistently been reported in great apes and humans (Barth and Call, 2006; Collier-Baker et al., 2006).

Metamemory

Metamemory defines the ability to monitor one's own memory. Smith (2009) compared the discrimination or memory capacities of visual or acoustic pattern. In a delayed matching sample task rhesus monkeys revealed high accuracies in chosen memory tests with an escape option, but not in forced memory tests. Rhesus macaques are able to monitor their performance on a series of perceptual tasks and to transfer that ability to a qualitatively different task based on working memory (WM). Macaques also can apply control strategies to correct deficiencies in their knowledge (Kornell et al., 2007). In one of two tested capuchin monkeys, matching accuracy on chosen memory tests decreased more slowly as a function of delay length. The subject was able to recognize the strength of his own memory trace, but not the content of the memory (Fujita, 2009). When tested in a delayed position discrimination task with an escape option, orangutans made more frequent choices for escape when they were ignorant of the baited location (Suda-King, 2008). It appears that the capacity for metamemory is widespread among primates.

Theory of mind

ToM is the ability to understand and take into account another individual's mental state (Premack and Woodruff, 1978). In humans, ToM and the understanding that a person can hold a false belief develop between the ages of 3–4years and is fully developed only at the age of 5. O'Connell and Dunbar (2003) studied chimpanzees, a group of autistic children (assumed to lack ToM) and children at ages between 3 and 6 years. "False belief" was tested using nonverbal tests. The chimpanzees performed better than autistic and 3-year-old normal children; they were equal to 4–5-year-old and inferior to 6-year-old children. This would corroborate the idea that chimpanzees exhibit at least some aspects of ToM. At present, the capability of ToM in nonhuman primates remains controversial. Call and Tomasello (2008) report that chimpanzees understand the goals and intentions of others as well as the perception and knowledge of others but found no evidence for understanding false beliefs, while Penn and Povinelli (2007) argue that there is no evidence that nonhuman animals possess anything remotely resembling ToM.

Syntactical–grammatical language

Most linguists agree that human language strongly differs from languages found in nonhuman animals in the sense that (1) there is no relationship between a sound or sign and its meaning, (2) human language is syntactical having complex rules and principles for the construction of sentences to create meaning, (3) language can be used to communicate ideas about things not present, and (4) a finite number of units can be used to create an indefinitely large number of utterances. Sentences consisting of up to three words appear to be understood and used by chimpanzees, gorillas, and dolphins. Savage-Rumbaugh et al. (1993) demonstrated that the 8-year-old bonobo *Kanzi* who was raised in a language environment similar to that of children shows linguistic capabilities typical of a 2-year-old girl, but *Kanzi* does not go beyond these abilities despite long-lasting training. There is a controversy about the roots from which human language evolved (vocal, mostly affective-emotional communication vs. visual communication/gestures, mimic, or a combination of both) (Goldin-Meadow, 1999). It is also argued that speech and gesture develop in parallel phylogenetically and ontogenetically (Kelly et al., 2002). Accordingly, the ability of humans to use language without accompanying gestures would only be a further specialization, because under normal conditions humans use both components (Kelly et al., 2002). The debate of the dominance of either gestural or vocal precursors in the evolution of speech continues (Cartmill and Byrne, 2010; Seyfarth et al., 2005; see also Chapter 22), although some researchers argue that both gesture and vocal precursors might have been there from the start and worked synergistically (Balter, 2010).

In summary, our survey of recent data suggests that intelligent behavior is distributed across nonhuman primates in a much more overlapping manner than previously thought (Fig. 1). Prosimians exhibit manipulatory, perceptual, and cognitive capacities, although often only as a basic ability. New World monkeys possess moderate capacities in various cognitive domains that partially overlap with those of Old World monkeys. The behavior of the latter shares characteristics with apes, although great apes clearly outperform the other nonhuman primate taxa in most respect, with humans on top.

Neural correlates of intelligence in primates

Neuroscientists generally agree that differences in intelligence of animals including humans are related to properties of their brains. These may either regard *general* properties such as absolute or relative size of the entire brain or of the cortex, or *specialties* in anatomy or physiology of the brains and cortices that may determine "information processing capacity" (IPC), and finally centers or functions that are found in some groups or only in one group of primates (e.g., humans).

Absolute brain size

Absolute brain size (ABS, cm^3 or g) is the most fundamental brain trait. Many authors assume that animals with larger brains are more intelligent (cf. Gibson et al., 2001; Jerison, 1973). ABS mostly depends on body size. Figures 2 and 3 demonstrate that in vertebrates including mammals, ABS increases with body size at the exponential function of $E=kP^{\alpha}$, in which E and P are brain and body weights or volumes, respectively, and k and α are constants. In double-logarithmic transformation, this becomes the linear equation $\log E=\log k-\alpha \log P$, where k is the intercept with the y-axis and α is the slope of the line. The exact value of α is still a matter of debate; for

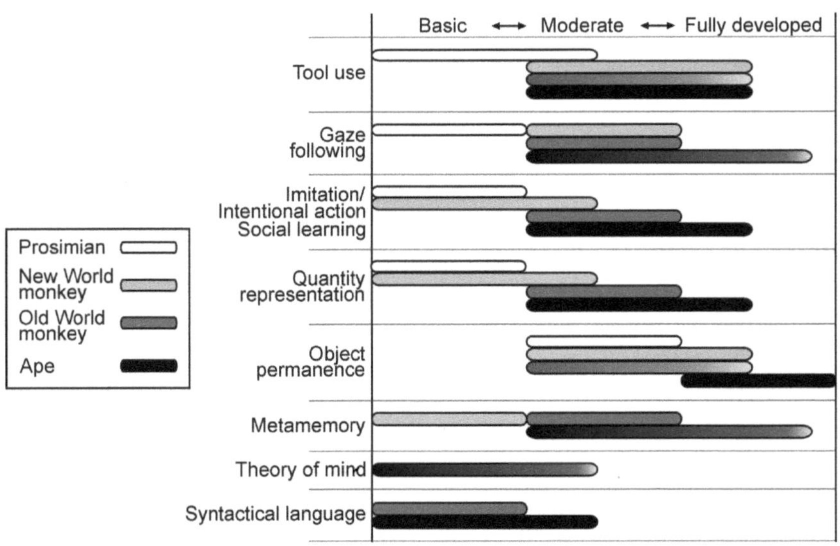

Fig. 1. Cognitive-intelligent competences in nonhuman primates. Recent data on the performance of species reported in the wild, in captivity, and after training in captivity are included. Shading of horizontal bars indicates that the range of performance is either debated or not investigated.

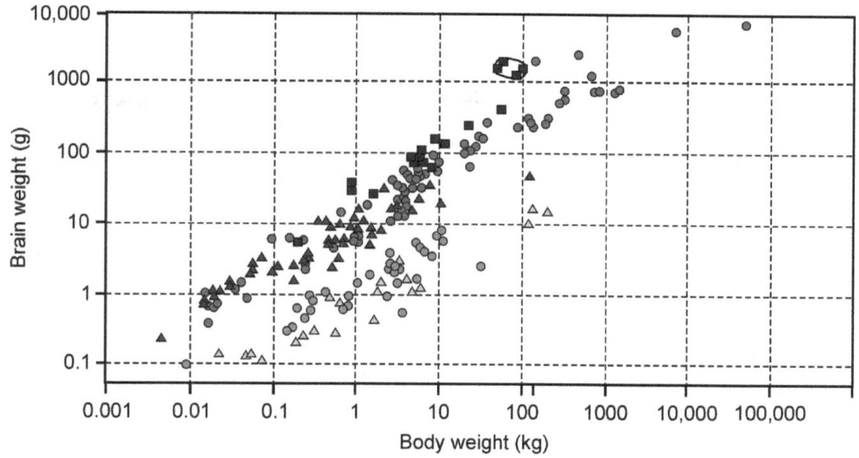

Fig. 2. The relationship between brain size and body size in vertebrates. Double-logarithmic scaling. Purple circles: bony fishes; yellow triangles: reptiles; red triangles: birds; blue circles: mammals except primates; green squares: primates; encircled green squares: *Homo sapiens*. After Jerison (1973). (For interpretation of the references to color in this figure legend, the reader is referred to the Web version of this chapter.)

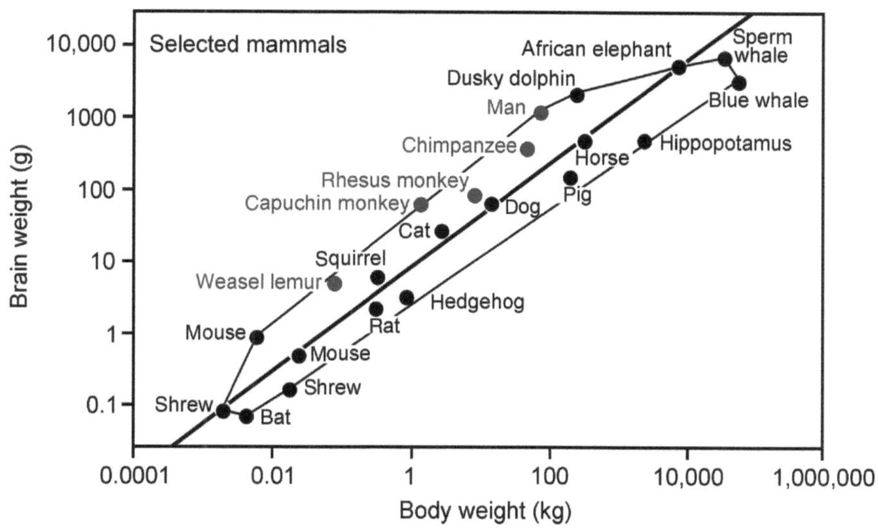

Fig. 3. Relationship between brain weight (g) and body weight (kg) in selected mammals in double-logarithmic scaling. Primates are shown in green. Two different shrew and mice species are shown. Data points above and below the regression line, respectively, indicate the positive and negative deviation (residuals) of brain sizes from average mammalian brain–body ratio. From van Dongen (1998), modified.

vertebrates in general, von Bonin (1937) and Jerison arrived at a value of 2/3, Martin (1996) of 3/4. Thus, in most vertebrates and most orders of mammals, brain size increases at a slower pace than body size. This is called *negative brain allometry*. In primates, however, an *isometric* growth of brain size (with $\alpha=1$) is found (Herculano-Houzel et al., 2007). Finally, in extinct hominins plus living *Homo sapiens* α amounts to 1.73 (Pilbeam and Gould, 1974), which is the steepest increase in size during the entire brain evolution.

Mammals and birds have about 10 times larger brains than bony fishes, amphibians, and reptiles of the same body size (Fig. 2; Jerison, 1973). Within mammals, primates, with the exception of prosimians, generally have larger brains than the other orders with the same body size (Fig. 3). In primates, ABS ranges from 1.67g in the prosimian mouse lemur *Microcebus* to 1350g in *H. sapiens* (Table 1). Generally, prosimians and tarsiers have relatively small brains with a range of 1.67–12.9g (average 6.7g), followed by New World monkeys with a range of 9.5–118g (average 45g) and Old World monkeys with a range of 36–222g (average 115g), with the largest monkey brains found in baboons. Among apes, gibbons have brain sizes (105–135g) within the range of Old World monkeys, while the large apes, that is, orangutans, gorillas, and chimpanzees, have brain weights between 330 and 570g (males). Thus, in extant primates, we recognize five non-overlapping or only slightly overlapping groups with respect to brain size: (1) prosimians and tarsiers, (2) New World monkeys, (3) Old World monkeys and hylobatids, (4) the large apes, and (5) extant humans. The gap between nonhuman apes and humans is filled by brains of extinct australopithecines (e.g., *Australopithecus afarensis*, *A. africanus*) having reconstructed brain sizes of 343–515cc, *Homo habilis* of 600–780cc, and *H. erectus* of 909–1149cc (Jerison, 1973). The largest hominine brain, that of *H. neanderthalensis*, had a mean weight of 1487cc (Falk, 2007).

Whether ABS is the appropriate measure for determining the level of "encephalization" and intelligence is a matter of debate, because brain size mostly depends on body size (cf. Hofman, 2001; Jerison, 1973).When we compare across primate taxa average ABS with the levels of intelligence described above, we get a reasonably good fit: the lowest level of intelligence is found in the small-brained prosimians, while the more intelligent New World and Old World monkeys have significantly larger brains. Great apes are more intelligent than monkeys, and their brains are significantly larger. Finally, humans, undoubtedly much smarter than the great apes, have much larger brains. Thus, it seems that larger brains mean higher intelligence or that "bigger is better" (Gibson et al., 2001).

A closer look at the distribution of ABS *within* primate taxa reveals remarkable deviations from that general conclusion. There is an essential overlap in intelligence between New and Old World monkey, but the latter has a much larger ABS on average. The intelligence of baboons does not significantly exceed that of the other monkeys, while their brains are twice as large as those of macaques. Among the great apes, gorillas have considerably larger brains (570g) than orangutans (395g) or chimpanzees (440g), but their cognitive performance is somewhat lower than that of the other two, which exhibit roughly equal intelligence. These inconsistencies occur, when primates have brain–body relationship that strongly deviates from average primate brain–body relationship. Thus, body size has to be taken into account as a "confounding variable," when studying the relationship between brain and intelligence.

Encephalization and corrected relative brain size

Jerison (1973) tried to account for this fact by calculating the "encephalization quotient" $EQ=E_a/E_e$, which indicates the extent to which the brain size of a given species E_a deviates from the expected brain size E_e. Within primates, humans have the highest EQ of 7.4–7.8 meaning that the human brain is seven to eight times larger than an average mammal of the same body size (Table 1). They are followed by *Cebus* and *Saimiri* with EQs of 4.8 and 2.8, respectively, while chimpanzees and orangutans have low (1.7 and 1.9, respectively) and gorillas very low EQs (1.5). Evidently, such EQ ranking is inconsistent with the above reported levels of intelligence.

Table 1. Brain and body sizes, encephalization quotient (EQ), and "extra" neurons (N_c) in primates

Primate taxa/species	Body size (kg)[a]	Brain size (g)[a]	EQ[a]	N_c ($\times 10^9$)[a]
Prosimians				
Microcebus murinus	65	1.67	0.86	
Lepilemur ruficaudatus	550	7.15	0.88	
Tarsiiformes				
Tarsius spectrum	120	3.76	1.29	
Prosimians and tarsiers			0.6–1.39	
Simiiformes, Platyrrhini				
Alouatta villosa	7824	66	1.38	0.88
Cebus albifrons	1640	80	4.79	1.28
Cebus capucinus	3765	74	2.54	1.10
Cebinae			2.54–4.79	
Saimiri sciureus	630	25	2.81	0.54
Callicebus moloch	670	18	1.92	0.40
Ateles paniscus	7400	106	2.33	1.39
Atelinae			2.33–2.48	
New World monkeys (n=15)			0.88–4.79	0.81
Simiiformes, Catarrhini				
Cercopithecini				
Cercopithecus aethiops	4819	73	2.14	1.06
Cercopithecus talapoin	1380	41	2.76	0.76
Cercopithecus (n=8)			1.66–2.18	1.03
Macaca mulatta	8719	106	2.09	1.36
Macaca fascicularis	7080	80.5	1.81	1.09
Papio cynocephalus	22,220	213	2.24	2.19
Papio anubis	35,000	222	1.73	2.12
Papio (n=5)			1.73–2.35	2.03
Cercopithecoidea			1.66–2.76	1.37
Old World monkeys (n=24)				
Hominoidea				
Hylobates lar	5700	105	2.74	1.42
Symphalangus s. (Siamang)	12,744	133	2.03	1.57
Hylobatidae (n=3)			2.35	1.30
Pongo pygmaeus (male)	90,720	395	1.63	3.07
Pan troglodytes (male)	56,690	440	2.48	3.62
Gorilla gorilla (male)	172,370	570	1.53	3.86
Nonhuman hominids (males)			1.88	3.22
Australopithecines	31,000	455	3.82	3.91
Homo sapiens (male)	55,500	1361	7.79	8.83
Homo sapiens (female)	51,500	1228	7.39	8.21

[a]Data from Jerison (1973).

Partly for these reasons, Jerison (1973) distinguished between brain parts necessary for the maintenance and control of the body (E_v) and those associated with improved cognitive capacities (E_c), in mammals mostly cortex, which Jerison called "extra neurons" (N_c). Calculating the number of such "extra neurons" removes some striking inconsistencies in the EQ list (cf. Table 1).

For example, while *Cebus albifrons* and *C. apella* have unusually high EQs compared even with large-brained apes, their N_c is much lower than that of the latter and even lower than that of the Old World monkeys. As with EQ, there is a huge gap between the great apes (3.2) and humans (8.8 for male *H. sapiens*), which, however, can be filled by an average N_c of 3.9 in australopithecines (Jerison, 1973).

Gibson et al. (2001) compared ABS, EQ, and N_c in a number of primate species with their mental flexibility defined as the ability for reversal in reward learning or "switching" and expressed as "transfer index" (TI). This TI can achieve positive values, if animals are easily capable to switch, or negative values, if they tend to stay with a once learned reward value. The TIs obtained in 12 primate species correlated significantly with body size, brain size or cranial capacity, and extra neurons. No significant correlation was found between TI and EQ: the New World monkeys with very high EQs had low TI, while the gorilla with a low EQ had the highest TI among all nonhuman primates. In summary, prosimians performed poorly on the test, monkeys exhibited intermediate TIs, and the great apes showed the best performance. The authors conclude that bigger brains are better for intelligence, because they contain more "extra neurons."

In a meta-analysis, Deaner et al. (2007) tested ABS and cortex size, cortex-to-brain ratio, EQ, and corrected relative brain size (cRBS) against global cognitive capacities such as reversal learning, delayed response, invisible displacement, detour behavior, object discrimination learning, and tool use. They found that ABS and neocortex size highly predicted cognitive ability, followed by neocortex-to-brain ratio, while EQ or cRBS did not yield significant results (see also Chapter 19). After correcting for common phylogeny, only correlations between ABS and neocortex size on the one hand and global cognition on the other remained significant.

It appears that in primates ABS and neocortex size, as well as N_c, correlate fairly well with global cognitive ability. However, there still remain some inconsistencies with N_c. For example, while the surprisingly low EQ of 1.9 found in the gorilla is now "adjusted" to a N_c of 3.9, this value is much higher than that found in the orangutan (3.0) and even higher than in the chimpanzee (3.6). Also, baboons have much higher N_c than *Cebus* or *Macaca*, while their cognitive performance does not exceed that of these species. These inconsistencies are largely due to the fact that the calculation of N_c was based on the assumption that the number of cortical neurons increases steadily with cortical surface, which in this generality is incorrect, because it does not consider the variability in cortical thickness and cortical neuron density (see below).

Social and ecological intelligence and brain

Dunbar and colleagues argue for a close correlation between the ratio of cortex to the rest of the brain on the one hand and social group size as well as complexity of social relationships on the other. Recently, Dunbar and Shultz (2007) confirmed that in monkeys, apes, and humans relative neocortex size correlates significantly with mean social group size. They argue that social cohesion requires a high level of behavioral flexibility. In another recent article, they compared cRBS with an index of sociality in primates, carnivores, and ungulates (Pérez-Barberia et al., 2007). This index depends on whether or not animals form regular associations with a minimum number of two adult conspecifics or spend most of the year alone. Carnivores and ungulates exhibit a significant correlation between cRBS and sociality, but in primates this holds only when applying a more complex index with several levels of sociality.

There is less evidence for a significant correlation between ecological intelligence and increase in brain or cortex size in primates. Deaner et al. (2000) reported a correlation between cRBS and home range, and the same appears to be true for tool use frequency (Lefebvre et al., 2004;

Reader and Laland, 2002; Chapter 19). Walker et al. (2006) emphasize that in primates, diet, group size, and life span significantly correlate with cRBS before correction for phylogeny, whereas after such a correction only home range remains. In primates, a correlation between ABS or cRBS on the one hand and either ecological or social intelligence remains controversial, when data are corrected for phylogeny, with the possible exception of home range.

Information processing capacity

It has been argued that differences in intelligence are most intimately connected to the IPC of brains, which, in turn, underlies both social and ecological intelligence (Hofman, 2001; Chapter 19). Brain IPC is largely determined by (1) *storage capacity or memory*, which in turn largely depends on the number of neurons and synapses; (2) *information processing speed*, which depends on interneuronal distance, axonal conduction velocity, and speed of synaptic transmission; and (3) *connectivity pattern* of neurons. These three factors may influence IPC at least partially independent of each other, that is, small and large networks with low or high storage capacities, respectively, may have low or high processing speed, and they may have a "favorable" or "unfavorable" connectivity pattern.

The number of cortical neurons depends on cortical gray matter volume and neuronal packing density (NPD). Cortical gray matter volume increases with cortical surface and thickness at exponents of 0.66–0.75 in most mammals and about 1 in primates (Herculano-Houzel et al., 2007). In "insectivores," small rodents and cetaceans, the cortex is relatively thin (0.5–1.5 mm), whereas primates generally have thick cortices. The thickest cortices of 2–4 mm are found in great apes including humans. With increasing cortical volume, NPD tends to decrease with an exponent of −0.33 (Changizi, 2001). However, data by Haug (1987) reveal high variability. Among primates, the prosimian mouse lemur and the New World marmoset have the highest NPD with about 75,000 neurons/mm^3, followed by the New World squirrel monkey and baboons with about 60,000 neurons/mm^3. Macaques, talapoins, and chimpanzees have about 40,000 neurons/mm^3, spider monkeys, woolly monkeys, gorillas, and humans have about 25–30,000 neurons/mm^3. Thus, in primates, NPD does not correlate with brain or cortex size.

Due to their large cortex volume and their high NPD, primates have much more cortical neurons than expected on the basis of ABS. Table 2 gives estimates calculated on the basis of data from Haug (1987). The relatively small squirrel monkey has 430, the larger white-fronted capuchin 600–700, and again the larger rhesus monkey 500–800, the gorilla about 4000, the chimpanzee about 6000, and humans have 11–14,000 million cortical neurons. Such estimates depend on the methods applied. Herculano-Houzel et al. (2007), using their isotropical fractionator method, report 1100 million cortical neurons for the rhesus monkey, which appear to be too high based on the data by Haug on cortex volume and NPD in that species. Estimates in humans likewise vary widely from 10 to 22,000 million, the latter being reported by Pakkenberg and Gundersen (1997), which likewise appear to be too high even when calculated on the basis of the highest measured human NPD. Herculano-Houzel (2009) reports 16,000 million cortical neurons in humans.

The number of synapses per cortical neuron is assumed to increase with cortical volume and neuron size with an exponent of 0.33 (Changizi, 2001). Thus, larger cortical neurons should have larger numbers of synapses, but this increase in number of synapses is believed to be compensated by the decrease in NPD, so that in mammals cortical synapse density would remain constant. Unfortunately, exact data on number of synapses are largely lacking. The number of synapses per neuron in the human cortex is controversial; Cherniak (1990) reports 1000–10,000, Rockland (2002) nearly 30,000 synapses per neuron on average.

Table 2. Brain weight, encephalization quotient, and number of cortical neurons in selected mammals

Animal taxa	Brain weight (g)[a]	Encephalization quotient[b,c]	Number of cortical neurons (millions)[d]
Whales	2600–9000	1.8	10,500
False killer whale	3650		
African elephant	4200	1.3	11,000
Man	1200–1450[e]	7.4–7.8	11–14,000
Bottlenose dolphin	1350	5.3	5800
Walrus	1130	1.2	
Camel	762	1.2	
Ox	490	0.5	
Horse	510	0.9	1200
Gorilla	430[e]–570	1.5–1.8	4300
Chimpanzee	330–440[e]	2.2–2.5	6200
Lion	260	0.6	
Sheep	140	0.8	
Old World monkeys	41–122	1.7–2.7	500–800
Rhesus monkey	88–106	2.1	
Gibbon	88–105	1.9–2.7	
Capuchin monkeys	26–80	2.4–4.8	600–700
White-fronted capuchin	57–80	4.8	
Dog	64	1.2	160
Fox	53	1.6	
Cat	25	1.0	300
Squirrel monkey	23	2.3	430
Rabbit	11	0.4	
Marmoset	7	1.7	
Opossum	7.6	0.2	27
Squirrel	7	1.1	
Hedgehog	3.3	0.3	24
Rat	2	0.4	15
Mouse	0.3	0.5	4

[a]Data from Haug (1987).
[b]Indicates the deviation of brain size of a species from brain size expected on the basis of average brain–body relationship in the same taxon.
[c]Data after Jerison (1973).
[d]Calculated using data from Haug (1987).
[e]Basis for calculation of neuron number.

The number of cortical neurons and synapses is assumed to largely determine the storage capacity or long-term memory, which would be highest in species with the maximum number of neurons and synapses. Processing speed is considered to be an independent variable depending on interneuronal distance, axonal conduction velocity, and synaptic transmission speed. Axonal conduction velocity depends on axon diameter. In mammals, it varies little from 0.5 μm in the mouse to 1 μm in monkeys (Schüz, 2001). Apes are reported to have thicker axons than other mammals. The average interneuronal distance is trivially larger in cortices with low NPD, while the speed of synapse transmission is assumed to be constant among mammals and primates (exact data are lacking). As a consequence, small primate brains with fewer, but densely packed neurons and reasonably thick axons would have a higher and larger primate brains with lower NPD and accordingly larger interneuronal distances would have a lower processing speed, if this is not compensated by thicker axon diameters.

An important organizational principle of the mammalian cortex is the parcellation into functionally different (sensory, motor, integrative) areas. In small mammalian brains, the number of such areas is around 10 primary, sensory, and motor areas, without signs of integrative–associative areas (cf. Kaas, 2007) and increases with cortex volume at an exponent of 0.33 in most mammals and all primates (but not in elephants or cetaceans, see below). At the same time, the relative sizes of cortical areas are supposed to decrease. However, the number of connections between areas is assumed to be constant, such that the number of area connections increases with cortical volume. The human cortex is assumed to possess 150 areas and 60 connections per area resulting in 9000 area–area connections (Changizi and Shimojo, 2005). This has been interpreted as tendency to maintain an optimal connectivity at increasing cortical volume and number of neurons and areas, which is realized via the principle of *dense local* (within cortical areas and columns) and *sparse global* connections (across cortical areas) (Hofman, 2001; Chapter 18).

Prefrontal cortex, working memory, and intelligence

There are other structural and functional cortical properties possibly relevant for IPC and intelligence. It has been claimed that the frontal or prefrontal cortex (PFC) as "seat" of intelligence is exceptionally large in humans (Deacon, 1990), and this was assumed to essentially contribute to the alleged superiority of human intelligence. However, recent studies argue that the human frontal or PFC is *not* disproportionally large compared to other primates (Semendeferi et al., 2002). While among primates, humans have the largest frontal cortex, its percentage (36–39%) is equal to that found in the orangutan (37–39%) and only slightly larger than in gorillas (35–38%) and chimpanzees (32–38%). At the same time, the relative size of the frontal cortex of hylobatids and all monkeys is significantly smaller (28–32%). These data correlate with a higher intelligence of the great apes compared to hylobatids and monkeys but would not explain the superiority of human intelligence over that of the great apes.

A number of authors relate primate intelligence to WM, which in turn is considered to be highly related to general intelligence and here especially to "fluid" intelligence. WM is defined as the ability for online maintenance and manipulation of information needed to do complex tasks in the context of reasoning, comprehension, and learning that is not directly available in the current environment. It is assumed that in humans and other primates WM is based on a distributed frontoparietal system (cf. Colom et al., 2007).

Only primates possess a "granular" PFC with a layer IV containing densely packed small neurons, whereas in nonprimate mammals the PFC is agranular. In primates, PFC lesioning leads to strong cognitive deficits, whereas in nonprimate mammals it does not (cf. Wise, 2008). The central function of the primate PFC appears to be integration of sensory information from different modalities and temporal organization of behavior (Fuster, 2008; Kolb, 2007). Thus, the evolution of a granular PFC could be regarded as basis for increased intelligence in primates.

Comparison of primate brains with those of other mammals

In nonprimate mammals, brain weight ranges from less than 0.2 g to 9–10 kg and cortex volume from less than 0.1 cm^3 in small "insectivores" to more than 2000 cm^3 in large whales (Haug, 1987; Fig. 2 and Table 2). The EQ of nonprimate mammals ranges from a very low 0.2 in the opossum to 5.3 in the bottlenose dolphin, which is the highest EQ next to that of humans. Based on data by Haug, the lowest number of cortical neurons is 3–4 million in small "insectivores" and rodents. A dog has about 160, a horse 1200 and the bottlenose dolphin 5800 million cortical neurons. The largest number of cortical neurons in nonprimate mammals is found in the false killer whale with 10,500 and the African elephant with 11,000 millions. Although data by Herculano-Houzel (2009) and Chapter 15 partly differ, they confirm that primates have substantially more cortical neurons than nonprimate mammals with equal or even larger brain and cortex size, as is the case in squirrel monkey versus cat, gorilla versus horse, and humans versus bottlenose dolphin, whales, or elephants (cf. Table 2).

The finding that whales have a lower number of cortical neurons than humans despite their much larger brain is due to the fact that their cortex is much thinner (1–1.2 mm) and has a much lower NPD of 10–15,000 compared to 25–30,000 neurons/mm^3 in humans. The cortex of whales is agranular because of the absence of layer IV (Hof and van der Gucht, 2007; Oelschläger et al., 2008). In addition, in cetaceans relative cortical volume is somewhat lower compared to humans amounting in dolphins to 42% of forebrain mass compared to 47.5% in humans. In the frontal, temporal, and parietal cortex of cetaceans, one finds very large auditory cortical areas, normal visual and somatosensory and motor areas, but only few integrative areas. Their frontal cortex is regarded not homologous to the primate PFC (Oelschläger et al., 2008). Further, in cetaceans, the corpus callosum is extremely thin, which possibly could explain the large hemispheric independence found in cetaceans (cf. Marino, 2007). The very low NPD, combined with a large but thin cortex, leads to large interneuronal distance, especially with respect to interhemispheric information transfer. All this strongly reduces IPC and could explain, why the cognitive abilities of cetaceans and even dolphins are not comparable to that of great apes or even some monkeys despite their highly developed "social intelligence" (Connor, 2007).

The elephant cortex is thicker than that of cetaceans, but still thinner than that of humans, although data differ in literature (Haug, 1987; Shoshani et al., 2006), and its NPD is the lowest among mammals with about 5000 neurons/mm^3 resulting in about 11,000 million cortical neurons. In elephants, like in cetaceans, long transcortical global connections dominate, and cortical areas are mostly devoted to sensorimotor functions (Hart and Hart, 2007). Experts have always been surprised to find that elephants are not as smart as expected given their large brain. Their memory is legendary, as is their navigation and communication capacity, but their cognitive capacities including insight behavior are remarkably restricted (Hart and Hart, 2007; Rensch and Altevogt, 1955).

There is constant search for other cortical specialties that could contribute to the superiority of primates and particularly humans among mammals regarding intelligence. One is morphological (and probably functional) diversity of cortical neurons (i.e., number of subtypes) which is particularly high in primates (reviewed in DeFelipe et al., 2002). In layer 5b of the anterior cingulate cortex of great apes and humans, spindle-shaped cells called "von Economo cells" are found that are four times larger than average pyramidal cells and have widespread connections with other brain parts (Nimchinsky et al., 1999). However, they have been found recently in some cetaceans and in elephants as well, but not consistently in all large-brained mammals (Hof and van der Gucht, 2007). Whether this is due to independent evolution or secondary loss, is unclear, as is their specific significance for cognition (Sherwood et al., 2008).

Comparison with other intelligent vertebrates

Most nonmammalian groups of vertebrates are assumed to be much less intelligent than mammals, let alone primates. Astonishingly high intelligence has been found in a number of bird taxa, particularly in parrots and corvids. Domestic pigeons are capable of categorization, concept formation, transitivity, and numerosity, and often reach levels typical of monkeys (Delius et al., 2001). Corvid birds reveal even higher abilities regarding tool use, flexibility, action planning, mirror self-recognition and may even rival the great apes in some respect (Emery and Clayton, 2004; Prior et al., 2008). While pigeon brains have a volume of about 2.5 cm^3, corvids have brains that are about four times larger and telencephala that occupy up to 80% of brain mass, and the relative size of those parts believed to be analogous to the PFC of mammals (the nidopallium caudolaterale) is much larger than in other birds except parrots (Iwaniuk and Hurd, 2005). However, corvid brains, with up to 10 g, are rather small in absolute terms. Unfortunately, exact data about the number of neurons, synapses, and connections in their brain or pallium are lacking. Since birds generally have much smaller cells than mammals (Olmo, 1983), this, together with higher NPD, might result in 300 million meso-nidopallial neurons in large corvids, which would be about five times more per volume than in the cortex of a monkey brain. Their IPC may be exceptionally high given the small interneuronal distances.

General conclusions

Primates turn out to be, at least on average, more intelligent than other mammals, with great apes and ultimately humans leading. Primates generally have larger brains than other mammals of the same body size. However, there is only a moderate fit between differences in ABS and intelligence. This is predominantly due to considerable variation in the brain–body relationship, that is, some primates have much larger and others much smaller brains and cortices than expected, and there is additional variation in NPD independent of brain and cortex size. Cortex volume and packing density determine the number of cortical neurons, which largely determines memory capacity as one decisive factor for intelligence. Primates in general have much more cortical neurons than other mammals of the same brain size. The second important factor for intelligence is information processing speed, which is generally high in primates due to their relatively high NPD and resulting short interneuronal distance and high axon conduction velocity. The third factor is extensive parcellation of the cortex according to the rule of intense local and sparse global connectivity. Thus, IPC rather than ABS yield the best correlate with intelligence. The human brain combines a large cortex with a relatively high NPD, high conduction velocity, and high parcellation, which together result in the highest IPC and intelligence among animals.

Cetaceans and the elephants have larger to much larger brains and cortices than even humans, but less cortical neurons due to the fact that their NPD is much lower, interneuronal distance much larger, and axonal conduction velocity lower. Finally, cortex parcellation seems to be poorly developed. This appears to strongly impair IPC and could explain, why whales, dolphins, and elephants are not nearly as intelligent as one would predict on the basis of brain size. The opposite seems to happen in corvid birds (and presumably parrots) with very small brains, but high NPD and processing speed, which could explain why these animals reveal an intelligence comparable to monkeys and even great apes with much larger brains.

In most respects, the human brain fits the general trends found in primates and mammals. With one remarkable exception, we find no aspects of human intelligence that is not present at least in rudimentary form in nonhuman primates and at least in some mammals or vertebrates. The exception appears to be syntactical language, which presumably has evolved about 100,000years ago, and the neuroanatomical correlate is the formation of the Broca speech center. Syntactical language can be regarded as a very potent "intelligence amplifier," as was later the case with the invention of scripture and eventually computers.

References

Amici, F., Aureli, F., Visalberghi, E., & Call, J. (2009). Spider monkeys (*Ateles geoffroyi*) and capuchin monkeys (*Cebus apella*) follow gaze around barriers: Evidence for perspective taking? *Journal of Comparative Psychology*, *123*, 368–374.

Balter, M. (2010). Animal communication helps reveal roots of language. *Science*, *328*, 969–971.

Barth, J., & Call, J. (2006). Tracking the displacement of objects: A series of tasks with great apes (*Pan troglodytes*, *Pan paniscus*, *Gorilla gorilla*, and *Pongo pygmaeus*) and young children (*Homo sapiens*). *Journal of Experimental Psychology. Animal Behavior Processes*, *32*, 239–252.

Bates, L. A., & Byrne, R. W. (2010). Imitation: What animal imitation tells us about animal cognition. *Wiley Interdisciplinary Reviews: Cognitive Science*, *1*, 685–695.

Beran, M. J. (2007). Rhesus monkeys (*Macaca mulatta*) enumerate large and small sequentially presented sets of items using analog numerical representations. *Journal of Experimental Psychology. Animal Behavior Processes*, *33*, 42–54.

Bräuer, J., Call, J., & Tomasello, M. (2005). All great ape species follow gaze to distant locations and around barriers. *Journal of Comparative Psychology*, *119*, 145–154.

Burkart, J. M., & Heschl, A. (2007). Understanding visual access in common marmosets, *Callithrix jacchus*: Perspective taking or behaviour reading? *Animal Behaviour*, *73*, 457–469.

Byrne, R. (1995). *The thinking ape. Evolutionary origins of intelligence.* New York: Oxford University Press.

Call, J., Hare, B., Carpenter, M., & Tomasello, M. (2004). 'Unwilling' versus 'unable': Chimpanzees' understanding of human intentional action. *Developmental Science*, *7*, 488–498.

Call, J., & Tomasello, M. (2008). Does the chimpanzee have a theory of mind? 30 years later. *Trends in Cognitive Sciences*, *12*, 187–192.

Cartmill, E. A., & Byrne, R. W. (2010). Semantics of primate gestures: Intentional meanings of orangutan gestures. *Animal Cognition, 13*, 793–804.
Cattell, R. B. (1963). Theory of fluid and crystallized intelligence: A critical experiment. *Journal of Educational Psychology, 54*, 1–22.
Changizi, M. A. (2001). Principles underlying mammalian neocortical scaling. *Biological Cybernetics, 84*, 207–215.
Changizi, M. A., & Shimojo, S. (2005). Parcellation and area-area connectivity as a function of neocortex size. *Brain, Behavior and Evolution, 66*, 88–98.
Cherniak, C. (1990). The bounded brain: Toward quantitative neuroanatomy. *Journal of Cognitive Neuroscience, 2*, 58–66.
Collier-Baker, E., Davis, J. M., Nielsen, M., & Suddendorf, T. (2006). Do chimpanzees (*Pan troglodytes*) understand single invisible displacement? *Animal Cognition, 9*, 55–61.
Colom, R., Jung, R. E., & Haier, R. J. (2007). General intelligence and memory span: Evidence for a common neuroanatomic framework. *Cognitive Neuropsychology, 24*, 867–878.
Connor, R. C. (2007). Dolphin social intelligence: Complex alliance relationships in bottlenose dolphins and a consideration of selective environments for extreme brain size evolution in mammals. *Philosophical Transactions of the Royal Society of London. Series B, Biological Sciences, 362*, 587–602.
Deacon, T. W. (1990). Rethinking mammalian brain evolution. *American Zoologist, 30*, 629–705.
Deaner, R. O., Isler, K., Burkhart, J., & van Schaik, C. (2007). Overall brain size, and not encephalization quotient, best predicts cognitive abilities across non-human primates. *Brain, Behavior and Evolution, 70*, 115–124.
Deaner, R. O., Nunn, C. L., & van Schaik, C. P. (2000). Comparative tests of primate cognition: Different scaling methods produce different results. *Brain, Behavior and Evolution, 55*, 44–52.
DeFelipe, J., Alonso-Nanclares, L., & Arellano, J. I. (2002). Microstructure of the neocortex: Comparative aspects. *Journal of Neurocytology, 31*, 299–316.
Delius, J. M., Siemann, J., Emmerton, L., & Xia, L. (2001). Cognition of birds as products of evolved brains. In G. Roth & M. Wullimann (Eds.), *Brain evolution and cognition* (pp. 451–490). New York: Wiley and Sons.
Deppe, A. M., Wright, P. C., & Szelistowski, W. A. (2009). Object permanence in lemurs. *Animal Cognition, 12*, 381–388.
Dunbar, R. I. M., & Shultz, S. (2007). Evolution in the social brain. *Science, 317*, 1344–1347.
Emery, N. J., & Clayton, N. S. (2004). The mentality of crows: Convergent evolution of intelligence in corvids and apes. *Science, 306*, 1903–1907.
Evans, T. A., Beran, M. J., Harris, E. H., & Rice, D. F. (2009). Quantity judgments of sequentially presented food items by capuchin monkeys (*Cebus apella*). *Animal Cognition, 12*, 97–105.
Falk, D. (2007). Evolution of the primate brain. In W. Henke & I. Tattersall (Eds.), *Handbook of paleaanthropology. Primate evolution and human origins* (Vol. 2, pp. 1133–1162). Berlin: Springer.
Fichtel, C., & Kappeler, P. M. (2010). Human universals and primate symplesiomorphies: Establishing the lemur baseline. In P. M. Kappeler & J. Silk (Eds.), *Mind the gap. Tracing the origins of human universals* (pp. 395–426). Berlin: Springer.
Fujita, K. (2009). Metamemory in tufted capuchin monkeys (*Cebus apella*). *Animal Cognition, 12*, 575–585.
Fuster, J. M. (2008). *The prefrontal cortex* (4th ed.). London: Academic Press.
Genty, E., & Roeder, J. J. (2011). Can lemurs (*Eulemur fulvus* and *E. macaco*) use abstract representations of quantities to master the reverse-reward contingency task? *Primates, 52*, 253–260. doi:10.1007/s10329-011-0245-7.
Gibson, K. R., Rumbaugh, D., & Beran, M. (2001). Bigger is better: Primate brain size in relationship to cognition. In D. Falk & K. R. Gibson (Eds.), *Evolutionary anatomy of the primate cerebral cortex* (pp. 79–97). Cambridge: Cambridge University Press.
Goldin-Meadow, S. (1999). The role of gesture in communication and thinking. *Trends in Cognitive Sciences, 3*, 419–429.
Goossens, B. M. A., Dekleva, M., Reader, S. M., Sterck, E. H. M., & Bolhuis, J. J. (2008). Gaze following in monkeys is modulated by observed facial expressions. *Animal Behaviour, 75*, 1673–1681.
Gould, J. L. (2003). Animal cognition. *Current Biology, 14*, 372–375.
Hanus, D., & Call, J. (2007). Discrete quantity judgments in the great apes (*Pan paniscus, Pan troglodytes, Gorilla gorilla, Pongo pygmaeus*): The effect of presenting whole sets versus item-by-item. *Journal of Comparative Psychology, 121*, 241–249.
Hart, B. L., & Hart, L. A. (2007). Evolution of the elephant brain: A paradox between brain size and cognitive behavior. In J. H. Kaas (Ed.), *Evolution of Nervous Systems* (Vol. 3, pp. 491–497). Oxford: Academic Press.
Haug, H. (1987). Brain sizes, surfaces, and neuronal sizes of the cortex cerebri: A stereological investigation of man and his variability and a comparison with some mammals (primates, whales, marsupials, insectivores, and one elephant). *The American Journal of Anatomy, 180*, 126–142.
Herculano-Houzel, S. (2009). The human brain in numbers: A linearly scaled-up primate brain. *Frontiers in Human Neuroscience, 3*, 31.
Herculano-Houzel, S., Collins, C. E., Wong, P., & Kaas, J. H. (2007). Cellular scaling rules for primate brains. *Proceedings of the National Academy of Sciences of the United States of America, 104*, 3562–3567.

Hof, P. R., & van der Gucht, E. (2007). Structure of the cerebral cortex of the humpback whale, *Megaptera novaeangliae* (Cetacea, Mysticeti, Balaenopteridae). *Anatomical Record (Hoboken)*, *290*, 1–31.

Hofman, M. A. (2001). Evolution and complexity of the human brain: Some organizing principles. In G. Roth & M. F. Wullimann (Eds.), *Brain evolution and cognition* (pp. 501–521). New York: Wiley and Sons.

Horner, V., & Whiten, A. (2005). Causal knowledge and imitation/emulation switching in chimpanzees (*Pan troglodytes*) and children (*Homo sapiens*). *Animal Cognition*, *3*, 164–181.

Iriki, A., & Sakura, O. (2008). The neuroscience of primate intellectual evolution: Natural selection and passive and intentional niche construction. *Philosophical Transactions of the Royal Society of London. Series B, Biological Sciences*, *363*, 2229–2241.

Iwaniuk, A. N., & Hurd, P. L. (2005). The evolution of cerebrotypes in birds. *Brain, Behavior and Evolution*, *65*, 215–230.

Jerison, H. J. (1973). *Evolution of the brain and intelligence.* New York: Academic Press.

Kaas, J. H. (2007). Reconstructing the organization of neocortex of the first mammals and subsequent modifications. In J. H. Kaas (Ed.), *Evolution of Nervous Systems* (Vol. 3, pp. 27–48). Oxford: Academic Press.

Kelly, S. D., Iverson, J. M., Terranova, J., Niego, J., Hopkins, M., & Goldsmith, L. (2002). Putting language back into the body: Speech and gesture on three time frames. *Developmental Neuropsychology*, *22*, 323–349.

Kendal, R. L., Custance, D. M., Kendal, J. R., Vale, G., Stoinski, T. S., Rakotomalala, N. L., et al. (2010). Evidence for social learning in wild lemurs (*Lemur catta*). *Learning & Behavior*, *38*, 220–234.

Kolb, B. (2007). Do all mammals have a prefrontal cortex? In J. H. Kaas & L. A. Krubitzer (Eds.), *Evolution of nervous systems. A comprehensive review* (Vol. 3, pp. 443–450). Amsterdam: Academic Press (Elsevier).

Kornell, N., Son, L. K., & Terrace, H. S. (2007). Transfer of metacognitive skills and hint seeking in monkeys. *Psychological Science*, *18*, 64–71.

Kuroshima, H., Kuwahata, H., & Fujita, K. (2008). Learning from others' mistakes in capuchin monkeys (*Cebus apella*). *Animal Cognition*, *11*, 599–609.

Lefebvre, L., Reader, S. M., & Sol, D. (2004). Brains, innovations and evolution in birds and primates. *Brain, Behavior and Evolution*, *63*, 233–246.

Marino, L. (2007). Cetacean brain evolution. In J. H. Kaas (Ed.), *Evolution of Nervous Systems* (Vol. 3, pp. 261–266). Oxford: Academic Press.

Martin, R. D. (1996). Scaling of the mammalian brain: The maternal energy hypothesis. *News in Physiological Sciences*, *11*, 149–156.

McGrew, W. C. (2010). Evolution. Chimpanzee technology. *Science*, *328*, 579–580.

Mendes, N., Hanus, D., & Call, J. (2007). Raising the level: Orangutans use water as a tool. *Biology Letters*, *3*, 453–455.

Neiworth, J. J., Steinmark, E., Basile, B. M., Wonders, R., Steely, F., & DeHart, C. (2003). A test of object permanence in a new-world monkey species, cotton top tamarins (*Saguinus oedipus*). *Animal Cognition*, *6*, 27–37.

Nimchinsky, E. A., Gilissen, E., Allman, J. M., Perl, D. P., Erwin, J. M., & Hof, P. R. (1999). A neuronal morphologic type unique to humans and great apes. *Proceedings of the National Academy of Sciences of the United States of America*, *96*, 5268–5273.

O'Connell, S., & Dunbar, R. I. M. (2003). A test for comprehension of false belief in chimpanzees. *Evolution and Cognition*, *9*, 131–140.

Oelschläger, H. H., Haas-Rioth, M., Fung, C., Ridgway, S. H., & Knauth, M. (2008). Morphology and evolutionary biology of the dolphin (*Delphinus* sp.) brain—MR imaging and conventional histology. *Brain, Behavior and Evolution*, *71*, 68–86.

Olmo, E. (1983). Nucleotype and cell size in vertebrates: A review. *Basic and Applied Histochemistry*, *27*, 227–256.

Osvath, M., & Osvath, H. (2008). Chimpanzee (*Pan troglodytes*) and orangutan (*Pongo abelii*) forethought: Self-control and pre-experience in the face of future tool use. *Animal Cognition*, *11*, 661–674.

Ottoni, E. B., & Izar, P. (2008). Capuchin monkey tool use: Overview and implications. *Evolutionary Anthropology: Issues, News, and Reviews*, *17*, 171–178.

Pakkenberg, B., & Gundersen, H. J. G. (1997). Neocortical neuron number in humans: Effect of sex and age. *The Journal of Comparative Neurology*, *384*, 312–320.

Penn, D. C., & Povinelli, D. J. (2007). On the lack of evidence that non-human animals possess anything remotely resembling a 'theory of mind'. *Philosophical Transactions of the Royal Society of London. Series B, Biological Sciences*, *362*, 731–744.

Pérez-Barberia, F. J., Shultz, S., & Dunbar, R. I. M. (2007). Evidence for coevolution of sociality and relative brain size in three orders of mammals. *Evolution*, *61*, 2811–2821.

Phillips, W., Barnes, J. L., Mahajan, N., Yamaguchi, M., & Santos, L. R. (2009). 'Unwilling' versus 'unable': Capuchin monkeys' (*Cebus apella*) understanding of human intentional action. *Developmental Science*, *12*, 938–945.

Piaget, J. (1954). *The construction of reality in the child.* New York: Basic Books.

Pilbeam, D., & Gould, S. J. (1974). Size and scaling in human evolution. *Science*, *186*, 892–901.

Premack, D., & Woodruff, G. (1978). Does the chimpanzee have a theory of mind? *The Behavioral and Brain Sciences*, *4*, 515–526.

Prior, H., Schwarz, A., & Güntürkün, O. (2008). Mirror-induced behavior in the magpie (*Pica pica*): Evidence of self recognition. *PLoS Biology*, *6*, 1642–1650.

Reader, S. M., & Laland, K. N. (2002). Social intelligence, innovation, and enhanced brain size in primates. *Proceedings of the National Academy of Sciences of the United States of America*, *99*, 4436–4441.

Rensch, B., & Altevogt, R. (1955). Zähmung und Dressurleistungen indischer Arbeitselefanten. *Zeitschrift für Tierpsychologie*, *11*, 497–510.

Rizzolatti, G., & Craighero, L. (2004). The mirror-neuron system. *Annual Review of Neuroscience*, *27*, 169–192.

Rockland, K. S. (2002). Non-uniformity of extrinsic connections and columnar organization. *Journal of Neurocytology*, *31*, 247–253.

Roth, G., & Dicke, U. (2005). Evolution of the brain and intelligence. *Trends in Cognitive Sciences*, *9*, 250–257.

Ruiz, A., Gómez, J. C., Roeder, J. J., & Byrne, R. W. (2009). Gaze following and gaze priming in lemurs. *Animal Cognition*, *12*, 427–434.

Sanz, C. M., & Morgan, D. B. (2009). Flexible and persistent tool-using strategies in honey-gathering by wild chimpanzees. *International Journal of Primatology*, *30*, 411–427.

Savage-Rumbaugh, E. S., Murphy, J., Sevcik, R. A., Brakke, K. E., Williams, S. L., & Rumbaugh, D. M. (1993). Language comprehension in ape and child. *Monographs of the Society for Research in Child Development*, *58*, 1–222.

Schüz, A. (2001). What can the cerebral cortex do better than other parts of the brain? In G. Roth & M. Wullimann (Eds.), *Brain evolution and cognition* (pp. 491–500). New York: Wiley and Sons.

Semendeferi, K., Lu, A., Schenker, N., & Damasio, H. (2002). Humans and great apes share a large frontal cortex. *Nature Neuroscience*, *5*, 272–276.

Seyfarth, R. M., Cheney, D. L., & Bergman, T. J. (2005). Primate social cognition and the origins of language. *Trends in Cognitive Sciences*, *9*, 264–266.

Shepherd, S. V., & Platt, M. L. (2008). Spontaneous social orienting and gaze following in ringtailed lemurs (*Lemur catta*). *Animal Cognition*, *11*, 13–20.

Sherwood, C. C., Subiaul, F., & Zawidzki, T. W. (2008). A natural history of the human mind: Tracing evolutionary changes in brain and cognition. *Journal of Anatomy*, *212*, 426–454.

Shoshani, J., Kupsky, W. J., & Marchant, G. H. (2006). Elephant brain Part. I: Gross morphology, functions, comparative anatomy, and evolution. *Brain Research Bulletin*, *70*, 124–157.

Smith, J. D. (2009). The study of animal metacognition. *Trends in Cognitive Sciences*, *13*, 389–396.

Spearman, C. (1904). 'General intelligence' objectively determined and measured. *The American Journal of Psychology*, *15*, 201–293.

Subiaul, F., Cantlon, J. F., Holloway, R. L., & Terrace, H. S. (2004). Cognitive imitation in rhesus macaques. *Science*, *305*, 407–410.

Suda-King, C. (2008). Do orangutans (*Pongo pygmaeus*) know when they do not remember? *Animal Cognition*, *11*, 21–42.

Tomasello, M., Hare, B., Lehmann, H., & Call, J. (2007). Reliance on head versus eyes in the gaze following of great apes and human infants: The cooperative eye hypothesis. *Journal of Human Evolution*, *52*, 314–320.

van Dongen, P. A. M. (1998). Brain size in vertebrates. In R. Nieuwenhuys, H. J. ten Donkelaar & C. Nicholson (Eds.), *The central nervous system of vertebrates* (Vol. 3, pp. 2099–2134). Berlin: Springer.

Visalberghi, E., Addessi, E., Truppa, V., Spagnoletti, N., Ottoni, E., Izar, P., et al. (2009). Selection of effective stone tools by wild bearded capuchin monkeys. *Current Biology*, *19*, 213–217.

von Bonin, G. (1937). Brain weight and body weight in mammals. *The Journal of General Psychology*, *16*, 379–389.

Walker, R., Burger, O., Wagner, J., & Von Rueden, C. R. (2006). Evolution of brain size and juvenile periods in primates. *Journal of Human Evolution*, *51*, 480–489.

Wise, S. P. (2008). Forward frontal fields: Phylogeny and fundamental function. *Trends in Neurosciences*, *31*, 599–608.

M. A. Hofman and D. Falk (Eds.)
Progress in Brain Research, Vol. 195
ISSN: 0079-6123

CHAPTER 21

Evolution of human emotion: A view through fear

Joseph E. LeDoux*

Center for Neural Science, New York University, New York, NY, USA

Abstract: Basic tendencies to detect and respond to significant events are present in the simplest single cell organisms and persist throughout all invertebrates and vertebrates. Within vertebrates, the overall brain plan is highly conserved, though differences in size and complexity also exist. The forebrain differs the most between mammals and other vertebrates. The classic notion that the evolution of mammals led to radical changes such that new forebrain structures (limbic system and neocortex) were added has not held up nor has the idea that so-called limbic areas are primarily involved in emotion. Modern efforts have focused on specific emotion systems, like the fear or defense system, rather than on the search for a general purpose emotion systems. Such studies have found that fear circuits are conserved in mammals, including humans. Animal work has been especially successful in determining how the brain detects and responds to danger. Caution should be exercised when attempting to discuss other aspects of emotion, namely subjective feelings, in animals since there are no scientific ways of verifying and measuring such states except in humans.

Keywords: emotion; feelings limbic system; fear; vertebrate; mammal; amygdala; hippocampus; neocortex.

Introduction

The topic of emotion and evolution typically brings to mind Darwin's classic treatise, Emotions in Man and Animals (Darwin, 1872). In this book, Darwin sought to extend his theory of natural selection beyond the evolution of physical structures and into the domain of mind and behavior by exploring how emotions too might have evolved. Particularly important to his argument was the fact that certain emotions are expressed similarly in people around the world, including in isolated areas where there had been little contact with the outside world and thus little opportunity for emotional expressions to have been learned and culturally transmitted. This suggested to him that there must be a strong heritable component to emotions in people. Also important was his observation that certain emotions are expressed similarly across species, especially closely related species, further

*Corresponding author.
Tel.: +1-212-998-3930; Fax: +1-212-995-4704
E-mail: joseph.ledoux@nyu.edu

DOI: 10.1016/B978-0-444-53860-4.00021-0

suggesting that these emotions are phylogenetically conserved.

With the rise of experimental brain research in the late nineteenth century, emotion was one of the key topics that early neuroscientists sought to relate to the brain (see LeDoux, 1987). The assumption was that emotion circuits are conserved across mammalian species, and that it should be possible to understand human emotions by exploring emotional mechanisms in the nonhuman mammalian brain.

In this chapter, I will first briefly survey the history of ideas about the emotional brain and especially ideas that have attempted to explain the emotional brain in terms of evolutionary principles. This will lead to a discussion of fear, since this is the emotion that has been studied most thoroughly in terms of brain mechanisms. The chapter will conclude with a reconsideration of what the term emotion refers to, and specifically which aspects of emotion can be studied in animals and which must be studied in humans.

A brief history of the emotional brain: The rise and fall of the limbic system theory

All organisms, even single cell organisms, must have the capacity to detect and respond to significant stimuli in order to survive. Bacteria, for example, approach nutrients and avoid harmful chemicals (Macnab and Koshland, 1972). With the evolution of multicellular, metazoan organisms with specialized systems, particularly a nervous system, the ability to detect and respond to significant events increases in sophistication (Shepherd, 1983).

Invertebrates, the oldest and largest group of multicellular organisms, exhibit a wide variety of types of nervous systems. However, all vertebrates share a common basic brain plan consisting of three broad zones (hindbrain, midbrain, and forebrain) with conserved basic circuits (Butler and Hodos, 2005; Nauta and Karten, 1970; Striedter, 2005; Swanson, 2002). In spite of this overall similarity, differences in size and complexity exist. For example, the forebrain differs the most between mammals and reptiles. On the basis of such differences, the classic view of forebrain evolution emerged in the first half of the twentieth century (e.g., Ariëns Kappers et al., 1936; Herrick, 1933; MacLean, 1949, 1952; Papez, 1937; Smith, 1924). According to this view, with the emergence of mammals, the forebrain plan underwent radical changes in which new structures, especially cortical structures, were added. These were layered over and covered the reptilian forebrain, which mainly consisted of the basal ganglia. First came "primitive" cortical regions in early mammals. In these organisms, the basic survival functions related to feeding, defense, and procreation were taken care of by fairly undifferentiated (weakly laminated) cortical regions (primitive cortex, including the hippocampus and cingulate cortex) and related subcortical areas (such as the amygdala) that were closely tied to the olfactory system. Later mammals added highly novel, laminated cortical regions (neocortex) that made possible enhanced non-olfactory sensory processing and cognitive functions (including learning and memory, reasoning, and planning capacities, and in humans, language).

The basic principle that equated cognition with evolutionarily new cortex (neocortex) and emotion with older cortex and related subcortical forebrain regions culminated in Paul MacLean's limbic system theory of emotion (1949, 1952, 1970). The term limbic was first used by the French anatomist Paul Broca as a structural designation for a rim of cortex in the medial wall of the hemisphere. Broca called this rim the limbic lobe (*le grande lobe limbique*) (limbic is from the Latin word for rim, *limbus*). MacLean built on the classic findings of comparative anatomists such as Herrick and Papez, and experimental findings from Walter Cannon, Phillip Bard, and Henrich Kluver and Paul Bucy (Bard, 1928; Cannon, 1929; Kluver and Bucy, 1937) to transform the limbic lobe into an emotion system, the limbic system. The limbic system was defined anatomically as the primitive medial cortical areas and interconnected subcortical nuclei (including the amygdala and septum).

MacLean called the limbic system the paleomammalian brain (since it was said to have emerged with the evolution of early mammals) and contrasted it with the reptilian brain (basal ganglia and brainstem). In more recent mammals, the neocortex, also called the neomammalian brain, was said by MacLean to increases in size and complexity at the expense of the limbic system. The decrease in the limbic system reduced the dependence of humans on base emotions, and the increase in the neocortex allowed humans greater control over remaining emotional circuits as well as greater cognitive capacities.

The limbic system concept stimulated much research in the 1950s, 1960s, and 1970s. However, it has been criticized on a number of grounds and has been rejected by many scientists (Butler and Hodos, 2005; Kotter and Meyer, 1992; LeDoux, 1987, 1991, 1996; Swanson, 1983). Because the limbic system concept continues to be referred to in some scientific circles (e.g., Panksepp, 1998, 2005) and persists in many lay accounts of the brain, it is worth considering why it is not acceptable.

First, the theory presumes that the neocortex and limbic system are unique mammalian specializations. Neither of these ideas is accepted by contemporary comparative neuroanatomists (Butler and Hodos, 2005; Nauta and Karten, 1970; Northcutt and Kaas, 1995; Striedter, 2005). Birds and reptiles, for example, have been shown to have structures that correspond with both mammalian neocortex and MacLean's cortical and subcortical limbic areas (hippocampus, amygdala). Second, MacLean argued the architecture of limbic areas is ill-suited for cognitive processes. However, the hippocampus, viewed by MacLean as the centerpiece of the limbic system and a central structure for emotional functions, is recognized as one of the key areas related to higher cognitive functions, such as declarative or explicit memory (Eichenbaum, 2002; Squire, 1987) and spatial cognition (O'Keefe and Nadel, 1978). Third, efforts to define the system have failed. Connectivity with old cortex is a flawed criterion if old cortex is itself an unjustified notion. Connectivity with the hypothalamus once seemed plausible, since that was a way of distinguishing relevant and irrelevant cortical areas (Isaacson, 1982). However, as anatomical techniques improved, areas from the neocortex were also found to be connected with the hypothalamus, as were areas of the spinal cord, potentially extending the limbic system across the entire central nervous system. Finally, and perhaps most important, there is no evidence that the limbic system, however defined, functions as an integrated system in the mediation of emotion. While specific areas of the limbic system contribute to some emotional functions, these areas do not do so because they belong to a limbic system that evolved to perform emotional functions. Indeed, relatively few limbic areas have been shown to contribute to emotional functions. As noted above, the hippocampus, the centerpiece of the limbic system theory of emotion, has been strongly implicated in cognitive functions, but the evidence for a role in emotion is far less impressive.

The limbic system theory attempted to explain all emotions within a single anatomical concept. Contemporary researchers are more inclined to focus on tasks designed to study the brain systems of specific emotions. As we will see, this has been a more profitable empirical approach.

Contributions of the amygdala to avoidance conditioning: An early approach to linking emotional behavior to the limbic system

Why, then, has the limbic system concept persisted for so long given that it proved questionable on evolutionary, structural, and functional grounds? The key reason can be summarized in the term "guilt by association." One limbic area, the amygdala, has consistently been implicated in emotional behavior. Because the amygdala was part of the limbic system concept, its involvement vindicated the whole concept. This does not mean that the amygdala is the only structure involved in emotion, but instead that the amygdala is one area that has been extensively implicated in emotion, in part because of the behavioral tasks that have most often been used to study emotion.

The amygdala was first implicated in emotion through studies of the Kluver–Bucy syndrome, a set of unusual behaviors observed in monkeys after removal of large areas of the temporal lobe (Kluver and Bucy, 1937). Monkeys with such lesions attempted to eat inappropriate items and copulate with inappropriate partners and lost their fear of snakes and humans. It was concluded that the animals had psychic blindness, an inability to appreciate the significance or value of visual stimuli. Weiskrantz (1956) attempted to localize the effects within the temporal lobe using a behavioral task where behavior was guided by stimulus value. Specifically, he used an avoidance conditioning paradigm where monkeys learned to use a cue to signal when to perform a behavioral response in order to avoid receiving a painful shock. Such a paradigm was viewed as especially useful in assessing the role of the amygdala in processing threats that lead to fear. Damage targeted to the amygdala disrupted performance, leading Weiskrantz to conclude that the amygdala was responsible for the inability of animals with temporal lobe damage to use stimulus value to guide behavior, and thus that an important function of the amygdala was to ascertain stimulus value. Specifically, Weiskrantz proposed that the amygdala processes the rewarding and punishing consequences of events. However, the data were essentially about aversive or punishing events since avoidance conditioning is a fear-based paradigm.

Subsequently, throughout the 1960s and 1970s, avoidance conditioning paradigms were used to study the contribution of the amygdala to emotion, an especially to fear. The bulk of the evidence was consistent, in general, with the idea that the amygdala is a key structure in avoidance conditioning, and by implication, in processing the value of emotional stimuli. Such findings were treated as evidence in support of the limbic system theory of emotion since the amygdala was part of the limbic system.

By the mid-1980s, 30years of research on the brain mechanisms of avoidance had been conducted. While it seemed clear that the amygdala was somehow involved, there was considerable confusion as exactly what its role might be (Sarter and Markowitsch, 1985). There are several explanations likely for this unsettled state of affairs. First, there was little appreciation of the anatomical complexity of the amygdala, a brain region with a dozen or so nuclei, each with subnuclei (Amaral et al., 1992; LeDoux, 2007; Pitkänen et al., 1997). Failure to recognize this anatomical complexity may have led to confusion. Indeed, more recent work has shown that different nuclei and subnuclei have different functions (LeDoux, 2007; Repa et al., 2001). Second, the behavioral complexity of avoidance conditioning itself was not fully appreciated. Avoidance tasks can be constructed in various ways (active, passive, signaled, unsignaled), and each involves the learning of a Pavlovian association and an instrumental association (Amorapanth et al., 2000; Cain and LeDoux, 2007; Cain et al., 2010; Choi et al., 2010; Lazaro-Munoz et al., 2010; LeDoux et al., 2009). In retrospect, failure to separate these components also probably played a role in adding confusion to efforts to understand the brain mechanisms of avoidance.

Contribution of the amygdala to fear: Studies of aversive Pavlovian conditioning

During the 1960s, researchers began using Pavlovian conditioning to pursue the cellular and molecular mechanisms of learning in invertebrates (e.g., Kandel and Spencer, 1968; Kandel et al., 1986). The success of this approach, together with the fact that avoidance conditioning was stuck in a rut, led mammalian researchers to turn to Pavlovian conditioning as well (Kapp et al., 1979; LeDoux et al., 1984; Thompson, 1986; Tischler and Davis, 1983).

As mentioned already, Pavlovian conditioning is the initial phase of avoidance conditioning. After the subjects rapidly undergo Pavlovian conditioning, they then slowly learn to perform avoidance responses using the conditioned stimulus (CS) as a warning signal. Indeed, the emotional learning that occurs in avoidance conditioning occurs during the Pavlovian phase. Pavlovian conditioning is thus a more direct means of studying emotional processing. Perhaps Pavlovian conditioning would be easier to understand.

In Pavlovian fear conditioning, the subject receives a neutral CS, usually a tone, followed by an aversive unconditioned stimulus (US), typically footshock. After one or at most a few pairings, the CS comes to elicit innate emotional responses that naturally occur in the presence of threatening stimuli, such as predators. For example, after conditioning a CS elicits defensive freezing behavior and associated autonomic and endocrine responses that support the behavior (Blanchard and Blanchard, 1969; Bolles and Fanselow, 1980; LeDoux et al., 1984). The subject does not have to learn to perform these responses. The responses are innate. What is learned is an association that allows a novel stimulus, a warning of danger, to elicit the defensive responses in anticipation of the actual danger.

With the simpler approach provided by fear conditioning, as opposed to avoidance, much progress was made in mapping the circuitry, including the regions in the brain where the CS and US converge to form the associations and the regions involved in the control of emotional responses by the CS in animals (see Davis, 1992; Davis et al., 1997; Fanselow and Poulos, 2005; Johansen et al., 2011; LeDoux, 2000; Maren, 2001, 2005; Pape and Pare, 2010; Rodrigues et al., 2004) and humans (Phelps, 2006; Phelps and LeDoux, 2005; Sehlmeyer et al., 2009; Whalen and Phelps, 2009). In brief, CS and US convergence occurs in the lateral nucleus of the amygdala (LA), and specifically in the dorsal subregion of the LA. This convergence leads to synaptic plasticity and the formation of a CS–US association. Damage to LA, inactivation of LA, or manipulation of a variety of molecular pathways in LA prevents fear conditioning. A second important region is central nucleus of the amygdala (CE). Manipulations of the region also disrupt conditioning. LA and CE are connected directly and by way of various intra-amygdala pathways. Once the CS–US association is formed, later exposure to the CS results in the retrieval of the learned association formed by CS–US convergence during conditioning. Information then flows from LA to CE, which then connects to hypothalamic and brainstem areas that control behavioral, autonomic, and hormonal responses that help the organism cope with the threat. Plasticity also occurs in CE, and in CS processing regions and motor control regions. This simplified description omits many details.

Much has been learned about the molecular mechanisms in LA that make fear conditioning possible (Blair et al., 2001; Fanselow and Poulos, 2005; Johansen et al., 2011; Maren, 2001, 2005; Pape and Pare, 2010; Rodrigues et al., 2004; Sah et al., 2008; Schafe et al., 2001). In brief, the CS input synapses undergo plasticity when the LA neurons they connect with are depolarized by the shock US. As a result, the ability of the CS to activate the LA cell is potentiated. Plasticity is triggered when the depolarizing US allows calcium to flow into the cell via NMDA receptors and voltage-sensitive calcium channels. The elevated calcium activates a number of protein kinases that ultimately lead to phosphorylation of transcription factors such as CREB that lead to gene expression and protein synthesis. The newly synthesized proteins then stabilize the synaptic potentiation, allowing the CS to strongly activate the LA cell for over long periods of time. It is particularly interesting that many of the molecular changes that underlie fear conditioning in mammals have also been shown to be important for Pavlovian conditioning in invertebrates, showing the conserved nature of the molecular mechanisms of learning and memory.

The advances made in understanding Pavlovian fear conditioning made it possible to revisit the neural system of avoidance and related aversive instrumental behaviors (Amorapanth et al., 2000; Cain and LeDoux, 2007; Cain et al., 2010; Choi et al., 2010; Lazaro-Munoz et al., 2010; LeDoux et al., 2009). This work showed that as in fear conditioning, the LA is essential for forming the CS–US association. But in contrast to fear conditioning, in avoidance information flows from the LA to the basal amygdala (not to the CE). Extrapolating from appetitive conditioning finding (Cardinal et al., 2002; Everitt et al., 1989, 1999), it has been proposed that connections from the basal amygdala to the ventral striatum allow the CS–US association to control aversively motivated instrumental behavior (Fig. 1).

Pavlovian fear and avoidance conditioning circuits

Avoidance responses are not emotional responses *per se*. They are simply responses. An animal can learn to avoid harm by running, climbing, pressing, swimming, or even remaining stationary. The animal learns to do what it needs to do to attain safety. But the same responses could be used to obtain food if the animal is hungry and those responses are a way to gain access to food. In contrast, in Pavlovian fear conditioning the CS elicits specific emotional responses, fear, or defense responses. Researchers were much more inclined to discuss Pavlovian conditioning results specifically in terms of fear/defense circuits.

Comparative observations

Amygdala areas have been implicated in fear conditioning in a variety of mammals, including rats, mice, rabbits, and monkeys (see Davis, 1992; Davis et al., 1997; Fanselow and Poulos, 2005; Johansen et al., in review; LeDoux, 1996, 2000; Maren, 2001, 2005; Pape and Pare, 2010; Rodrigues et al., 2004). This suggests strong conservation of the circuitry within mammals, including humans. Indeed, a large body of work implicates the human amygdala in fear conditioning and in instrumental responses like avoidance (Bechara et al., 1995; Damasio, 1994; Delgado et al., 2009; Dolan and Vuilleumier, 2003; Gianaros et al., 2008; Labar, 2003; Ousdal et al., 2008; Phelps, 2006; Phelps and LeDoux, 2005; Whalen and Phelps, 2009; Whalen et al., 2004). Thus, damage to the amygdala in humans prevents fear conditioning from occurring, and functional imaging studies show that activity increases in the amygdala during fear conditioning. Additionally, a number of studies have found amygdala activation in response to angry or fearful faces, considered to be unconditioned threat stimuli (Adolphs, 2008). Thus, findings involving both lesion studies and functional imaging suggest strong correspondence with the animal literature, at least at a gross anatomical level. Techniques available for studying the human brain do not allow precise localization of specific nuclei, though some progress is being made in this regard (Bach et al., 2011; Davis et al., 2011).

An important question concerns the nature of the amygdala in nonmammals and the role of the homologous structure in fear conditioning. According to classic view, areas such as the amygdala, being paleomammalian structures, should not exist in reptiles. However, in the 1970s, Cohen (1975) claimed to have indentified the amygdala in avian species and found that lesions of this regions disrupted of Pavlovian fear conditioning in pigeons. More recently, there has been much debate about what constitutes the amygdala, and specifically individual amygdala nuclei, in reptiles and birds (Bruce and Neary, 1995; Karten, 1997; Lanuza et al., 1998; Martinez-Garcia et al., 2002; Moreno and Gonzalez, 2007). Using connectivity patterns established in

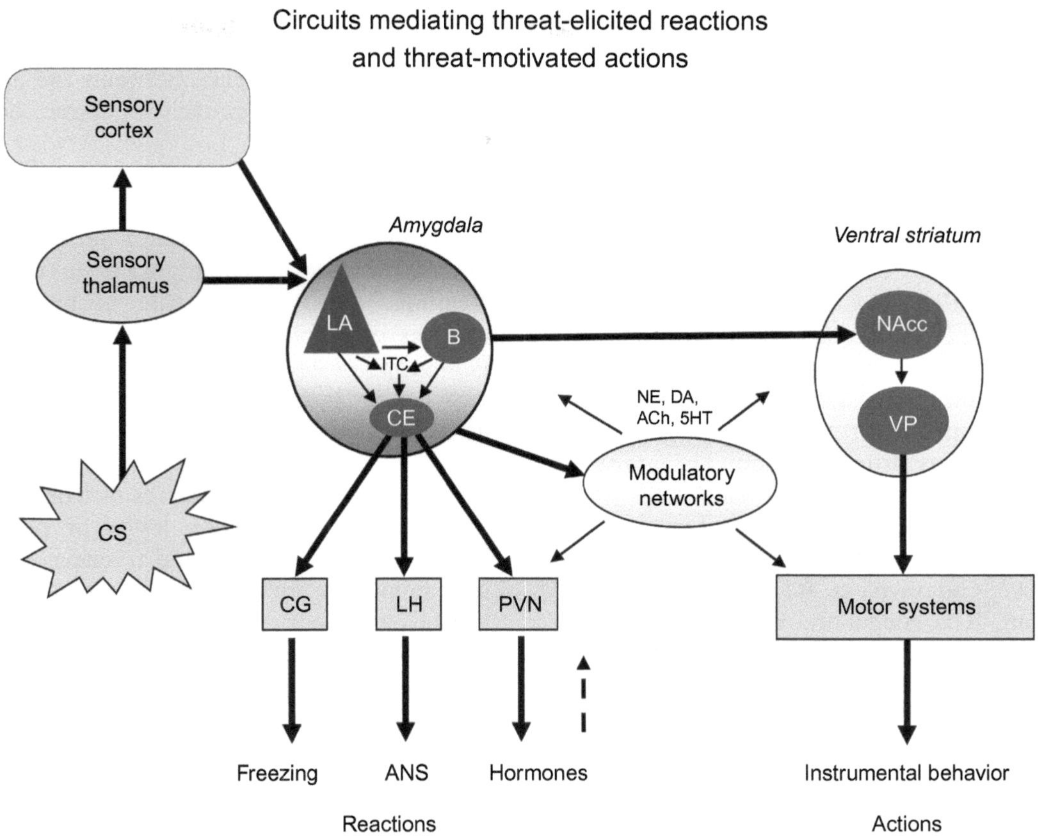

Fig. 1. Fear is the emotion most thoroughly understood in the brain. Much of the progress made has involved studies of Pavlovian fear conditioning in rats. During conditioning the conditioned stimulus (CS), usually a tone, and the unconditioned stimulus (US), usually a footschock, converge in the lateral nucleus of the amygdala (LA) to induce synaptic plasticity of the CS inputs (CS–US convergence not shown). The CS is then able to flow through amygdala circuits to the central nucleus (CE) to control the expression of hard-wired, automatic, defensive reactions (freezing behavior, autonomin nervous system, ANS, activity, and hormonal release). CE outputs also activate networks in that control the release of neuromodulators, such as norepinephrine (NE), dopamine (DA), acetylcholine (ACh), and serotonin (5HT) throughout the brain. These, like hormonal feedback, help add intensity to and prolong the duration of the aroused state. In addition to these various automatic responses controlled by CE, the LA also sends information, via the basal nucleus (B) to the ventral striatum, especially the nucleus accumbens. The latter connections are likely to be involved in the invigoration of goal-directed behaviors that allow the organism to act in certain ways on the basis of past instrumental learning or on-the-spot decisions about how to cope with the threat. Other abbreviations: ITC, intercalated nuclei of the amydala; CG, central gray; LH, lateral hypothalamus; PVN, paraventricular nucleus of the hypothalamus; VP, ventral pallidum.

mammals, areas believed to be homologous to the lateral and central nucleus have been identified in lizards (Lanuza et al., 1998; Martinez-Garcia et al., 2002). When threatened, these animals undergo tonic immobility, and damage to the CE homologue interferes with this defensive response (Davies et al., 2002).

Much more work is needed to resolve what constitutes the amygdala in nonmammalian vertebrates and to determine whether the functions of the amygdala known to exist in mammals have some relation to the function of the homologous regions in the vertebrate ancestors of mammals.

Amygdala contributions to other emotions

While the contribution of the amygdala to fear has been most thoroughly studied, it is clear that the amygdala contributes to other emotional states as well. A relatively large body of research has focused on the role of the amygdala in processing of rewards and the use of rewards to motivate and reinforce behavior (Cardinal et al., 2002; Everitt et al., 1999; Holland and Gallagher, 2004; Murray, 2007; Nishijo et al., 2008; Salzman et al., 2007). As with aversive conditioning, the lateral, basal, and central amygdala have been implicated in different aspects of reward learning and motivation, as well as drug addiction. The amygdala has also been implicated in emotional states associated with aggressive, maternal, sexual, and ingestive (eating and drinking) behaviors (Bahar et al., 2003; Galaverna et al., 1993; Miczek et al., 2007; Pfaff, 2005; Siegel and Edinger, 1981). Less is known about the detailed circuitry involved in these emotional states than is known about fear.

Emotional evolution in perspective

There is no shortage of theories that have speculated about the relation of emotion circuits to brain evolution. In the tradition of Darwin, basic emotions theorists have proposed that certain emotions are innate, in part because they are expressed the same in people around the world (Buck, 1981; Ekman, 1977, 1992; Izard, 1971, 1992; Plutchik, 1980; Tomkins, 1962). These innate emotions are said to be mediated by affect programs in the brain. An affect program, in effect, is psychological description of a dedicated neural circuit. Some neuroscientist have adopted the basic emotions idea and have proposed specific circuits for different basic emotions (Panksepp, 1980, 1998, 2005), though the basic emotions discussed do not completely correspond with those proposed in the psychological theories.

The above discussion of the amygdala and its role in fear and defense might be construed as a mini-version of basic emotions theory, a version focused on one basic emotion. However, there is a fundamental difference between the approach I take and the approach of basic emotions theorists.

The goal of basic emotions theories is to understand subjective states of conscious experience that humans label with emotion words (fear, love, sadness, joy, etc.). Their goal is to understand "feelings." This is also true of brain science theories of emotion focused on basic emotions. Panksepp (1980, 1998, 2005), for example, searches for brain systems in animals that underlie feelings in the animals as a way of understanding the brain systems that underlie human feelings. Vocalizations that result from tickling a rat are ways of indexing joyful or pleasurable feelings in the rat brain, and freezing, flight, and fight behaviors are markers of fearful feelings.

The approach I take is quite different (LeDoux, 1984, 1996, 2002, 2008). I use emotional behavior as a means of indexing circuits that have evolved to allow organisms to deal with challenges and opportunities in their environments. I make no assumption about what an animal is feeling, since I believe it is not possible to scientifically measure, and thus not possible to research, feelings in animals other than humans. I do not deny that other animals may have feelings. I simply question whether these can be studied using scientific methods. Beyond this methodological barrier, I am also critical of attempts to equate feelings in humans and other animals for other reasons. First, most studies that have explored conscious experience in humans have found that when information (including emotional information) reaches awareness the dorsolateral prefrontal cortex is active, and if information is experimentally prevented from reaching awareness this area is not active (for summary, see LeDoux, 2008). The dorsolateral granular prefrontal cortex is a unique primate specialization (Preuss, 1995; Wise, 2008) and has features in the human brain that are lacking in other primates (Semendeferi et al., 2011). If human conscious experience depends on these unique features of brain organization, we should be cautious about attributing the kinds of mental states made possible by these features to animals that lack the feature or the brain region. Second, language is a unique human capacity, and conscious experience, including emotional experience, is influenced by language. The once disputed idea that language, and the cognitive processes required to support language functions, add complexity to human experience, has regained respect (Lakoff, 1987; Lucy, 1997). In the absence of language, experience cannot be partitioned in the same way—English speakers can partition fear and anxiety into more than 30 categories (Marks, 1987). The diversity with which nonverbal organisms can conceptualize the world and their experiences in it is thus likely constricted by the absence of language.

Conclusion

In sum, basic tendencies to detect and respond to significant events are present in the simplest single cell organisms and persist throughout all invertebrates and vertebrates. Within vertebrates, the overall brain plan is highly conserved, though differences in size and complexity also exist. The forebrain differs the most between mammals and other vertebrates, though the old notion that the evolution of mammals led to radical changes such that new forebrain structures were added has not held up. Thus, the idea that mammalian evolution is characterized by the addition of a limbic system (devoted to emotion) and a neocortex (devoted to cognition) is flawed. Modern efforts to understand the brain mechanisms of emotion have made more progress by focusing on specific emotion systems, like the fear or defense system, rather than on efforts to find a single brain system devoted to emotion. Also, progress has been made in animal studies by focusing on emotion in terms of brain circuits that contribute to behaviors related to survival functions. Efforts to use scientific methods identify circuits in animals that might correspond to circuits in the human brain that are responsible for conscious feelings are not likely to succeed since we have no way of scientifically measuring feelings in animals. Conscious feelings are an important topic, but one that is best pursued through studies of humans.

References

Adolphs, R. (2008). Fear, faces, and the human amygdala. *Current Opinion in Neurobiology*, *18*, 166–172.

Amaral, D. G., Price, J. L., Pitkänen, A., & Carmichael, S. T. (1992). Anatomical organization of the primate amygdaloid complex. In J. P. Aggleton (Ed.), *The amygdala: Neurobiological aspects of emotion, memory, and mental dysfunction* (pp. 1–66). New York: Wiley-Liss, Inc.

Amorapanth, P., LeDoux, J. E., & Nader, K. (2000). Different lateral amygdala outputs mediate reactions and actions elicited by a fear-arousing stimulus. *Nature Neuroscience*, *3*, 74–79.

Ariëns Kappers, C. U., Huber, C. G., & Crosby, E. C. (1936). *The comparative anatomy of the nervous system of vertebrates, including man*. New York: Macmillan Company.

Bach, D. R., Talmi, D., Hurlemann, R., Patin, A., & Dolan, R. J. (2011). Automatic relevance detection in the absence of a functional amygdala. *Neuropsychologia*, *49*, 1302–1305.

Bahar, A., Samuel, A., Hazvi, S., & Dudai, Y. (2003). The amygdalar circuit that acquires taste aversion memory differs from the circuit that extinguishes it. *European Journal of Neuroscience*, *17*, 1527–1530.

Bard, P. (1928). A diencephalic mechanism for the expression of rage with special reference to the sympathetic nervous system. *American Journal of Physiology*, *84*, 490–515.

Bechara, A., Tranel, D., Damasio, H., Adolphs, R., Rockland, C., & Damasio, A. R. (1995). Double dissociation of conditioning and declarative knowledge relative to the amygdala and hippocampus in humans. *Science*, *269*, 1115–1118.

Blair, H. T., Schafe, G. E., Bauer, E. P., Rodrigues, S. M., & LeDoux, J. E. (2001). Synaptic plasticity in the lateral amygdala: A cellular hypothesis of fear conditioning. *Learning and Memory*, *8*, 229–242.

Blanchard, R. J., & Blanchard, D. C. (1969). Crouching as an index of fear. *Journal of Comparative and Physiological Psychology*, *67*, 370–375.

Bolles, R. C., & Fanselow, M. S. (1980). A perceptual-defensive-recuperative model of fear and pain. *The Behavioral and Brain Sciences*, *3*, 291–323.

Bruce, L. L., & Neary, T. J. (1995). The limbic system of tetrapods: A comparative analysis of cortical and amygdalar populations. *Brain, Behavior, and Evolution, 46*, 224–234.

Buck, R. (1981). The evolution and development of emotion expression and communication. In S. S. Brehm, S. M. Kassin & F. X. Gibbons (Eds.), *Developmental social psychology* (pp. 127–151). New York: Oxford University Press.

Butler, A. B., & Hodos, W. (2005). *Comparative certebrate neuroanatomy: Evolution and adaptation* (2nd ed.). Hoboken: John Wiley & Sons, Inc.

Cain, C. K., Choi, J. S., & LeDoux, J. E. (2010). Active avoidance and escape learning. In G. Koob, R. Thompson & M. Le Moal (Eds.), *Encyclopedia of behavioral neuroscience.* New York: Elsevier.

Cain, C. K., & LeDoux, J. E. (2007). Escape from fear: A detailed behavioral analysis of two atypical responses reinforced by CS termination. *Journal of Experimental Psychology. Animal Behavior Processes, 33*, 451–463.

Cannon, W. B. (1929). *Bodily changes in pain, hunger, fear, and rage* (Vol. 2). New York: Appleton.

Cardinal, R. N., Parkinson, J. A., Hall, J., & Everitt, B. J. (2002). Emotion and motivation: The role of the amygdala, ventral striatum, and prefrontal cortex. *Neuroscience and Biobehavioral Reviews, 26*, 321–352.

Choi, J. S., Cain, C. K., & LeDoux, J. E. (2010). The role of amygdala nuclei in the expression of auditory signaled two-way active avoidance in rats. *Learning and Memory, 17*, 139–147.

Cohen, D. H. (1975). Involvement of the avian amygdalar homologue (archistriatum posterior and mediale) in defensively conditioned heart rate change. *The Journal of Comparative Neurology, 160*, 13–35.

Damasio, A. (1994). *Descarte's error: Emotion, reason, and the human brain.* New York: Gosset/Putnam.

Darwin, C. (1872). *The expression of the emotions in man and animals.* London: Fontana Press.

Davies, D. C., Martinez-Garcia, F., Lanuza, E., & Novejarque, A. (2002). Striato-amygdaloid transition area lesions reduce the duration of tonic immobility in the lizard Podarcis hispanica. *Brain Research Bulletin, 57*, 537–541.

Davis, M. (1992). The role of the amygdala in fear and anxiety. *Annual Review of Neuroscience, 15*, 353–375.

Davis, F. C., Johnstone, T., Mazzulla, E. C., Oler, J. A., & Whalen, P. J. (2011). Regional response differences across the human amygdaloid complex during social conditioning. *Cerebral Cortex, 20*, 612–621.

Davis, M., Walker, D. L., & Lee, Y. (1997). Amygdala and bed nucleus of the stria terminalis: Differential roles in fear and anxiety measured with the acoustic startle reflex. *Philosophical Transactions of the Royal Society of London. Series B, Biological Sciences, 352*, 1675–1687.

Delgado, M. R., Jou, R. L., LeDoux, J. E., & Phelps, E. A. (2009). Avoiding negative outcomes: Tracking the mechanisms of avoidance learning in humans during fear conditioning. *Frontiers in Behavioral Neuroscience, 3*, 33.

Dolan, R. J., & Vuilleumier, P. (2003). Amygdala automaticity in emotional processing. *Annals of the New York Academy of Sciences, 985*, 348–355.

Eichenbaum, H. (2002). *The cognitive neuroscience of memory.* New York: Oxford University Press.

Ekman, P. (1977). Biological and cultural contributions to body and facial movement. In J. Blacking (Ed.), *The anthropology of the body* (pp. 39–84). London: Academic Press.

Ekman, P. (1992). An argument for basic emotions. *Cognition and Emotion, 6*, 169–200.

Everitt, B. J., Cador, M., & Robbins, T. W. (1989). Interactions between the amygdala and ventral striatum in stimulus- reward associations: Studies using a second-order schedule of sexual reinforcement. *Neuroscience, 30*, 63–75.

Everitt, B. J., Parkinson, J. A., Olmstead, M. C., Arroyo, M., Robledo, P., & Robbins, T. W. (1999). Associative processes in addiction and reward. The role of amygdala-ventral striatal subsystems. *Annals of the New York Academy of Sciences, 877*, 412–438.

Fanselow, M. S., & Poulos, A. M. (2005). The neuroscience of mammalian associative learning. *Annual Review of Psychology, 56*, 207–234.

Galaverna, O. G., Seeley, R. J., Berridge, K. C., Grill, H. J., Epstein, A. N., & Schulkin, J. (1993). Lesions of the central nucleus of the amygdala. I: Effects on taste reactivity, taste aversion learning and sodium appetite. *Behavioural Brain Research, 59*, 11–17.

Gianaros, P. J., Sheu, L. K., Matthews, K. A., Jennings, J. R., Manuck, S. B., & Hariri, A. R. (2008). Individual differences in stressor-evoked blood pressure reactivity vary with activation, volume, and functional connectivity of the amygdala. *Journal of Neuroscience, 28*, 990–999.

Herrick, C. J. (1933). The functions of the olfactory parts of the cerebral cortex. *Proceedings of the National Academy of Sciences of the United States of America, 19*, 7–14.

Holland, P. C., & Gallagher, M. (2004). Amygdala-frontal interactions and reward expectancy. *Current Opinion in Neurobiology, 14*, 148–155.

Isaacson, R. L. (1982). *The limbic system.* New York: Plenum Press.

Izard, C. E. (1971). *The face of emotion.* New York: Appleton-Century-Crofts.

Izard, C. E. (1992). Basic emotions, relations among emotions, and emotion-cognition relations. *Psychological Review, 99*, 561–565.

Johansen, J., Cain, C., Ostroff, L., & LeDoux, J. E. (2011). Molecular mechanisms of fear learning and memory. *Cell, 147*, 509–524.

Kandel, E. R., Klein, M., Castellucci, V. F., Schacher, S., & Goelet, P. (1986). Some principles emerging from the study of short- and long-term memory. *Neuroscience Research, 3*, 498–520.

Kandel, E. R., & Spencer, W. A. (1968). Cellular neurophysiological approaches to the study of learning. *Physiological Reviews, 48*, 65–134.

Kapp, B. S., Frysinger, R. C., Gallagher, M., & Haselton, J. R. (1979). Amygdala central nucleus lesions: Effect on heart rate conditioning in the rabbit. *Physiology and Behavior, 23*, 1109–1117.

Karten, H. J. (1997). Evolutionary developmental biology meets the brain: The origins of mammalian cortex. *Proceedings of the National Academy of Sciences of the United States of America, 94*, 2800–2804.

Kluver, H., & Bucy, P. C. (1937). "Psychic blindness" and other symptoms following bilateral temporal lobectomy in rhesus monkeys. *American Journal of Physiology, 119*, 352–353.

Kotter, R., & Meyer, N. (1992). The limbic system: A review of its empirical foundation. *Behavioural Brain Research, 52*, 105–127.

LaBar, K. S. (2003). Emotional memory functions of the human amygdala. *Current Neurology and Neuroscience Reports, 3*, 363–364.

Lakoff, G. (1987). *Women, fire, and dangerous things: What categories reveal about the mind.* Chicago: The University of Chicago Press.

Lanuza, E., Belekhova, M., Martinez-Marcos, A., Font, C., & Martinez-Garcia, F. (1998). Identification of the reptilian basolateral amygdala: An anatomical investigation of the afferents to the posterior dorsal ventricular ridge of the lizard Podarcis hispanica. *European Journal of Neuroscience, 10*, 3517–3534.

Lazaro-Munoz, G., LeDoux, J. E., & Cain, C. K. (2010). Sidman instrumental avoidance initially depends on lateral and Basal amygdala and is constrained by central amygdala-mediated Pavlovian processes. *Biological Psychiatry, 67*, 1120–1127.

LeDoux, J. E. (1984). Cognition and emotion: Processing functions and brain systems. In M. S. Gazzaniga (Ed.), *Handbook of cognitive neuroscience* (pp. 357–368). New York: Plenum Publishing Corp.

LeDoux, J. E. (1987). Emotion. In F. Plum (Ed.), *Handbook of physiology. 1: The nervous system: Vol. V. Higher functions of the brain* (pp. 419–460). Bethesda: American Physiological Society.

LeDoux, J. E. (1991). Emotion and the limbic system concept. *Concepts in Neuroscience, 2*, 169–199.

LeDoux, J. E. (1996). *The emotional brain.* New York: Simon and Schuster.

LeDoux, J. E. (2000). Emotion circuits in the brain. *Annual Review of Neuroscience, 23*, 155–184.

LeDoux, J. E. (2002). *Synaptic self: How our brains become who we are.* New York: Viking.

LeDoux, J. (2007). The amygdala. *Current Biology, 17*, R868–R874.

LeDoux, J. E. (2008). Emotional colouration of consciousness: How feelings come about. In L. Weiskrantz & M. Davies (Eds.), *Frontiers of consciousness: Chichele lectures* (pp. 69–130). Oxford: Oxford University Press.

LeDoux, J. E., Sakaguchi, A., & Reis, D. J. (1984). Subcortical efferent projections of the medial geniculate nucleus mediate emotional responses conditioned to acoustic stimuli. *Journal of Neuroscience, 4*, 683–698.

LeDoux, J. E., Schiller, D., & Cain, C. (2009). Emotional reaction and action: From threat processing to goal-directed behavior. In M. S. Gazzaniga (Ed.), *The cognitive neurosciences.* (4th ed., pp. 905–924). Cambridge: MIT Press.

Lucy, J. A. (1997). Linguistic relativity. *Annual Review of Anthropology, 26*, 291–312.

MacLean, P. D. (1949). Psychosomatic disease and the "visceral brain": Recent developments bearing on the Papez theory of emotion. *Psychosomatic Medicine, 11*, 338–353.

MacLean, P. D. (1952). Some psychiatric implications of physiological studies on frontotemporal portion of limbic system (visceral brain). *Electroencephalography and Clinical Neurophysiology, 4*, 407–418.

MacLean, P. D. (1970). The triune brain, emotion and scientific bias. In F. O. Schmitt (Ed.), *The neurosciences: Second study program* (pp. 336–349). New York: Rockefeller University Press.

Macnab, R. M., & Koshland, D. E. Jr. (1972). The gradient-sensing mechanism in bacterial chemotaxis. *Proceedings of the National Academy of Sciences of the United States of America, 69*, 2509–2512.

Maren, S. (2001). Neurobiology of Pavlovian fear conditioning. *Annual Review of Neuroscience, 24*, 897–931.

Maren, S. (2005). Synaptic mechanisms of associative memory in the amygdala. *Neuron, 47*, 783–786.

Marks, I. (1987). *Fears, phobias, and rituals: Panic, anxiety and their disorders.* New York: Oxford University Press.

Martinez-Garcia, F., Martinez-Marcos, A., & Lanuza, E. (2002). The pallial amygdala of amniote vertebrates: Evolution of the concept, evolution of the structure. *Brain Research Bulletin, 57*, 463–469.

Miczek, K. A., de Almeida, R. M., Kravitz, E. A., Rissman, E. F., de Boer, S. F., & Raine, A. (2007). Neurobiology of escalated aggression and violence. *Journal of Neuroscience, 27*, 11803–11806.

Moreno, N., & Gonzalez, A. (2007). Evolution of the amygdaloid complex in vertebrates, with special reference to the anamnio-amniotic transition. *Journal of Anatomy, 211*, 151–163.

Murray, E. A. (2007). The amygdala, reward and emotion. *Trends in Cognitive Sciences, 11*, 489–497.

Nauta, W. J. H., & Karten, H. J. (1970). A general profile of the vertebrate brain, with sidelights on the ancestry of cerebral cortex. In F. O. Schmitt (Ed.), *The neurosciences: Second study program* (pp. 7–26). New York: The Rockefeller University Press.

Nishijo, H., Hori, E., Tazumi, T., & Ono, T. (2008). Neural correlates to both emotion and cognitive functions in the monkey amygdala. *Behavioural Brain Research*, *188*, 14–23.

Northcutt, R. G., & Kaas, J. H. (1995). The emergence and evolution of mammalian neocortex. *Trends in Neurosciences*, *18*, 373–379.

O'Keefe, J., & Nadel, L. (1978). *The hippocampus as a cognitive map.* Oxford: Clarendon Press.

Ousdal, O. T., Jensen, J., Server, A., Hariri, A. R., Nakstad, P. H., & Andreassen, O. A. (2008). The human amygdala is involved in general behavioral relevance detection: Evidence from an event-related functional magnetic resonance imaging Go-NoGo task. *Neuroscience*, *156*, 450–455.

Panksepp, J. (1980). Hypothalamic integration of behavior: Rewards, punishments, and related psychological processes. In P. J. Morgane & J. Panksepp (Eds.), *Handbook of the hypothalamus: Vol. 3. Behavioral studies of the hypothalamus* (pp. 289–431). New York: Marcel Dekker.

Panksepp, J. (1998). *Affective neuroscience.* New York: Oxford U. Press.

Panksepp, J. (2005). Affective consciousness: Core emotional feelings in animals and humans. *Consciousness and Cognition*, *14*, 30–80.

Pape, H. C., & Pare, D. (2010). Plastic synaptic networks of the amygdala for the acquisition, expression, and extinction of conditioned fear. *Physiological Reviews*, *90*, 419–463.

Papez, J. W. (1937). A proposed mechanism of emotion. *Archives of Neurology and Psychiatry*, *79*, 217–224.

Pfaff, D. (2005). Hormone-driven mechanisms in the central nervous system facilitate the analysis of mammalian behaviours. *Journal of Endocrinology*, *184*, 447–453.

Phelps, E. A. (2006). Emotion and cognition: Insights from studies of the human amygdala. *Annual Review of Psychology*, *57*, 27–53.

Phelps, E. A., & LeDoux, J. E. (2005). Contributions of the amygdala to emotion processing: From animal models to human behavior. *Neuron*, *48*, 175–187.

Pitkänen, A., Savander, V., & LeDoux, J. E. (1997). Organization of intra-amygdaloid circuitries in the rat: An emerging framework for understanding functions of the amygdala. *Trends in Neurosciences*, *20*, 517–523.

Plutchik, R. (1980). *Emotion: A psychoevolutionary synthesis.* New York: Harper & Row.

Preuss, T. M. (1995). Do rats have a prefrontal cortex? The Rose-Woolsey-Akerty program reconsidered. *Journal of Cognitive Neuroscience*, *7*, 1–24.

Repa, J. C., Muller, J., Apergis, J., Desrochers, T. M., Zhou, Y., & LeDoux, J. E. (2001). Two different lateral amygdala cell populations contribute to the initiation and storage of memory. *Nature Neuroscience*, *4*, 724–731.

Rodrigues, S. M., Schafe, G. E., & LeDoux, J. E. (2004). Molecular mechanisms underlying emotional learning and memory in the lateral amygdala. *Neuron*, *44*, 75–91.

Sah, P., Westbrook, R. F., & Luthi, A. (2008). Fear conditioning and long-term potentiation in the amygdala: What really is the connection? *Annals of the New York Academy of Sciences*, *1129*, 88–95.

Salzman, C. D., Paton, J. J., Belova, M. A., & Morrison, S. E. (2007). Flexible neural representations of value in the primate brain. *Annals of the New York Academy of Sciences*, *1121*, 336–354.

Sarter, M. F., & Markowitsch, H. J. (1985). Involvement of the amygdala in learning and memory: A critical review, with emphasis on anatomical relations. *Behavioral Neuroscience*, *99*, 342–380.

Schafe, G. E., Nader, K., Blair, H. T., & LeDoux, J. E. (2001). Memory consolidation of Pavlovian fear conditioning: A cellular and molecular perspective. *Trends in Neurosciences*, *24*, 540–546.

Sehlmeyer, C., Schoning, S., Zwitserlood, P., Pfleiderer, B., Kircher, T., Arolt, V., et al. (2009). Human fear conditioning and extinction in neuroimaging: A systematic review. *PLoS One*, *4*, e5865.

Semendeferi, K., Teffer, K., Buxhoeveden, D. P., Park, M. S., Bludau, S., Amunts, K., et al. (2011). Spatial organization of neurons in the frontal pole sets humans apart from great apes. *Cerebral Cortex*, *21*, 1485–1497.

Shepherd, G. (1983). *Neurobiology.* New York: Oxford University Press.

Siegel, A., & Edinger, H. (1981). Neural control of aggression and rage behavior. In P. J. Morgane & J. Panksepp (Eds.), *Handbook of the hypothalamus: Vol. 3. Behavioral studies of the hypothalamus* (pp. 203–240). New York: Marcel Dekker.

Smith, G. E. (1924). *The evolution of man.* Oxford: Oxford University Press.

Squire, L. (1987). *Memory and brain.* Oxford: New York.

Striedter, G. J. (2005). *Principles of brain evolution.* Sunderland: Sinauer Associates.

Swanson, L. W. (1983). The hippocampus and the concept of the limbic system. In W. Seifert (Ed.), *Neurobiology of the hippocampus* (pp. 3–19). London: Academic Press.

Swanson, L. W. (2002). *Brain architecture: Understanding the basic plan.* Oxford: Oxford University Press.

Thompson, R. F. (1986). The neurobiology of learning and memory. *Science*, *233*, 941–947.

Tischler, M. D., & Davis, M. (1983). A visual pathway that mediates fear-conditioned enhancement of acoustic startle. *Brain Research*, *276*, 55–71.

Tomkins, S. S. (1962). *Affect, imagery, consciousness.* New York: Springer.

Weiskrantz, L. (1956). Behavioral changes associated with ablation of the amygdaloid complex in monkeys. *Journal of Comparative and Physiological Psychology*, *49*, 381–391.

Whalen, P. J., Kagan, J., Cook, R. G., Davis, F. C., Kim, H., Polis, S., et al. (2004). Human amygdala responsivity to masked fearful eye whites. *Science*, *306*, 2061.

Whalen, P. J., & Phelps, E. A. (2009). *The human amygdala.* New York: Guilford Press.

Wise, S. P. (2008). Forward frontal fields: Phylogeny and fundamental function. *Trends in Neurosciences*, *31*, 599–608.

M. A. Hofman and D. Falk (Eds.)
Progress in Brain Research, Vol. 195
ISSN: 0079-6123

CHAPTER 22

Evolution of brain and language

P. Thomas Schoenemann*

Department of Anthropology, Indiana University, Bloomington, IN, USA

Abstract: In this chapter evolutionary changes in the human brain that are relevant to language are reviewed. Most of what is known involves assessments of the relative sizes of brain regions. Overall brain size is associated with some key behavioral features relevant to language, including complexity of the social environment and the degree of conceptual complexity. Prefrontal cortical and temporal lobe areas relevant to language appear to have increased disproportionately. Areas relevant to language production and perception have changed less dramatically. The extent to which these changes were a consequence specifically of language versus other behavioral adaptations is a good question, but the process may best be viewed as a complex adaptive system, whereby cultural learning interacts with biology iteratively over time to produce language. Overall, language appears to have adapted to the human brain more so than the reverse.

Keywords: coevolution; conceptual complexity; communication; cortex; Broca's area; Wernicke's area; comparative primate; brain scaling.

Introduction

Among all the behavioral changes made possible by human brain evolution, language is arguably the most critical to defining the human condition. Other animals communicate, but none do so with the richness and complexity of human language. This means that there must be some important differences between the brains of humans and other animals in areas relevant to communication. Comparative studies of the brains of humans and other animals, combined with an understanding of the different functions of specific brain regions, and considered within a realistic evolutionary perspective, allow a reasonable sketch of the evolution of brain and language.

Languages must be learnable by the brains of children in each generation. Thus, language change (a form of cultural evolution) is constrained by the existing abilities of brains in each generation. Because language is critical to an individual's adaptive fitness, language also likely had a fundamental influence on brain evolution. Humans are particularly socially interactive

*Corresponding author.
Tel.: +1-812-8558800; Fax: +1-812-8554358
E-mail: toms@indiana.edu

DOI: 10.1016/B978-0-444-53860-4.00022-2

creatures, which makes communication central to our existence. Two interrelated evolutionary processes therefore occurred simultaneously: language adapted to the human brain (cultural evolution), while the human brain adapted to better subserve language (biological evolution). This coevolutionary process resulted in language and brain evolving to suit each other (Christiansen, 1994; Christiansen and Chater, 2008; Deacon, 1992).

The coevolution of language and brain can be understood as a complex adaptive system (Beckner et al., 2009). Complex adaptive systems are characterized by interacting sets of agents (which can be individuals, neurons, etc.), where each agent behaves in an individually adaptive way to local conditions, often following very simple rules. The sum total of these interactions nevertheless leads to various kinds of emergent, system-wide order. With respect to the coevolution of brain and language, a number of language-relevant neural systems interact with and influence each other in important ways. Syntax depends fundamentally on the structure of semantics, since the function of syntax is to code higher-level semantic information. Semantics in turn depends on the structure of conceptual understanding, which—as will be reviewed below—is a function of brain structure. These structures are in turn the result of biological adaptation: circuits that result in conceptual understanding that is useful to a given individual's environmental realities will be selected for.

In general, the selective environment for primate species is largely a social one. The adaptiveness (reproductive benefit) of an individual's particular behavior at any given moment in time depends crucially on the flexible responses of others in the group, who are also attempting to behave in an adaptive manner in response. Understanding language evolution ultimately involves understanding how the repeated complex communicative interactions of individuals result in cultural change in languages, and how these changes in turn influence biological change in the long term. The evolution of brain circuits cannot be understood independent of the evolution of language, and vice versa.

Because the evolutionary benefits of language to an individual would always have been dependent on the preexisting cognitive abilities of others, language evolution is inherently constrained. New genetic variants enhancing the perception of linguistically relevant signals would have been favored only to the extent that they increase the individual's ability to perceive and rapidly process the acoustic signals *already used by others for language*. Similarly, changes affecting the production of linguistically relevant signals would be favored only to the extent that they could be understood by the preexisting perceptual abilities of others. Signals too complicated or subtle for others to process would not be adopted, and hence mutations influencing them would not likely spread.

It is possible for some new variant to be adaptive strictly at the individual level (and therefore spread) even if it was not immediately useful for language, but this could only be true if they were beneficial for some other reason. In this case, however, it would not be a "language" variant. If it spread wide enough (for nonlinguistic reasons), it might later be co-opted for language. This would not result in the evolution of highly language-specific circuits.

For these reasons, any adaptive changes in language circuitry occurring during a given sequence of the evolutionary process will be biased toward slight modifications of preexisting circuits, and away from major changes in the ways communication is processed by the brain (Schoenemann, 2005). Language circuits should show extensive homologies with preexisting systems in closely related animals. These hijacked circuits would, by definition, be domain general (contra Tooby and Cosmides, 1992).

Inferences about evolutionary changes in the brain relevant to language are derived from knowledge of how language is processed in the brain, combined with knowledge of how our brains are different from those of our closest evolutionary relatives. To the extent that a particular area known to be relevant to language appears to have also changed significantly, we are justified in inferring that this area was important for language evolution, though whether it increased specifically for language will be difficult to determine. Evolutionary inferences will also involve thinking about the interplay of different behavioral abilities over our history. Both an evolutionary perspective and a complex adaptive systems approach predict that language evolution occurred hand in hand with the evolution of other aspects of cognition. Language processing depends heavily on the integration of a large number of abilities that are processed in widely dispersed circuits across the brain (Damasio and Damasio, 1992; Mueller, 1996). Assessing the coevolution of language and brain therefore requires a broad focus on a number of brain regions.

Evolutionary changes in the brain relevant to language

Knowing how different language-relevant brain areas have changed over our evolutionary history is central to understanding the coevolution of brain and language. However, it is not clear what counts as significant change: increases relative to brain size, body size, or simply absolute size independent of either brain or body? Because of the evolutionary costs to increasing the absolute numbers of neurons (Hofman, 1983), changes in absolute size of an area *independent* of body or brain increases are likely to be behaviorally relevant. Further, there are many examples of changes that appear to be unrelated to either body or brain size increases. The olfactory bulb (responsible for sense of smell) has actually decreased in size (being half that found in chimpanzees; Stephan et al., 1981), the primary motor cortex (Brodmann area 4) has apparently not increased at all in absolute size (Blinkov and Glezer, 1968), and premotor cortex (Brodmann area 6) and primary visual cortex (Brodmann area 17) appear to have lagged significantly behind the increase in overall brain size (Deacon, 1997; Schoenemann, 2006). Thus, brain evolution is remarkably plastic over the long term (contra Finlay et al., 2001). It is true that brain size correlates with body size across major groups of animals, which in turn has led to the use of relative brain size measures when comparing species (e.g., the encephalization quotient or EQ; Jerison, 1973; see also Chapter 20). However, this association does not require that relative brain size is behaviorally relevant. It likely just represents an inevitable tradeoff between the utility of brains and their disproportionate metabolic costs: larger brains may always be useful, but only larger animals can afford to pay for them metabolically (e.g., Martin, 1981). In fact, absolute brain size is empirically a much better predictor of species differences in behavior than is relative brain size (Deaner et al., 2007; Gibson et al., 2001; for a discussion see Chapter 15). For this reason, any changes in brain anatomy are therefore potentially of behavioral importance.

One caveat that needs to be stated; however, we do not know the details of exactly how different areas of the brain actually process information. We lack the neural equivalent of a circuit diagram for anything larger than the 302 neuron brain of the worm *Caenorhabditis elegans* (White et al., 1986; Chapter 17). Because of this, we cannot be sure that a fourfold increase in one area has greater behavioral implications than a twofold increase in some other area. Nevertheless, any increase would seem to be important, given the costs of maintaining excess neurons.

Overall brain size

Human brains are about three times larger than that found in our closest relatives, the African apes (even taking body size into account; Jerison, 1973).

Exactly what behavioral implications this has is a matter of debate, but given the importance of language to the human condition, it is reasonable to believe that at least part of this increase is due to language. The idea that the evolution of brain size and language are related is an old one going back at least to Darwin (1882), who argued there was likely a "...relation between the continued use of language and the development of the brain..." (p. 87). Because different parts of the brain have changed in different ways, focusing solely on overall brain size is an oversimplification. However, there are some interesting correlates of overall brain size that are likely relevant to language evolution.

First, as pointed out above, absolute brain size is empirically behaviorally relevant. Overall brain size also correlates strongly with both length of maturation (Harvey and Clutton-Brock, 1985) and overall lifespan (Allman et al., 1993; Hofman, 1993). This means that the bigger the brain, the greater the potential for behavioral learning to be a central part of the organism's behavioral repertoire. Larger brained animals rely on learning more than do smaller brained animals (Deacon, 1997), and larger brained primates do better at a variety of experimental learning tasks (Deaner et al., 2007).

A great deal of modern human behavior (including language) depends critically on learning. While learning can be biased in particular ways by evolved innate influences, human behavioral evolution is better characterized by increasing behavioral flexibility rather than greater numbers of hardwired, innate circuits. Learning language obviously depends on being able to understand changing, fluid contingencies between constituents and meaning. The increasing behavioral flexibility and reliance on learning made possible by the increase in brain size, therefore, made language increasingly possible, if not inevitable.

Primates as a group are particularly interactively social, and interactive sociality is a particularly complicated niche (Holloway, 1975; Humphrey, 1984). The size of the typical social group has been shown to be associated with various measures of brain size (including absolute brain size) in primates (Fig. 1; Dunbar, 1995; Reader and Laland, 2002). Social group size is a reasonable proxy for the complexity of one's social existence. Human social networks appear to be particularly complex, and given that language is an inherently social activity, the selective value of language is likely greatest for humans.

The size of the neocortex, which plays a key role in conscious awareness generally as well as mediating a number of complex cognitive functions including language, appears to be strongly associated with overall brain size (Hofman, 1985). The neocortex makes up over 80% of the entire human brain, which is the highest value among all primates. The corresponding values for apes (who have the next largest brains among primates) range from ~76% to 73%, while particularly small brained monkeys range down to 59%, and the smallest brained primate of all, the mouse lemur (*Microcebus murinus*), has a neocortex that takes up only 44% of its brain (Hofman, 1985, 1988;

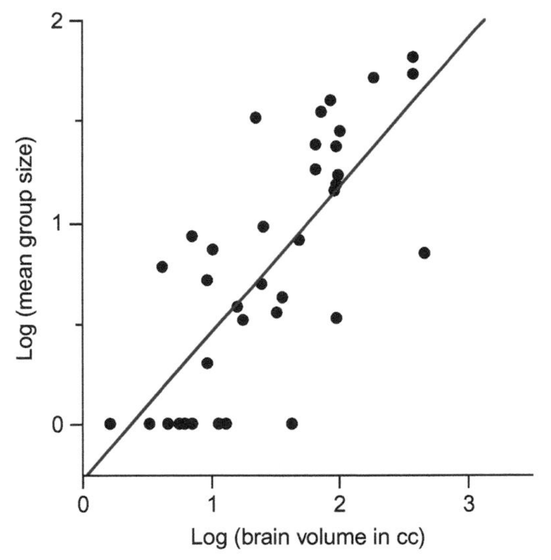

Fig. 1. The relationship between brain volume and mean group size in primate species. N=36, r=0.75, p<0.0001. Data from Dunbar (1995).

Stephan et al., 1981). This would suggest that as brains get larger, conscious behavior becomes increasingly important.

In addition, as the neocortex increases, areas of it that are devoted specifically to integrating different types of information (so-called association areas) increase disproportionately, at the expense of areas devoted either to the processing of sensory information from a single modality (such as the primary visual or primary auditory cortices) or to the conscious control of muscle movement (Fig. 2). The larger these "association areas" are, the greater the likely potential for increasingly complex types of integrative processing (Schoenemann, 2010).

Further, as these association areas expand, they appear to evolve increasing numbers of relatively specialized processing areas. Larger brains have greater numbers of identifiably distinct cortical areas (Changizi and Shimojo, 2005; Northcutt and Kaas, 1995). This turns out to be a predictable consequence of increasing brain size: any given area of the neocortex will tend to be less directly connected to other areas in larger brains compared to smaller brains (Ringo, 1991). This means that areas are able to carry out tasks increasingly independent of each other, leading predictably to increasing functional localization.

An fundamentally important consequence of this that it allows for the formation of richer, more complex, and more subtle conceptual understanding (Gibson, 2002; Schoenemann, 1999, 2005). Much of the brain appears to be relevant to concept formation (Barsalou, 2008; Damasio and Damasio, 1992; Schoenemann, 2005). When a subject imagines an object that is not actually present, similar areas of their brain are activated as when the object *is* being viewed (Damasio et al., 1993; Kosslyn et al., 1993). Different kinds of basic sensory input—visual, auditory, olfactory, taste, and somatosensory (touch, temperature, pain, body position)—are processed in different areas. While some basic concepts involve only a single sensory modality (e.g., [green] or [smooth (texture)]), most concepts require the integration of more than one sense. For example, the concept "coffee" typically invokes not just a particular taste but also a smell, a visual image of a mug, the sensation of warmth, and so forth (Damasio and Damasio, 1992). For these sensory impressions to be bound in some way into the concept "coffee," the different areas that process these impressions must be connected. A complete list of areas that are relevant to just the basic features of conceptual awareness would be very long, involving all the visual (color, shape, motion, etc.), spatial, auditory, temporal organization, olfactory, taste, somatosensory, and limbic system (emotion) areas. These are processed using extensive regions of the parietal, occipital, and temporal lobes (Fig. 3b).

Given that conceptual awareness forms the very foundation of language (Hurford, 2003a), and given that larger brains appear to give rise to more complex conceptual universes (and hence more interesting things to communicate about), and given that humans are intensely socially interactive, increasing brain size itself should be seen as an excellent proxy for language evolution (Gibson, 2002; Schoenemann, 1999, 2005).

Classical language areas

Broca's and Wernicke's areas were the first cortical regions to be associated with specific linguistic abilities. Broca's aphasics display nonfluent, effortful, and agrammatical speech, whereas Wernicke's aphasics display grammatical but meaningless speech in which the wrong words (or parts of words) are used (Bear et al., 2007; Damasio and Damasio, 1992). Broca's area is located in the posterior–inferior frontal convexity of the neocortex, while Wernicke's area is localized to the general area where parietal, occipital, and temporal lobes meet (Fig. 3d). For most people, these areas are functional for language primarily in the left hemisphere. However, it turns out that Broca's and Wernicke's aphasias (the specific types of

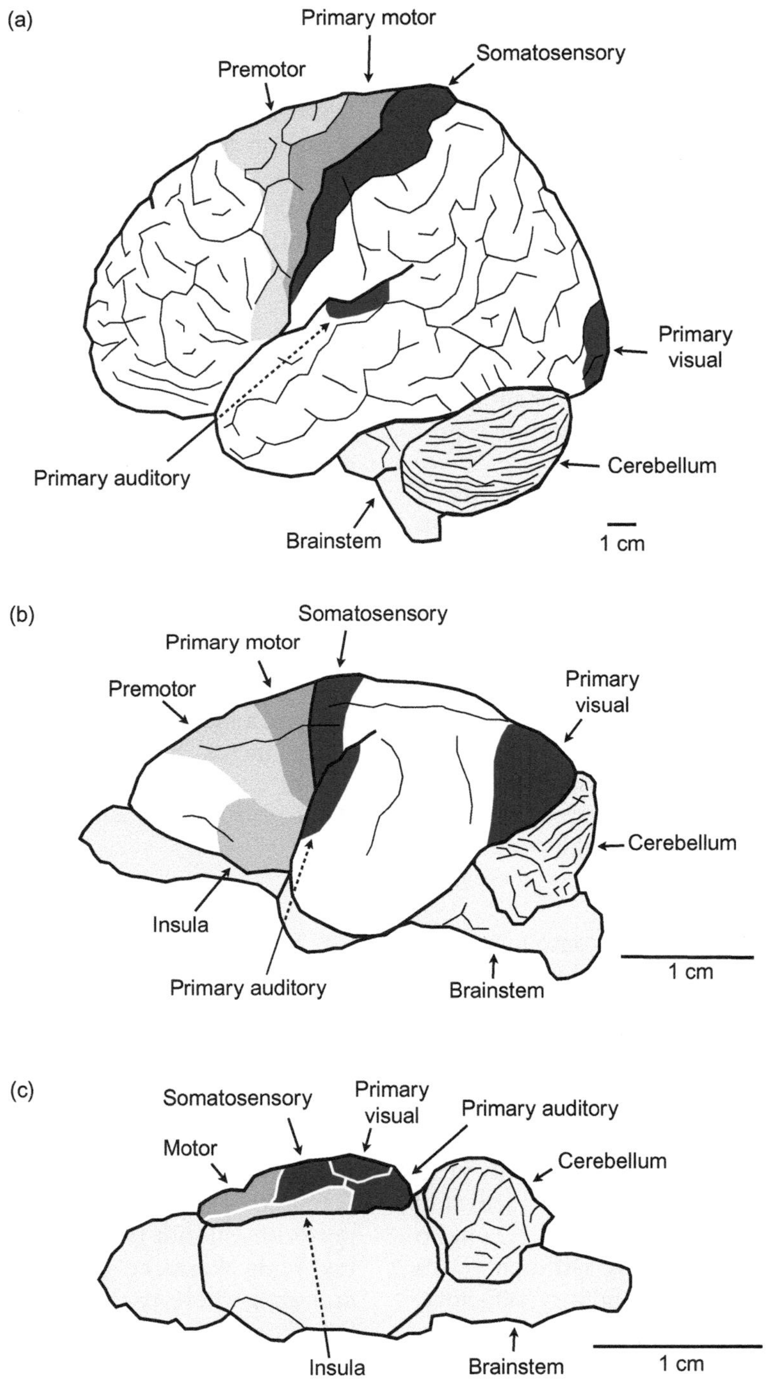

Fig. 2. Size of "association" cortex in mammals of different brain size. Lateral line drawings of the cortex of human (a), galago (prosimian primate) (b), and hedgehog (non-primate mammal) (c). Images are not to scale (note 1 cm scale bars for each image;

language deficits) are not exclusively associated with damage to Broca's and Wernicke's cortical areas (Dronkers, 2000). Damage to the caudate nucleus, putamen, and internal capsule (structures of the cerebral hemispheres that are deep to the cortex) also appear to play a role in Broca's aphasia, including aspects of syntactic processing (Lieberman, 2000). It is clear that a simple model of language being processed solely in Broca's and Wernicke's areas is too simplistic (Poeppel and Hickok, 2004), though these areas are clearly relevant.

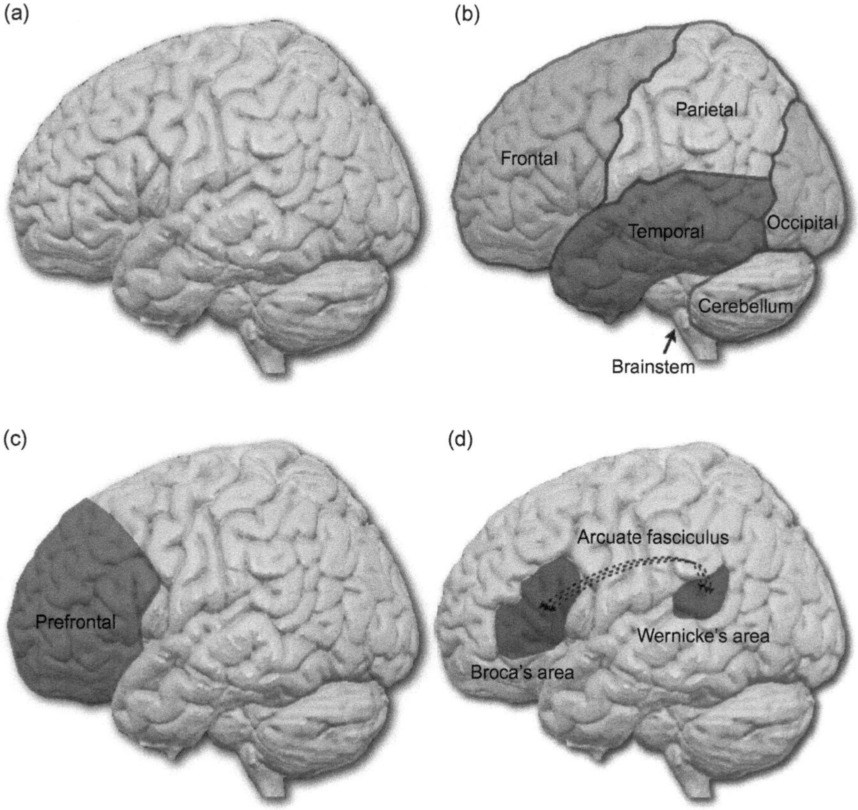

Fig. 3. Major regions of the brain. (a) Lateral view of the brain. (b) Lobes of the cortex, with cerebellum and brainstem indicated. (c) Prefrontal cortex. (d) Classical language areas. The arcuate fasciculus is a connective pathway deep to the cortex. The brain image is a 3D rendering of an average of 27 MRIs of the same individual from Holmes et al. (1998), used with permission.

---

corresponding volumes: hedgehog: ~3ml, galago: ~10ml, human: ~1350ml). The white regions represent cortical areas that are devoted to processing information *other* than primary sensory or motor (muscle movement) information. They function to integrate information in various interesting ways. Hedgehog brains have essentially no association cortex, whereas human brains have significantly more association cortex than other primates, both absolutely and proportionately. The human insula is not visible on the surface, being buried deep to the Sylvian fissure (which separates the temporal lobe from the frontal and anterior parietal lobes). The human brain was drawn from a 3D rendering of an average of 27 MRIs of the same individual (Holmes et al., 1998, used with permission). Galago and hedgehog brains were drawn from images at http://www.brains.rad.msu.edu, and http://brainmuseum.org, used with permission. Functional areas are mapped following Nieuwenhuys et al. (2008).

The evolutionary histories of Broca's and Wernicke's areas are quite intriguing, since homologs to both of these areas have been identified in nonhuman primate brains (Striedter, 2005). These species lack human language capabilities, of course, so these areas must have evolved originally for other purposes. An evolutionary perspective would predict that in nonhuman primates these areas likely process information in ways that would be useful to language (Schoenemann, 2005), with language then naturally making use of them (i.e., language adapting to the human brain: Christiansen and Chater, 2008). The presence of these "language" areas in nonlinguistic animals undermine models that imply the evolution of completely new kinds of circuits (e.g., Bickerton, 1990; Pinker, 1995). Some intriguing findings suggest that these areas do function in primates in ways that would predispose them to human language processing. The homolog of Broca's area in monkeys has been shown to contain neurons that fire both when a monkey performs a specific action as well as when it hears a sound related to that same action ("mirror neurons"; Kohler et al., 2002), which may form the basis for the ability to attach meaning to sounds. Stimulation of the Broca's area homolog in macaque monkeys results in orofacial movements (Petrides et al., 2005), which are foundational to human speech. Hearing species-specific calls has been shown to activate Broca's and Wernicke's areas in monkeys (Gil-da-Costa et al., 2006). In chimpanzees, communicative signaling (begging) has activated the homolog of Broca's area (Taglialatela et al., 2008). Further exploration of the function of these areas in nonhumans will allow a better idea of how and why they became co-opted for human language.

Detailed quantitative data on the size of these areas have been reported only for humans and chimpanzees so far. For the two areas that comprise Broca's area, one study reported that Brodmann area 44 in the left hemisphere is 6.6 times larger and in the right 4.1 times larger in humans as compared to chimpanzees, while Brodmann area 45 was 6.0 times larger on the left and 5.0 times larger on the right (Schenker et al., 2009). For comparison, overall brain size was 3.6 times larger for this sample, thus suggesting that there have been disproportionate increases in Broca's area—particularly, in the left hemisphere—during human evolution. Quantitative comparisons of Wernicke's area have not been reported, though it does appear that it is significantly bigger in both absolute and relative terms in humans as compared to macaque monkeys (Petrides and Pandya, 2002; Striedter, 2005).

Given that Broca's and Wernicke's areas play different but complementary roles in language processing, they must be connected in some way. A tract of nerve fibers known as the arcuate fasciculus (Fig. 3d) directly connects these areas (Geschwind, 1974). It tends to be larger on the left side than the right in humans, consistent with the lateralization of expressive language processing to the left hemisphere for most people (Nucifora et al., 2005). In addition, it appears to have been elaborated in human evolution. The homolog of Wernicke's area in macaque monkeys projects to prefrontal regions that are close to the homolog of Broca's area, but apparently not directly to it (Aboitiz and Garcia, 1997). Instead, projections directly to their homolog of Broca's area originate from a region just adjacent to their homolog of Wernicke's area (Aboitiz and Garcia, 1997). This would suggest that there has been an elaboration and extension of projections to more closely connect Broca's and Wernicke's areas over the course of human (or ape) evolution. Recent work using diffusion tensor imaging (which delineates approximate white matter axonal tracts *in vivo*) suggests that both macaques and chimpanzees have tracts connecting areas in the vicinity of Wernicke's area to regions in the vicinity of Broca's area (Rilling et al., 2007). However, connections between Broca's area and the middle temporal regions (important to semantic processing—see below) are only clear in chimpanzees and humans, and are even more extensive in humans (Rilling et al., 2007). These changes are certainly relevant to language evolution, though knowing whether they evolved specifically *for* language, rather than to support more general cognitive operations involving conceptual understanding, is unknown.

Prefrontal cortex

Areas in the prefrontal cortex (in addition to Broca's area; Fig. 3c) appear to be involved in a variety of linguistic tasks, including various semantic aspects of language (Gabrieli et al., 1998; Kerns et al., 2004; Luke et al., 2002; Maguire and Frith, 2004; Noppeney and Price, 2004; Thompson-Schill et al., 1998), syntax (Indefrey et al., 2004; Novoa and Ardila, 1987), and higher level linguistic processing, such as understanding the reasoning underlying a conversation (Caplan and Dapretto, 2001).

There appears to have been a significant elaboration of the prefrontal cortex during human evolution, with cytoarchitectural data pointing to an approximately twofold increase for the entire prefrontal cortex over what would be predicted for a primate brain as large as ours (Brodmann, 1909; Deacon, 1997). MRI studies generally support these conclusions though some debate remains (reviewed in Schoenemann, 2006; see also Smaers et al., 2011). Using a proxy for prefrontal cortex, we found that connective tracts (white matter areas composed mostly of axons) seem to account for a greater portion of the increase (Schoenemann et al., 2005). This makes sense given that prefrontal areas generally have an oversight role, modifying activity in other posterior areas of the brain.

Because prefrontal areas mediate a number of important behaviors besides language, language evolution may not be the primary driving force behind these changes. A variety of higher-order behavioral abilities that were likely crucial for human evolution are known to be mediated there, including planning, maintaining behavioral goals, processing social information, temporary storage/manipulation of information (working memory), memory for serial order and temporal information, and attention (references in Schoenemann, 2006). Teasing apart the relative contributions of these various behavioral abilities to the evolution of prefrontal areas will likely be very difficult.

The prefrontal cortex itself has many components, not all of which have changed to the same extent. Broca's area has increased disproportionately, particularly on the left side, as discussed above (Schenker et al., 2009). Brodmann area 13, located in the posterior orbitofrontal cortex (posteroinferior frontal cortex, above the eyes), seems to have lagged behind the increase in overall brain size, being only ~1.5 times larger than the average ape (Semendeferi et al., 1998). This area is involved in assessing emotional aspects of social interactions. Given the increasing importance of the social context during human evolution, its increase might seem relatively small (though not trivial). However, its specific relevance to language is unclear, except insofar as it presumably contributes to conceptual understanding of social relationships, which form the basis for caring about communication in the first place (see discussion below about conceptual understanding and language evolution).

Brodmann area 10, by contrast, is ~6.6 times larger than the corresponding areas in pongids (Semendeferi et al., 2001; Chapter 9). This increase is actually close to what one would expect given the peculiar way in which area seems to scale with overall brain size in primates (Holloway, 2002). Nevertheless, because this area is specifically active in linguistic tasks that require selection of appropriate words given a specific semantic context (Gabrieli et al., 1998; Luke et al., 2002), it seems likely its increase is relevant to language evolution.

Concepts and semantic processing

As discussed above, language depends critically on a foundation of conceptual understanding of the world, which in turn appears to depend on a wide network of many different areas of the brain. Humans are particularly biased toward visual information, which, as a consequence, forms an important component of conceptual understanding for most people (blind people being an exception). Visual information processing starts in the retina of the eye and is transferred through intermediate nuclei to the primary visual cortex, located in the occipital lobe (Figs. 2a and 3b), where it becomes available to conscious awareness (Bear et al., 2007). From here, visual information is processed along two major pathways: the dorsal stream (extending up into the parietal lobe), which processes information regarding the location and motion of an object, and the ventral stream (extending to the anterior tip of the temporal lobe), which processes information regarding the characteristics of the objects themselves (e.g., shape, color, etc.; Bear et al., 2007). The dorsal stream can therefore be thought of as the "where" pathway, and the ventral stream as the "what" pathway (Bear et al., 2007). These two pathways consequently correspond at least broadly to the networks involved in conceptualizing objects (which get mapped as nouns) versus actions/orientations/directions (which are central to concepts generally mapped as verbs) (cf., Hurford, 2003b).

The understanding of proper nouns appears to depend on anterior and medial areas of the temporal lobe, whereas understanding common nouns appears to depend on the lateral and inferior temporal lobes (Damasio and Damasio, 1992). In a comparative perspective, the human temporal lobe as a whole is 23% larger than predicted given our brain size, or about four times larger in absolute terms (though this is similar to overall brain size difference in this sample; data from Rilling and Seligman, 2002). The human disproportion in the temporal lobe appears to be greatest with respect to white matter axonal tracts, which represent long distance connections with other parts of the brain. This suggests a significant increase in the dense network of interconnectivity underlying conceptual understanding.

Areas involved in auditory processing—the primary auditory cortex of the temporal lobe (Brodmann areas 41 and 42) and immediately adjacent auditory association areas—appear to be slightly larger than predicted based on overall brain size (primary auditory by ~6%, auditory association by ~17%; Deacon, 1997). In absolute terms, these areas would still be more than approximately three to four times larger than the equivalent area in apes, suggesting the increases reflect enhancements with respect to auditory processing.

In addition to areas involved in auditory processing, and conceptual and semantic processing, the temporal lobe also contains the hippocampus (which plays a key role in memory formation) and the amygdala (which is central to marking the emotional valence of events). Both of these structures lie on the medial surface of the temporal lobes (hidden from lateral view). In absolute volume, the hippocampus is about 2.7 times larger in humans compared to chimpanzees (N=1 for each species, however; Stephan et al., 1981). In Stephan et al.'s (1987) data, the human amygdala was 3.7 times larger than their chimpanzee's (also N=1), which may indicate that marking the emotional content and/or importance of events was particularly important.

The planum temporale, located just posterior to the primary auditory cortex, has been well studied comparatively. The planum temporale is larger on the left side in humans (Geschwind and Levitsky, 1968; Sommer et al., 2008), which was assumed to reflect a functional anatomical correlate of language evolution. However, apes show a similar asymmetry in this region (Gannon et al., 1998), suggesting that the lateralized size differences are not language specific. Details of neuronal cell spatial organization have been shown to be asymmetric in the planum temporale of humans but not other primates, however, with the left hemisphere displaying wider minicolumns than the right (Buxhoeveden et al., 2001). The functional significance of this cytoarchitectural asymmetry is unknown, though presumably it has some behavioral relevance. It may be that the planum temporale has a role in auditory information for communication generally (not just for language). If so, it would be yet another example of language co-opting preexisting language-friendly circuits.

With respect to the "where" pathway, which likely grounds concepts central to most verbs, there is some indication of an evolutionary expansion of this area during human evolution, based on analyses of brain endocasts of fossil hominins (Bruner, 2004). Detailed comparative neuroanatomical studies of the parietal lobe have not been reported, however. The semantic generation of verbs (the actual words themselves) seems to also involve Broca's area (Damasio and Damasio, 1992; Posner and Raichle, 1994) which also appears to have evolved disproportionately (see above).

Right hemisphere

Although the cortical language areas discussed so far are localized to the left hemisphere in most people, the right hemisphere also appears to be involved in language. The right hemisphere understands short words (Gazzaniga, 1970) and also entertains alternative possible meanings for particular words (Beeman and Chiarello, 1998), suggesting it is involved in interpreting multiple intended meanings of a given linguistic communication. The right hemisphere also plays a greater role in spatial processing in most people (Tzeng and Wang, 1984; Vallar, 2007), thus presumably grounding the semantics of spatial terms. The right frontal lobe mediates aspects of prosody (Alexander et al., 1989), which is critically important to understanding intended meaning.

As discussed above, the planum temporale has been shown to be asymmetric in both apes and humans (Gannon et al., 1998), but only humans seem to show differences in neuronal spatial organization, favoring the left (Buxhoeveden et al., 2001). Humans also show asymmetries in Broca's area, again favoring the left (Gannon et al., 1998; Schenker et al., 2009). The left prefrontal overall has been shown to be particularly large relative to the right, in humans compared to other primates (Smaers et al., 2011). Although this appears to be largely predicted by primate scaling trends for these regions, it nevertheless likely has behavioral implications. Curiously, in humans, the whole right hemisphere is actually very slightly *larger* overall than the left ($\leq$1%; Allen et al., 2002). The author is not aware of studies comparing the volumes of entire hemispheres in other primates, however. Given that the right hemisphere is very close to the left in overall size in humans, it has probably increased approximately threefold along with the cortex as a whole. Also, since Broca's area (at least) is larger on the left hemisphere, by definition other areas of the right must therefore be larger than their corresponding areas on the left, given that the right hemisphere is slightly larger as a whole. Certainly, there is nothing to suggest that the right hemisphere has particularly lagged behind the left overall during human evolution, though additional detailed comparative asymmetry studies remain to be done.

Basal ganglia

Although the cortex is heavily involved in language processing, a group of interconnected nuclei deep to the cortex, collectively known as the basal ganglia, appear to play important roles also. They participate in an important circuit loop that functions in the selection and initiation of willed movements generally (Bear et al., 2007). A variety of studies have implicated these circuits not just in language production but also in language comprehension (see references in Hochstadt et al., 2006). Important symptoms of Parkinson's and Huntington's diseases, which affect the basal ganglia in particular, include problems understanding complicated syntax (e.g., center-embedded clauses), as well as processing semantic information (Hochstadt et al., 2006). Broca's aphasia typically requires disruption to areas other than Broca's area, likely including circuits involving the basal ganglia (Lieberman, 2002).

Comparative studies of the relative size of the basal ganglia in humans suggest that these nuclei are only about 65% as large as predicted for a primate brain as large as ours (Schoenemann, 1997; Stephan et al., 1981). However, they are still about twice as large in absolute terms as predicted based on body size. Since humans do not appear to have significantly more sophisticated willed movements than apes (except possibly those related to vocalization and hand manipulation), it seems possible these increases indicate an important role supporting higher cortical functions including language.

Cerebellum

The cerebellum is not only involved in modulating motor signals from the cortex (Bear et al., 2007) but has also been implicated in higher cognitive functions, including goal organization and planning, aspects of memory and learning, attention, visuospatial processing, modulating emotional responses, and language (Baillieux et al., 2008). The cerebellum appears to play a role in speech production and perception, as well as both semantic and grammatical processing (Ackermann et al., 2007; Baillieux et al., 2008; De Smet et al., 2007). The cerebellum also seems to play a role in timing mechanisms generally (Ivry and Spencer, 2004). Given the importance temporal information plays in language production and perception, this may explain why the cerebellum is implicated in studies of language brain function.

The human cerebellum is slightly smaller than one would predict based on brain size (Rilling and Insel, 1998) but is ~2.9 times larger based on body size (the largest increase of all brain regions outside the neocortex). The higher cognitive functions appear to be localized to the lateral hemispheres of the cerebellum (MacLeod et al., 2003), which have undergone a significant increase in relative proportion in both apes and humans. With respect to overall body weight, human lateral cerebellar hemispheres are ~2.9 times larger than predicted (MacLeod et al., 2003) and therefore are not explained by any possible differences in muscle mass. Given its role in language processing, this increase may therefore be relevant to language evolution.

Vocalization

The muscles responsible for vocalization are directly innervated by nuclei in the brainstem, which relay signals from midbrain and higher cortical areas. The muscles of the vocal folds, responsible for changing the pitch of the primary vocal signal, are innervated by the nucleus ambiguous. The vocal signal is filtered in various ways by manipulations of the tongue, lower jaw, and lips, thereby producing the sound changes we interpret as speech (Denes and Pinson, 1963). The tongue is innervated by the hypoglossal nucleus and nucleus ambiguous, the lower jaw by the trigeminal nucleus, and the lips by the facial motor nucleus (all in the brain stem). This whole system depends on the maintenance and manipulation of air pressure in the lungs, using the muscles of the chest and abdomen, innervated by anterior horn areas of the spinal cord (Carpenter and Sutin, 1983).

Even though vocal production is clearly important to language, these brainstem nuclei do not appear to be particularly enlarged in humans. The hypoglossal nucleus in humans is large but shows substantial overlap with apes, while the human trigeminal and facial motor nuclei completely overlap in size with those of the great apes (Sherwood et al., 2005). Thus, there is little evidence that language evolution substantially modified these nuclei.

Because conscious muscle movement originates in the cerebral cortex, intentional speech utilizes direct connections from cortex to the relevant brainstem nuclei. Humans also have two indirect pathways, the first involving connections routed through the reticular formation of the brainstem (for tongue, lower jaw, and lip movement) and the nucleus retroambiguus (for muscles of respiration; Deacon, 1997; Striedter, 2005), and a second through the cingulate gyrus of the limbic system (Jürgens, 1994), which mediates involuntary vocal responses to pain or strong emotions (Deacon, 1997; Striedter, 2005).

Comparative studies show that nonhuman primates have these same indirect pathways, allowing them a variety of emotionally mediated vocalizations, but have only weak direct connections controlling the tongue and respiration muscles, and appear to completely lack direct connections to the larynx (Jurgens, 2002; Jurgens and Alipour, 2002). This suggests that at least some new direct cortical pathways to the brainstem evolved for deliberate conscious vocalization in humans.

Auditory perception

Conscious awareness of sound depends on signals from the cochlea (where sound is translated into neural signals) reaching the primary auditory cortex located in the temporal lobe through a series of intermediate nuclei in the brainstem and midbrain. Comparative data suggest that primates have somewhat smaller intermediate auditory nuclei than expected for their brain weights (though the primate sample was small; Glendenning and Masterton, 1998). Both in absolute terms and in relation to body size, human auditory nuclei appear to be reasonably large, though not dramatically so. Overall, this suggests only modest evolutionary changes in these nuclei.

Conclusion

Our current understanding of brain/language coevolution is based on comparative gross anatomical assessments of differences in parts of the brain that are relevant to language processing. The increase in overall brain size paved the way for language both by encouraging localized cortical specialization and by making possible increasingly complicated social interactions, which provided the central usefulness for language, thereby driving its evolution. Specific brain areas relevant to language increased disproportionately: the prefrontal cortex (areas relevant to semantics and syntax) and the temporal lobe (particularly, areas relevant to connecting words to meanings and concepts). Broca's and Wernicke's areas, and the arcuate fasciculus connecting them, were not specially evolved for language but do appear to have been elaborated. Other areas that participate in language processing, including the basal ganglia and cerebellum, are larger than predicted based on overall body weight, though they have not increased as much relative to overall brain size. Finally, the auditory processing pathways do not appear to have changed much, though pathways allowing for enhanced conscious control of speech have been either added (to the larynx) or strengthened.

These make excellent starting points for future research, which hopefully will clarify exactly how circuitry in these areas support language and may have been modified by language evolution. A better understanding of the complex interactions between areas will allow us to test some of the more interesting speculative ideas in the literature (e.g., Deacon, 1997). Because language was not the only behavioral change in human evolution, unraveling the interplay between behavioral domains will be critical. Assessing nonlanguage functions of "language" areas will also be crucial. All these findings are consistent with the view that language adapted to the brain more than the reverse (Christiansen and Chater, 2008).

Three major factors seem to have conspired to drive the evolution of language: first, the general elaboration of—and increasing focus on—the importance of learned behavior; second, a significant increase in the complexity, subtlety, and range of conceptual understanding that was possible; and third, an increasingly complex, socially interactive existence. Because language itself facilitates thinking and conceptual awareness, language evolution would have been a mutually reinforcing process: increasingly complicated brains led to increasingly rich and varied thoughts, driving the evolution of increasingly complicated language, which itself facilitated even more complex conceptual world that these brains would then want to communicate (Savage-Rumbaugh and Rumbaugh, 1993; Schoenemann, 2009a,b). This process highlights the usefulness of thinking about language evolution as a complex adaptive system. The extent to which increasing conceptual complexity itself might have driven language evolution represents an intriguing research question for the future.

Acknowledgments

This review has been adapted from Schoenemann (2009a), which grew out of a contribution to the IIAS International Seminar on Language, Evolution, and the Brain held in Kyoto, Japan in April 2007, organized by William Wang and Junjiro Kanamori, as well as from Schoenemann (2009b), which grew out of a workshop on language evolution at the Santa Fe Institute in March 2007, organized by John Holland. I wish to thank Michel Hofman for thoughtful comments and for pointing out errors in an earlier draft. Any errors that remain are of course my own. My thinking on language evolution has benefited from various discussions with William Wang, James Minett, Vince Sarich, Jim Hurford, and Morten Christiansen. The galago and hedgehog brains in Fig. 2a and b were adapted with permission from images at http://www.brains.rad.msu.edu, and http://brainmuseum.org, supported by the US National Science Foundation.

References

Aboitiz, F., & Garcia, V. R. (1997). The evolutionary origin of the language areas in the human brain. A neuroanatomical perspective. *Brain Research. Brain Research Reviews*, *25*(3), 381–396.

Ackermann, H., Mathiak, K., & Riecker, A. (2007). The contribution of the cerebellum to speech production and speech perception: Clinical and functional imaging data. *Cerebellum (London, England)*, *6*(3), 202–213.

Alexander, M. P., Benson, D. F., & Stuss, D. T. (1989). Frontal lobes and language. *Brain and Language*, *37*, 656–691.

Allen, J. S., Damasio, H., & Grabowski, T. J. (2002). Normal neuroanatomical variation in the human brain: An MRI-volumetric study. *American Journal of Physical Anthropology*, *118*, 341–358.

Allman, J., McLaughlin, T., & Hakeem, A. (1993). Brain weight and life-span in primate species. *Proceedings of the National Academy of Sciences of the United States of America*, *90*(1), 118–122.

Baillieux, H., De Smet, H. J., Paquier, P. F., De Deyn, P. P., & Marien, P. (2008). Cerebellar neurocognition: Insights into the bottom of the brain. *Clinical Neurology and Neurosurgery*, *110*(8), 763–773.

Barsalou, L. W. (2008). Grounded cognition. *Annual Review of Psychology*, *59*, 617–645.

Bear, M. F., Connors, B. W., & Paradiso, M. A. (2007). *Neuroscience: Exploring the brain* (3rd ed.). Baltimore; Philadelphia, PA: Lippincott Williams & Wilkins.

Beckner, C., Blythe, R., Bybee, J., Christiansen, M. H., Croft, W., Ellis, N. C., et al. (2009). Language is a complex adaptive system: Position paper. *Language Learning*, *59*(Suppl. 1), 1–26.

Beeman, M. J., & Chiarello, C. (1998). Complementary right- and left-hemisphere language comprehension. *Current Directions in Psychological Science*, *7*(1), 2–8.

Bickerton, D. (1990). *Language and species*. Chicago: University of Chicago Press.

Blinkov, S. M., & Glezer, I. I. (1968). *The human brain in figures and tables (B. Haigh, Trans.)*. New York: Plenum Press.

Brodmann, K. (1909). *Vergleichende lokalisationslehre der grosshirnrinde in ihren prinzipien dargestellt auf grund des zellenbaues*. Leipzig: Johann Ambrosius Barth Verlag.

Bruner, E. (2004). Geometric morphometrics and paleoneurology: Brain shape evolution in the genus homo. *Journal of Human Evolution*, *47*, 279–303.

Buxhoeveden, D. P., Switala, A. E., Litaker, M., Roy, E., & Casanova, M. F. (2001). Lateralization of minicolumns in human planum temporale is absent in nonhuman primate cortex. *Brain, Behavior and Evolution*, *57*, 349–358.

Caplan, R., & Dapretto, M. (2001). Making sense during conversation: An fMRI study. *Neuroreport: An International*

Journal for the Rapid Communication of Research in Neuroscience, *12*(16), 3625–3632.

Carpenter, M. B., & Sutin, J. (1983). *Human neuroanatomy* (8th ed.). Baltimore, MD: Williams & Wilkins.

Changizi, M. A., & Shimojo, S. (2005). Parcellation and area-area connectivity as a function of neocortex size. *Brain, Behavior and Evolution*, *66*(2), 88–98.

Christiansen, M. H. (1994). *Infinite languages, finite minds: Connectionism, learning and linguistic structure.* Unpublished PhD dissertation, University of Edinburgh, Edinburgh, Scotland.

Christiansen, M. H., & Chater, N. (2008). Language as shaped by the brain. *The Behavioral and Brain Sciences*, *31*, 489–509.

Damasio, A. R., & Damasio, H. (1992). Brain and language. *Scientific American*, *267*(3), 89–95.

Damasio, H., Grabowski, T. J., Damasio, A., Tranel, D., Boles-Ponto, L., Watkins, G. L., et al. (1993). Visual recall with eyes closed and covered activates early visual cortices. *Society for Neuroscience Abstracts*, *19*, 1603.

Darwin, C. (1882). *The descent of man and selection in relation to sex* (2nd ed.). London: John Murray.

De Smet, H. J., Baillieux, H., De Deyn, P. P., Marien, P., & Paquier, P. (2007). The cerebellum and language: The story so far. *Folia Phoniatrica et Logopaedica*, *59*(4), 165–170.

Deacon, T. W. (1992). Brain-language coevolution. In J. A. Hawkins & M. Gell-Mann (Eds.), *The evolution of human languages* (Vol. XI, pp. 49–83). Redwood City, CA: Addison-Wesley.

Deacon, T. W. (1997). *The symbolic species: The co-evolution of language and the brain.* New York: W.W. Norton.

Deaner, R. O., Isler, K., Burkart, J., & van Schaik, C. (2007). Overall brain size, and not encephalization quotient, best predicts cognitive ability across non-human primates. *Brain, Behavior and Evolution*, *70*(2), 115–124.

Denes, P. B., & Pinson, E. N. (1963). *The speech chain.* Garden City, NY: Anchor Press/Doubleday.

Dronkers, N. F. (2000). The pursuit of brain-language relationships. *Brain and Language*, *71*(1), 59–61.

Dunbar, R. I. M. (1995). Neocortex size and group size in primates: A test of the hypothesis. *Journal of Human Evolution*, *28*, 287–296.

Finlay, B. L., Darlington, R. B., & Nicastro, N. (2001). Developmental structure in brain evolution. *The Behavioral and Brain Sciences*, *24*, 263–308.

Gabrieli, J. D., Poldrack, R. A., & Desmond, J. E. (1998). The role of left prefrontal cortex in language and memory. *Proceedings of the National Academy of Sciences of the United States of America*, *95*(3), 906–913.

Gannon, P. J., Holloway, R. L., Broadfield, D. C., & Braun, A. R. (1998). Asymmetry of chimpanzee planum temporale: Humanlike pattern of Wernicke's brain language area homolog. *Science*, *279*(5348), 220–222.

Gazzaniga, M. S. (1970). *The bisected brain.* New York: Appleton-Century-Crofts.

Geschwind, N. (1974). *Selected papers on language and the brain.* Boston: Reidel Publishing Co.

Geschwind, N., & Levitsky, W. (1968). Human brain: Left-right asymmetries in temporal speech region. *Science (New York, NY)*, *161*(837), 186–187.

Gibson, K. R. (2002). Evolution of human intelligence: The roles of brain size and mental construction. *Brain, Behavior and Evolution*, *59*(1–2), 10–20.

Gibson, K. R., Rumbaugh, D., & Beran, M. (2001). Bigger is better: Primate brain size in relationship to cognition. In D. Falk & K. Gibson (Eds.), *Evolutionary anatomy of the primate cerebral cortex* (pp. 79–97). Cambridge: Cambridge University Press.

Gil-da-Costa, R., Martin, A., Lopes, M. A., Munoz, M., Fritz, J. B., & Braun, A. R. (2006). Species-specific calls activate homologs of Broca's and Wernicke's areas in the macaque. *Nature Neuroscience*, *9*(8), 1064–1070.

Glendenning, K. K., & Masterton, R. B. (1998). Comparative morphometry of mammalian central auditory systems: Variation in nuclei and form of the ascending system. *Brain, Behavior and Evolution*, *51*(2), 59–89.

Harvey, P. H., & Clutton-Brock, T. H. (1985). Life history variation in primates. *Evolution*, *39*, 559–581.

Hochstadt, J., Nakano, H., Lieberman, P., & Friedman, J. (2006). The roles of sequencing and verbal working memory in sentence comprehension deficits in Parkinson's disease. *Brain and Language*, *97*(3), 243–257.

Hofman, M. A. (1983). Energy metabolism, brain size, and longevity in mammals. *The Quarterly Review of Biology*, *58*, 495–512.

Hofman, M. A. (1985). Size and shape of the cerebral cortex in mammals. I. The cortical surface. *Brain, Behavior and Evolution*, *27*, 28–40.

Hofman, M. A. (1988). Size and shape of the cerebral cortex in mammals. II. The cortical volume. *Brain, Behavior and Evolution*, *32*, 17–26.

Hofman, M. A. (1993). Encephalization and the evolution of longevity in mammals. *Journal of Evolutionary Biology*, *6*, 209–227.

Holloway, R. L. (1975). *The role of human social behavior in the evolution of the brain.* New York: American Museum of Natural History.

Holloway, R. L. (2002). Brief communication: How much larger is the relative volume of area 10 of the prefrontal cortex in humans? *American Journal of Physical Anthropology*, *118*(4), 399–401.

Holmes, C. J., Hoge, R., Collins, L., Woods, R., Toga, A. W., & Evans, A. C. (1998). Enhancement of mr images using registration for signal averaging. *Journal of Computer Assisted Tomography*, *22*(2), 324–333.

Humphrey, N. (1984). The social function of intellect. In *Consciousness regained* (pp. 14–28). Oxford: Oxford University Press.

Hurford, J. R. (2003a). The language mosaic and its evolution. In M. H. Christiansen & S. Kirby (Eds.), *Language evolution* (pp. 38–57). Oxford: Oxford University Press.

Hurford, J. R. (2003b). The neural basis of predicate-argument structure. *The Behavioral and Brain Sciences, 26*(3), 261–283 discussion 283–316.

Indefrey, P., Hellwig, F., Herzog, H., Seitz, R. J., & Hagoort, P. (2004). Neural responses to the production and comprehension of syntax in identical utterances. *Brain and Language, 89*(2), 312–319.

Ivry, R. B., & Spencer, R. M. (2004). The neural representation of time. *Current Opinion in Neurobiology, 14*(2), 225–232.

Jerison, H. J. (1973). *Evolution of the brain and intelligence.* New York: Academic Press.

Jurgens, U. (2002). Neural pathways underlying vocal control. *Neuroscience and Biobehavioral Reviews, 26*(2), 235–258.

Jürgens, U. (1994). The role of the periaqueductal grey in vocal behaviour. *Behavioural Brain Research, 62*(2), 107–117.

Jurgens, U., & Alipour, M. (2002). A comparative study on the cortico-hypoglossal connections in primates, using biotin dextranamine. *Neuroscience Letters, 328*(3), 245–248.

Kerns, J. G., Cohen, J. D., Stenger, V. A., & Carter, C. S. (2004). Prefrontal cortex guides context-appropriate responding during language production. *Neuron, 43*(2), 283–291.

Kohler, E., Keysers, C., Umilta, M. A., Fogassi, L., Gallese, V., & Rizzolatti, G. (2002). Hearing sounds, understanding actions: Action representation in mirror neurons. *Science, 297*(5582), 846–848.

Kosslyn, S. M., Alpert, N. M., Thompson, W. L., Maljkovic, V., Weise, S. B., Chabris, C. F., et al. (1993). Visual mental imagery activates topographically organized visual cortex: Pet investigations. *Journal of Cognitive Neuroscience, 5*, 263–287.

Lieberman, P. (2000). *Human language and our reptilian brain: The subcortical bases of speech, syntax, and thought.* Cambridge, MA: Harvard University Press.

Lieberman, P. (2002). On the nature and evolution of the neural bases of human language. *American Journal of Physical Anthropology, 45*, 36–62.

Luke, K. K., Liu, H. L., Wai, Y. Y., Wan, Y. L., & Tan, L. H. (2002). Functional anatomy of syntactic and semantic processing in language comprehension. *Human Brain Mapping, 16*(3), 133–145.

MacLeod, C. E., Zilles, K., Schleicher, A., Rilling, J. K., & Gibson, K. R. (2003). Expansion of the neocerebellum in hominoidea. *Journal of Human Evolution, 44*(4), 401–429.

Maguire, E. A., & Frith, C. D. (2004). The brain network associated with acquiring semantic knowledge. *NeuroImage, 22*(1), 171–178.

Martin, R. D. (1981). Relative brain size and basal metabolic rate in terrestrial vertebrates. *Nature, 293*, 57–60.

Mueller, R.-A. (1996). Innateness, autonomy, universality? Neurobiological approaches to language. *The Behavioral and Brain Sciences, 19*(4), 611–675.

Nieuwenhuys, R., Voogd, J., & van Huijzen, C. (2008). *The human central nervous system* (4th ed.). Berlin: Springer-Verlag.

Noppeney, U., & Price, C. J. (2004). Retrieval of abstract semantics. *NeuroImage, 22*, 164–170.

Northcutt, R. G., & Kaas, J. H. (1995). The emergence and evolution of mammalian neocortex. *Trends in Neurosciences, 18*(9), 373–379.

Novoa, O. P., & Ardila, A. (1987). Linguistic abilities in patients with prefrontal damage. *Brain and Language, 30*, 206–225.

Nucifora, P. G., Verma, R., Melhem, E. R., Gur, R. E., & Gur, R. C. (2005). Leftward asymmetry in relative fiber density of the arcuate fasciculus. *Neuroreport, 16*(8), 791–794.

Petrides, M., Cadoret, G., & Mackey, S. (2005). Orofacial somatomotor responses in the macaque monkey homologue of Broca's area. *Nature, 435*(7046), 1235–1238.

Petrides, M., & Pandya, D. N. (2002). Comparative cytoarchitectonic analysis of the human and the macaque ventrolateral prefrontal cortex and corticocortical connection patterns in the monkey. *The European Journal of Neuroscience, 16*(2), 291–310.

Pinker, S. (1995). Facts about human language relevant to its evolution. In J.-P. Changeux & J. Chavaillon (Eds.), *Origins of the human brain* (pp. 262–283). Oxford: Clarendon.

Poeppel, D., & Hickok, G. (2004). Towards a new functional anatomy of language. *Cognition, 92*(1–2), 1–12.

Posner, M. I., & Raichle, M. E. (1994). *Images of mind.* New York: W.H. Freeman.

Reader, S. M., & Laland, K. N. (2002). Social intelligence, innovation, and enhanced brain size in primates. *Proceedings of the National Academy of Sciences of the United States of America, 99*(7), 4436–4441.

Rilling, J. K., Glasser, M. F., Preuss, T. M., Ma, X., Zhang, X., Zhao, T., et al. (2007). A comparative diffusion tensor imaging (DTI) study of the arcuate fasciculus language pathway in humans, chimpanzees and rhesus macaques. *American Journal of Physical Anthropology, 132*(S44), 199–200.

Rilling, J. K., & Insel, T. R. (1998). Evolution of the cerebellum in primates: Differences in relative volume among monkeys, apes and humans. *Brain, Behavior and Evolution, 52*(6), 308–314.

Rilling, J. K., & Seligman, R. A. (2002). A quantitative morphometric comparative analysis of the primate temporal lobe. *Journal of Human Evolution, 42*(5), 505–533.

Ringo, J. L. (1991). Neuronal interconnection as a function of brain size. *Brain, Behavior and Evolution, 38*, 1–6.

Savage-Rumbaugh, E. S., & Rumbaugh, D. M. (1993). The emergence of language. In K. R. Gibson & T. Ingold (Eds.), *Tools, language and cognition in human evolution* (pp. 86–108). Cambridge: Cambridge University Press.

Schenker, N. M., Hopkins, W. D., Spocter, M. A., Garrison, A. R., Stimpson, C. D., Erwin, J. M., et al. (2009). Broca's area homologue in chimpanzees (Pan troglodytes): Probabilistic mapping, asymmetry, and comparison to humans. *Cerebral Cortex, 20*(3), 730–742.

Schoenemann, P. T. (1997). *An MRI study of the relationship between human neuroanatomy and behavioral ability.* Unpublished dissertation, University of California, Berkeley, Berkeley.

Schoenemann, P. T. (1999). Syntax as an emergent characteristic of the evolution of semantic complexity. *Minds and Machines, 9*, 309–346.

Schoenemann, P. T. (2005). Conceptual complexity and the brain: Understanding language origins. In W. S.-Y. Wang & J. W. Minett (Eds.), *Language acquisition, change and emergence: Essays in evolutionary linguistics* (pp. 47–94). Hong Kong: City University of Hong Kong Press.

Schoenemann, P. T. (2006). Evolution of the size and functional areas of the human brain. *Annual Review of Anthropology, 35*, 379–406.

Schoenemann, P. T. (2009a). Brain evolution relevant to language. In J. Minett & W. S.-Y. Wang (Eds.), *Language, evolution, and the brain.* Hong Kong: City University of Hong Kong Press.

Schoenemann, P. T. (2009b). Evolution of brain and language. *Language Learning, 59*(Suppl. 1), 162–186.

Schoenemann, P. T. (2010). The meaning of brain size: The evolution of conceptual complexity. In *Human brain evolving: Papers in honor of Ralph Holloway.* Bloomington, Indiana: Stone Age Institute Press.

Schoenemann, P. T., Sheehan, M. J., & Glotzer, L. D. (2005). Prefrontal white matter volume is disproportionately larger in humans than in other primates. *Nature Neuroscience, 8*(2), 242–252.

Semendeferi, K., Armstrong, E., Schleicher, A., Zilles, K., & Van Hoesen, G. W. (1998). Limbic frontal cortex in hominoids: A comparative study of area 13. *American Journal of Physical Anthropology, 106*(2), 129–155.

Semendeferi, K., Armstrong, E., Schleicher, A., Zilles, K., & Van Hoesen, G. W. (2001). Prefrontal cortex in humans and apes: A comparative study of area 10. *American Journal of Physical Anthropology, 114*(3), 224–241.

Sherwood, C. C., Hof, P. R., Holloway, R. L., Semendeferi, K., Gannon, P. J., Frahm, H. D., et al. (2005). Evolution of the brainstem orofacial motor system in primates: A comparative study of trigeminal, facial, and hypoglossal nuclei. *Journal of Human Evolution, 48*(1), 45–84.

Smaers, J. B., Steele, J., Case, C. R., Cowper, A., Amunts, K., & Zilles, K. (2011). Primate prefrontal cortex evolution: Human brains are the extreme of a lateralized ape trend. *Brain, Behavior and Evolution, 77*, 67–78.

Sommer, I. E., Aleman, A., Somers, M., Boks, M. P., & Kahn, R. S. (2008). Sex differences in handedness, asymmetry of the planum temporale and functional language lateralization. *Brain Research, 1206*, 76–88.

Stephan, H., Frahm, H., & Baron, G. (1981). New and revised data on volumes of brain structures in insectivores and primates. *Folia Primatologica, 35*, 1–29.

Stephan, H., Frahm, H. D., & Baron, G. (1987). Comparison of brain structure volumes in insectivora and primates. VII. Amygdaloid components. *Journal für Hirnforschung, 28*(5), 571–584.

Striedter, G. F. (2005). *Principles of brain evolution.* Sunderland, MA: Sinauer Associates.

Taglialatela, J. P., Russell, J. L., Schaeffer, J. A., & Hopkins, W. D. (2008). Communicative signaling activates 'Broca's' homolog in chimpanzees. *Current Biology, 18*(5), 343–348.

Thompson-Schill, S. L., Swick, D., Farah, M. J., D'Esposito, M., Kan, I. P., & Knight, R. T. (1998). Verb generation in patients with focal frontal lesions: A neuropsychological test of neuroimaging findings. *Proceedings of the National Academy of Sciences of the United States of America, 95*(26), 15855–15860.

Tooby, J., & Cosmides, L. (1992). The psychological foundations of culture. In J. H. Barkow, L. Cosmides & J. Tooby (Eds.), *The adapted mind: Evolutionary psychology and the generation of culture* (pp. 19–136). New York: Oxford University Press.

Tzeng, O. J. L., & Wang, W. S.-Y. (1984). Search for a common neurocognitive mechanism for language and movements. *American Journal of Physiology. Regulatory, Integrative and Comparative Physiology, 246*(15), R904–R911.

Vallar, G. (2007). Spatial neglect, Balint-Homes' and Gerstmann's syndrome, and other spatial disorders. *CNS Spectrums, 12*(7), 527–536.

White, J. G., Southgate, E., Thomson, J. N., & Brenner, S. (1986). The structure of the nervous system of the nematode Caenorhabditis elegans. *Philosophical Transactions of the Royal Society of London B, Biological Sciences, 314*(1165), 1–340.

Subject Index

Note: Page numbers followed by "*f*" indicate figures, and "*t*" indicate tables.

Volume 181: Neuroendocrinology: The Normal Neuroendocrine System, by Luciano Martini, George P. Chrousos, Fernand Labrie, Karel Pacak and Donald W. Pfaff (Eds.) – 2010, 978-0-444-53617-4.

Volume 182: Neuroendocrinology: Pathological Situations and Diseases, by Luciano Martini, George P. Chrousos, Fernand Labrie, Karel Pacak and Donald W. Pfaff (Eds.) – 2010, 978-0-444-53616-7.

Volume 183: Recent Advances in Parkinson's Disease: Basic Research, by Anders Björklund and M. Angela Cenci (Eds.) – 2010, 978-0-444-53614-3.

Volume 184: Recent Advances in Parkinson's Disease: Translational and Clinical Research, by Anders Björklund and M. Angela Cenci (Eds.) – 2010, 978-0-444-53750-8.

Volume 185: Human Sleep and Cognition Part I: Basic Research, by Gerard A. Kerkhof and Hans P.A. Van Dongen (Eds.) – 2010, 978-0-444-53702-7.

Volume 186: Sex Differences in the Human Brain, their Underpinnings and Implications, by Ivanka Savic (Ed.) – 2010, 978-0-444-53630-3.

Volume 187: Breathe, Walk and Chew: The Neural Challenge: Part I, by Jean-Pierre Gossard, Réjean Dubuc and Arlette Kolta (Eds.) – 2010, 978-0-444-53613-6.

Volume 188: Breathe, Walk and Chew; The Neural Challenge: Part II, by Jean-Pierre Gossard, Réjean Dubuc and Arlette Kolta (Eds.) – 2011, 978-0-444-53825-3.

Volume 189: Gene Expression to Neurobiology and Behaviour: Human Brain Development and Developmental Disorders, by Oliver Braddick, Janette Atkinson and Giorgio M. Innocenti (Eds.) – 2011, 978-0-444-53884-0.

Volume 190: Human Sleep and Cognition Part II: Clinical and Applied Research, by Hans P.A. Van Dongen and Gerard A. Kerkhof (Eds.) – 2011, 978-0-444-53817-8.

Volume 191: Enhancing Performance for Action and perception: Multisensory Integration, Neuroplasticity and Neuroprosthetics: Part I, by Andrea M. Green, C. Elaine Chapman, John F. Kalaska and Franco Lepore (Eds.) – 2011, 978-0-444-53752-2.

Volume 192: Enhancing Performance for Action and Perception: Multisensory Integration, Neuroplasticity and Neuroprosthetics: Part II, by Andrea M. Green, C. Elaine Chapman, John F. Kalaska and Franco Lepore (Eds.) – 2011, 978-0-444-53355-5.

Volume 193: Slow Brain Oscillations of Sleep, Resting State and Vigilance, by Eus J.W. Van Someren, Ysbrand D. Van Der Werf, Pieter R. Roelfsema, Huibert D. Mansvelder and Fernando H. Lopes da Silva (Eds.) – 2011, 978-0-444-53839-0.

Volume 194: Brain Machine Interfaces: Implications for Science, Clinical Practice and Society, by Jens Schouenborg, Martin Garwicz and Nils Danielsen (Eds.) – 2011, 978-0-444-53815-4.

Kolibri 360 GmbH
Ferdinand-Jühlke-Straße 7
99095 Erfurt, Deutschland
produktsicherheit@kolibri360.de